Just Enough Nitrogen

Mark A. Sutton · Kate E. Mason ·
Albert Bleeker · W. Kevin Hicks ·
Cargele Masso · N. Raghuram ·
Stefan Reis · Mateete Bekunda
Editors

Just Enough Nitrogen

Perspectives on how to get there for regions with too much and too little nitrogen

Springer

Editors
Mark A. Sutton
UK Centre for Ecology & Hydrology (UKCEH)
Edinburgh Research Station
Penicuik, Midlothian, UK

Kate E. Mason
UK Centre for Ecology & Hydrology (UKCEH)
Edinburgh Research Station
Penicuik, Midlothian, UK

Albert Bleeker
National Institute for Public Health and Environment (RIVM)
Bilthoven, The Netherlands

W. Kevin Hicks
Stockholm Environment Institute (SEI)
Department of Environment and Geography
University of York
York, UK

Cargele Masso
International Institute of Tropical Agriculture (IITA)
Yaoundé Research Station
Yaoundé, Cameroon

N. Raghuram
University School of Biotechnology
Guru Gobind Singh Indraprastha University
New Delhi, India

Stefan Reis
UK Centre for Ecology & Hydrology (UKCEH)
Edinburgh Research Station
Penicuik, Midlothian, UK

Mateete Bekunda
International Institute of Tropical Agriculture (IITA)
Arusha, Tanzania

ISBN 978-3-030-58064-3 ISBN 978-3-030-58065-0 (eBook)
https://doi.org/10.1007/978-3-030-58065-0

This Springer imprint is published by the registered company Springer Nature Switzerland AG
The registered company address is: Gewerbestrasse 11, 6330 Cham, Switzerland

Preface

The International Nitrogen Initiative—or INI for short—is a global network that focuses on bringing together scientific evidence to inform the development of policies and practices for better nitrogen management. In doing so, INI highlights a dual global challenge: that some regions of the world have excess nitrogen input, leading to major losses to the environment, while other regions have insufficient nitrogen input, constraining food production and exacerbating soil degradation including nitrogen depletion. It is therefore highly appropriate that the present volume seeks to bring these challenges together. It provides evidence to help local, national and global communities on how to get there with 'Just Enough Nitrogen'.

The setting for this discussion is also appropriate, as the volume represents the final outcomes from the 6th International Nitrogen Conference organized by the INI and hosted at the Speke Resort in Kampala, Uganda (24–27 November 2013). This was the first time that the International Nitrogen Conference was hosted in sub-Saharan Africa, as all previous conferences in the series had been organized in regions typified by excess nitrogen (Europe, North America, East Asia, Latin America, South Asia). The Kampala Conference therefore brought the INI community to a region where fertilizer inputs are typically less than one tenth of the rates per hectare in other world regions. With food security being one of the top issues for sub-Saharan Africa, and nitrogen supply being critical for food security, the conference therefore addressed a core challenge of the region. At the same time, the discussions set Africa in the global context, making the comparison with lessons learned in other regions.

The present volume reports the emerging messages. It builds on papers presented to the conference, including additional chapters that have been specifically developed after the conference as a result of the emerging discussions. The products include the conference declaration agreed by the delegates, representing a wide range of science and stakeholder interests, the *Kampala Statement-for-Action on Reactive Nitrogen in Africa and Globally*, together with the results of primary scientific studies and syntheses from local to regional and global scales. The subsequent analysis has also led to a chapter that assesses the impact of advance planning to halve meat intake, as compared with the reference intake for such a conference. The findings demonstrate how this demitarian approach greatly reduced the nitrogen footprint and environmental impact associated with the conference.

In launching this volume, we take the opportunity to thank all those who helped make the Kampala Conference such a success. In particular, we thank John Stephen Tenywa, Giregon Olupot, Patrick Musingusi, Peter Ebanyat, Trust Tumwesigye and colleagues in the local organizing committee. We thank the INI Coordination Team for its ongoing support, including Clare Howard, Will Brownlie, Agnieszka Becher and Sarah Blackman, together with the valuable support from Susan Greenwood-Etienne of the Scientific Committee on Problems of Environment (SCOPE).

We gratefully acknowledge the funding support from a wide range of sponsors for the conference, without whose support the endeavour would not have been possible. Together with the contributions-in-kind of many networks, we are grateful for conference funding from the Alliance for a Green Revolution in Africa (AGRA), the International Fertilizer Industry Association (IFA), the International Plant Nutrition Institute (IPNI), Africa Research in Sustainable Intensification for the Next Generation (Africa RISING), the National Agricultural Research Organisation of Uganda (NARO), the Department of Agricultural Production of Makerere University, the Global Partnership on Nutrient Management (GPNM) in cooperation with the Scientific Committee on Problems of Environment (SCOPE), the UK Centre for Ecology & Hydrology (UK CEH), the European Commission Joint Research Centre (JRC), the N2Africa project and the International Centre for Tropical Agriculture (CIAT).

We would also like to thank the Speke Resort, Kampala, for the additional work of sharing data on food supplies, comparison with a reference conference, and willingness to halve the normal amount of meat for the Nitrogen Conference. Last but not least, we thank all the authors, co-editors and chapter reviewers for their unstinting efforts at bringing the volume to such a high standard.

The outcomes provide serious food for thought. The volume shows how our food system is impacting all aspects of the global nitrogen cycle, contributing to climate change, air pollution, water pollution, and threatening human health, ecosystems and biodiversity. The chapters highlight how we need and benefit from nitrogen supply for food, yet the inefficient use of nitrogen—the majority of which is lost as pollution—is threatening our global environment. At the same time, rapidly increasing nitrogen oxide emissions from combustion sources in developing parts of the world point to a fast-growing threat, unless action is taken.

Together with the Kampala Statement-for-Action and an accompanying Special Issue of Environmental Research Letters, the present volume brings to completion the reporting of the Kampala Conference. Nevertheless, it is obvious that the challenge of solving the nitrogen challenge is just beginning. How indeed can we reach 'Just Enough Nitrogen'? With nitrogen cutting across most of the UN Sustainable Development Goals, humanity faces a global systemic challenge in the disruption of the world's nitrogen cycle. This systemic alteration points to the need for an equally transformational change to a global 'nitrogen circular economy', where efficient nitrogen use, food, health, wellbeing, environment and profit all go hand in hand.

Edinburgh, UK

Mark A. Sutton
Chair, International Nitrogen Initiative (2012–2018)

Yaoundé, Cameroon

Cargele Masso
Director, African Centre of the International
Nitrogen Initiative (2014–2019)

Arusha, Tanzania

Mateete Bekunda
Conference Convener, 6th International Nitrogen Conference
Director, African Centre of the International
Nitrogen Initiative (until 2014)

Contents

Part XI Conclusions and Outlook

Contributors

Aliyu A. Abdullahi Department of Soil Science, Faculty of Agriculture, Institute for Agricultural Research, Ahmadu Bello University, Zaria, Nigeria;
Centre for Rhizobium Studies, Murdoch University, Perth, WA, Australia

Y. P. Abrol South Asian Nitrogen Centre, Indian Nitrogen Group, Sustainable India Trust and Society for Conservation of Nature, New Delhi, India

T. K. Adhya South Asian Nitrogen Centre, Indian Nitrogen Group, Sustainable India Trust and Society for Conservation of Nature, New Delhi, India;
School of Biotechnology, Kalinga Institute of Industrial Technology (KIIT), Bhubaneswar, Odisha, India

Rocío Alonso Ecotoxicology of Air Pollution, CIEMAT, Madrid, Spain

Alice Amoding Department of Agricultural Production, College of Agricultural and Environmental Sciences, Makerere University, Kampala, Uganda

Juliette Anglade Sorbonne Université, CNRS, EPHE, UMR 7619 Metis, Paris, France

Pedro M. Aparicio-Tejo Institute for Multidisciplinary Research in Applied Biology (IMAB), Edificio Jerónimo de Ayanz, Campus Arrosadia, Pamplona, Navarra, Spain

Nataly L. Ascarrunz Instituto Boliviano de Investigación Forestal (IBIF), Santa Cruz de La Sierra, Bolivia

Godfrey Asea National Crop Resources Research Institute (NaCRRI), Kampala, Uganda

Amy T. Austin University of Buenos Aires and IFEVA, Buenos Aires, Argentina

Jill S. Baron U.S. Geological Survey, Colorado State University, Fort Collins, CO, USA

Iskander Barrena Department of Plant Biology and Ecology, University of the Basque Country UPV/EHU, Bilbao, Spain

William J. Bealey UK Centre for Ecology and Hydrology, Edinburgh Research Station, Penicuik, Midlothian, UK

Mateete Bekunda International Institute of Tropical Agriculture (IITA), Arusha, Tanzania

Victoria Bermejo Ecotoxicology of Air Pollution, CIEMAT, Madrid, Spain

George Bigirwa Alliance for a Green Revolution in Africa, Nairobi, Kenya

Gilles Billen Sorbonne Université, CNRS, EPHE, UMR 7619 Metis, Paris, France

Albert Bleeker National Institute of Public Health and Environment (RIVM), Bilthoven, The Netherlands

Lex Bouwman Department of Earth Sciences—Geochemistry, Faculty of Geosciences, Utrecht University, Utrecht, The Netherlands;
PBL Netherlands Environmental Assessment Agency, Bilthoven, The Netherlands

Jean D. Brender Epidemiology and Biostatistics, Texas A&M University, School of Public Health, College Station, TX, USA

Aleksandr Briukhanov Federal State Budgetary Scientific Institution "Federal Scientific Agroengineering Center VIM", Branch "Institute for Engineering and Environmental Problems in Agricultural Production (IEEP)", Saint Petersburg, Russian Federation

Steen Brock Department of Culture and Society, Aarhus University, Aarhus C, Denmark

Will J. Brownlie UK Centre for Ecology & Hydrology, Edinburgh Research Station, Bush Estate, Penicuik, Midlothian, UK;
School of Geosciences, University of Edinburgh, King's Buildings, Edinburgh, UK

Mercedes M. Bustamante Departamento de Ecologia, Universidade de Brasília, Brasília, CEP, Brazil

María Eréndira Calleja-Cervantes Institute for Multidisciplinary Research in Applied Biology (IMAB), Edificio Jerónimo de Ayanz, Campus Arrosadia, Pamplona, Navarra, Spain

Héctor Calvete-Sogo Ecotoxicology of Air Pollution, CIEMAT, Madrid, Spain

Lynne A. Carpenter-Boggs Department of Crop and Soil Sciences, Washington State University, Pullman, WA, USA

Cláudia M. d. S. Cordovil Universidade de Lisboa, Instituto Superior de Agronomia, CEF, Lisboa, Portugal

Christopher Cox United Nations Environment Programme, Nairobi, Kenya

Gisleine Cunha-Zeri Earth System Science Center (CCST), National Institute for Space Research (INPE), São José dos Campos, SP, Brazil

Tommy Dalgaard Department of Agroecology, Aarhus University, Tjele, Denmark

Eric A. Davidson University of Maryland Center for Environmental Science, Appalachian Laboratory, Frostburg, MD, USA

Delegates of the 6th International Nitrogen Conference, Kampala c/o UK Centre for Ecology & Hydrology, Edinburgh Research Station, Bush Estate, Penicuik, Midlothian, UK

Wim de Vries Wageningen University and Research (WUR), Wageningen, The Netherlands

Enzai Du State Key Laboratory of Earth Surface Processes and Resource Ecology, School of Natural Resources, Faculty of Geographical Science, Beijing Normal University, Beijing, China

Christian Dupraz INRAE, UMR, ABSys, University of Montpellier, Montpellier, France

Peter Ebanyat Department of Agricultural Production, Makerere University, Kampala, Uganda;
International Institute of Tropical Agriculture (IITA), Kampala, Uganda

Susana Elvira Ecotoxicology of Air Pollution, CIEMAT, Madrid, Spain

Jan Willem Erisman Louis Bolk Institute for International Advice and Research on Sustainable Agriculture, Nutrition and Health Care, Driebergen, The Netherlands;
VU University, Amsterdam, The Netherlands;
Institute of Environmental Sciences (CML), Leiden University, Leiden, The Netherlands

A. O. Esilaba Kenya Agricultural and Livestock Research Organisation, Nairobi, Kenya

José María Estavillo Department of Plant Biology and Ecology, University of the Basque Country UPV/EHU, Bilbao, Spain

Idowu B. Famuwagun Department of Crop, Soil and Pest Management, Federal University of Technology, Akure, Nigeria

Maria C. Forti Earth System Science Center (CCST), National Institute for Space Research (INPE), São José dos Campos, SP, Brazil

David Fowler UK Centre for Ecology & Hydrology, Edinburgh Research Station, Bush Estate, Penicuik, Midlothian, UK

Tamon Fumoto Institute for Agro-Environmental Sciences, National Agriculture and Food Research Organization (NARO), Tsukuba, Ibaraki, Japan

James N. Galloway Environmental Sciences Department, University of Virginia, Charlottesville, VA, USA

Corinne Galy-Lacaux Laboratoire d'aérologie, UMR 5560, Université Toulouse III Paul Sabatier, CNRS, Toulouse, France

Héctor García-Gómez Ecotoxicology of Air Pollution, CIEMAT, Madrid, Spain

Josette Garnier Sorbonne Université, CNRS, EPHE, UMR 7619 Metis, Paris, France

Patricia M. Glibert University of Maryland Center for Environmental Science, Horn Point Laboratory, Cambridge, MD, USA

Vitória Gonçalves Universidade de Lisboa, Instituto Superior de Agronomia, Lisboa, Portugal

Ignacio González-Fernández Ecotoxicology of Air Pollution, CIEMAT, Madrid, Spain

Carmen Gonzalez-Murua Department of Plant Biology and Ecology, University of the Basque Country UPV/EHU, Bilbao, Spain

Morten Graversgaard Department of Agroecology, Aarhus University, Tjele, Denmark

Bruna Grizzetti Sorbonne Université, CNRS, EPHE, UMR 7619 Metis, Paris, France;
European Commission, Joint Research Centre (JRC), Ispra, Italy

Birgitte Hansen Geological Survey of Denmark and Greenland (GEUS), Højbjerg, Denmark

Steffen C. B. Hansen Global Environment Facility, GEF Secretariat—Programs Unit, Washington, DC, USA

Toshihiro Hasegawa Tohoku Agricultural Research Center, National Agriculture and Food Research Organization (NARO), Shimokuriyagawa, Morioka, Iwate, Japan

Fatemeh Hashemi Department of Agroecology, Aarhus University, Tjele, Denmark

Berit Hasler Department of Environmental Sciences, Aarhus University, Roskilde, Denmark

Kentaro Hayashi Institute for Agro-Environmental Sciences, National Agriculture and Food Research Organization (NARO), Tsukuba, Ibaraki, Japan

Patrick Heffer Agriculture Service, International Fertilizer Association, Paris, France

Ole Hertel Department of Environmental Sciences, Aarhus University, Roskilde, Denmark

Jonathan E. Hickman NASA Goddard Institute for Space Studies, Broadway, NY, USA;
Earth Institute at Columbia University, New York, USA

W. Kevin Hicks Stockholm Environment Institute (SEI), Department of Environment and Geography, University of York, York, UK

Arjen Y. Hoekstra Twente Water Centre, University of Twente, Enschede, The Netherlands

Judith Hoelzemann Universidade Federal do Rio Grande do Norte, s/n Natal, RN, Brazil

Clare M. Howard UK Centre for Ecology & Hydrology, Edinburgh Research Station, Bush Estate, Penicuik, Midlothian, UK

John Howieson Centre for Rhizobium Studies, Murdoch University, Perth, WA, Australia

Ximena Huerfano Department of Plant Biology and Ecology, University of the Basque Country UPV/EHU, Bilbao, Spain

Nicholas J. Hutchings Department of Agroecology, Aarhus University, Tjele, Denmark

Brian H. Jacobsen Department of Food and Resource Economics, University of Copenhagen, Frederiksberg C, Denmark

Annie P. Jangam University School of Biotechnology, Guru Gobind Singh Indraprastha University, New Delhi, India

Victor J. Jaramillo Universidad Nacional Autónoma de México, Centro de Investigaciones en Ecosistemas, Morelia, Michoacán, Mexico

Lars Stoumann Jensen Department of Plant and Environmental Sciences, University of Copenhagen, Frederiksberg C, Denmark

Xiaotang Ju Chinese Agricultural University, College of Resources and Environmental Sciences, Beijing, China

David R. Kanter New York University, Department of Environmental Studies, New York, NY, USA

George Y. Kanyama-Phiri Lilongwe University of Agriculture and Natural Resources, Lilongwe, Malawi

Genki Katata Ibaraki University, Mito, Ibaraki, Japan

Nobuko Katayanagi Institute for Agro-Environmental Sciences, National Agriculture and Food Research Organization (NARO), Tsukuba, Ibaraki, Japan

P. Kathuli Kenya Agricultural and Livestock Research Organization, Katumani, Machakos, Kenya;
National Dry-Land Research Centre, Katumani, Nairobi, Kenya

Job Kihara International Center for Tropical Agriculture (CIAT), c/o ICIPE Duduville Campus, Nairobi, Kenya

Joseph Kikafunda National Crop Resources Research Institute (NaCRRI), Kampala, Uganda

Chris Kjeldsen Department of Agroecology, Aarhus University, Tjele, Denmark

Natalia Kozlova Federal State Budgetary Scientific Institution "Federal Scientific Agroengineering Center VIM", Branch "Institute for Engineering and Environmental Problems in Agricultural Production (IEEP)", Saint Petersburg, Russian Federation

Richard G. Kraaijvanger Hogeschool van Hall Larenstein, University of Applied Sciences, Velp, The Netherlands;
College of Dryland Agriculture, Mekelle University, Mekelle, Ethiopia

Brian Kronvang Department of Bioscience, Aarhus University, Silkeborg, Denmark

Luis Lassaletta Sorbonne Université, CNRS, EPHE, UMR 7619 Metis, Paris, France;
CEIGRAM/Department of Agricultural Production, Universidad Politécnica de Madrid, Madrid, Spain

Gerry Lawson UK Centre for Ecology and Hydrology, Edinburgh Research Station, Penicuik, Midlothian, UK

Allison M. Leach University of New Hampshire, The Sustainability Institute at UNH, Durham, NH, USA

L. Michael Lege School of the Environment, Washington State University, Pullman, WA, USA

Adrian Leip European Commission, Joint Research Centre (JRC), Directorate D—Sustainable Resources, Ispra, VA, Italy

Sergei M. Lukin All-Russian Scientific Research Institute of Organic Fertilizers and Peat—Branch of the Federal Budget Scientific Institution "Verkhnevolzsky Federal Agrarian Scientific Center", Vladimir, Russian Federation

John M. Maingi Department of Biochemistry, Microbiology and Biotechnology, Kenyatta University, Nairobi, Kenya

Dmitry Maksimov Federal State Budgetary Scientific Institution "Federal Scientific Agroengineering Center VIM", Branch "Institute for Engineering and Environmental Problems in Agricultural Production (IEEP)", Saint Petersburg, Russian Federation

Roxane Maranger Groupe de Recherche Interuniversitaire en Limnologie et en Environnement Aquatique (GRIL), Département de Sciences Biologiques, Université de Montréal, Montréal, Québec, Canada

Luiz A. Martinelli CENA-Universidade de São Paulo, Piracicaba, SP, Brazil

Kate E. Mason UK Centre for Ecology & Hydrology, Edinburgh Research Station, Penicuik, Midlothian, UK

Cargele Masso International Institute of Tropical Agriculture (IITA), Yaoundé, Cameroon

Miwa Y. Matsushima Chiba University, Chiba, Japan

Mesfin M. Mekonnen Robert B. Daugherty Water for Food Global Institute, University of Nebraska, Lincoln, NE, USA;
Department of Civil, Construction and Environmental Engineering, University of Alabama, Tuscaloosa, AL, USA

Sergio Menéndez Department of Plant Biology and Ecology, University of the Basque Country UPV/EHU, Bilbao, Spain

Wezi G. Mhango Department of Crop and Soil Sciences, Lilongwe University of Agriculture and Natural Resources, Lilongwe, Malawi

Fredrick Mhina Mngube Lake Victoria Basin Commission, Kisumu, Kenya

Dative Mukaminega Faculty of Applied Sciences, Kigali Institute of Science and Technology (KIST), Kigali, Rwanda

John V. O. Muoma Department of Biological Sciences, School of Natural Sciences, Masinde Muliro University of Science and Technology (MMUST), Kakamega, Kenya;
Centre for African Medicinal and Nutritional Flora and Fauna (CAMNFF), Masinde Muliro University of Science and Technology, Kakamega, Kenya

Patrick Musinguzi Department of Agricultural Production, Makerere University, Kampala, Uganda

Lizzy A. Mwamburi Department of Biological Sciences, University of Eldoret, Eldoret, Kenya

Hirofumi Nakamura Taiyo Keiki Co. Ltd, Tokyo, Kita, Japan

Victoria Naluyange Department of Agriculture and Land Use Management, School of Agriculture, Veterinary Sciences and Technology, Masinde Muliro University of Science and Technology (MMUST), Kakamega, Kenya

S. N. Nguluu Kenya Agricultural and Livestock Research Organization, Katumani, Machakos, Kenya;
South Eastern Kenya University (SEKU), Kitui, Kenya

Graham O'Hara Centre for Rhizobium Studies, Murdoch University, Perth, WA, Australia

Dennis M. W. Ochieno Department of Biological Sciences, School of Natural Sciences, Masinde Muliro University of Science and Technology (MMUST), Kakamega, Kenya;
Centre for African Medicinal and Nutritional Flora and Fauna (CAMNFF), Masinde Muliro University of Science and Technology, Kakamega, Kenya

Martins Odendo Socio-Economics and Statistics Division, Kenya Agricultural and Livestock Research Organization (KALRO), Kakamega, Kenya

Oene Oenema Alterra, Wageningen University and Research Centre (WUR), Wageningen, The Netherlands

Titilayo Oladitan Rufus Giwa Polytechnic, Owo, Ondo State, Nigeria

Jørgen E. Olesen Department of Agroecology, Aarhus University, Tjele, Denmark

Giregon Olupot Department of Agricultural Production, Makerere University, Kampala, Uganda

Omwoyo Ombori Department of Plant Sciences, Kenyatta University, Nairobi, Kenya

Jean P. Ometto Earth System Science Center (CCST), National Institute for Space Research (INPE), São José dos Campos, SP, Brazil

Geoffrey Onaga National Crop Resources Research Institute (NaCRRI), Kampala, Uganda

Keisuke Ono Institute for Agro-Environmental Sciences, National Agriculture and Food Research Organization (NARO), Tsukuba, Ibaraki, Japan

J. W. Onyango National Agricultural Research Laboratories, Nairobi, Kenya

Felipe Pacheco Earth System Science Center (CCST), National Institute for Space Research (INPE), São José dos Campos, SP, Brazil

Cheryl A. Palm Institute for Sustainable Food Systems, University of Florida, Gainesville, FL, USA

H. Pathak South Asian Nitrogen Centre, Indian Nitrogen Group, Sustainable India Trust and Society for Conservation of Nature, New Delhi, India;
ICAR-National Institute of Abiotic Stress Management, Baramati, Maharashtra, India

Peter Pepó Faculty of Agricultural and Food and Environmental Sciences, Institute of Crop Sciences, University of Debrecen, Debrecen, Hungary

Cecilia Perez Instituto de Ecología Y Biodiversidad (IEB), Universidad de Chile, Ñuñoa, Santiago, Chile

Tibisay Perez Instituto Venezolano de Investigaciones Cientificas, Lab. Química Atmosférica, Caracas, Venezuela

Isaura Rábago Ecotoxicology of Air Pollution, CIEMAT, Madrid, Spain

N. Raghuram University School of Biotechnology, Guru Gobind Singh Indraprastha University, New Delhi, India;
South Asian Nitrogen Centre, Indian Nitrogen Group, Sustainable India Trust and Society for Conservation of Nature, Pusa, New Delhi, India

Ramesh Ramachandran National Centre for Sustainable Coastal Management, Ministry of Environment Forest & Climate Change, Government of India, Anna University Campus, Chennai, India

K. P. C. Rao International Crop Research Institute for Semi-arid Tropics, Nairobi, Kenya

Stefan Reis UK Centre for Ecology & Hydrology, Edinburgh Research Station, Penicuik, Midlothian, UK

Javier Sanz Ecotoxicology of Air Pollution, CIEMAT, Madrid, Spain

Jan K. Schjørring Department of Plant and Environmental Sciences, University of Copenhagen, Frederiksberg C, Denmark

Ekaterina Shalavina Federal State Budgetary Scientific Institution "Federal Scientific Agroengineering Center VIM", Branch "Institute for Engineering and Environmental Problems in Agricultural Production (IEEP)", Saint Petersburg, Russian Federation

Torben Sigsgaard Department of Public Health, Aarhus University, Aarhus C, Denmark

Vimlendu Bhushan Sinha Department of Biotechnology, School of Engineering and Technology, Sharda University, Greater Noida, Uttar Pradesh, India;
University School of Biotechnology, Guru Gobind Singh Indraprastha University, New Delhi, India

Ute M. Skiba UK Centre for Ecology and Hydrology, Edinburgh Research Station, Penicuik, Midlothian, UK

Sieglinde Snapp Department of Crop and Soil Sciences, Lilongwe University of Agriculture and Natural Resources, Lilongwe, Malawi;
Department of Plant, Soils and Microbial Sciences, Michigan State University, East Lansing, MI, USA

Clifford S. Snyder Conway, AR, USA

Daniel J. Sobota Environmental Solutions Division, Oregon Department of Environmental Quality, Portland, OR, USA

Ariel Stein Earth Resources and Technology on assignment to NOAA's Air Resources Laboratory, Silver Spring, MD, USA

Igor Subbotin Federal State Budgetary Scientific Institution "Federal Scientific Agroengineering Center VIM", Branch "Institute for Engineering and Environmental Problems in Agricultural Production (IEEP)", Saint Petersburg, Russian Federation

Mark A. Sutton UK Centre for Ecology & Hydrology, Edinburgh Research Station, Bush Estate, Penicuik, Midlothian, UK

John S. Tenywa Department of Agricultural Production, Makerere University, Kampala, Uganda

Jason Terpolilli Centre for Rhizobium Studies, Murdoch University, Perth, WA, Australia

Chaopu Ti State Key Laboratory of Soil and Sustainable Agriculture, Institute of Soil Science, Chinese Academy of Sciences, Nanjing, China

M. K. Tiwari South Asian Nitrogen Centre, Indian Nitrogen Group, Sustainable India Trust and Society for Conservation of Nature, New Delhi, India

Ravi Tiwari Centre for Rhizobium Studies, Murdoch University, Perth, WA, Australia

Takeshi Tokida Institute for Agro-Environmental Sciences, National Agriculture and Food Research Organization (NARO), Tsukuba, Ibaraki, Japan

Sakae Toyoda Tokyo Institute of Technology, Yokohama, Kanagawa, Japan

Trust Tumwesigye Department of Agricultural Production, Makerere University, Kampala, Uganda

Katrine Turner Department of Agroecology, Aarhus University, Tjele, Denmark

Fernando Valiño Ecotoxicology of Air Pollution, CIEMAT, Madrid, Spain

Isabelle Van der Beck United Nations Environment Programme, Ecosystems Division (International Waters), Washington, DC, USA

Hans J. M. van Grinsven PBL Netherlands Environmental Assessment Agency, The Hague, The Netherlands

Bernard Vanlauwe International Institute of Tropical Agriculture (IITA), Ibadan, Oyo State, Nigeria

Amarilis de Varennes Universidade de Lisboa, Instituto Superior de Agronomia, Lisboa, Portugal

Eduard Vasilev Federal State Budgetary Scientific Institution "Federal Scientific Agroengineering Center VIM", Branch "Institute for Engineering and Environmental Problems in Agricultural Production (IEEP)", Saint Petersburg, Russian Federation

Henrik Vejre Department of Geosciences and Natural Resource Management, University of Copenhagen, Frederiksberg C, Denmark

Tom Veldkamp Faculty of Geo-Information Science and Earth Observation (ITC), Twente University, Enschede, The Netherlands

Maren Voss Leibniz Institute for Baltic Sea Research, Warnemuende, Germany

Philip Wandahwa Department of Agriculture and Land Use Management, School of Agriculture, Veterinary Sciences and Technology, Masinde Muliro University of Science and Technology (MMUST), Kakamega, Kenya

Henk Westhoek PBL Netherlands Environmental Assessment Agency, Bilthoven, The Netherlands

Irene A. Wiborg Knowledge Centre for Agriculture, Aarhus N, Denmark

Wilfried Winiwarter International Institute for Applied Systems Analysis (IIASA), Laxenburg, Austria

Xiaoyuan Yan State Key Laboratory of Soil and Sustainable Agriculture, Institute of Soil Science, Chinese Academy of Sciences, Nanjing, China

Midori Yano Institute for Agro-Environmental Sciences, National Agriculture and Food Research Organization (NARO), Tsukuba, Ibaraki, Japan

Ado A. Yusuf Department of Soil Science, Faculty of Agriculture, Institute for Agricultural Research, Ahmadu Bello University, Zaria, Nigeria

A. R. Zaharah Department of Soil Science, Universiti Putra Malaysia (UPM), Serdang, Selangor, Malaysia

Shamie Zingore International Plant Nutrition Institute (IPNI), c/o ICIPE Duduville Campus, Nairobi, Kenya

Acronyms and Abbreviations

AE	Agronomic Efficiency. Calculated in units of yield increase per unit of applied nutrient, such as nitrogen
AEZ	Agro-ecological Zone
AMF	Arbuscular Mycorrhizal Fungi
AUE	Agronomic Use Efficiency
BAT	Best Available Technique
BNF	Biological Nitrogen Fixation
C	Carbon
CAP	Common Agricultural Policy of the European Union
CBD	Convention of Biological Diversity
CCAC	Climate and Clean Air Coalition
CE	Capture Efficiency, i.e., the amount of a nutrient in the harvested product compared with the total nutrient uptake by the crop
CH_4	Methane
CL	Critical Loads
CO_2	Carbon dioxide
CPR	Committee of Permanent Representatives of the United Nations Environment Programme
CRIN	Cacao Research Institute of Nigeria
CT	Conventional Tillage
DAP	Di-ammonium phosphate, used as a mineral fertilizer
DM	Dry Matter
DMPP	3,4-dimethylpyrazole phosphate—a nitrification inhibitor
DNMARK	Danish Nitrogen Mitigation Assessment
DON	Dissolved Organic Nitrogen
EANET	Acid Deposition Monitoring Network in East Asia
EC	European Commission
[eCO_2]	Elevated CO_2 levels
EMEP	European Monitoring and Evaluation Programme

ENA	European Nitrogen Assessment
EPNB	Expert Panel on Nitrogen Budgets of the TFRN
ES	Ecosystem Services
EU	European Union
EU27	European Union 27 Member States
EU-NEP	European Union Expert Nitrogen Panel
FACE	Free Air Carbon dioxide Enrichment ecosystem manipulation system
FAO	Food and Agriculture Organization of the United Nations
FYM	Farmyard Manure
GAW	Global Atmospheric Watch
GDP	Gross Domestic Product
GEF	Global Environment Facility
GHG	Greenhouse Gas
GIS	Geographic Information Systems
GMO	Genetically Modified Organism
GN	Groundnut
GPA	Global Programme of Action for the Protection of the Marine Environment
GPNM	UNEP Global Partnership on Nutrient Management
GS	Glutamine Synthetase
HAB	Harmful Algal Bloom
HB	Haber-Bosch
HI	Harvest Index
HLPF	High-Level Political Forum on Sustainable Development
IITA	International Institute of Tropical Agriculture
INC	Internal Nitrogen Cycle
INE	Internal Nutrient Efficiency
ING-SCON	Indian Nitrogen Group under the Society for Conservation of Nature
INI	International Nitrogen Initiative
INMS	International Nitrogen Management System
INS	Indigenous Nutrient Supply
IOC-UNESCO	International Oceanographic Commission of of the United Nations Educational, Scientific and Cultural Organization
IPBES	Intergovernmental Platform on Biodiversity and Ecosystem Services
IPCC	Intergovernmental Panel on Climate Change
ISFM	Integrated Soil Fertility Management
K	Potassium
KAP	Knowledge, Attitudes, Practices
LRTAP	UNECE Convention on Long-range Transboundary Air Pollution (informally the 'UNECE Air Convention')
m.a.s.l	Metres above sea level
miRNAs	Micro-Ribonucleic Acids

N	Nitrogen
N_2	Di-nitrogen, a colourless and odourless diatomic gas, forming about 78% of Earth's atmosphere
N_2O	Nitrous oxide—a powerful greenhouse gas
NB	(Partial) Nitrogen Balance, i.e., the difference between inputs (e.g., fertilizer, biological nitrogen fixation, manure) and outputs (crop harvest and other removed residues). May be defined at field, farm and regional scales
NBPT	N-(n-butyl) thiophosphoric triamide—a urease inhibitor that slows the conversion of urea to NH_x
NCE	Nitrogen Capture Efficiency, the amount of nitrogen taken up or 'captured' by a crop as a fraction of the N added as input to the soil (i.e., 'availability') from external supply and internal supply (mineralization)
NEWS India-UK	Indo-UK Virtual Nitrogen Centre on Nitrogen Efficiency in Whole-cropping Systems
NF_3	Nitrogen trifluoride
NGO	Non-governmental Organisation
NH_3	Ammonia—an air and water pollutant and the primary nitrogen form in biological systems.
NH_4^+	Ammonium—present in biological systems and soils, while forming a pollutant in atmospheric PM and aquatic systems
NH_x	Total ammoniacal nitrogen sometimes referred to as TAN
NI	Nitrification Inhibitor
Nnet	Nitrogen Human Environment Network
NO	Nitric oxide—a tropospheric air pollutant
NO_2	Nitrogen dioxide—a tropospheric air pollutant
NO_3^-	Nitrate—present as a secondary pollutant in atmospheric PM and a eutrophying pollutant of aquatic systems
NO_x	Nitrogen oxides—a combination of NO and NO_2
NPK	Nitrogen, Phosphorus and Potassium in combination
N_r	Reactive Nitrogen, a term used for a variety of nitrogen compounds that support growth directly or indirectly, as opposed to N_2 which is inert
NUE	Nitrogen Use Efficiency. Typically defined as the ratio of N in outputs divided by the N in inputs. May be defined for different systems such as crops, livestock, food chain and the whole economy
O_3	Ozone
OECD	Organization for Economic Co-operation and Development
P	Phosphorus
PM	Particulate Matter, which includes NH_4^+ and NO_3^- as major components; PM_{10} and $PM_{2.5}$ refer to atmospheric particulate matter (PM) that has a diameter of less than 10 and 2.5 μm respectively. PM_2 is also known as Fine Particulate Matter

POM	Particulate Organic Matter
PP	Pigeonpea
PUE	Phosphorus Use Efficiency
RCBD	Randomized Complete Block Design
RE	Recovery Efficiency, i.e., mass increase of nutrients in harvested crop as a fraction of the mass of nutrients applied
RF	Rain-Fed
RNA	Ribonucleic acid
SACEP	South Asia Co-operative Environment Programme
SANC	South Asian Nitrogen Centre
SDGs	Sustainable Development Goals
SI	Supplemental Irrigation
SOC	Soil Organic Carbon
SOM	Soil Organic Matter
SSA	Sub-Saharan Africa
SSC	Soil Supply Capacity, i.e., ability of the soil system to replenish a given plant nutrient in the soil solution for plant uptake
STFR	Soil Testing and Fertilizer Recommendation
TAN	Total Ammoniacal Nitrogen
TFRN	Task Force on Reactive Nitrogen of the UNECE Convention on Long-range Transboundary Air Pollution
TN	Total Nitrogen
TR	Tied Ridges
TSP	Triple Super Phosphate
UN	United Nations
UNDP	United Nations Development Programme
UNDSD	United Nations Division for Sustainable Development
UNEA	United Nations Environment Assembly
UNECE	United Nations Economic Commission for Europe
UNEP	United Nations Environment Programme
UNFCCC	United Nations Framework on Climate Change
UNIDO	United Nations Industrial Development Organization
USD	United States Dollars (US$)
VCR	Value-Cost Ratio, i.e., the ratio of the price of additional yield (e.g., crop yield increment) following application of inputs (e.g., fertilizer, but excluding seeds) to the cost of the inputs
WHO	World Health Organization
WMO	World Meteorological Organization

Chapter 1
Just Enough Nitrogen: Summary and Synthesis of Outcomes

Mark A. Sutton, Kate E. Mason, Albert Bleeker, W. Kevin Hicks, Cargele Masso, N. Raghuram, Stefan Reis, and Mateete Bekunda

Abstract Food production and power generation have increased to feed growing populations and to keep pace with economic development, leading to major human alteration of the global nitrogen (N) cycle. The result is a global challenge, with many regions having 'too much' or 'too little' nitrogen. As di-nitrogen (N_2) in the atmosphere, nitrogen is one of the most abundant elements, but which cannot be used

---

M. A. Sutton (✉) · K. E. Mason · S. Reis
UK Centre for Ecology & Hydrology, Edinburgh Research Station, Bush Estate, Penicuik, Midlothian, EH26 0QB, UK
e-mail: ms@ceh.ac.uk

K. E. Mason
e-mail: katson@ceh.ac.uk

S. Reis
e-mail: srei@ceh.ac.uk

A. Bleeker
National Institute of Public Health and Environment (RIVM), 1, 3720 BA Bilthoven, The Netherlands
e-mail: albert.bleeker@rivm.nl

W. K. Hicks
Stockholm Environment Institute (SEI), Department of Environment and Geography, University of York, York, YO10 5DD, UK
e-mail: kevin.hicks@york.ac.uk

C. Masso
International Institute of Tropical Agriculture (IITA), 1st Main IRAD Road, Nkolbisson, P.O. Box 2008 (Messa), Yaoundé, Cameroon
e-mail: C.Masso@cgiar.org

N. Raghuram
University School of Biotechnology, Guru Gobind Singh Indraprastha University, Dwarka Sec. 16C, New Delhi 110078, India
e-mail: raghuram@ipu.ac.in; raghuram98@hotmail.com

South Asian Nitrogen Centre, Indian Nitrogen Group, Sustainable India Trust and Society for Conservation of Nature, F-4, A-Block, NASc Complex, DPS Marg, Pusa, New Delhi 110 012, India

M. A. Sutton et al. (eds.), *Just Enough Nitrogen*,
https://doi.org/10.1007/978-3-030-58065-0_1

by most organisms. Conversely, reactive nitrogen (N_r) is essential for organisms, but is mostly in short supply for natural ecosystems. Human activities have polarized the differences in N_r flows between different world regions, leading to major sustainability challenges, with implications for food security, adverse impacts on health and ecosystems, and the need to develop tools and policies for better management. In developed regions, abundant use of manufactured fertilizers, crop biological nitrogen fixation and inadvertent formation of nitrogen oxides via combustion processes are leading to a plethora of environmental problems. These threaten air quality, water quality, soil quality, greenhouse gas balance, stratospheric ozone levels, biodiversity and human health. At the same time, in many developing regions, insufficient access to reactive nitrogen is leading to degradation of agricultural soils including N depletion, making it vital to reduce losses and recycle available nitrogen stocks. Nitrogen emissions as a result of agricultural practices and combustion for energy represent a major economic loss. Adding up all N losses in the world (excluding emissions from oceans) amounts to a lost agricultural fertilizer resource worth around $200 billion USD annually. The societal costs to human health, ecosystems and climate are even larger at $400–4000 billion USD annually. Knowledge of these figures can help motivate society to optimize with 'just enough' nitrogen. This chapter provides an overview of results from the 6th International Nitrogen Conference, Kampala (Uganda), which considered the question of how to optimize practices for 'just enough' nitrogen both internationally and specifically for the African Continent. From experimental trials to scenario analysis, the contributions demonstrate the approaches being used. The messages in very different regions often turn out to be surprisingly similar. They encompass all aspects of society: optimizing the use of available fertilizer and manure resources (both under excess and under scarcity conditions), improving nitrogen use efficiency, developing landscape integration, and optimizing our food choices by prior planning that can also reduce food waste. Together, such nitrogen-related strategies will have major benefits for global environmental sustainability.

Keywords Nitrogen · Environment · Nitrogen use efficiency · Regional assessment · Environmental economics · Pollution mitigation strategies

M. Bekunda
International Institute of Tropical Agriculture (IITA), Arusha, Tanzania
e-mail: M.Bekunda@cgiar.org

1.1 Introduction

The 6th International Nitrogen Conference was hosted in Kampala, Uganda (24–27th November 2013). This was the first time that the International Nitrogen Initiative (INI) held its triannual conference in Africa and provided a major opportunity to see a perspective that especially emphasized the situation of "too little" nitrogen (N). Across most of sub-Saharan Africa (SSA) access to mineral nitrogen fertilizers is limited, with the result that fertilizer rates are typically of the order of 8–10 kg N ha^{-1} $year^{-1}$ as compared with ten times this amount in many developed countries (Vitousek et al. 2009; Hickman et al. 2020). The meeting provided the opportunity to bring these issues together, especially as all the previous International Nitrogen Conferences had taken place in regions with too much nitrogen. In particular, it was asked: to what extent are messages for regions with too little nitrogen different to those with too much nitrogen?

While agricultural nitrogen flows provide the starting point for many of the science activities across the INI community, it has always been recognized by INI that the nitrogen challenge covers all nitrogen flows and all sectors of society (e.g., Galloway et al. 2008). In this way, all sources of nitrogen supply are considered relevant, including biological nitrogen fixation and manure recycling, as well as in the form of nitrogen oxides (NO_x) from combustion sources and recycling of nitrogen in food waste streams (Sutton et al. 2013). This implies that there are opportunities for all sectors of society to contribute to improved nitrogen management. Nevertheless, as footprint analysis has shown, including an analysis conducted for the Kampala conference (Leip et al. 2014), the overall food chain is especially relevant for nitrogen. It means that our dietary choices have a key impact on nitrogen needs, as well as the resulting nitrogen losses and pollution effects.

It was with this in mind that previous INI activities led to the establishment of the Barsac Declaration on Environmental Sustainability and the Demitarian Diet (Sutton et al. 2009), which pointed to the option to reduce meat consumption by half. Analysis by Westhoek et al. (2014, 2015) had shown that such a scenario in Europe (including the halving of dairy and egg intake) would lead to reductions in nitrogen pollution of 40% for ammonia and 25–40% for nitrate and nitrous oxide, while freeing up substantial agricultural land, currently used for growing livestock feed, for other agricultural products (including for bio-energy production).

In SSA, however, the issues were surely different. Here, many people do not have access to sufficient nutritious food, so how can one have a conversation about *reducing* meat and dairy consumption? In advance of the INI conference this seemed to be an extremely sensitive topic of discussion. In practice, the topic turned out to be an important conversation during the conference and one which could interact constructively with the science of how to improve agricultural nitrogen management, as well as how to reduce food waste. Following the experience from previous INI associated conferences, the approach taken was to work with the conference hotel and chef in advance to broadly halve the amount of meat intake compared with a typical conference at the same venue (the Speke Resort, Kampala). The outcomes of this experience have been analysed in detail by Tumwesigye et al. (2020), as summarized later in this chapter.

The main body of this volume is naturally dedicated to chapters describing results presented at the 6th International Nitrogen Conference. In addition, the opportunity has also been taken to summarize how the science is feeding into the regional activities of INI. This led to the development of six chapters on the global nitrogen challenge as it applies to each of the INI regions: Africa, East Asia, South Asia, Latin America, North America and Europe. The comparison of these chapters clearly illustrates the major challenges faced in these different regions of too much and too little nitrogen.

In the following sections, we summarize the major findings of the conference as detailed in the subsequent chapters of this book. The broad themes include: Food and Agriculture, Nitrogen Impacts on Health and Ecosystems, Management Tools and Assessment. Finally, we finish with the analysis of Tumwesigye et al. (2020) on consequences of dietary change during the conference, together with the overall conclusions of the INI conference as encapsulated in the Kampala Statement-for-Action on Reactive Nitrogen in Africa and Globally (Sutton et al. 2020b).

With these calls to action, the present volume completes the reporting of the INI Kampala Conference, which complements a parallel Special Issue of Environmental Research Letters, the results of which are summarized by Reis et al. (2016).

1.2 Food and Agriculture

1.2.1 Nitrogen and Food Security

We open the presentation of science findings with the chapter of Lassaletta et al. (2014, 2020) who assess how yields and nitrogen use efficiency (NUE) have changed across the world since 1961. They use statistics from the UN Food and Agriculture Organization (FAO) to assess trajectories in the relationship between nitrogen inputs to cropland and yields. Importantly, they identify how these trajectories can be characterized in four groups. In the first group, increasing N inputs increased yield, but substantially decreased NUE, substantially exacerbating pollution risks (e.g., India, China, Egypt). In the second group, increased N inputs provided higher return in yield, matched with stable or decreasing NUE (e.g., USA, Brazil, Bangladesh). In a third group (e.g., France, Netherlands, Greece) initially increasing N inputs were later reduced, while at the same time substantially increasing yield and NUE. Lastly, a fourth group showed less clear relationships, and was especially related to countries with low N inputs (e.g., Morocco, Benin, Nigeria), where other constraints such as inter-annual fluctuations in meteorology affected yield. The comparison of these four groups provides important lessons to share between countries, pointing at how to meet food security goals and reduce the polluting effects of N surplus. At the same time, the results show the continuing risk to sub-Saharan countries of soil N mining, where soils are used for crop production without soil nutrients being replenished after harvest. Further analysis of such datasets is needed to explore combinations of practices and N input regimes that find the balance between too much and too little N. In this regard, a key observation of Lassaletta et al. (2020) is that around 100 Tg N is excreted by the world's livestock each year, while only 25% of this is actually received as a N input by croplands.

The opportunity for improving nitrogen management practices related to improved yields is examined by Onyango et al. (2020), Famuwagun and Oladitan (2020), and Hickman et al. (2020).

Onyango et al. (2020) established an experiment to assess the performance of tied ridges (or furrow diking) as an adaptation strategy to improve resilience to climate change by reducing water loss. In this approach, each crop row is set on a ridge, with the intervening furrows between ridges blocked (or 'tied') at regular (1–2 m) intervals, so as to hold rain where it falls. The results of Onyango et al. (2020) offer a note of caution, highlighting how a careful targeting of climate adaptation measures is needed to the situations where they are most needed. In the case of tied ridges, the extra work involved and the lack of yield benefit argue against the practice in moist climates, unless it can have benefits as an 'insurance policy' for years where the rainy season fails.

Famuwagan and Oladitan (2020) emphasize the importance of different forms of fertilizer application, considering the effects of combined nutrition with nitrogen, phosphorus and potassium (NPK) on the performance of Cacao seedlings in Nigeria. The authors compared different rates for mineral NPK, when applied to the soil surface or foliage, with the effects of a proprietary organo-mineral fertilizer and of poultry manure. Considering a range of indicators, including seedling height, number of leaves, stem girth and root parameters, they found a rather poor reward for the mineral NPK treatments, as compared with the organic sources. The authors suggest that this may be due to added benefits of micronutrients, especially in the poultry litter, highlighting the importance of improving fertilizer formulation and application methods. While the authors suspected that significant losses from volatilization and leaching may be part of the cause of reduced yields, they also recognized the need for further tests that improve comparability of dosage rates.

Hickman et al. (2020) summarize the current state of knowledge of N dynamics in agricultural systems in SSA and assess the evidence for potential synergies and trade-offs from nitrogen use in Africa, stressing that fertilizer use in African agriculture is currently extremely low, with just 4% of global fertilizer use, and an annual average of only 8 kg nutrients ha^{-1} in SSA. They warn that increased N inputs may also be accompanied by a decline in N use efficiency, causing increased N losses to the environment with potential impacts on water and air quality, soil pH and biodiversity. As there have been fewer studies examining N cycling and losses in Africa compared with other parts of the world, they highlight the need for coordinated efforts to assess the implications of increased N use in Africa and fill information gaps. On the positive side, increased inputs can be expected to increase crop productivity in most soils, and may also increase soil organic matter, particularly when added with organic inputs or crop residues, which, in turn, can lead to increased water use efficiency. There is also potential for N losses from sewage effluent and agricultural activities to affect water quality in African coastal and aquatic environments and the potential for increased reactive nitrogen (N_r) emissions and deposition to terrestrial and aquatic ecosystems.

Van Grinsven et al. (2015, 2020) discuss the case to consider ‘sustainable extensification’ as an alternative strategy to the more commonly discussed paradigm of ‘sustainable intensification’ (e.g., Garnett and Godfrey, 2012). They conclude that, in Europe specifically, while extensification of agriculture may come at a cost of reduced yields, it could be a relevant strategy with positive environmental and biodiversity benefits, especially if it were combined with changes in diets leading to reduced meat and dairy intake. Such changes in consumption patterns as a consequence of reduced animal protein intake due to adopting a demitarian diet, may amplify or weaken these effects. Building on the work of Westhoek et al. (2014), these authors considered a demitarian scenario, where European meat and dairy intake were halved, linking this also with potential health benefits associated with avoidance of excessive intake.

1.2.2 Nitrogen Intensification and Biological Nitrogen Fixation in Low-Input Systems

Insufficient access to, and therefore limited use of manufactured fertilizers in developing countries has led to the identification of cheaper alternative sources of N_r, notably biological nitrogen fixation (BNF), the benefits of which from legume cropping have been recognized for centuries. It has been well established that BNF and consequent productivity of grain legumes and associated crops can be improved by use of rhizobium inoculants.

Abdullahi et al. (2020) present a timeline on research processes aimed at promoting use of rhizobia inoculants as a cheaper, easier and safer option to improve the N_2-fixation and productivity of grain legumes, using Nigeria as a case study. Initial studies were conducted on “US type” Soybean (*Glycine max* (L.) Merrill), which required specific inoculation with *Bradyrhizobim japonicum* for optimum productivity. During the 1980s, research introduced promiscuous soybean cultivars (Tropical Glycine Cross, TGx), which nodulated freely with the indigenous rhizobium population, fixed large amounts of atmospheric nitrogen and produced higher grain yields than the local genotypes. However, some experiments indicated up to 40–45% increases in yield by some of the TGx cultivars following inoculation. Hence, the ultimate BNF solution still remains the development of inoculants using highly effective indigenous rhizobia strains for legume crops. Studies with cowpea (*Vigna unguiculata* (L.) Walp), bambara groundnut (*Vigna subterranea* (L.) Verdc.), peanut (*Arachis hypogaea* L.) and common bean (*Phaseolus vulgaris* L.), showed responses to inoculation, mostly with imported inoculants.

In Malawi, Mhango et al. (2020) demonstrated the main drivers of BNF in sole or intercropped pigeon pea and groundnut as being plant density, inorganic phosphorus (P) and interspecific competition. The proportion of N derived from the atmosphere (22–99%) was influenced by soil P status across seasons and crop species, but not by the sole crop or intercrop system. Total N fixed, on the other hand, differed with cropping system in the dry year, where intercropping was associated with low levels of N fixed by pigeon pea (15 kg N ha^{-1}) compared to sole pigeon pea (32 kg N ha^{-1}). A short rainfall season could not support biomass production of pigeon pea and this has negative implications for relying on BNF to drive productivity on rain-fed smallholder farms.

Using the BNF technology as an example, Lege and Carpenter-Boggs (2020) give evidence on why nutrient intensification in low input systems should assess community knowledge, attitudes, practices (KAP) and resources alongside agronomic research. They argue that this is essential to improve the potential for long-term adoption of new production methods and technologies. In a collaborative study with ongoing development of improved bean varieties and inoculants for BNF, a KAP survey of farmers regarding bean production identified that farmers viewed crop growth and yield in different areas and over time as indicators of soil health, but considered crop pests, diseases and weather as greater limiters of bean production than soil health. When deciding which bean varieties to plant, market price and yields were the most important determining factors. By identifying current practices, beliefs, desires, and concerns of producers, research and extension on N_r intensification and crop improvement can become more effective.

The potential for BNF to be part of the solution to address wider challenges is investigated by Naluyange et al. (2020). These authors highlight a double challenge for the Lake Victoria basin, where soil nutrient resources are simultaneously being depleted, while Lake Victoria itself is increasingly suffering eutrophication problems from the run-off and atmospheric deposition of excessive nutrients. One of the features of eutrophication of Lake Victoria is the rapid growth of water hyacinth. The authors therefore established experiments to investigate whether harvesting water hyacinth from Lake Victoria could simultaneously contribute to lake restoration, while acting as a useful compost to improve crop yields (e.g., Naluyange et al. 2014). The results that they report in this volume focus on trials using the Pinto sugars bean (*Mwitemania*), where they combined a water hyacinth compost with inoculation by Rhizobium to promote BNF. They also co-added either cattle manure culture or effective microbes and compared performance with the use of diammonium phosphate fertilizer. While they found Rhizobium inoculation to increase root nodulation when grown with water hyacinth and effective microbes as compared with a control, there was no indication that water hyacinth increased key crop outcomes (flowering, number of pods, bean yield). The authors suggest that better results could be found for the most unfertile soils. Nevertheless, the findings highlight the difficulty of translating an aspirational approach into a viable economic practice. As the authors note, water hyacinth has a high water content and low nutrient content, while large scale harvesting and transport could be rather costly.

1.2.3 Improving Nitrogen Management in Fertilizers and Manures

Crop NUE is a key factor for improving N management and is crucial for sustainable food security, as well as for a sustainable environment. Sinha et al. (2020) consider the biological determinants of crop NUE and biotechnological avenues for improvement. They show how advances in our understanding of the functional biology of N-response and NUE in the last decade have led to the discovery of several candidate genes that could improve NUE through transgenic approaches. These genes would also be amenable to non-transgenic approaches for crop NUE improvement. However, they advise that despite these exciting developments, it may take a decade or more before the biologically improved crop varieties with high-NUE become widely available or used in most parts of the world. Thus, crop biotechnology may deliver in the medium to long term, but until then, improved fertilizer formulations and crop management practices for integrated nutrient management coupled with matching extension services to the farmers is the way forward.

The question of integrating mineral and organic manures is addressed by experiments reported by Lukin (2020). The author examined nitrogen budgets and used a lysimeter approach to evaluate nitrogen losses for different treatments to a grain rotation on light podzolic soils in the Russian Federation. Lukin considered different rates of N fertilizer addition and of organic manure addition and of several combinations. The highest rate of loss was associated with a combined treatment of Farmyard Manure (FYM) (10 tonne ha^{-1}) plus 100 kg N ha^{-1} of mineral fertilizer (ammonium nitrate), where over 60% of the input N was estimated to be lost. By examining the seasonal dynamics of soil nitrogen levels the author offers a basis for improved application timing with smaller losses. Annual nitrogen losses measured by the lysimeters (mainly as NO_3^-) ranged from 15 to 112 kg N ha^{-1}, indicating the potential for substantial tuning to reduce these losses. Overall Lukin found that a combination of mineral fertilizer and manure was associated with lower N losses than use of mineral fertilizer alone, while giving better yields than using FYM alone.

Musinguzi et al. (2020) evaluated the contribution of soil organic carbon (SOC) to enhance the recovery efficiency of applied N for sorghum production in various agro-ecological zones of Uganda. They found the agronomic efficiency of applied N was high under medium and high SOC conditions, whereas there was significant spatial variability of yield response to N rates irrespective of the SOC levels. At SOC levels >1.2%, increased sorghum response to applied N was much more noticeable. Further investigations to inform N rates based on SOC would be crucial to optimize recovery efficiency and agronomic efficiency of applied N inputs.

Kraaijvanger and Veldkamp (2020) demonstrate how different indicators available to evaluate efficiency and sustainability of fertilizer use illustrate complex feedbacks when using N and P fertilizers. They applied a suite of indicators including Agronomic Use Efficiency (AUE), Value-Cost-Ratio (VCR), Recovery Efficiency (RE), Capture Efficiency (CE), Soil Supply Capacity (SSC) and (partial) Nutrient Balances (NB) for Ethiopian wheat, teff and hanfets crops. They found significant differences between crops for four out of the six indicators, with wheat appearing to be the most 'nutrient extractive crop': i.e., the one that it is most effective at taking nutrients out of the soil. Correlations between SSC and N-total, and RE and N-uptake were significant for all crops. For both nitrogen and phosphorus, NB correlated significantly with SSC for wheat and teff. Interactions between SSC, RE and NB demonstrated a significant trend for wheat: soils with higher SSC had lower NB and higher RE than soils with lower SSC. These observations point to dynamic trade-offs existing between soil nutrient stock sustainability and nutrient use efficiency, and to the need for Integrated Soil Fertility Management (ISFM) strategies. Such ISFM strategies look for complementarity between the use of all on-farm nutrient resources to arrive at a sound balance between efficiency and sustainability of fertilizer use. They note that such an approach may go some way to addressing a concern expressed by Ethiopian farmers that use of mineral fertilizers creates 'an addiction', whereby, once started, fertilizer inputs need to be continued to maintain yields.

Onaga et al. (2020) report on field experiments to establish upland rice crop nutrient requirements in Uganda using the popular rice NERICA 4 and 36 combinations of NPK fertilizers under supplementary irrigation (SI) and rain-fed (RF) conditions. They applied a factorial design by partially employing the nutrient omission technique. Their data show that applying N beyond 80 kg N ha^{-1} at the current P and K recommendation of 40 kg ha^{-1} for upland rice is less profitable in RF and SI conditions and may contribute to excess nitrogen with negative environmental consequences. However, the low agronomic efficiency and harvest index at higher NPK rates suggests the need for improvement of NERICA 4 or deployment of rice cultivars with higher agronomic efficiency (> 25 kg^{-1}).

Kathuli et al. (2020) assessed alternative sources of N that would represent a significant contribution to plant N nutrition in low-input farming systems in World regions like sub-Saharan Africa. They found that leaf prunings of the tree legume Quickstick (*Gliricidia sepium*) mineralized very fast and could supply N to Sweetcorn (maize) in a timely way, provided that sufficiently frequent application of the prunings was made. Sweet corn yield and dry matter were significantly enhanced using the *Gliricidia sepium* pruning. However, the recovery efficiency of N from prunings was low (< 10%), and further investigation on good agronomic management of the prunings would be required to improve the NUE derived from *Gliricidia sepium* prunings. They conclude that understanding of the release pattern of N from the prunings under variable agro-ecological zones would be crucial to inform the application frequency and rate to optimize the agronomic use efficiency, while minimizing environment losses, such as those related to residual N from mineralization of the prunings.

1.3 Nitrogen Impacts on Health, Ecosystems and Climate

1.3.1 Nitrogen Impacts on Health and Ecosystems

In contrast to a steadily growing demand for N fertilizers and the extreme 'leakiness' of nitrogen use in agriculture, global phosphorus (P) losses to the environment have been seen to level off as industrialized nations reduced P use in detergents and upgraded sewage treatment processes in the mid-1980s and 1990s. Glibert et al. (2014, 2020) relate this increase in N:P ratio to the occurrence and proliferation of harmful algal blooms (HABs) in water bodies including lakes, rivers and coastal waters bringing about large negative economic and ecological impacts. As an example, Glibert et al. (2014, 2020) point to how fertilizer use in China has increased from 0.5 Mt in the 1960s to 42 Mt in 2010, with the use of urea increasing five-fold in the last two decades (IFA 2014). They show how this has led to nitrogen export during this period increasing from 500 to 1,200 kg N km^{-2} in the Yangtze River catchment, with an increase from 400 to >1,200 kg N km^{-2} in the Zhujiang (Pearl) River catchment (Ti and Yan 2013).

1.3.1.1 Human Health

Brender (2020) reviews human exposure to nitrogen pollution via nitrate in food and water and nitrogen oxides in ambient air. Recent data indicate that nitrogen dioxide inhalation may reduce lung function in children, increase risk of myocardial infarction, increase morbidity and mortality in persons with heart failure, and exacerbate symptoms in children with asthma. Higher intake of nitrate/nitrite via food and water has been associated with methemoglobinemia in infants and young children and altered thyroid function in children and adults. The review shows that harmful health effects of nitrogen exposure may be contingent on the presence of other substances and may partly be due to other co-pollutants rather than nitrogen. The review highlights that harmful effects might occur in people with deficient intake of vitamins and other nutrients, such as vitamin C, and that further research on the impact of too much nitrogen on human health must recognize that nitrogen is essential for life and, in the right amounts as part of a balanced diet, is beneficial to health.

1.3.1.2 Coastal Seas

Du (2020) synthesizes data from various studies to assess the status of N deposition to China's coastal seas and its ecological impacts. Overall, the average total N deposition to China's coastal seas during the 2000s was around 20 g N ha^{-1} $year^{-1}$, leading to an estimate of total atmospheric inputs of 4.56 Tg N $year^{-1}$, with two thirds contributed by wet deposition. The effects of N deposition on primary production in China's seas differ by region due to variations in background nutrient status and nitrogen deposition can also affect phytoplankton composition via regulating competitive interactions. However, the chapter shows that large uncertainty still exists in the magnitudes, patterns, trends and ecological impacts of N deposition and recommends that a long-term network be established to monitor wet and dry N deposition and that modelling tools are needed to combine site-observed data at regional scale. The author proposes that novel experiments are needed to assess the integrative impacts of N deposition and that critical loads should be determined for each coastal region to improve the marine ecosystem management.

1.3.1.3 Freshwater

In a high-resolution global study of anthropogenic nitrogen loads to fresh water systems, Mekonnen and Hoekstra (2020) report that, at a spatial resolution of 5 by 5 arcminute, N load from diffuse and point sources in the period 2002–2010 was 32.6 million tonnes per year. China is estimated to contribute about 45%, followed by the USA (7%), Russia (6%) and India (5%) to this amount. Three quarters of the N loads came from diffuse sources in agriculture, 23% from domestic point sources and 2% from industrial point sources. Among the crops, production of cereals had the largest contribution to the N loads (18%, of which 7% was from wheat and 6% maize production), followed by vegetables (15%) and oil crops (11%). Globally, the authors estimate that 18% of the total N input on crop fields in the form of artificial fertilizer and manure leaches to freshwater systems.

1.3.1.4 Terrestrial Ecosystems

The importance of nitrogen deposition as the third most important driver affecting biodiversity after land-use and climate change (Sala et al. 2000) is highlighted in the chapter of García-Gómez et al. (2020), who focus their attention on Mediterranean conditions in Spain. They point out that little is known about nitrogen deposition effects on such ecosystems, despite the very high biodiversity of this region. While NO_x emissions have reduced in the region, NH_3 emissions from Spain have increased in the last decade. The authors highlight the importance of dry deposition compared with wet deposition of N in Spanish forests, such as for *Quercus ilex* and *Pinus halapensis*, while the quantitative estimation remains uncertain. Nevertheless, inputs exceed estimated critical loads and the authors report an ongoing increase in nitrophilous species, as well as in N content of moss species. In parallel, they find an increase in NO_3^- concentrations in headwater streams, where there is little agricultural activity, which therefore is thought to be related to atmospheric N deposition. Overall, they find natural grasslands to be the most threatened habitat from N deposition in this region.

1.3.2 Nitrogen, Climate Change and Trace Gas Enrichment

Hayashi et al. (2020) open the topic of climate change interactions in this volume by summarizing an ambitious programme to address how future atmospheric CO_2 levels may alter the nitrogen balance of rice paddy systems. They outline the main features of a Free Air CO_2 Enrichment (FACE) system and compare the effects with those measured under ambient conditions. The results presented by Hayashi et al. (2020) demonstrate the complexity of the challenge to measure the major nitrogen flows as well as to address the system responses. One of these challenges is that the most reliable estimates of N flows require field-scale micrometeorological measurements, such as those for ammonia (NH_3) fluxes reported by the authors. These show both substantial emissions (following fertilization) and substantial deposition (during fallow periods), with deposition rates up to 0.06 $\mu g\ m^{-2}\ s^{-1}$. Considering the matching NH_3-N concentrations of c. 3.5 $\mu g\ m^{-3}$, their results imply deposition velocities of up to 18 mm s^{-1}, which are characteristic of rates implying substantial uptake of NH_3 directly on to leaf surfaces (i.e., not limited by stomatal uptake).

Such detailed dynamics apply for each of the nitrogen compounds, illustrating the ambition of the authors to provide a complete picture, as well as developing methods to measure the component fluxes at the plot scale. For example, the authors describe the NH_3 compensation point as an indicator (i.e., the NH_3 concentration below which emission occurs and above which deposition occurs, which may differ at leaf and canopy scales). They found that this term was not affected by elevated CO_2 concentration, even though CO_2 did decrease photo-respiratory production of NH_3, relevant for cell metabolism. In the case of N_2O and denitrification fluxes the authors applied labelled isotope methods to discriminate effects. They also used a resistance modelling approach to simulate NH_3 fluxes, pointing to the potential for relationships with canopy structure to be used to minimize emissions to the atmosphere.

Calleja-Cervantes et al. (2020) have estimated nitrous oxide (N_2O) emissions from forests, grasslands and agricultural soils in northern Spain, including forests soils (*Pinus radiata* and *Fagus sylvatica*) at different growth stages and grassland and agricultural soils in the edapho-climatic conditions of northern Spain. They show that N_2O losses from forests in northern Spain are limited by atmospheric N deposition, resulting in lower losses than temperate forest in the rest of Europe. The results suggest that an emission factor of around 0.1–0.3% (depending on crop type, time of year and fertilizer type) should be applied for this area instead of the default value of 1% suggested by the Intergovernmental Panel on Climate Change (IPCC). The authors also suggest that the nitrification inhibitor 3,4-dimethylpyrazole phosphate (DMPP) can potentially be used to mitigate N_2O emissions, both in grasslands and in crops. Sewage sludge application should also be further studied, but the authors conclude that significant reduction in N_2O emission is possible with DMPP.

Pepó (2020) reports that climate change and crop year exert strong effects on the yield and fertilizer N use efficiency in winter wheat production in Hungary. Using studies on chernozem soil in long-term experiments in 2007 and 2008, the author reports that the crop year and the climatic factors (mainly the water supply:

quantity of rainfall and its distribution) exert strong effects on the natural nutrient utilization, yield surpluses of N-fertilization, the maximum yield and the optimum NPK fertilizer doses of different winter wheat genotypes. Pepó demonstrates that by harmonizing the optimal agrotechnical factors (irrigation, crop rotation, fertilization) yields can be protected even in unfavourable, dry crop years, though this needs extremely intensive agro-technical management and high inputs. The efficiency of fertilization was strongly modified by crop year, while the varied response of the tested wheat genotypes indicates the importance of variety-specific fertilization in sustainable wheat production.

1.4 Management Tools and Assessment

This part of the book explores examples of different nitrogen management and assessment approaches. Concerning developing approaches to nitrogen management, the contributions include a national approach from Denmark that considers all nitrogen sources, a case study from North West Russia on the challenges of ensuring effective manure utilization, and a discussion on the potential contribution of agroforestry systems for improved nitrogen capture and utilization. The subsequent chapters then contrast global and national assessments, followed by overviews for each of the main world regions.

1.4.1 Implementing Nitrogen Management Policies

Dalgaard et al. (2020) demonstrate the case for developing a more integrated approach to nitrogen management for Denmark, implemented through the *dNmark* initiative. They take a multi-actor approach linking science, business and government as a basis to assess past regulatory performance through the analysis of long-term trends, combined with indicator development and evaluation of approaches for improved nitrogen management. They took the European Nitrogen Assessment (ENA, Sutton et al. 2011) as a starting point for recommendations to improve nitrogen management, and use this to assess both national scale and landscape flows. Their overall nitrogen budget for Denmark illustrates the opportunities that could be achieved through more efficient nitrogen cycling in production chains, landscape differentiation of the possible measures incorporating spatial planning, and the results of changed consumption patterns (including food choice and waste reduction). The chapter provides an overview of the *dNmark* process, where further ongoing findings are presented by Dalgaard et al. (2017).

1.4.2 Manure Management

Analysis of statistical data and field studies have revealed the problems associated with agricultural manure utilization in the North-West Federal District of Russia (Briukhanov et al. 2020). Large-scale livestock farms face a difficult task as the vast majority of them do not have sufficient agricultural land to utilize the amount of animal and poultry manure produced. This surplus results in accumulation of animal and poultry manure near the farms and increases the risk of nutrient discharge into water sources and ammonia emissions into the atmosphere. The results suggest that animal/poultry manure processing and use as organic fertilizer is unprofitable unless financial support systems are in place. Such financial support would help catalyse the introduction of agricultural technologies with minimal nitrogen and other nutrient loss. According to preliminary estimates, if such financial support were to be put in place, the input of nitrogen into the environment may be reduced by up to 30%. Ultimately, the goal would be for such 'circular economy' approaches to become financially self-standing, but it is clear that much greater investment is needed in the first instance.

1.4.3 Multi-species and Agroforestry Systems

Multi-species systems include herbaceous mixtures (cover cropping, living mulches, intercropping), woody mixtures and herbaceous-woody mixtures, i.e., "agroforestry", and have the potential to improve N availability to both the tree and herbaceous components, and to increase N-conservation and recycling. Lawson et al. (2020) show how this can be achieved through factors such as deeper and more extensive root distributions, greater root and shoot turnover, N-fixation and mycorrhizal associations, greater interception of light and water resources, improved soil structure and organic matter content, and control of erosion and leaching. Ammonia recapture by tree-foliage from livestock emissions reduces N deposition on nearby sensitive ecosystems, and may contribute to reductions in net N_2O emissions by providing an additional biomass nitrogen sink. Trees, especially if they are not N-fixers, are also likely to reduce soil N_2O emissions by drying the soil, increasing soil-aeration, and removing excess nitrate concentrations throughout most of the year. Agroforestry therefore is identified by the authors as one of the best options for climate change mitigation and adaptation. At the same time, they recognize that more measurements are required on fluxes of N_2O and CH_4 in different types and ages of agroforestry system. Another issue that requires further consideration is the impact of agroforestry systems on soil nitric oxide (NO) emissions, especially as Butterbach-Bahl et al. (2011) concluded that emission factors for NO from forested systems are higher than had previously been thought. Such soil sources of NO are becoming more relevant in Europe, where they are contributing to an increasing share of total NO_x emission as policies reduce NO_x emissions from combustion sources (Sutton et al. 2017).

1.5 National, Regional and Global Nitrogen Assessment

The global flows of nitrogen and phosphorus pollution are assessed by Grizzetti et al. (2020). Humans have altered the natural nitrogen and phosphorus biogeochemical cycles by the massive input of fertilizers to the agricultural system to boost production. As a result a large amount of nitrogen and phosphorus have been mobilized and delivered to the environment, creating threats to aquatic and terrestrial ecosystem functioning and human health. Based on recent studies published in the literature, Grizzetti et al. (2020) describe and estimate the main fluxes of the current global nitrogen and phosphorus cycles (representative of the period 2000–2010). Overall, the efficiency of nutrient use is very low: considering the full chain, on average over 80% of nitrogen and 25–75% of phosphorus that are consumed end up lost to the environment (when not temporarily stored in agricultural soils). This causes pollution through emissions of the greenhouse gas nitrous oxide (N_2O) and ammonia (NH_3) to the atmosphere, and losses of nitrates (NO_3^-), phosphate and organic N and P compounds to water.

Cordovil et al. (2020) quantify the nitrogen footprint for the city of Lisbon and for Portugal, based on the N-Calculator developed by the University of Virginia USA, which transforms data on food and energy consumption in kg nitrogen lost to the environment by year. In order to calculate an average individual's N footprint in the area of Lisbon, 1,000 surveys were conducted. Based on the survey results, the average N footprint for Portugal was estimated at 25.5 kg N capita^{-1} year^{-1}, while the average personal nitrogen footprint for the Lisbon area was calculated to be 24.7 kg N capita^{-1} year^{-1}. Overall, the average nitrogen footprint for men and women was found to be similar. The average individual´s N footprint the age group of over 65 was 23.5 kg N capita^{-1} year^{-1}, while for a younger demographic (people 21–30 years of age) a higher value of 25.8 kg N capita^{-1} year^{-1} was found. As a general conclusion, the authors note that the losses of N to the environment in Portugal were higher for food consumption than for energy use.

1.5.1 International Nitrogen Initiative Regional Centres

Under the International Nitrogen Initiative (INI) six regional centres have been established. These centres facilitate an enhanced cooperation and integration among researchers, policy makers and practitioners on environmental issues related to reactive nitrogen in the different regions. They have all been working on developing nitrogen assessments to various extents. The following chapters describe these regional activities, followed by an overarching summary of the global activities ongoing under INI.

The chapter of Winiwarter et al. (2020) characterizes Europe as a region of agronomic and environmental challenges posed by high population density and an associated large food demand. The authors point to it as being largely an area of excess nitrogen, a fact that is increasingly being recognized by stakeholders and environmental policy. INI-Europe has been instrumental in awareness building and provision of scientific information, which has facilitated implementation of measures to reduce environmental nitrogen loads and associated impacts. This has been accomplished by stimulating, inventorying, reviewing and synthesizing work in a number of platforms including through EUROSTAT and OECD, the Task Force on Reactive Nitrogen (TFRN) and its expert panels operating under the 'Geneva Air Convention'—the UNECE Convention on Long-Range Transboundary Air Pollution (LRTAP). An important outcome has been the publication of the first European Nitrogen Assessment (Sutton et al. 2011), in which much of the European knowledge on all aspects of nitrogen has been brought together.

While INI-Europe represents a region with too much nitrogen in the environment, INI-Africa is addressing nitrogen management more from a perspective of too little nitrogen. Masso et al. (2020) explain how insufficient reactive nitrogen inputs in food production have been associated with chronic food insecurity and malnutrition. At the same time, there is nevertheless still too much nitrogen load in water bodies, which is mainly due to excessive soil erosion, leaching, and limited nitrogen recovery from wastewater. In addition, the authors highlight that atmospheric nitrogen deposition contributes to eutrophication in some areas, reflective of the losses and dispersal of N from agricultural N volatilization and biomass burning. While significant research has been conducted to improve N use for production, the authors point out that the 'too much' N pollution remains an important issue in Africa, where much more scientific attention is needed. Masso et al. explain how INI-Africa is aiming at addressing the current research gaps, as well as the operationalization of supportive policies, related to the nitrogen paradox of 'too little' and 'too much'. According to the authors, innovation platforms involving key stakeholders in Africa are required to address the full chain of nitrogen use efficiency.

Raghuram et al. (2020) describe the way in which INI-South Asia has been working on the implementation of a comprehensive assessment of reactive nitrogen in India (Abrol et al. 2017) and showcase the local and regional capacity for science-led policies for sustainable N management across South Asia. Furthermore, INI-South Asia has been instrumental in mainstreaming the importance of reactive nitrogen in India and South Asia through advocacy, workshops and publication. A scoping study on nutrient management for the South Asian Cooperative Environment Programme (SACEP) has led to the intergovernmental recognition of sustainable management of nitrogen and other nutrients at the ministerial level (SACEP 2014). Over the last few years, INI-South Asia has contributed as a core partner to the development of the International Nitrogen Management System (INMS).

Yan et al. (2020) describe how East Asia is one of the most densely populated regions in the world, and is also a region with intensive nitrogen loads. Nitrogen-related environmental problems, such as surface water eutrophication and atmospheric smog are of common concern in the region. Efforts are being made by INI-East Asia to investigate how a balance can be found between the beneficial effects of nitrogen for food security and its adverse environmental impacts. One of the important messages of Yan et al. (2020) is that there are similar challenges faced by China and Japan, for which they give a special focus of comparison. They highlight the cross cutting importance of nitrogen for air pollution, water pollution and climate. For example, drawing on an analysis of Gu et al. (2012), the authors point out that estimated health-damage related to atmospheric nitrogen alone reached \$19 to \$62 billion USD in 2008, representing around 1% of China's gross domestic product. Although figures are not available for Japan, major economic losses are also expected. Even if nitrogen air pollution levels in Japan are half those of China, the impacts in China and Japan for 2016 would be valued at around \$110 billion and \$25 billion USD, respectively.

Baron and Davidson (2020) describe the main activities of INI in North America, a region where large challenges exist in reducing nitrogen loss from all parts of the North American food production and supply chain, including consumers. As with Europe, the nitrogen challenge in North America is one of excess. The authors emphasize how a technologically advanced modern lifestyle and protein rich diets cause more reactive nitrogen to be lost to the North American environment than is produced in usable goods and services. To address the different nitrogen related challenges, INI-North America is working on three main topics: (1) conducting assessments on nitrogen flows within North America and the consequences for human health, water resources, biodiversity, and greenhouse gas emissions; (2) facilitating efforts to develop solutions to the problem of excess nitrogen in agricultural, institutional, and natural resource management sectors; and (3) presenting these results to policy makers. Baron and Davidson (2020) explain how INI-North America is working with producers, trade groups, universities and supply chains to develop effective practices for minimizing the loss of reactive nitrogen to the environment. Furthermore, INI-North America is helping public land management and regulatory agencies prepare effective policy approaches toward minimizing ecological damage from atmospheric reactive nitrogen deposition.

Ometto et al. (2020) describe the situation in Latin America, where a general lack of information on the nitrogen cycle is seen as a serious impediment to evaluate how human activity is altering nitrogen pools and turnover at regional and global scales. Empirical measurements of nitrogen deposition and other nitrogen cycle processes are also scarce in Latin America (Austin et al. 2013). Data feeding into global and regional circulation models lack spatial distribution information for this region. The authors explain how the data and information scarcity is something that INI-Latin America is addressing. Trying to overcome this, INI-Latin America is working with the Nitrogen Human Environment Network (Nnet), which aims to gather scientific information, acquire new data and inform the policy processes on the nitrogen budget and nutrient management in a broad region. Overall, the purpose of Nnet is to examine human impact on natural and modified ecosystems across a wide range of climates, ranging from direct measurements to regional modelling, aspiring to build a greater understanding of how nitrogen excess or shortage affects ecosystem processes and how it relates to biodiversity, food production and environmental pollution.

Sutton et al. (2020a) show how these regional contributions are contributing to the overarching global programme of the International Nitrogen Initiative. Specifically, they show how INI is developing a coordinated approach of science support for international policy development through the International Nitrogen Management System (INMS), which has recently been established in partnership with UN Environment and with the support of the Global Environment Facility. One of the main challenges emphasized by Sutton et al. (2020a) is the multi-dimensional nature of the global nitrogen challenge. The authors highlight how nitrogen pollution is relevant for many multi-lateral environmental agreements for water, air pollution, climate, biodiversity and stratospheric ozone depletion, as well of course for food and energy security. In this way, they highlight how the challenges of too much and too little nitrogen cut across most of the UN Sustainable Development Goals.

The authors point to the development of the multi-actor platform provided by INMS, working to deliver guidance on practices for better nitrogen management, and developing the first International Nitrogen Assessment as a means to better understand the scientific synergies and catalyse change for better nitrogen management (Sutton et al. 2020a). By acting in synergy with the regional contributions of INI described above, a mutual exchange between global, regional and local action is anticipated. One way that the authors emphasize the urgency of this challenge is to highlight the value of nitrogen in economic terms. For example, they estimate that emissions from agriculture alone represent a lost agricultural fertilizer resource worth around $200 billion USD annually. A goal to halve nitrogen waste would therefore save $100 billion USD annually. The societal costs to human health, ecosystems and climate are estimated to be even bigger, at $400–4000 billion USD annually. It remains to be seen which of these valuations will provide the most powerful driver for change. The authors finish by reporting the recent adoption of a first ever Resolution on Sustainable Nitrogen Management by the United Nations Environment Assembly (UNEP/EA.4/L.16). This is now setting a path for the next steps including examination of the case to establish an Interconvention Nitrogen Coordination Mechanism.

1.6 Conclusions and Outlook

This volume closes its author contributions with the contribution of Tumwesigye et al. (2020). As outlined in the introduction to the present chapter (Sect. 1.1), these authors took the 6th International Nitrogen Conference itself as their study material. The challenge was to see what it would mean to organize a major international event in Africa following the principles of the Barsac Declaration i.e., making available nutritious food to the delegates, but doing this with half the usual local amount of meat products. What would this mean in terms of reduced nitrogen footprint? And, just as importantly, how would the delegates react in African context, where many lack a sufficient balanced diet?

The findings of Tumwesigye et al. (2020) are just as impressive as the reaction from the Speke Resort hosting the conference. The hotel chef also engaged enthusiastically, welcoming the interest shown by the conference in food choice. According to the calculations of Tumwesigye et al. (2020), the average delegate in the Nitrogen Conference consumed 118 g meat per day (dry weight basis), representing 50% of the baseline conference intake at the Speke Resort. Milk consumption was 75 g per person per day (fresh weight), which was 47% of the baseline conference, while fruit and vegetable intake correspondingly increased by 48% and 36%, respectively. Taking account of the increased replacement in plant-based foods, the overall nitrogen footprint of the conference was reduced by 40%. It was particularly interesting that the Nitrogen Conference showed an increase in uneaten meat compared with the baseline conference, even though the overall food supply was lower than the baseline conference. This was attributed to prior awareness raising of the delegates to the food discussion, implying that the overall meat supply could have been reduced by even more than half, while fully satisfying the delegates' appetite.

The authors conclude by emphasizing the importance of moderating dietary intake of meat products by the most affluent in developing societies, given the contribution of high meat intake to a wide range of nitrogen pollution problems and human health effects. Their findings highlight the importance of demography: the challenge is not just to manage average intake by citizens, but to understand the differences between different sectors of society, especially as this affects their aspirations for a better life.

The volume closes with the conclusions as agreed in in the final plenary of the conference. These are summarized in the Kampala Statement for Action on Nitrogen in Africa and Globally (Sutton et al. 2020b). The actual conclusions are relatively short, emphasizing the importance of improving nitrogen efficiency and decreasing pollution, and setting the special challenges for sub-Saharan Africa in the global context. The opportunity was also taken to incorporate a summary of previous international statements on nitrogen, providing a longer background to the Kampala Statement for Action. We let the conclusions speak for themselves, as they provide an important reference point, especially to ensure that future development in Africa finds a way to improve overall food and energy security for its citizens, while not exacerbating the already significant nitrogen pollution and other sustainable development challenges.

As the global community works to address the UN Sustainable Development Goals, this volume emphasizes that nitrogen cuts across most of the challenges. In this way, finding the solutions for 'Just Enough Nitrogen' will be at the heart of making substantial progress toward sustainable development.

Acknowledgements Financial support for the 6th International Nitrogen Conference, Kampala, was generously provided through the Alliance for a Green Revolution in Africa (AGRA), the International Fertilizer Industry Association (IFA), the International Plant Nutrition Institute (IPNI), Africa Research in Sustainable Intensification for the Next Generation (Africa RISING), the National

Agricultural Research Organisation of Uganda (NARO), the Department of Agricultural Production of Makerere University, the Global Partnership on Nutrient Management (GPNM) in cooperation with the Scientific Committee on Problems of Environment (SCOPE), the UK Centre for Ecology & Hydrology (CEH), the European Commission Joint Research Centre (JRC), the N2Africa project, and the International Centre for Tropical Agriculture (CIAT). We gratefully acknowledge the subsequent associated funding for INMS, firstly through the UK Natural Environment Research Council (International Opportunities Fund) and subsequently from the Global Environment Facility through UN Environment, together with the Global Challenges Research Fund (GCRF) South Asian Nitrogen Hub, which have contributed to the preparation of this chapter.

References

Abdullahi, A. A., Howieson, J., O'Hara, G., Terpolilli, J., Tiwari, R., Yusuf, A. A. (2020). History of Rhizobia inoculants use for improving performance of grain legumes based on experience from Nigeria. In M. A. Sutton, K. E. Mason, A. Bleeker, W. K. Hicks, C. Masso, S. Reis, M. Bekunda (Eds.), *Just enough nitrogen. Perspectives on how to get there for regions with too much and too little nitrogen (Proceedings of the 6th International Nitrogen Conference).* Chapter 7 (this volume). Springer.

Abrol, Y. P., Adhya, T. K., Aneja, V. P., Raghuram, N., Pathak, H., Kulshrestha, U., Chhemendra, S., et al. (Eds.). (2017). *The Indian Nitrogen Assessment: Sources of reactive nitrogen, environmental and climate effects, management options, and policies.* Elsevier.

Austin, A. T., Bustamante, M. M. C., Nardoto, G. B., Mitre, S. K., Pérez, T., Ometto, J. P. H. B., et al. (2013). Latin America's nitrogen challenge. *Science, 340*(6129), 149.

Baron, J. S., & Davidson, E. A. (2020). The INI North American Regional Nitrogen Center: 2011–2015 nitrogen activities in North America. In M.A. Sutton, K.E. Mason, A. Bleeker, W.K. Hicks, C. Masso, S. Reis, M. Bekunda (Eds.), *Just enough nitrogen. Perspectives on how to get there for regions with too much and too little nitrogen (Proceedings of the 6th International Nitrogen Conference).* Chapter 34 (this volume). Springer.

Brender, J. D. (2020). Human health effects of exposure to nitrate, nitrite, and nitrogen dioxide. In M. A. Sutton, K. E. Mason, A. Bleeker, W. K. Hicks, C. Masso, S. Reis, M. Bekunda, (Eds.), *Just enough nitrogen. Perspectives on how to get there for regions with too much and too little nitrogen (Proceedings of the 6th International Nitrogen Conference).* Chapter 18 (this volume). Springer.

Briukhanov, A., Vasilev, E., Kozlova, N., Maksimov, D., Shalavina, E., Subbotin, I. (2020). Farm level assessment of nitrogen use efficiency as part of environmental management. In M. A. Sutton, K. E. Mason, A. Bleeker, W. K. Hicks, C. Masso, S. Reis, M. Bekunda (Eds.), *Just enough nitrogen. Perspectives on how to get there for regions with too much and too little nitrogen (Proceedings of the 6th International Nitrogen Conference).* Chapter 26 (this volume). Springer.

Butterbach-Bahl, K., Nemitz, E., Zaehle, S., Billen, G., Boeckx, P., Erisman, J. W., et al. (2011). Nitrogen as a threat to the European greenhouse gas balance. In M. A. Sutton, C. M. Howard, J. W. Erisman, G. Billen, A. Bleeker, P. Grennfelt, H. van Grinsven, B. Grizzetti (Eds.), *The European Nitrogen Assessment. Sources, effects and policy perspectives* (Chapter 19, pp. 434–462). Cambridge University Press.

Calleja-Cervantes, M. E., Huerfano, X., Barrena, I., Estavillo, J. M., Aparicio-Tejo, P. M., Gonzalez-Murua, Menéndez, S. (2020). Nitrous oxide (N_2O) emissions from forests, grasslands and agricultural soils in northern Spain. In M. A. Sutton, K. E. Mason, A. Bleeker, W. K. Hicks, C. Masso, S. Reis, M. Bekunda (Eds.), *Just enough nitrogen. Perspectives on how to get there for regions with too much and too little nitrogen (Proceedings of the 6th International Nitrogen Conference).* Chapter 23 (this volume). Springer.

Cordovil, C. M. d. S., Gonçalves, V., de Varennes, A., Leach, A. M., Galloway, J. M. (2020). A first approach to the calculation of nitrogen footprint in Lisbon, Portugal. In M.A. Sutton, K. E. Mason, A. Bleeker, W. K. Hicks, C. Masso, S. Reis, M. Bekunda (Eds.), *Just enough nitrogen. Perspectives on how to get there for regions with too much and too little nitrogen (Proceedings of the 6th International Nitrogen Conference).* Chapter 29 (this volume). Springer.

Dalgaard, T., Olesen, J. E., Schjørring, J. K., Jensen, L. S., Vejre, H., Andersen, P. S., Gundersen, P., et al. (Eds.) (2017). *Innovative solutions for sustainable management of nitrogen. Proceedings from the International Conference, Aarhus, Denmark, 25–28 June 2017, and the Following United Nations Economic Commission for Europe Task Force on Reactive Nitrogen Meeting (TFRN-12, 29–30 June 2017.* Denmark: Aarhus University and the DNMARK Research Alliance.

Dalgaard, T., Brock, S., Hansen, B., Hasler, B., Hertel, O., Hutchings, N. J., Jacobsen, B., et al. (2020). DNMARK: Danish nitrogen mitigation assessment: Research and know-how for a sustainable, low-nitrogen food production. In M. A. Sutton, K. E. Mason, A. Bleeker, W. K. Hicks, C. Masso, S. Reis, M. Bekunda (Eds.), *Just enough nitrogen. Perspectives on how to get there for regions with too much and too little nitrogen (Proceedings of the 6th International Nitrogen Conference).* Chapter 25 (this volume). Springer.

Du, E. (2020). Nitrogen deposition to China's coastal seas: Status and ecological impacts. In M. A. Sutton, K. E. Mason, A. Bleeker, W. K. Hicks, C. Masso, S. Reis, M. Bekunda (Eds.), *Just enough nitrogen. Perspectives on how to get there for regions with too much and too little nitrogen (Proceedings of the 6th International Nitrogen Conference).* Chapter 19 (this volume). Springer.

Famuwagun, I. B., & Oladitan, T. (2020). Influence of varying rates of fertilizers on the performance of cacao (*Theobroma cacao*) seedlings in the nursery. In M. A. Sutton, K. E. Mason, A. Bleeker, W. K. Hicks, C. Masso, S. Reis, M. Bekunda (Eds.), *Just enough nitrogen. Perspectives on how to get there for regions with too much and too little nitrogen (Proceedings of the 6th International Nitrogen Conference).* Chapter 4 (this volume). Springer.

Galloway, J. N., Townsend, A. R., Erisman, J. W., Bekunda, M., Cai, Z., Freney, J. R., et al. (2008). Transformation of the nitrogen cycle: Recent trends, questions and potential solutions. *Science, 320*(5878), 889–892.

García-Gómez, H., Calvete-Sogo, H., González-Fernández, I., Rábago, I., Bermejo, V., Valiño, F., Sanz, J., et al. (2020). Atmospheric nitrogen deposition in Spain: Emission and deposition trends, critical load exceedances and effects on terrestrial ecosystems. In M. A. Sutton, K. E. Mason, A. Bleeker, W. K. Hicks, C. Masso, S. Reis, M. Bekunda (Eds.), *Just enough nitrogen. Perspectives on how to get there for regions with too much and too little nitrogen (Proceedings of the 6th International Nitrogen Conference).* Chapter 21 (this volume). Springer.

Garnett, T., & Godfray, C. (2012). *Sustainable intensification in agriculture. Navigating a course through competing food system priorities.* Food Climate Research Network and the Oxford Martin Programme on the Future of Food, UK: University of Oxford.

Glibert, P. M., Maranger, R., Sobota, D. J., & Bouwman, L. (2014). The Haber Bosch–harmful algal bloom (HB–HAB) link. *Environmental Research Letters, 9*(10), 105001.

Glibert, P., Maranger, R., Sobota, D. J., Bouwman, L. (2020). Further evidence of the Haber-Bosch—harmful algal bloom (HB-HAB) link and the risk of suggesting HAB control through phosphorus reductions only. In M. A. Sutton, K. E. Mason, A. Bleeker, W. K. Hicks, C. Masso, S. Reis, M. Bekunda (Eds.), *Just enough nitrogen. Perspectives on how to get there for regions with too much and too little nitrogen (Proceedings of the 6th International Nitrogen Conference).* Chapter 17 (this volume). Springer.

Grizzetti, B., Billen, G., Davidson, E. A., Winiwarter, W., de Vries, W., Fowler, D., Howard, C. M., et al. (2020). Global nitrogen and phosphorus pollution. In M. A. Sutton, K. E. Mason, A. Bleeker, W. K. Hicks, C. Masso, S. Reis, M. Bekunda (Eds.), *Just enough nitrogen. Perspectives on how to get there for regions with too much and too little nitrogen (Proceedings of the 6th International Nitrogen Conference).* Chapter 28 (this volume). Springer.

Gu, B. J., Ge, Y., Ren, Y., Xu, B., Luo, W. D., Jiang, H., et al. (2012). Atmospheric reactive nitrogen in China: Sources, recent trends, and damage costs. *Environmental Science and Technology, 46*(17), 9420–9427.

Hayashi, K., Ono, K., Tokido, T., Matsushima, M. Y., Yano, M., Toyoda, S., Katata, G., et al. (2020). Nitrogen aspects of the Free-Air CO_2 Enrichment (FACE) study for paddy rice ecosystems. In M. A. Sutton, K. E. Mason, A. Bleeker, W. K. Hicks, C. Masso, S. Reis, M. Bekunda (Eds.), *Just enough nitrogen. Perspectives on how to get there for regions with too much and too little nitrogen (Proceedings of the 6th International Nitrogen Conference).* Chapter 22 (this volume). Springer.

Hickman, J., Zingore, S., Galy-Lacaux, C., Kihara, J., Bekunda, M., Palm, C. (2020). Assessing synergies and trade-offs from nitrogen use in Africa. In M. A. Sutton, K. E. Mason, A. Bleeker, W. K. Hicks, C. Masso, S. Reis, M. Bekunda (Eds.), *Just enough nitrogen. Perspectives on how to get there for regions with too much and too little nitrogen (Proceedings of the 6th International Nitrogen Conference).* Chapter 5 (this volume). Springer.

IFA. (2014). *International Fertilizer Association 2014 Database.* http://www.fertilizer.org/ifa/Home-Page/STATISTICS.

Kathuli, P., Zaharah, A. R., Nguluu, S. N. (2020). Contribution of *Gliricidia sepium* pruning and fallow to sweet corn (*Zea mays L. var. rugosa*) yield, nitrogen uptake, release pattern and use efficiency in a humid tropical environment of Malaysia. In M. A. Sutton, K. E. Mason, A. Bleeker, W. K. Hicks, C. Masso, S. Reis, M. Bekunda (Eds.), *Just enough nitrogen. Perspectives on how to get there for regions with too much and too little nitrogen (Proceedings of the 6th International Nitrogen Conference).* Chapter 16 (this volume). Springer.

Kraaijvanger, R. G., & Veldkamp, T. (2020). Evaluating resource use efficiency and stock balances of nitrogen and phosphorus fertilizer inputs: The effect of soil supply capacity in Tigray (Ethiopia). In M. A. Sutton, K. E. Mason, A. Bleeker, W. K. Hicks, C. Masso, S. Reis, M. Bekunda (Eds.), *Just enough nitrogen. Perspectives on how to get there for regions with too much and too little nitrogen (Proceedings of the 6th International Nitrogen Conference).* Chapter 14 (this volume). Springer.

Lassaletta, L., Billen, G., Grizzetti Anglade, J., & Garnier, J. (2014). 50 year trends in nitrogen use efficiency of world cropping systems: The relationship between yield and nitrogen input to cropland. *Environmental Research Letters, 9*(10), 105011.

Lassaletta, L., Billen, G., Grizzetti, B., Anglade, J., Garnier, J. (2020). Long term trends in agronomical and environmental performances of world cropping systems: The relationship between yield and nitrogen input to cropland at the country and regional scales. In M. A. Sutton, K. E. Mason, A. Bleeker, W. K. Hicks, C. Masso, S. Reis, M. Bekunda (Eds.), *Just enough nitrogen. Perspectives on how to get there for regions with too much and too little nitrogen (Proceedings of the 6th International Nitrogen Conference).* Chapter 2 (this volume). Springer.

Lawson, G., Bealey, B., Dupraz, C., Skiba, U. (2020). Agroforestry and opportunities for improved nitrogen management. In M. A. Sutton, K. E. Mason, A. Bleeker, W. K. Hicks, C. Masso, S. Reis, M. Bekunda (Eds.), *Just enough nitrogen. Perspectives on how to get there for regions with too much and too little nitrogen (Proceedings of the 6th International Nitrogen Conference).* Chapter 27 (this volume). Springer.

Lege, L. M., & Carpenter-Boggs, L. A. (2020). Producer knowledge, attitudes, and practices for dry beans and biological nitrogen fixation in Kamuli district, Uganda. In M. A. Sutton, K. E. Mason, A. Bleeker, W. K. Hicks, C. Masso, S. Reis, M. Bekunda (Eds.), *Just enough nitrogen. Perspectives on how to get there for regions with too much and too little nitrogen (Proceedings of the 6th International Nitrogen Conference).* Chapter 8 (this volume). Springer.

Leip, A., Leach, A., Musinguzi, P., Tumwesigye, T., Olupot, G., Tenywa, J. S., et al. (2014). Nitrogen-neutrality: A step towards sustainability. *Environmental Research Letters, 9*(11), 115001.

Lukin, S. M. (2020). Nitrogen loss when using organic and mineral fertilizers on soddy podzolic sandy-loan soil in central Russia. In M. A. Sutton, K. E. Mason, A. Bleeker, W. K. Hicks, C. Masso, S. Reis, M. Bekunda (Eds.), *Just enough nitrogen. Perspectives on how to get there for regions with too much and too little nitrogen (Proceedings of the 6th International Nitrogen Conference).* Chapter 12 (this volume). Springer.

Masso, C., Bekunda, M., Mngube, F. M. (2020). The INI African regional nitrogen centre: Challenges and opportunities in Africa. In M. A. Sutton, K. E. Mason, A. Bleeker, W. K. Hicks,

C. Masso, S. Reis, M. Bekunda (Eds.), *Just enough nitrogen. Perspectives on how to get there for regions with too much and too little nitrogen (Proceedings of the 6th International Nitrogen Conference).* Chapter 31 (this volume). Springer.

Mekonnen, M. M., & Hoekstra, A. Y. (2020). Anthropogenic nitrogen loads to fresh water: A high-resolution global study. In M. A. Sutton, K. E. Mason, A. Bleeker, W. K. Hicks, C. Masso, S. Reis, M. Bekunda (Eds.), *Just enough nitrogen. Perspectives on how to get there for regions with too much and too little nitrogen (Proceedings of the 6th International Nitrogen Conference).* Chapter 20 (this volume). Springer.

Mhango, W. G., Snapp, S., Kanyama-Phiri, G. Y. (2020). Biological nitrogen fixation of pigeonpea and groundnut: Quantifying response across 18 farm sites in northern Malawi. In M. A. Sutton, K. E. Mason, A. Bleeker, W. K. Hicks, C. Masso, S. Reis, M. Bekunda (Eds.), *Just enough nitrogen. Perspectives on how to get there for regions with too much and too little nitrogen (Proceedings of the 6th International Nitrogen Conference).* Chapter 10 (this volume). Springer.

Musinguzi, P., Ebanyat, P., Tenywa, J. S., Bekunda, M. (2020). Sorghum response to nitrogen in organic carbon-categorized Ferralsol and Andosol in Uganda. In M.A. Sutton, K. E. Mason, A. Bleeker, W. K. Hicks, C. Masso, S. Reis, M. Bekunda (Eds.), *Just enough nitrogen. Perspectives on how to get there for regions with too much and too little nitrogen (Proceedings of the 6th International Nitrogen Conference).* Chapter 13 (this volume). Springer.

Naluyange, V., Ochieno, D. M. W., Maingi, J. M., Ombori, O., Mukaminega, D., Amoding, A., et al. (2014). Compatibility of Rhizobium inoculant and water hyacinth compost formulations in Rosecoco bean and consequences on Aphis fabae and Colletotrichum lindemuthianum infestations. *Applied Soil Ecology, 76*(4), 66–77.

Naluyange, V., Ochieno, D. M. W., Wandahwa, P., Maingi, J. M., Ombori, O., Mukaminega, D., Amoding, A., et al,. (2020). Performance of Mwitemania bean under the influence of nitrogen-fixing *Rhizobium* inoculant, water hyacinth composts and DAP fertilizer in a field infested with *Aphis fabae* and *Colletotrichum lindemuthianum.* In M. A. Sutton, K. E. Mason, A. Bleeker, W. K. Hicks, C. Masso, S. Reis, M. Bekunda (Eds.), *Just enough nitrogen. Perspectives on how to get there for regions with too much and too little nitrogen (Proceedings of the 6th International Nitrogen Conference).* Chapter 9 (this volume). Springer.

Ometto, J. P., Ascarrunz, N. L., Austin, A. T., Bustamante, M. M., Cunha-Zeri, G., Forti, M.C., Hoelzemann, J., et al. (2020). The latin America regional nitrogen centre: Concepts and recent activities. In M. A. Sutton, K. E. Mason, A. Bleeker, W. K. Hicks, C. Masso, S. Reis, M. Bekunda (Eds.), *Just enough nitrogen. Perspectives on how to get there for regions with too much and too little nitrogen (Proceedings of the 6th International Nitrogen Conference).* Chapter 35 (this volume). Springer.

Onaga, G., Kikafunda, J., Bigirwa, G., Asea, G., Mwamburi, L. A. (2020). Rice response to nitrogen and supplemental irrigation under low phosphorus and potassium in upland production systems in East Africa. In M. A. Sutton, K. E. Mason, A. Bleeker, W. K. Hicks, C. Masso, S. Reis, M. Bekunda (Eds.), *Just enough nitrogen. Perspectives on how to get there for regions with too much and too little nitrogen (Proceedings of the 6th International Nitrogen Conference).* Chapter 15 (this volume). Springer.

Onyango, J. W., Esilaba, A. O., Rao, K. P. C., Kathuli, P. (2020). A critique of combining tillage practices and nitrogen for enhanced maize production on a humic nitisol in Kenya. In M. A. Sutton, K. E. Mason, A. Bleeker, W. K. Hicks, C. Masso, S. Reis, M. Bekunda (Eds.), *Just enough nitrogen. Perspectives on how to get there for regions with too much and too little nitrogen (Proceedings of the 6th International Nitrogen Conference).* Chapter 3 (this volume). Springer.

Pepó, P. (2020). Effect of climate change and crop-year on the yield and nitrogen fertilizer efficiency in winter wheat (*Triticum aestivum* L.) production. In M. A. Sutton, K. E. Mason, A. Bleeker, W. K. Hicks, C. Masso, S. Reis, M. Bekunda (Eds.), *Just enough nitrogen. Perspectives on how to get there for regions with too much and too little nitrogen (Proceedings of the 6th International Nitrogen Conference).* Chapter 24 (this volume). Springer.

Raghuram, N., Abrol, Y. P., Pathak, H. (2020). The INI South Asian regional nitrogen centre: Capacity building for regional nitrogen assessment and management. In M. A. Sutton, K. E.

Mason, A. Bleeker, W. K. Hicks, C. Masso, S. Reis, M. Bekunda (Eds.), *Just enough nitrogen. Perspectives on how to get there for regions with too much and too little nitrogen (Proceedings of the 6th International Nitrogen Conference)*. Chapter 32 (this volume). Springer.

Reis, S., Bekunda, M., Howard, C. M., Karanja, N., Winiwarter, W., Yan, X., et al. (2016). Synthesis and review: Tackling the nitrogen management challenge: From global to local scales. *Environmental Research Letters, 11*(12), 120205.

SACEP. (2014). *Nutrient loading and eutrophication of coastal waters of the South Asian seas—A scoping study*. South Asia Cooperative Environment Program (SACEP). Available online at: http://www.sacep.org/pdf/Reports-Technical/2014.05.21–22-Scoping-study-on-Nutrient-loading-in-SAS-Region.pdf. Accessed September 2019.

Sala, O. E., Chapin, F. S., Armesto, J. J., Berlow, E., Bloomfield, J., Dirzo, R., et al. (2000). Global biodiversity scenarios for the year 2100. *Science, 287*, 1770–1774.

Sinha, V. B., Jangam, A. P., Raghuram, N. (2020). Biological determinants of crop nitrogen use efficiency and biotechnological avenues for improvement. In M. A. Sutton, K. E. Mason, A. Bleeker, W. K. Hicks, C. Masso, S. Reis, M. Bekunda (Eds.), *Just enough nitrogen. Perspectives on how to get there for regions with too much and too little nitrogen (Proceedings of the 6th International Nitrogen Conference)*. Chapter 11 (this volume). Springer.

Sutton, M. A, Bleeker, A., Stevens, C., et al. (2009). *The Barsac Declaration: Environmental sustainability and the demitarian diet*. Available online at: http://www.initrogen.org/publications/barsac-declaration. Accessed 11 September 2018.

Sutton, M. A., Howard, C., Erisman, J. W., Billen, G., Bleeker, A., Grennfelt, P., van Grinsven, H., et al. (Eds.). (2011). *The European Nitrogen Assessment: Sources, effects and policy perspectives*. Cambridge University Press.

Sutton, M. A., Bleeker, A., Howard, C. M., Bekunda, M., Grizzetti, B., de Vries, W., van Grinsven, H. J. M., et al. (2013). *Our Nutrient World: The challenge to produce more food and energy with less pollution*. Global Overview of Nutrient Management. Edinburgh: Centre for Ecology and Hydrology, on behalf of the Global Partnership on Nutrient Management and the International Nitrogen Initiative.

Sutton, M. A., Howard, C. M., Brownlie, W. J., Skiba, U., Hicks, W. K., Winiwarter, W., van Grinsven, H., et al. (2017). The European nitrogen assessment 6 years after: What was the outcome and what are the future research challenges?. In T. Dalgaard, J. E. Olesen, J. K. Schjørring, L. S. Jensen, H. Vejre, P. S. Andersen, P. Gundersen, et al. (Eds.) (2017). *Innovative Solutions for Sustainable Management of Nitrogen. Proceedings from the International conference, Aarhus, Denmark, 25–28 June 2017, and the following United Nations Economic Commission for Europe Task Force on Reactive Nitrogen Meeting (TFRN-12, 29–30 June 2017* (pp. 40–49). Denmark: Aarhus University and the DNMARK Research Alliance.

Sutton, M. A., Howard, C. M., Brownlie, W. J., Kanter, D. R., de Vries, W., Adhya, T. K., Ometto, J. P., et al. (2020a). Global challenges for nitrogen science-policy interactions: Towards the International Management System (INMS) and improved coordination between multi-lateral environmental agreements. In M. A. Sutton, K. E. Mason, A. Bleeker, W. K. Hicks, C. Masso, S. Reis, M. Bekunda (Eds.), *Just enough nitrogen. Perspectives on how to get there for regions with too much and too little nitrogen (Proceedings of the 6th International Nitrogen Conference)*. Chapter 36 (this volume). Springer.

Sutton, M. A., Ebanyat, P., Raghuram, N., Bekunde, M., Tenywa, J. S., Winiwarter, W., Bleeker, A., et al. (2020b). The kampala statement-for-action on reactive nitrogen in Africa and globally. In M. A. Sutton, K. E. Mason, A. Bleeker, W. K. Hicks, C. Masso, S. Reis, M. Bekunda (Eds.), *Just enough nitrogen. Perspectives on how to get there for regions with too much and too little nitrogen (Proceedings of the 6th International Nitrogen Conference)*. Chapter 38 (this volume). Springer.

Ti, C., & Yan, X. (2013). Spatial and temporal variations of river nitrogen exports from major basins in China. *Environmental Science Pollution, 20*(9), 6509–6520.

Tumwesigye, T., Olupot, G., Musinguzi, P., Leip, A., Bekunda, M., Sutton, M. A. (2020). Pre-informed consumers on a pre-adjusted menu had smaller nitrogen footprints during the N2013

Conference, Kampala, than those on a conventional menu. In M. A. Sutton, K. E. Mason, A. Bleeker, W. K. Hicks, C. Masso, S. Reis, M. Bekunda (Eds.), *Just enough nitrogen. Perspectives on how to get there for regions with too much and too little nitrogen (Proceedings of the 6th International Nitrogen Conference).* Chapter 37 (this volume). Springer.

Van Grinsven, H. J. M., Erisman, J. W., de Vries, W., & Westhoek, H. (2015). Potential of extensification of European agriculture for a more sustainable food system, focusing on nitrogen. *Environmental Research Letters, 10*(2), 025002.

Van Grinsven, H. J. M., Erisman, J. W., de Vries, W., Westhoek, H., Lassaletta, L. (2020). Potential of extensification of European and Dutch agriculture for a more sustainable food system focusing on nitrogen and livestock. In M. A. Sutton, K. E. Mason, A. Bleeker, W. K. Hicks, C. Masso, S. Reis, M. Bekunda (Eds.), *Just enough nitrogen. Perspectives on how to get there for regions with too much and too little nitrogen (Proceedings of the 6th International Nitrogen Conference).* Chapter 6 (this volume). Springer.

Vitousek, P. M., Naylor, R., Crews, T., David, M. B., Drinkwater, L. E., Holland, E., et al. (2009). Nutrient imbalances in agricultural development. *Science, 324*(5934), 1519–1520.

Westhoek, H., Lesschen, J. P., Rood, T., Wagner, S., De Marco, A., Murphy-Bockern, D., et al. (2014). Food choices, health and environment: Effects of cutting Europe's meat and dairy intake. *Global Environmental Change, 26,* 196–205.

Westhoek, H., Lesschen, J. P., Leip A., Rood, T., Wagner, S., De Marco, A., Murphy-Bokern, D., et al. (2015). *Nitrogen on the table: The influence of food choices on nitrogen emissions and the European environment. (European Nitrogen Assessment Special Report on Nitrogen and Food).* Edinburgh: Centre for Ecology and Hydrology.

Winiwarter, W., de Vries, W., Grizzetti, B., Hicks, W. K., van Grinsven, H. J. M., Voss, M. (2020). The INI European Regional nitrogen centre: Concepts and vision. In M. A. Sutton, K. E. Mason, A. Bleeker, W. K. Hicks, C. Masso, S. Reis, M. Bekunda (Eds.), *Just enough nitrogen. Perspectives on how to get there for regions with too much and too little nitrogen (Proceedings of the 6th International Nitrogen Conference).* Chapter 30 (this volume). Springer.

Yan, X., Ti, C., Hayashi, K. (2020). The INI East Asia regional nitrogen centre: Balancing food production and the environment—nitrogen-related research and management in East Asia. In M. A. Sutton, K. E. Mason, A. Bleeker, W. K. Hicks, C. Masso, S. Reis, M. Bekunda (Eds.), *Just enough nitrogen. Perspectives on how to get there for regions with too much and too little nitrogen (Proceedings of the 6th International Nitrogen Conference).* Chapter 33 (this volume). Springer.

Part I
Food and Agriculture: Nitrogen and Food Security

Chapter 2
Long Term Trends in Agronomical and Environmental Performances of World Cropping Systems: The Relationship Between Yield and Nitrogen Input to Cropland at the Country and Regional Scales

Luis Lassaletta, Gilles Billen, Bruna Grizzetti, Juliette Anglade, and Josette Garnier

Abstract Nitrogen (N) is one of the main production factors in agricultural systems and has to be properly managed to sustain crop yields. Nowadays more than half of the N added to cropland is lost to the environment, wasting the resource, producing threats to air, water, soil and biodiversity, and generating greenhouse gas emissions. Based on FAO data of the past ~ 50 years (1961–2009 period), we have reconstructed

---

This chapter is an abridged version of Lassaletta et al. (2014a)(https://iopscience.iop.org/article/10.1088/1748-9326/9/10/105011), 50 year trends in nitrogen use efficiency of world cropping systems: the relationship between yield and nitrogen input to cropland, *Environmental Research Letters*, 9(10), 105011, https://doi.org/10.1088/1748-9326/9/10/105011, re-used here in modified form with the permission of IOP Publishing Ltd. under Creative Commons license CC BY 3.0 (https://creativecommons.org/licenses/by/3.0). The paper has been shortened and edited, the literature updated, and includes the addition of some new sentences and a new figure (Fig. 2.6). It is complemented with new regional results of Lassaletta et al. (2016), (https://iopscience.iop.org/article/10.1088/1748-9326/11/9/095007) Nitrogen use in the global food system: past trends and future trajectories of agronomic performance, pollution, trade, and dietary demand, *Environmental Research Letters*, 11(9), 095007, doi:0.1088/1748-9326/11/9/095007 (under Creative Commons license CC BY 3.0, (https://creativecommons.org/licenses/by/3.0).

L. Lassaletta (✉) · G. Billen · B. Grizzetti · J. Anglade · J. Garnier
Sorbonne Université, CNRS, EPHE, UMR 7619 Metis, 4 place Jussieu, 75005 Paris, France
e-mail: luis.lassaletta@upm.es

G. Billen
e-mail: gilles.billen@upmc.fr

B. Grizzetti
e-mail: Bruna.GRIZZETTI@ec.europa.eu

J. Anglade
e-mail: juliettealice.anglade@gmail.com

J. Garnier
e-mail: josette.garnier@upmc.fr

M. A. Sutton et al. (eds.), *Just Enough Nitrogen*,
https://doi.org/10.1007/978-3-030-58065-0_2

the trajectory followed by 124 countries in terms of crop yield (expressed in nitrogen) and total nitrogen inputs to cropland (synthetic fertilizer, manure, symbiotic fixation and atmospheric deposition). During the past five decades the response of cropping systems to increased nitrogen fertilization has evolved differently between world countries. While some countries have improved their agro-environmental performances, in others the increased fertilization has produced low agronomical benefits and higher environmental losses. Despite very significant improvements in N use efficiency in several countries, N losses above recommendable levels are still frequently observed. Our data also suggests that, in general, those countries using a higher proportion of N inputs from symbiotic N fixation rather than from synthetic fertilizer have a better N use efficiency. By evidencing the long-term response of N inputs to the soil in terms of production and potential losses to the environment, this chapter provides a summarised and comprehensive diagnosis of the effective changes in agronomical and environmental performances of the cropping systems of 124 countries, as well as of the corresponding aggregated values for 12 macro-regions of the world.

Keywords Nitrogen use efficiency · Country scales · Global scale · Cropping systems · Crop yields · Nitrogen pollution

2.1 Introduction

Over the past half century, world agriculture has succeeded in increasing its production of vegetal proteins by a factor of 3 (Lassaletta et al. 2014a, 2016). This has been made possible by changes in cropping systems generally referred to as the Green Revolution, based on the adoption of improved crop varieties, use of pesticides, expansion of irrigated land and increased application of synthetic fertilizers, among which nitrogen (N) was by far the most crucial (Tilman et al. 2002; Mueller et al. 2012; Sinclair and Rufty 2012). Improper management of N leads to environmental N loss significantly contributing to the degradation of air, water and soil quality, threatens biodiversity, and boosts climate change, with high economic costs for the society (Billen et al. 2013; Sutton et al. 2013; van Grinsven et. al. 2013; Glibert et al. 2014; Sobota et al. 2015). From this perspective, strongly contrasting situations exist in the different countries and regions of the world, linked to the disparity of their agro-food-system (Billen et al. 2014; Lassaletta et al. 2016).

It is the purpose of this chapter to describe the relationship between yield and nitrogen inputs to cropland (an expression of the agronomical performances of cropping systems at the national or regional scale), and the related potential losses of

L. Lassaletta
CEIGRAM/Department of Agricultural Production, Universidad Politécnica de Madrid, Madrid 28040, Spain

B. Grizzetti
European Commission, Joint Research Centre (JRC), Ispra, Italy

reactive nitrogen (N_r) to the hydrosphere and atmosphere (an indicator of their environmental performances). We established the long term trajectory of yield and total N inputs to cropland for 124 countries based on the analysis of the data available in the FAO database from 1961 to 2009 (www.fao.org/faostat). From these data an assessment of the changing trends of agronomical and environmental performances of world cropping systems can be delivered.

2.2 Methods

2.2.1 Yield Fertilization Relationship

In this chapter we use the concept of the yield-fertilization (Y, F) relationship in an original way compared to the concept commonly used, relating here the mean yield integrated over the entire crop rotation to the total fertilization of the cropland soils of a given territory. While the yield-fertilization relationship is normally used in conventional agronomy as a tool to predict the yield increase of a given crop that could be expected from increasing fertilization in a given pedo-climatic context, we consider the integrative values of Y and F as overall indicators of the agronomical and environmental performances of a regional cropping system: the Y/F ratio is a measure of its nitrogen use efficiency (NUE), while the F–Y difference is the regional N surplus (or N balance) representing the potential for hydrological or gaseous losses of nitrogen to the environment.

Several mathematical formulations of the yield-fertilization relationship in a given pedo-climatic and technical-agronomical context have been proposed in the agronomical literature, most of them involving negative exponential or hyperbolic functions (de Wit 1992; Llewely and Featherstone, 1997; Harmsen 2000; Nijland et al. 2008). Since we express both output and input in exactly the same unit (kg N ha^{-1} $year^{-1}$) and because we look for a simple long-term integrative theoretical relationship, we make use of the simplest possible function obeying the three following properties: (i) the function intercept should be zero; (ii) the slope of the function should be 1 (unity) at low fertilization; (iii) the function should reach a plateau at high fertilization. The first two properties reflect the fact that, in the long run, harvest cannot exceed N restitutions to the soil, and that the effect of low fertilization in strongly N-limited systems is characterized by a NUE close to 1. The third property expresses the classical law of diminishing return and the fact that, in constant technical-agronomical context, some other limiting factor will always impose a ceiling to production at saturating N availability. Two mathematical functions with only one parameter obey both conditions: a hyperbolic function of the form:

$$Y = Ymax \cdot F/(F + Ymax) \tag{2.1}$$

and a negative exponential function such as:

$$Y = Y_{max}[1 - \exp(-F/Y_{max})] \quad (2.2)$$

We observed that the hyperbolic function (Eq. 2.1) generally provides the best fit to the data. Ymax represents the yield value reached at saturating N fertilization, and also characterizes the value of fertilization at which half the maximum yield is reached. Ymax can thus be viewed as a lumped parameter characterizing the cropping system of a given region with a given technical and pedo-climatic context, from the angle of its response to N fertilization. In support of this view, Anglade et al. (2015a, b) recently demonstrated by careful analysis of a large variety of crop rotations in the central Paris Basin (France), including conventional and organic specialized cropping systems, as well as organic mixed crop and livestock systems, widely differing in their level of fertilization. They showed that all these systems can be described by one and the same yield-total soil N input relationship.

2.2.2 Long Term National Trajectories in Terms of Yield and Fertilization

Our approach consists of the calculation of the various components of the arable soil budget of 124 countries and, most importantly, on the description of the trajectory drawn from 1961 to 2009 by these countries in terms of their total crop production (Y, expressed in harvested vegetal N contained in the protein, kg N ha^{-1} $year^{-1}$) and the total N inputs onto cropland, excluding permanent grassland, as synthetic fertilizers, manure, symbiotic fixation and atmospheric deposition (F, also in kg N ha^{-1} $year^{-1}$). Together these 124 countries represent 99.2% of the world population and 99.6% of the cropland surface in 2009.

The detailed methodology can be found in the original paper (Lassaletta et al. 2014a) and its supplementary material. Total annual crop production by each country was calculated taking into account the yearly harvested yield of 178 primary crops and their N content, as reported in Lassaletta et al. (2014a). The cropland surface was estimated by summing up the surfaces of all individual crops. Total fertilization of cropland was defined as the total N input as synthetic fertilizers, symbiotic N fixation, manure application and atmospheric deposition onto cropland, excluding permanent grassland. We focused our analysis on cropland because the fate of the agricultural surplus (excess N input over N export by plant harvest) strongly differs between cropland and permanent grassland, particularly with respect to the relative proportions of NH_3 volatilization, denitrification, leaching and storage in the soil organic pool (Velthof et al. 2009; Billen et al. 2013). Note that temporary grasslands, included within crop rotations, are considered as cropland. Yearly data on synthetic N fertilizer application, provided from the Resources module of the FAOstat database, refer to total synthetic fertilizer use in agriculture without distinction between arable and grassland. We therefore had to subtract the proportion used for grassland fertilization, which in some European countries such as Ireland and the Netherlands accounts for a significant proportion. We have estimated the proportion of synthetic fertilizers to grasslands at the country scales using information compiled from different sources (see Lassaletta et al. 2014a).

To estimate the crop biological nitrogen fixation by fixing crops we used a yield-based approach:

$$\text{Nfixed} = \%\text{Ndfa} * \frac{\text{Y}}{\text{NHI}} * \text{BGN} \tag{2.3}$$

where %Ndfa is the percentage of N uptake derived from N fixation, Y is the yield (expressed in kg N ha^{-1} $year^{-1}$), NHI is the N harvest index, defined as the ratio of the harvested material to the total above-ground N production, and BGN is a multiplicative factor expressing the total N_2 fixation including below-ground contributions associated with roots, nodules and rhizo-deposition via exudates and decaying root cells and hyphae. These parameters have been obtained from different sources (Herridge et al. 2008; Salvagiotti et al. 2008; Laberge et al. 2009; Kombiok and Buak, 2013; Álvarez et al. 2014; Anglade et al. 2015a). We applied a regional %Ndfa for soybean N fixation. For sugar cane, rice, paddy and forage products, we applied a constant rate of biological fixation per hectare, as suggested by Herridge et al. (2008).

To estimate the animal excretion factors, we have followed the methodology of Sheldrick et al. (2003) assuming that excretion rates, within a given livestock category, are proportional to the slaughtered animal weights, using a particular excretion factor for each type of animal, country and year. The proportion of N excreted that is finally used as manure applied onto cropland was taken from the estimates of Sheldrick et al. (2003) at the regional level for each type of animal. It was considered that 30% of the available manure is lost during management and storage before reaching the crop, as proposed by Oenema et al. (2007) for Europe and close to the value estimated by Liu et al. (2010). We finally discounted the amount of N that is applied to permanent grasslands by applying the proportions provided by regions, and in some cases at the country scale, by Liu et al. (2010).

Deposition of oxidised and reduced nitrogen compounds onto croplands over the period was estimated using data from Dentener et al. (2006), Bouwman et al. (2009) and Seitzinger et al. (2010).

2.3 Results and Discussion

2.3.1 Yield Versus Fertilization Trajectories of World Agriculture

The trajectory followed from 1961 to 2009 by a number of countries in terms of crop yield and total N inputs into cropland is shown in Fig. 2.1. The results for all countries of the FAO database are provided in the Supplementary Material of Lassaletta et al. (2014a).[1]

Over the 1961–2009 period, certain countries that we will call 'type I', such as China, Egypt and India, present a simple trajectory with regularly increasing fertilization and gradual reduction in the crop yield response, following a consistent and unique Y versus F relationship (Fig. 2.1a). Other countries (called 'type II'), such as the USA, Brazil and Bangladesh, show a historical trajectory with first a regularly increasing fertilization and yield, fitting the Y versus F relationship with a definite Ymax, then a turning point with a shift of the trajectory to another relationship with a significantly higher Ymax. This likely reflects improved agronomical practices in terms of production factors other than nitrogen and changes in the crop mix (Zhang et al. 2015, together with the pursuit of increasing fertilization. The turning point seems to have occurred in the 1980s or later depending on the country (Fig. 2.1b). The case of the USA, for example, is consistent with a slowdown in the increase of synthetic fertilizers inputs from the 1980s parallel to a moderate increase in the yields of the most important crops (Howarth et al. 2002; Alston et al. 2010; van Grinsven et al. 2015a).

In most European countries (see the example of France, the Netherlands and Greece in Fig. 2.1c), the trajectory also shows a bi-phasic pattern, describing a regular increase in both fertilization and yield during the 1960–1975 period, followed by a shift towards improved yields without further increasing fertilization and even decreasing fertilization from the 1980s on (type III). The case of the Netherlands is the most spectacular, as in this country, which has always used very high rates of fertilization, the level applied in recent years has been reduced to the same as in the 1960s, however with yields doubled. This trend is related to the reduction of N inputs prescribed by European environmental policies and regulations (van Grinsven et al. 2012; van Grinsven et al. 2016), which interestingly did not prevent significant yield increases. These changes match with the three Kuznets curves described by Zhang et al. (2015). It should be noted that, despite the increase of NUE and decrease in N surpluses, the nitrogen surplus emitted to the environment in many cases remains much higher than that of other countries belonging to types I and II and therefore the environmental impact is still above acceptable limits (Romero et al. 2016; van Grinsven et al. 2016) In these countries, not only sustainable intensification but also sustainable extensification could conciliate productive and societal demands (van Grinsven et al. 2015b).

[1] http://iopscience.iop.org/1748-9326/9/10/105011/media/erl502906suppdata2annex.pdf.

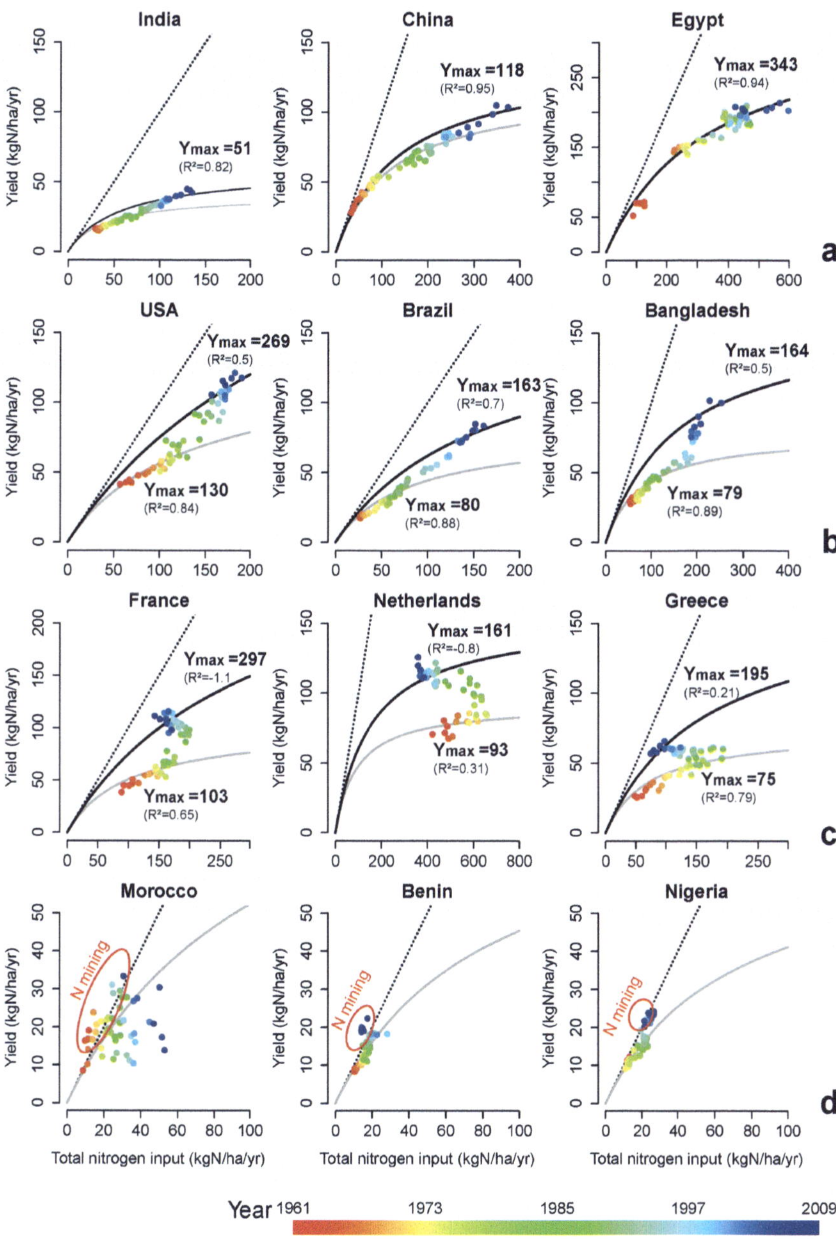

India
Ymax =51
(R²=0.82)
China
Ymax =118
(R²=0.95)
Egypt
Ymax =343
(R²=0.94)
a.
USA
Ymax =269
(R²=0.5)
Ymax =130
(R²=0.84)
Brazil
Ymax =163
(R²=0.7)
Ymax =80
(R²=0.88)
Bangladesh
Ymax =164
(R²=0.5)
Ymax =79
(R²=0.89)
b.
France
Ymax =297
(R²=-1.1)
Ymax =103
(R²=0.65)
Netherlands
Ymax =161
(R²=-0.8)
Ymax =93
(R²=0.31)
Greece
Ymax =195
(R²=0.21)
Ymax =75
(R²=0.79)
c.
Morocco
N mining
Benin
N mining
Nigeria
N mining
d.
Yield (kgN/ha/yr)
Total nitrogen input (kgN/ha/yr)
Year 1961 1973 1985 1997 2009

◂**Fig. 2.1** Examples of trajectories followed by countries in the Yield vs Fertilization (Y versus F) diagram: **a** examples of type I trajectories, **b** examples of type II trajectories, **c** examples of type III trajectories, and **d** examples of type IV trajectories. "R^2" stands here for the coefficient of determination of Nash-Sutclife, defined as: "R^2" $= 1 - [\Sigma\ (obs_i\text{-}calc_i)^2/\Sigma\ (obs_i - meanobs)^2]$, where obs_i are the observed values of yield, $calc_i$ the yield value calculated with the relationship and meanobs is the average value of the observed yields over the period considered. Negative values of the coefficient indicate poor fit of the relationship on the observed values. This is often the case for the most recent period of type III trajectories because of still evolving agronomical conditions. (Reproduced from Lassaletta et al. 2014a © IOP Publishing Ltd., under CC BY 3.0, https://creativecommons.org/licenses/by/3.0)

Finally, there is a small group of countries, such as Morocco, Benin and Nigeria, whose trajectory does not display any consistent Y versus F relationship (type IV). In all cases these countries have low inputs and yields with high inter-annual variation. Very often, their trajectory in the Y versus F diagram crosses the 1:1 line, indicating higher yield than fertilization. The 'negative' (or very small) nitrogen balance displayed in the Y versus F diagram can represent the signature of an unsustainable nitrogen mining of agricultural soils.

For type I to III countries, we were able to define the Ymax values providing the best fit of the hyperbolic relationship (Eq. 2.1) to the points corresponding to the 1961–1980 period or later, and another Ymax for the most recent 10–15 years. Comparison of the two periods shows a significant increase of Ymax in 45 countries (type II and III trajectories). For a large number (55) of others, however, nearly the same relationship holds over the 50-year period (type I trajectory), as is the case for China, Egypt, Turkey, Chile, India and many others. Possible N mining is indicated by higher crop yield than fertilization for 18 countries such as Canada, Morocco, Algeria, Iraq and Mozambique in the 1960–1980 period. In recent years, N mining has continued in 10 African countries, as well as in former Soviet Union countries, Afghanistan and Paraguay. Nitrogen mining has been observed in Argentina for the entire studied period. The severe problem of nutrient mining and loss of soil fertility in African countries has been frequently highlighted (Vitousek et al. 2009; Liu et al. 2010). In these countries, yields are among the lowest in the world, but apparently have potential for improvement through better fertilization practices, including an increasing use of legumes in crop rotations and better management of manure (Vanlauwe et al. 2014). In the former Soviet Union, after the abrupt changes which occurred from 1989 onwards, crops may have benefitted from nutrient legacies.

2.3.2 *Agronomical Performances: Trends in Nitrogen Use Efficiency*

The described trajectories can be translated in terms of changes in the NUE of the cropping system in the different countries (Fig. 2.2). Type I countries display a regularly decreasing trend of NUE. In type II countries, the shift in the trajectory

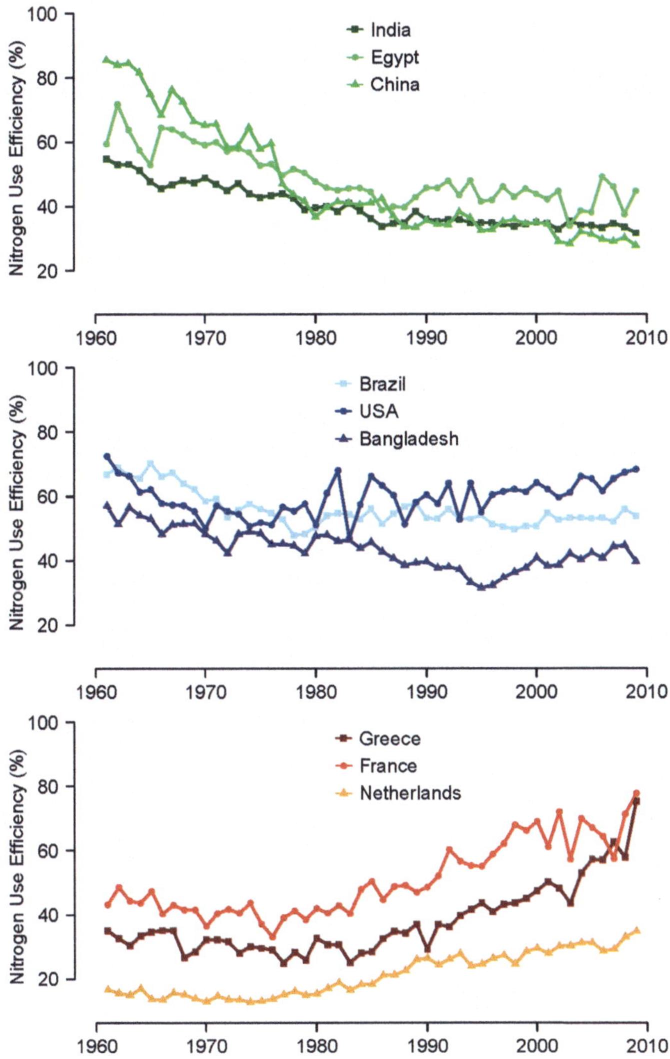

Fig. 2.2 50 years trends in nitrogen use efficiency of the cropping system of selected countries (Reproduced from Lassaletta et al. 2014a © IOP Publishing Ltd., under CC BY 3.0, https://creativecommons.org/licenses/by/3.0)

toward an improved Ymax results in the stabilization or in the increase of NUE. In type III countries, the reduction of N inputs in recent years with no drop in yield corresponds to increasing NUE. In addition to negative budgets, very high values of NUE observed in some African countries could indicate a high risk of soil mining and subsequent loss of fertility. The EU Nitrogen Expert Panel (2015) indicates that NUE values in excess of 90% constitute a higher risk of N mining.

2.3.3 Environmental Performances: Nitrogen Losses

Using N surpluses as an indicator of potential N pollution, our data show the global distribution of environmental N losses from agricultural soils. Losses are over 50 kg N ha^{-1} $year^{-1}$ in most of Europe, the Middle East, the USA and Central America, India and China. They remain on average below 25 kg N ha^{-1} $year^{-1}$ in most of sub-Saharan Africa, the former Soviet Union countries and Australia. High surplus values are associated with low NUE or very high inputs (even with acceptable values of NUE), as is the case for some European countries.

Total fertilization is mainly the sum of synthetic fertilizers, manure application and symbiotic nitrogen fixation because atmospheric N deposition generally contributes a much smaller share. The proportion of the three former N inputs to overall fertilization is very different in the world cropping countries. Our data show that NUE is generally higher (and the N surplus relatively lower) for agricultural systems with higher proportion of N inputs derived from symbiotic N fixation and lower for a higher proportion of synthetic fertilizers in total fertilization (Fig. 2.3).

The higher NUE associated with nitrogen fixation is probably explained by a higher efficiency in the incorporation by legumes of their self-supplied nitrogen (Herridge and Peoples 1990). This higher efficiency of cropping systems relying

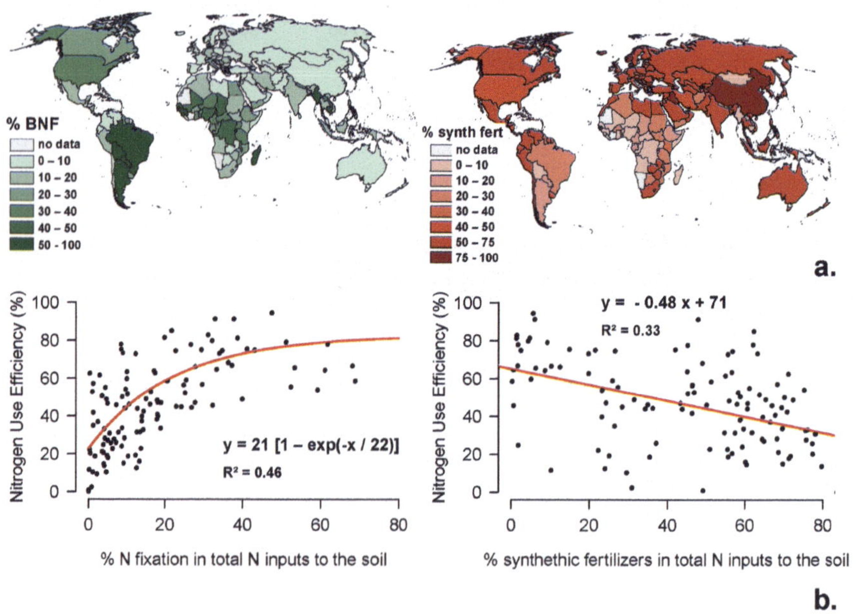

Fig. 2.3 **a** Distribution of the share of symbiotic fixation and synthetic fertilizers in total N inputs to cropland by countries in 2000–2009; **b** Observed relationship between NUE and the proportion of symbiotic fixation, or of synthetic fertilizers in total N inputs to cropland in the period 2000–2009 (Reproduced from Lassaletta et al. 2014a © IOP Publishing Ltd., under CC BY 3.0, https://creativecommons.org/licenses/by/3.0)

largely on biological N fixation is observed for the largest soybean producers of South America as noted by Liu et al. (2010), as well as for less productive countries in Africa and Asia with significant production of rice, groundnuts and beans.

2.3.4 Global Trends

At the global scale, aggregating all cropping systems of the world, a type II Y/F trajectory is observed, with a shift during the 1980s from one Y/F relationship characterized by a Ymax of 73 kg N ha^{-1} $year^{-1}$ to an improved one with Ymax of 110 kg N ha^{-1} $year^{-1}$ (Fig. 2.4a). The overall observed global trend is a decrease of NUE in the 19611980 period (from 68% to 45%), followed by a stabilization during the last 30 years around 47%) (Fig. 2.4b). The share of the different sources of N in the total inputs to cropland, depicted in Fig. 2.5, change considerably during the last 50 years, with synthetic fertilizers now being the largest sources. Even though the total rate of N excreted by livestock is equivalent to synthetic fertilizer application, the rate of manure that finally reaches the crops is much lower and nowadays is lower than crop biological N fixation (Fig. 2.5). The concentration of intensive landless livestock systems that has occurred during the last decades produces a misuse of the animal excreta reducing the country NUEs and exacerbating pollution problems (Lassaletta et al. 2014b, b; Garnier et al. 2016; Strokal et al. 2016).

These global trends are masking important regional disparities which can be highlighted by grouping the different countries into 12 macro-regions, as proposed by Lassaletta et al. (2014a). The yield-fertilization relationship of cropland from these 12 regions, as presented and discussed in Lassaletta et al. (2016) in the context of their whole agro-food system, are shown in Fig. 2.6.

2.4 Conclusions

Currently, only 47% of the reactive nitrogen added globally onto cropland is converted into harvested products, compared to 68% in the early 1960s, while synthetic N fertilizer input increased by a factor of 9 over the same period. This means that more than half of the nitrogen used for crop fertilization is currently lost into the environment, where it contributes to severe problems (e.g., air, water, biodiversity). Even though a significant improvement in NUE occurred in many countries after the 1980s, the present results suggest that a further increase of nitrogen fertilization would result in a disproportionately low increase of crop production with further environmental alterations, unless significant changes occur in cropping systems towards better efficiency. In that respect, improvement and development of agronomical practices and proper application of environmental policies have been demonstrated to be efficient strategies (van Grinsven et al. 2015a, b, 2016; Romero et al. 2016). A better integration of crop and livestock systems can also contribute

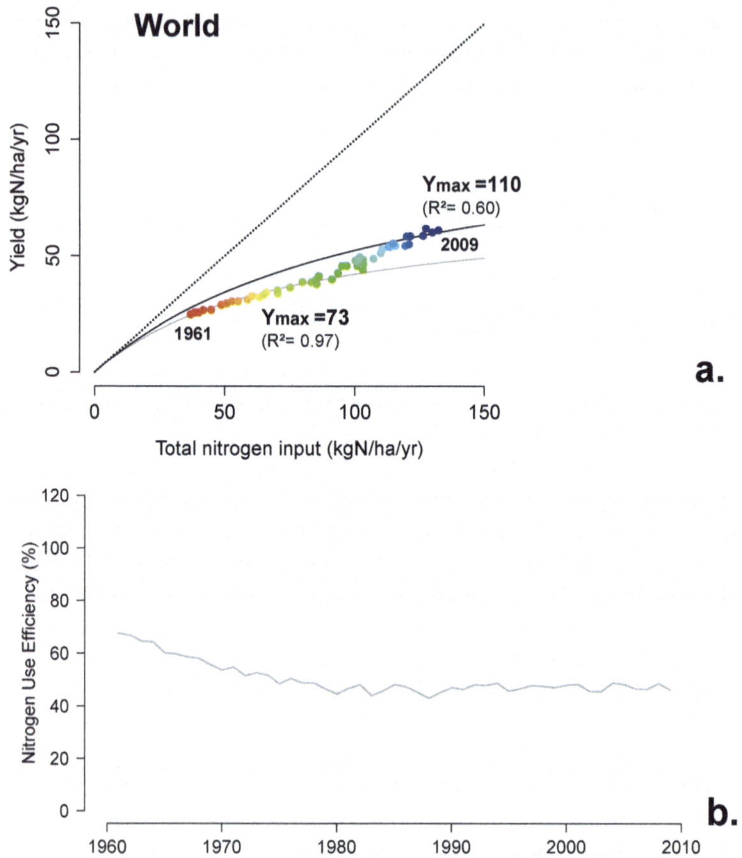

Fig. 2.4 The global cropping system: **a** Trajectory followed by global world cropping systems in the Y versus F diagram (Y: crop yield in protein harvested, kg N ha^{-1} $year^{-1}$; F: total N inputs to the cropland soil, kg N ha^{-1} $year^{-1}$); **b** Trends in nitrogen use efficiency of the global cropping system (Reproduced from Lassaletta et al. 2014a © IOP Publishing Ltd., under CC BY 3.0, https://creativecommons.org/licenses/by/3.0)

to increasing NUE at the local and global scale (Herrero et al. 2010; Lassaletta et al. 2014a; Bonaudo et al. 2014; Soussana and Lemaire 2014; Garnier et al. 2016; Billen et al. 2019). In addition to sustainable intensification, sustainable extensification could represent an appropriate alternative agricultural strategy in very polluted regions (van Grinsven et al. 2015a). Moreover, our data suggest that an increase in the contribution of symbiotic N fixation would result in increasing NUE. Peoples et al. (2009) and Anglade et al. (2015a) have stressed the large potential of symbiotic nitrogen fixation to improve the efficiency of agroecosystems. The benefits of N-fixing crops can be achieved by including more legume crops in rotations, or by the introduction of short-duration legume green manures or 'catch crops' (Blesh and Drinkwater 2013).

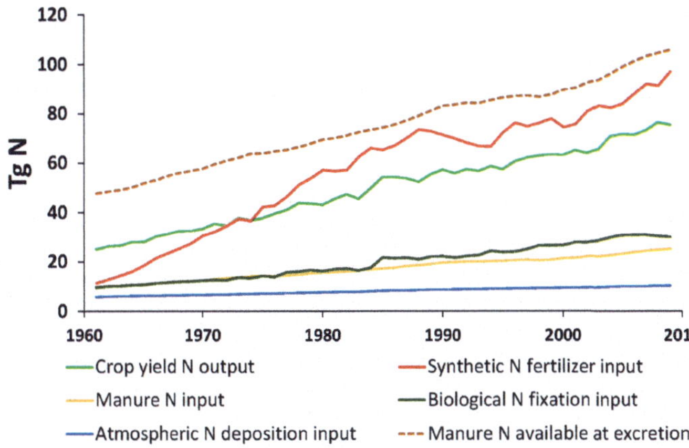

Fig. 2.5 Evolution of the components of the global cropping system budget (Reproduced from Lassaletta et al. 2014a © IOP Publishing Ltd., under CC BY 3.0, https://creativecommons.org/licenses/by/3.0)

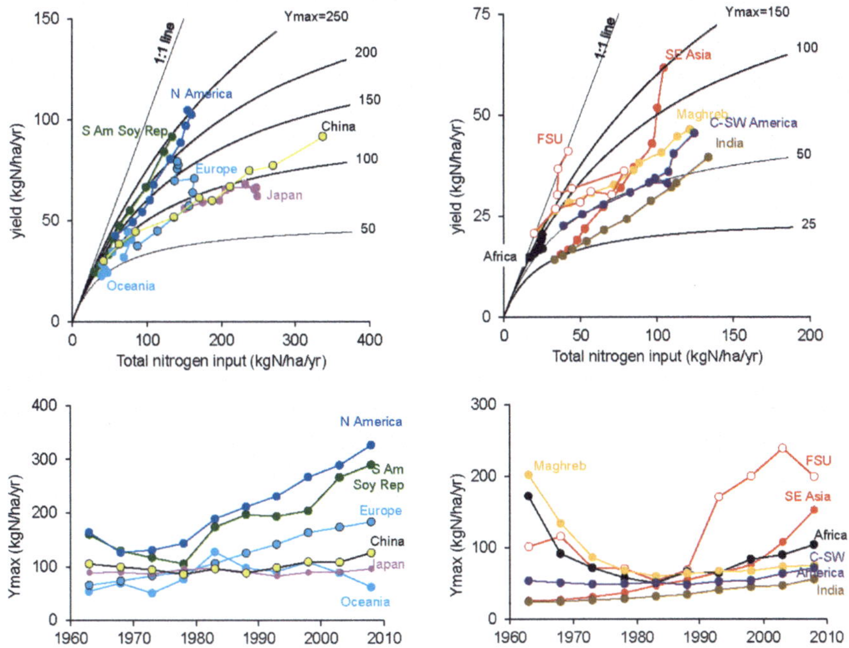

Fig. 2.6 Trajectories followed by the cropping system of 12 macro-regions of the world in the Y versus F diagram (Y: crop yield in protein harvested, kg N ha^{-1} $year^{-1}$; F: total N inputs to the cropland soil, kg N ha^{-1} $year^{-1}$), and time variation of the corresponding Ymax parameter (Values are shown for 5 years' averages over the period 1961–2009). See Lassaletta et al. 2016 for the dataset that this figure is based upon

An improved knowledge of the relationship between yield and nitrogen fertilization of cropland at the national, macro-regional and global scales, as provided by the present study, allows the implementation of better strategies and concerted actions for improving the agronomical and environmental performances of agro-food systems (Lassaletta et al. 2016).

Acknowledgements This study was partly supported by the SeasERA EMoSEM project (ANR-12-SEAS-0005–01). We wish to thank the FIRE (Fédération Ile de France de Recherche environnement, CNRS and UPMC). We thank Javier Castrillo who developed several computer routines for the data management and Angel Udias for his help in preparing the figures. This chapter is an edited version of Lassaletta et al. (2014a), Environmental Research Letters, 9(10), 105011, https://doi.org/10.1088/1748-9326/9/10/105011, re-used here in modified form with the permission of IOP Publishing Ltd. under Creative Commons license CC BY 3.0 (https://creativecommons.org/licenses/by/3.0) and complemented with new regional results of Lassaletta et al. (2016), *Environmental Research Letters*, 11(9), 095007, doi:0.1088/1748-9326/11/9/095007 under Creative Commons licence CC BY 3.0 (https://creativecommons.org/licenses/by/3.0). The paper has been shortened and edited, the literature updated, and includes the addition of some new sentences and a new figure (Fig. 2.6).

References

Alston, J. M., Babcock, B. A., & Pardey, P. G. (2010). *The shifting patterns of agricultural production and productivity worldwide*. CARD Books, 2. Iowa State University, Ames, Iowa: Midwest Agribuisiness Trade Research and Information Center.

Alvarez, R., Steinbach, H. S., & De Paepe, J. L. (2014). A regional audit of nitrogen fluxes in pampean agroecosystems. *Agriculture, Ecosystems & Environment, 184,* 1–8.

Anglade, J., Billen, G., Makridis, T., Garnier, J., Puech, T., & Tittel, C. (2015a). Nitrogen soil surface balance of organic vs conventional cash crop farming in the Seine watershed. *Agricultural Systems, 139,* 82–92.

Anglade, J., Billen, G., Garnier, J. (2015b). Relationships for estimating N_2 fixation in legumes: Incidence for N balance of legume-based cropping systems in Europe. *Ecosphere*, *6*(3), Article 37. doi. 10.1890/es14–00353.1.

Billen, G., Garnier, J., Lassaletta, L. (2013). The nitrogen cascade from agricultural soils to the sea: modelling N transfers at regional watershed and global scales. *Philosophical Transactions of the Royal Society B, 368*(1621), 20130123. https://doi.org/10.1098/rstb.2013.0123.

Billen, G., Garnier, J., & Lassaletta, L. (2014). A biogeochemical view of the global agro-food system: Nitrogen flows associated with protein production, consumption and trade. *Global Food Security, 3*(3–4), 209–219.

Billen, G., Lassaletta, L., Garnier, J., Le Noë, J., Aguilera, E., Sanz-Cobena, A. (2019). Opening to distant markets or local reconnection of agro-food systems? Environmental consequences at regional and global scales. In G. Lemaire, P. Carvalho, S. Kronberg, S. Recous (Eds.), *Agroecosystem Diversity. Reconciling Contemporary Agriculture and Environmental Quality* (Section VI, chapter 25, pp. 391–413). Elsevier.

Blesh, J., & Drinkwater, L. E. (2013). The impact of nitrogen source and crop rotation on nitrogen mass balances in the Mississippi River Basin. *Ecological Applications, 23*(5), 1017–1035.

Bonaudo, T., Bendahan, A. B., Sabatier, R., Ryschawy, J., Bellon, S., Leger, F., et al. (2014). Agroecological principles for the redesign of integrated crop–livestock systems. *European Journal of Agronomy, 57,* 43–51.

Bouwman, A. F., Beusen, A. H. W., Billen, G. (2009). Human alteration of the global nitrogen and phosphorus soil balances for the period 1970–2050. *Global Biogeochemical Cycles, 23*(4) GB0A04. https://doi.org/10.1029/2009gb003576.

Dentener, F., Drevet, J., Lamarque, J. F., Bey, I., Eickhout, B., Fiore, A. M., et al. (2006). Nitrogen and sulfur deposition on regional and global scales: A multi-model evaluation. *Global Biogeochemical Cycles, 20*(4), B4003. https://doi.org/10.1029/2005GB002672.

de Wit, C. T. (1992). Resource use efficiency in agriculture. *Agricultural Systems, 40*(1–3), 125–151.

EU Nitrogen Expert Panel. (2015). *Nitrogen Use Efficiency (NUE)—an indicator for the utilization of nitrogen in agriculture and food systems*. Wageningen, Netherlands: Wageningen University. Available online at: http://www.inms.international/inms-meeting-lisbon/NExpertPanelNUESession1.pdf. Accessed September 2019.

Garnier, J., Anglade, J., Benoit, M., Billen, G., Puech, T., Ramarson, A., et al. (2016). Reconnecting crop and cattle farming to reduce nitrogen losses to river water of an intensive agricultural catchment (Seine basin, France): Past, present and future. *Environmental Science & Policy, 63*, 76–90.

Glibert, P. M., Maranger, R., Sobota, D. J., & Bouwman, L. (2014). The Haber Bosch–harmful algal bloom (HB–HAB) link. *Environmental Research Letters, 9*, 105001. https://doi.org/10.1088/1748-9326/9/10/105001.

Harmsen, K. (2000). A modified Mitscerlich equation for rainfed crop production in semi-arid areas: 1. Theory. *NJAS—Wageningen Journal of Life Sciences, 48*(3), 237–250.

Herrero, M., Thornton, P. K., Notenbaert, A. M., Wood, S., Msangi, S., Freeman, H. A., et al. (2010). Smart investments in sustainable food production: Revisiting mixed crop-livestock systems. *Science, 327*(5967), 822–825.

Herridge, D. F., & Peoples, M. B. (1990). Ureide assay for measuring nitrogen-fixation by nodulated soybean calibrated by ^{15}N methods. *Plant Physiology, 93*(2), 495–503.

Herridge, D. F., Peoples, M. B., & Boddey, R. M. (2008). Global inputs of biological nitrogen fixation in agricultural systems. *Plant and Soil, 311*(1), 1–18.

Howarth, R. W., Boyer, E. W., Pabich, W. J., & Galloway, J. N. (2002). Nitrogen use in the United States from 1961–2000 and potential future trends. *Ambio, 31*(2), 88–96.

Kombiok, J., & Buah, S. (2013). Tillage depth effects on nodulation, nitrogen fixation and yield of three soybean varieties in the Northern Savanna zone of Ghana. *African Journal of Agricultural Research, 8*(20), 2340–2345.

Laberge, G., Franke, A. C., Ambus, P., & Høgh-Jensen, H. (2009). Nitrogen rhizodeposition from soybean (Glycine max) and its impact on nutrient budgets in two contrasting environments of the Guinean savannah zone of Nigeria. *Nutrient Cycling in Agroecosystems, 84*(1), 49–58.

Lassaletta, L., Billen, G., Grizzetti, B., Anglade, J., Garnier, J. (2014a). 50 year trends in nitrogen use efficiency of world cropping systems: the relationship between yield and nitrogen input to cropland. *Environmental Research Letters, 9*(10), 105011. doi.org/https://doi.org/10.1088/1748-9326/9/10/105011. https://iopscience.iop.org/article/10.1088/1748-9326/9/10/105011.

Lassaletta, L., Billen, G., Grizzetti, B., Garnier, J., Leach, A. M., & Galloway, J. N. (2014b). Food and feed trade as a driver in the global nitrogen cycle: 50-year trends. *Biogeochemistry, 118*(1), 225–241.

Lassaletta, L., Billen, G., Romero, E., Garnier, J., & Aguilera, E. (2014c). How changes in diet and trade patterns have shaped the N cycle at the national scale: Spain (1961–2009). *Regional Environmental Change, 14*(2), 785–797.

Lassaletta, L., Billen, G., Garnier, J., Bouwman, L., Velazquez, E., Mueller, N. D., Gerber, J. S. (2016). Nitrogen use in the global food system: Past trends and future trajectories of agronomic performance, pollution, trade, and dietary demand. *Environmental Research Letters, 11*(9), 095007. doi: 0.1088/1748-9326/11/9/095007. https://iopscience.iop.org/article/10.1088/1748-9326/11/9/095007.

Liu, J.-G., You, L.-Z., Amini, M., Obersteiner, M., Herrero, M., Zehnder, A. J. B., et al. (2010). A high-resolution assessment on global nitrogen flows in cropland. *Proceedings of the National Academy of Sciences of the United States of America (PNAS), 107*(17), 8035–8040.

Llewelyn, R. V., & Featherstone, A. M. (1997). A comparison of crop production functions using simulated data for irrigated corn in western Kansas. *Agricultural Systems, 54*(4), 521–538.

Mueller, N. D., Gerber, J. S., Johnston, M., Ray, D. K., Ramankutty, N., & Foley, J. A. (2012). Closing yield gaps through nutrient and water management. *Nature, 490,* 254–257.

Nijland, G. O., Schouls, J., & Goudriaan, J. (2008). Integrating the production functions of Liebig, Michaelis-Menten, Mitscherlich and Liebscher into one system dynamics model. *NJAS—Wageningen Journal of Life Sciences, 55*(2), 199–224.

Oenema, O., Oudendag, D., & Velthof, G. L. (2007). Nutrient losses from manure management in the European Union. *Livestock Science, 112*(3), 261–272.

Peoples, M., Brockwell, J., Herridge, D. F., Rochester, I. J., Alves, B. J. R., Urquiaga, S., et al. (2009). The contributions of nitrogen-fixing crop legumes to the productivity of agricultural systems. *Symbiosis, 48*(1), 1–17.

Romero, E., Le Gendre, R., Garnier, J., Billen, G., Fisson, C., Silvestre, M., et al. (2016). Long-term water quality in the lower Seine: Lessons learned over 4 decades of monitoring. *Environmental Science & Policy, 58,* 141–154.

Salvagiotti, F., Cassman, K. G., Specht, J. E., Walters, D. T., Weiss, A., & Dobermann, A. (2008). Nitrogen uptake, fixation and response to fertilizer N in soybeans: A review. *Field Crops Research, 108*(1), 1–13.

Seitzinger, S. P., Mayorga, E., Bouwman, A. F., Kroeze, C., Beusen, A. H. W., Billen, G., Van Drecht, G., et al. (2010). Global river nutrient export: A scenario analysis of past and future trends. *Global Biogeochemical Cycles, 24*(4), GB0A08. https://doi.org/10.1029/2009gb003587.

Sheldrick, W., Syers, J. K., & Lingard, J. (2003). Contribution of livestock excreta to nutrient balances. *Nutrient Cycling in Agroecosystems, 66*(2), 119–131.

Sinclair, T. R., & Rufty, T. W. (2012). Nitrogen and water resources commonly limit crop yield increases, not necessarily plant genetics. *Global Food Security, 1*(2), 94–98.

Sobota, D. J., Compton, J. E., McCrackin, M. L., & Singh, S. (2015). Cost of reactive nitrogen release from human activities to the environment in the United States. *Environmental Research Letters, 10*(2), 025006. https://doi.org/10.1088/1748-9326/10/2/025006.

Soussana, J.-F., & Lemaire, G. (2014). Coupling carbon and nitrogen cycles for environmentally sustainable intensification of grasslands and crop-livestock systems. *Agriculture, Ecosystems & Environment, 190,* 9–17.

Strokal, M., Ma, L., Bai, Z., Luan, S., Kroeze, C., Oenema, O., et al. (2016). Alarming nutrient pollution of Chinese rivers as a result of agricultural transitions. *Environmental Research Letters, 11,* 024014. https://doi.org/10.1088/1748-9326/11/2/024014.

Sutton, M. .A., Bleeker, A., Howard, C. M., Bekunda, M., Grizetti, B., de Vries, W., van Grinsven, H. J. M., et al. (2013). *Our Nutrient World: The challenge to produce more food and energy with less pollution.* Global Overview of Nutrient Management. Edinburgh: Centre for Ecology & Hydrology on behalf of the Global Partnership on Nutrient Management and the International Nitrogen Initiative.

Tilman, D., Cassman, K. G., Matson, P. A., Naylor, R., & Polasky, S. (2002). Agricultural sustainability and intensive production practices. *Nature, 418,* 671–677.

Vanlauwe, B., Wendt, J., Giller, K. E., Corbeels, M., Gerard, B., & Nolte, C. (2014). A fourth principle is required to define Conservation Agriculture in sub-Saharan Africa: The appropriate use of fertilizer to enhance crop productivity. *Field Crops Research, 155,* 10–13.

van Grinsven, H. J. M., ten Berge, H. F. M., Dalgaard, T., Fraters, B., Durand, P., Hart, A., et al. (2012). Management, regulation and environmental impacts of nitrogen fertilization in northwestern Europe under the Nitrates Directive; a benchmark study. *Biogeosciences, 9,* 5143–5160.

van Grinsven, H. J. M., Holland, M., Jacobsen, B. H., Klimont, Z., Sutton, M. A., Willems, W. J., et al. (2013). Costs and benefits of nitrogen for Europe and implications for mitigation. *Environmental Science and Technology, 47,* 3571–3579.

van Grinsven, H. J. M., Erisman, J. W., Vries, W., & Westhoek, H. (2015a). Potential of extensification of European agriculture for a more sustainable food system, focusing on nitrogen. *Environmental Research Letters, 10,* 025002. https://doi.org/10.1088/1748-9326/10/2/025002.

van Grinsven, H. J. M., Bouwman, L., Cassman, K. G., van Es, H. M., McCrackin, M. L., & Beusen, A. H. W. (2015b). Losses of ammonia and nitrate from agriculture and their effect on nitrogen recovery in the European Union and the United States between 1900 and 2050. *Journal of Environmental Quality, 44,* 356–367.

van Grinsven, H. J. M., Tiktak, A., & Rougoor, C. W. (2016). Evaluation of the Dutch implementation of the nitrates directive, the water framework directive and the national emission ceilings directive. *NJAS—Wageningen Journal of Life Sciences, 78,* 69–84.

Vitousek, P. M., Naylor, R., Crews, T., David, M. B., Drinkwater, L. E., Holland, E., et al. (2009). Nutrient imbalances in agricultural development. *Science, 324*(5934), 1519–1520.

Velthof, G. L., Oudendag, D., Witzke, H. P., Asman, W. A., Klimont, Z., & Oenema, O. (2009). Integrated assessment of nitrogen losses from agriculture in EU-27 using MITERRA-EUROPE. *Journal of Environmental Quality, 38*(2), 402–417.

Zhang, X., Davidson, E. A., Mauzerall, D. L., Searchinger, T. D., Dumas, P., & Shen, Y. (2015). Managing nitrogen for sustainable development. *Nature, 528,* 51–59.

Chapter 3
A Critique of Combining Tillage Practices and Nitrogen for Enhanced Maize Production on a Humic Nitisol in Kenya

J. W. Onyango, A. O. Esilaba, K. P. C. Rao, and P. Kathuli

Abstract A study was carried out in the short rainy season of 2012/2013 and the long rainy season of 2013 to compare the effects of a combination of two tillage practices and three fertilizer levels on maize production in a warm and wet environment of Kabete in Kenya. This was part of a larger experiment to test water use efficiency and viability of various climate adaptation strategies in east and southern Africa. The trial consisted of a split-plot design experiment of conventional tillage (CT) and tied ridges (TR) as main plots and three fertilizer levels of 0, 20 and 40 kg nitrogen (N) ha^{-1} as sub-plots. Air-dried grain harvested at physiological maturity was separated from the cob and comparative analysis done of the grain yields under the various fertility rates, tillage categories and their combinations. In the short rains of 2012–2013, TR had higher grain yield compared to CT across all the N levels, while in the long rains of 2013 the reverse was found. In 2012–2013, increasing fertilizer from 0 to 20 kg N ha^{-1} increased the yields by 20.9%, but increasing from 20 to 40 kg N ha^{-1} decreased yields by 4.9% across both the tillage categories. In the long rains of 2013, the reductions of the yields were 0.09% and 10.3% for the respective fertilizer increments. Under CT, increasing fertilizer from 0 to 20 kg N ha^{-1} increased the yields by 56.2%, but increasing from 20 to 40 kg N ha^{-1} reduced it by 8.0% in 2012–2013, while in the long rains of 2013 the yields were reduced by 8.2% and by

J. W. Onyango (✉)
National Agricultural Research Laboratories, P.O. Box 14733-00800, Nairobi, Kenya
e-mail: joabwamari@yahoo.com

A. O. Esilaba
Kenya Agricultural and Livestock Research Organisation, Nairobi, Kenya
e-mail: aesilaba@gmail.com

K. P. C. Rao
International Crop Research Institute for Semi-arid Tropics, Nairobi, Kenya
e-mail: k.p.rao@cgiar.org

P. Kathuli
Kenya Agricultural and Livestock Research Organization, Katumani, P.O. Box 340-90100, Machakos, Kenya
e-mail: peterkathuli@yahoo.com

National Dry-Land Research Centre, Katumani, Nairobi, Kenya

M. A. Sutton et al. (eds.), *Just Enough Nitrogen*,
https://doi.org/10.1007/978-3-030-58065-0_3

2.7% for the respective fertilizer increments. Under TR, increasing fertilizer from 0 to 20 kg N ha^{-1} increased yields by 0.08% and by 0.03% for 20 to 40 kg N ha^{-1} in 2012–2013, while in 2013 the increment was by 15.1% for the first fertilizer increment but a reduction by 23.6% for the second fertilizer treatment. Neither of the two tested tillage practices, fertilizer levels, nor their interactions, however, gave any significant grain yield difference at $p < 0.05$ for both seasons. Since TR is more labor intensive than CT, the later would be preferable and convenient for farmers in the agro-climatic conditions similar to those at the experimental site, whereas fertilizer application should be based on better understanding of the initial soil fertility.

Keywords Nitrogen fertilizer · Tied ridges · Conventional tillage · Climate change · Water use efficiency

3.1 Introduction

It is now acknowledged that climate change and variability will negatively impact on agricultural production through reduction of crop yields especially under subsistence farming in developing countries where impacts on food security can be devastating (Bochiolo et al. 2013). Climate change alone already adversely impacts on 175.4 million hectares of rain-fed agriculture accounting for some 440.8 MT of production losses in sub-Saharan Africa alone (Calzadilla et al. 2009; Ewbank 2012). Experience has also shown that these possible impacts can be dealt with by integrating a wide range of adaptation strategies especially into national development planning (Parry et al. 2007; Deressa et al. 2008). Some positive adaptation measures reported include seed choices and planting dates for the likely changes in climatic conditions to which significant increases in yields have been attributed (Campbell et al. 2009). Given that small-scale farmers globally use 60% of the world's land to produce half the world's food (Ewbank 2012), introducing adaptive capacity to prevent eminent hunger and reduce poverty in the face of impeding climate change is pertinent (Van Ardenne-van der Hoeven et al. 2002).

In Kenya effectiveness of tillage practices to improve rainfall water utilization is significantly influenced by soil and climatic conditions (Sijali and Kamoni 2005). Gichangi et al. (2007) proposed that whatever in-situ rainwater conservation technologies are used, organic matter (farmyard manure) should be an integral part of the technology for increased soil moisture conservation and utilization by crops for increased crop production and food security in eastern Kenya. Itabari et al. (2004) and Gichangi et al. (2007) have also explained that crop yield increments would arise from use of a combination of fertilizers and/or manures along with in-situ soil moisture conservation such as tied ridges (TR) through improved water use efficiency by crops planted in semi-arid regions of eastern Kenya. Kathuli and Itabari (2010, internal communication) reported that tied ridging when used with fertilizer, manure or their combination has the potential to increase crop yields by up to 100–300%. Manure has also been shown to increase soil moisture profile irrespective of whether

Table 3.1 Rainfall received at Kabete during experimental period

Season	Short rains	Long rains
Trial period	2012/2013	2013
Seasonal total	365.0 (mm)	570.0 (mm)
Mean monthly	91.0 (mm)	114.0 (mm)

it is used with in-situ soil moisture conservation technology or otherwise (Gichangi et al. 2007). There is therefore a need to complement these findings by implementing analysis of temporal rainfall and soil moisture distribution. This particular experiment was part of using analogue locations in east and southern Africa to test adaption strategies to climate change using three varieties for each of four crops, two tillage practices and two planting dates in various agro-ecological zones. The aim of the study was to assess whether the adoption of climate adaption practices would result in yield increase as this is the main primary benefit and driver for smallholder farmers to adopt new technologies or practices.

3.2 Materials and Methods

3.2.1 Area of Study

Experiments were carried out at Kabete University Campus Field station, Kenya, which represents a warm/wet agro-ecological zone (Upper Midland subzone 3) and lies at 36° 45' E 01° 16' S at an altitude of 1820 m above mean sea level. The site lies on a humic nitisol with nitrogen (N) content of 0.66% and organic carbon (C) content of 2.84% and receives about 970 mm of rainfall. It has a mean annual temperature of 18.2 °C and a bimodal pattern of rainfall with two cropping seasons per year (Kenya Meteorological Department 1983; Karuku et al. 2012). The results presented are of grain yields attributes for the short rains (October–January) season of 2012 and 2013 and the long rains (March–July) season of 2013 (Table 3.1).

3.2.2 The Experiment

The experiment was laid out in a split plot design consisting of two tillage practices namely conventional tillage (CT) and tied ridges (TR) as main plots and three fertilizer levels of 0, 20 and 40 kg N ha^{-1} in the sub plots. The ridges were tied at 1 m and 40 kg ha^{-1} basal phosphate (P_2O_5) fertilizer in the form of Triple Super phosphate (TSP) applied at planting to all the plots. Nitrogen in the form of Calcium Ammonium Nitrate (CAN) was applied as top dressing at 20–40 kg ha^{-1} in the N treated plots. Maize variety DK8031 was used as the test crop and planted at spacings of 75 cm × 30 cm in 5 m × 6 m plots. This medium maturing maize variety requires between 600 and 950 mm of well distributed rainfall and has a yield potential of 9 t ha^{-1} (Jaetzold and Schmidt 1983). The Portable GenStat version 12.1.0.3278 was used to carry out the analysis of variance of the grain yields.

3.3 Results

3.3.1 Responses of Maize in the Short Rainy Season

During the short rains of 2012–2013 TR had 15.5% higher grain yield compared to CT across all the N levels. Increasing fertilizer from 0 to 20 kg N ha^{-1} increased the yields by 20.9% but from 20 to 40 kg N ha^{-1} decreased this by 4.9% across both the tillage categories. Under CT increasing fertilizer from 0 to 20 kg N ha^{-1} increased the yields by 56.2% but from 20 to 40 kg N ha^{-1} reduced it by 8.0%. Under TR increasing fertilizer from 0 to 20 kg N ha^{-1} and 20 to 40 kg N ha^{-1} did not affect the yield compared to 0 kg N ha^{-1} (Figs. 3.1, 3.2 and 3.3).

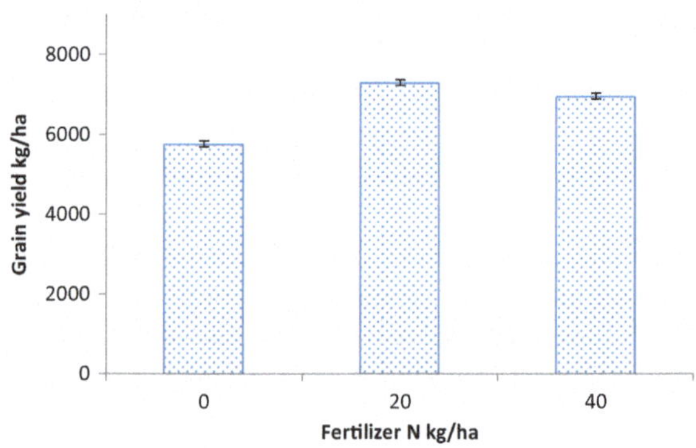

Fig. 3.1 Fertilizer responses in 2012/2013 short rains season across tillage categories

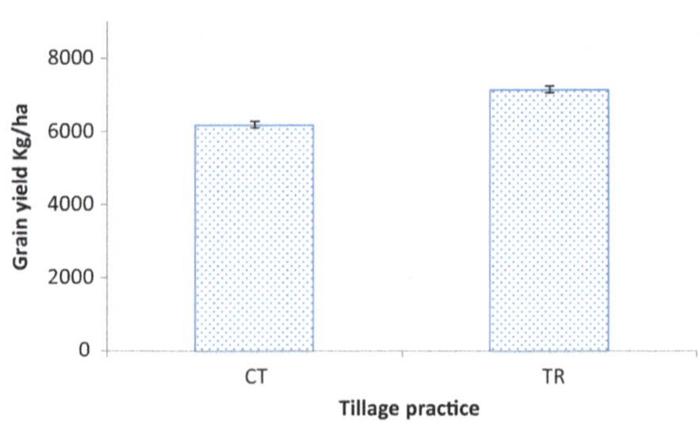

Fig. 3.2 Tillage responses in 2012/2013 short rains season across fertilizer rates. CT: conventional tillage, TR: tied ridges

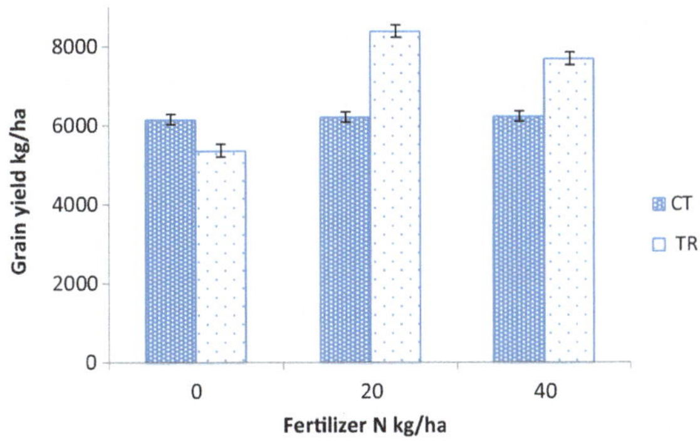

Fig. 3.3 Fertilizer and tillage responses in 2012/2013 short rains season. CT: conventional tillage, TR: tied ridges

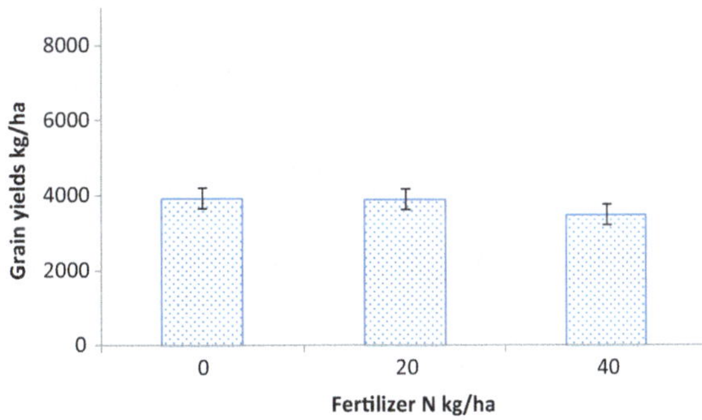

Fig. 3.4 Fertilizer responses in 2013 long rains season across tillage categories

3.3.2 Responses of Maize in Long Rains Season

In the long rains of 2013 TR had 50% lower grain yield compared to CT across all the N levels. Increasing fertilizer from 0 to 20 kg N ha^{-1} reduced the yields by 0.09%, while the reduction was by 10.3% when N was increased from 20 to 40 kg ha^{-1} across both the tillage categories. Under CT increasing fertilizer from 0 to 20 and 20 to 40 kg N ha^{-1} reduced the yields by 8.2–2.7%. Under TR increasing fertilizer from 0 to 20 kg N ha^{-1} increased yields by 15.1%, while from 20 to 40 kg N ha^{-1} reduced this yields by 23.6% (Figs. 3.4, 3.5 and 3.6).

3.4 Discussion

While CT had higher yields in the 2012/2013 short rains season, the trend was reversed in the 2013 long rains season. In the 2012/2013 short rains season, high yields were observed under the 20 N applications and especially under TR but this was not statistically significant. Under TR, increasing fertilizer from 20 to 40 kg N ha^{-1} even slightly reduced yields. However, the differences in yields were not significant

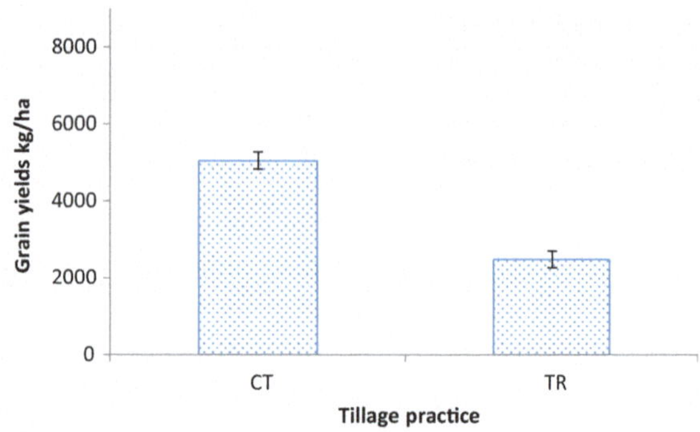

Fig. 3.5 Tillage responses in 2013 long rains season across fertilizer rates. CT: conventional tillage, TR: tied ridges

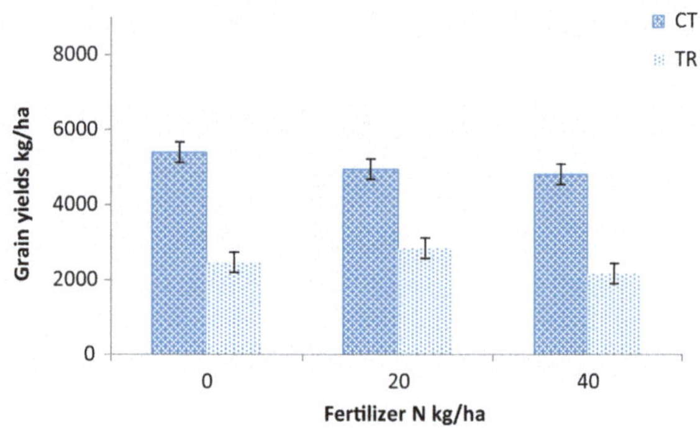

Fig. 3.6 Fertilizer and tillage responses in 2013 long rains season

at ($p < 0.05$) suggesting that there was no need for investing in labor to institute TR in the short rains.

In the long rains of 2013, no interaction responses were found between fertilizers and tillage practices, while enhanced fertilizer levels actually even appeared to decrease yields. The total rainfall amounts received in the long rains season were higher than those received during the short rains season and could have been responsible for these results (Table 3.1). This variability in yield variations between the seasons may implicate aspects of variations in rainfall amounts and or distribution. In humid areas like Kabete, which receive high amounts of rainfall such as those observed during the long rains, the tied ridging technology may not be necessary since it requires additional labor to establish but does not lead to significantly enhanced grain yields. Sijali and Kamoni (2005) reported that TR yielded more maize crop dry matter (1.18 Mg ha^{-1}) than flat tillage (1.04 Mg ha^{-1}) when seasonal rainfall was 222 mm. This emphasizes that water use efficiency is improved under TR as the rainfall amounts decrease and explains why no significant changes were registered in Kabete. Kathuli and Itabari (2010, internal communication) and Miriti et al. (2012) explained that increment in yields under tied ridges would arise from increased plant available moisture content in the dry-land areas. Hence, there may be a threshold rainfall level above which TR will not outperform CT. Future studies on climatic adaptation should determine such a threshold value.

3.5 Conclusions

Since the yields were not significant between fertilizer levels tested, tillage practices, nor their interaction at $p < 0.05$, there would be no need to invest in labor to establish Tied Ridges when there is (apparently) adequate rainfall. There is a need to compliment these findings through analysis of actual rainfall and/or soil moisture distributions. The lack of significant response to the nitrogen (N) fertilizer could be related to the high level of N in the experimental soil, and demonstrates the need to tailor fertilizer recommendations on the initial soil fertility in order to improve N and agronomic use efficiency.

Acknowledgements We acknowledge project funding provided by the Federal Ministry for Economic Cooperation and Development of Germany and the conducive working environment provided by the International Crops Research Institute for the Semi-Arid Tropics, the former Kenya Agricultural Research Institute (currently the Kenyan Agriculture and Livestock Research Organization) and University of Nairobi Kabete campus field office. The assistance of Kenneth Mungai and Phillip Ikholi during data collection is also greatly acknowledged. Dr Cargele Masso also gave invaluable comments which are greatly appreciated.

References

Bochiolo, D., Nana, E., & Soncini, A. (2013). Impact of climate scenarios on crop yields and water footprints of maize in the Po valley, Italy. *Agricultural Water Management, 116*, 50–60.

Calzadilla, A., Zhu, T., Rehdanz, K., Tol, R.S.J., & Ringler, C. (2009). *Economy wide impacts of climate change on agriculture in sub Saharan Africa.* IFPRI Research Brief 15–15. Washington DC, USA: International Food Policy Research Institute (IFPRI).

Campbell, A., Kapos, V., Scharlemann, J.P.S., Bubb, P., Chenery, A., Coad, L., Dickson, B., Donswald, N., Khan, M.S.I., Kershaw, F., & Rashid, M. (2009). *Review of literature on the links between biodiversity and climate change: Impacts, Adaptation and mitigation.* Secretariat of the Convention on Biological Diversity, Montreal, Technical Series No. 42. Montreal, Canada: Secretariat of the Convention on Biological Diversity, UNEP.

Deressa, T.T., Hassan, R.M., Ringler, C. Alemu, T., & Yesuf, M. (2008). *Analysis of the determinants of farmers' choices of adaptation. Methods and perceptions of climate change in the Nile basin of Ethiopia.* IFPRI Discussion paper. Washington DC: IFPRI.

Ewbank, R. (2012). *Forecasts benefit farmers at risk from weather extremes.* Feature article, Climate and Development Knowledge Network. Available at: https://cdkn.org/2012/12/features-forecasts-benefit-farmers-at-risk-from-weather-extremes/?loclang=en_gb. Accessed July 2019.

Gichangi, E. M., Njiru, E. N., Itabari, J. M., Wambua, J. M., Wambua, J. N., & Karuku, A. (2007). Assessment of improved soil fertility and water harvesting technologies through community based on-farm trials in the ASALs of Kenya. In A. Batiano, B. Waswa, J. Kihara, & J. Kimetu (Eds.), *Advances in integrated soil fertility management in sub-Saharan Africa: Challenges and opportunities* (pp. 759–765). Netherlands: Springer.

Itabari, J.K., Nguluu, S.N., Gichangi, E.M., Karuku, A.M., Njiru, E.N., Wambua, J.M., Maina, J.N., & Gachimbi, L.N. (2004). Managing land and water resources for sustainable crop production in dry areas. A case study of small-scale farms in semi-arid areas of eastern, central, and rift valley provinces of Kenya. In: Crissman, L. (Ed.) *Agricultural research and development for sustainable resource management and food security in Kenya*. Proceedings of end of Programme Conference, KARI, 11–12 November 2003 (pp. 31–42).

Jaetzold, R., & Schmidt, H. (1983). *Farm management handbook of Kenya. Vol. II—Natural conditions and farm management information. Part II/B: Central Kenya*. 2nd Edition. Kenya: Ministry of Agriculture.

Karuku, G. N., Gachene, C. K. K., Karanja, N., Cornelis, W., Verplancke, H., & Kironchi, G. (2012). Soil hydraulic properties of a nitisol in Kabete, Kenya. *Tropical and Subtropical Agroecosystems, 15*, 595–609.

Kathuli, P., & Itabari, J.K. (2010). *In-situ soil moisture conservation: utilization and management of rainwater for crop production in the arid and semi-arid lands of Kenya*. KARI-Katumani, Box 340–90100, Machakos (Internal communication).

Kenya Meteorological Department. (1983). *Climatological statistics for Kenya*. Nairobi, Kenya: Kenya Meteorological Department.

Miriti, J. M. G., Kironchi, G., Esilaba, A. O., Heng, L. K., Gachene, C. K. K., & Mwangi, D. M. (2012). Yield and water use efficiencies for maize and cowpeas as affected by tillage and cropping systems in semi-arid eastern Kenya. *Agricultural Water Management, 115*, 148–155.

Parry, M. L., Canziani, O. F., Palutikof, J. P., van der Linden, P. J., & Hansen, C. E. (Eds.). (2007). *Climate change 2007: Impacts, adaptation and vulnerability. contribution of working group II to the fourth assessment report of the intergovernmental panel on climate change*. Cambridge, UK: Cambridge University Press.

Sijali, V.I., & Kamoni, P.T. (2005). Optimization of water and nutrient use in rain-fed semi-arid farming through integrated soil-water-nutrient management practices. In *Nutrient and water management practices for increasing crop production in arid/semi-arid areas*. Proceedings of a Coordinated Research Projects. IAEA TECDOC-1468 (pp. 209–216). Vienna, Austria: IAEA Publications.

Van Ardenne-van der Hoeven, A., Benn, H., Brown, M.M., Chino, T.., Johnson, D.J., Kabbaj, O., Nielson, P., Töpfer, K., Wieizorel-Zeul, H., & Zhang, S. (2002). *Poverty and climate change: Reducing vulnerability of the poor through adaptation*. Report delivered to the 8th Conference of the UNFCC in New Delhi. Collaboration report by Asian Development Bank, DFID, Governments of Netherlands and Germany, EU, OECD, UNDP, UNEP and World Bank. World Bank.

Chapter 4
Influence of Varying Rates of Fertilizers on the Performance of Cacao (*Theobroma cacao*) Seedlings in the Nursery

Idowu B. Famuwagun and Titilayo Oladitan

Abstract The effects of the application of poultry manure, organo-mineral fertilizer, and mineral fertilizer including nitrogen (N), phosphorus (P) and potassium (K) applied at varying rates to the soil or foliage of cacao seedlings were studied in the nursery between January–May 2011 and December to April 2012. Poultry manure added at 40, 50 and 60 g per plant significantly influenced the height and number of leaves of cacao seedlings over other treatments. The effect of organo-mineral fertilizer application at 50 and 60 g per plant on the number of leaves was also significantly better than NPK applied to soil or foliage, starting from 12 weeks after treatment application. The stem girth development following application of poultry manure at 50 and 60 g per plant application had the best performance, followed by poultry manure at 40 g per plant. There were no significant differences on the stem girth of seedlings in almost all the treatments at five weeks after sowing. The treatments with NPK led to the longest tap root length, while poultry manure led to the largest average number of lateral roots, followed by the organo-mineral treatments. The results thereby indicated that the use of poultry manure at 40–50 g per plant and organo-mineral fertilizer as sources of nutrients had significant beneficial effects on cacao seedling quality.

Keywords Cacao · Fertilizer products · Fertilizer rate · Growth · Seedling quality

---

I. B. Famuwagun (✉)
Department of Crop, Soil and Pest Management, Federal University of Technology, Akure, Nigeria
e-mail: ibfamuwagun@futa.edu.ng

T. Oladitan
Rufus Giwa Polytechnic, Owo, Ondo State, Nigeria
e-mail: titipraise@yahoo.com

M. A. Sutton et al. (eds.), *Just Enough Nitrogen*,
https://doi.org/10.1007/978-3-030-58065-0_4

4.1 Introduction

Cacao (*Theobroma cacao* L.) is a tropical woody species, which belongs to the family Malvaceae (Alverson et al. 1999). The geographical origin of cacao is South America (Motamayor et al. 2002), where several wild populations can be found in the Amazon and Guyanian regions. It is considered one of the most important perennial crops with an estimated world output of 3.5 million tonnes in 2006 (ICCO 2007). It is one of the tropical crops that requires the raising of seedlings in a nursery before transplanting to the permanent field (Adenikinju et al. 1989). In cacao seedling production, adequate nutrients supply is essential for optimum growth and development both at the nursery and in the field after transplanting (Egbe 1968). In recent years, several reports revealed a decrease in the establishment percentage of transplanted cacao seedlings due to poor seedling development from the nursery, poor farm management practices and soil and air moisture stresses (Famuwagun and Agele 2010). These problems were also compounded by deficiencies of both macro and micronutrients (Ayanlaja 2002), which had resulted in a drastic reduction in the yield of such plantations. There is also a need to supply adequate nutrients to cacao seedling at the nursery stage to aid root and shoot development (Famuwagun and Agele 2010). Many field reports give an average of less than 40% survival in most cases, which has resulted in significant losses for farmers (Ibiremo et al. 2012). Effective management of cacao seedlings in the nursery using appropriate agronomic practices like fertilization, dry season irrigation, and appropriate self-multiplying bio-fertilizer like Arbuscular Mycorrhizal Fungi (AMF) to enhance root development will enhance seedling growth and allow optimum field establishment (Smith 1994). However, very few studies have tried to determine the appropriate fertilizer products and the application rates in the nursery in the Nigerian context.

The objective of the study was therefore to assess proper fertilization of cacao in the nursery to raise healthy and vigorous seedlings so as to provide good transplantable seedlings that can adapt well to field conditions and facilitate early establishment and growth of the transplanted seedlings. Various fertilizer materials were used at variable rates to determine the suitable product and rates.

4.2 Materials and Methods

An experiment was conducted between January and May 2011 and repeated in 2012 in the Teaching and Research Farm of the Federal University of Technology, Akure, Nigeria within the rainforest zone to investigate the effects of fertilizer application on cacao seedling growth and development in the nursery. The treatment groups were as follows: (a) Nitrogen (N), Phosphorus (P) and Potassium (K) in the ratio N:P:K of 15:15:15 applied to the soil at 15, 20 and 25 g per seedling; (b) N:P:K (15:15:15) applied to the foliage at 40, 50 and 60 ml per litre of water; (c) "Sunshine" organo-mineral fertilizer applied at 40, 50 and 60 g per plant; and (d) poultry manure applied at 40, 50, 60 g per plant. The forms of the NPK in the fertilizers were ammonium nitrate for N, single superphosphate (SSP) 20% P_2O_5 for P, and potassium chloride (KCl) for K. The Sunshine organo-mineral used was an organic fertilizer that was fortified with essential nutrients to meet various crop demands with 3.5% N, 2.5% P_2O_5 and 1.5% K_2O, while the poultry manure contained 3.2% N. The application rates used in this study were not adjusted to supply the equivalent amount of nutrients in each treatment.

Black polythene pots of 4.5 × 10 cm were filled with top soil from a well-fallowed forest. The planting materials (cacao bean from freshly harvested pods) from the Cocoa Research Institute of Nigeria (CRIN), Owena sub-station, Akure, Ondo State were sown at one seed per pot. The experiments were set up in a randomized complete block design (RCBD) with each treatment having five replicates. Watering was carried out at two-day intervals throughout the period of the experiment. The fertilizer treatments were applied at two weeks after sowing at the specified rates for the soil application and foliar fertilizers in a two split application at two and 12 weeks after sowing to enhance timely availability and reduce loss due to leaching and oxidation of essential nutrients.

Data collection from the raised seedlings was scheduled at intervals every two weeks after the application of the fertilizers; however, only data at four weeks intervals are presented. Data collected include plant height, number of leaves and stem girth as well as root parameters. The root parameters were taken at 16 weeks after sowing and include tap root, length and number of lateral roots. The data were subjected to analysis of variance (ANOVA) using the statistical package "SPSS" and the means were separated using Duncan's New Multiple Range Test (DNMRT).

4.3 Results

4.3.1 Seedling Height

The application of poultry manure positively influenced the development of cacao seedlings in term of plant height, which was significantly higher compared with other fertilizer treatments across the various rates of application in the 2011 experiment. In the 2012 experiment, a similar trend was observed, except for the data sets collected at eight weeks after sowing. The results for foliar applied NPK at 40 ml per litre, soil applied NPK at 20 g per plant and organo-mineral fertilizer at 40 g per plant were not significantly different from those recorded with poultry manure applied at 40–50 g per plant at eight weeks after sowing, as shown in Table 4.1. Organo-mineral fertilizer application significantly enhanced cacao seedlings plant height compared with foliar applied NPK, soil applied NPK and the control, but were significantly lower compared with those recorded in poultry manure treated seedlings (Table 4.1).

Table 4.1 Effects of varying rates of fertilizer application on plant height (cm) of cacao seedlings in the nursery. The superscripted letters indicate statistical significance between treatments (by column); shared letters imply no significance difference

Treatments	Weeks after sowing in 2011			Weeks after sowing in 2012		
	8	12	16	8	12	16
NPK foliar applied @ 40 ml	22.0^{cde}	27.0^{cdef}	31.0^{bcd}	24.2^{a}	26.0^{bc}	32.5^{cdef}
NPK foliar applied @ 50 ml	19.7^{def}	24.4^{efg}	28.4^{de}	21.3^{bc}	25.1^{cd}	31.6^{def}
NPK foliar applied @ 60 ml	16.90^{f}	31.4^{abc}	33.4^{bcd}	22.0^{b}	25.7^{cd}	32.4^{cdef}
NPK soil applied @ 15 g/plant	20.4^{de}	23.8^{efg}	29.8^{bcde}	21.3^{bc}	26.0^{bc}	34.0^{cde}
NPK soil applied @ 20 g/plant	20.9^{cde}	22.6^{fg}	23.8^{e}	25.0^{a}	28.4^{b}	36.2^{bc}
NPK soil applied @ 25 g/plant	22.7^{bcd}	26.4^{def}	28.8^{cde}	21.2^{bc}	25.7^{bcd}	33.4^{cde}
Organo-mineral fertilizer @ 40 g/plant	20.2^{de}	30.4^{abcd}	35.6^{b}	24.0^{ab}	28.5^{b}	37.5^{bc}
Organo-mineral fertilizer @ 50 g/plant	22.0^{bcde}	27.8^{bcde}	35.8^{b}	20.0^{bc}	25.3^{cd}	37.8^{bc}
Organo-mineral fertilizer @ 60 g/plant	20.1^{de}	24.8^{efg}	29.6^{bcde}	21.4^{bc}	27.5^{b}	37.0^{bc}
Poultry Manure @ 40 g/plant	26.6^{ab}	33.8^{a}	42.6^{a}	24.8^{a}	30.3^{a}	43.6^{ab}
Poultry Manure @ 50 g/plant	24.6^{ab}	32.0^{ab}	48.0^{a}	25.1^{a}	32.6^{a}	44.2^{ab}
Poultry Manure@ 60 g/plant	24.6^{ab}	21.6^{g}	31.6^{bcd}	22.3^{b}	33.3^{a}	46.7^{a}
Control	23.9^{abc}	27.8^{bcde}	31.8^{bcd}	21.9^{bc}	26.0^{bc}	32.5^{cdef}

4.3.2 *Number of Leaves of the Seedlings*

Poultry manure treated seedlings produced a significantly higher number of leaves in the two trials compared with other fertilizer treatments (Table 4.2). However, no marked difference was found among the three rates of fertilizer application in terms of number of leaves produced for soil and foliar applied NPK, poultry manure, and the organo-mineral fertilizers.

4.3.3 *Seedling Stem Girth*

In the 2011 experiment, foliar applied NPK showed a significantly thicker seedling stem girth compared with the other treatments, while in the 2012 experiment, the stem girth development was significantly higher in seedlings treated with poultry manure (40, 50 and 60 g per plant), NPK foliar applied (40 g per litre), and organo-mineral fertilizer (40 and 50 g per plant) compared with the other treatments (Table 4.3).

Table 4.2 Effects of varying rates of fertilizer application on number of leaves. The superscripted letters indicate statistical significance between treatments (by column); shared letters imply no significance difference

Treatments	Weeks after sowing in 2011			Weeks after sowing in 2012		
	8	12	16	8	10	16
NPK folia applied @ 40 ml	5.8^{bcde}	7.0^{bcd}	8.4^{def}	6.2^{bc}	7.5^{bcd}	8.8^{def}
NPK folia applied @ 50 ml	6.8^{abcd}	8.0^{abcd}	9.0^{cdef}	6.1^{bcd}	7.3^{bcd}	9.4^{cdef}
NPK folia applied @ 60 ml	5.2^{de}	8.0^{abcd}	9.0^{cdef}	6.0^{bcd}	7.6^{bcd}	8.9^{def}
NPK soil applied @ 15 g/plant	6.2^{bcde}	8.4^{abcd}	11.1^{abcd}	6.2^{bc}	7.8^{abcd}	10.6^{cde}
NPK soil applied @ 20 g/plant	7.6^{abc}	9.0^{abc}	8.2^{def}	6.3^{bc}	8.2^{abc}	11.6^{bc}
NPK soil applied @ 25 g/plant	7.8^{a}	9.0^{abc}	9.0^{cdef}	7.5^{a}	8.6^{a}	11.3^{bc}
Organo-mineral fertilizer @ 40 g/plant	6.6^{abcd}	7.8^{abcd}	10.2^{cdef}	6.0^{bcd}	8.1^{abc}	11.2^{bcd}
Organo-mineral fertilizer @ 50 g/plant	6.6^{abcd}	8.4^{abcd}	10.8^{bcde}	6.7^{ab}	8.6^{a}	11.0^{bcd}
Organo-mineral fertilizer @ 60 g/plant	5.8^{bcde}	6.8^{cd}	6.8^{f}	6.1^{bcd}	7.7^{bcd}	10.8^{cde}
Poultry Manure @ 40 g/plant	7.2^{abc}	9.6^{ab}	14.0^{ab}	6.5^{abc}	8.4^{ab}	12.6^{ab}
Poultry Manure @ 50 g/plant	8.0^{a}	9.8^{a}	14.4^{a}	6.8^{abc}	8.3^{ab}	13.5^{a}
Poultry Manure @ 60 g/plant	6.2^{abcd}	6.0^{d}	12.4^{abc}	6.5^{abc}	7.8^{abcd}	12.4^{ab}
Control	6.8^{abcd}	7.4^{abcd}	8.2^{def}	7.0^{ab}	8.1^{abc}	8.9^{def}

Table 4.3 Effects of varying rates of fertilizer application on stem girth (cm). The superscripted letters indicate statistical significance between treatments (by column); shared letters imply no significance difference

Treatments	Weeks after sowing in 2011			Weeks after sowing in 2012		
	8	12	16	8	12	16
NPK foliar applied @ 40 ml	0.62^{a}	0.82^{a}	0.90^{b}	0.72^{a}	0.87^{ab}	0.99^{a}
NPK foliar applied @ 50 ml	0.46^{de}	0.65^{cde}	0.76^{b}	0.58^{b}	0.75^{bcd}	0.82^{abc}
NPK foliar applied @ 60 ml	0.42^{def}	0.64^{cde}	1.18^{a}	0.58^{a}	0.78^{abcd}	0.86^{abc}
NPK soil applied @ 15 g/plant	0.48^{def}	0.58^{de}	0.89^{b}	0.62^{a}	0.74^{bcd}	0.81^{abc}
NPK soil applied @ 20 g/plant	0.42^{def}	0.56^{b}	0.62^{b}	0.59^{a}	0.58^{e}	0.62^{d}
NPK soil applied @ 25 g/plant	0.52^{bcd}	0.66^{cde}	0.76^{b}	0.64^{a}	0.76^{bcd}	0.79^{cd}
Organo-mineral fertilizer @ 40 g/plant	0.42^{def}	0.71^{bc}	0.83^{b}	0.56^{a}	0.78^{abcd}	0.85^{abc}
Organo-mineral fertilizer @ 50 g/plant	0.52^{bcd}	0.71^{bc}	0.83^{b}	0.66^{a}	0.78^{abcd}	0.90^{ab}
Organo-mineral fertilizer @ 60 g/plant	0.40^{ef}	0.58^{de}	0.70^{b}	0.58^{a}	0.66^{de}	0.78^{cd}
Poultry Manure @ 40 g/plant	0.58^{ab}	0.70^{bc}	0.88^{b}	0.64^{a}	0.84^{abc}	0.93^{ab}
Poultry Manure @ 50 g/plant	0.48^{cde}	0.70^{bc}	1.13^{b}	0.63^{a}	0.85^{abc}	0.99^{a}
Poultry Manure @ 60 g/plant	0.36^{f}	0.68^{bcd}	1.12^{b}	0.52^{a}	0.90^{a}	1.0^{a}
Control	0.56^{abc}	0.77^{ab}	0.96^{b}	0.67^{a}	0.90^{a}	0.75^{cd}

4.3.4 Seedling Root System

In both the 2011 and 2012 experiments, tap root length was enhanced significantly through the application of organo-mineral fertilizer at 60 g per plant, with soil applied NPK at 15 g per plant and the control compared with other treatments. No significant difference in the tap root length was found among the poultry manure and organo-mineral treated seedlings (Table 4.4). Lateral root numbers were significantly higher with treatments involving organo-mineral fertilizer at 60 g per plant and poultry manure at 60 g per plant as compared with other treatments.

4.4 Discussion

The significantly higher plant height obtained as a result of poultry manure application at 40, 50 and 60 g per plant was attributed to the fact that poultry manure contains a reasonable amount of nitrogen and other nutrients that are essential for plant growth and development (Parr et al. 1994). In particular, poultry manure contains reasonable amounts of calcium and magnesium, which increase the elemental content of the soil (Ahp 1979). Wilson (1999) and Famuwagun and Agele (2010) reported seedling vigour as a key factor in cacao seedling quality, which is determined by plant height, stem girth, number of leaves, and the root system. The non-significant effect on seedlings plant height under soil applied NPK may be due to the effects of the watering method, which imposed a leaching effect on the elemental nutrients supplied by the mineral fertilizer. This was in line with the findings of Agele et al. (2004) that top watering in potted plants enhances increased leaching losses. Foliar applied NPK has the lowest performance in term of plant height; this may be a result of the volatile nature of the nitrogen, which reduced its availability to the seedlings, or due to minimum intake of the foliar applied nutrients. Poultry manure may also present a competitive advantage by allowing slow release of nutrients, which would minimize losses during watering compared with soil and foliar applied NPK. This was in conformity with the earlier findings that organic manures are usually slow in mineralization (Ayeni 2011). Soil applied and foliar applied NPK were found to be associated with a reduced number of leaves. This may be accounted for by the ammonium ion present as the nitrogen source, which may reduce soil pH thereby causing soil acidity (Ojeniyi 1991; Famuwagun and Adekayode 2010). This assumption seems reasonable as the control plants performed better than the NPK treatments, although further studies may be required to validate the observation. Changes in soil pH (i.e., through acidification), might cause nutrient unavailability to the seedling and consequently affect the seedling quality.

The performance of poultry manure on seedling stem girths in this study was in line with the report of Ewulo (2005) that poultry manure contains high percentages of nitrogen, phosphorus and potassium. The gradual release of nutrients in poultry manure and organo-minerals fertilizer allows for uniform growth and development

Table 4.4 Effects of fertilizers application on root parameters of cacao seedlings. The superscripted letters indicate statistical significance between treatments (by column); shared letters imply no significance difference

Treatments	2011 season			2012 season		
	Tap root length (cm)	Number of lateral root	Average length of lateral root (cm)	Tap root length (cm)	Number of lateral root	Average length of lateral root (cm)
NPK foliar applied @ 40 ml	18.3^{abc}	54.0^{d}	11.3^{d}	16.2^{abc}	59.0^{d}	13.4^{d}
NPK foliar applied @ 50 ml	13.3^{f}	65.7^{c}	13.3^{abcd}	12.5^{f}	61.6^{c}	13.7^{abcd}
NPK foliar applied @ 60 ml	20.3^{a}	63.7^{c}	12.0^{cd}	22.1^{a}	73.7^{c}	11.6^{cd}
NPK soil applied @ 15 g/plant	18.3^{abc}	44.0^{e}	8.3^{e}	18.0^{abc}	55.0^{e}	9.2^{e}
NPK soil applied @ 20 g/plant	17.0^{bcd}	52.3^{d}	12.7^{bcd}	17.3^{bcd}	57.3^{d}	10.3^{bcd}
NPK soil applied @ 25 g/plant	16.7^{bc}	52.0^{d}	12.3^{bcd}	18.6^{bcde}	65.0^{d}	11.1^{bcd}
Organo-mineral fertilizer @ 40 g/plant	16.0^{cd}	61.0^{c}	12.2^{cd}	17.2^{de}	72.3^{c}	13.5^{cd}
Organo-mineral fertilizer @ 50 g/plant	14.7^{ef}	52.3^{d}	12.2^{cd}	16.7^{ef}	76.5^{d}	13.8^{cd}
Organo-mineral fertilizer @ 60 g/plant	16.3^{bc}	85.0^{a}	15.7^{a}	17.3^{a}	87.0^{a}	15.6^{a}
Poultry Manure @ 40 g/plant	15.3^{cde}	63.3^{c}	12.0^{cd}	15.5^{def}	83.3^{c}	12.5^{cd}
Poultry Manure @ 50 g/plant	16.3^{bc}	74.3^{b}	11.7^{cd}	16.7^{bcde}	78.5^{b}	11.4^{cd}
Poultry Manure @ 60 g/plant	16.2^{bc}	82.7^{a}	15.0^{ab}	17.7^{cde}	86.7^{a}	15.3^{ab}
Control	18.5^{abc}	65.0^{c}	14.2^{abc}	18.0^{ab}	45.0^{c}	12.0^{abc}

in terms of an increased number of lateral roots, both at the lower and the upper part of the tap root, and a higher average length of the lateral roots.

As noted in the methods (see Sect. 4.2 of this chapter), fertilizer application rates were not adjusted in this study to deliver the same rates of nutrients such as N, P, or K. Hence, there is a need to cross-validate the findings of this study with a similar assessment where the fertilizer rates are adjusted for at least one of the major nutrients.

This would be useful in order to fine-tune the fertilizer recommendations for cacao seedlings. This is particularly important as the best performance was obtained with poultry manure whose quality may be affected by seasonal and feeding variability, in addition to the manure management practice prior to its use as a fertilizer. An economic analysis of the various fertilizer products, when applied at the rate adjusted based on the nutrient content of at least one of the major nutrients, would also be necessary to determine the most cost-effective treatment.

4.5 Conclusion

Application of different types of fertilizer was found to enhance the development of cacao seedlings in the nursery with appreciable effects. Based on the study conditions, poultry manure and the organo-mineral fertilizer were found to perform better than inorganic NPK treatment (applied either to the soil surface or foliage) for the production of high quality seedlings based on height, leaf numbers, stem girth, and root system. Application of poultry manure at 50 g/seedling or 50 g of organo-mineral fertilizer, or 25 g of NPK 15:15:15 was proved to enhance seedling performance in cacao production and establishment.

Acknowledgements Our sincere appreciation goes to Professor Samuel O. Agele, Dr M. A. Awodun and Mrs D. Y. Famuwagun for their support towards the success of this research study. This chapter was first published as a paper in *International Letters of Natural Sciences*, 58, pp. 54–59, doi:10.18052/www.scipress.com/ILNS.58.54 (Famuwagun and Oladitan (2016). It is re-used here in slightly edited form (including a new sentence added to the conclusion) under Creative Commons license CC BY 4.0 (http://creativecommons.org/licenses/by/4.0).

References

Adenikinju, S. A., Esan, E. B., & Adeyemi, A. A. (1989). *Nursery techniques, propagation and management of cacao, kola, coffee, cashew and tea. In progress in tree crop research: A commemorative book to mark the 25th Anniversary of CRIN* (2nd ed.). Ibadan: Cocoa Research Institute of Nigeria.

Agele, S. O., Adeosun, S., & Oluwadare, D. (2004). A lysimeter study of nutrient release characteristics, leaching losses and effects on growth response of Amaranthus resulting from application of inorganic and organic nitrogen sources. *Journal of Food, Agriculture and the Environment, 2*(2), 301–306.

Ahp, P. M. (1979). *A text book on West Africa soil*. Oxford: Oxford University Press.

Alverson, W. S., Whitelock, B. A., Nyffeler, R., Bayer, C., & Baum, D. A. (1999). Phylogeny of the core Malvales: Evidence from ndhF sequence data. *American Journal of Botany, 86,* 1474–1486.

Ayanlaja, S. A. (2002). Soil nutrient management research in the humid forest zone of Nigeria for cacao production between 1940 and 2000. *Tropical Agriculture (Trinidad), 64,* 237–240.

Ayeni, L. S. (2011). Effect of sole and combined cocoa pod ash, poultry manure and NPK 20:10:10 fertilizer on soil organic carbon, available phosphorus and forms of nitrogen on alfisols in southwestern Nigeria. *International Research Journal of Agricultural and Soil Science, 1,* 77–82.

Egbe, N.E. (1968). Fertilizer on growth and yield of cocoa. In *CRIN Annual Report* (pp. 41–43). Ibadan: Cocoa Research Institute of Nigeria.

Ewulo, B. S. (2005). Effect of poultry dung and cattle manure on chemical properties of clay and sandy clay loamy soil. *Journal of Animal and Veterinary Advancement, 4,* 839–841.

Famuwagun, I. B., & Adekayode, F. O. (2010). Effect of wood ash, poultry manure and NPK fertilizer on growth, yield and seed quality of soya bean. *Applied Tropical Agriculture, 15,* 143–146.

Famuwagun, I. B., & Agele, S. O. (2010). Effects of sowing methods and plant population densities on root development of cacao seedlings in the nursery. *International Journal of Agricultural Research, 5*(7), 445–452.

Famuwagun, I. B., & Oladitan, T. O. (2016). Influence of varying rates of fertilizers on the performance of cacao (Theobroma cacao) seedlings in the nursery. *International Letters of Natural Sciences, 58,* 54–59.

Ibiremo, O. S., Oduwole, O. O., Fagbayide, J. A., & Nduka, B. A. (2012). Effect of root mat management and phosphate fertilizer application on the field establishment of cocoa in Ondo State, Nigeria. *International Journal of Plant, Animal and Environmental Sciences, 2*(2), 18–25.

ICCO. (2007). The World Cocoa Market. In *International Cocoa Organization Annual Report 2006/2007* (Chapter 3, pp. 19–22). London: International Cocoa Organization (ICCO). Available online at: https://www.icco.org/about-us/international-cocoa-agreements/cat_view/1-annual-report/23-icco-annual-report-in-english.html. Accessed September 2019.

Motamayor, J. C., Risterucci, A. M., Lopez, P. A., Ortiz, C. F., Moreno, A., & Lanaud, C. (2002). Cocoa domestication I: The origin of cacao cultivated by Mayas. *Heredity, 89,* 380–386.

Ojeniyi, S. O. (1991). Effect of bush clearing and tillage methods on soil physical and chemical properties of humid alfisol. *Soil Tillage Research, 15,* 269–277.

Parr, J.F., Hornick, S.B., & Kaufman, D.D. (1994). Use of microbial inoculants and organic fertilizers in agricultural production. *Extension Bulletin* No. 394. Taipei, Taiwan: Food & Fertilizer Technology Center (ASPC/FFTC). Available online at: http://www.agnet.org/library.php?func=view&id=20110722114739. Accessed September 2019.

Smith, J. N. (1994). Cocoa production systems, option and constraints. *Cocoa Growers' Bulletin, 47,* 20–26.

Wilson, C.K. (1999). *Coffee, cocoa and tea.* Crop production science in horticulture series, number 8. Wallingford: CABI Publishing.

Chapter 5
Assessing Synergies and Trade-Offs from Nitrogen Use in Africa

Jonathan E. Hickman, Shamie Zingore, Corinne Galy-Lacaux, Job Kihara, Mateete Bekunda, and Cheryl A. Palm

Abstract Fertilizer use in African agriculture is extremely low—just 4% of global fertilizer use, with an annual average of only 8 kg nutrients ha^{-1} in sub-Saharan Africa (SSA). A major focus of a new African Green Revolution is increasing inputs of nitrogen (N) to help restore soil fertility to soils that have experienced decades of nutrient depletion. These increased inputs can be expected to increase crop productivity in most soils, and may also increase soil organic matter, particularly when added with organic inputs or crop residues, which, in turn, can lead to increased water use efficiency. The increased N inputs may also be accompanied by a decline in N use efficiency, causing increased N losses to the environment with potential

J. E. Hickman (✉)
NASA Goddard Institute for Space Studies, 2880 Broadway, New York, NY 10025, USA
e-mail: jonathan.e.hickman@nasa.gov

Earth Institute at Columbia University, New York, NY 10025, USA

S. Zingore
International Plant Nutrition Institute (IPNI), c/o ICIPE Duduville Campus, Kasarani, P.O. Box 30772–00100, Nairobi, Kenya
e-mail: szingore@ipni.net

C. Galy-Lacaux
Laboratoire d'aérologie, UMR 5560, Université Toulouse III Paul Sabatier, CNRS, 31400 Toulouse, France
e-mail: corinne.galy-lacaux@aero.obs-mip.fr

J. Kihara
International Center for Tropical Agriculture (CIAT), c/o ICIPE Duduville Campus, Kasarani, Nairobi, Kenya
e-mail: j.kihara@cgiar.org

M. Bekunda
International Institute of Tropical Agriculture (IITA), Arusha, Tanzania
e-mail: M.Bekunda@cgiar.org

C. A. Palm
Institute for Sustainable Food Systems, University of Florida, Frazier Rogers Hall, PO Box 110570, Gainesville, FL 32611, USA
e-mail: cpalm@ufl.edu

M. A. Sutton et al. (eds.), *Just Enough Nitrogen*,
https://doi.org/10.1007/978-3-030-58065-0_5

impacts on water and air quality, soil pH and biodiversity. Here we briefly summarize the current state of knowledge of N dynamics in agricultural systems in SSA and the potential synergies and trade-offs that may result from increasing N inputs. As there have been fewer studies examining N cycling and losses in Africa compared to other parts of the world, it is difficult to predict confidently the magnitude of any environmental impacts associated with increased N inputs, and computer models are limited by data for model parameterization and operation. Given the potential of N losses from sewage effluent and agricultural activities to affect water quality in African coastal and aquatic environments and the potential for increased reactive N emissions and deposition to terrestrial and aquatic ecosystems, coordinated efforts should be made to fill these gaps.

Keywords Fertilizer · Soil fertility · Soil organic matter · Trade-offs · African Green Revolution

5.1 Introduction

The African Green Revolution launched by former UN secretary general Kofi Annan (Bationo and Bürkert, 2001; Annan 2004) is taking off to varying degrees within sub-Saharan Africa (SSA), most notably in Malawi (Annan 2004; Denning et al. 2009; Nziguheba et al. 2010; Minot & Benson 2009). The original Green Revolution starting in the 1960s tripled food production in many countries in 30 years (Hazell & Wood 2008), halved the number of rural poor, and reduced the proportion of undernourished from 30 to 18% (UN Millennium Project 2005). These benefits, which have long been seen as the initial key steps to economic transformation (Johnston & Kilby 1975), were not realized in SSA (Hazell and Wood 2008; Sanchez 2010). This failure of a green revolution in SSA was in part due to soil nutrient depletion (Sanchez 2002, 2010), a result of decades of nutrient removal from crop harvest, leaching, erosion and other loss pathways that exceeded nutrient inputs to the soil through mineral fertilizers, manures, biological nitrogen (N) fixation and atmospheric deposition. Annual N depletion rates averaged 22 kg N ha^{-1} over 30 years in 37 sub-Saharan countries (Stoorvogel and Smaling 1990), though nutrient depletion of agricultural soils is not universal in SSA (Cobo et al. 2010). In general, N deficiency is prevalent in smallholder agriculture in SSA, with an average annual application of only 8 kg of nutrients (not just N) per hectare in comparison with 100 kg N ha^{-1} in the US and India and 220 in China (Vitousek et al., 2009).

Sub-Saharan Africa represents the clearest example of the ongoing challenges created by "too little N", and the African Green Revolution emphasizes the challenge of overcoming soil nutrient depletion. Research and development projects have doubled or tripled maize yields with applications of 80 to 100 kg N ha^{-1} (Sanchez 2010; Nziguheba et al. 2010). At a national level, Malawi increased production from 0.8 to 2.0 t ha^{-1} of maize following the implementation of a national input subsidy program that included vouchers for two 50 kg bags of fertilizers (equivalent to 60–70 kg N ha^{-1}) (Denning et al. 2009). This "Malawi Miracle" increased interest in subsidy programs (Jayne and Rashid 2013), and many countries in SSA now give high priority to increasing fertilizer inputs. In addition, the World Bank has established a funding mechanism for sub-Saharan countries to scale up the African Green Revolution and is co-financing subsidy programs with multiple African nations (Jayne and Rashid 2013).

Subsidy programs have been associated with greater use of organic inputs (Holden and Lunduka 2012) and greater adoption of high-yielding varieties (Fisher and Kandiwa 2014; Chibwana et al. 2012). However, many economists criticize fertilizer subsidy programs as being an inefficient means for increasing production and reducing poverty (Jayne and Rashid 2013). This is because subsidies can provide fewer benefits to the poorest households than to those with greater resources (Jayne and Rashid 2013; Lunduka et al. 2013; Mason et al. 2013), reduce the diversity of crops cultivated (Chibwana et al. 2012), and potentially crowd out commercial fertilizer (Jayne and Rashid 2013). Nevertheless, critics acknowledge that fertilizer and seed subsidy programs have become a popular policy in SSA for increasing nutrient inputs and agricultural production, and as a result a lot of focus has shifted towards improving subsidy program targeting and efficiency (Jayne and Rashid 2013).

The main objective of increasing N inputs is to increase food security to reduce hunger and under-nutrition. Improved nutrition is linked to improved health status including reduction in child and maternal mortality and increased immunity to diseases (Bhutta et al. 2008). Increased production could also result in higher incomes if it enables farmers to grow surplus crops, and more importantly, shift to high value cash crops supported by accessible markets. A second potential positive impact is that increases in crop yields, due to the addition of sufficient levels of N (and other nutrients), may lead to increases in carbon (C) concentrations and moisture retention in some soils.

Limiting losses of reactive nitrogen (N_r) compounds to the environment will be critical for sustainable intensification of agriculture. Nitrate leaching, nutrient runoff, and erosion could potentially lead to contamination of surface and subsurface waters with N_r, as well as the eutrophication of aquatic and coastal ecosystems. Nitrous oxide (N_2O) is a greenhouse gas with 298 times the warming potential of carbon dioxide, while nitric oxide (NO) and ammonia (NH_3) can contribute to the formation of aerosols and particulates in the atmosphere, rain acidification, and N deposition; NO_x is also a precursor to tropospheric ozone, which is a powerful greenhouse gas and harmful pollutant. The key question is how to achieve an African Green Revolution while avoiding the N pollution seen ubiquitously in other developed and developing countries with intensified agriculture. In particular, adoption of appropriate management of N and other nutrients by African farmers will have to make an essential contribution to ensuring a sustainable Green Revolution in Africa. This chapter reviews the literature to summarize the current state of knowledge of N dynamics in agricultural systems in SSA and the potential synergies and trade-offs that may result from increasing N inputs.

5.2 Crop Yields and Nitrogen Use Efficiency

Crop yields have shown strong responses to N fertilizer additions, with yields doubling and tripling under recent subsidy schemes (e.g., Denning et al. 2009), although the use of improved varieties and management have also been critical drivers of increasing yields.

In a review of yield response data from across SSA, Chikowo et al. (2009) found that N application rates explained a substantial part of the variation in maize yields in the reviewed experiments, while there was no relationship in the in an estimation of nutrient capture efficiency (NCE). In contrast, increased N additions are not a universal panacea for crop productivity. More generally, the maximum yields achievable with mineral N fertilizers vary substantially and are a function of genetic potential, soil constraints and available moisture. Soil organic carbon (SOC) and clay content appear to be two important factors determining yield responses to inorganic fertilizer in SSA.

In a study by Zingore et al. (2007), differences in soil N, soil phosphorus (P), and SOC within soil types led to large differences in baseline yields, yield responses to fertilizer additions, and nutrient use efficiencies. Losses of SOC result from imbalances in losses of organic matter through factors such as erosion, decomposition, and inputs of organic matter (such as manure or plant roots, stems, and leaves). Soil degradation from continuous cultivation without nutrient or organic inputs, or sustained interruptions in additions of fertilizer or organic matter, can reduce or eliminate productivity responses to fertilizer additions (Tittonell and Giller 2013). These shifts in productivity may represent relatively stable—or at least highly resilient—states. Restoration of non-responsive soils can take years and large investments in manure and fertilizer (Zingore et al. 2007; Tittonell and Giller 2013), and productivity can still fall short of adjacent fields that did not experience an interruption in inputs (Tittonell and Giller 2013). Some soils may be degraded beyond the point where they can be fully restored (Morris et al. 2007; Sanchez 2010; Nziguheba et al. 2010; Tittonell and Giller 2013).

Currently there are not many data on the level of SOC needed to ensure a yield response to nutrient inputs. Yields responded to fertilizer at some sites in West Africa when SOC concentrations were as little as 1.7 to 3.5 g C kg^{-1} soil (Bationo et al. 2007; Mtambanengwe and Mapfumo 2005), but did not respond when levels were 4.4 g C kg^{-1} soil at another site (Mtambanengwe and Mapfumo 2005), or at 3 g C kg^{-1} soil in Zimbabwe (Zingore et al. 2007). Understanding multiple site-specific factors constraining N response, and the spatial distribution of soil characteristics in particular, is therefore critical for appropriately managing N to maximize yield or economic returns, as well as to minimize losses and off-site effects.

Chikowo et al. (2009) found no relationship between N inputs and nitrogen use efficiency (NUE), presumably due to the wide variability in soils, differences in limitation of other nutrients, climate, and crop varieties between sites evaluated. In a global model of agricultural N flows, Liu et al. (2010) estimated recovery of fertilizer N in harvested grain and residue to be 59% in SSA, compared to over

70% in the United States and Europe, in spite of an order of magnitude difference in fertilizer additions. However, some initial estimates of N recovery based on yield estimates for maize grown at recommended rates of fertilizer addition on smallholder farms in several countries in SSA were comparable to those in the United States (Nziguheba et al. 2010). Increasing crop N recovery through the use of improved varieties and improved management should lead to improved yields and could limit losses associated with increased fertilizer application in the future.

5.3 Effects of Increasing Nitrogen Inputs on Soil Organic Carbon

There are clear arguments that fertilizer use in intensive agricultural production can lead to decreases in SOC (Khan et al. 2007), but in the low-input, low-yield farming systems in SSA, increased additions of N can be essential for increasing soil organic matter through increases in root biomass (Bationo et al. 2007) and increases in stover (leaves and stalks of field crops) production that can help build SOC when added to the soil (Bationo and Bürkert 2001; Bationo et al. 2007; Zingore et al. 2005). However, there are exceptions to the positive benefits of N input on SOC in the low-input systems of SSA (Pichot et al. 1981). Regardless, relying exclusively on organic sources of nutrients may not be a sustainable agricultural strategy in the region. Livestock densities are too low to supply sufficient manure to maintain soil fertility in existing cropland (Tittonell and Giller 2013); a case study from Zimbabwe found that the carrying capacity of the region's grasslands limited potential manure production to 1/3 of what would be required for farmers' fields (Zingore et al. 2011).

The boost in biomass production from increased mineral N inputs can limit SOC losses in many agricultural systems in SSA. Comparison of long-term changes in soil organic matter after forest clearance in Zimbabwe revealed that equilibrium contents of soil C under high N input commercial agriculture were 32 t C ha^{-1}, nearly twice as much as under low-input smallholder agriculture (Zingore et al. 2005). The long-term SOC differences were attributed to a six-fold gap in biomass production between commercial and smallholder farms (8 t ha^{-1} and 1.2 t ha^{-1}, respectively), which is the result of large differences in fertilizer use (~ 150 kg N ha^{-1}, 30 kg P ha^{-1}, and 30 kg K ha^{-1} in commercial farms, but little to no fertilizer use by smallholders). In another example, long-term trials in West Africa have shown that use of mineral N in combination with P can contribute to substantial increases in SOC contents (Bationo et al. 2007).

In many cases, inorganic N inputs may require organic inputs to maintain SOC and yields. In an example from Niger, incorporation of crop residues along with inorganic fertilizer resulted in SOC levels about twice as high as in plots receiving no inputs after 12 years, although it was still lower than in adjacent fallows (Bationo and Bürkert 2001). Without other sources of organic matter, an estimated 2–5 t of crop residues are required every year to maintain the SOC contents necessary to support the productive capacity of semi-arid West African clay soils, and, consequently, help to maintain yield responses (Bationo et al. 2007; Rufino et al. 2011). When organic and other resources are limiting, targeting organic applications to less fertile soils can produce substantially higher yields and larger increases in SOC at a landscape scale (Rufino et al. 2011). However, these positive effects of N inputs on SOC are not always observed. In some instances, fertilizer use accompanied by crop residue mulching lead to increases in soil respiration (Bationo and Bürkert 2001), and greater losses in SOC (Pieri 1995). In addition, organic resources have value outside of their use in soil fertility restoration or maintenance. In Morocco, residues represent 25% of the value of a crop in normal years, and up to 75% in drought years (Magnan et al. 2012). Competing uses of organic resources may make management of fields for SOC and yield a challenging factor particularly for smallholder farmers in areas where production is not high enough to produce a substantial surplus of residues above the amounts required for livestock feed or fuel (Valbuena et al. 2012).

5.4 Water Use Efficiency

Crop productivity in SSA can be so poor that only 10 to 15% of total rainfall is taken up (Heng and Nguyen 2006). Increased N additions may increase crop water use efficiency by increasing the amount of water that flows through the transpiration stream. Any improvements in soil organic matter and soil structure resulting from N additions will be accompanied by increased water retention within the soil. In addition, shading of the soil by increased aboveground biomass reduces evaporation and increases the proportion of soil water passing through the transpiration stream (Rockstrom 2003). The increased N also increases the size of rooting systems in the soil, with added benefits for reducing erosion. These effects of N applications can increase the proportion of soil water accessed by the crop (Rockstrom 2003).

5.5 NO_3^- Leaching and N_2O Losses

Africa is currently estimated to be responsible for 18% of global anthropogenic N_2O emissions, overwhelmingly from agriculture and biomass burning (Hickman et al. 2011). The majority of agricultural emissions are from unmanaged livestock excretions (Fig. 5.1), although imbedded assumptions about high N excretion rates by African livestock may make these estimates too high (Hickman et al. 2011). Additional studies of losses from livestock systems will be important for refining estimates of N flows in African agriculture (e.g., Schlecht et al. 2004). Even under the low-input regimes prevalent in SSA, 20, 14 and 12% of current N inputs are estimated to be lost through erosion, gas emissions, and leaching, respectively (Liu et al. 2010). Using Millennium Ecosystem Assessment scenarios for 2050, in which fertilizer use increases by a factor of 1.5 to 6, N_2O emissions from African agriculture

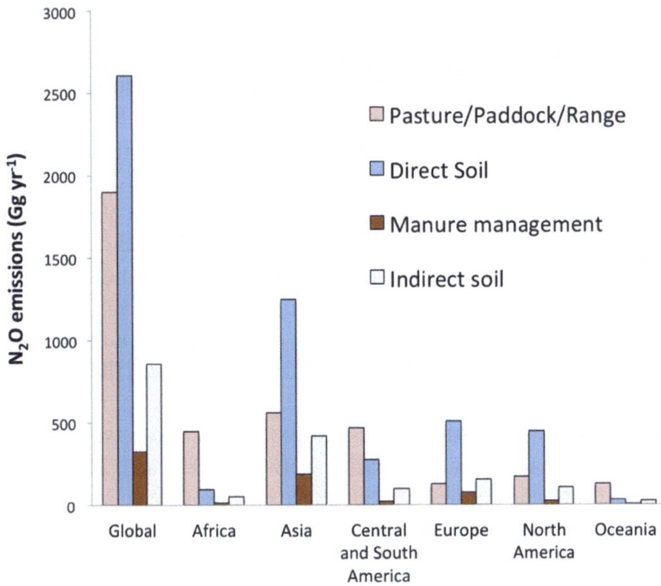

Fig. 5.1 Continental estimates of agricultural N_2O emissions for 2005, not including emissions from biomass burning (EDGAR v. 4.1). Yr = year

are expected to roughly double, in keeping with estimates from the IMAGE model (Bouwman et al. 2009).

Direct measurements of N losses in leaching or gas emissions from agricultural soils are relatively scarce in SSA. Roughly 20 studies have been published examining N_2O or NO emissions from African agriculture, including both field and laboratory measurements. Fluxes from unfertilized agricultural soils in SSA tend to be lower than global averages (a mean of 670 g N ha^{-1} in SSA compared to a global average of roughly 1 kg N ha^{-1}). Emissions of N_2O have been seen to increase by a factor of more than 10 with additions of inorganic fertilizers in SSA (Hickman et al. 2011). However, the overall emission losses from fertilizer are generally small in published studies from SSA, with emission factors (the percentage of fertilizer N lost as N_2O) largely well below the IPCC default of 1% for direct emissions (Hickman et al. 2014).

A larger range of losses is reported for leaching losses with fertilizer additions. Over two seasons with seasonal inputs of 60 kg N ha^{-1} to potassium-limited soils at a Togo experimental station, more N was lost through leaching than was added in fertilizer (Poss 1992). In contrast, leaching losses of ^{15}N-enriched ammonium sulphate in a loamy sand with 0.8% SOC in Zimbabwe never exceeded 0.25% of the applied N (about 2% of the total N leached) (Kamukondiwa and Bergström 1994), but much of the labeled N was retained within soils, and may be leached in subsequent years. If increased production is accompanied by increased residue applications, immobilization of N may reduce leaching losses, particularly in regions where crop residue does not decompose quickly (Bationo and Bürkert 2001).

Losses of N may be characterized by important thresholds of N application, above which losses may be expected to increase rapidly. For example, hydrologic NO_3^- exports increase exponentially with increasing rates of N application, presumably when N is no longer limiting to plant growth (Andraski et al. 2000; Bergström and Brink 1986; Power et al. 2000; Færge and Magid 2004). N_2O emissions have exhibited similar exponential responses to incremental increases in N additions (Hoben et al. 2010; McSwiney and Robertson 2005; Van Groenigen et al. 2010; Shcherbak et al. 2014). In addition to identifying how widespread these thresholds may be in sub-Saharan agro-ecosystems, additional studies can be stratified across soils and agro-ecological zones to identify how climate and the physical and chemical characteristics of soils affect these critical thresholds, and to identify nutrient input ranges that allow for maximizing yields while minimizing the risks to the environment.

Currently, land clearing for agriculture is a large source of agriculturally-related greenhouse gas emissions in the tropics; tropical agriculture was responsible for 98% of CO_2 emissions from land clearing between 2000 and 2009 (DeFries and Rosenzweig 2010). Intensification of agriculture may provide an avenue for avoiding these emissions. To date, increasing productivity of agricultural land in the tropics globally has resulted in emissions reductions of 161 Gt CO_{2eq} between 1961 and 2005 (Burney et al. 2010). This mitigating effect is expected to continue in the future (Tilman et al. 2011), with the largest potential in developing countries with relatively high food supplies (Ewers et al. 2009). Simple landscape models found that in East Africa agricultural intensification using mineral fertilizer or agro-forestry resulted in reducing emissions by 1 to 6.5 t CO_{2eq} ha^{-1} (Palm et al. 2010).

5.6 Other Reactive Nitrogen Losses and Deposition

Deposition and emissions of other N_r species are poorly quantified components of N budgets in SSA. Future projections in Asia and in SSA suggest that these regions can expect substantial increases in N_r deposition over coming decades (Rufino et al. 2011; Bationo and Bürkert 2001; Dentener et al. 2006; Bationo et al. 2007; Dentener et al. 2014), due to the combination of urban development and agriculture extensification and intensification (Bouwman et al. 2009). A useful way to assess the impacts of these increases on ecosystems is the critical load concept, which is centered on the idea that there are threshold rates above which N deposition will begin to have substantial negative effects on ecosystem health. Although N deposition exceeding critical loads has been a greater issue for Europe and North America, several tropical regions in Latin America and SSA are likely to become vulnerable to the expected increases in N deposition rates (Bobbink et al. 2010).

Quantification of the processes contributing to changing levels of N emissions and deposition is restricted by a very limited capacity for monitoring N deposition in SSA. A global assessment of precipitation chemistry and deposition has been carried out under the direction of the World Meteorological Organization (WMO) Global Atmosphere Watch (GAW) Scientific Advisory Group for Precipitation Chemistry

(SAG-PC) (Vet et al. 2014). The INDAAF project (International Network to study Deposition and Atmospheric Chemistry in AFrica) has provided budgets of wet and dry N deposition at a regional ecosystem scale that have been estimated for western and southern Africa, using direct measurements made through the INDAAF network (https://indaaf.obs-mip.fr; Bationo and Bürkert 2001; Galy-Lacaux et al. 2014; Adon et al. 2013; Delon et al. 2014; Pieri 1995; Williams et al. 2009; Delon et al. 2010). A new initiative for the development of a similar network for eastern and central Africa—the Equatorial African Deposition Network—is currently in the planning stages (Odada 2012).

Figure 5.2 shows the estimated N deposition budget in West and Central Africa through the comparison of results obtained in a 2000 ensemble mean global modelling study (Dentener et al. 2006) with INDAAF deposition fluxes estimated over a ten-year period (1997-2008, Galy-Lacaux et al. 2014). Measured and modelled N deposition fluxes in West and Central Africa are not negligible, ranging between 5.6 and 8.7 kg N ha^{-1} year^{-1} in dry savannah, 7 to 10 kg N ha^{-1} year^{-1} in wet savannah, and about 12.6 kg N ha^{-1} year^{-1} respectively in forested ecosystems. Modelled and observed results are in good agreement for wet savannahs and forests. In contrast, the modelling study underestimates N deposition fluxes in dry savannahs by a factor of 2. This

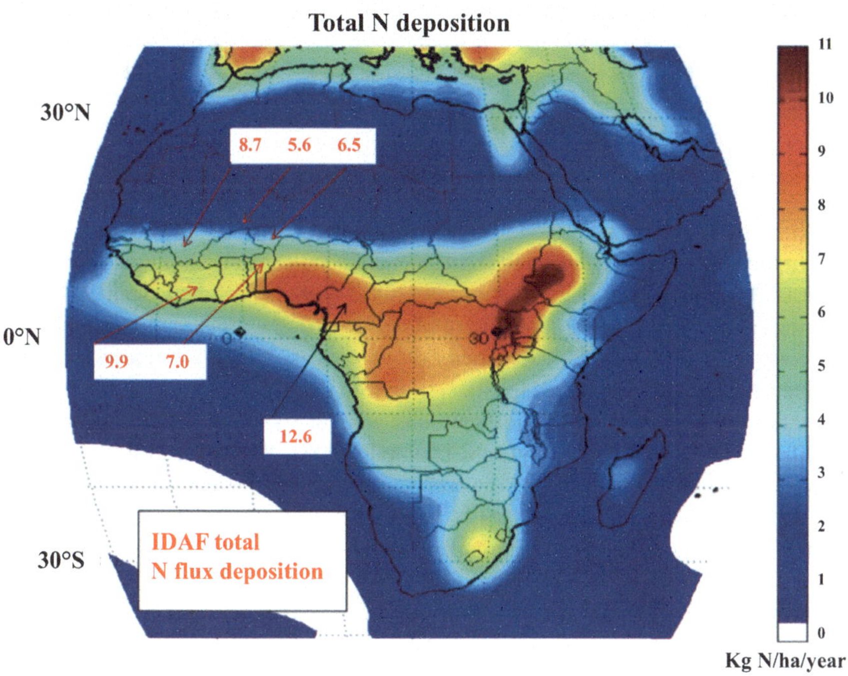

Fig. 5.2 2000 ensemble-mean pattern of $N_{oxidized} + N_{reduced}$ Wet + Dry Deposition with INDAAF observations superimposed (in kg N ha^{-1} year^{-1}). The observations in west central Africa represent the estimated 10-year-average annual $N_{oxidized} + N_{reduced}$ Wet + Dry Deposition fluxes for the period 1997–2008

result is likely the consequence of the underestimation of soil biogenic NO and NH_3 emissions resulting from agro-pastoralism.

Measurements of emissions of N_r from agricultural soils are even more limited than measurements of deposition. Fluxes of nitric oxide (NO) from agricultural soils have been measured for three months at one site in Zimbabwe, but increased by a factor of 3 in response to modest fertilizer additions of 16 kg N ha^{-1} (Meixner et al. 1997). Modelling and remote sensing studies have found that NO fluxes from soils represent a larger proportion of the continental flux of NO than previously believed, suggesting that agricultural emissions of NO as N fertilizer use increases in SSA may represent a substantial new source of this reactive gas (Jaeglé et al. 2004; Delon et al. 2008, 2010, 2012).

Several new tools and databases for the African continent have emerged from the African Monsoon Multidisciplinary Analysis (AMMA) to better address deposition and emission rates and inform other modelling efforts and assessments. Satellite data have improved emission inventories (https://eccad3.sedoo.fr/; Assamoi and Liousse 2010), vegetation maps (Tchuente et al. 2010), and our understanding of regional meteorology (Boone et al. 2009). Efforts such as the global precipitation chemistry assessment (http://wdcpc.org) and the African Soil Information Service (AfSIS) digital soil mapping project (Sanchez et al. 2009) are providing improved datasets and tools for evaluating emissions and deposition. In combination, these maps, inventories, and models may be used to conduct spatially-explicit evaluations of how increased fertilizer use may affect both reactive N emissions and deposition.

5.7 Soil Acidification

Even in the absence of external perturbation, acidic soils are common in the tropics; a quarter of tropical soils exhibit aluminum toxicity (Sanchez and Logan 1992). In addition, many soils used for agriculture in SSA, with low organic matter and high sand content, have low buffering capacity (Ngatunga et al. 2001). In these soils, the application of ammonium-based fertilizers can lead to decreases in soil pH. Seasonal application of urea can lead to acidification of soils in SSA at high application rates (> 100 kg N ha^{-1} $year^{-1}$, Lungu and Dynoodt 2008). Without additions of crop residues, inorganic N and P inputs lowered soil pH in a three-year study on weakly-buffered Sahelian soils (Bationo 1993). Studies of continuous cultivation of cotton in Nigeria (Bache 1969), of maize and maize/cassava intercrops in a Nigerian kaolinitic Alfisol at 120 kg N ha^{-1} as urea (Juo et al. 1995), of maize in Ultisols in Burkino Faso under 90 kg N ha^{-1} as urea (Bado et al. 2004), and of millet in Niger (Bationo and Bürkert 2001) suggest that soil pH decreases with the use of mineral fertilizers, although in the latter studies, the degree of acidification was reduced when fertilizer additions were accompanied by stover and manure, respectively.

5.8 Biodiversity

Documented impacts of N inputs on biodiversity in SSA are uncommon. There are many ways in which agricultural intensification with increased N inputs can be expected to affect biodiversity. Major pathways include: (1) through the direct effects of N loading in terrestrial ecosystems, (2) through the nutrient enrichment (and possible eutrophication) of aquatic systems, and (3) by reducing the amount of natural habitat converted to agriculture (i.e., land sparing). Effects in terrestrial systems appear largely unstudied, but in general, increasing N inputs tends to reduce terrestrial biodiversity (Robertson and Vitousek 2009; Tilman and Lehman 2001), with potentially serious consequences as N inputs exceed critical loads (Bobbink et al. 2010). A few studies have found links between N inputs and degradation of aquatic ecosystems. In Lake Victoria, N is seen as one of multiple factors (of which the introduction of the Nile Perch is particularly notable) contributing to a dramatic decline in diversity of endemic cichlid species (Hecky et al. 1994; Verschuren et al. 2002; Njiru et al. 2010). Several cichlid species that are rare or extinct from Lake Victoria are present in surrounding ox-bow lakes, but there are no policies for their protection (Katunzi et al. 2010). In Lake Malawi, the highest loads of nutrients and sediment are associated with high agricultural activity, and the clearing of woodland for agriculture may have increased nutrient loading to the lake by 50% (Hecky et al. 1994).Vanlauwe and Giller (2006) argue that because fertilizers are not used in excessive quantities in SSA, agricultural systems are unlikely to be a major cause of aquatic eutrophication, suggesting instead that untreated sewage waste is the most likely cause of increased nutrient inputs to Lake Victoria. Less than 30% of sewage is treated in SSA, and untreated sewage may be responsible for most coastal and lake eutrophication in the region (Nyenje et al. 2010). The effects of sewage are clear in Ghana, where coastal lagoons have become nutrient enriched downstream of high population densities (Nixon et al. 2007). Without new practices and policies to protect water quality, inputs to Ebrie lagoon near Abidjan, Ivory Coast are expected to quintuple by 2050 (Scheren et al. 2004). While it may vary regionally, intensification should result in substantially reduced conversion of other habitat to agriculture in SSA (Palm et al. 2010). Given the high diversity of natural habitat in SSA (particularly forested ecosystems), any sparing of natural lands resulting from agricultural intensification should have net positive benefits for biodiversity.

5.9 Conclusions

A new African Green Revolution is underway, and will involve substantial increases in N inputs to African environments. Currently, there is not enough data collected for SSA to make robust conclusions regarding the goal of using increased N inputs to increase agricultural productivity without repeating the severe environmental degradation experienced in parts of Asia during the original Green Revolution (Table 5.1).

Table 5.1 Preliminary conclusions regarding current state of knowledge and the magnitude of changes likely with increased inputs of nitrogen (N) for different responses

Effect of increased N inputs	Preliminary conclusion
N_2O	N inputs appear to have small effects, but few studies have been conducted
NO emissions	Potential large source but few studies have been conducted
N_r deposition	N deposition currently 5–12 kg N ha^{-1} $year^{-1}$ in West Africa; deposition expected to increase and may cause negative effects in future. Deposition measurement network expanding
NO_3^-	Data on leaching rates vary widely, from less than 1% to over 100% of added N lost; few studies have been conducted
SOC	Organic inputs can help build SOC and increase yields; inorganic N inputs may reduce rates of SOC losses over time and may increase SOC in some low-SOC soils
Soil acidification	Seen to accompany inorganic fertilizer use; effects often ameliorated with organic additions
Biodiversity	Limited data on current effects. On-farm biodiversity likely to decline, but intensification may provide benefits if it results in land-sparing. Policies may be needed to protect aquatic ecosystems and endemic aquatic species

It appears likely that the greatest benefits to soils and long-term soil fertility will result from the thoughtful use of both organic and inorganic resources. Moreover, with careful management of N inputs, there appears to be potential for obtaining benefits while avoiding excessive losses of N to the environment. Emissions of N_2O are currently low from agriculture in SSA, and while increased N inputs will increase these emissions, current data suggest that these increases may be small in the near-term. However, data are still very limited, and if emissions increase over time, approaches for mitigation that do not adversely affect local livelihoods may become a higher priority. It is even more difficult to draw strong conclusions regarding risks to groundwater and aquatic systems, and additional data on leaching losses and watershed-scale nutrient transport are needed for policy makers to have the information they need to maintain the integrity of aquatic ecosystems and groundwater quality.

The robustness of these conclusions is severely limited by a body of research on N flows and cycling that is not as extensive as in many other parts of the world. Given the urgent need for practicable knowledge, the most effective way forward may require an emphasis on models and controlled field studies. The development of a digital soil map for SSA by AfSIS should provide an improved basis for many modelling activities, particularly when used in combination with improved datasets of cropping areas and management (e.g., Potter et al. 2010; Ramankutty et al. 2008); modelling activities supported by direct measurements can provide a powerful approach for evaluating emissions and deposition (Vet et al. 2014).

Characterizations of land cover and weather, based on remote sensing observations, have improved in recent years (e.g., Boone et al. 2009; Tchuente et al. 2010). The compilation of existing datasets on cropping area, yield, fertilizer use, and varieties from district and regional sources (such as government agricultural offices) is time-consuming and challenging, but could provide important data. More detailed—and more frequently published—data from field trials would help improve simulations of production and losses under different crops and varieties. It is critical that simulations include links to socio-economic models to more reliably assess the impacts of pathways for agricultural intensification. Complementing modelling exercises, field studies examining both N and C dynamics under increasing fertilizer inputs will be essential for identifying important environmental thresholds. Ultimately, extension agents and farmers must have access to detailed information on local soil characteristics and best management practices informed by these research efforts.

As N inputs increase towards recommended rates in SSA, identifying the critical thresholds, trade-offs, and synergies that may accompany different approaches to agricultural intensification in SSA represents the first critical step towards avoiding the mistakes made during the Asian Green Revolution. These new experiments, datasets, and modelling studies can provide decision makers, extension agents, farmers, and other members with the information they need to implement sound policy and management practices that can help ensure food security, alleviate poverty, and maintain the health of the environment.

Acknowledgements The INDAAF (International Network to study Deposition and Atmospheric chemistry in Africa) long-term project (http://indaaf.obs-mip.fr) is supported by the CNRS/INSU (Centre National de la Recherche Scientifique/ Institut National des Sciences de l'Univers), by the ACTRIS-FR research infrastructure and by the IRD (Institut de Recherche pour le Developpement). The Editors would also like to thank Dr. R. P. Raju (School of Public and Environmental Health, Hawassa University, Ethiopia) and Dr. Konka Bheemaligeswara (Department of Earth Sciences, Mekelle University, Ethiopia) for additional valuable comments and suggestions.

References

Adon, M., Galy-Lacaux, C., Delon, C., Yoboue, V., Solmon, F., & Kaptue Tchuente, A. T. (2013). Dry deposition of nitrogen compounds (NO_2, HNO_3, NH_3), sulfur dioxide and ozone in west and central African ecosystems using the inferential method. *Atmospheric Chemistry and Physics, 13,* 11351–11374.

Andraski, T. W., Bundy, L. G., & Brye, K. R. (2000). Crop management and corn nitrogen rate effects on nitrate leaching. *Journal of Environment Quality, 29,* 1095–1103.

Annan, K. (2004). *Secretary-General calls for "Uniquely African Green Revolution" in 21st century, to end continent's plague of hunger, in Addis Ababa remarks.* Press Release. Available online at: https://www.un.org/press/en/2004/sgsm9405.doc.htm. Accessed Aug 2019.

Assamoi, E.-M., & Liousse, C. (2010). Atmospheric environment. *Atmospheric Environment, 44,* 3985–3996.

Bache, B. W. (1969). Long-term effects of fertilizers and manure on soil and leaves of cotton in nigeria. *Experimental Agriculture, 5,* 241–247.

Bado, B. V., Sedogo, M. P., & Lompo, F. (2004). Long term effects of mineral fertilizers, phosphate rock, dolomite and manure on the characteristics of an ultisol and maize yield in Burkina Faso. In A. Bationo (Ed.), *Managing nutrient cycles to sustain soil fertility in sub-Saharan Africa* (pp. 77–88). Nairobi: Academy Science Publishers.

Bationo, A. (1993). The effect of crop residue and fertilizer use on pearl millet yields in Niger. *Fertilizer Research, 34,* 251–258.

Bationo, A., & Bürkert, A. (2001). Soil organic carbon management for sustainable land use in Sudano-Sahelian West Africa. *Nutrient Cycling in Agroecosystems, 61,* 131–142.

Bationo, A., Kihara, J., Vanlauwe, B., Waswa, B., & Kimetu, J. (2007). Soil organic carbon dynamics, functions and management in West African agro-ecosystems. *Agricultural Systems, 94,* 13–25.

Bergström, L., & Brink, N. (1986). Effects of differentiated applications of fertilizer N on leaching losses and distribution of inorganic N in the soil. *Plant and Soil, 93,* 333–345.

Bhutta, Z. A., Ahmed, T., Black, R. E., Cousens, S., Dewey, K., Giugliani, E., et al. (2008). What works? Interventions for maternal and child undernutrition and survival. *The Lancet, 371,* 417–440.

Bobbink, R., Hicks, K., Galloway, J., Spranger, T., Alkemade, R., Ashmore, M., et al. (2010). Global assessment of nitrogen deposition effects on terrestrial plant diversity: a synthesis. *Ecological Applications, 20,* 30–59.

Boone, A., Decharme, B., Guichard, F., de Rosnay, P., Balsamo, G., Beljaars, A., et al. (2009). The AMMA land surface model intercomparison project (ALMIP). *Bulletin of the American Meteorological Society, 90,* 1865–1880.

Bouwman, A. F., Beusen, A. H. W., & Billen, G. (2009). Human alteration of the global nitrogen and phosphorus soil balances for the period 1970–2050. *Global Biogeochemical Cycles,* 23(4), GB0A04. https://doi.org/10.1029/2009gb003576.

Burney, J. A., Davis, S. J., & Lobell, D. B. (2010). Greenhouse gas mitigation by agricultural intensification. *Proceedings of the National Academy of Sciences, 107*(26), 12052–12057.

Chibwana, C., Fisher, M., & Shively, G. (2012). Cropland allocation effects of agricultural input subsidies in malawi. *World Development, 40,* 124–133.

Chikowo, R., Corbeels, M., Mapfumo, P., Tittonell, P., Vanlauwe, B., & Giller, K. E. (2009). Nitrogen and phosphorus capture and recovery efficiencies, and crop responses to a range of soil fertility management strategies in sub-Saharan Africa. *Nutrient Cycling in Agroecosystems, 88,* 59–77.

Cobo, J. G., Dercon, G., & Cadisch, G. (2010). Nutrient balances in African land use systems across different spatial scales: A review of approaches, challenges and progress. *Agriculture, Ecosystems & Environment, 136,* 1–15.

DeFries, R., & Rosenzweig, C. (2010). Toward a whole-landscape approach for sustainable land use in the tropics. *Proceedings of the National Academy of Sciences, 107,* 19627–19632.

Delon, C., Galy-Lacaux, C., Adon, M., Liousse, C., Boone, A., Serca, D., Diop, B., et al. (2014). Interannual variability of the atmospheric nitrogen budget in West African dry savannas. In M.A. Sutton, K.E. Mason, L.J. Sheppard, H. Sverdrup, R. Haeuber, W.K. Hicks (Eds.), *Nitrogen deposition, critical loads and biodiversity (Proceedings of the International Nitrogen Initiative workshop, linking experts of the Convention on Long-range Transboundary Air Pollution and the Convention on Biological Diversity* (Chapter 11, pp. 93–106). Dordrecht: Springer.

Delon, C., Galy-Lacaux, C., Adon, M., Liousse, C., Serca, D., Diop, B., et al. (2012). Nitrogen compounds emission and deposition in West African ecosystems: Comparison between wet and dry savanna. *Biogeosciences, 9,* 385–402.

Delon, C., Galy-Lacaux, C., Boone, A., Liousse, C., Serca, D., Adon, M., et al. (2010). Atmospheric nitrogen budget in Sahelian dry savannas. *Atmospheric Chemistry and Physics, 10,* 2691–2708.

Delon, C., Reeves, C. E., Stewart, D. J., Serca, D., Dupont, R., Mari, C., et al. (2008). Biogenic nitrogen oxide emissions from soils–impact on CNO_X and ozone over West Africa during AMMA

(African Monsoon Multidisciplinary Experiment): modelling study. *Atmospheric Chemistry and Physics, 8,* 2351–2363.

Denning, G., Kabambe, P., Sanchez, P., Malik, A., Flor, R., Harawa, R., et al. (2009). Input subsidies to improve smallholder maize productivity in malawi: Toward an african green revolution. *PLoS Biology, 7*(1), e1000023. https://doi.org/10.1371/journal.pbio.1000023.

Dentener, F., Drevet, J., Lamarque, J. F., Bey, I., Eickhout, B., Fiore, A. M., Hauglustaine, D., et al. (2006). Nitrogen and sulfur deposition on regional and global scales: A multimodel evaluation. *Global Biogeochemical Cycles, 20,* GB4003. https://doi.org/10.1029/2005gb002672.

Dentener, F., Vet, R., Dennis, R.L., Du, E., Kulshrestha, U.C., & Galy-Lacaux, C. (2014). Progress in Monitoring and Modelling Estimates of Nitrogen Deposition at Local, Regional and Global Scales. In M. A. Sutton, K. E. Mason, L. J. Sheppard, H. Sverdrup, R. Haeuber, & W. K. Hicks (Eds.), *Nitrogen deposition, critical loads and biodiversity (Proceedings of the International Nitrogen Initiative workshop, linking experts of the Convention on Long-range Transboundary Air Pollution and the Convention on Biological Diversity* (Chapter 2, pp. 7–22). Dordrecht: Springer.

Ewers, R. M., Scharlemann, J. P. W., Balmford, A., & Green, R. E. (2009). Do increases in agricultural yield spare land for nature? *Global Change Biology, 15,* 1716–1726.

Færge, J., & Magid, J. (2004). Evaluating NUTMON nutrient balancing in sub-Saharan Africa. *Nutrient Cycling in Agroecosystems, 69,* 101–110.

Fisher, M., & Kandiwa, V. (2014). Can agricultural input subsidies reduce the gender gap in modern maize adoption? Evidence from Malawi. *Food Policy, 45,* 101–111.

Galy-Lacaux, C., Delon, C., Solmon, F., Adon, M., Yoboue, V., Mphepya, J., Pienaar, J.J., et al. (2014). Dry and wet atmospheric nitrogen deposition in West Central Africa. In M. A. Sutton, K. E. Mason, L. J. Sheppard, H. Sverdrup, R. Haeuber, W. K. Hicks (Eds.), *Nitrogen deposition, critical loads and biodiversity (Proceedings of the International Nitrogen Initiative workshop, linking experts of the convention on long-range transboundary air pollution and the convention on biological diversity* (Chapter 10, pp. 83–92). Dordrecht: Springer.

Hazell, P., & Wood, S. (2008). Drivers of change in global agriculture. *Philosophical Transactions of the Royal Society B: Biological Sciences, 363,* 495–515.

Hecky, R. E., Bugenyi, F. W. B., Ochumba, P., Talling, J. F., Mugidde, R., Gophen, M., et al. (1994). Deoxygenation of the deep water of Lake Victoria, East Africa. *Limnology and Oceanography, 39,* 1476–1481.

Heng, L., & Nguyen, L. (2006). *Water use efficiency in agriculture: The role of nuclear and isotopic techniques.* Vienna: International Atomic Energy Agency. Available online at: http://www-naweb.iaea.org/nafa/swmn/public/Nguyen-NTR-2007-WUE-27–10-2006. pdf. Accessed February 2017.

Hickman, J. E., Havlikova, M., Kroeze, C., & Palm, C. A. (2011). Current and future nitrous oxide emissions from African agriculture. *Current Opinion in Environmental Sustainability, 5,* 370–378.

Hickman, J., Scholes, R. J., Rosenstock, T. S., Pérez García-Pando, C., & Nyamangara, J. (2014). Assessing non-CO_2 climate-forcing emissions and mitigation in sub-Saharan Africa. *Current Opinion in Environmental Sustainability, 9–10,* 65–72.

Hoben, J. P., Gehl, R. J., Millar, N., Grace, P. R., & Robertson, G. P. (2010). Nonlinear nitrous oxide (N_2O) response to nitrogen fertilizer in on-farm corn crops of the US Midwest. *Global Change Biology, 17,* 1140–1152.

Holden, S., & Lunduka, R. (2012). Do fertilizer subsidies crowd out organic manures? The case of Malawi. *Agricultural Economics, 43,* 303–314.

Jaeglé, L., Martin, R. V., Chance, K., Steinberger, L., Kurosu, T. P., Jacob, D. J., et al. (2004). Satellite mapping of rain-induced nitric oxide emissions from soils. *Journal of Geophysical Research, 109,* D21310. https://doi.org/10.1029/2004jd004787.

Jayne, T. S., & Rashid, S. (2013). Input subsidy programs in sub-Saharan Africa: A synthesis of recent evidence. *Agricultural Economics, 44,* 547–562.

Johnston, B. F., & Kilby, P. (1975). *Agriculture and structural transformation: Economic strategies in late developing countries*. New York: Oxford University Press.

Juo, A. S., Franzluebbers, K., Dabiri, A., & Ikhile, B. (1995). Changes in soil properties during long-term fallow and continuous cultivation after forest clearing in Nigeria. *Agriculture, Ecosystems & Environment, 56,* 9–18.

Kamukondiwa, W., & Bergström, L. (1994). Nitrate leaching in field lysimeters at an agricultural site in Zimbabwe. *Soil Use and Management, 10,* 118–124.

Katunzi, E. F. B., Mbonde, A., Waya, R., & Mrosso, H. D. J. (2010). Minor water bodies around southern Lake Victoria—a replica of lost biodiversity. *Aquatic Ecosystem Health & Management, 13,* 277–283.

Khan, S. A., Mulvaney, R. L., Ellsworth, T. R., & Boast, C. W. (2007). The myth of nitrogen fertilization for soil carbon sequestration. *Journal of Environment Quality, 36,* 1821–1832.

Liu, J., You, L., Amini, M., Obersteiner, M., Herrero, M., Zehnder, A. J., et al. (2010). A high-resolution assessment on global nitrogen flows in cropland. *Proceedings of the National Academy of Sciences, 107,* 8035–8040.

Lunduka, R., Ricker-Gilbert, J., & Fisher, M. (2013). What are the farm-level impacts of Malawi's farm input subsidy program? A critical review. *Agricultural Economics, 44,* 563–579.

Lungu, O. I., & Dynoodt, R. F. (2008). Acidification from long-term use of urea and its effect on selected soil properties. *African Journal of Food, Agriculture, Nutrition and Development, 8,* 63–76.

Magnan, N., Larson, D. M., & Taylor, J. E. (2012). Stuck on stubble? The non-market value of agricultural byproducts for diversified farmers in Morocco. *American Journal of Agricultural Economics, 94,* 1055–1069.

Mason, N. M., Jayne, T. S., & Mofya-Mukuka, R. (2013). Zambia's input subsidy programs. *Agricultural Economics, 44,* 613–628.

McSwiney, C. P., & Robertson, G. P. (2005). Nonlinear response of N_2O flux to incremental fertilizer addition in a continuous maize (Zea mays L.) cropping system. *Global Change Biology, 11,* 1712–1719.

Meixner, F. X., Fickinger, T., Marufu, L., Serca, D., Nathaus, F. J., Makina, E., et al. (1997). Preliminary results on nitric oxide emission from a southern African savanna ecosystem. *Nutrient Cycling in Agroecosystems, 48,* 123–138.

Minot, N., & Benson, T. (2009). *Fertilizer subsidies in Africa: Are vouchers the answer?*. Washington, DC: International Food Policy Research Institute.

Morris, M., Kelly, V. A., Kopicki, R. J., & Byerlee, D. (2007). *Fertilizer use in African agriculture*. Washington DC: The World Bank.

Mtambanengwe, F., & Mapfumo, P. (2005). Organic matter management as an underlying cause for soil fertility gradients on smallholder farms in Zimbabwe. *Nutrient Cycling in Agroecosystems, 73,* 227–243.

Ngatunga, E. L., Cools, N., Dondeyne, S., Deckers, J. A., & Merckx, R. (2001). Buffering capacity of cashew soils in Southeastern Tanzania. *Soil Use and Management, 17,* 155–162.

Nixon, S. W., Buckley, B. A., Granger, S. L., Entsua-Mensah, M., Ansa-Asare, O., White, M.J., McKinney, R.A., et al. (2007). Anthropogenic enrichment and nutrients in some tropical lagoons of Ghana, West Africa. *Ecological Applications, 17*(5), Supplement, S144–64.

Njiru, M., Mkumbo, O. C., & van der Knaap, M. (2010). Some possible factors leading to decline in fish species in Lake Victoria. *Aquatic Ecosystem Health & Management, 13,* 3–10.

Nyenje, P. M., Foppen, J. W., Uhlenbrook, S., Kulabako, R., & Muwanga, A. (2010). Eutrophication and nutrient release in urban areas of sub-Saharan Africa—A review. *Science of the Total Environment, 408,* 447–455.

Nziguheba, G., Palm, C., Berhe, T., Denning, G., Dicko, A., Diof, O., et al. (2010). The African Green Revolution: Results from the Millennium Villages Project. *Advances in Agronomy, 109,* 75–115.

Odada, E. (2012). The 1st inaugural workshop of the equatorial africa deposition network (EADN), *Imperial Hotel*, Kisumu-Kenya: IGAC News, *46*, 32–33 5–9 December 2011.

Palm, C. A., Smukler, S. M., Sullivan, C. C., Mutuo, P. K., Nyadzi, G. I., & Walsh, M. G. (2010). Identifying potential synergies and trade-offs for meeting food security and climate change objectives in sub-Saharan Africa. *Proceedings of the National Academy of Sciences, 107,* 19661–19666.

Pichot, J., Sedogo, M. P., Poulain, J. F., & Arrivets, J. (1981). Fertility evolution in a tropical ferruginous soil under the effect of organic manure and inorganic fertilizer applications. *Agronomie Tropicale, 36,* 122–133.

Pieri, C. (1995). Long-term soil management experiments in semiarid Francophone Africa. In R. Lal & B. A. Stewart (Eds.), *Soil management: Experimental basis for sustainability and environmental quality* (pp. 225–266). Boca Raton: CRC Press.

Poss, R. (1992). Leaching of nitrate, calcium and magnesium under maize cultivation on an oxisol in Togo. *Fertilizer Research, 33,* 123–133.

Potter, P., Ramankutty, N., Bennett, E. M., & Donner, S. D. (2010). Characterizing the spatial patterns of global fertilizer application and manure production. *Earth Interactions, 14*(2), 1–22.

Power, J. F., Wiese, R., & Flowerday, D. (2000). Managing nitrogen for water quality—Lessons from management systems evaluation area. *Journal of Environment Quality, 29,* 355–366.

Ramankutty, N., Evan, A. T., Monfreda, C., Foley, J. A. (2008). Farming the planet: 1. Geographic distribution of global agricultural lands in the year 2000. *Global Biogeochemical Cycles,* 22(1), GB1003. https://doi.org/10.1029/2007gb002952.

Robertson, G. P., & Vitousek, P. M. (2009). Nitrogen in agriculture: balancing the cost of an essential resource. *Annual Review of Environment and Resources, 34,* 97–125.

Rockstrom, J. (2003). Water for food and nature in drought-prone tropics: Vapour shift in rainfed agriculture. *Philosophical Transactions of the Royal Society B: Biological Sciences, 358,* 1997–2009.

Rufino, M. C., Dury, J., Tittonell, P., van Wijk, M. T., Herrero, M., Zingore, S., et al. (2011). Competing use of organic resources, village-level interactions between farm types and climate variability in a communal area of NE Zimbabwe. *Agricultural Systems, 104,* 175–190.

Sanchez, P. A. (2002). Soil fertility and hunger in Africa. *Science, 295,* 2019–2020.

Sanchez, P. A. (2010). Tripling crop yields in tropical Africa. *Nature Geoscience, 3,* 299–300.

Sanchez, P. A., & Logan, T. J. (1992). Myths and science about the chemistry and fertility of soils in the tropics. In R. Lal & P. A. Sanchez (Eds.), *Myths and science of soils of the Tropics* (pp. 35–46). Madison: Soil Science Society of America.

Sanchez, P. A., Ahamed, S., Carre, F., Hartemink, A. E., Hempel, J., Huising, J., et al. (2009). Digital soil map of the world. *Science, 325,* 680–681.

Scheren, P. A. G. M., Kroeze, C., Janssen, F. J. J. G., Hordijk, L., & Ptasinski, K. J. (2004). Integrated water pollution assessment of the Ebrié Lagoon, Ivory Coast, West Africa. *Journal of Marine Systems, 44,* 1–17.

Schlecht, E., Hiernaux, P., Achard, F., & Turner, M. D. (2004). Livestock related nutrient budgets within village territories in western Niger. *Nutrient Cycling in Agroecosystems, 68,* 199–211.

Shcherbak, I., Millar, N., & Robertson, G. P. (2014). Global metaanalysis of the nonlinear response of soil nitrous oxide (N2O) emissions to fertilizer nitrogen. *Proceedings of the National Academy of Sciences, 111,* 9199–9204.

Stoorvogel, J. J., & Smaling, E. (1990). *Assessment of soil nutrient depletion in sub-saharan Africa: 1983–2000. Volume 1: Main report.* Wageningen, The Netherlands: Winand Staring Centre for Integrated Land, Soil and Water Research.

Tchuente, A. T. K., Roujean, J.-L., & Faroux, S. (2010). Remote Sensing of Environment. *Remote Sensing of Environment, 114,* 961–976.

Tilman, D., & Lehman, C. (2001). Human-caused environmental change: impacts on plant diversity and evolution. *Proceedings of the National Academy of Sciences, 98,* 5433–5440.

Tilman, D., Balzer, C., Hill, J., & Befort, B. L. (2011). Global food demand and the sustainable intensification of agriculture. *Proceedings of the National Academy of Sciences, 108,* 20260–20264.

Tittonell, P., & Giller, K. E. (2013). When yield gaps are poverty traps: The paradigm of ecological intensification in African smallholder agriculture. *Field Crops Research, 143,* 76–90.

United Nations Millennium Project. (2005). *Halving hunger: It can be done*. London: UN Task Force on Hunger.
Valbuena, D., Erenstein, O., Tui, S. H.-K., Abdoulaye, T., Claessens, L., Duncan, A. J., et al. (2012). Conservation Agriculture in mixed crop–livestock systems: Scoping crop residue trade-offs in sub-Saharan Africa and South Asia. *Field Crops Research, 132,* 175–184.
Van Groenigen, J. W., Velthof, G. L., Oenema, O., Van Groenigen, K. J., & Van Kessel, C. (2010). Towards an agronomic assessment of N2O emissions: A case study for arable crops. *European Journal of Soil Science, 61,* 903–913.
Vanlauwe, B., & Giller, K. (2006). Popular myths around soil fertility management in sub-Saharan Africa. *Agriculture, Ecosystems & Environment, 116,* 34–46.
Verschuren, D., Johnson, T. C., Kling, H. J., Edgington, D. N., Leavitt, P. R., Brown, E. T., et al. (2002). History and timing of human impact on Lake Victoria, East Africa. *Proceedings of the Royal Society B: Biological Sciences, 269,* 289–294.
Vet, R., Artz, R. S., Carou, S., Shaw, M., Ro, C.-U., Aas, W., et al. (2014). A global assessment of precipitation chemistry and deposition of sulfur, nitrogen, sea salt, base cations, organic acids, acidity and pH, and phosphorus. *Atmospheric Environment, 93,* 3–100.
Vitousek, P.M., Naylor, R., Crews, T., David, M.B., Drinkwater, L.E., Holland, E., Johnes, P.J., et al. (2009). Nutrient imbalances in agricultural development. Science, 324, 1519–1520
Williams, J., Scheele, M., Van Velthoven, P., Cammas, J., Thouret, V., Galy-Lacaux, C., et al. (2009). The influence of biogenic emissions from Africa on tropical tropospheric ozone during 2006: A global modeling study. *Atmospheric Chemistry and Physics, 9,* 5729–5749.
Zingore, S., Manyame, C., Nyamugafata, P., & Giller, K. E. (2005). Long-term changes in organic matter of woodland soils cleared for arable cropping in Zimbabwe. *European Journal of Soil Science, 56,* 727–736.
Zingore, S., Murwira, H. K., Delve, R. J., & Giller, K. E. (2007). Influence of nutrient management strategies on variability of soil fertility, crop yields and nutrient balances on smallholder farms in Zimbabwe. *Agriculture, Ecosystems & Environment, 119,* 112–126.
Zingore, S., Tittonell, P., Corbeels, M., van Wijk, M. T., & Giller, K. E. (2011). Managing soil fertility diversity to enhance resource use efficiencies in smallholder farming systems: A case from Murewa District, Zimbabwe. *Nutrient Cycling in Agroecosystems, 90,* 87–103.

Chapter 6
Potential of Extensification of European and Dutch Agriculture for a More Sustainable Food System Focusing on Nitrogen and Livestock

Hans J. M. van Grinsven, Jan Willem Erisman, Wim de Vries, Henk Westhoek, and Luis Lassaletta

Abstract Most global strategies for future food security focus on sustainable intensification of production of food and involve an increase of nitrogen (N) fertilizer use, livestock production and risk of N pollution. In this chapter, we explore the potential of sustainable extensification for agriculture in the European Union (EU) and the Netherlands by analyzing cases and scenario studies focusing on reducing N inputs and livestock densities. Benefits of extensification to society include higher local

---

This chapter is an edited version of van Grinsven et al. 2015, Potential of extensification of European agriculture for a more sustainable food system, focusing on nitrogen, *Environmental Research Letters*, 10(2), doi:10.1088/1748-9326/10/2/025002 (https://iopscience.iop.org/article/10.1088/1748-9326/10/2/025002/meta). It is re-used here in modified form with the permission of IOP Publishing Ltd., under Creative Commons license CC BY 3.0 (https://creativecommons.org/licenses/by/3.0). The modifications include edited text (including the removal of the original Scope and Objective section), a slight rewording of the definition of sustainable intensification and extensification, and additional references.

H. J. M. van Grinsven (✉)
PBL Netherlands Environmental Assessment Agency, Bezuidenhoutseweg 30, 2594 AV The Hague, The Netherlands
e-mail: Hans.vanGrinsven@pbl.nl

J. W. Erisman
Institute of Environmental Sciences (CML), Leiden University, Leiden, The Netherlands
e-mail: j.w.erisman@cml.leidenuniv.nl

Louis Bolk Institute for International Advice and Research on Sustainable Agriculture, Nutrition and Health Care, Driebergen, The Netherlands

VU University, Amsterdam, The Netherlands

W. de Vries
Wageningen University and Research (WUR), Droevendaalsesteeg 3, 6708 PB Wageningen, The Netherlands
e-mail: wim.devries@wur.nl

H. Westhoek
PBL Netherlands Environmental Assessment Agency, P.O. Box 302, 3720 AH Bilthoven, The Netherlands
e-mail: henk.westhoek@pbl.nl

M. A. Sutton et al. (eds.), *Just Enough Nitrogen*,
https://doi.org/10.1007/978-3-030-58065-0_6

biodiversity and less environmental pollution. Societal costs of N losses in the EU from agriculture are substantial and for 2008 are here estimated at 0.3–1.9% of the gross domestic product (GDP). Extensification also has risks such as a reduction of yields and therewith a decrease of both GDP and farm income. This also implies a smaller contribution to global food production and, potentially, an increase of global demand for land. For N-intensive agricultural systems in northwest EU, a reduction of N fertilization rate and livestock densities of up to 30% would reduce the external cost of N pollution to such an extent that society as a whole would benefit. However, compensation would be needed for net loss of farm income, e.g., by price premiums for cleaner production or improved animal welfare. Extensification scenarios with > 30% decrease of livestock production would require adjustment of human diets. A 2030 scenario for the EU halving consumption and production of animal products (demitarian diet) is here estimated to reduce N pollution by 10%, benefits human health and would transform the EU from a feed importer to a food exporter.

Keywords Agriculture · External costs · Sustainable intensification · Extensification · Nitrogen pollution · Fertilizer · Livestock density

6.1 Introduction

6.1.1 Changing Demands on Food Production Systems

Until now, food production has kept pace with population growth. Increased use of synthetic fertilizer has been a major driver in increasing crop yields. Agricultural policies have kept and still keep a focus on increasing agricultural productivity by increasing external inputs, in both the developing and the developed world. In the developing economies, agriculture is a dominant economic sector, in some African countries contributing up to 50% to the gross domestic product (GDP), while in developed countries agriculture generally contributes less than 5% to GDP (World Bank, 2014). In the Netherlands, the agro-food sector contributes 20% to the export of products and 10% to GDP (Eurostat 2014).

Together with the enormous boost in global agricultural production, land productivity and the associated improvement of human nutrition and well-being, the efficiency of the use of critical natural resources has decreased. This includes resources such as biodiversity, phosphate and fossil fuel in the global food production system (the system that feeds the global population). As a consequence, impacts on environment, climate and human health have increased. The efficiency in terms of what

L. Lassaletta
CEIGRAM/Department of Agricultural Production, Universidad Politécnica de Madrid, Madrid 28040, Spain
e-mail: luis.lassaletta@upm.es

Sorbonne Université, CNRS, EPHE, UMR 7619 Metis, 4 place Jussieu, Paris, France eventually is consumed per use of these resources has decreased because of: (i) food waste and losses in the whole chain, which currently reach 30–40% (Gustavsson et al. 2011), (ii) increased production and consumption of animal products, and (iii) the increased use of agricultural land for the production of luxurious products, such as tea, coffee, sugar, etc. Food production is estimated to contribute about 25% to all greenhouse gas emissions (Vermeulen et al. 2012), is responsible for 60% of loss of land-based biodiversity (inferred from PBL 2010) and is the major source of global eutrophication (Sutton et al. 2013).

The increasing world population in combination with a more affluent diet is projected to demand an increase in crop production of 50% and in livestock production of 70% by 2050 (FAO 2012). According to Tilman et al. (2011), the food industry may have to double their productivity to satisfy people's demand by 2050. The larger part of the increase in demand for cereals would be necessary to feed livestock to meet the increase in diets that are rich in animal protein. Many projections indicate that such an increase can be met by yield-gap closure (e.g., Foley et al. 2011; Mueller et al. 2012). However, at the global scale the annual yield increase for staple food crops is currently slowing down (Grassini et al. 2012). This risks an expansion of agriculture area (notably in developing countries). For the first time since 1980, harvested areas for wheat and corn have started to increase (Grassini et al. 2013). This increase in part has been caused by an increased demand for biofuels. Such an extension of agricultural land is a direct threat to biodiversity (Alkemade et al. 2009; PBL 2012). Without additional efforts, current agricultural practices will also increase global emissions of greenhouse gases and lead to higher losses of nitrogen (N) and phosphorus to the environment (Garnett and Godfray 2012; Tilman et al. 2011). The main challenge is therefore to guarantee future food security, while reducing environmental pollution and biodiversity loss (Sutton et al. 2013).

6.1.2 Sustainable Intensification and Sustainable Extensification

Currently, there is an ongoing debate in society on how to accommodate increasing demand for livestock products and global food security for a larger and wealthier population with more sustainable agriculture (Lang and Heasman 2004; Roberts 2008). The scientific community applies different definitions of sustainable intensification and holds different views on the potential of alternatives for *conventional* agricultural production systems to minimize environmental externalities, while maintaining food security and socio-economic viability (Petersen and Snapp 2015). Here we define:

- Sustainable intensification of agriculture as "*Increasing food security without compromising the environment and depleting natural resources*". This generally translates to increasing the production per hectare by increasing external inputs, such as N, while decreasing resource losses to the environment;

- Sustainable extensification of agriculture as *"Decreasing the depletion of natural resources and the environmental impacts without compromising food security and farm income"*. This translates to reducing external inputs per hectare, such as N, and livestock densities, while minimizing food loss and reduction of production per hectare.

Both sustainable intensification and extensification require improved management of nutrients, such as nitrogen and phosphorus, as well as of water, pesticides and the diversity of seed, plant and livestock. The aim in both cases includes to increase resource efficiencies. While both strategies focus on agricultural production, they also have implications for natural resource efficiencies of food processing and consumption. Sustainable intensification with increased food production provides more room for resource-demanding food choices, with a larger share of animal products, and for food loss in retail and consumption to satisfy consumer preferences. When accompanied by decreasing food production, sustainable extensification requires less resource-demanding food choices, different diets and minimization of food wastes.

Sustainable intensification appears to be a good strategy for improving food security in areas where there is a large yield gap, such as in Africa (Garnett and Godfray 2012), under the condition that smallholders profit from it. According to Phalan et al. (2011), intensification is also the best global strategy to spare land and halt biodiversity loss. The current dominant strategy for closure of crop yield-gaps is intensification of land use by the combination of increasing external inputs and use of high-yielding crop and animal varieties (van Grinsven et al. 2014). This form of intensification may be regarded as unsustainable in view of risks for the environment, such as soil degradation and losses of nutrients and pesticides, which pose threats for biodiversity and human health (e.g., Garnett et al. 2013; Sutton et al. 2011). These threats are particularly manifest in countries or regions where current inputs are already high. In situations with currently low crop yields and low levels of fertilization, as in various regions in Africa, the external cost of land extension for increased crop production would probably outweigh the cost of intensification for biodiversity and health. However, many African smallholder farmers lack resources to purchase fertilizers, pesticides and technology. For Europe, sustainable intensification is currently the dominant strategy to meet the increasing demand for agricultural commodities given the small potential to increase the area of agricultural land. In view of concerns about nutrient losses to air and water, one could however, hypothesize that sustainable extensification could also be a strategy for Europe, enabling it to meet both the global and regional demand for food and biodiversity.

A necessary condition for both strategies would be to maintain farm income, or, in the case of poor smallholder farmers, to increase income. Extensification holds the risk of loss of income because of reduced yield (at constant market prices). Conversely, intensification holds the risk of income loss due to a decrease of commodity prices, combined with increased cost of fertilizers, pesticides, seeds, machinery etc. This risk is currently manifest in the EU dairy sector. In anticipation of the abolishment of the EU milk quota system in 2015 and the prospect of increasing export to China and Russia, dairy farmers in the Netherlands and other EU member states made large investments to increase dairy production. This included building new stables, installing automatic milking systems and sometimes buying land. However, as export demand did not increase according to expectations, conventional Dutch milk prices dropped from over 0.40 euro/litre in 2014 to 0.25 euro/litre in 2016, while prices for organic milk remained between 0.45 and 0.50 euro/l.

Both intensification and extensification can create risks and opportunities for biodiversity, farmers and consumers. The challenge is to find, for each world region, the optimal combination and trajectory of agricultural intensification and extensification, which is needed to reduce risks of environmental pollution, resource depletion and food insecurity at regional and global scales. Achieving this optimal combination is facilitated by a globally operating food system with intensive shipping of agricultural products and resources across the globe (Galloway et al. 2007). In Europe, and particularly in the Netherlands, there is room to improve the balance between agricultural production and environmental pressure.

6.1.3 Characteristics of Low Input/Organic Farming and Impacts on Crop Yield

To meet the objectives of sustainable intensification it is necessary to close the yield- and harvest-gap, while minimizing losses of nutrients and pesticides to the environment. The yield-gap is the difference between the highest possible yield using best management and all the technology and inputs available relative to the current yield (Foley et al. 2011), whereas the harvest-gap is the difference between current and maximum cropping frequency (Ray and Foley 2013). Yield-gaps depend on crop type, genotype, management, and also on regional edaphic physical and environmental conditions (van Grinsven et al. 2013). Yield-gaps are largest for poorly managed and low input farms. For developed agricultural countries, yield-gaps are largest for organic production methods and amount up to 20–30% (de Ponti et al. 2012; Seufert et al. 2012; Ponisio et al. 2015). A problem is to establish good reference cases, both for organic and conventional methods. Yield-gaps show a large range, with several organic farms showing even-higher yields than conventional farms. The global harvest-gap is estimated at 57% (Ray and Foley 2013), as compared with a global yield-gap of 45–70% for major cereals (Mueller et al. 2012). While there is huge potential to close the harvest-gap in the tropics, the harvest-gap in European countries is generally less than 10%.

Organic agriculture is a special case of extensification, which does not allow inputs like synthetic fertilizers and pesticides, nor allow the use of genetically modified organisms (GMOs). There is no consensus on the relative contribution of various factors to the yield-gap between organic and conventional production methods, but lack of nitrogen fertilizer is claimed to be an important reason (Smil 2002). Most agricultural systems are nitrogen-limited and therefore nitrogen is important when closing the yield-gap. Both conventional and organic systems show nitrogen leakages. Although organic farmers are forced to manage nutrients better than conventional farmers, they may occasionally over-use organic sources of N to boost production of valuable crops. A main critique of organic farming is that the amount of naturally available nitrogen, including biological nitrogen fixation, is not sufficient to sustain and increase production to feed the world. Smil (2002) estimated that synthetic nitrogen fertilizer feeds about 40% of the world's population; Erisman et al. (2008) estimated that nitrogen fertilizers fed 48% of the world's population in 2008. Arable organic systems depend on N-fixing crops and animal manure to provide nutrients, which would effectively mean that about one third of the available arable land would not be available to produce food directly (Schröder and Sørensen 2011). The yield-gap, such as estimated by Seufert et al. (2012) and de Ponti et al. (2012), reflects the combined effect of the use of synthetic fertilizer, pesticides, GMOs, as well as improved management of nutrients, soils and crops. This implies that the effect of synthetic N fertilizer on cereal yields would be less than 40%. Ponisio et al. (2015) found a yield-gap of 9% between organic and conventional treatments when N inputs were similar, and a gap of 30% when taking into account the effect of higher N input in conventional systems. This implies that, by adopting practices from organic agriculture, the effect of synthetic N fertilizer on global yields could be reduced to 20%, so that agriculture could feed a much-larger number of people without synthetic N fertilizer than estimated by Smil (2002) and Erisman et al. (2008).

The challenge for both conventional and organic systems is to use nitrogen/nutrients and other resources, such as water, as efficiently as possible. Organic farming focuses on using natural processes to increase the production of quality food, with sustainable soil management in a clean environment (e.g., Tomich et al. 2011). Strict organic production systems demand sophisticated management skills to conserve nutrients and control pests and diseases. Organic arable agriculture typically requires more drought-resistant systems with higher soil organic matter and more biodiversity. The organic approach forces the community to create resilient systems while boosting production, in that sense aiming for sustainable intensification, as in conventional systems (e.g., Mäder et al. 2002). Conventional farms using best management practices to reduce external input and to control disease aim for the same, with similar results (Oenema et al. 2011). Conventional farming also depends on manure to maintain organic carbon in the soil, especially when crop residues are not returned to the field. However, intensification through conventional farming has been shown to reduce soil quality, while increasing concerns about resistance of pests and diseases to antibiotics and pesticides.

6.2 Examples of Sustainable Extensification in the Netherlands and Their Impacts on Crop Yields, Animal Welfare, Income and Environment

6.2.1 Impacts of Low Input Organic Dairy Farming in the Netherlands on Milk Yields and Farm Income

Milk yield data for organic and conventional dairy farming suggest a smaller yield-gap than for arable farming (Offerman and Nieberg 2000). Dairy farming systems based on grazing are more suitable for extensification than arable systems because the net export of minerals embedded in dairy products per hectare is lower than in arable systems, due to more efficient recycling of manure N. Furthermore, as a perennial crop, grassland is more efficient in conserving nutrients and organic matter than arable crops. Several authors, among which Alan Savory is perhaps the most prominent (as discussed by Holecheck et al. 2000), have suggested that rapid rotation grazing is more resource efficient than continuous/season long grazing systems. Holecheck et al. (2000) could not find proof of superior functioning of rapid rotation grazing in semi-arid regions in the USA and Mexico, in terms of forage production, resource (water and nutrients) efficiency and financial return. In spite of this lack of proof, rapid rotation grazing is becoming increasingly popular in intensive systems in countries with maritime climates like New Zealand, Ireland, as well as the Netherlands. This system might increase resource efficiency through a combination of deeper rooting and an increase of effective photosynthesizing area. In addition, the fatty acid composition and antioxidant content of meat and milk from grass fed systems are more favourable for human health than from grain-fed systems (Daley et al. 2010). The downside of the system is the increased demand for labour.

Peer reviewed information on environmental and economic performance of extensive dairy farming applying strip grazing, which is a form of rapid rotational grazing, is scarce. Provisional data for two organic farms, with no use of synthetic fertilizer and applying rapid rotation grazing, show an annual milk production per hectare of land used for feed of 10,000 litres as compared to 13,000 litres for conventional Dutch dairy (unpublished data). In this system, milk production per cow is 30–40% lower than for conventional farming, but production life of cows is longer. The income of both farms is higher than for conventional farms, and is representative for such farms in the Netherlands, including 50 dairy farms applying low input and intensive grazing and 300 organic dairy farms.

6.2.2 *Impact of Improved Animal Welfare Pig Production Systems in the Netherlands on Farm Income*

A system of improved welfare was developed by the Dutch organization for animal protection, in cooperation with the pig-meat processing industry and retail (Dierenbescherming 2014). Pigs raised under the one—and two-star welfare production system have a 25 and 50% increase of floor space, respectively, and are supplied with distraction material and bedding material. Since systems with improved welfare do not require a major adjustment of housing, conventional pig farmers changing to this system create extra living space by decreasing their livestock numbers. This will have an immediate negative effect on their fixed costs per unit of production. Additional production costs in this system include costs for: bedding material, additional fuel for heating in winter (because a smaller stock produces less body heat), and loss of discount on feed and other material in view of the smaller stock.

In the Netherlands, the share of sales of pork from fattening systems with increased welfare in 2013 was nearly 30% (LEI 2014). Extensification measures in pig production decrease livestock numbers, while extending the finishing-period for the purpose of:(a) improving welfare and (b) complying with conditions for a price bonus. Average Dutch consumers are willing to pay a premium of up to 10% for improved animal welfare (Carabain and Spitz 2012), and increasingly buy one-star pork. In 2012, about 15% of pork consumption in the Netherlands was produced on farms with increased pig-welfare systems. Production costs for pig farmers in the Netherlands that provide 25% more floor space and distraction material are about 6% higher than in a conventional system. For systems with 50% more floor space, additional costs are about 25%, mainly due to increased feed requirements. The welfare star-system also has export potential. More than half of Dutch pork production is exported. There are high production costs associated with manure disposal, amounting to 5% of total production costs of Dutch conventional pig farming (Willems et al. 2016). If there were a change to a more stringent system (e.g., tighter animal welfare requirements), then pig numbers would be expected to reduce and be accompanied by a reduction in pig manure production. This would increase the price that arable farmers pay for pig manure and, by that means, decrease the cost of manure disposal per pig. In addition, there would be potential demand for pork from improved welfare systems in surrounding countries like Germany and the UK, with consumers having similar preferences as Dutch consumers. Therefore, we conclude that systems with improved welfare and smaller livestock populations could maintain or can even improve farm income.

6.2.3 *Impacts of Improved Nutrient Cycling Approaches in the Netherlands on Farm Income and Environment*

Several dairy farms in the Netherlands aim to reduce their environmental impact by improving the internal nutrient cycle (INC) on their farm by innovative farm management approaches. Such farms targeted to improved INC, focus on optimizing use of on-farm available resources, including nutrients from manure and home-grown feed production, thus reducing purchased feed and fertilizer, while maintaining a sufficient income in the long term (van Hees et al. 2009). Nitrate concentrations in the upper groundwater at INC farms are relatively low compared to dairy farms with regular farming practices on similar soils (e.g., Sonneveld et al. 2010), while soil organic matter contents are relatively high (e.g., van Apeldoorn et al. 2011). Furthermore, N losses through ammonia volatilization are lower when application takes place under the same approach as conventional farming (e.g., Sonneveld et al. 2008). Recently, Dolman et al. (2014) showed that INC farms had a lower non-renewable energy use per kg FPCM (fat-and-protein-corrected milk), higher soil organic carbon content and received higher annual payments for agri-environmental measures, whereas economic and other environmental-societal indicators were not significantly different. De Boer et al. (2012) showed that at landscape level the calculated N losses to air and water would be on average 5–10% lower if INC farming were to be implemented for the whole region.

6.3 Effects of Extensification of the Agricultural Sector on Environment, Economic Welfare and Food Security for the Netherlands and Europe

6.3.1 *Effects of Extensification of the Dutch Livestock Sector on External Cost and Economic Welfare*

The Dutch livestock sector is perhaps the most intensive in the world. It is probably also the most efficient livestock sector regarding feed conversion and environmental emissions per unit of livestock product. Dutch citizens are concerned about the quality of the local environment and effects related to ammonia and manure, about the impacts in South America of large import of soy product for feed, about zoonotic diseases (particularly after the recent outbreak of Q-fever), about biodiversity loss and about animal welfare. Most of these problems are inextricably related to intensification by modern industrial scale livestock farming. To assess the consequences of extensification for environment and economy, a scenario with 50% reduction of pig and poultry livestock, 20% reduction of dairy livestock and 40% reduction of the use of synthetic N fertilizer was analysed (van Grinsven et al. 2012). Calculation of external costs was limited to nitrogen emissions and based on the unit cost approach, as described in van Grinsven et al. (2013). The extensification scenario reduces N-fertilizer use by 42%, N-excretion by 35%, ammonia emission by 37% and nitrate leaching by 58%. In this scenario, there would be no exceedance of the 50 mg/litre EU standard for nitrate (NO_3) in the upper groundwater (also in sandy soils), and the Netherlands would comply with the EU Nitrates Directive. Furthermore, current exceedance of critical N-deposition loads on more than 60% of the Dutch area of natural ecosystems would decrease. Ammonia emissions from agricultural sources contribute about 30% to the mean N deposition of 22 kg N/ha/year. The emission reduction in the extensification scenario would lower deposition by about 10% and decrease the area with exceedance of critical loads from 60% to an area between 20 and 40%.

The benefits for society of reducing of N emissions were quantified by the decrease of the external cost of nitrogen pollution. These costs decreased by 0.2–2.2 billion euro, a decrease of 40% relative to 2008. External costs estimates of N pollution are highly uncertain in view of various problems in valuation of impacts (van Grinsven et al. 2013). The decrease of external cost outweighs the loss of added value (0.6 billion euro from primary production), but not the loss of added value of 2.5 billion euro in the full supply and processing chain resulting from agricultural products. One possibility to create net benefits is to transfer the reduction of external cost of agricultural product to price premiums, as for organic products. External N costs offset the benefits of N fertilization for crop yields considerably. This would alter current N recommendations for Europe, which are based on what is economically optimal for farmers. van Grinsven et al. (2013) estimated that the N rate for winter wheat in northwest Europe for creating the highest societal benefits is about 50 kg/ha/year (25–30%) lower than the economic optimum for the farmers.

6.3.2 *Global and Regional Effects of Extensification of Agricultural Production*

As productive land is running short, land extension to compensate for productivity loss will include the taking of less productive land. This implies that the relative increase in land demand due to a transition to less intensive systems will exceed the relative decrease in land productivity, provided that food demands increase as predicted by FAO (2012). Important factors determining the impact of extensification scenarios in affluent regions on land use and food production are:

(1) Yield-gap between conventional and more extensive and/or organic agriculture.
(2) The effect of the increase of the share of extensive (organic) production methods on crop and food waste.
(3) The change in food choice by consumers shifting to products from more extensive, or, ultimately organic products.

(4) The regional correlation between the increase of the share of extensive (organic) production in total production and the increase of the share of alternative consumers of these products.

A common criticism to the option of saving biodiversity by extensification of agricultural production in Europe, or affluent countries in general, is that this will increase the demand for land in view of yield loss per hectare. A transition to, for example, organic farming in the USA or the EU might create a food security problem if not combined with limiting food waste and a transition to diets that require less land or resources in general. Such diets often infer consuming less animal products, but also reducing amounts of pre-prepared food and sugar (Brandt et al. 2011). Consumers shifting to organic products will indeed tend to consume less animal products for both ethical and economic reasons (Honkanen et al. 2006), usually going along with a much healthier lifestyle (Kesse-Guyot et al. 2012). As a result of the more labour-intensive production methods associated with organic agriculture, a food basket of organic products in the USA is typically 50% more expensive than the equivalent for conventional products (Brown and Sperow 2005).

Westhoek et al. (2011) analysed effects of scenarios in 2030 assessing effects of low input farming, as well as of dietary changes and increased use efficiencies. A partial shift to organic production in the 27 Member States of the European Union (EU27) would hardly increase arable land use in EU27, but would increase land use in the rest of the world (RoW) by 16 million ha. A demitarian diet, with a 50% reduction of livestock products, only leads to small changes in land demand for feed crops (−6 million ha) and food crops (+ 4 million ha) in the EU27, according to Westhoek et al. (2011). However, the major effect would be a large reduction in land demand in the RoW both for food crops (47 million ha) and for grassland and fodder crops (60 million ha), compensating more than six times the increased land demand for organic farming. In this scenario the EU would become a net exporter of cereals. A shift to a healthy diet, following WHO recommendations for consumption of red meat and saturated fats, would have a similar impact on global land use and in the EU27 as the demitarian diet. These results show that in the EU27, diets that are good for health also demand less land. Clearly this type of diet change would have large economic impacts on the livestock sector, on farms, and the feed and meat processing industry, and on the environmental and societal costs (Westhoek et al. 2014). In summary, this study shows that the slight increase in land that is needed when changing to organic production can be more than compensated by dietary changes and improved efficiencies.

6.4 Discussion and Conclusion

Both intensification and extensification of agricultural production, as a strategy for building a more sustainable future food system, hold risks and opportunities. Both the environment, as well as human health, would greatly benefit from a combination of extensification, decreasing food waste and a change to diets with less animal products. Less consumption of saturated (animal) fats and red meats (and sugars) can greatly reduce common and increasing public health concerns about food related cardiovascular disease, cancers, obesity and diabetes. The last two health problems, however, are also related to a general overconsumption of calories and lack of physical exercise (McAllister et al. 2009). Governments appear to be reluctant to promote diets with less animal products. Experiments with introducing a tax on unhealthy consumption in Denmark in 2011 were not successful (Smed 2012). Important conditions for a general adjustment towards healthier and less resource-demanding diets are increasing public awareness and agreements with the food and retail sectors to adjust marketing and labelling of particular food items which are not considered to fit within a healthy diet.

To make extensification go beyond niche solutions, establishing adequate and stable price premiums requires more cooperation between the various stakeholders in the food chain, particular those in processing and retail. Currently, primary producers are "squeezed" between the market power of suppliers, processers and retail, in part due to a lack of organization. Although a fair proportion of primary production sectors in the EU is organized in cooperatives (e.g., cereals 35%, dairy over 50%, pork 25%), these cooperatives increasingly act similarly to normal multinationals, focusing on cost price reduction rather than sustainability (Bijman et al. 2012). An important step is to create premium certificates for intermediate production systems and products between conventional and more sustainable products with regard to animal welfare and the use of chemical inputs (Paarlberg 2013). Modest and targeted inputs of synthetic fertilizer, pesticides and antibiotics are very effective in increasing the average productivity of land and animals, reducing the risks of loss of production and farm income in individual years. Theoretically, the cumulative premium for extensification should not exceed the conservative estimates of reduction of external costs of environmental pollution. As reduction of external costs is a public concern, government should play a more pronounced role in communicating the need for price premiums for food products from less polluting and also less resource demanding production systems.

Extensification and intensification are not silver bullets, but examples of various possible strategies to create a more sustainable food system (Garnett and Godfray 2012). Sustainable intensification may be the way forward in regions with low crop yields, and may be the best option for efficient production of staple food. Both intensification and extensification are not sustainable without a more integrated view on how to change to healthier and more resource efficient diets, reduction of food waste and on fairer sharing of costs and benefits of food production between players in the food chain. All these possible strategies not only create opportunities to increase sustainability and synergy, but also create risks of trade-offs, and need "smart" operationalization to make progress. In essence, the food system should be demand driven and make the consumer the primary agent of change instead of relying solely on actions by industry, retail and government. However, to play this role, consumers in affluent countries need to regain their sovereignty in the food system by making informed choices on the food they buy. Currently, the majority of consumers are not very well informed and interested and rely on NGOs and government to make the food system more sustainable.

Acknowledgements This chapter is an edited version of van Grinsven et al. 2015, *Environmental Research Letters*, 10(2), https://doi.org/10.1088/1748-9326/10/2/025002, re-used here in modified form with the permission of IOP Publishing Ltd., under Creative Commons license CC BY 3.0 (https://creativecommons.org/licenses/by/3.0). The modifications include edited text (including the removal of the original Scope and Objective section), a slight rewording of the definition of sustainable intensification and extensification, and additional references.

References

Alkemade, R., van Oorschot, M., Miles, L., Nellemann, C., Bakkenes, M., & ten Brink, B. (2009). GLOBIO3: A framework to investigate options for reducing global terrestrial biodiversity loss. *Ecosystems,12,* 374–390.

Bijman, J., Iliopoulos, C., Poppe, K. J., Gijselinckx, C., Hagedorn, K., Hanisch, M., et al. (2012). *Support for farmers' cooperatives*. European Commission: Final Report.

Brandt, K., Leifert, C., Sanderson, R., & Seal, C. J. (2011). Agroecosystem management and nutritional quality of plant foods: The case of organic fruits and vegetables. *Critical Reviews in Plant Sciences,30,* 177–197.

Brown, C., & Sperow, M. (2005). Examining the cost of an all-organic diet. *Journal of Food Distribution Research,36,* 20–27.

Carabain, C., & Spitz, G. (2012). *Nederlanders & voedselzekerheid. Jonge en oudere Nederlanders over voedsel*. Amsterdam: NCDO (in Dutch).

Daley, C. A., Abbott, A., Doyle, P. S., Nader, G. A., Larson, S. (2010). A review of fatty acid profiles and antioxidant content in grass-fed and grain-fed beef. *Nutrition Journal, 9*(10). https://doi.org/10.1186/1475-2891-9-10.

De Boer, H. C., Dolman, M. A., Gerritsen, A. L., Kros, J., Sonneveld, M. P. W., Stuiver, M., Termeer, C. J. A. M., et al,. (2012). *Effecten van kringlooplandbouw op ecosysteemdiensten en milieukwaliteit—Een integrale analyse op People, Planet & Profit, effecten op gebiedsniveau, en potentie voor zelfsturing, met de Noardlike Fryske Wâlden als inspirerend voorbeeld*. Wageningen Livestock Research Report. The Netherlands: Wageningen University & Research (in Dutch).

De Ponti, T., Rijk, B., & van Ittersum, M. K. (2012). The crop yield gap between organic and conventional agriculture. *Agricultural Systems,108,* 1–9.

Dierenbescherming (Dutch organization for animal protection). (2014). http://beterleven.dierenbescherming.nl/voor-bedrijven/deelnemers. Accessed August 2019.

Dolman, M. A., Sonneveld, M. P. W., Mollenhorst, H., & de Boer, I. J. M. (2014). Benchmarking the economic, environmental and societal performance of Dutch dairy farms aiming at internal recycling of nutrients. *Journal of Cleaner Production,73,* 245–252.

Erisman, J. W., Sutton, M. A., Galloway, J., Klimont, Z., & Winiwarter, W. (2008). How a century of ammonia synthesis changed the world. *Nature Geoscience,1,* 636–639.

Eurostat. (2014). http://epp.eurostat.ec.europa.eu/portal/page/portal/statistics/search_database. Accessed 5 June 2014.

FAO. (2012). *Statistical yearbook 2013: World food and agriculture*. Rome: FAO (Food and Agriculture Organization of the United Nations).

Foley, J. A., Ramankutty, N., Brauman, K. A., Cassidy, E. S., Gerber, J. S., Johnston, M., et al. (2011). Solutions for a cultivated planet. *Nature,478,* 337–342.

Galloway, J. N., Burke, M., Bradford, G. E., Naylor, R., Falcon, W., Chapagain, A. K., et al. (2007). International trade in meat: the tip of the pork chop. *Ambio,36*(8), 622–629.
Garnett, T., & Godfray, C. (2012). *Sustainable intensification in agriculture. Navigating a course through competing food system priorities.* Food Climate Research Network and the Oxford Martin Programme on the Future of Food, University of Oxford, UK.
Garnett, T., Appleby, M. C., Balmford, A., Bateman, I. J., Benton, T. G., Bloomer, P., et al. (2013). Sustainable intensification in agriculture: premises and policies. *Science,341*(6141), 33–34.
Grassini, P., Eskridge, K. M., Cassman, K. G. (2013). Distinguishing between yield advances and yield plateaus in historical crop production trends. *Nature Communications, 4,* 2918. 10.1038/ncomms3918.
Gustavsson, J., Cederberg, C., Sonesson, U., van Otterdijk, R., Mybeck, A. (2011). *Global food losses and food waste. Extent, causes and prevention.* Rome, Italy: UN Food and Agriculture Organization (FAO).
Holechek, J. L., Gomes, H., Molinar, F., Galt, D., & Valdez, R. (2000). Short-duration grazing: The facts in 1999. *Rangelands,22,* 18–22.
Honkanen, P., Verplanken, B., & Olsen, S. O. (2006). Ethical values and motives driving organic food choice. *Journal of Consumer Behaviour,5*(5), 420–430.
Kesse-Guyot, E., Péneau, S., Méjean, C., Szabo de Edelenyi, F., Galan, P., Hercberg, S., et al. (2013). Profiles of organic food consumers in a large sample of French adults: Results from the Nutrinet-Santé Cohort Study. *PLoS ONE,8*(10), e76998. https://doi.org/10.1371/journal.pone.0076998.
Lang, T., & Heasman, M. (2004). *Food wars: The global battle for mouths, minds and markets.* London: Earthscan.
LEI. (2014). *Monitor Duurzaam Voedsel 2013.* Agricultural Economics Research Institute (LEI), Wageningen University and Research Centre. The Hague, The Netherlands: Ministry of Economic Affairs (in Dutch).
Mäder, P., Flieβbach, A., Dubois, D., Gunst, L., Fried, P., & Niggli, U. (2002). Soil fertility and biodiversity in organic farming. *Science,296*(5573), 1694–1697.
McAllister, E. J., Dhurandhar, N. V., Keith, S. W., Aronne, L. J., Barger, J., Baskin, M., et al. (2009). Ten putative contributors to the obesity epidemic. *Critical Reviews in Food Science and Nutrition,49*(10), 868–913.
Mueller, N. D., Gerber, J. S., Johnston, M., Ray, D. K., Ramankutty, N., & Foley, J. A. (2012). Closing yield gaps through nutrient and water management. *Nature,490,* 254–257.
Oenema, J., van Keulen, H., Schils, R. L. M., & Aarts, H. F. M. (2011). Participatory farm management adaptations to reduce environmental impact on commercial pilot dairy farms in the Netherlands. *NJAS-Wageningen Journal of Life Sciences,58,* 39–48.
Offermann, F., & Nieberg, H. (2000). *Economic performance of organic farms in Europe.* Organic farming in Europe: economics and policy, vol. 5. Stuttgart, Germany: University of Hohenheim.
Paarlberg, R. (2013). *Food politics: What everyone needs to know.* Oxford University Press.
PBL Netherlands Environmental Assessment Agency. (2010). *Rethinking global biodiversity strategies: Exploring structural changes in production and consumption to reduce biodiversity loss.* The Netherlands: The Hague/Bilthoven,
PBL Netherlands Environmental Assessment Agency. (2012). *Roads from Rio + 20. Pathways to achieve global sustainability goals by 2050.* The Hague, The Netherlands: PBL Netherlands Environmental Assessment Agency.
Petersen, B., & Snapp, S. (2015). What is sustainable intensification? Views from experts. *Land Use Policy,46,* 1–10.
Phalan, B., Onial, M., Balmford, A., & Green, R. E. (2011). Reconciling food production and biodiversity conservation: Land sharing and land sparing compared. *Science,333*(6047), 1289–1291.
Ponisio, L. C., M'Gonigle, L. K., Mace, K. C., Palomino, J., de Valpine, P., & Kremen, C. (2015). Diversification practices reduce organic to conventional yield gap. *Proceedings of the Royal*

Society of London B: Biological Sciences,282(1799), 20141396. https://doi.org/10.1098/rspb.2014.1396.

Ray, D. K., & Foley, J. A. (2013). Increasing global crop harvest frequency: recent trends and future directions. *Environmental Research Letters,8*(4), 044041. https://doi.org/10.1088/1748-9326/8/4/044041.

Roberts, P. (2008). *The end of food: The coming crisis in the world food industry*. London: Bloomsbury.

Schröder, J. J., & Sørensen, P. (2011). *The role of mineral fertilisers in optimising the use efficiency of manure and land*. International Fertiliser Society Proceedings 701. Leek, UK: International Fertilizer Society.

Seufert, V., Ramankutty, N., & Foley, J. A. (2012). Comparing the yields of organic and conventional agriculture. *Nature,485,* 229–232.

Smed, S. (2012). Financial penalties on foods: The fat tax in Denmark. *Nutrition Bulletin,37*(2), 142–147.

Smil, V. (2002). Nitrogen and food production: Proteins for human diets. *Ambio,31,* 126–131.

Sonneveld, M. P. W., Schröder, J. J., de Vos, J. A., Monteny, G. J., Mosquera, J., Hol, J. M., et al. (2008). A whole-farm strategy to reduce environmental impacts of nitrogen. *Journal of Environmental Quality,37*(1), 186–195.

Sonneveld, M. P. W., Brus, D. J., & Roelsma, J. (2010). Nitrate concentrations in the upper groundwater on dairy farms with different N strategies: Validation of regression models for sandy soils. *Environmental Pollution,158,* 92–97.

Sutton, M. A., Oenema, O., Erisman, J. W., Leip, A., van Grinsven, H. J. M., & Winiwarter, W. (2011). Too much of a good thing. *Nature,472,* 159–161.

Sutton, M. A., Bleeker, A., Howard, C. M., Bekunda, M., Grizzetti, B., de Vries, W., van Grinsven, H. J. M., et al. (2013). *Our Nutrient World: The challenge to produce more food and energy with less pollution.* Global Overview of Nutrient Management. Centre for Ecology and Hydrology, Edinburgh on behalf of the Global Partnership on Nutrient Management and the International Nitrogen Initiative.

Tilman, D., Balzer, C., Hill, J., Befort, B. L. (2011). Global food demand and the sustainable intensification of agriculture. *PNAS: Proceedings of the National Academy of Sciences, 108*(50), 20260–20264.

Tomich, T. P., Brodt, S., Ferris, H., Galt, R., Horwath, W. R., Kebreab, E., et al. (2011). Agroecology: A review from a global-change perspective. *Annual Review of Environment and Resources,36,* 193–222.

Van Apeldoorn, D. F., Sonneveld, M. P. W., & Kok, K. (2011). Landscape asymmetry of soil organic matter as a source of agro-ecosystem resilience. *Agriculture, Ecosystems & Environment,140,* 401–410.

Van Grinsven, H. J. M., van Dam, J., van der Sluis, S., Willems, J. (2012). *Welke veestapel past in Nederland? Achtergrondrapport scenarioberekeningen.* PBL report 500245002. Bilthoven, The Netherlands: PBL Netherlands Environmental Assessment Agency (in Dutch).

Van Grinsven, H. J. M., Holland, M., Jacobsen, B. H., Klimont, Z., Sutton, M. A., & Willems, W. J. (2013). Costs and benefits of nitrogen for Europe and implications for mitigation. *Environmental Science and Technology, 47,* 3571–3579.

Van Grinsven, H. J. M., Spiertz, J. H. J., Westhoek, H. J., Bouwman, A. F., Erisman, J. W. (2014). Nitrogen use and food production in European regions from a global perspective *The Journal of Agricultural Science, 152*(S1), 9–19. 10.1017/S0021859613000853.

Van Grinsven, H. J. M, Erisman, J. W., de Vries, W.,Westhoek, H. (2015). Potential of extensification of European agriculture for a more sustainable food system, focusing on nitrogen. *Environmental Research Letters, 10*(2), 025002. https://doi.org/10.1088/1748-9326/10/2/025002. https://iopscience.iop.org/article/10.1088/1748-9326/10/2/025002/meta.

Van Hees. E, Van der Schans, F., Aarts, F., Verhoeven, F., Takens. P. (2009). *Van top-down naar bodem-up: Review van de kringlooplandbouw in de melkveehouderij.* CLM-rapport 703. Utrecht, The Netherlands: CLM Onderzoek en advies (in Dutch).

Vermeulen, S. J., Campbell, B. M., & Ingram, J. S. I. (2012). Climate change and food systems. *Annual Review of Environment and Resources,37,* 195–222.

World Bank. (2014). World Development Indicators. http://wdi.worldbank.org/table/4.2# (Accessed August 2019).

Westhoek, H. J., Rood, G. A., van den Berg, M., Janse, J. H., Nijdam, D. S., Reudink, M. A., et al. (2011). *The protein puzzle: The consumption and production of meat, dairy and fish in the European Union.* The Hague, The Netherlands: PBL Netherlands Environmental Assessment Agency.

Westhoek, H. J., Lesschen, J. P., Rood, T., Wagner, S., de Marco, A., Murphy-Bokern, D., et al. (2014). Food choices, health and environment: Effects of cutting Europe's meat and dairy intake. *Global Environmental Change,26,* 196–205.

Willems, J., van Grinsven, H. J. M., Jacobsen, B. H., Jensen, T., Dalgaard, T., Westhoek, H., et al. (2016). Why danish pig farms have far more land and pigs than dutch farms? implications for feed supply, manure recycling and production costs. *Agricultural Systems,144,* 122–132.

Part II
Food and Agriculture: Nitrogen Intensification and Biological Nitrogen Fixation in Low-Input Systems

Chapter 7
History of Rhizobia Inoculants Use for Improving Performance of Grain Legumes Based on Experience from Nigeria

Aliyu A. Abdullahi, John Howieson, Graham O'Hara, Jason Terpolilli, Ravi Tiwari, and Ado A. Yusuf

Abstract The use of rhizobium inoculants for improvement in nitrogen-fixation and productivity of grain legumes has been well established in developed countries. However, the practice is still under-utilized in Nigeria. Nitrogen (N) is the most frequently deficient nutrient for crop production, while nitrogen fertilizers are costly, inadequate, and may not be timely in supply. These make rhizobia inoculants a cheaper, easier and safer option to improve the N_2-fixation and productivity of grain legumes. Inoculant use in Nigeria was initiated in the 1970s, but still remains very limited. Studies conducted on inoculant use were initially on "US

This chapter is an edited and modified version of Abdullahi et al. (2013), *History of Rhizobia inoculants use for grain legumes improvement in Nigeria—the journey so far* (N2013 Kampala Conference paper). Abdullah et al. (2013) includes figures and additional tables and is available online at the N2Africa project website and the Tropical Soybean Information Portal, https://n2africa.org/sites/default/files/History%20of%20rhizobium%20inoculants%20%20use%20in%20Nigeria-%20Aliyu%20Anchau%20N2013%20paper%2018-22%20Nov.pdf (N2Africa Project website), and https://tropicalsoybean.com/history-rhizobia-inoculants-use-grain-legumes-improvement-nigeria-journey-so-far (Tropical Soybean Information Portal).

A. A. Abdullahi (✉) · A. A. Yusuf
Department of Soil Science, Faculty of Agriculture, Institute for Agricultural Research, Ahmadu Bello University, Zaria, Nigeria
e-mail: aanchau@yahoo.com

A. A. Yusuf
e-mail: adamu_99@yahoo.co.uk

A. A. Abdullahi · J. Howieson · G. O'Hara · J. Terpolilli · R. Tiwari
Centre for Rhizobium Studies, Murdoch University, South Street, Perth, WA 6150, Australia
e-mail: J.Howieson@murdoch.edu.au

G. O'Hara
e-mail: G.Ohara@murdoch.edu.au

J. Terpolilli
e-mail: J.Terpolilli@murdoch.edu.au

R. Tiwari
e-mail: R.Tiwari@murdoch.edu.au

M. A. Sutton et al. (eds.), *Just Enough Nitrogen*,
https://doi.org/10.1007/978-3-030-58065-0_7

type" Soybean (*Glycine max* (L.) Merrill), which has been found to require specific inoculation with *Bradyrhizobim japonicum* for optimum productivity. Studies were also conducted on inoculation of cowpea (*Vigna unguiculata* (L.) Walp), but rarely on bambara groundnut (*Vigna subterranea* (L.) Verdc.) and groundnut or peanut (*Arachis hypogaea* L.). In the 1980s, the International Institute of Tropical Agriculture (IITA) Ibadan, Nigeria, introduced promiscuous soybean cultivars; Tropical Glycine Cross (TGx). These genotypes nodulate freely with the indigenous rhizobium population, fix large amount of atmospheric nitrogen and produce higher grain yields than the local genotypes. However, some experiments indicated up to 40–45% increases in yield by some of the genotypes on inoculation. Hence, the ultimate solution remains the development of inoculants using highly effective indigenous rhizobia strains for particular crops. The recent efforts of the project "Putting Nitrogen fixation to work for smallholder farmers in Africa (N2Africa)" towards the promotion of inoculants technology are highly welcomed in the country.

Keywords Legumes · Nitrogen-fixation · Nitrogen · Nodulation · Rhizobia inoculants

7.1 Introduction

Legumes have been a source of food since before mankind first tilled the soil. They were recognized from very early times as 'soil improvers', but encountered many problems (GRDC 2013). Most of the soils on which legumes are grown in Africa, particularly in Nigeria, are relatively unproductive because of poor nutrient status, especially with regard to nitrogen (N) (Machido et al. 2011). In addition, many soils do not have adequate populations of native rhizobia in terms of number, quality or effectiveness to enhance biological nitrogen fixation (FAO 1984). This situation calls for the provision of an external source of rhizobia to enable effective nodulation and N_2 fixation, known as inoculation. The benefits of rhizobia inoculants for legume improvement have been known for over a century; production and utilization of legume inoculants is a well-established practice in many developed countries (Nelson 2004). However, production and use of inoculants in Africa are relatively underdeveloped. Initiatives to establish local inoculant production industries in many African countries only started in the 1980s and 1990s, led by organizations such as Nitrogen Fixation in Tropical Agricultural Legumes (NifTAL), Food and Agriculture Organization (FAO) and the Regional Microbiological Resources Centres (MIRCENs). However, only a few initiatives led to a large scale in production (Bala et al. 2011). Moreover, most of these inoculants are for soybean.

The introduction of soybean to Nigeria in 1910 (Shurtleff and Aoyagi 2009) led to the initiation of studies on the use of inoculants in the country. The earliest introduced varieties were the "US-type" soybean that requires inoculation with specific rhizobia strains of (*Bradyrhizobium japonicum*) (Osunde et al. 2003). In Nigeria, a series of field experiments conducted in 1978 to screen strains of *Bradyrhizobium japonicum* that were efficient in N_2 fixation, showed grain yield of the American soybean cultivars to increase by as much as 100% (Chianu et al. 2011, for the Bossier and TGm 294–4 varieties). This benefit was limited by difficulties faced by farmers in obtaining inoculants as a result of unavailability or lack of access (Mpepereki et al. 2000). A breeding programme at the International Institute of Tropical Agriculture (IITA) Ibadan, Nigeria, later developed 'promiscuous' soybean cultivars that naturally nodulate freely with indigenous rhizobia, thus obviating the need to inoculate with *B. japonicum* (Pulver et al. 1985; Mpepereki et al. 2000). However, some studies still show responses of the promiscuous soybeans to additional inoculation with rhizobia and other commercial products (Yusuf et al. 2012b; N'cho et al. 2013; Sanginga et al. 2000; Osunde et al. 2003). A few other studies focused on groundnuts, cowpea and bambara groundnut which showed sporadic responses to inoculation due to their long adaptation and compatibility with indigenous rhizobia that naturally nodulate them (Yusuf et al. 2012b).

This chapter reviews the history of the use of inoculants in Nigeria with a view to highlighting problems and opportunities for future improvement of inoculant production and use for higher N_2 fixation, grain yield and soil fertility.

7.2 Rhizobia Inoculants and Their Use up to the 1970s

The discovery of symbiotic nitrogen fixation between rhizobia and legumes in the late nineteenth century resulted in a wealth of research confirming its potential as an alternative to inorganic N-fertilizers in agriculture (Bala 2011; N'cho et al. 2013). Rhizobia inoculants have been on the commercial market for over 100 years in many developed countries (Nelson 2004; Giller 2008). Inoculating legumes, with rhizobia achieved substantial increases in legume nodulation, grain and biomass yield, nitrogen fixation and post-crop soil nitrate levels. These benefits are usually highest when inoculated legumes are grown in nil-rhizobia or low-rhizobia soils, but tend to be marginal in soils already containing high numbers of compatible and effective rhizobia (GRDC 2013).

Generally, inoculation activities have been on-going in sub-Saharan Africa since the 1950s, mostly on soybean and forage legumes; however, the adoption of inoculation on a commercial scale has not been appreciable, except in a few countries, such as Zimbabwe and South Africa, where commercial farms dominate the agricultural sector (Bala 2011). Kang (1975) and Rhoades and Nanju (1979) confirmed the importance of soybean inoculation with *B. japonicum* strains in Nigeria during this period. Indigenous *Bradyrhizobium japonicum* strain IRj 2180A has been isolated and used for inoculation of soybean since 1979 (Muhammad 2010; Yusuf et al. 2012b). A breeding programme to introduce promiscuous soybean was initiated in 1977 at IITA and later developed to produce genotypes for release in the 1980s (Sanginga et al. 2000). During this period, cowpea rhizobia were also confirmed to be fairly well distributed in most Nigerian soils and native rhizobia seemed as effective as inoculated rhizobia strains tested at IITA (Balasubramanian and Annadi 1978).

7.3 Research on the Use of Rhizobia Inoculants in the 1980s

Inoculant research during this period was dominated by the introduction of promiscuous soybean cultivars, Tropical Glycine Cross (TGx) by IITA. These genotypes nodulate freely with indigenous rhizobia, and were therefore widely adopted by farmers in different parts of Nigeria (Pulver et al. 1985; Osunde et al. 2003). Soon, the need to inoculate these genotypes became obvious. Thereafter, studies on inoculation in the country turned to evaluation of the response of the promiscuous soybean along with other legumes to inoculation. This was because of the impossibility of having a genotype that nodulates effectively with indigenous rhizobia in soils at all locations (Sanginga et al. 2000; Osunde et al. 2003). Several studies showed positive responses to inoculation by the promiscuous genotypes (Sanginga et al. 2000; Yusuf et al. 2012b). For example, Olufajo and Adu (1992) conducted three experiments on inoculating promiscuous soybean cultivars in relation to SAMSOY 1 and 2 of Malaysian origin between 1982 to 1988 in the northern Guinea savanna zone of Nigeria, They observed significant increases in nodule number, nodule weight and grain yield, with a rhizobia strain (IRj 2123) superseding other inoculants. Oloke and Odeyemi (1988) also conducted a study on an imported peat based *B. japonicum* inoculant in relation to local inoculants (prepared by incorporating three *Bradyrhizobium* strains—Ife CR9, Ife CR15 and *Bradyrhizobium japonicum*—each into a different carrier material) on three cowpea varieties of TVU 1190, IT82E-60 and Ife brown. The local inoculants, with lignite, sub-bituminous coal and cow manure as carriers, increased cowpea yield by 72, 54 and 10% respectively, relative to uninoculated plants, while the imported peat based inoculant increased the yield only by 25%. The authors attributed this to better adaptation of the local inoculants to tropical conditions as compared with the imported inoculant.

7.4 Studies on Rhizobia Inoculants in Nigeria from 1990s to Date

Studies on rhizobia inoculants in this period continued to evaluate the response of promiscuous cultivars of soybean to inoculation and phosphorus or other nutrients application. Results showed higher yields in response to inoculation, depending on the cultivars and level of nutrient deficiency of the soils, especially in relation to phosphorus deficiency, with on-farm yield increases of up to 3 t/ha in amended soils (Abaidoo et al. 2013). Okogun et al. (2004) reported significantly higher root biomass (20%), nitrogen fixation (35%) and grain yield (6%) of TGx 1448–2E over SAMSOY-2 associated with inoculation by R25B and IRj 2180A rhizobia strains in the northern Guinea savanna of Nigeria. Turner et al. (2011) also observed varying responses of different promiscuous soybean varieties to rhizobia inoculation, phosphorus and nitrogen application in Kano State, Nigeria. Inoculants were found to fix varying levels of atmospheric nitrogen, based on their effectiveness as confirmed by inoculation of TGx 1485 with four rhizobia inoculants; R25B, IRj 2180A, IRc 461 and IRc 291 and application of 60 kg N ha^{-1} which gave yield increases of 23, 38, 19, 17 and 22% respectively, over the control in the southern Guinea savanna of Nigeria (Muhammad 2010). Osunde et al. (2003) in a two seasons experiment, observed a response of up to 40% increase in yield in the first season, though there was a decline in the second season, with TGx1456-2E and TGx1660-19F, when using two inoculants R25B and IRj 2180A in the same southern Guinea savanna zone of Nigeria. A mixture of *Bradyrhizobia* isolates R25B and IRj 2180A was found to increase the grain yield of 1456–2E by 30 and 25% in the northern and southern Guinea savanna of Nigeria, respectively. This study indicated compatibility synergistic effect and effectiveness of the indigenous strains (Sanginga et al. 2000).

Studies involving rhizobial inoculants, other microbial inoculants and foliar fertilizers application on promiscuous soybean were also conducted during the period. *Bradyrhizobium* spp. (RACA 6), arbuscular mycorrhizal fungi (Rhizatech) and *Trichodermaharzianum* (Eco-T), Agroleaf high P, triple superphosphate; TSP (30 kg P_2O_5 ha^{-1}) and their various combinations gave yield increases of between 1–52% (N'cho et al. 2013), when tested under smallholder farmers' conditions on TGx 1448–2E in the northern Guinea savanna of Nigeria. This indicates how amelioration of the deficiency of other essential nutrients using other products affect rhizobia inoculation of the promiscuous soybean genotypes. Screening of 15 commercial and laboratory rhizobium inoculants in on TGx 1448–2E and a Malaysian genotype (SAMSOY-2) at Kadawa (Sudan savanna) and Samaru (northern Guinea savanna) identified 1495 MAR, USDA 4675, USDA 110, TSBF 531 and TSBF 560 as good inoculants with significant increase in grain yield of 35%, on average, in Samaru, Zaria, Nigeria (Yusuf et al. 2012b).

Inoculation of existing legumes other than soybean was also investigated during this period. In the Sudano-Sahelian zone of Nigeria, the influence of rhizobia inoculation on N-fixation by cowpea (*Vigna unguiculata* (L.) Walp.), groundnut (*Arachis hypogaea* L.) and bambara groundnut (*Vigna subterranea* L.Verdc.) with rhizobia strains isolated from the same crops in the previous year, increased the amount of N-fixed by up to 46% (Yakubu et al. 2010). Cowpea, groundnut and bambara groundnut fixed 42.68, 27.19 and 32.53 kg N ha^{-1}, respectively.

Cowpea has higher potential for N fixation. In contrast, when Yusuf et al. (2012a) evaluated the effect of two rhizobia inoculants; Biofix (Rhizobium spp.), obtained from the MIRCEN project, based in University of Nairobi, Kenya and Vault (various rhizobia spp. and *Bacillus subtilis*), from Becker Underwood, USA and their mixture on groundnut in the northern Guinea savanna of Nigeria, the results showed significantly higher pod (20%) and haulm (28%) yields of the uninoculated over the inoculated. The inoculants' poor performance was attributed to a long history of groundnut cultivation, and lack of acclimatization of the foreign rhizobia to the environment, necessitating the need to conduct research on indigenous rhizobia to identify and isolate more efficient strains. Inoculation trials by Aliyu et al. (2013) on two soil types (Eutric Cambisols and Rhodic Nitisols) and three crops (promiscuous soybean genotype (TGx 1448–2E), cowpea (IT90K-277–2) and groundnut (SAMNUT 21)) using different rhizobial inoculants (MAR 1495, TSBF Mixture, Legumefix, HiStick and IRj 2180A) showed that rhizobia strains MAR 1495 and TSBF mixture had a similar ability to improve the productivity of soybean and groundnut.

These findings indicate varying responses of the cowpea, groundnut and bamabara groundnut to inoculation, which could be attributed to their free nodulation with existing indigenous strains; hence sometimes introduced inoculants fail to perform as expected.

7.5 Reasons Why Research on Inoculants and Their Use Is Mostly on Soybean

A number of reasons explain the observed domination of inoculants research on soybean in Nigeria. Firstly, more than 90% of rhizobial inoculants worldwide are used on soybean, because of its 'specific' requirements in terms of the type of rhizobia able to form nodules on its roots and actively fix nitrogen (Giller 2008). Secondly, soybean positively responds to inoculation more frequently, relative to other legumes, such as groundnut and cowpea grown Nigeria (Bala et al. 2011). Hence, the greatest success in raising legume yields with biological nitrogen fixation in the country was experienced with soybean crops that respond synergistically to inoculation with rhizobia and improved fertilization (Woomer et al. 2013). Thirdly, a highly efficient soybean—*Bradyrhizobium* symbiosis can lead to an increase in yield and fixation of large amounts of nitrogen under good conditions, with up to 300 kg N ha^{-1} (Keyser and Li 1992). Moreover, soybean is becoming one of the most cultivated grain legumes in sub-Saharan Africa, due to its great potential for producing cheap food protein and other essential nutrients for farm households (Rao and Reddy 2010), among other vital products.

Other legumes such as cowpea, groundnut and bamabra groundnut are highly permissive in their interaction with indigenous rhizobia and hardly respond to inoculation in most Nigerian and West African soils (Bala et al. 2011). Hence, reports of sporadic positive responses to rhizobia inoculation by cowpea, and less evidence of responses with groundnut are an indication of a requirement for high quality inoculants in terms of both strains and formulations (Giller 2012).

Generally, the use of inputs (inoculants and/or fertilizers) for beans, cowpea and groundnut in Nigeria is considerably less than with soybean, leading to lower yield increases in these crops attributed to better agronomic management in the absence of inoculants (Woomer et al. 2013). Identification of effective strains for different legume crops from the natural indigenous pool of rhizobia strains in the Nigerian soils or other effective imported inoculants could be a good solution to the problem.

7.6 Factors Responsible for Poor Adoption of Inoculant Research Results by Nigerian Farmers

Many factors are responsible for the low level of adoption of the existing promising results from inoculant studies by farmers. The information reported in this chapter is mostly research conducted by scientists at different institutions or managed by scientist on-farm. The data shown in Table 7.1 provide a clear indication of the low level of adoption of inoculants by Nigerian legume farmers. In Nigeria, the current market is too small, the skill base for scientists and technicians trained in rhizobiology for inoculant production and quality control is weak, there are no adequately tested carriers and there is only one season for production (Giller 2012). There are networks of agricultural dealers in Nigeria, but they do not market inoculants, because inoculants are not readily available or commonly used in legume production, and even fertilizers are mostly used only for cereal production (Woomer et al. 2013). Moreover, most countries in the whole sub-Saharan Africa, including Nigeria, do not have factories to produce viable rhizobia inoculants. In addition, the inoculants are not available at affordable prices or smallholder farmers are faced by difficulties in accessing them (Keyser and Li 1992; Mpepereki et al. 2000; Bala 2011). As a result, supply and adoption of inoculants and inoculant technology are lacking in Nigeria (IITA/CGIAR 2015b).

Large-scale commercial production of soybean, intensive livestock industry and the incentive for inoculant adoption among farmers were further discouraged by the introduction of promiscuous soybean cultivars (Bala et al. 2011). Hence, the use of rhizobia and other microbial inoculants by smallholder farmers in Nigeria, like other West African countries, is not popular (N'cho et al. 2013). The percentage use of inoculants and/or phosphorus on three important legumes among Nigerian farmers shows only 11% of Nigerian farmers use inoculants and on soybean alone, while up to 26% apply inoculants in combination with phosphorus fertilizers (Table 7.1).

Table 7.1 The percentage of farmers using specific legume inputs in their own fields, based on data from the N2Africa impact survey in Nigeria

Crops	Without use of inputs	+ Inoculant only	+ P fertilizer	+ P fertilizer and inoculant
Soybean	6	11	57	26
Cowpea	18	–	82	–
Groundnut	24	–	76	–

Source Huising and Franke (2013)

7.7 Recent Efforts of N2Africa Project Towards Production and Use of Inoculants

The N2Africa project is a research-and-development partnership programme that aims to develop, disseminate, and promote appropriate N_2-fixation technologies for smallholder farmers, focusing on major grain legumes (Abaidoo et al. 2013). In particular, the project focuses on delivery and dissemination of legume inoculant technologies to farmers throughout sub-Saharan Africa and creating sufficient demand to reinforce private sector investment (Giller 2012). The project supports an initiative to establish inoculant production facilities in Nigeria, for soybean and other legumes (Giller 2012), while a public-private partnership approached is used to achieve its objectives.

N2Africa collaborates with institutions and research partners in the country such as: the Institute for Agricultural Research (IAR); Ahmadu Bello University, Samaru, Zaria, Nigeria; Bayero University, Kano; Federal University of Technology, Minna; and Federal University, Dutsin Ma. Each of these are involved in rhizobiology and agronomy research in different legume crops in the savanna of northern Nigeria to address the increasing demand of the major grain legumes in the country, as shown in Table 7.2. N2Africa also partners with another non-governmental organization, Sasakawa Global (SG2000), to conduct training, demonstration and dissemination of the use of inoculants and phosphorus fertilizer to increase the yield of legumes, soil fertility and N_2 fixation (N2Africa 2014a). In addition, it partners with Borno State government of Nigeria in the north-eastern part of the country (N2Africa 2014b). Capacity building, in terms of training African M.Sc. and Ph.D. students from various countries on inoculant technology is also a focus of the project.

The history of implementation of inoculant production in eight N2Africa project operating countries is shown in Table 7.3. While Kenya, Malawi and Zimbabwe have reasonably high inoculant production rates as of 2000; there were no production records in Nigeria (Table 7.3). In Zimbabwe, the production is most successful, partially due to long-term production and experience since 1964 (Giller 2008; Bala 2011) and partially due to the production of highly determinate, specifically-nodulating soyabean varieties, highly responsive to inoculants in commercial farms being inoculated, even with locally produced inoculants (Mpepereki et al. 2000).

Table 7.2 Projected production and demand for cowpea, groundnut and soybean in Nigeria

Commodity	Production (1000 MT)			Demand (1000 MT)		
	2010	2015	2020	2010	2015	2020
Cowpea	2,761	3,364	4,097	4,092	5,273	6,906
Groundnut	3,275	3,563	3,784	3,335	3,497	3,726
Soybean	633	709	793	643	748	869
Total	6,669	7,536	8,674	8,070	9,518	11,501

Source Tropical legumes II (TLII 2011)

Table 7.3 Partner laboratories involved in the rhizobiology activities of N2Africa

S/No	Country	Laboratory	Location	Year Established	Inoculant Production	Quantity of inoculant produced in 2010
1	DRC	Soil Microbiology Laboratory	Bukavu		No	None
2	Ghana	Soil Research Institute	Kumasi		No	None
3	Kenya	MIRCEN/University of Nairobi	Nairobi	1977 (1981)	Yes	ND
4	Kenya	MEA Fertilizer Ltd.	Nakuru	1977 (2008)	Yes	25,000
5	Malawi	Chitedze Research Station	Chitedze, Lilongwe	1964	Yes	15,000
6	Nigeria	Institute for Agricultural Research (IAR)	Samaru, Zaria	1922	No	None
7	Rwanda	Institut des Sciences Agronomique du Rwanda (ISAR)	Robena	1984	Yes	ND
8	Mozambique	IIAM	Nampula	Under construction	No	None
9	Zimbabwe	Soil Productivity Research Laboratory (SPRL)	Marondera	1964	Yes	80,000

Year in parenthesis is the year inoculant production commenced; ND = Not determined
Source: Bala (2011)

This led to higher production and demand for the inoculants in that country. Nigeria could emulate these practices as well as engage in bioprospecting for new elite strains for major grain legumes such as soybean, cowpea, and groundnut. This would be complemented by work to characterize elite strains using molecular tools, make them available to inoculant producers to scale up the technology and to tailor dissemination to local settings. In particular, there is a need to focus on 'last mile' delivery networks (Giller and Vanlauwe 2013).

7.8 Future Outlook on Inoculants Production and Use in Nigeria

Establishment of effectively nodulated legumes requires the use of effective strains of rhizobia. Machido et al. (2011) indicate the need for a holistic approach to improve the entire cropping system, which calls for researchers on biotechnology in Nigeria to include the selection of more competitive and efficient indigenous rhizobia that could serve as local inoculants. This is because introduced commercial inoculants often fail to show a significant increase in the yields of inoculated crops, due to environmental factors. This needs to be complemented by establishing a database on the occurrence, abundance, distribution and composition of indigenous populations of rhizobia in soils by researchers. Work should also identify crop combinations, sequences and rotations that would take maximum advantage of the N_2-fixing potentials of the existing legumes under continuous cropping systems in Nigeria. Involving stakeholders, farmers, private sector and policy makers will surely improve the situation. A new legume inoculant, NoduMax, recently developed at IITA Ibadan for soybean, compares favourably with inoculants produced in other countries, and production is scheduled to commence in 2015 (IITA/CGIAR 2015a). The aim is to produce inoculants for sale, demonstrate their economic viability to private investors and provide incentives and training for future operations (Woomer et al. 2013). More such developments are in the pipeline.

7.9 Conclusions

The use of rhizobia inoculants for improvement in N-fixation and productivity of grain legumes has been established in developed countries over a long time, but is still in the development stage in Nigeria, where research on biological nitrogen fixation started only in the 1970s. The introduction of promiscuous soybean cultivars (TGx) by IITA, Ibadan, Nigeria in the 1980s significantly impacted on inoculants research. Only a few studies have been conducted on cowpea, groundnut and bambara groundnut, showing sporadic responses to inoculation, mostly with imported inoculants. Results of these inoculant studies mostly end in the hand of researchers with little uptake by farmers. However, current and future studies need to aim at breaking grounds by identifying effective rhizobia strains for making local inoculants specific to particular legumes grown in Nigeria. This effort, when matched with awareness by farmers and other stakeholders, will lead to improved symbiotic N-fixation, with increased productivity of the legumes, food security and soil fertility. The intervention of the N2Africa project in this direction is thus highly welcomed.

Acknowledgements The authors acknowledge the sponsorship of Aliyu Anchau Abdullahi's Ph.D. studies by the N2Africa project and Murdoch University, Western Australia in the course of which this review has been made. This chapter is an edited and modified version of Abdullahi et al. (2013), *History of Rhizobia inoculants use for grain legumes improvement in Nigeria—the journey so far* (N2013 Kampala Conference paper). Abdullahi et al. (2013) includes figures and additional tables and is available online at the N2Africa project website and the Tropical Soybean Information Portal.

References

Abaidoo, R., Boahen, S., Turner, A., Dianda, M. (2013). Bridging the grain legume gap through Agronomy. *IITA Research for Development (R4D) Review, 9,* 31 January 2013. Available online at: http://r4dreview.iita.org/index.php/2013/01/31/bridging-the-grain-legume-yield-gap-through-agronomy. Accessed 7 August 2018.

Abdullahi, A. A., Howieson, J., O'Hara, G., Tepolilli, J., Vivas-Marfisi, A., Yusuf, A. A. (2013). History of Rhizobia inoculants use for grain legumes improvement in Nigeria—the journey so far. *N2Africa Project*. N2013 Kampala Conference paper. Available online at: https://n2africa.org/sites/default/files/History%20of%20rhizobium%20inoculants%20%20use%20in%20Nigeria-%20Aliyu%20Anchau%20N2013%20paper%2018–22%20Nov.pdf and https://tropicalsoybean.com/history-rhizobia-inoculants-use-grain-legumes-improvement-nigeria-journey-so-far. Accessed December 2019.

Aliyu, I. A., Yusuf, A. A., & Abaidoo, R. C. (2013). Response of grain legumes to rhizobial inoculation in two savanna soils of Nigeria. *African Journal of Microbiology Research, 7*(15), 1332–1342.

Bala, A. (2011). Emerging challenges in cross-border movement of inoculants in sub-Saharan Africa. *N2Africa Podcaster,* 8, 3–4. Available online at: http://www.n2africa.org/sites/n2africa.org/files/N2Africa%20Podcaster%208.pdf. Accessed 7 August 2018.

Bala, A., Karanja, N., Murwira, M., Lwimbi, L., Abaidoo, R., Giller K. (2011) Production and use of Rhizobial inoculants in Africa. Milestone reference 3.4.1. *N2Africa.* Available online at: http://www.n2africa.org/sites/n2africa.org/files/images/images/N2Africa_Production%20and%20use%20of%20rhizobial%20inoculants%20in%20Africa.pdf. Accessed 7 August 2018.

Balasubramanian, V., & Annadi, L. A. (1978). Nitrogen nutrition of cowpea (*Vigna unguiculata* (L.) Walp.) and its relationship to-soil fertility. *Ghana Journal of Agricultural Science, 11,* 5–10.

Chianu, J. N., Nkonya, E. M., Mairura, F. S., Chianu, J. N., & Akinnifesi, F. K. (2011). Biological nitrogen fixation and socioeconomic factors for legumes production in sub-Saharan Africa. A review. *Agronomy for Sustainable Development, 31*(1), 139–154.

FAO. (1984). *Legume inoculants and their use.* A pocket manual jointly prepared by Nitrogen Fixation for Tropical Agricultural Legumes (NifTAL) Project, USA and FAO Fertilizer and Plant Nutrition Service Land and Water Development Division in association with FAO Crop and Grassland Production Service. Rome: Food and Agriculture Organization of the United Nations. Available online at: http://www.ctahr.hawaii.edu/bnf/Downloads/Training/Legume%20use/Title.PDF. Accessed 7 August 2018.

Giller, K. E. (2008). The successful intensification of smallholder farming in Zimbabwe. *LEISA Magazine, 24*(2), 30–31. Available online at: http://edepot.wur.nl/58722. Accessed 7 August 2018.

Giller, K. E. (2012). Establishment of rhizobial inoculant production factories in Africa. *N2Africa Podcaster, 12,* 3–4. http://www.n2africa.org/sites/n2africa.org/files/N2Africa%20Podcaster%2012.pdf. Accessed 7 August 2018.

Giller, K. E., & Vanlauwe B. (2013). N2Africa Phase II. *N2Africa Podcaster, 22,* 2. Available online at: http://www.n2africa.org/sites/n2africa.org/files/images/N2Africa%20Podcaster%2022.pdf. Accessed 7 August 2018.

GRDC. (2013). *Rhizobial inoculants fact sheet.* Australia: Grains Research and Development Corporation (GRDC). Available online at: https://grdc.com.au/__data/assets/pdf_file/0014/117221/rhizobial-inoculants-fact-sheet.pdf.pdf. Accessed 7 August 2018.

Huising, J., & Franke L. (2013). Early impacts of the N2Africa project. *N2Africa Podcaster*, 22, 4. Available online at: http://www.n2africa.org/content/early-impacts-n2africa-project. Accessed 7 August 2018.

IITA/CGIAR (2015a). CGIAR top managers visit IITA Business Incubation Platform. *The Bulletin,* 2265(2–6 March 2015). IITA. Available online at: https://issuu.com/iita/docs/bulletin_2265. Accessed 7 August 2018.

IITA/CGIAR (2015b). *N2Africa revitalizes legume production in Nigeria.* Press release. Available online at: http://grainlegumes.cgiar.org/n2africa-revitalizes-legume-production-in-nigeria. Accessed 7 August 2018.

Kang, B. T. (1975). Effect of inoculation and nitrogen fertilizers on soybeans in western Nigeria. *Experimental Agriculture, 11,* 23–33.

Keyser, H. H., & Li, F. (1992). Potential for increasing biological nitrogen fixation in soybean. *Plant and Soil, 141,* 119–135.

Machido, D. A., Olufajo, O. O., Yakubu, S. E., & Yusufu, S. S. (2011). Enhancing the contribution of the legumes to the N fertility of soils of the semi-arid zone of Nigeria. *African Journal of Biotechnology, 10*(10), 1848–1853.

Mpepereki, S., Javaheri, F., Davis, P., & Giller, K. E. (2000). Soybeans and sustainable agriculture, promiscuous soybean in Southern Africa. *Field Crops Research, 65,* 137–149.

Muhammad, A. (2010). Response of a promiscuous soybean cultivar to rhizobial inoculation and phosphorus in Nigeria's southern Guinea savanna alfisol. *Nigerian Journal of Basic and Applied Sciences, 18*(1), 79–82.

N'cho, C. O., Yusuf, A. A., Ama-Abina, J. T., Jemo, M., Abaidoo, R. C., Savane, I. (2013). Effects of commercial microbial inoculants and foliar fertilizers on Soybean nodulation and yield in the northern Guinea Savannah of Nigeria. *International Journal of Advance Agricultural Science, 1,* 66–73.

N2Africa. (2014a). SG2000 partnership with N2Africa in Nigeria. http://www.n2africa.org/content/sg2000-partnership-n2africa-nigeria. Accessed 7 August 2018.

N2Africa. (2014b). N2Africa ventures into Borno state Nigeria. http://www.n2africa.org/content/n2africa-ventures-borno-state-nigeria. Accessed 7 August 2018.

Nelson, L. M. (2004). Plant growth promoting rhizobacteria (PGPR): Prospects for new inoculants. *Crop Management, 3*(1). https://doi.org/10.1094/cm-2004-0301-05-rv.

Okogun, J. A., Otuyemi, B. T., & Sanginga, N. (2004). Soybean yield determinants and response to rhizobial inoculation in on-farm trial in the northern Guinea savanna of Nigeria. *West African Journal of Applied Ecology, 6,* 30–39.

Oloke, J. K., & Odeyemi, O. (1988). Effect of some *Bradyrhizobium* inoculants on the performance of three cowpea (Vignaunguiculata) cultivars. *Biology and Fertility of Soils, 6,* 178–182.

Olufajo, O. O., & Adu, J. K. (1992). Response of soybean to inoculation with *Bradyrhizobiumjaponicum* in the Northern Guinea savanna of Nigeria. In K. Mulongoy, M. Gueye, & D. S. C. Spencer (Eds.), *Biological nitrogen fixation and sustainability of tropical agriculture* (pp. 147–154). London: Wiley-Sayce.

Osunde, A. O., Gwam, S., Bala, A., Sanginga, N., & Okogun, J. A. (2003). Responses to rhizobial inoculation by two promiscuous soybean cultivars in soils of the Southern Guinea savanna zone of Nigeria. *Biology and Fertility of Soils, 37,* 274–279.

Pulver, E. L., Kueneman, E. A., & Ranarao, V. (1985). Identification of promiscuous nodulating soybean efficient in N_2 fixation. *Crop Science, 25*(4), 660–663.

Rao, A. S., & Reddy, K. S. (2010). Nutrient management in soybean. In G. Singh (Ed.), *The soybean: Botany, production and uses* (Chapter 8). Centre for Agriculture and Biosciences International (CABI).

Rhoades, E. R., & Nanju, D. (1979). Effects of pelleting soybean seed with fertilizer dust. *Experimental Agriculture, 15,* 27–32.

Sanginga, N., Thottappilly, G., & Dashiell, K. (2000). Effectiveness of rhizobia nodulating recent promiscuous soybean selections in the moist savanna of Nigeria. *Soil Biology & Biochemistry, 32,* 127–133.

Shurtleff, W., & Aoyagi, A. (2009). *History of soybean and soy foods in Africa (1857–2009): Extensively annotated bibliography and sourcebook.* Soyinfo Center. Available online at: http://www.soyinfocenter.com/pdf/134/Afri.pdf. Accessed 7 August 2018.

TLII (Tropical Legumes II). (2011). Grain legumes in Nigeria. *Bulletin of Tropical Legumes*, 12 (December 2011), 1–4. Available online at: http://www.icrisat.org/TropicalLegumesII/pdfs/BTL 12-2011223.pdf. Accessed 7 August 2018.

Turner, A., Bala, A., Abaidoo, R., Boahen, S., de Wolf, J., Irenge, Y., Baijukya, F., et al. (2011). *Annual country reports 2011*. N2Africa. Available online at: http://www.n2africa.org/sites/n2africa.org/files/images/images/N2Africa_Annual%20country%20reports%202011.pdf. Accessed 7 August 2018.

Woomer, P., Husing, J., Giller K. (2013). *N2Africa: Final report of the first phase 2009–2013*. Available online at: http://www.n2africa.org/sites/n2africa.org/files/images/N2Africa%20Final%20Report%20of%20the%20First%20Phase%20%20-%20%202009-2013.pdf. Accessed 7 August 2018.

Yakubu, H., Kwari, J. D., & Ngala, A. L. (2010). N_2 fixation by grain legume varieties as affected by Rhizobia inoculation in the sandy loam soil of Sudano-Sahelian zone of north eastern Nigeria. *Nigerian Journal of Basic and Applied Science, 18*(2), 229–236.

Yusuf, A. A., Jemo, M., Abaidoo, R. C. (2012a). Effects of rhizobium inoculation and phosphorus sources on the productivity of groundnut in Samaru, northern Guinea savanna of Nigeria. In M. K. A. Adeboye, A. J. Odofin, A. O. Osunde, A. Bala, S. O. Ojeniyi (Eds). *Soil resources management, global climate change and food security*. Proceedings of the 35th Annual Conference of the Soil Science Society of Nigeria, March 7–11 2011, Federal University of Technology, Minna, Niger State, Nigeria (pp. 191–198). Soil Science Society of Nigeria (SSSN).

Yusuf, A. A., Jemo, M., Nwoke, O. C., Abaidoo, R. C. (2012b). Evaluation of commercial and laboratory rhizobium inoculants on nodulation and yield of promiscuous soybean in the Nigerian savanna. In M. K. A. Adeboye, A. J. Odofin, A. O. Osunde, A. Bala, S. O. Ojeniyi (Eds). *Soil resources management, global climate change and food security*. Proceedings of the 35th Annual Conference of the Soil Science Society of Nigeria, March 7–11 2011, Federal University of Technology, Minna, Niger State, Nigeria (pp. 199–214). Soil Science Society of Nigeria (SSSN).

Chapter 8
Producer Knowledge, Attitudes, and Practices for Dry Beans and Biological Nitrogen Fixation in Kamuli District, Uganda

L. Michael Lege and Lynne A. Carpenter-Boggs

Abstract Past and current research toward increasing crop yield, nitrogen (N) use efficiency, and food security in east-central Africa includes the development of common beans varieties, inoculants and systems that improve biological N fixation (BNF) . Improved inoculants and bean varieties could play a role in increasing yields for subsistence farmers if the causes of, and limitations to maximum BNF efficiency can be identified and controlled through genetics, breeding, and improved management. However, adoption of new crop varieties and related technologies in developing regions is often poor due to agronomic, infrastructural, and socioeconomic factors. Community knowledge, attitudes, practices (KAP) and resources should be assessed alongside agronomic research in order to improve the potential for long-term adoption of new production methods and technologies. In collaboration with ongoing development of improved bean varieties and inoculants for BNF, we conducted a KAP survey of farmers regarding bean production. The study found that farmers viewed crop growth and yield in different areas and over time as indicators of soil health. Crop pests, diseases and weather were viewed as greater limiters of bean production than was soil health. Most growers (54%) were farming without inputs and 32% reported using green manure. Only 4% were currently using mineral fertilizer; however, 64% said that they would prefer to use fertilizer over other agricultural inputs. When deciding which bean varieties to plant, market price and yields were the most important determining factors. By identifying current practices, beliefs, desires, and concerns of producers, research and extension on soil and crop improvement can become more effective. Such KAP data as these can be used to improve the relevance, dissemination, and adoption of agronomic and genetic research outputs.

Keywords Legumes · Common bean · Rhizobium · Biological nitrogen fixation · KAP

L. M. Lege
School of the Environment, Washington State University, Pullman, WA, USA
e-mail: llege@wsu.edu

L. A. Carpenter-Boggs (✉)
Department of Crop and Soil Sciences, Washington State University, Pullman, WA, USA
e-mail: lcboggs@wsu.edu

M. A. Sutton et al. (eds.), *Just Enough Nitrogen*,
https://doi.org/10.1007/978-3-030-58065-0_8

8.1 Introduction

New crop varieties and agricultural products, even when proven effective, are not necessarily adopted broadly or quickly. The prevailing agricultural system of a given place, including producer decision-making, can present barriers and opportunities for technology adoption. In order to improve the potential for adoption, the prevailing system should be understood and potential barriers addressed before broad scale dissemination is attempted.

The 2009 International Assessment of Agricultural Knowledge, Science and Technology for Development (IAASTD) Report for sub-Saharan Africa (McIntyre et al. 2009) cites soil depletion and land degradation as main barriers to agricultural productivity in sub-Saharan Africa. With about 80% of employment in Uganda coming from the agricultural sector (Buyinza 2010), soil management strategies must be a priority in research and implementation. McIntyre et al. (2009) suggest several means for doing this, including the use of chemical fertilizers, as a part of an integrated soil fertility management (ISFM) protocol. However, chemical fertilizers are usually not an economically feasible option for resource poor farmers in East Africa (Bationo 2004; Nkonya et al. 2004). Therefore, more cost-effective and accessible solutions are needed to improve soil fertility and productivity.

Legume crops are able to supply a portion of their own nitrogen (N) requirement, as well as contribute to that of the following crop, by hosting rhizobium bacteria capable of biological nitrogen fixation (BNF) (Giller 2001). BNF can be inhibited by many soil quality issues including low pH, phosphorus, boron, or molybdenum, or high available N. Still, BNF is very important to plant N supply and yield in N depleted soil conditions.

Across east-central Africa, low N conditions predominate and either N fertilization or inoculation with effective rhizobia commonly increase productivity of common bean (*Phaseolus vulgaris* L.) (Woomer et al. 1997). Similarly, research is identifying legume varieties and breeding lines capable of greater BNF (Biabani et al. 2011). Yet, use of inoculants and new varieties is far from consistent due to multiple socioeconomic, technical, and environmental barriers. Many studies in sub-Saharan Africa have shown the benefits of different soil management protocols in field trials but note social and ecological barriers (access to markets, poor infrastructure, soil acidity or nutrient deficiencies, etc.) which limit adoption and diffusion of these technologies and require further study (Bohlool et al. 1992; Buyinza 2010; Nkonya et al. 2011; Ojiem et al. 2006; Snapp et al. 1998; Snapp et al. 2010).

Effectiveness of such research may be improved by identifying and understanding the socioeconomic and agricultural context in greater depth before or concurrent to field trials. Ojiem et al. (2006) assert that introducing improved legume technologies requires an understanding of the socio-ecological niche that they may fill in a given system. They suggest, like Hall and Clark (1995) that successful adoption requires engagement with farmers and a more holistic understanding of the niches in society or ecology that may be addressed by new technologies and varieties.

In order to engage with farmers and identify niches for BNF technologies in a community being studied for development interventions, members of the farming communities in Butansi Sub-County, Kamuli District, Uganda were interviewed about their current knowledge, attitudes, and practices (KAP) about bean seeds and fertilizers. KAP studies are used to establish baseline data for the development of extension programs in healthcare and agriculture (Machila et al. 2003 and Salameh et al.. 2004). They provide useful insight into the current perceptions of a particular process within a larger socioeconomic and ecological context, and can point to misinformation or solutions to problems in a system (Machila et al. 2003 and Salameh et al. 2004). Farmer interviews can yield KAP information that provides insights into producers' perceived constraints and decision-making criteria, which can then inform and improve research and extension efforts.

8.2 Methods

Ten farmer groups with ten members each were involved in a USAID funded BNF project in Butansi Sub-county, Kamuli District, Uganda. These groups were formed by extension agents of a cooperating non-governmental organization (NGO). The farmers were exposed to BNF technology through a one-season demonstration using plots managed by the farmers in cooperation with the NGO extension agents.

Half of the farmers who had participated in the BNF demonstrations (50) were interviewed using a semi-structured interview process. The interviews were conducted in October and November of 2011 at centrally located demonstration gardens and farmer homesteads. In compliance with standard Institutional Review Board procedures, farmers gave informed consent to participate verbally prior to each interview. Thirty-four of the farmers interviewed were female and 16 were male.

An interview guide was used to provide similar structure to each interview and to focus discussion on farmers' KAP about beans and BNF. Topics included current input use, desired input use, criteria for bean seed selection, yield-limiting factors, as well as farmers' perceived level and sources of access to agronomic information. Questions were asked in an open-ended format to avoid restricting answers to preconceived options. The interviews were conducted in Lusoga and Luganda, both being local vernacular languages. NGO extension agents with experience in conducting interviews translated the questions to farmers and farmer responses to the researcher. Interviews were recorded and responses were coded for percentage and correlation analyses.

8.3 Results and Discussion

8.3.1 Knowledge and Attitudes About Soil

Of the 50 farmers interviewed, 48 described their soil quality during the interview. Forty-four percent of those described their soil as unhealthy, 35% reported having healthy soil, and 21% reported having mixed healthy and unhealthy soils (Fig. 8.1). Farmers used crop performance as the indicator of soil health. One woman farmer said: "I think that the soil is healthy because I see that the crops look good" (Interview 28). Her neighbor, a 30 year old woman, said: "Part of the soil is good, part is not. On one part the plants grow well and in good condition, but not on the other side" (Interview 29). Another woman said of her farm: "The soil is not healthy, it is hard pan" (Interview 35).

Farmers who reported unhealthy soils cited poor yields or declining crop growth as evidence. For example, one man said: "I can tell the land is depleted of nutrients because of the plants; every year the yields are decreasing… I would consider using fertilizers, if possible, but fertilizers are expensive" (Interview 26). Another farmer said: "My soils are not healthy. I have cultivated them for a very long period of time and over time I see that there is reduction in both plant vigor and yield" (Interview 17). Although 42% of farmers considered their soil to be "unhealthy," only 14% of the interviewees named soil quality as a yield-limiting factor. Eighty-six percent of farmers that reported soil quality to be a yield limiting factor also reported having

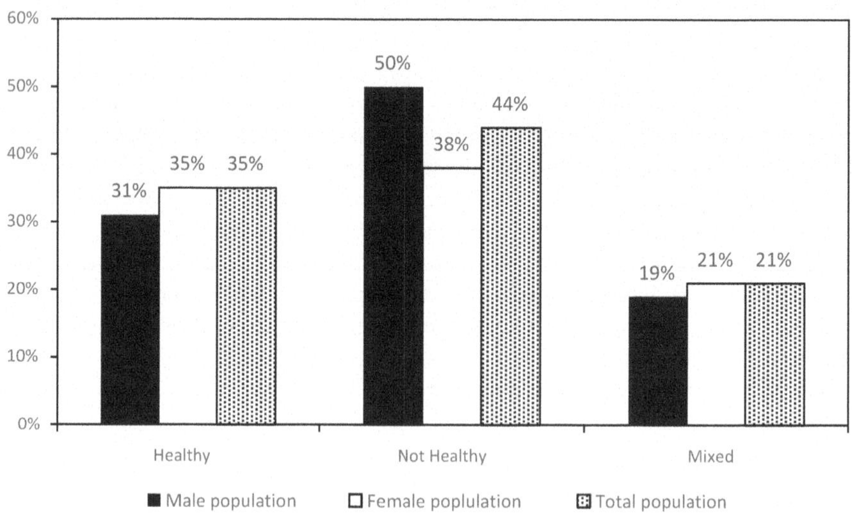

Fig. 8.1 Farmer perceptions of soil health. The black bars represent the percentage of the male population that described the soil on their farms as healthy, unhealthy, or a mixture of healthy and unhealthy; the white bars represent the percentage of the female population, and the grey bars indicate the percentage of the total interviewed population using these soil descriptions

unhealthy soil, but only 29% of the farmers that reported having poor soil quality also mentioned soil quality as a yield-limiting factor. This suggests that the other 71% of farmers reporting poor soil quality consider factors other than soil quality to be more impactful to their yields.

8.3.2 Attitudes and Practices About Inputs

BNF can be inhibited by many soil quality issues including low pH, phosphorus, boron, or molybdenum. Solutions to poor soil nutrition that consider farmer concerns as well as scientific goals are more likely to be broadly adopted (Chianu et al. 2011; Hall and Clark 1995). Although 65% of producers recognized at least some areas of unhealthy soil or poor crop growth on their farms, only 46% of the farmers reported using any inputs on their crops. The most common input was "green manure" (plant residue), which was used by 32% of farmers. All of the farmers using green manure applied it to non-bean crops like matoke and maize, but 44% of those using green manure also applied it to beans. Farmyard animal manure was also used by 14% of farmers. Green manures and farmyard manures were mentioned as being the easiest inputs to obtain but were also reported to be bulky and labor intensive to apply. Two individuals reported using more than one input, both used green manure and farmyard manure on their farms. Only two farmers in the study used mineral fertilizer (Fig. 8.2).

Mineral fertilizer use was rarely reported in this community, but farmers would prefer to use fertilizers rather than their current practice. One farmer said: "I don't use fertilizers, they are expensive… I only use manure compost on the bananas, but no other crop. I would like to use fertilizer. I have seen other people have better yields with fertilizer" (Interview 26). Sixty-four percent of the farmers interviewed said that they would prefer to use fertilizer if they could have any agricultural input. Of these, 72% reported receiving information about agriculture through government extension programs. Among all interviewees, 62% reported receiving farming information through government extension. Thus, more farmers receiving information from government sources favored the use of fertilizers. Farmers in Kamuli District would like to use mineral fertilizers because they have been told of, or seen the positive benefits of fertilizers. However, these fertilizers must be located and purchased. Both accessibility and cost of fertilizers were seen as being prohibitive: "I have never used fertilizers… I don't have access, and no money" (Interview 35). Many farmers adopted "green manure" as a means of managing soil fertility because it is easy to obtain, and is not cost prohibitive (Fig. 8.3). One farmer explained "I like it [green manure], because the plant grows well… I would like to use fertilizer but I have no access." (Interview 28)

Agricultural development interventions should continue to work with farmers to show the positive impacts of BNF, but also appeal to the attitudes and realities of rural farmers. As this community's relationship with fertilizer shows, technologies will not

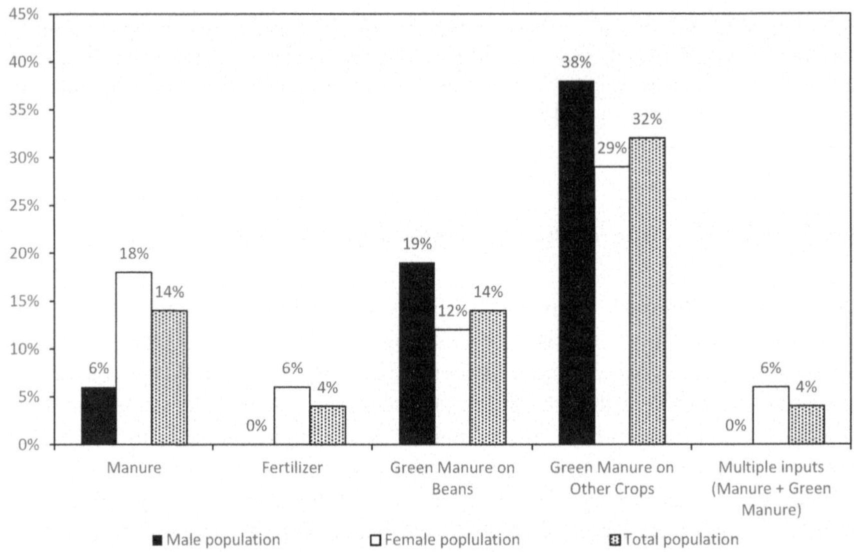

Fig. 8.2 Current input use by farmers. The black bars represent the percentage of the male population, white bars represent the percentage of the female population, and the grey bars indicate the percentage of the total population using each agricultural input

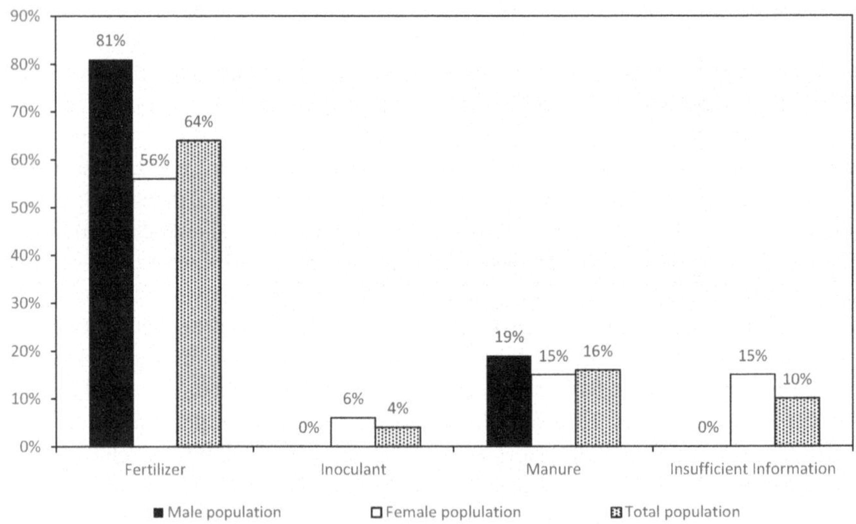

Fig. 8.3 Farmer input preferences. The black bars represent the percentage of the male population, white bars represent the percentage of the female population, and the grey bars indicate the percentage of the total population that would prefer to use each input type

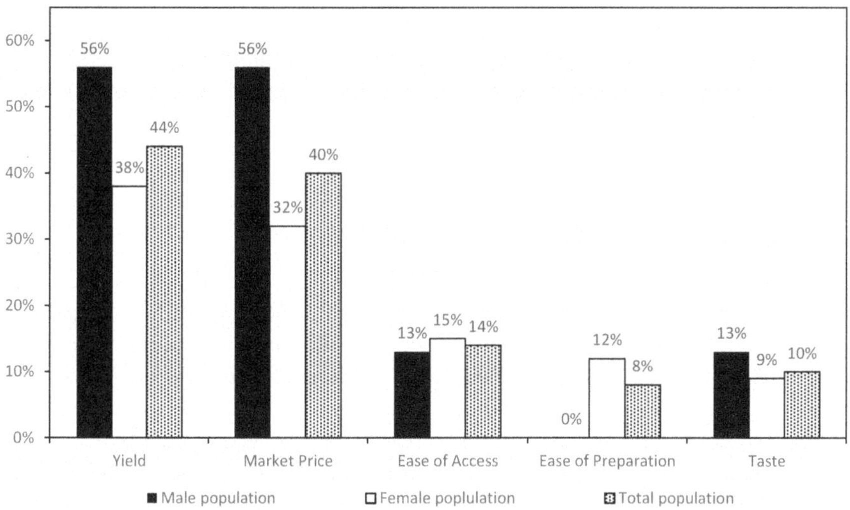

Fig. 8.4 Farmer self-selected bean choice criteria. The black bars represent the percentage of the male population, white bars represent the percentage of the female population, and the grey bars indicate the percentage of the total population using these criteria to choose what bean varieties to plant

be widely adopted if research and demonstrations are not paired with socioeconomic, business, and policy environments that provide affordable access to new technologies.

8.3.3 Attitudes and Practices About Bean Seed

Farmers interviewed were mostly concerned about yield (47%) and market price (44%) when making decisions about which varieties to plant (Fig. 8.4). These concerns must be considered when offering improved bean varieties. Bean varieties selected for enhanced BNF must maintain or improve overall yield and market price in order for broad dissemination to be successful.

Technologies and practices that improve soil quality can increase yields; but only address some of the many problems limiting yields in Kamuli District. Other yield limiting factors most commonly mentioned by farmers were "pests" (78%), "disease" (28%), and "weather/climate" (26%) (data not shown). Forty-six percent of producers mentioned two or more significant yield limiters. Improved inoculants and bean varieties can play a role in improving yields for subsistence farmers if the causes of and limitations to maximum BNF efficiency can be identified and controlled through soil management, genetics and breeding; and, if other traits that address producer concerns are also addressed in the new, high-fixing varieties. In a greenhouse trial conducted as part of the cooperating research to this KAP study, 51 varieties of common bean were able to fix between 48% and 77% of their total N requirement, indicating good potential for improved BNF varieties. Pest and disease resistance and drought tolerance are also important crop traits for improving yields and these qualities must be maintained in varieties that are developed or chosen for enhanced BNF (Bationo 2004; Hall and Clark 1995).

8.4 Conclusions for Using KAP Information

Biological nitrogen fixation is a useful trait in East Africa, where N is frequently yield limiting but chemical fertilizers are cost prohibitive. Identifying or creating bean varieties with high BNF is not sufficient, however, because people must be willing to adopt them (and effective rhizobia) in order for enhanced BNF technology to make a difference in the target community's food or economic security.

Agricultural extension providers could improve the dissemination of many technologies, including enhanced BNF, by improving farmers' access to information (Bationo 2004; Hall and Clark 1995). Farmers in Butansi would prefer to use fertilizer because they have received effective education about its efficacy and low labor requirement. Increasing the information available to farmers about BNF, inoculants, and improved BNF beans may likewise increase their popularity.

The farmers in this study were all personally engaged with a demonstration project yet showed little interest in obtaining BNF inoculants or in choosing bean varieties based on BNF or other rotational benefits. Clearly, the approach of the extension efforts should change if the goal is to increase producer interest in the products of BNF research. Demonstrations must be designed to highlight the benefits of the new technology, and address other limitations in realistic ways. Since farmers would prefer to use fertilizer as an input if it was available and affordable, inoculants might be introduced as a biological fertilizer or as an alternate cheaper source of nitrogen in order to better appeal to farmer interests. Rhizobial inoculants are not heavy like manures or as expensive as fertilizers, and may become a more popular option if farmers understand these advantages, can see the benefits of their use, and can access them locally and affordably.

Acknowledgements We would like to acknowledge Dr. Michael Ugen of the National Agricultural Research Organization (NARO) in Uganda for his local support of the project and participation. We would also like to acknowledge Volunteer Efforts for Development Concerns (VEDCO) Uganda and the Center for Sustaining Rural Livelihoods (CSRL) at Iowa State University for their logistical support.

References

Bationo, A. (2004). *Managing nutrient cycles to sustain soil fertility in sub-saharan Africa*. Nairobi, Kenya: Academy Science Publishers.

Biabani, A., Carpenter-Boggs, L., Coyne, C. J., Taylor, L., Smith, J. L., & Higgins, S. (2011). Nitrogen fixation potential in global chickpea mini-core collection. *Biology and Fertility of Soils, 47*(6), 679–685.

Bohlool, B. B., Ladha, J. K., Garrity, D. P., & George, T. (1992). Biological nitrogen fixation for sustainable agriculture: A perspective. *Plant and Soil, 141,* 1–11.

Buyinza, M. (2010). Is soil degradation to blame for the rural poverty in Southeastern Uganda? *Research Journal of Soil and Water Management, 1*(2), 27–33.

Chianu, J. N., Nkonya, E. M., Mairura, F. S., Chianu, J. N., & Akinnifesi, F. K. (2011). Biological nitrogen fixation and socioeconomic factors for legume production in sub-Saharan Africa: A review. *Agronomy and Sustainable Development, 31,* 139–154.

Giller, K. E. (2001). *Nitrogen Fixation in Tropical Cropping Systems,* 2nd edn. Wallingford, and New York: CABI Publishing.

Hall, A., & Clark, N. (1995). Coping with change, complexity and diversity in agriculture—the case of rhizobium inoculants in Thailand. *World Development, 23*(9), 1601–1614.

Machila, N., Wanyangu, S. W., McDermott, J., Welburn, S. C., Maudlin, I., & Eisler, M. C. (2003). Cattle owners' perceptions of African bovine trypanosomiasis and its control in Busia and Kwale Districts of Kenya. *Acta Tropica, 86,* 25–34.

McIntyre, B. D., Herren, H. R., Wakhungu, J., & Watson, R. T. (Eds.). (2009). *International assessment of agricultural knowledge, science and technology for development: Sub-saharan Africa report.* Washington DC: Island Press.

Nkonya, E., Pender, J., Jagger, P., Sserunkuuma, D., Kaizzi, C., Ssali, H. (2004). *Strategies for sustainable land management and poverty reduction in Uganda.* International Food Policy Research Institute (IFPRI) Research Report 133. Washington, DC: IFPRI.

Nkonya, E., Place, F., Pender, J., Mwanjololo, M., Okhimamhe, A., Kato, E., Crespo, S., et al,. (2011). *Climate risk management through sustainable land management in sub-saharan Africa.* International Food Policy Research Institute (IFPRI). Discussion Paper 01126. Washington DC: IFPRI.

Ojiem, J. O., de Ridder, N., Vanlauwe, B., & Giller, K. E. (2006). Socio-ecological niche: A conceptual framework for integration of legumes in smallholder farming systems. *International Journal of Agricultural Sustainability, 4*(1), 79–93.

Salameh, P. R., Baldi, I., Brochard, P., & Saleh, B. A. (2004). Pesticides in Lebanon: a knowledge, attitude, and practice study. *Environmental Research, 94,* 1–6.

Snapp, S. S., Mafongoya, P. L., & Waddington, S. (1998). Organic matter technologies for integrated nutrient management in smallholder cropping systems of southern Africa. *Agriculture, Ecosystems & Environment, 71,* 185–200.

Snapp, S. S., Blackie, M. J., Gilbert, R. A., Bezner-Kerr, R., & Kanyama-Phiri, G. Y. (2010). Biodiversity can support a greener revolution in Africa. *Proceedings of the National Academy of Sciences of the United States of America, 107*(48), 20840–20845.

Woomer, P. L., Karanja, N. K., Mekki, E. I., Mwakalombe, B., Tembo, H., Nkwiine, C., et al. (1997). Indigenous populations of rhizobia, legume response to inoculation and farmer awareness of inoculants in East and Southern Africa. *African Crop Science Conference Proceedings, 3,* 297–308.

Chapter 9
Performance of Mwitemania Bean Under the Influence of Nitrogen-Fixing *Rhizobium* Inoculant, Water Hyacinth Composts and DAP Fertilizer in a Field Infested with *Aphis fabae* and *Colletotrichum lindemuthianum*

Victoria Naluyange, Dennis M. W. Ochieno, Philip Wandahwa, John M. Maingi, Omwoyo Ombori, Dative Mukaminega, Alice Amoding, Martins Odendo, and John V. O. Muoma

Abstract The production of beans in the Lake Victoria basin of East Africa has been declining while the water hyacinth has been invasively spreading in the lake. These are understood to be effects of nitrogen (N) loss among other nutrients from land and their deposition into the lake resulting in eutrophication. To mitigate these problems, bean seeds are being inoculated with *Rhizobium* inoculants to fix nitrogen as an

---

V. Naluyange (✉) · P. Wandahwa
Department of Agriculture and Land Use Management, School of Agriculture, Veterinary Sciences and Technology, Masinde Muliro University of Science and Technology (MMUST), P.O. Box 190-50100, Kakamega-Webuye Road, Kakamega, Kenya
e-mail: vicluy@gmail.com

P. Wandahwa
e-mail: Wandahwa2002@yahoo.com

D. M. W. Ochieno · J. V. O. Muoma
Department of Biological Sciences, School of Natural Sciences, Masinde Muliro University of Science and Technology (MMUST), P.O. Box 190-50100, Kakamega-Webuye Road, Kakamega, Kenya

Centre for African Medicinal and Nutritional Flora and Fauna (CAMNFF), Masinde Muliro University of Science and Technology, P.O. Box 190-50100, Kakamega, Kenya
e-mail: dochieno@mmust.ac.ke

J. V. O. Muoma
e-mail: jmuoma@mmust.ac.ke

J. M. Maingi
Department of Biochemistry, Microbiology and Biotechnology, Kenyatta University, P.O. Box 43844-00100, Nairobi, Kenya
e-mail: maingijohn@gmail.com

O. Ombori
Department of Plant Sciences, Kenyatta University, P.O. Box 43844-00100, Nairobi, Kenya
e-mail: omwoyo.ombori@ku.ac.ke

M. A. Sutton et al. (eds.), *Just Enough Nitrogen*,
https://doi.org/10.1007/978-3-030-58065-0_9

alternative to fertilizers in maintaining soil nitrogen; while water hyacinth biomass is processed into composts to enrich soil fertility. The present study evaluates the growth and yield of *Rhizobium* inoculated seeds when grown without fertilizer (control), with diammonium phosphate fertilizer (DAP), or supplied with water hyacinth composts containing cattle manure culture (H + CMC) or effective microbes (H + EM). Infestations by the black bean aphid *Aphis fabae* and the anthracnose disease infestations caused by *Colletotrichum lindemuthianum* on these bean plants were also assessed. Root nodulation was found to be higher in *Rhizobium* inoculated plants grown with H + EM (20 nodules) and those without fertilizer (19 nodules) when compared to controls (7 nodules). In DAP treated plants, the germination percentage was low, growth rate slow, with few root nodules, flowers and pods, resulting in reduced yield. The differences between treatments for anthracnose disease incidence were similar to those of root nodules; while aphid populations did not vary between the treatments. The results of the present study do not offer a reliable basis for the application of water hyacinth composts and *Rhizobium* in improving the yields of Mwitemania beans. This is very likely because the soils were relatively fertile; N-fixation by *Rhizobium* is more efficient in unfertile soils. Better results may be achieved in soils of low fertility especially those deficient in nitrogen, hence conducive for N-fixation.

Keywords *Aphis fabae* · Compost · Nitrogen · *Rhizobium* · Water hyacinth

9.1 Introduction

Common beans are very important crops in Africa, especially within the Lake Victoria basin. Apart from their importance as food crops, common beans have an environmental role of fixing nitrogen (N) in soil through symbiotic *Rhizobium* species (Fischer 1994; Gage 2004; Masson-Boivin et al. 2009; De Lajudie et al. 2019). Nitrogen is a very important macronutrient for all plant growth and development functions (Vance 2001), and helps boost tolerance to pests and diseases (Ochieno 2010; Naluyange et al. 2014). However, effects of pests such as the black bean aphid *Aphis fabae* and the anthracnose pathogen *Colletotrichum lindemuthianum* have become severer (Kharinda 2013; Naluyange et al. 2014; Naluyange et al. 2016;

D. Mukaminega (Deceased)
Faculty of Applied Sciences, Kigali Institute of Science and Technology (KIST), P.O. Box 3900, Kigali, Rwanda

A. Amoding
Department of Agricultural Production, College of Agricultural and Environmental Sciences, Makerere University, P.O. Box 7062, Kampala, Uganda
e-mail: amodinga@yahoo.com

M. Odendo
Socio-Economics and Statistics Division, Kenya Agricultural and Livestock Research Organization (KALRO), P.O. Box 169-50100, Kakamega, Kenya
e-mail: Odendos@yahoo.com

Karoney et al. 2020). These effects have been complicated by loss of soil fertility (Vanlauwe et al. 2008; Ayuke et al. 2011), drought and other effects associated with climate change (Adger et al. 2003). Therefore, the production of common beans and other crops within the Lake Victoria basin has been declining (Katungi et al. 2009). Conserving nitrogen within the agricultural soils of the Lake Victoria basin is a primary move towards enhancing the production of common beans and other food crops.

The Lake Victoria Basin is highly populated with intensive food production activities like agriculture, animal husbandry and fishing (Odada et al. 2009). These create high demand for nitrogen. At the same time, there has been rapid environmental degradation, which manifests in forms such as loss of vegetative cover, erosion of soil nutrients and their deposition into Lake Victoria (Machiwa 2003). This has been a double-edged problem, as soils have lost fertility (Vanlauwe et al. 2008), while Lake Victoria undergoes extreme eutrophication (Lung'ayia et al. 2001). Nitrogen is among the most important nutrients whose movement from land into the lake has been linked to such problems (Swallow et al. 2009). Rates and sources of nutrient movement into Lake Victoria have not been conclusively researched (Ong and Orego 2002), although it has been estimated that 90% of the nitrogen input into this lake is associated with atmospheric deposition and land runoff (Scheren et al. 2000). The land runoff of nutrients is mainly contributed by soil erosion (Leip et al. 2014; Mwanuzi et al. 2005; World Bank 1996). The contribution of agricultural input in the Lake Victoria Basin from commercial fertilizers is very low as farmers rarely use them (Leip et al. 2014; Mwanuzi et al. 2005; World Bank 1996). As a result of nutrient loss through soil erosion and runoff, crop production is thought to have declined on land, while the deposition of nutrients in the lake through runoff and effluents has resulted into the invasive growth of water hyacinth impeding fishing activities.

There have been efforts to conserve the waters and soils of Lake Victoria basin. Among the strategies, water hyacinth with total nitrogen estimated between 1.11% and 1.33% fresh weight is being developed into compost formulations for replenishing nitrogen and other nutrients in soil (Gunnarsson and Petersen 2007; Singh and Kalamdhad 2015). The water hyacinth composts are being tested for compatibility with *Rhizobium*-inoculated beans that fix nitrogen in the soil (Naluyange et al. 2014). Currently, water hyacinth composts formulated with cattle manure culture (H + CMC) and effective microbes (H + EM) have been developed and found to have some positive effects on Rosecoco bean production (Naluyange et al. 2014). The present work evaluated these two composts and diammonium phosphate fertilizer (DAP) on the performance of Pinto sugars bean (Mwitemania) inoculated with nitrogen-fixing *Rhizobium* inoculant in the presence of aphids and anthracnose disease.

9.2 Materials and Methods

9.2.1 Experimental Design

A field experiment was conducted at the Masinde Muliro University of Science and Technology (MMUST) farm (N 00 17.104′, E 034° 45.874′; altitude 1,561 m a.s.l.). The land had been fallow and colonized by the African couch grass *Digitaria scalarum* (Schweinf.) Chiov. (Poaceae) for over five years. Soils in this region have been classified as Dystro-mollic Nitisols (FAO 1974; Rota et al. 2006). Nutrient composition for the soil was; total phosphorus (18.9 ppm), total nitrogen (0.26%), organic carbon (2.5%), potassium (0.41 $cmol_c\ kg^{-1}$), sodium (0.1 $cmol_c\ kg^{-1}$), calcium (2.3 $cmol_c\ kg^{-1}$), magnesium (0.8 $cmol_c\ kg^{-1}$), zinc (1.9 ppm) and iron (0.37 ppm), with acidic pH of 4.2 (Naluyange et al. 2014). The experiment was laid out in a randomized block design comprising of a 2 × 4 factorial treatment with *Rhizobium* inoculum factor having two levels (with or without inoculation) and fertilizer factor with four levels, i.e., no fertilizer (Control), diammonium phosphate fertilizer (DAP), water hyacinth compost + cattle manure culture (H + CMC), and water hyacinth compost + effective microbes (H + EM). Each of the resulting eight treatment combinations (plots) had 25 plants (n) in three blocks (i.e., N = 600). Each plot was in form of a row containing the 25 plants spaced at 20 cm, with a distance of 40 cm between the plots, without border rows. The treatment rows were completely randomized to minimize non-experimental bias in sampling for natural infestations of aphids and anthracnose disease on bean plants. This experiment was conducted during the long rain season between 20th April and 30th July 2012, and then repeated between 30th May and 31st August 2012.

9.2.2 Water Hyacinth Composts

Water hyacinth composts were produced as described by Naluyange et al. (2014). The H + CMC compost had a density of 58 g/100 cm^3, with a nutrient concentration of total phosphorus (375 ppm), total nitrogen (1.1%), organic carbon (13.4%), potassium (21 $cmol_c\ kg^{-1}$), sodium (1.9 $cmol_c\ kg^{-1}$), calcium (22.3 $cmol_c\ kg^{-1}$), magnesium (12 $cmol_c\ kg^{-1}$), zinc (2 ppm) and iron (1.9 ppm), with alkaline pH of 8.1 (Naluyange et al. 2014). The H + EM compost had a density of 62 g/100 cm^3, with the nutrient concentration of total phosphorus (270 ppm), total nitrogen (1%), organic carbon (13.5%), potassium (24.5 $cmol_c\ kg^{-1}$), sodium (1.7 $cmol_c\ kg^{-1}$), calcium (27.5 $cmol_c\ kg^{-1}$), magnesium (15.3 $cmol_c\ kg^{-1}$), zinc (4 ppm) and iron (1.7 ppm), with alkaline pH of 8.4 (Naluyange et al. 2014).

9.2.3 *Seed Inoculation and Planting*

The experiment utilized Mwitemania common bean variety (GLPx92) (Kenya Seed Company Ltd). This common bean variety grows well in altitudes ranging between 900 and 1600 m a.s.l. It matures between 2 and 3 months with potential grain yields ranging between 1.2 and 1.5 t/ha (Kenya Plant Health Inspectorate Service (KEPHIS) 2015). The seeds were inoculated with BIOFIX®(MEA Ltd., Kenya) containing Rhizobium tropici CIAT899 (Balume et al. 2013) as per manufacture's directions. The seeds (250 g) were mixed in gum Arabic solution (0.5 gum Arabic/5 mL of sterile lukewarm water). The gum Arabic-coated seeds (250 g) were mixed with the *Rhizobium* inoculant powder (1 g). Controls were coated with the gum Arabic solution only.

Planting holes of ~200 cm^3 volume (i.e., ~5 cm diameter and ~10 cm deep) were dug using a shovel. The water hyacinth composts were applied using containers of 150 mL volumes per hole (~90 g) as per the respective treatments and mixed with soil. Therefore, each planting hole received approximately 0.03 g phosphorus and 0.99 g nitrogen for the H + CMC compost; or 0.02 g phosphorus and 0.90 g nitrogen in case of the H + EM compost. For the DAP treatment, one levelled teaspoon (4.7 g) was mixed with soil in the planting hole (Naluyange et al. 2014). DAP fertilizer contains nitrogen (18%) and phosphorus pentoxide P_2O_5 (46%), with phosphorus (P) constituting 20% of the total mass. Hence, every planting hole in the DAP treatment received 0.94 g phosphorus and 0.85 g nitrogen. One bean seed was sown in every planting hole at a depth of ~2 cm.

9.2.4 *Plant Growth, Root Nodulation and Yields*

Data were collected as described by Naluyange et al. (2014). The emergence date of every seedling was recorded independently, and used to determine the duration for germination. The number of seedlings that germinated out of the total number of seeds that were planted was used to determine the germination percentage within 20 days from the planting date. When the first trifoliate leaves were fully formed in ~80% of the seedlings, plant height (stem base to petiole), length of the middle leaf (base to apex) and its width (widest part) were recorded. The date when the first flower of every plant appeared was recorded and used to calculate the duration for flowering in days from the date of planting. The number of flowers on each plant was recorded every three days for a period of three weeks. Ten days from the onset of flowering, five bean plants were randomly selected from each treatment per block for the estimation of number of root nodules associated with *Rhizobium* colonization. The bean plants were dug out from the soil into plastic bags, and the number of root nodules per plant was recorded using a tally counter in the laboratory.

The date of ripening of the first pod per plant was recorded and used to calculate the duration to maturity in days from the date of planting. Plants were harvested independently and the number of harvested pods per plant recorded. The harvested pods from every plant were packed in separate paper packets and sun dried for a period of five days. The weight of all seeds per packet was recorded as seed weight per plant.

9.2.5 Aphid and Anthracnose Incidences

Aphid infestations on bean plants were recorded at the vegetative and flowering stage of bean plants. Three screw-capped containers each containing 10 mL of 70% ethanol were placed on every treatment row of 25 plants. Aphids from every eight plants per row were collected into each container using a camel hair brush from leaves and stems. The collected aphids were identified under a dissection microscope (Model Z45E, Leica Inc., USA) at $\times 10$ magnifications using the features described in Martin (1983) and Holman (1998), and their absolute counts recorded using a tally counter. At the vegetative stage, the bean plants were also scored for anthracnose disease incidence i.e., the proportion of plants having anthracnose symptoms, characterized by dark brown to black lesions on leaves (Hagedorn and Inglis 1986; Buruchara et al. 2010).

9.2.6 Statistical Analysis

Statistical analysis was conducted using SAS 9.1 software (SAS Institute Inc.) at $p <$ 0.05 confidence level. Descriptive statistics such as means were generated using proc means, while frequencies (percentages) were generated using proc freq. Data on plant growth was checked for normality using proc univariate; while proc transreg was used to find appropriate Box-Cox power transformations for normalization of data. Proc mixed was used for analyses of variance (ANOVA) among the treatments; and means were separated using Ls-means when the effects of treatments were significant. Data on aphid populations and root nodule counts were analyzed using proc genmod (χ^2 test; Poisson) and the means separated using proc multtest. Anthracnose disease incidences and germination percentages were analyzed by proc genmond (χ^2 test; binomial) and percentages compared using proc multtest.

9.3 Results and Discussion

Rhizobium inoculation was associated with high number of root nodules, particularly in plants receiving water hyacinth compost with effective microbes (H + EM) and those without fertilizer (Fig. 9.1). Enhancement of root nodulation by *Rhizobium* species is desirable for nitrogen fixation (Mandri et al. 2012; Salvagiotti et al. 2013).

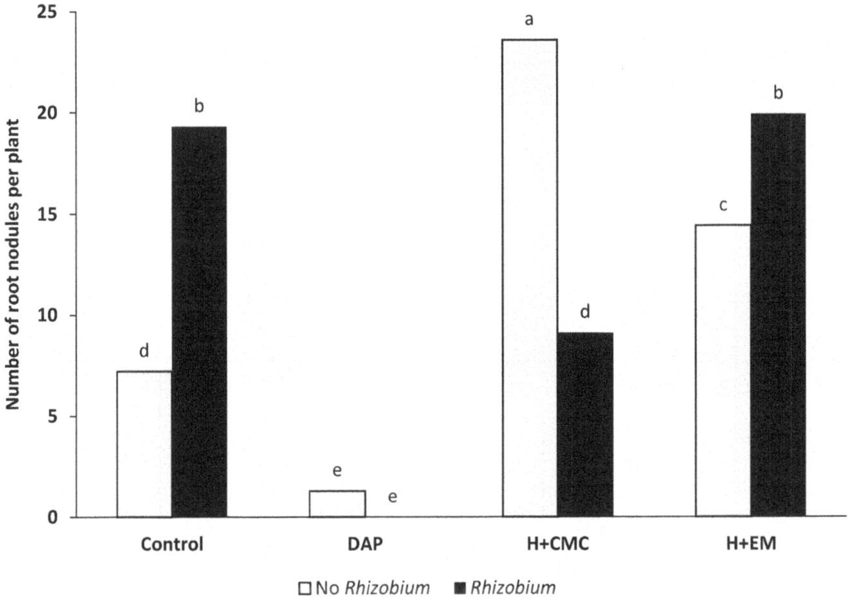

Fig. 9.1 Number of root nodules in Pinto sugars (Mwitemania) bean plants as affected by commercial *Rhizobium* inoculant and soil fertility amendments, i.e., without fertilizer (Control), diammonium phosphate fertilizer (DAP), water hyacinth compost + cattle manure culture (H + CMC) and water hyacinth compost + effective microbes (H + EM). The black bars represent *Rhizobium* inoculated plants while the white bars are for the non-inoculated plants. Bars with the same letter(s) are not significantly different (Proc Genmod, $\chi 2$ test, $p > 0.05$)

However, water hyacinth compost with cattle manure culture (H + CMC) appeared to favour root nodulation by native *Rhizobium* species, since the number of root nodules was higher in non-inoculated plants (Fig. 9.1). Native rhizobia are known to function differently from introduced ones (Dean et al. 2009). In a similar study that utilized Rosecoco bean variety, both H + EM and H + CMC enhanced root nodulation by the *Rhizobium* inoculant (Naluyange et al. 2014). The two types of composts contain different microbial profiles, with H + EM having *Rhodopseudomonas palustris*, lactic acid bacteria (*Lactobacillus plantarum* and *L. casei*) and yeast (*Saccharomyces cerevisae*) (Naluyange et al. 2014), while cattle manure culture constitutes different microbial profiles (Tiquia 2005). This implies that bean varieties vary in response towards the *Rhizobium* inoculant and different microbes in water hyacinth composts. This points to the need for further in-depth studies on relationships between these two common bean cultivars and *Rhizobium* strains.

There have been reports that the population of *A. fabae* and other aphid species increases in *Rhizobium*-inoculated legumes (Dean et al. 2009, 2014; Naluyange et al. 2014). This was not evident in the present study. It has also been observed that *Rhizobium* inoculation in Rosecoco bean was associated with an increase in anthracnose incidences (Naluyange et al. 2014); this was not the case in the present study with

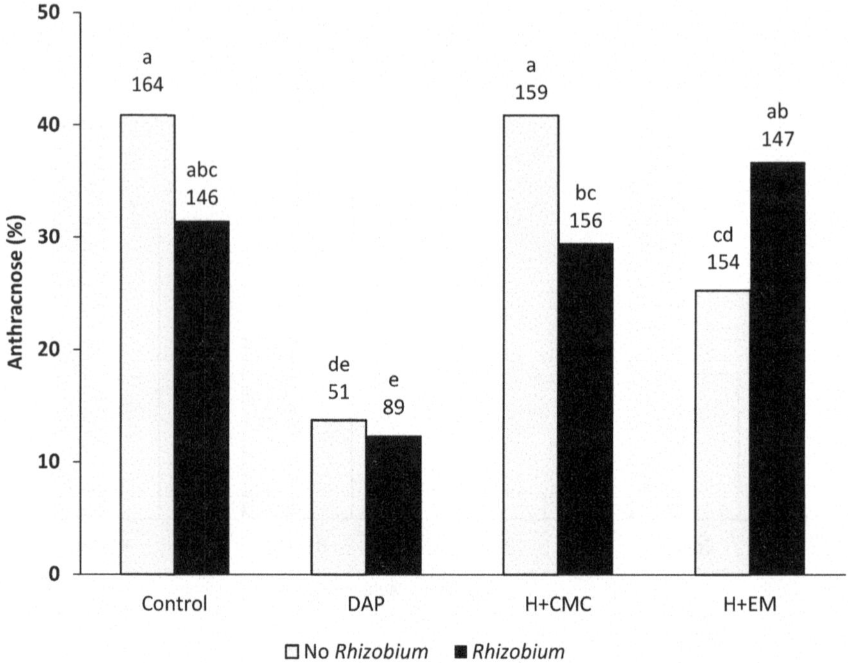

Fig. 9.2 Anthracnose incidences in Pinto sugars (Mwitemania) bean plants as affected by commercial *Rhizobium* inoculant and soil fertility amendments in the second trial, i.e., without fertilizer (Control), diammonium phosphate fertilizer (DAP), water hyacinth compost + cattle manure culture (H + CMC) and water hyacinth compost + effective microbes (H + EM). The black bars represent *Rhizobium* inoculated plants while the white bars are for the non-inoculated plants. Numbers on top of bars represent sample sizes. Bars with the same letter(s) are not significantly different (Proc Genmod, $\chi 2$ test, $p > 0.05$)

Mwitemania bean when trends in Figs. 9.1 and 9.2 are compared. Figure 9.2 shows that while *Rhizobium* inoculation was associated with reduced anthracnose incidence for the control and H + CMC treatments, it was associated with increased anthracnose incidence for the H + EM treatment. The DAP treatment had the lowest anthracnose incidence.

Plant height, leaf length and width did not vary due to *Rhizobium* inoculation and fertilizer treatments ($p > 0.05$), with the overall means being 4.7 ± 0.25 cm, 6.53 ± 0.23 cm and 4.3 ± 0.20 cm, respectively. This was also the case for the post-emergence developmental time in days to flowering (39 ± 0.8) and days to harvest (74 ± 0.7), which did not vary between the treatments ($p > 0.05$). However, bean seeds grown with DAP took more days to emerge (9.7 ± 0.5) compared to the controls (7.3 ± 0.6), H + CMC (6.8 ± 0.4) and H + EM (7.2 ± 0.6), while their germination percentage was low (Fig. 9.3). The emerging bean plants also had very low number of root nodules when grown with DAP (Fig. 9.1). Consequently, the DAP treated bean plants produced few flowers and pods (Fig. 9.4), resulting in low seed weight per

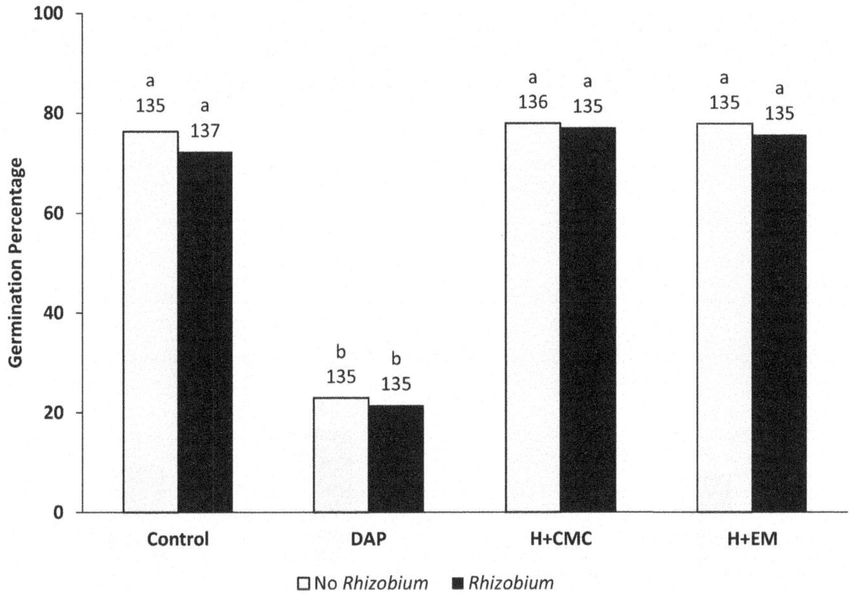

Fig. 9.3 Germination percentage of Pinto sugars (Mwitemania) bean seeds as affected by commercial *Rhizobium* inoculant and soil fertility amendments, i.e., without fertilizer (Control), diammonium phosphate fertilizer (DAP), water hyacinth compost + cattle manure culture (H + CMC) and water hyacinth compost + effective microbes (H + EM). The black bars represent Rhizobium inoculated plants while the white bars are for the non-inoculated plants. Numbers on top of bars represent sample sizes. Bars with the same letter(s) are not significantly different (Proc Genmod, $\chi 2$ test, $p > 0.05$)

hectare (Fig. 9.5). These effects were also observed in the Rosecoco bean (Naluyange et al. 2014), and have been linked to phytotoxicity of phosphate fertilizers (Zhang and Rengel 2002; Kabir et al. 2010; Salvagiotti et al. 2013). Although phosphorus (P) is necessary for root nodulation by rhizobia (Mandri et al. 2012), DAP may not have been the suitable source of P for the bean cultivar, *Rhizobium* inoculant and the type of soil in the present study. The concentration of DAP, application method, seed placement and the type of soil may have contributed to the low plant population. For example, lower rates of DAP addition may have different effects if phytotoxicity is an issue. Grain yields that were attained without any amendment (0.4 t/ha) were generally below the potential for Mwitemania cultivar, which ranges between 1.2 and 1.5 t/ha. This may partly be attributed to insufficient nutrient acquisition by the bean plants.

Inoculation of the Mwitemania bean seeds with the *Rhizobium* inoculant and the application of water hyacinth composts influenced root nodulation, but more likely due to the added microbes. There was no clear evidence that *Rhizobium* inoculant and the water hyacinth composts enhanced plant growth and yields. This implies that the influenced root nodulation did not translate to nitrogen fixation. Inoculation

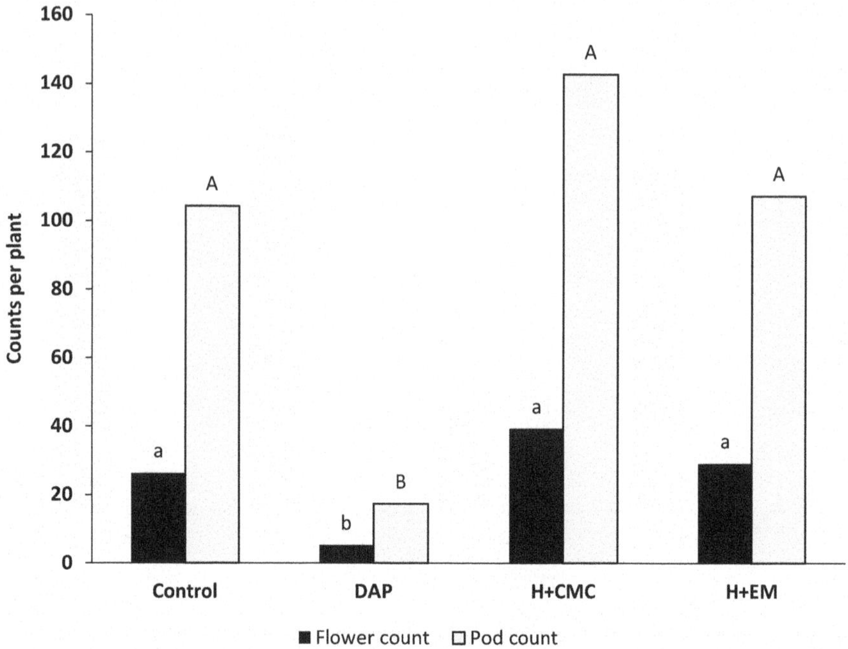

Fig. 9.4 Number of flowers and pods in Pinto sugars (Mwitemania) bean plants as affected by soil fertility amendments, i.e., without fertilizer (Control), diammonium phosphate fertilizer (DAP), water hyacinth compost + cattle manure culture (H + CMC) and water hyacinth compost + effective microbes (H + EM). Bars with the same letter(s) [*capital letters for pod counts*] are not significantly different (F test, $p > 0.05$)

of beans and other legumes with the symbiotic bacterium does not always result in improved plant growth (Graham 1981; Graham and Vance 2000). Although mean yield of H + CMC was 50% higher than the control (Fig. 9.5), the means were not significantly different at the $p = 0.05$ level. Therefore the results of the present study do not offer reliable basis for the application of water hyacinth composts and *Rhizobium* in improving the yields of Mwitemania beans. This is very likely because the soils were relatively fertile and yet N-fixation by *Rhizobium* is more efficient in unfertile soils. Better results may be achieved in soils of low fertility especially those deficient in nitrogen, hence conducive for N-fixation. The fact that a 50% difference in the case of H + CMC was not significantly different from the control also points to high variability between the replicates indicating a lack of consistent response.

It had been proposed that using water hyacinth in crop production would reduce its invasive spread on Lake Victoria (Naluyange et al. 2014). However, many questions have been raised on this issue. Results in the present study have not demonstrated that the compost is a viable soil fertility amendment. Furthermore, water hyacinth is succulent with very low dry matter content, therefore huge amounts may be required to replenish soils, bearing in mind the fact that the invasive weed grows very fast

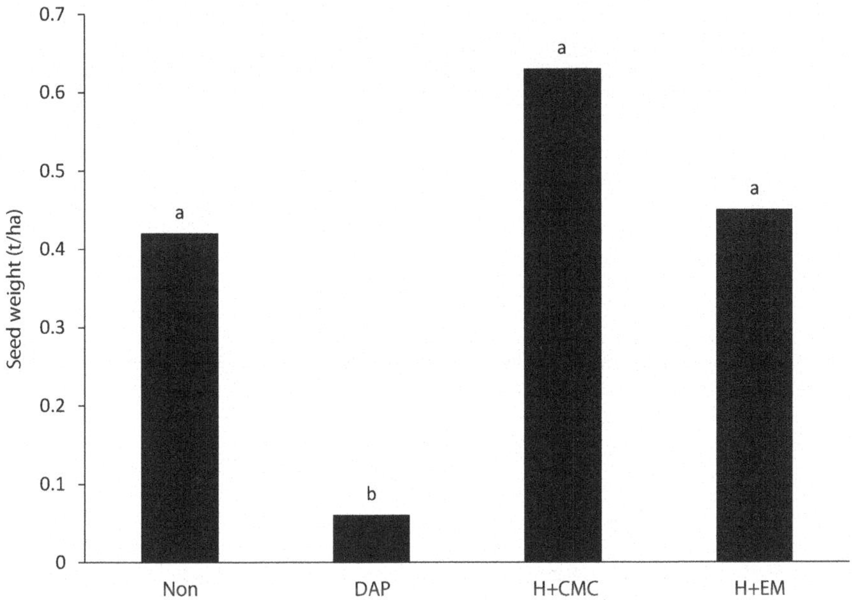

Fig. 9.5 Seed weight in Pinto sugars (Mwitemania) bean plants as affected by soil fertility amendments, i.e., without fertilizer (Control), diammonium phosphate fertilizer (DAP), water hyacinth compost + cattle manure culture (H + CMC) and water hyacinth compost + effective microbes (H + EM). Bars with the same letter(s) are not significantly different (F test, $p > 0.05$)

and drifts in water. It may be very expensive to harvest any meaningful quantities and hence not be economically viable. Therefore, despite composting being well known to improve soil fertility and organic matter, water hyacinth may not be the best material for this process.

Acknowledgements This work was conducted under the sponsorship of the Lake Victoria Research (VicRes) Initiative, a regional collaborative research programme of the Inter-University Council for East Africa (IUCEA). Research funds were provided by the Government of Sweden through the Swedish International Development Cooperation Agency (Sida), under the framework of the Lake Victoria Development Partnership (LVDP) Programme.

References

Adger, W. N., Huq, S., Brown, K., Conway, D., & Hulme, M. (2003). Adaptation to climate change in the developing world. *Progress in Development Studies, 3*(3), 179–195.

Ayuke, F. O., Brussaard, L., Vanlauwe, B., Six, J., Lelei, D. K., Kibunja, C. N., et al. (2011). Soil fertility management: Impacts on soil macrofauna, soil aggregation and soil organic matter allocation. *Applied Soil Ecology, 48*(1), 53–62.

Balume, I. K., Keya, S. O., Karanja, N. K., & Woomer, P. L. (2013). Improving shelf life of legume inoculants in East Africa. In *Joint Proceedings of the 27th Soil Science Society of East Africa and the 6th African Soil Science Society Conference*.

Buruchara, R., Mukankusi, C., & Ampofo, K. (2010). *Bean disease and pest identification and management*. CIAT publication no. 371. Handbooks for small-scale seed producers, No. 04. Kampala, Uganda: International Center for Tropical Agriculture (CIAT); Pan-Africa Bean Research Alliance (PABRA).

De Lajudie, P. M., Andrews, M., Ardley, J., Eardly, B., Jumas-Bilak, E., Kuzmanović, N., Lassalle, F., Lindström, K., Mhamdi, R., Martínez-Romero, E., & Moulin, L. (2019). Minimal standards for the description of new genera and species of rhizobia and agrobacteria. *International Journal of Systematic and Evolutionary Microbiology, 69*(7),1852–1863.

Dean, J. M., Mescher, M. C., & De Moraes, C. M. (2009). Plant-rhizobia mutualism influences aphid abundance on soybean. *Plant and Soil, 323*(1–2), 187–196.

Dean, J. M., Mescher, M. C., & De Moraes, C. M. (2014). Plant dependence on rhizobia for nitrogen influences induced plant defenses and herbivore performance. *International Journal of Molecular Sciences, 15*(1), 1466–1480.

FAO. (1974). *Soil map of the world 1:5,000,000* (Vol. 1). FAO/UNESCO: Legend.

Fischer, H. M. (1994). Genetic regulation of nitrogen fixation in rhizobia. *Microbiology Reviews, 58*(3), 352–386.

Gage, D. J. (2004). Infection and invasion of roots by symbiotic, nitrogen-fixing rhizobia during nodulation of temperate legumes. *Microbiology and Molecular Biology Reviews, 68*(2), 280–300.

Graham, P. H. (1981). Some problems of nodulation and symbiotic nitrogen fixation in *Phaseolus vulgaris* L.: A review. *Field Crops Research, 4*, 93–112.

Graham, P. H., & Vance, C. P. (2000). Nitrogen fixation in perspective, an overview of research and extension needs. *Field Crops Research, 65*(2–3), 93–106.

Gunnarsson, C. C., & Petersen, C. M. (2007). Water hyacinths as a resource in agriculture and energy production: A literature review. *Waste Management, 27*(1), 117–129.

Hagedorn, D. J., & Inglis, D. A. (1986). *Handbook of bean diseases*. University Wisconsin-Madison Extension Bulletin, A3374. USA: University of Wisconsin-Madision.

Holman, J. (1998). Species of the genus *Aphis* (Sternorrhyncha: Aphidoidea) living on *Hieracium* (Asteraceae: Cichorieae). *European Journal of Entomology, 95*(3), 383–394.

Kabir, E., Bell, R. W., & Johansen, C. (2010). Triple superphosphate placement affects early growth of chickpea. In *Proceedings of the 19th World Congress of Soil Science, Soil Solutions for a Changing World* (pp. 276–279), 1–6 August 2010, Brisbane, Australia. Wien, Austria: International Union of Soil Sciences (IUSS).

Karoney, E. M., Ochieno, D. M. W., Baraza, D. L., Muge, E. K., Nyaboga, E. N., & Naluyange, V. (2020). Rhizobium improves nutritive suitability and tolerance of Phaseolus vulgaris to *Colletotrichum lindemuthianum* by boosting organic nitrogen content. *Applied Soil Ecology , 149*, 103534. https://doi.org/10.1016/j.apsoil.2020.103534

Katungi, E., Farrow, A., Chianu, J., Sperling, L., & Beebe, S. (2009). *Common bean in Eastern and Southern Africa: A situation and outlook analysis*. International Centre for Tropical Agriculture.

KEPHIS (Kenya Plant Health Inspectorate Service). (2015). *National crop variety list*. http://www.kephis.org/images/Uploads/UPNVLIST.pdf. Accessed December 2018.

Kharinda, B. T. M. (2013). *Phenotypic characterization of common bean genotypes for resistance to the pathogen Colletotrichum lindemuthianum constraining bean production in Western Kenya*. M.Sc. thesis, Kakamega, Kenya: Masinde Muliro University of Science and Technology.

Leip, A., Leach, A., Musinguzi, P., Tumwesigye, T., Olupot, G., Tenywa, J. S., et al. (2014). Nitrogen-neutrality: A step towards sustainability. *Environmental Research Letters, 9*, 115001. https://doi.org/10.1088/1748-9326/9/11/115001.

Lung'ayia, H., Sitoki, L., & Kenyanya, M. (2001). The nutrient enrichment of Lake Victoria (Kenyan waters). *Hydrobiologia*, 458(1–3), 75–82.

Machiwa, P. K. (2003). Water quality management and sustainability: The experience of Lake Victoria Environmental Management Project (LVEMP)-Tanzania. *Physics and Chemistry of the Earth, 28,* 1111–1115.

Mandri, B., Drevon, J.-J., Bargaz, A., Oufdou, K., Faghire, M., Plassard, C., et al. (2012). Interactions between common bean genotypes and rhizobia strains isolated from Moroccan soils for growth, phosphatase and phytase activities under phosphorus deficiency conditions. *Plant Nutrition, 35*(10), 1477–1490.

Martin, J. H. (1983). The identification of common aphid pests of tropical agriculture. *Tropical Pest Management, 29*(4), 395–411.

Masson-Boivin, C., Giraud, E., Perret, X., & Batut, J. (2009). Establishing nitrogen-fixing symbiosis with legumes: How many *Rhizobium* recipes? *Trends in Microbiology, 17*(10), 458–466.

Mwanuzi, F. L., Abuodha, J. O. Z., Muyodi, F. J., Hecky, R. E. (Eds.). (2005). *Lake Victoria regional water quality synthesis report.* Lake Victoria Environment Management Project (LVEMP) water quality and ecosystem status.

Naluyange, V., Ochieno, D. M. W., Maingi, J. M., Ombori, O., Mukaminega, D., Amoding, A., et al. (2014). Compatibility of *Rhizobium* inoculant and water hyacinth compost formulations in Rosecoco bean and consequences on *Aphis fabae* and *Colletotrichum lindemuthianum* infestations. *Applied Soil Ecology, 76*(4), 66–77.

Naluyange, V., Ochieno, D. M., Wandahwa, P., Odendo, M., Maingi, J. M., Amoding, A., & Muoma, J. (2016). Belowground influence of Rhizobium inoculant and water hyacinth composts on yellow bean infested by Aphis fabae and *Colletotrichum lindemuthianum* under field conditions. *Journal of Plant Studies, 5,* 32–41. https://doi.org/10.5539/jps.v5n2p32

Ochieno, D. M. W. (2010). *Endophytic control of Cosmopolites sordidus and Radopholus similis using Fusarium oxysporum V5w2 in tissue culture banana.* Ph.D. thesis, Wageningen University, The Netherlands.

Odada, E. O., Ochola, W. O., & Olago, D. O. (2009). Drivers of ecosystem change and their impacts on human well-being in Lake Victoria basin. *African Journal of Ecology, 47*(s1), 46–54.

Ong, C., & Orego, F. (2002). Links between land management, sedimentation, nutrient flows and smallholder irrigation in the Lake Victoria basin. In H. G. Blank, M. M. Clifford, & M. R. Hammond (Eds.), *The changing face of irrigation in Kenya: Opportunities for anticipating change in eastern and southern Africa* (pp. 135–154). Colombo, Sri Lanka: International Water Management Institute.

Rota, J. A., Wandahwa, P., & Sigunga, D. O. (2006). Land evaluation for soybean (*Glycine max* L. Merrill) production based on kriging soil and climate parameters for the Kakamega district Kenya. *Journal of Agronomy, 5*(1), 142–150.

Salvagiotti, F., Barraco, M., Dignani, D., Sanchez, H., Bono, A., Vallone, P., et al. (2013). Plant stand, nodulation and seed yield in soybean as affected by phosphate fertilizer placement, source and application method. *European Journal of Agronomy, 51,* 25–33.

Scheren, P. A. G. M., Zanting, H. A., & Lemmens, A. M. C. (2000). Estimation of water pollution sources in Lake Victoria, East Africa: Application and elaboration of the rapid assessment methodology. *Journal of Environmental Management, 58*(4), 235–248.

Singh, J., & Kalamdhad, A. S. (2015). Assessment of compost quality in agitated pile composting of water hyacinth collected from different sources. *International Journal of Recycling of Organic Waste in Agriculture, 4*(3), 175–183.

Swallow, B. M., Sang, J. K., Nyabenge, M., Bundotich, D. K., Duraiappah, A. K., & Yatich, T. B. (2009). Tradeoffs, synergies and traps among ecosystem services in the Lake Victoria basin of East Africa. *Environmental Science & Policy, 12*(4), 504–519.

Tiquia, S. M. (2005). Microbial community dynamics in manure composts based on 16S and 18S rDNA T-RFLP profiles. *Environmental Technology, 26*(10), 1101–1113.

Vance, C. P. (2001). Symbiotic nitrogen fixation and phosphorus acquisition. Plant nutrition in a world of declining renewable resources. *Plant Physiology, 127*(2), 390–397.

Vanlauwe, B., Kanampiu, F., Odhiambo, G. D., De Groote, H., Wadhams, L. J., & Khan, Z. R. (2008). Integrated management of *Striga hermonthica*, stemborers, and declining soil fertility in western Kenya. *Field Crops Research, 107*(2), 102–115.

World Bank. (1996). *Staff Appraisal Report. The Republic of Kenya, United Republic of Tanzania, and the Republic of Uganda for the Lake Victoria Environmental Management Project*. Report No. 15541-AFR. Agriculture and Environment Operations Division, Eastern Africa Department, Africa Region. Washington: Global Environment Division, World Bank.

Zhang, X. K., & Rengel, Z. (2002). Temporal dynamics of gradients of phosphorus, ammonium, pH, and electrical conductivity between a di-ammonium phosphate band and wheat roots. *Australian Journal of Agricultural Research, 53*(8), 985–992.

Chapter 10
Biological Nitrogen Fixation of Pigeonpea and Groundnut: Quantifying Response Across 18 Farm Sites in Northern Malawi

Wezi G. Mhango, Sieglinde Snapp, and George Y. Kanyama-Phiri

Abstract The global nitrogen (N) cycle is markedly, and increasingly, influenced by anthropogenic inputs. A large unknown remains the quantity of biological N fixation (BNF) inputs derived from agriculture. This leads to major uncertainties in modelling reactive N interactions with climate change, and understanding N biogeochemical processes. Understanding N dynamics is central to enhancing productivity in cropping systems. To fill this gap, we used measurements of natural abundance of the ^{15}N isotope to quantify BNF and yield of groundnut and pigeonpea at 18 on-farm sites in Ekwendeni, Northern Malawi. The study was conducted over the 2007/08 (2008) and 2008/09 (2009) cropping seasons under farmer management, for a range of edaphic environments. Overall, the soils are largely sandy with low to moderate organic carbon (0.12–1.56%), pH (5.5–6.5), and very low to moderately high inorganic phosphorus (P) (3–85 mg kg^{-1}). The main drivers of BNF were plant density, inorganic P and interspecific competition. The proportion of N derived from the atmosphere (22–99%) was influenced by soil P status across seasons and crop species, but not by cropping system. The mean proportion of nitrogen derived from

This chapter is an edited version of Mhango et al. (2017), Biological nitrogen fixation and yield of pigeonpea and groundnut: quantifying response on smallholder farms in northern Malawi, *African Journal of Agricultural Research*, 12(16), 1385–1394, doi:10.5897/AJAR2017.12232 (https://academicjournals.org/journal/AJAR/article-abstract/637190263880). It is re-used here in edited form under Creative Commons license CC BY 4.0 (https://creativecommons.org/licenses/by/4.0).

W. G. Mhango (✉) · S. Snapp
Department of Crop and Soil Sciences, Lilongwe University of Agriculture and Natural Resources, Bunda Campus, P. O. Box 219, Lilongwe, Malawi
e-mail: wezzi2002@yahoo.com

S. Snapp
e-mail: snapp@msu.edu

S. Snapp
Department of Plant, Soils and Microbial Sciences, Michigan State University, East Lansing, MI 48824, USA

G. Y. Kanyama-Phiri
Lilongwe University of Agriculture and Natural Resources, Bunda Campus, P. O. Box 219, Lilongwe, Malawi
e-mail: gykphiri@gmail.com

M. A. Sutton et al. (eds.), *Just Enough Nitrogen*,
https://doi.org/10.1007/978-3-030-58065-0_10

atmosphere (Ndfa) was high in both groundnut (75%) and pigeonpea (76%). Total N fixed by groundnut and pigeonpea differed between cropping system in the dry year, where intercropping was associated with low levels of N fixed by pigeonpea (15 kg N ha^{-1}) compared to sole pigeonpea (32 kg N ha^{-1}). A short rainfall season could not support biomass production of pigeonpea and this has negative implications for relying on BNF to drive productivity on smallholder farms.

Keywords Groundnut · Pigeonpea · Intercropping · Biological nitrogen fixation · Smallholder farms

10.1 Introduction

Nitrogen (N) deficiency is a major factor limiting productivity of maize based systems on smallholder farms in Southern Africa. The major sources of N in agro-ecosystems are inorganic fertilizers, livestock manures, compost manures and legumes. Legumes fix atmospheric N biologically into inorganic forms that can be used by plants (Giller 2001). Biological N fixation is important for small holder farmers as it is a relatively cheaper source of N compared to inorganic fertilizers, and less prone to losses through leaching and denitrification. In Malawi, farmers grow a diversity of legumes including groundnut, common beans, soybean, pigeonpea, cowpea and green manures such as velvet bean, fish bean and a number of agroforestry species. These legumes are planted in sole stands or intercropped with cereals or other legumes, and rotated with cereals. Intercropping is the growing of two or more crops simultaneously on the same piece of land (Hauggaard-Nielsen et al. 2008). This helps to optimize the use of resources such as land, nutrients, labor and water. The traditional intercrop system consists of maize-legume intercrops at low density. The relative proportion of each component species depends on the main crop of interest to the farmers, complementarity in growth habits and use of resources. Productivity of intercrops can be maximized with careful selection of component crops and appropriate agronomic practices (Casper and Jackson 1997; Szumigalski and Van Acker 2008).

Legumes improve soil quality through biological nitrogen fixation (BNF) and crop residue incorporation, and increase crop productivity. The amount of N fixed by legumes and residual benefits varies with plant species, agronomic practices and environmental factors. Groundnut can fix 32–206 kg N ha^{-1} (Giller et al. 1987; Unkovich and Pate 2000) and a net N contribution of 13–100 kg ha^{-1} if crop residues are incorporated in the soil (Toomsan et al. 1995). Pigeonpea can fix 69–100 kg N ha^{-1} (Kumar Rao et al. 1987), and a net N contribution of 2–60 kg ha^{-1} depending on the genotype and environmental factors (Myaka et al. 2006; Egbe et al. 2007).

An intercrop of a legume and a legume in one field has rarely been considered in research, in contrast to an intercrop of a legume and a cereal, which is widely grown and has a nutritional complementarity (Snapp and Silim 2002). This includes the nitrogen fixation capacity of the legume and the high nitrogen requirement of the cereal. Complementary growth forms can also be the basis for a successful intercrop, such as a slow-growing deep-rooted perennial grown in mixtures with a fast-growing and shallow-rooted annual. In India, some farmers intercrop pigeonpea with groundnut at low density (Willey et al. 1981), and this arrangement was tested in Zimbabwe (Natarajan and Mafongoya 1992) as an example of a doubled up legume system. In this study, BNF of intercropped groundnut and pigeonpea legumes was investigated. We hypothesized that N fixation rate per area basis will be higher under groundnut—pigeonpea intercrop (GNPP) than if either is sole cropped.

10.2 Materials and Methods

10.2.1 Site Description

Participatory on-farm researcher designed farmer managed trials were conducted in Ekwendeni area of Mzimba District, northern Malawi (33°53′ E and 11°20′ S; altitude 1200 m) in the 2008 and 2009 growing seasons. Note that the growing season usually starts around November of the previous year but we refer to the season in terms of the year in which the majority of the season occurs. The annual precipitation is 800–1200 mm, with a unimodal distribution from November/December to March/April. Ekwendeni soils are classified as ferruginous latosols (Young and Brown 1962).

In the 2008 crop season, rainfall was considerably below average at 669 mm, and most of it (~45–50%) occurred between the last half of December 2007 to end of January which caused sheet erosion on some farm sites depending on the slope of the field (personal observation), and could have resulted into leaching of some nutrients. Another concern was poor seedling development at some sites possibly due to saturated soil. From the second half of February 2008, the area experienced a dry spell and this coincided with grain and pod filling growth stages in maize and groundnut respectively while pigeonpea was still at vegetative stage. The scattered rain showers during March 2008 were not adequate to support optimum growth of the crops. In the 2009 season (November 2008–April 2009), the area received 826 mm of well distributed rainfall which was adequate for growth of legumes and maize.

10.2.2 Experiment Description and Data Collection

The treatments for the cropping seasons 2008 and 2009 consisted of five cropping systems (treatments): three sole crops, maize (*Zea mays*), groundnut (*Arachis hypogaea*) and pigeonpea (*Cajanus cajan*); and two intercrops, maize-pigeonpea and groundnut-pigeonpea. The varieties grown were CG7 groundnut, ICEAP00040 pigeonpea and ZM621 maize. CG7 groundnut is characterized by bunch growth habit and high oil content averaging 48% and a yield potential of 2500 kg ha^{-1} (Malawi

Government Ministry of Agriculture and Food Security (MoAFS) 2012). Pigeonpea is a semi perennial legume that grows with a deep root system (Werner 2005). It has a deep root system and initial growth rates are slower compared to other crops such as groundnut or maize. ICEAP00040 is a long duration indeterminate variety that takes 7–9 months to mature, and a yield potential of 1500 kg ha^{-1} (Malawi Government, MoAFS 2012). ZM621 maize is an open pollinated variety that matures within 4–5 months (Malawi Government MoAFS 2005).

The experiment was laid out as a randomized complete block design with each treatment replicated 18 times, one replicate per farm. Treatment plot size was 10 m by 10 m, and consisted of 11 rows (aligned on a ridge following farmer practice in Malawi), each 10 m long and spaced at 0.90 m. The net plot used for measurements of grain and biomass consisted of the interior 8 m of 7 centrally located ridges. The goal for the planting pattern for intercrops was based on maximizing the plant population of the main crop for all cropping systems. The legumes were seeded at a 0.20 and 0.90 m within row spacing for groundnut and pigeonpea respectively to achieve 43,210 plants ha^{-1} (0.90 × 0.2 × 1) and 55,555 plants ha^{-1} (0.90 × 0.20 × 1) for intercropped and sole groundnut respectively, with an additional 37,000 plants ha^{-1} of maize or pigeonpea in the intercrops. Maize and pigeonpea were seeded alternately in the intercrop, along rows in stations of three plants each, spaced at 0.45 m intervals. The planting pattern for maize and pigeonpea was an additive design, sole crop and intercrop all planted at 37,000 plants ha^{-1} density for both crops. Planting was done in December 2007 and 2008. All plots received a uniform basal application of 10 kg N ha^{-1} at one week after planting based on observations that soils were highly N deficient, to improve uniformity of plant stands and early vigor while remaining consistent with the limited use of external inputs which can be afforded by smallholder farmers in Malawi (Snapp et al. 2002). All field management practices were conducted by participating farmers. In the 2009 crop season, year one treatments were replicated in time by planting adjacent plots. The varieties and planting pattern were the same as described for year one.

10.2.3 Soil Sampling and Analysis

At planting in December of 2007, composite soil samples (8–10 subsamples) were collected from each farm using a Z-scheme to ensure random collection. Two depths were sampled, 0–15 cm and 15–30 cm, for site characterization. These were air dried and sieved through a 2 mm sieve. Soil texture was determined using the hydrometer method (Anderson and Ingram 1996). Particulate organic matter (POM) was analyzed on ungrounded soil samples using a modification of the light-large POM fractionation method described by Cambardella and Elliott (1992, 1993). Sodium polytungstate was recycled according to Six et al. (1999) after POM extraction and weighing, the sample was ground into powder with a clean mortar and pestle. POM-C and POM-N were determined using a dry combustion C and N Analyzer (Costech ECS 4010, Costech Analytical Technologies, and Valencia, CA). The remaining soil samples were ground and sent to A and L Great Lakes Lab in Fort Wayne, Indiana, USA for analysis of the following variables: pH in a 1:1 ratio in H_2O, inorganic phosphorus (Bray P), and Mehlich 3 extraction of Ca, K and Mg (Mehlich 1984).

10.2.4 Plant Sampling and Analysis

10.2.4.1 Assessment of Nitrogen Fixation by Legumes

Plant sampling for N fixation measurements: Two 1 m × 1 m quadrants were demarcated in each plot for BNF measurements. At harvest, the plants were separated into grain and leafy biomass. Dried grain samples were ground into fine powder using a Wiley mill to pass a sieve size of 1 mm, then carefully sub-sampled and weighed into capsules before ^{15}N and ^{14}N mass spectrophotometer analysis conducted at the University of California Davis, USA. BNF was determined using ^{15}N natural abundance method. The proportion of N derived from atmosphere (Ndfa) was calculated according to Shearer and Khol (1986) and Peoples et al. (1989) as follows:

$$\%Ndfa = 100\frac{\left(\delta^{15}Nref - \delta^{15}Nlegume\right)}{\delta^{15}Nref - B} \tag{10.1}$$

where $\delta^{15}N_{ref}$ is the ^{15}N natural abundance of grain of the reference plant (maize) grown on same soil as the legume; $\delta^{15}N_{legume}$ is the ^{15}N natural abundance of the grain of the legume crop; B is the $\delta^{15}N$ of the test legume where the only N source is atmospheric N. The lowest $\delta^{15}N$ for each legume was used as B value (Hansen and Vinther 2001).

10.2.4.2 Plant Biomass

Plant biomass at early vegetative stage and at harvest were determined from the net plot of 57.6 m^2 (8 middle ridges × 8 m × 0.90 m). Groundnut and maize were harvested in June whilst pigeonpea was harvested in September after full physiological maturity. Grain yields were adjusted for quadrant samples used to measure N fixation. Grain moisture was determined by net weight basis of oven dried sub sample of grain dried at 70 °C for 48 h. Grain yields were reported on an adjusted basis at 8% and 15% moisture content for groundnut and pigeonpea, respectively.

10.2.5 *Statistical Analysis*

Soil nutrient and physical properties were analyzed using a one-way anova for location. Data on plant growth and yield were analyzed as a RCBD (Randomised Complete Block Design) using SAS Proc mixed procedure for a two-way model, with cropping system by year as factors (SAS Institute 2001). Where variances were not homogenous, data were analyzed with unequal variances assumption. Significant differences were determined at $p = 0.05$.

10.3 Results

10.3.1 *Soil Characterization*

Table 10.1 shows results on soil chemical properties and texture. Soil pH is slightly acidic to neutral and within the recommended range (5.5–6.5) for most production of most arable crops. The soils are largely light textured with low organic carbon (OC) and medium levels of total N. Inorganic P was highly variable ranging from low to high. Exchangeable cations (calcium, potassium and magnesium) were adequate for growth of the maize, groundnut and pigeonpea. The mean cation exchange capacity (CEC) was 5.8 ± 1.6 and 4.9 ± 1.4 cmol kg^{-1} at 0–15 cm and 15–30 cm soil depths respectively.

Table 10.1 Soil chemical properties and texture of on-farm experimentation fields. Baseline analysis sampled early December, 2008. N = 18 (Mhango et al. 2017)

Variable	0–15 cm		15–30 cm	
	Mean	Range	Mean	Range
pH (in H_2O)	5.8 ± 0.3	5.5–6.5	5.9 ± 0.4	5.1–6.9
OC (g kg^{-1})	6.5 ± 2.1	3–11	5.0 ± 2.2	2–11
Total N (g kg^{-1})	0.5 ± 0.1	0.4–0.8	0.5 ± 0.2	0.3–1.1
POM-C (g kg^{-1})	0.4 ± 0.01	0.20–0.94	0.2 ± 0.12	0.03–0.45
POM-N (g kg^{-1})	0.02 ± 0.01	0.008–0.04	0.01 ± 0.002	0.003–0.04
Bray P (mg kg^{-1})	10 ± 8.4	3–85	3 ± 1.9	1–66
Sand (%)	74 ± 9.8		72 ± 10.7	
Clay (%)	18 ± 8.2		21 ± 8.8	

Key: POM-C = particulate organic matter carbon; POM-N = particulate organic matter nitrogen, units for POM-C and POM-N are g per kg POM. OC = organic carbon

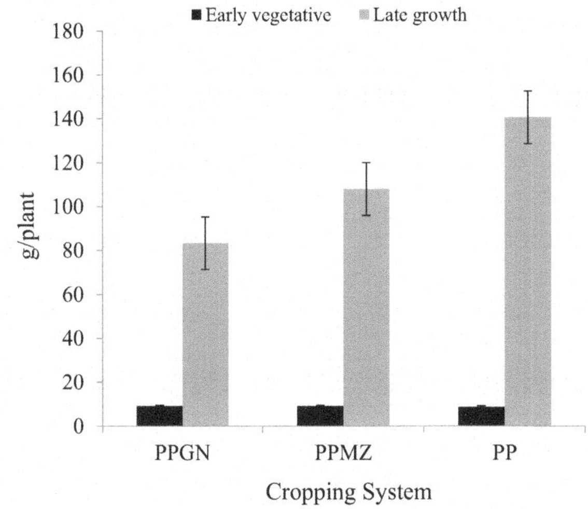

Fig. 10.1 Pigeonpea growth at vegetative stage (8.5 weeks after planting) and at harvest in 2008. Key: PP = sole pigeonpea; PPGN = pigeonpea intercropped with groundnut; PPMZ = pigeonpea intercropped with maize. Standard error presented as error bar (Reproduced from Mhango et al. 2017, under CC BY 4.0, https://creativecommons.org/licenses/by/4.0)

10.3.2 Plant Growth in 2008 and 2009

Dry matter accumulation by pigeonpea and groundnut at early vegetative stages averaged 8.9 ± 1.7 g $plant^{-1}$ and 17 ± 3.4 g $plant^{-1}$, respectively, with no effect observed of cropping system. The mean biomass/plant of sole and intercropped groundnut were 63 ± 20 g and 53 ± 20 g respectively. However, intercropping reduced pigeonpea biomass by 30–60%, $p = 0.0023$ (Fig. 10.1). Late season growth in pigeonpea in 2009 was similar to 2008.

10.3.3 Biological Nitrogen Fixation of Groundnut and Pigeonpea

The %Ndfa, total N fixed by legumes and correlation matrix between N fixed and selected variables are shown in Tables 10.2 and 10.3. Total N fixed was positively correlated to crop N. In groundnut, the %Ndfa ranged from 29–99% with a mean of 78%. For sole groundnut, the %Ndfa was positively correlated to POM (r = 0.68), inorganic P (r = 0.59) and plant density (r = 0.65) (Fig. 10.2). The relationship between total N fixed (kg ha^{-1}) and plant density was described by following fitted regression model:

$$Total\,N\,fixed = -93.404 + 0.003\,plant\,density \tag{10.2}$$

Table 10.2 Biological nitrogen fixed (proportion and total) in grain and leafy biomass by sole and intercropped groundnut and pigeonpea in 2008 and 2009 (Mhango et al. 2017)

Season	Cropping system	%Ndfa	N fixed in grain (kg/ha)	N fixed in leafy biomass (kg/ha)	Total N fixed (kg/ha)	Range, Total N fixed (kg/ha)	Estimated N fixed in defoliated PP leaves* (kg/ha)
2008	Groundnut (GN)	78	17b	21c	50c	21–102	–
	Pigeonpea (PP)	76	2a	15b	31b	11–64	8.6
	GNPP	–	16b	19bc	42bc	23–69	6.2
	Pr > F		0.0001	0.0005	< 0.0002		
2009	Groundnut (GN)	73	28b	33	62	21–96	–
	Pigeonpea (PP)	75	8a	34	53	21–86	13.4
	GNPP	–	23b	39	72	34–148	6.5
	Pr > F		< 0.001	0.726	0.238		

2008 = 2007/08 season; 2009 = 2008/09 season; GNPP = groundnut intercropped with pigeonpea
B values obtained from lowest ^{15}N of legume (Hansen and Vinther 2001). B values in 2008 are −0.45, −0.38 and −0.80 for sole GN, GNPP and sole PP. B values in 2009 are −0.26 and −0.21 for sole GN and GNPP; and −0.83, −0.74 for sole PP and PP intercropped with GN. Means in a column by year category followed by same letter are not statistically significant at $p = 0.05$
*The estimated N fixed in defoliated leaves calculated based on determined proportion of defoliation in pigeonpea at harvest, 41 and 57% for intercropped and sole cropped ICEAP00040 pigeonpea

Table 10.3 Correlation matrix of nitrogen fixation with %Ndfa, Bray P and crop N of sole and intercropped pigeonpea and groundnut, 2008 and 2009 seasons (Mhango et al. 2017)

Season	Variable	N fixed by pigeonpea			N fixed by groundnut	
		Sole PP	PPGN	PPMZ	Sole GN	GNPP
2008	N fixed	1.000	1.000	1.000	1.000	1.000
	Bray P	**0.780****	0.175	**0.857*****	**0.587***	0.352
	%Ndfa	**0.731***	0.488	**0.676***	0.428	0.296
	Crop N	**0.911*****	**0.92*****	**0.599***	**0.98*****	**0.93*****
2009	N fixed	1.000	1.000	nd	1.000	1.000
	Bray P	**0.646***	**0.590***	nd	0.263	0.170
	%Ndfa	0.598	0.566*	nd	0.337	**0.868*****
	Crop N	**0.894*****	**0.904*****	nd	**0.614***	**0.981*****

2008 = 2007/08 season; 2009 = 2008/09 season;
Values in bold are significant. Level of significance $^{*}p = 0.05$; $^{**}p = 0.01$; $^{***}p = 0.0001$
GN—groundnut; PP—pigeonpea; PPGN—pigeonpea intercropped with groundnut; GNPP—groundnut intercropped with pigeonpea; PPMZ—pigeonpea intercropped with maize; nd—no data for PPMZ in 2009

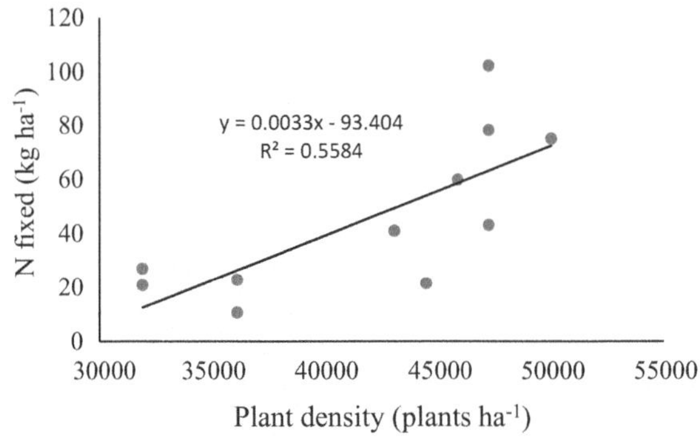

Fig. 10.2 Relationship between nitrogen (N) fixed by sole cropped groundnut and plant density, 2008 season (Reproduced from Mhango et al. 2017, under CC BY 4.0, https://creativecommons.org/licenses/by/4.0)

However, no linear relationship was observed between the total N fixed per unit area and inorganic P or plant density when groundnut was intercropped with pigeonpea.

In pigeonpea, the %Ndfa ranged from 41 to 99%, mean = 76 ± 20, and no differences were observed between sole and intercrop. In 2008, the %Ndfa was positively correlated with total N fixed, r = 0.73 and 0.68 for sole pigeonpea and pigeonpea intercropped with maize (PPMZ) respectively. Similar findings for %Ndfa were observed under sole pigeonpea and pigeonpea intercropped with groundnut (PPGN) in 2009. Inorganic P was positively correlated with total N fixed, r = 0.86 for PPMZ in 2008, r = 0.59 for PPGN (Table 10.3), r = 0.78 and 0.65 for sole pigeonpea in 2008 and 2009 respectively (Table 10.3; Fig. 10. 3).

On area basis, total N fixed by groundnut and pigeonpea was variable. Overall, there were no differences in total N fixed by sole groundnut and GNPP. A trend of higher total N fixed under GNPP than PP was observed ($p = 0.094$). Total N fixed in aboveground leafy biomass of pigeonpea was twice as much in 2009 season than 2008. Of the total N fixed across the two seasons, 34% and 11% was fixed in the grain of sole groundnut and sole pigeonpea respectively. Defoliation in senesced leaves was estimated at 41 and 57% for PPGN and sole PP. The estimates of N fixed in defoliated leaves are included in Table 10.2.

10.4 Discussion

10.4.1 Soil Fertility

Soil fertility was low and variable among smallholder farms and this is consistent with findings from earlier studies (Snapp 1998; Mhango et al. 2013). Inorganic P

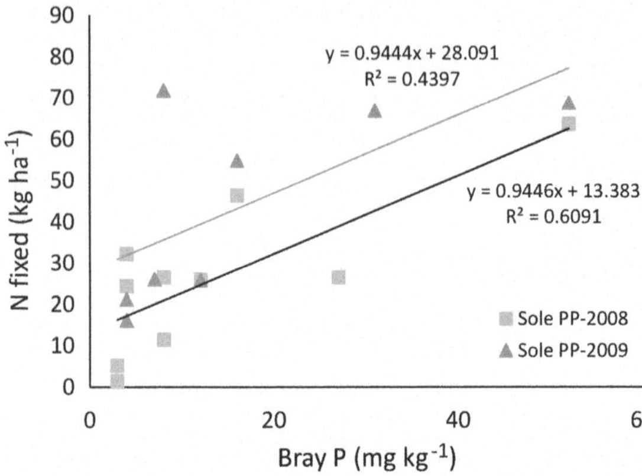

Fig. 10.3 Relationship between nitrogen fixed by sole cropped pigeonpea and inorganic soil phosphorus content, 2008 and 2009 seasons. Key: PP-2008 = pigeonpea, 2008 season; PP-2009 = pigeonpea, 2009 season (Reproduced from Mhango et al. 2017, under CC BY 4.0, https://creativecommons.org/licenses/by/4.0)

was variable and low and yet this is important for BNF (Jemo et al. 2006). The high correlation of inorganic P and total N fixed by pigeonpea in maize-pigeonpea intercrops (Table 10.3) suggests interspecific competition and that the two crops were accessing P from same pools during part of their growth cycles. In contrast, there is evidence that groundnut and pigeonpea access P from different pools (Shibata and Yano 2003). Phosphorus (P) is important for root development and growth of legume species. Positive correlations observed between inorganic P and N fixed by pigeonpea could be related effects of P on nodulation, biomass production and N fixation process.

10.4.2 Biomass Production of Sole and Intercropped Legumes

Early growth was not altered by crop system, indicating that competition was minimal early in the growing season. This is not surprising as pigeonpea has a slow growth pattern, and had 50% of groundnut dry matter accumulation during early vegetative growth. This may have also been due to relatively low plant population densities, which follows farmer practice. Surprisingly, groundnut biomass was not affected in late growth stages and this could be due to differences in growth rates relative to semi-perennial pigeonpea. A follow up study in 2012 showed that different densities of pigeonpea, from 12,350 to 37,000 plants per hectare did not alter the growth and grain yield of CG7 groundnut (Mhango and Chanza, unpublished data). In contrast, the late season growth competition demonstrated by low pigeonpea biomass could be related to inadequate soil moisture due to early cessation of rain in 2008 which inhibited vegetative biomass production.

10.4.3 Biological Nitrogen Fixation of Sole and Intercropped Groundnut and Pigeonpea, and Sustainability of Cropping Systems

Interspecific competition is one of the determinants of intercrop productivity. It is an important finding that %Ndfa was not affected by cropping system. This is consistent with Katayama et al. (1995), but differ in that they reported higher %Ndfa in pigeonpea intercropped with cereals (84%) than sole or doubled up legumes (52–70%).

Groundnut met 78% of its N requirement from BNF. This was higher than 22–67% reported by Katayama et al. (1995) and Phoomthaisong et al. (2003) and this could be due to low soil N (Table 10.1). The total amount of N fixed per area was lower than that reported by Ojiem et al. (2007) for CG7 groundnut probably due to low inorganic soil P (Table 10.3), plant density and biomass production. Ojiem et al. (2007) reported 115–124 kg ha^{-1} as N fixed with application of inorganic P fertilizer and higher plant densities of approximately 1.5 times than the density in this study. Since crop N was positively correlated with amount of N fixed by legume, inadequate soil moisture during reproductive growth stage of groundnut may have limited pod formation and grain filling consequently reducing total plant biomass. The positive correlation between inorganic P and total N fixed by intercropped pigeonpea may suggest that the two crops were accessing the same P pools (Makumba et al. 2009). However, in a short rainfall season (2008), the lack of a correlation between inorganic P and N fixed by PPGN is probably due to poor growth of pigeonpea with inadequate soil moisture and hence less competition for nutrients.

Crop genotype and duration are some of the factors that influence BNF (Giller 2001). Long duration and indeterminate pigeonpea are expected to fix more N and produce higher biomass than the early maturing varieties (Kumar-Rao and Dart 1987). In this study, pigeonpea fixed less N than groundnut in a short rainfall season probably because of inadequate soil moisture. The average total N fixed by pigeonpea is lower compared to 46–118 kg ha^{-1} reported in earlier studies for ICEAP00040 variety in Malawi (Adu-Gyamfi et al. 2007). Sole pigeonpea fixed 30 and 53 kg ha^{-1} in a dry and wet season respectively and this is within 20–60 kg N ha^{-1}, values reported for the same variety on selected sites in Tanzania. These findings can be attributed to low inorganic P (Table 10.1) and inadequate soil moisture to support biomass production following a dry spell that occurred when pigeonpea was still at early vegetative stage in 2008 season.

Legumes have been promoted in farming systems as an alternative strategy to improving soil N and productivity of cereals. In Malawi, the recommended N rate for maize on most smallholder farms is 92 kg N ha^{-1}. The proportion of N requirement met by sole and intercropped legume systems is 12–50%. This implies that on low fertility soils (< 15 g kg^{-1} soil organic matter, SOM), legume based cropping systems alone cannot sustain maize productivity and hence the need for Integrated Soil Fertility Management (ISFM) approaches (Snapp et al. 1998).

10.5 Conclusion

This study evaluated biological nitrogen fixation (BNF) and yield of sole and intercropped groundnut and pigeonpea on smallholder farms. Soil phosphorus (P) availability was not related to general soil properties such as soil organic matter (SOM) and texture, yet it was an important determinant of BNF in these legume diversified cropping systems. This indicates that it may be possible to support greater legume growth without building SOM to higher levels, rather the emphasis should be on judicious use of P-fertilizer and other P amendments such as compost. Intercropping pigeonpea with groundnut or maize can help to improve crop productivity, maximize use of limited land and labor. The results have demonstrated that the drivers of N fixation are inorganic P, plant density and interspecific competition. In a short rainfall season interspecific competition may limit vegetative growth of semi-perennial pigeonpea and this has negative implications for BNF. The findings from this study also suggest that different legume cropping systems should be recommended for farmers, sole cropping for grain maximization; and intercropping for smallholder farmers interested in multiple benefits.

Acknowledgements The authors would like to acknowledge the funding from The Mc Knight Foundation Collaborative Crop Research Program through the Legume Best Bets project. We are also grateful for the financial support from CS Mott Fellowship for Sustainable Agriculture of Michigan State University (MSU), University of Malawi—Bunda College of Agriculture, and MSU International Predissertation Travel Award. Lastly, we are grateful to the staff and farmers working with the SFHC project of Ekwendeni Mission Hospital for their support. This chapter is an edited version of Mhango et al. (2017), Biological nitrogen fixation and yield of pigeonpea and groundnut: quantifying response on smallholder farms in northern Malawi, *African Journal of Agricultural Research*, 12(16), 1385–1394, https://doi.org/10.5897/ajar2017.12232. It is re-used here in edited form under Creative Commons license CC BY 4.0 (https://creativecommons.org/licenses/by/4.0).

References

Adu-Gyamfi, J. J., Myaka, F. A., Sakala, W. D., Odgaard, R., Vesterager, J. M., & Høgh-Jensen, H. (2007). Biological nitrogen fixation and nitrogen and phosphorus budgets in farmer-managed intercrops of maize-pigeonpea in semi-arid southern and eastern Africa. *Plant and Soil, 295*(1–2), 127–136.

Anderson, J. M., & Ingram, J. S. J. (1996). *Tropical soil biology and fertility. A handbook of methods.* UK: CAB International.

Cambardella, C. A., & Elliot, E. T. (1992). Particulate soil organic matter changes across a grassland cultivation sequence. *Soil Science Society of America Journal, 56,* 777–783.

Cambardella, C. A., & Elliot, E. T. (1993). Carbon and nitrogen distribution in aggregates from cultivated and native grassland soils. *Soil Science Society of America Journal, 57,* 1071–1076.

Casper, B. B., & Jackson, R. B. (1997). Plant competition underground. *Annual Review of Ecology and Systematics, 28,* 545–570.

Egbe, O. M., Idoga, S., & Idoko, J. A. (2007). Preliminary investigation of residual benefits of pigeonpea genotypes intercropped with maize in Southern Guinea Savanna of Nigeria. *Journal of Sustainable Development in Agriculture & Environment, 3,* 58–75.

Giller, K. E., Nambiar, P. T. C., Srinivasa Rao, B., Dart, P. J., & Day, J. M. (1987). A comparison of nitrogen fixation in genotypes of groundnut (*Arachis hypogaea* L.) using ^{15}N isotope dilution. *Biology and Fertility of Soils, 5,* 23–25.

Giller, K. (2001). *Nitrogen fixation in tropical cropping systems* (2nd ed.). UK: CABI Publishing.

Hansen, J. P., & Vinther, F. P. (2001). Spatial variability of symbiotic N_2 fixation in grass-white clover pastures estimated by the ^{15}N isotope dilution method and the natural ^{15}N abundance method. *Plant and Soil, 230,* 257–266.

Hauggaard-Nielsen, H., Jørnsgaard, B., Kinane, J., & Jensen, E. S. (2008). Grain legume–cereal intercropping: The practical application of diversity, competition and facilitation in arable and organic cropping systems. *Renewable Agriculture and Food Systems, 23*(1), 3–12.

Jemo, M., Abaidoo, R. C., Nolte, C., Tchienkoua, M., Sanginga, N., & Horst, W. J. (2006). Phosphorus benefits from grain legume crops to subsequent maize grown on an acid soil of southern Cameroon. *Plant and Soil, 284,* 385–397.

Katayama, K., Ito, O., Matsunanga, R., Adu-Gyamfi, J. J., Rao, T. R., Anders, M. M., et al. (1995). Nitrogen balance and root behavior in four pigeonpea-based intercropping systems. *Fertilizer Research, 42,* 315–319.

Kumar-Rao, J. V. D. K., & Dart, P. J. (1987). Nodulation, nitrogen fixation and nitrogen uptake in pigeonpea (*Cajanus cajan* (L.) Mill sp.) of different maturity groups. *Plant and Soil, 99,* 255–266.

Kumar Rao, J. V. D. K., Thomson, J. A., Sastry, P. V. S. S., Giller, K. E., & Day, J. M. (1987). Measurement of N_2 fixation in field-grown pigeonpea (*Cajanus cajan* (L.) Mill sp.) using ^{15}N-labelled fertilizer. *Plant and Soil, 101,* 107–113.

Malawi Government, Ministry of Agriculture and Food Security. (2005). *Guide to Agriculture Production and Natural Resource Management in Malawi.* Lilongwe, Malawi: Agriculture Communication Branch, Department of Agricultural Extension Services.

Malawi Government, Ministry of Agriculture and Food Security. (2012). *Guide to Agriculture Production and Natural Resource Management in Malawi.* Lilongwe, Malawi: Agriculture Communication Branch, Department of Agricultural Extension Services.

Makumba, W., Akinnifesi, F. K., & Janssen, B. H. (2009). Spatial rooting patterns of *Gliricidia,* pigeonpea and maize intercrops and effect on profile soil N and P distribution in southern Malawi. *African Journal of Agricultural Research, 4*(4), 278–288.

Mehlich, A. (1984). Mehlich no.3 extractant: A modification of Mehlich no.2 extractant. *Communications in Soil Science and Plant Analysis, 15,* 1409–1416.

Mhango, W. G., Snapp, S. S., & Kanyama-Phiri, G. Y. (2013). Opportunities and constraints to legume diversification for sustainable maize production on smallholder farms in Malawi. *Renewable Agriculture and Food Systems, 28,* 234–244.

Mhango, W. G., Snapp, S., & Kanyama-Phiri, G. Y. (2017). Biological nitrogen fixation and yield of pigeonpea and groundnut: quantifying response on smallholder farms in northern Malawi. *African Journal of Agricultural Research, 12*(16), 1385–1394.

Myaka, F. M., Sakala, W. D., Adu-Gyamfi, J. J., Kamalongo, D., Ngwira, A., Odgaard, R., et al. (2006). Yields and accumulation of N and P in farmer-managed intercrops of maize-pigeonpea in semi-arid Africa. *Plant and Soil, 285,* 207–220.
Natarajan, M., & Mafongoya, P. L. (1992). A study on intercropping of pigeonpea (*Cajanun Cajun* L. Mill sp) with maize, sunflower and groundnut in Zimbabwe. *Zimbabwe Journal of Agricultural Research, 30,* 163–171.
Ojiem, J. O., Vanlauwe, B., de Ridder, N., & Giller, K. E. (2007). Niche-based of contributions of legumes to the nitrogen economy of Western Kenya smallholder farms. *Plant and Soil, 292,* 119–135.
Peoples, M. B., Faizah, A. W., Rerkasem, B., & Herridge, D. F. (1989). *Methods for evaluating nitrogen fixation by nodulated legumes in the field.* ACIAR Monograph. No. 11. Canberra, Australia: Australian Centre for International Agricultural Research.
Phoomthaisong, J., Toomsan, B., Limpinuntana, V., Cadisch, G., & Patanothai, A. (2003). Attributes affecting residual benefits of N_2 fixing mungbean and groundnut cultivars. *Biology and Fertility of Soils, 39,* 16–24.
SAS Institute. (2001). *SAS statistics users' guide.* Release 8.2. Cary: SAS Institute.
Shearer, G., & Kohl, D. H. (1986). Nitrogen fixation in field settings estimations based on natural ^{15}N abundance. *Australian Journal of Plant Physiology, 13*(6), 699–756.
Shibata, R., & Yano, K. (2003). Phosphorus acquisition from non-labile sources in peanut and pigeonpea with mycorrhizal interaction. *Applied Soil Ecology, 24*(2), 133–141.
Snapp, S. S. (1998). Soil nutrient status of smallholder farms in Malawi. *Communication in Soil and Plant Analysis, 29*(17&18), 2571–2588.
Snapp, S. S., Mafongoya, P. L., & Waddington, S. (1998). Organic matter technologies for integrated nutrient management in smallholder cropping systems of southern Africa. *Agriculture, Ecosystems & Environment, 71,* 185–200.
Snapp, S. S., & Silim, S. N. (2002). Farmer preferences and legume intensification for low nutrient environments. *Plant and Soil, 245,* 181–192.
Snapp, S., Kanyama-Phiri, G. Y., Kamanga, B., Gilbert, R., & Wellard, K. (2002). Farmer and researcher partnerships in Malawi: Developing soil fertility technologies for the near-term and far-term. *Experimental Agriculture, 38,* 411–431.
Szumigalski, A. R., & Van Acker, R. C. (2008). Land equivalent ratios, light interception, and water use in annual intercrops in the presence or absence of in-crop herbicides. *Agronomy Journal, 100*(4), 1145–1154.
Toomsan, B., McDonagh, J. F., Limpinuntana, V., & Giller, K. E. (1995). Nitrogen fixation by groundnut and soybean and residual nitrogen benefits to rice in farmers' fields in Northeast Thailand. *Plant and Soil, 175,* 45–56.
Unkovich, M. J., & Pate, J. S. (2000). An appraisal of recent measurements of symbiotic nitrogen fixation by annual legumes. *Field Crops Research, 65,* 211–228.
Werner, D. (2005). Production and biological nitrogen fixation of tropical legumes. In D. Werner & W. E Newton (Eds.), *Nitrogen fixation in agriculture, ecology and the environment* (pp. 1–13). Netherlands: Springer.
Willey, R. W., Rao, M. R., & Natarajan, M. (1981). Traditional cropping systems in pigeonpea and their improvement. In Y. L. Nene, & V. Kumble (Eds.), *Proceedings of the International Workshop on Pigeonpeas, Volume 1, 15–19 December 1980, Patancheru, A.P., India* (pp. 11–25). ICRISAT (International Crops Research Institute for the Semi-Arid Tropics).
Young, A., & Brown, P. (1962). *The physical environment on northern Nyasaland with special reference to soils and agriculture.* Zomba, Nyasaland, Malawi: Government Printer.

Further Reading

Six, J., Schultz, P. A., Jastrow, J. D., & Merckx, R. (1999). Recycling of sodium polytungstate used in soil organic matter studies. *Soil Biology & Biochemistry, 31*(8), 1193–1196.

Part III
Food and Agriculture: Improving Nitrogen Management in Fertilizers and Manures

Chapter 11
Biological Determinants of Crop Nitrogen Use Efficiency and Biotechnological Avenues for Improvement

Vimlendu Bhushan Sinha, Annie P. Jangam, and N. Raghuram

Abstract Crop nitrogen use efficiency (NUE) is crucial for sustainable food security as well as for a sustainable environment. It can be improved in the short term through improved fertilizer formulations and cropping practices under integrated nutrient management, but the inherent capacity of the plant to take up, retain and use the available nitrogen (N) has to be tackled biologically. The last decade has witnessed several major advances in our understanding on the biological determinants of N-response and N-use efficiency, which are opening up biotechnological opportunities for improvement in the medium to long term. This chapter highlights the various biological determinants including the uptake and assimilation of external N, remobilization of internal N, efflux or loss of N from plants. The emerging opportunities for NUE enhancement span a vast array of approaches including germplasm diversity, root architecture, molecular markers, phenomics, genomics and functional genomics, metabolomics and micro-Ribonucleic Acids (miRNAs). They are amenable to both transgenic, as well as non-transgenic selection/breeding options.

Keywords Efficiency · Nitrogen use efficiency · Nitrogen response · Breeding · Transgenics

---

V. B. Sinha
Department of Biotechnology, School of Engineering and Technology, Sharda University, Greater Noida, Uttar Pradesh 201306, India
e-mail: vimlendusinha@gmail.com

V. B. Sinha · A. P. Jangam · N. Raghuram (✉)
University School of Biotechnology, Guru Gobind Singh Indraprastha University, Dwarka Sec. 16C, New Delhi 110078, India
e-mail: raghuram98@hotmail.com; raghuram@ipu.ac.in

A. P. Jangam
e-mail: jangamannie@gmail.com

N. Raghuram
South Asian Nitrogen Centre, Indian Nitrogen Group, Sustainable India Trust and Society for Conservation of Nature, F-4, A-Block, NASc Complex, DPS Marg, Pusa, New Delhi 110012, India

M. A. Sutton et al. (eds.), *Just Enough Nitrogen*,
https://doi.org/10.1007/978-3-030-58065-0_11

11.1 Introduction

Nitrogen (N) is a major nutrient that is crucial for the survival of all living organisms and is therefore an important determinant of the biodiversity and productivity in every ecosystem (Vitousek et al. 2002). Plants cannot utilize gaseous N_2 in the air (except with the help of symbiotic N-fixing microbes hosted by legumes in their root nodules) and need dissolved forms of reactive nitrogen (N_r) such as nitrate, ammonium or urea. These plant nutrients, required to meet their nutritional requirements for their growth and development, are obtained either from natural processes in the soil or from inorganic or organic fertilizers. Plants generally take up N_r compounds through their roots, though they can also absorb from foliar application of fertilizers or deposition of N_r compounds by rain. Poor nitrogen use efficiency (NUE) often results from the limited capacity of the plant or excessive/inefficient use of fertilizers (Good and Beatty 2011). Poor agricultural NUE not only threatens sustainable food/feed/fiber security, but also has major environmental consequences. These include N-pollution of ground and surface waters, causing eutrophication, affecting fisheries, biodiversity and ecosystem services as well as air pollution through volatilized N_r forms, including nitrous oxide, which has 300 times more global warming potential than carbon dioxide (Galloway et al. 2008; Sutton et al. 2013).

The availability of N_r may vary in form and concentration, to which plants, being sessile, have to adjust and respond biologically (Andrews and Lea 2013). This opens avenues for biological improvements in crop NUE at the genetic level for sustainable agriculture and food security (Pathak et al. 2008; McAllister et al. 2012; Andrews and Lea 2013). Such approaches usually target either N uptake or internal retention/utilisation in order to produce the most desirable farm output (grain/fruit/leaf/root/tuber/fodder/fiber). This includes prevention of N-losses from the plant during cropping, but N-losses from the soil or during post-harvest processing are outside the purview of crop improvement and therefore not elaborated further in this chapter.

Crop NUE is defined in various ways such as N uptake efficiency (NUpE), N assimilation/utilization efficiency (NUtE), apparent N recovery rate (ANR), agronomic efficiency of fertilizer N (AE), N physiological use efficiency (NpUE), N transport efficiency (NTE), and N remobilization efficiency (NRE) (Xu et al. 2012). For most practical purposes, biologists consider physiological definitions of NUE that rely on N-uptake from external sources but internal remobilization is known to be important in cereals (Pathak et al. 2011). This review summarizes the various plant biological processes that could determine NUE, as well as the emerging strategies for improving NUE at the plant level.

11.2 Nitrogen Assimilation and Remobilization

Nitrifying soil bacteria convert organic N compounds or fertilizers like urea to ammonia and then to nitrate, making it the most predominant form of usable N in aerobic soils. Plants have families of transporters of different affinities in their roots to take up nitrate/ammonia/urea from the soil, as well as in the shoots for their internal translocation and remobilization between various tissues and organs (Sorgona et al. 2011; Xu et al. 2012). Plant root architecture has an important bearing on N use efficiency, due to the predominant role of roots in N-sensing and acquisition (Chakraborty and Raghuram 2011). The nitrate acquired is partly utilized in the roots but largely transported to the shoots that have the photo-synthetically accumulated reducing power to convert it into nitrite by nitrate reductase in the cytoplasm, followed by its translocation into plastids where it is converted into ammonium by nitrite reductase. The ammonium ions are finally assimilated into amino acids by plastidic glutamine synthase (GS2) and glutamate synthetase (GOGAT) cycle, with the help of 2-oxoglutarate from photosynthetic C-fixation, followed by further transformations through aminotransferases (Masclaux-Daubresse et al. 2006). This is called primary N metabolism or assimilation, which depends heavily on carbon–nitrogen (C–N) balance. The leaves of plants serve as a sink for N during the vegetative growth of the plants and become a source for remobilization of N for seed development in cereals and other plants that undergo senescence during seed setting (Okumoto and Pilot 2011). The cytosolic glutamine synthase (GS1) plays a more prominent role in this secondary N metabolism/remobilization (Goodall et al. 2013).

11.3 Nitrogen Efflux and Loss

Nitrogen losses may happen in agricultural and horticultural systems through: (a) ammonia volatilisation; (b) leaching or runoff (i.e., removal in drainage water) and (c) denitrification i.e., transformations into gaseous forms (Cameron et al. 2013). Not all these losses are from the plant alone, though plant biologists are primarily concerned with minimizing plant losses. At the root level, excess uptake of ammonium ions could lead to their efflux and inhibition of further development of roots (Li et al. 2010). At the shoot level, the volatilization and ammonia emission are related to photorespiration (Rachmilevitch et al. 2004). In rice, the ammonia emission is related directly to Glutamine Synthetase (GS) activity in photo-respiratory ammonium recycling (Kumagai et al. 2011). The main cause for N volatilization is the imbalance between N accumulation and assimilation, though its forms, rates and impact on the overall plant budget may vary. These need to be quantified in different crops, clearly accounting for plant and non-plant N losses to optimize biological interventions for minimizing N losses. In addition, there is a growing emphasis on integrated models that incorporate cultivar performance in the prediction of the interactions between genotype, environment and cropping system to define the breeding targets and/or

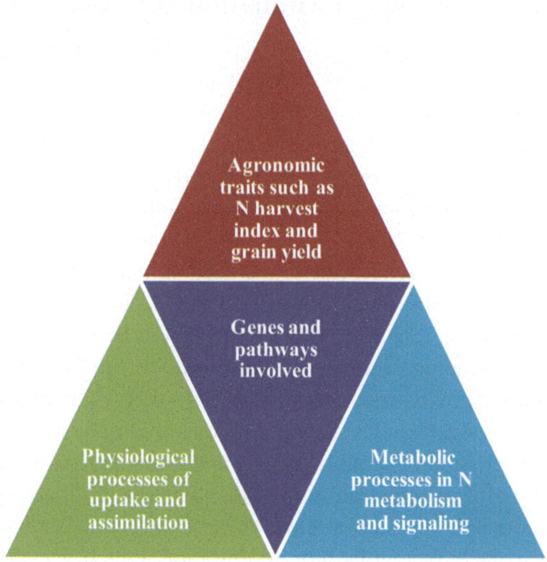

Fig. 11.1 Biological determinants of crop nitrogen use efficiency

identify the best cultivar to grow in a given situation (Jeuffroy et al. 2014). Such approaches can be applied for management/enhancement of NUE of crops.

11.4 Biological Strategies for Improving Nitrogen Use Efficiency

There are several agronomic crop management practices such as the timing, dosage, application method and formulation of fertilizer, which can be employed to enhance crop productivity and NUE in a given agro-ecosystem for each cultivar (Hawkesford 2012). However, the inherent biological capacity of the plant for NUE can only be improved genetically, by selection or breeding for better genotypes using all available biological avenues of improvement in an integrated manner (Parry and Hawkesford 2012). This requires the identification of the various genes/alleles involved in various physiological and metabolic processes that determine the various agronomic traits related to NUE (Fig. 11.1) and their improvement, which is the main focus of this review.

11.5 Germplasm Diversity Utilisation for Plant Breeding

A major challenge for improved crop NUE is the abysmally poor characterization of the available germplasm of most crops for genetic variation in NUE, not only among the wild varieties but even among the cultivated varieties. This challenge is further compounded by the fact that most of the current crop cultivars that accumulated various other important traits were chosen in fertilizer rich environments. Their allelic diversity for NUE-related traits could be very narrow at best. At worst, they may even lack the alleles for good agronomic performance in N-limited environments. Recently, there has been an attempt to address this problem by mapping the genomic regions associated with response to N-limitation using a global core collection of wheat (Bordes et al. 2013). While such studies are required on a wider scale on all important crops, they require huge scientific teams with access to seeds representing the entire germplasm, field facilities and financial resources. The role of international public sector research infrastructure becomes particularly important in this regard, regardless of the national differences in their approaches towards transgenics (Rothstein et al. 2014). Alternatively, innovative approaches for large scale phenotyping are needed, such as digital imaging of N-responses (Poiré et al. 2014). However, conventional evaluation of differences in NUE at different developmental stages of the plant/crop under different N regimes continues to be important to assist plant breeders in crop improvement (Coque and Gallais 2006; Igarashi et al. 2009; Chardon et al. 2010).

11.6 Molecular Markers for Selection/Breeding for High Nitrogen Use Efficiency

Over the years, several quantitative trait loci (QTLs) for NUE have been identified in barley, maize, rice, wheat, sorghum and *Arabidopsis*, as reviewed recently (Xu et al. 2012; Simons et al. 2014). In wheat, the various QTLs identified for NUE in different studies were subjected to a meta-analysis on the wheat genome map, providing an integrated view for functional validation (Quraishi et al. 2011). Recent studies have revealed more QTLs in wheat, barley and other cereals (Sun et al. 2013; Kindu et al. 2014; Li et al. 2014). The QTLs for NUE are conserved in genomes of maize, rice, wheat and sorghum, revealing concerted cereal genome evolution for NUE-related traits (Quraishi et al. 2011). These findings enhance the possibilities of finding common QTLs for NUE enhancement in multiple crops.

11.7 Improvement in Root Architecture

Most of the crop requirements for water and nutrients are met by uptake through the roots and therefore, the plant's ability to perform well under varying conditions of water and nutrient supply depends heavily on its root architecture. The potential for genetic improvements in root architecture for multiple input use efficiencies is being increasingly recognized (de Dorlodot et al. 2007; Giehl et al. 2014; Orman-Ligeza et al. 2014) and recently reviewed for specific crops like rice (Wu and Cheng 2014) and tubers (Villordon et al. 2014). Studies on nitrogen-responsive changes in root architecture in *Arabidopsis* and other model plants are increasingly demonstrating their potential for breeding crops for denser or deeper roots (Smith and De Smet 2012; Forde 2014). The identification of the genes and signalling pathways involved in the stimulation of lateral root growth by nitrate has been a major contribution in this area (Forde 2014). Crop breeding for root architecture could be combined with QTLs for NUE and other agronomic traits, such as yield and stress resistance, when their QTLs converge or lie in close proximity. Recent developments in imaging-based approaches for field scale root phenotyping (Bucksch et al. 2014) show the potential for rapid progress in phenotyping for root architecture and their use for selection/breeding. However, this is not to underestimate the importance of shoot biomass in N-response and NUE under low-N conditions, as recently demonstrated in maize (Kamiji et al. 2014).

11.8 Omics-Based Approaches for Molecular Breeding

There have been major strides in the use of high throughput screening methods at various levels in plant N response and NUE. Though collectively called 'omics', owing to their high throughput nature, they are very different in terms of their details, such as phenomics, genomics, transcriptomics and metabolomics, and their integration into networks (Fig. 11.2). Some recent developments in this area have been highlighted below:

Phenomics: This is the high throughput study of plant growth, performance and composition by using phenotypic tools. It combines genetics and physiology for revealing the molecular genetic basis for less understood plant processes (Furbank and Tester 2011). The identification of an N use efficient phenotype for different soil conditions can facilitate conventional breeding, as well as open up synergistic opportunities with transcriptomics in molecular breeding (Ruzicka et al. 2010). The development of imaging-based phenotyping approaches for NUE may prove to be very useful in this regard (Bucksch et al. 2014; Poiré et al. 2014).

Genomics: The availability of genome sequences of various crop species like rice, sorghum, maize, wheat, etc., and the advent of new generation sequencing technologies are increasingly aiding our understanding of genome evolution and genomic sequence diversity within and between crop species (Varshney et al. 2011). Further, progress in genome annotation has paved the way for functional genomics, gene discovery and analysis of regulatory sequences in relation to various traits, including NUE. This in turn has facilitated the integration of transcriptomic, proteomic and metabolomic data on a genomic scale into some model crops, including maize (Amiour et al. 2012; Simons et al. 2014). While these approaches are very powerful aids to unravelling the complete genotype for the multi-genic trait, considerable ground needs to be covered in every crop before a firm grip is obtained on the genotype to assist breeders.

Trancriptomics: Whole transcriptome studies in the presence or absence of N, high or low N, or different sources of N have produced a wealth of information in model plants/crops within the last decade (e.g., Wang et al. 2000; Wei et al. 2013). The study of N-responsive gene expression in the presence of different sources of N is often more complex than the study involving a single N source (Patterson et al. 2010). Nevertheless, transcriptomics have improved our understanding of the hundreds of N-responsive genes involved in dozens of processes in *Arabidopsis* and other species (Xu et al. 2012). They also aided the identification of signalling intermediates like kinases and other transcription factors involved in N response (e.g., Hu et al. 2009; Marchive et al. 2013), but many more candidate genes from many different processes are expected to be shortlisted from transcriptomic data in the coming years, as targets for genetic improvement of NUE alone, or in combination with other agronomically important traits like water stress (Humbert et al. 2013).

Proteomics: Protein profiling of plants under variable N source/supply has been extensively studied and reviewed recently (Liang et al. 2013; Simons et al. 2014). They include important crops like rice (Hakeem et al. 2012, 2013), maize (Amiour et al. 2012; Liao et al. 2012), and barley (Møller et al. 2011). As the proteome

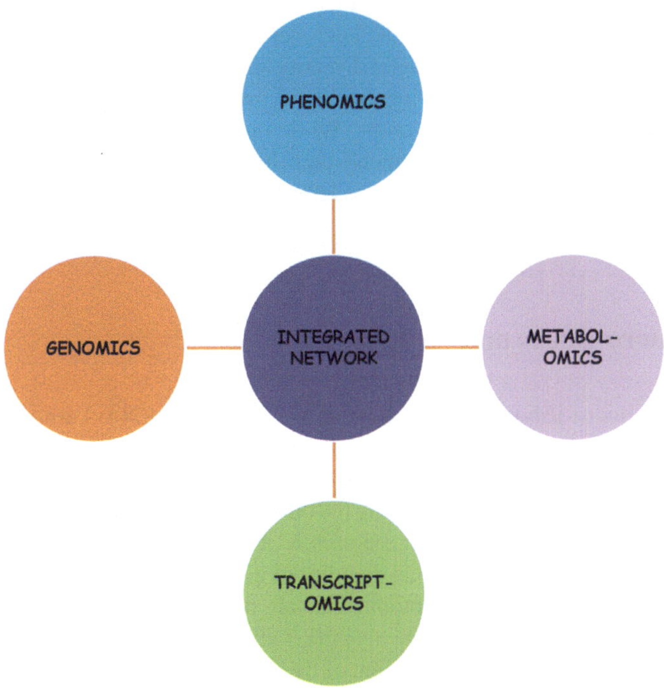

Fig. 11.2 Omics-based approaches to mine nitrogen use efficiency (NUE) genes and loci

datasets for NUE from both model plants and crop species reach levels comparable to those of transcriptomes, candidate proteins from critical processes and/or with critical functions in N-response and/or N-use efficiency will increasingly become available for crop improvement.

Metabolomics: Metabolite profiling of plant tissues under variable N source/supply has been one of the emerging approaches to understanding N-response and NUE (Kusano et al. 2011). There have been some studies on the N-responsive metabolome in maize, rice and *Arabidopsis* (Pavlík et al. 2010; Albinsky et al. 2010; Sato and Yanagisawa 2014). Such studies on crop species will be increasingly relevant in integrated omics-enabled identification of the NUE targets for crop improvement, as detailed below.

Integrated network: Accurate genotyping of quantitative, multigenic traits is a formidable challenge for plant biologists, especially when the phenotype itself is ill defined or equivocal, as is the case with NUE. While it is in the nature of omics-approaches to generate large data sets of N-responsive genes, proteins and metabolites, shortlisting the critical candidates for crop improvement can be a daunting task. Fortunately, integrative analyses of transcriptomic, proteomic and metabolomics data have already begun to address these issues from the systems network approach in single crops like maize (Amiour et al. 2012; Simons et al. 2014) or across multiple species (Fukushima and Kusano 2014). Such approaches could improve the sophistication and robustness of the shortlisting of targets for genetic improvement of NUE and reduce the uncertainty in the later steps of crop improvement. The shortlisted candidate genes can fuel genetic diversity studies and development of molecular markers for NUE, which in turn could aid marker-assisted selection/breeding and/or development of transgenic crops. While the omics approaches have not yet reached that stage of crop improvement, their high throughput nature will almost certainly enlarge the number of genetic avenues available for crop improvement to unprecedented levels.

11.9 Micro-RNAs and NUE

The roles of micro-Ribonucleic Acids (micro-RNAs or miRNAs) are increasingly becoming important in plant nutrient responses, including N, P, S, Cu and Fe (Zeng et al. 2014). These are small RNAs that participate in the post-transcriptional regulation of their target genes, thus controlling the outcome of gene expression. The potential of miRNA-controlled gene expression in NUE improvement has been reviewed recently (Fischer et al. 2013). The availability of gene chips and bioinformatic tools to study miRNAs in large transcriptomic data are contributing to the fast expanding list of miRNAs in N responsive gene network and potentially in NUE. They need to be shortlisted and validated for the extent of their contribution to the improvement in NUE in specific crops before being used widely in crop improvement programmes.

11.10 Transgenic Manipulations

Transgenic manipulation of gene expression has been used extensively, both for the identification/validation of genes potentially involved in NUE, as well as in enhancing NUE per se. Over-expression and silencing of genes under the control of constitutive or inducible promoters has been attempted in various plants like *Arabidopsis*, rice, maize, wheat, sorghum, barley, finger millet, potato, tobacco, etc. (Table 11.1) (Good and Beatty 2011; Pathak et al. 2011; Xu et al. 2012). These authors targeted enhanced N uptake, assimilation, remobilization, seed nitrogen/protein content, biomass nitrogen/protein content, growth under conditions such as N limiting environment (Table 11.2). However, their success has been varied in producing robust candidate genes for improvements in NUE, as well as in the extent of improvement achieved. The most successful case is of the gene alanine aminotransferase (AAT) that was manipulated in several crops and went through trials in multiple countries (Good et al. 2007), but two decades since it was developed, "its commercial relevance is still unclear" (Rothstein et al. 2014). Another example is the rice early nodulin gene that was found using transcriptomic studies of NUE and manipulated transgenically (Bi et al. 2009) at Syngenta, but its commercial potential is still under evaluation. Considering the large number of steps involved in developing transgenic products (Privalle et al. 2012) and the estimated time scale of over 22 years (McDougall 2011), it is not surprising that these and many other first generation transgenics for NUE are yet to reach widespread commercial cultivation in the farmer's's fields. In due course, it is expected that at least some of the dozens of candidate genes being currently used for transgenic improvement would become available as fully validated transgenic crops for cultivation.

Table 11.1 Various candidate genes modified to increase nitrogen use efficiency in different plants (Adapted from Rothstein et al. 2014)

Organism	Genes
Arabidopsis thaliana	Alanine Aminotransferase, Asparagine Synthetase, Aspartate Aminotransferase, Cytokinin Biosynthesis, Hexose transporter, Glutamate receptor, ANR1 MADS transcription Factor, Dof1 Transcription factor, GLB1 PII regulatory Protein, 14-3-3 regulatory protein, Yeast nitrate transporter 1
Zea mays	Yeast nitrate transporter 1, Cytokinin biosynthesis, NADP-dependent glutamate dehydrogenase
Oryza sativa	Peptide transporter/nitrate transporter, OsENOD93-1 Mitochondrial membrane protein, Cytokinin oxidase, Aspartate aminotransferase, Alanine aminotransferase
Nicotiana tabacum	Rubisco small subunit antisense gene, Cytokinin biosynthesis, NADP-dependent glutamate dehydrogenase, Aspartate aminotransferase, Asparagine synthetase
Solanum lycopersicum	NADP-dependent glutamate dehydrogenase
Brassica napus	Alanine Aminotransferase, Aspartate aminotransferase

Table 11.2 Nitrogen responsive genes and phenotypes and agronomic traits they manifest (Adapted from Rothstein et al. 2014)

Gene	Phenotype observed/agronomic traits
Alanine aminotransferase	Increased biomass and seed yield in the laboratory and field (for *B. napus*) under low N
Asparagine synthetase	Increased free asparagine, enhanced seeds protein, higher seed N yield, and improved nitrogen harvest index at high N
Aspartate aminotransferase (AspAT)	Increased AspAT activity, PEPC activity, greater seed amino acid and protein content
NADP-dependent glutamate dehydrogenase	Increased free amino acid, ammonium assimilation, biomass, and dry weight. Higher water potential during water deficit. Increased yield in the field
Cytokinin oxidase	More panicles and a 23–34% increase in grain numbers
Cytokinin biosynthesis	Delayed leaf senescence, increase in biomass, larger embryo and seed, higher seed protein content and flood tolerance
Ferredoxin NADP+ reductase	Enhanced root growth, ear size, seed weight
OsENOD93-1 Mitochondrial membrane protein	Higher concentration of total amino acids and total N in roots, increased dry biomass and seed yield
Hexose transporter	Improved growth, higher biomass and N use when provided with exogenous sugar
Amino acid permease	Seed size increased by 20–30%, increase in relative abundance of Asparagine, Aspartate, Glutamate, and Glutamine in the seed, higher seed storage protein content
Glutamate receptor	Reduced growth rate, impairs calcium utilization and sensitivity to ionic stress in transgenic plants
ANR1 MADS transcription factor	Significantly more lateral root growth after plants were treated with synthetic steroid dexamethasone
Dof1 Transcription factor	Enhanced growth rate under N-limiting conditions
GLB1 PII regulatory protein	Increased anthocyanin production under low N condition
14-3-3 regulatory protein regulates NR, post-translationally ATL31 ubiquitin ligase that degrades 14-3-3χ	Overexpression of 14-3-3 under N stress (low N relative to high C) resulted in hypersensitivity to the N stress and stunted growth Overexpression of ATL31 under N stress allowed for continued growth regardless of N stress conditions

(continued)

Table 11.2 (continued)

Gene	Phenotype observed/agronomic traits
Rubisco small subunit antisense gene	Total nitrogen (total nitrogen/total mass) increased. Increase in vacuolar nitrate
Yeast nitrate transporter 1	Increased yield under low or normal N fertility. Field yield trials
Peptide transporter/ nitrate transporter	Enhanced ammonium uptake, promotion of lateral root formation, and increased grain yield

11.11 Conclusions and Future Perspectives

The improvement of NUE is as much a biological problem as an agronomic or environmental problem. Advances in our understanding of the functional biology of N-response and N use efficiency in the last decade have revealed how N source and concentration can have genome-wide and organism-wide impacts involving thousands of genes belonging to dozens of critical biological processes. Several candidate genes have been found and some of them have been demonstrated to be useful to improve NUE through transgenic approaches and many more are in the pipeline. Once their NUE-enhancing potential is proven, these genes will also be amenable to non-transgenic approaches of crop improvement. Molecular markers are also being developed for selection/breeding, using conventional germplasm screening or mining genomic sequences as available. Despite these exciting developments, it may take a decade or more before the biologically improved crop varieties with high-NUE become widely available or used in most parts of the world. Thus, crop biotechnology may deliver in the medium to long term, but until then, short-term benefits can accrue from factors such as improved fertilizer formulations and crop management practices for integrated nutrient management coupled with matching extension services to the farmers.

Acknowledgements VBS and JAP are thankful to ICAR and CSIR (India) for their fellowships. This work was supported in part by grants to NR from ICAR under NICRA, through CRIDA, Hyderabad, India and the Indian Department of Biotechnology under the NEWS India-UK Virtual Joint Network.

References

Albinsky, D., Kusano, M., Higuchi, M., Hayashi, N., Kobayashi, M., Fukushima, A., et al. (2010). Metabolomic screening applied to rice FOX *Arabidopsis* lines leads to the identification of a gene-changing nitrogen metabolism. *Molecular Plant, 3*(1), 125–142.

Amiour, N., Imbaud, S., Clément, G., Agier, N., Zivy, M., Valot, B., et al. (2012). The use of metabolomics integrated with transcriptomic and proteomic studies for identifying key steps

involved in the control of nitrogen metabolism in crops such as maize. *Journal of Experimental Botany, 63*(14), 5017–5033.
Andrews, M., & Lea, P. J. (2013). Our nitrogen 'footprint': The need for increased crop nitrogen use efficiency. *Annals of Applied Biology, 163*(2), 165–169.
Bi, Y. M., Kant, S., Clark, J., Gidda, S., Ming, F., Xu, J., et al. (2009). Increased nitrogen-use efficiency in transgenic rice plants over-expressing a nitrogen-responsive early nodulin gene identified from rice expression profiling. *Plant, Cell & Environment, 32*(12), 1749–1760.
Bordes, J., Ravel, C., Jaubertie, J., Duperrier, B., Gardet, O., Heumez, E., et al. (2013). Genomic regions associated with the nitrogen limitation response revealed in a global wheat core collection. *Theoretical and Applied Genetics, 126*(3), 805–822.
Bucksch, A., Burridge, J., York, L. M., Das, A., Nord, E., Weitz, J. S., & Lynch, J. P. (2014). Image-based high-throughput field phenotyping of crop roots. *Plant Physiology, 166*(2), 470–486.
Cameron, K., Di, H., & Moir, J. (2013). Nitrogen losses from the soil/plant system: A review. *Annals of Applied Biology, 162*(2), 145–173.
Chakraborty, N., & Raghuram, N. (2011). Nitrate sensing and signaling in genomewide plant N response. In V. Jain & P. Anandakumar (Eds.), *Nitrogen use efficiency in plants* (pp. 45–62). New Delhi: New India Publishing Agency.
Chardon, F., Barthélémy, J., Daniel-Vedele, F., & Masclaux-Daubresse, C. (2010). Natural variation of nitrate uptake and nitrogen use efficiency in Arabidopsis thaliana cultivated with limiting and ample nitrogen supply. *Journal of Experimental Botany, 61*(9), 2293–2302.
Coque, M., & Gallais, A. (2006). Genomic regions involved in response to grain yield selection at high and low nitrogen fertilization in maize. *Theoretical and Applied Genetics, 112*(7), 1205–1220.
De Dorlodot, S., Forster, B., Pagès, L., Price, A., Tuberosa, R., & Draye, X. (2007). Root system architecture: Opportunities and constraints for genetic improvement of crops. *Trends in Plant Science, 12*(10), 474–481.
Fischer, J. J., Beatty, P. H., Good, A. G., & Muench, D. G. (2013). Manipulation of microRNA expression to improve nitrogen use efficiency. *Plant Science, 210,* 70–81.
Forde, B. G. (2014). Nitrogen signalling pathways shaping root system architecture: An update. *Current Opinion in Plant Biology, 21,* 30–36.
Fukushima, A., & Kusano, M. (2014). A network perspective on nitrogen metabolism from model to crop plants using integrated 'omics' approaches. *Journal of Experimental Botany, 65*(19), 5619–5630.
Furbank, R. T., & Tester, M. (2011). Phenomics–technologies to relieve the phenotyping bottleneck. *Trends in Plant Science, 16*(12), 635–644.
Galloway, J. N., Townsend, A. R., Erisman, J. W., Bekunda, M., Cai, Z., Freney, J. R., et al. (2008). Transformation of the nitrogen cycle: Recent trends, questions, and potential solutions. *Science, 320*(5878), 889–892.
Giehl, R. F., Gruber, B. D., & von Wirén, N. (2014). It's time to make changes: Modulation of root system architecture by nutrient signals. *Journal of Experimental Botany, 65*(3), 769–778.
Good, A. G., & Beatty, P. H. (2011). Fertilizing nature: A tragedy of excess in the commons. *PLoS Biology, 9*(8), e1001124. https://doi.org/10.1371/journal.pbio.1001124
Good, G., Johnson, S. J., De Pauw, M., Carroll, R. T., Savidov, N., Vidmar, J., Lu, Z., Taylor, G., & Stroeher, V. (2007). Engineering nitrogen use efficiency with alanine aminotransferase. *Canadian Journal of Botany, 85*(3), 252–262.
Goodall, A. J., Kumar, P., & Tobin, A. K. (2013). Identification and expression analyses of cytosolic glutamine synthetase genes in barley (*Hordeum vulgare* L.). *Plant and Cell Physiology, 54*(4), 492–505.
Hakeem, K., Chandna, R., Ahmad, A., Qureshi, M. I., & Iqbal, M. (2012). Proteomic analysis for low and high nitrogen-responsive proteins in the leaves of rice genotypes grown at three nitrogen levels. *Applied Biochemistry and Biotechnology, 168*(4), 834–850.

Hakeem, K., Mir, B., Qureshi, M. I., Ahmad, A., & Iqbal, M. (2013). Physiological studies and proteomic analysis for differentially expressed proteins and their possible role in the root of N-efficient rice (*Oryza sativa* L.). *Molecular Breeding, 32*(4), 785–798.

Hawkesford, M. J. (2012). Improving nutrient use efficiency in crops. In *eLS*. Chichester: Wiley. https://doi.org/10.1002/9780470015902.a0023734.

Hu, H. C., Wang, Y. Y., & Tsay, Y. F. (2009). AtCIPK8, a CBL-interacting protein kinase, regulates the low-affinity phase of the primary nitrate response. *The Plant Journal, 57*(2), 264–278.

Humbert, S., Subedi, S., Cohn, J., Zeng, B., Bi, Y.-M., Chen, X., et al. (2013). Genome-wide expression profiling of maize in response to individual and combined water and nitrogen stresses. *BMC Genomics, 14,* 3. https://doi.org/10.1186/1471-2164-14-3

Igarashi, D., Ishizaki, T., Totsuka, K., & Ohsumi, C. (2009). ASN2 is a key enzyme in asparagine biosynthesis under ammonium sufficient conditions. *Plant Biotechnology, 26*(1), 153–159.

Jeuffroy, M.-H., Casadebaig, P., Debaeke, P., Loyce, C., & Meynard, J.-M. (2014). Agronomic model uses to predict cultivar performance in various environments and cropping systems. A Review. *Agronomy for Sustainable Development, 34*(1), 121–137.

Kamiji, Y., Pang, J., Milroy, S. P., & Palta, J. A. (2014). Shoot biomass in wheat is the driver for nitrogen uptake under low nitrogen supply, but not under high nitrogen supply. *Field Crops Research, 165,* 92–98.

Kindu, G., Tang, J., Yin, X., & Struik, P. (2014). Quantitative trait locus analysis of nitrogen use efficiency in barley (Hordeum vulgare L.). *Euphytica, 199*(1–2), 207–221.

Kumagai, E., Araki, T., Hamaoka, N., & Ueno, O. (2011). Ammonia emission from rice leaves in relation to photorespiration and genotypic differences in glutamine synthetase activity. *Annals of Botany, 108*(7), 1381–1386.

Kusano, M., Fukushima, A., Redestig, H., & Saito, K. (2011). Metabolomic approaches toward understanding nitrogen metabolism in plants. *Journal of Experimental Botany, 62*(4), 1439–1453.

Li, Q., Li, B.-H., Kronzucker, H. J., & Shi, W.-M. (2010). Root growth inhibition by NH_4^+ in Arabidopsis is mediated by the root tip and is linked to NH_4^+ efflux and GMPase activity. *Plant, Cell & Environment, 33*(9), 1529–1542.

Li, X.-M., Chen, X.-M., Xiao, Y.-G., Xia, X.-C., Wang, D.-S., He, Z.-H., & Wang, H.-J. (2014). Identification of QTLs for seedling vigor in winter wheat. *Euphytica, 198*(2), 199–209.

Liang, C., Tian, J., & Liao, H. (2013). Proteomics dissection of plant responses to mineral nutrient deficiency. *Proteomics, 13*(3–4), 624–636.

Liao, C., Peng, Y., Ma, W., Liu, R., Li, C., & Li, X. (2012). Proteomic analysis revealed nitrogen-mediated metabolic, developmental, and hormonal regulation of maize (*Zea mays* L.) ear growth. *Journal of Experimental Botany, 63*(14), 5275–5288.

Marchive, C., Roudier, F., Castaings, L., Bréhaut, V., Blondet, E., Colot, V., et al. (2013). Nuclear retention of the transcription factor NLP7 orchestrates the early response to nitrate in plants. *Nature Communications, 4,* 1713. https://doi.org/10.1038/ncomms2650

Masclaux-Daubresse, C., Reisdorf-Cren, M., Pageau, K., Lelandais, M., Grandjean, O., Kronenberger, J., et al. (2006). Glutamine synthetase-glutamate synthase pathway and glutamate dehydrogenase play distinct roles in the sink-source nitrogen cycle in tobacco. *Plant Physiology, 140*(2), 444–456.

McAllister, C. H., Beatty, P. H., & Good, A. G. (2012). Engineering nitrogen use efficient crop plants: The current status. *Plant Biotechnology Journal, 10*(9), 1011–1025.

McDougall, P. (2011). *The cost and time involved in the discovery, development and authorisation of a new plant biotechnology derived trait.* A consultancy study for Crop Life International. Midlothian, UK: Phillips McDougall.

Møller, A. L. B., Pedas, P. A. I., Andersen, B., Svensson, B., Schjoerring, J. K., & Finnie, C. (2011). Responses of barley root and shoot proteomes to long-term nitrogen deficiency, short-term nitrogen starvation and ammonium. *Plant, Cell & Environment, 34*(12), 2024–2037.

Okumoto, S., & Pilot, G. (2011). Amino acid export in plants: A missing link in nitrogen cycling. *Molecular Plant, 4*(3), 453–463.

Orman-Ligeza, B., Civava, R., de Dorlodot, S., & Draye, X. (2014). Root system architecture. In A. Morte & A. Varma (Eds.), *Root engineering* (pp. 39–56). Berlin Heidelberg: Springer.

Parry, M. A., & Hawkesford, M. J. (2012). An integrated approach to crop genetic improvement. *Journal of Integrative Plant Biology, 54*(4), 250–259.

Pathak, R., Ahmad, A., Lochab, S., & Raghuram, N. (2008). Molecular physiology of plant nitrogen use efficiency and biotechnological options for its enhancement. *Current Science, 94*(11), 1394–1403.

Pathak, R., Lochab, S., & Raghuram, N. (2011). Improving nitrogen use efficiency. In M. Moo-Young (Ed.) *Comprehensive biotechnology. Volume 4: Agricultural and related biotechnologies* (pp. 209–218). Oxford: Elsevier.

Patterson, K., Cakmak, T., Cooper, A., Lager, I., Rasmusson, A. G., & Escobar, M. A. (2010). Distinct signalling pathways and transcriptome response signatures differentiate ammonium-and nitrate-supplied plants. *Plant, Cell & Environment, 33*(9), 1486–1501.

Pavlík, M., Pavlíková, D., & Vašíčková, S. (2010). Infrared spectroscopy-based metabolomic analysis of maize growing under different nitrogen nutrition. *Plant, Soil and Environment, 56*(11), 533–540.

Poiré, R., Chochois, V., Sirault, X. R., Vogel, J. P., Watt, M., & Furbank, R. T. (2014). Digital imaging approaches for phenotyping whole plant nitrogen and phosphorus response in Brachypodium distachyon. *Journal of Integrative Plant Biology, 56*(8), 781–796.

Privalle, L. S., Chen, J., Clapper, G., Hunst, P., Spiegelhalter, F., & Zhong, C. X. (2012). Development of an agricultural biotechnology crop product: Testing from discovery to commercialization. *Journal of Agricultural and Food Chemistry, 60*(41), 10179–10187.

Quraishi, U. M., Abrouk, M., Murat, F., Pont, C., Foucrier, S., Desmaizieres, G., et al. (2011). Cross-genome map based dissection of a nitrogen use efficiency ortho-metaQTL in bread wheat unravels concerted cereal genome evolution. *The Plant Journal, 65*(5), 745–756.

Rachmilevitch, S., Cousins, A. B., & Bloom, A. J. (2004). Nitrate assimilation in plant shoots depends on photorespiration. *Proceedings of the National Academy of Sciences of the United States of America, 101*(31), 11506–11510.

Rothstein, S. J., Bi, Y.-M., Coneva, V., Han, M., & Good, A. (2014). The challenges of commercializing second-generation transgenic crop traits necessitate the development of international public sector research infrastructure. *Journal of Experimental Botany, 65*(19), 5673–5682.

Ruzicka, D. R., Barrios-Masias, F. H., Hausmann, N. T., Jackson, L. E., & Schachtman, D. P. (2010). Tomato root transcriptome response to a nitrogen-enriched soil patch. *BMC Plant Biology, 10*(1), 75.

Sato, S., & Yanagisawa, S. (2014). Characterization of metabolic states of Arabidopsis thaliana under diverse carbon and nitrogen nutrient conditions via targeted metabolomic analysis. *Plant and Cell Physiology, 55*(2), 306–319.

Simons, M., Saha, R., Guillard, L., Clément, G., Armengaud, P., Cañas, R., Maranas, C. D., Lea, P. J., & Hirel, B. (2014). Nitrogen-use efficiency in maize (*Zea mays* L.): from 'omics' studies to metabolic modelling. *Journal of Experimental Botany, 65*(19), 5657–5671.

Smith, S., & De Smet, I. (2012). Root system architecture: Insights from Arabidopsis and cereal crops. *Philosophical Transactions of the Royal Society B: Biological Sciences, 367*(1595), 1441–1452.

Sorgona, A., Lupini, A., Mercati, F., Di Dio, L., Sunseri, F., & Abenavoli, M. R. (2011). Nitrate uptake along the maize primary root: An integrated physiological and molecular approach. *Plant, Cell & Environment, 34*(7), 1127–1140.

Sun, J.-J., Guo, Y., Zhang, G.-Z., Gao, M.-G., Zhang, G.-H., Kong, F.-M., et al. (2013). QTL mapping for seedling traits under different nitrogen forms in wheat. *Euphytica, 191*(3), 317–331.

Sutton, M. A., Bleeker, A., Howard, C., Bekunda, M., Grizzetti, B., de Vries, W., van Grinsven, H. J. M., Abrol, Y. P., Adhya, T. K., Billen, G., Davidson, E. A., Datta, A., Diaz, R., Erisman, J. W., Liu, X. J., Oenema, O., Palm, C., Raghuram, N., Reis, S., Scholz, R. W., Sims, T., Westhoek, H., & Zhang, F. S. (2013). *Our Nutrient World: The challenge to produce more food and energy with less pollution.* Global overview of Nutrient Management. Edinburgh: Centre for Ecology and Hydrology (CEH) on behalf of the Global Partnership on Nutrient Management and the International Nitrogen Initiative.

Varshney, R. K., Bansal, K. C., Aggarwal, P. K., Datta, S. K., & Craufurd, P. Q. (2011). Agricultural biotechnology for crop improvement in a variable climate: Hope or hype? *Trends in Plant Science, 16*(7), 363–371.

Villordon, A. Q., Ginzberg, I., & Firon, N. (2014). Root architecture and root and tuber crop productivity. *Trends in Plant Science, 19*(7), 419–425.

Vitousek, P., Cassman, K., Cleveland, C., Crews, T., Field, C., Grimm, N., et al. (2002). Towards an ecological understanding of biological nitrogen fixation. In E. Boyer & R. Howarth (Eds.), *The nitrogen cycle at regional to global scales* (pp. 1–45). Netherlands: Springer.

Wang, R., Guegler, K., LaBrie, S. T., & Crawford, N. M. (2000). Genomic analysis of a nutrient response in Arabidopsis reveals diverse expression patterns and novel metabolic and potential regulatory genes induced by nitrate. *The Plant Cell Online, 12*(8), 1491–1509.

Wei, H., Yordanov, Y. S., Georgieva, T., Li, X., & Busov, V. (2013). Nitrogen deprivation promotes Populus root growth through global transcriptome reprogramming and activation of hierarchical genetic networks. *New Phytologist, 200*(2), 483–497.

Wu, W., & Cheng, S. (2014). Root genetic research, an opportunity and challenge to rice improvement. *Field Crops Research, 165,* 111–124.

Xu, G., Fan, X., & Miller, A. J. (2012). Plant nitrogen assimilation and use efficiency. *Annual Review of Plant Biology, 63,* 153–182.

Zeng, H., Wang, G., Hu, X., Wang, H., Du, L., & Zhu, Y. (2014). Role of microRNAs in plant responses to nutrient stress. *Plant and Soil, 374*(1–2), 1005–1021.

Chapter 12
Nitrogen Loss When Using Organic and Mineral Fertilizers on Soddy Podzolic Sandy-Loam Soil in Central Russia

Sergei M. Lukin

Abstract This chapter presents information on the nitrogen (N) balance in the long-term field experiment and some supporting data from a lysimeter experiment. The long-term field experiment treatments included different rates of manure/fertilizer application and also were separated by the type of the incorporated nitrogen: N from the mineral fertilizer only, N from farmyard manure (FYM) only, N from a combination of the fertilizer and FYM. The lowest calculated average annual N loss in the long-term field experiment was in the treatment of FYM 10 t ha^{-1} (N loss of 12 kg ha^{-1} or 24% of the N input), while the highest loss was calculated for the treatment FYM 10 t ha^{-1} + N100P50K120 (N loss of 92 kg ha^{-1} or 61% of the N input). On average, the mineral N content in the soil layer 0–60 cm declined during the autumn-winter-early spring period by 12 kg N ha^{-1} for zero fertilizer treatment, by 9 kg N ha^{-1} for the treatment of 20 t FYM ha^{-1}, by 16 kg N ha^{-1} for the treatment 10 t FYM ha^{-1} + N50P25K60 and by 34 kg N ha^{-1} for the treatment N100P50K120. In the four-year lysimeter experiment, the highest concentrations of mineral N in the lysimeter waters were measured in autumn. On the average, for four years of the experiment, the lowest annual N loss with the lysimeter waters was 15.3 kg N ha^{-1} in the control treatment and the highest—111.7 kg N ha^{-1} in the treatment of 320 t FYM ha^{-1}. Most of the mineral N in the lysimeter waters was in the nitrate form (91–95%).

Keywords Soil management · Manure · Mineral nitrogen · Nitrogen balance · Nitrogen loss

12.1 Introduction

Soddy podzolic sandy and sandy loam soils are widespread in Russia. The total area of these soils is 130 million hectares and most of it is occupied by forests. Only 21.5

S. M. Lukin (✉)
All-Russian Scientific Research Institute of Organic Fertilizers and Peat—Branch of the Federal Budget Scientific Institution "Verkhnevolzsky Federal Agrarian Scientific Center", Vladimir, Russian Federation
e-mail: vnion@vtsnet.ru

M. A. Sutton et al. (eds.), *Just Enough Nitrogen*,
https://doi.org/10.1007/978-3-030-58065-0_12

million hectares (16.6%) are used for agriculture, of which 8.8 million ha (6.8%) is arable land (Romanenko et al. 1996). The largest proportion of light-textured soddy podzolic soils is located in the non-Chernozem zone of the country, where they form broad lateral stripes extending from the Pripyat river to the Northern Urals "almost parallel to the Northern boundary of chernozems" (Dokuchaev 1949).

Improving fertility of the soddy podzolic soils is associated with the use of complex measures in order to optimize the agrochemical, physico-chemical and biological properties of the soils. Among the most important measures is the use of fertilizers.

Since 1989, the use of manures in the agriculture of the Russian Federation decreased to 1/8, while the use of mineral fertilizers decreased to 1/5 of their previous values. The average annual nitrogen (N) input from mineral fertilizers and manures in Russia in 2013 was 29 kg ha^{-1} (ROSSTAT 2014). Despite the overall low level of manure and fertilizer use, large cattle-breeding complexes have continued to use large rates of manure, which have led to surface and groundwater pollution (Lukin et al. 2014).

Nitrogen compounds of manures and mineral fertilizers are transformed in soils due to chemical, physical, biochemical and microbiological processes (Kudeyarov 1989; Osipov and Sokolov 2001). The direction and intensity of these processes are determined by many factors: the level of soil productivity, the intensity of different soil treatments, hydrothermal conditions of the area, biological characteristics of crops, types of fertilizers, and methods of fertilizer application to soils. Due to the unstable soil water regime of the sandy and sandy-loam soddy podzolic soils, substantial losses of mineral N can be observed (Wild and Cameron 1980; Kutra et al. 2006; Buchkina et al. 2010; Shilnikov et al. 2012).

The purpose of this project was to determine the amount of mineral N loss associated with the use of manures and mineral fertilizers on the light-textured soddy podzolic soils.

12.2 Materials and Methods

Studies were performed at a long-term experimental site, established at the institute in 1968, and also in a separate lysimeter experiment conducted in 1998–2002. The experimental site is situated 10 km to the North-East from the town of Vladimir (56°03′ N, 40°29′ E). Geomorphologically the site can be characterized as a sandy (fluvioglacial) slightly undulated lowland (130–170 m above sea level) with sediments of lacustrine and glacial origin. The climate of the region is moderately continental. The average annual air temperature is +3.9 °C (for the last 50 years it was + 4.46 °C). Average annual rainfall is 599 mm (for the last 50 years it was 608 mm). Average rainfall for the growing season (May–September) is 281 mm (Table 12.1).

Table 12.1 Climate parameters at the long-term experimental site for the last 50 years

Years	Annual average air temperature, °C	Annual average precipitation, mm
1961–1970	3.80	585
1971–1980	4.21	610
1981–1990	4.37	648
1991–2000	4.59	607
2001–2010	5.35	589

The studied soil is the most typical soil of the area: soddy podzolic sandy-loam soil (Albeluvisols/Glacial Till, WRB 2006) containing 5% of clay (particles ≤ 1 μm), 21% of silt (particles 1–50 μm) and 74% of sand (particles > 50 μm) with 6–8 g of soil organic carbon (SOC) kg^{-1} soil (Fig. 12.1).

The long-term field experiment "Effect of long-term fertilization systems on the productivity of grain-row crop rotation and the soddy podzolic soil fertility" included 16 treatments with different rates of manure or mineral fertilizer or the combination of the two. Each plot was 161 m^2. The experiment was conducted in four replicates. Eight of the 16 treatments were used to study N balance and N loss for this chapter (where "N50P25K60" means an annual application of 50 kg N, 25 kg phosphorus (P_2O_5) and 60 kg potassium (K_2O) per hectare as a mineral fertilizer, "FYM 10 t ha^{-1}" means an annual application of 10 tons of farmyard manure (FYM) per hectare), as follows:

- Treatment 1: Zero inputs
- Treatment 6: FYM 10 t ha^{-1}
- Treatment 7: FYM 20 t ha^{-1}
- Treatment 8: N50P25K60
- Treatment 10: FYM 5 t ha^{-1} + N25P12K30
- Treatment 13: FYM 10 t ha^{-1} + N50P25K60
- Treatment 14: N100P50K120
- Treatment 16: FYM 10 t ha^{-1} + N100P50K120

All the treatments were separated not only by the rates of the manure/fertilizer but also by the type of the incorporated nitrogen: N from the mineral fertilizer only (treatments 8 and 14), N from the manure only (treatments 6 and 7), N from a combination of the fertilizer and manure (treatments 10, 13 and 16).

Nitrogen as a mineral fertilizer was applied in the form of ammonium nitrate. The four-year crop rotation consisted of annual lupine, winter wheat, potato and spring barley. For treatments 6, 7, 13 and 16, the farmyard manure was applied twice during the four-year crop rotation: for winter wheat and for potatoes in the rates either 20 (treatments 6, 13 and 16) or 40 t ha^{-1} (treatment 7). For treatment 10, the FYM was applied only for winter wheat. In the treatments' list the manure rates are given on the annual basis. An annual input of 10 t ha^{-1} of FYM was roughly equivalent to 50 kg N ha^{-1}.

a
b

Fig. 12.1 The long-term experimental site at the Research Institute for Organic Fertilizers and Peat (**a**), and the soil profile at the site (**b**)

Total N loss (emissions to the atmosphere and leaching), fertilizer/manure N accumulated by the crop and intensity of N balance (INB) were calculated:

$$N\,loss\left(\mathrm{kg\,N\,ha^{-1}}\right) = Input\ N - fertilizer/manure\ N accumulated\ by\ the\ crop \\ - Increase\ of\ the\ soil\ N\ content \quad (12.1)$$

$$Fertilizer/manure\ N\ accumulated\ by\ the\ crop\left(\mathrm{kg\,N\,ha^{-1}}\right) \\ = N\ accumulated\ by\ the\ crop\ in\ fertilizer/manure\ treatment \\ - N accumulated\ by\ the\ crop\ in\ the\ Zero\ treatment \quad (12.2)$$

$$Intensity\ of\ N\ balance(\%) = (Input\ N/N\ accumulated\ by\ the\ crop) * 100 \quad (12.3)$$

The mineral N content was measured in the 60 cm soil layer in spring, summer and autumn (after harvesting). Supplies of mineral N in the soil were calculated using the soil bulk density. The difference in the soil mineral N supplies in autumn and next spring allowed calculations of the soil N losses during the autumn-winter-early spring time.

In the lysimeter experiment, the N loss was calculated for the four-year crop rotation (fallow, winter wheat, potato, spring barley) for four treatments consisting of manure (one application in four years at the beginning of the experiment, at the rates of 0, 40, 80 and 320 t $\mathrm{ha^{-1}}$), and mineral fertilizer (two applications in four years: for winter wheat and potato in the form of ammonium nitrate (N60), superphosphate (P60) and potassium chloride (K60)). The lysimeter area was 0.17 m^2 (Shilova 1955).

Based on the results of the long-term experiment, the nitrogen use efficiency (NUE), and N losses were calculated:

$$NUE(\%) = 100 * (Fertilizer/manure\ N\ accumulated\ by\ the\ crop \\ + increase\ in\ soil\ N\ stock)/N\ input \quad (12.4)$$

$$N\,loss(\%) = 100 - NUE \quad (12.5)$$

"*Fertilizer/manure N accumulated by the crop*" here refers to the N in the harvested products.

The soil mineral N content, in the nitrate form, was measured using the potentiometric method, while the content of the soil mineral N in the ammonium form was measured by the fotokalorimetric method (GOST 26951-86 1986; GOST 26489-85 1985). The results of the measurements were analyzed using the ANOVA and STAT programs.

12.3 Results and Discussion

The total N content in the topsoil (20 cm layer) of the soddy podzolic sandy loam soil of the long-term field experiment before the beginning of the experiment was on average 0.082%, which was equivalent to 2.46 t N ha^{-1}. In the 0–60 cm soil layer the total N content was equal to 4.62–4.73 t N ha^{-1}.

Application of the fertilizer/manure in the long-term field experiment resulted in a significant increase of the crop yields and higher crop rotation productivity. An average yield increase of 71% in grain units was found under complex application of manure and mineral fertilizer (treatment 16, N input is equivalent to 150 kg N ha^{-1} annually), as compared to the zero fertilizer application (treatment 1, Table 12.2). The results on the crop yields also show that the mineral fertilizer (treatment 8) has about the same efficiency with the combination of manure and mineral fertilizer (treatment 10) at low N rates, but to achieve higher crop yields at higher N rates it is better to use a combination of manure and fertilizer (treatment 13) rather that the mineral fertilizer only (treatment 14).

According to the calculated N balance for the long-term field experiment (Table 12.3) the total N input (excluding non-symbiotic N fixation) for the first seven cycles of the crop rotation was equal to 1036–5311 kg N ha^{-1} (37.0–189.7 kg N ha^{-1} $year^{-1}$), depending on the type and the rates of the manure/fertilizer. Nitrogen

Table 12.2 Effect of different fertilization systems on the crop yields in the first seven cycles of the crop rotation

Treatment number and treatments	Crop yields, t ha^{-1}					Yield increase	
	Lupin	Winter wheat	Potato	Barley	Average, in t ha^{-1} G.U.*	t ha^{-1}	%
1. Zero input	22.2	1.80	11.5	1.30	2.39		
6. FYM 10 t ha^{-1}	24.3	2.39	17.4	1.86	3.21	0.82	34
7. FYM 20 t ha^{-1}	25.0	2.63	20.0	2.20	3.56	1.17	49
8. N50P25K60	25.0	2.57	17.8	2.66	3.56	1.17	49
10. FYM 5 t ha^{-1} + N25P12K30	24.9	2.63	18.3	2.35	3.53	1.14	48
13. FYM 10 t ha^{-1} +N50P25K60	25.4	2.71	21.7	2.94	3.98	1.59	67
14. N100P50K120	25.5	2.61	19.8	2.91	3.83	1.44	60
16. FYM 10 t ha^{-1} +N100P50K120	26.8	2.73	21.9	3.04	4.08	1.69	71
LSD	0.92	0.09	0.82	0.12	0.14		

LSD = least significant difference at a reference crop yield of 0.95 t ha^{-1}

*G.U. (grain unit) is a measure used to equivalent measure different types of crop production. The basic unit of measurement taking grain, food from other cultures is transferred into the grain units, which uses the various factors (grain of winter wheat and barley—1.0, potatoes—0.25, lupin—0.14, straw—0.20–0.25) (Ministry of Agriculture of the Russian Federation, 2017)

Table 12.3 Nitrogen (N) balance in the long-term field experiment (for the first seven cycles of the crop rotation), kg ha^{-1}

Treatment number and treatments	N input		N accumulated by the crop	N balance		INB,%
	Total	Symbiotic N only		Total	Annual	
1. Zero inputs	1036	728	1351	–315	−11.3	77
6. FYM 10 t ha^{-1}	2471	763	1772	699	25.0	139
7. FYM 20 t ha^{-1}	3872	764	1913	1959	70.0	202
8. N50P25K60	2467	759	2084	383	13.7	118
10. FYM 5 t ha^{-1} + N25P12K30	2483	765	1936	537	19.2	128
13. FYM 10 t ha^{-1} + N50P25K60	3879	771	2207	1672	59.7	176
14. N100P50K120	3865	757	2221	1644	58.7	174
16. FYM 10 t ha^{-1} + N100P50K120	5311	803	2346	2965	105.9	179

INB = intensity of nitrogen balance

accumulated by the crops for the same period amounted to 1351–2346 kg N ha^{-1} (48.3–83.8 kg N ha^{-1} year^{-1}). The N balance for the different treatments of the experiment changed from −11.3 kg N ha^{-1} year^{-1} (implying additional N inputs, e.g., atmospheric N deposition, non-symbiotic N fixation) to 105.9 kg N ha^{-1} year^{-1} (implying substantial N losses). As a result of N application with fertilizers and manure, the crop yields and soil N stocks increased, as well as the N balance became positive.

The intensity of N balance for the average annual N rate of 50 kg ha^{-1} was: for manure application (treatment 6): 139%; for the combined application of manure and fertilizer (treatment 10): 128%; for mineral fertilizer application (treatment 8): 118%. For an average annual N rate of 100 kg N ha^{-1}, the intensity of nitrogen balance was for manure application (treatment 7): 202%, for the combined application of manure and fertilizer (treatment 13): 176% and for mineral fertilizer application (treatment 14): 174% (Table 12.3).

Higher intensity of N balance for the manure treatments, as compared to the equivalent amount of the mineral fertilizer treatments, can be explained by the less efficient use of N from manure than from the mineral fertilizer, as well as by a lower content of N in the products.

The annual N losses in the long-term field experiment as well as the components of the overall N budget for the set of first seven cycles of the crop rotation are shown in Table 12.4. For treatment 6 (10 t ha^{-1} FYM), annual N losses were 12 kg ha^{-1}, equivalent to 24% loss of the N input. For treatment 8 (N50P25K60) the losses were 21 kg N ha^{-1} year^{-1}, equivalent to 43% loss of the N input. For treatment 10 (5 t

Table 12.4 Nitrogen (N) loss and the components of the N balance (for the first seven cycles of the crop rotation) in the long-term field experiment

Treatment number and treatments	Input N, kg ha^{-1}	Fertilizer/manure N accumulated by the crop, kg ha^{-1}	N content increase in soil, kg ha^{-1}	Total uptake, kg ha^{-1}	NUE, %	N loss, kg ha^{-1}	
						Total	Annual
6. FYM 10 t ha^{-1}	1400	421	638	1059	76	341	12
7. FYM 20 t ha^{-1}	2800	562	885	1447	52	1353	48
8. N50P25K60	1400	733	67	800	57	600	21
10. FYM 5 t ha^{-1} + N25P12K30	1400	585	375	960	69	440	16
13. FYM 10 t ha^{-1} + N50P25K60	2800	856	365	1221	44	1579	56
14. N100P50K120	2800	870	195	1065	38	1735	62
16. FYM 10 t ha^{-1} + N100P50K120	4200	995	623	1618	39	2582	92

NUE = nitrogen use efficiency

ha^{-1} FYM + N25P12K30) the losses were 16 kg N ha^{-1} year^{-1} or 31% loss of the N input. With the N application rate increasing two times, annual N losses increased to 48–62 kg N ha^{-1} (treatments 7, 13, 16), equivalent to a loss of 48–62% of the N input. With an annual N application of 150 kg ha^{-1} (treatment 14) N losses increased to 92 kg ha^{-1}, which was 61% of the N input (Table 12.4; Fig. 12.2). These results are also expressed in the Table 12.4 as the nitrogen use efficiency (NUE).

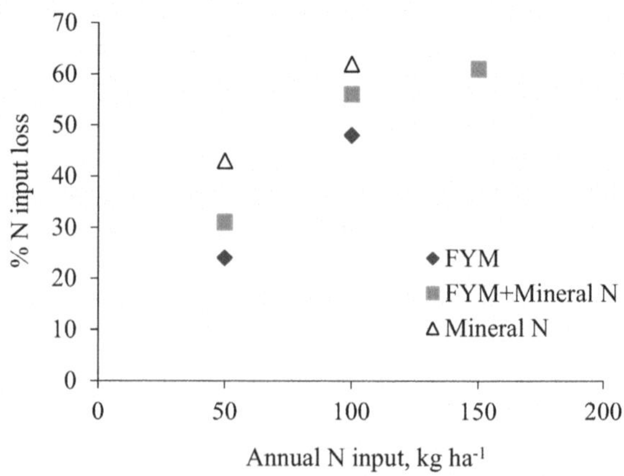

Fig. 12.2 Annual average (for the first 7 cycles of the crop rotation) nitrogen (N) input and N loss in the long-term field experiment for different rates of N application with fertilizer, manure or a combination of the two

The results on the dynamics of the mineral N content (average for the period from 1990 to 2005) have shown that long-term use of manures and fertilizers has caused significant changes in the mineral N content of the soddy podzolic sandy loam soil. In spring, the highest soil mineral nitrogen content (19 kg ha^{-1} in the 0–20 cm soil layer and 35 kg ha^{-1} in the 0–60 cm soil layer) was measured for treatment 7 (manure 20 t ha^{-1}). The soil mineral nitrogen content in the mineral fertilizer treatments (treatments 8 and 14) was 13–30% lower compared with the manure treatments and was almost the same as the control. For treatments 10, 13 and 16, where both manure and the mineral fertilizer were used, the soil mineral N content was lower than in the manure treatments, but higher than in the mineral fertilizer treatments.

At the beginning of the growing season, most of the mineral N in the topsoil was in the form of ammonium. With increasing temperature and the development of the nitrification processes, the ammonium N fraction in the mineral N decreased. In the second half of the summer, the soil mineral N was mainly in the nitrate form.

According to the data on the soil mineral N content in autumn and spring, quite high losses of mineral N were occurring in the autumn—winter—early spring periods. Comparison of the soil mineral N content in autumn and in spring shows that the soil mineral N content depends on the crop type and fertilizer/manure application. The highest soil mineral N loss was observed after the potato harvest when using mineral fertilization (treatments 14 and 16). The main reason could be the intense mineralization of the soil N in the potato crop and the accumulation of a high amount of nitrate N in the soil. Relatively high losses of mineral N were also observed after the barley harvest, which might be due to the early timing of the harvest (early August) and a small amount of plant residues with a more narrow carbon/nitrogen (C/N) ratio than that of winter crops.

The lowest losses of the soil mineral N in the autumn-winter-early spring period occurred under winter wheat. Relatively low loss of mineral N was also observed after the winter wheat harvest, which can be explained by the accumulation of large amounts of crop residues with a high C/N ratio and partial immobilization of mineral N during their decomposition. In 1995–2005, the average annual loss of the mineral nitrogen content in the soil layer 0–60 cm during the autumn–winter–early spring season in treatment 1 was 12 kg ha^{-1}, in treatment 7: 9 kg ha^{-1}, in treatment 13: 16 kg ha^{-1}, and in treatment 14: 34 kg ha^{-1}. Based on the differences in N stock in the soil in the autumn and spring, the relative loss of mineral N during autumn-winter-early spring time (Table 12.5) amounted to 18% of the initial N stocks for the organic fertilizer system (treatment 7), to 31% for the organic-mineral system (treatment 13), and to 53% for the mineral system (treatment 14).

Maximum concentrations of N were measured in the lysimeter water during autumn time. An average (for four years) N loss with the lysimeter water in the zero input treatment was 15.3 kg ha^{-1} $year^{-1}$, which is close to the N loss calculated for the same treatment in the long-term field experiment (11.3 kg ha^{-1} $year^{-1}$, Table 12.3). Higher rates of manure and mineral fertilizer in the lysimeter experiment

Table 12.5 Changes in the soil mineral nitrogen (N) content in the long-term field experiment (layer 0–60 cm) in autum-winter-spring period (average for 1990–2005)

Treatment number and treatments	Average soil mineral N, kg ha^{-1}		Average changes in autum-winter-spring period	
	Autumn	Spring	kg ha^{-1}	%
1. Zero inputs	43 ± 7.2	31 ± 9.9	−12 ± 4.4	28
7. FYM 20 t ha^{-1}	51 ± 11.0	42 ± 14.7	−9 ± 8.4	18
13. FYM 10 t ha^{-1} + N50P25K60	51 ± 9.6	35 ± 11.8	−16 ± 13.0	31
14. N100P50K120	64 ± 10.1	30 ± 9.3	−34 ± 13.3	53

resulted in higher N losses with the lysimeter water (Table 12.6). Most of the N in the lysimeter water (91–95%) was in the form of nitrates.

12.4 Conclusions

The main findings of the study are as follows:

- Various N sources had a different effect on the components of nitrogen balance in the soddy podzolic sandy-loam soil. When manure was used in the cereal crops, there was a shortage of mineral components, which could be eliminated only by the additional application of mineral fertilizers.
- The use of high rates of mineral fertilizers led to the accumulation of higher amounts of nitrate nitrogen in the soil which was associated with increased losses of mineral N in the non-vegetation period of the year.
- Application of manure in combination with mineral fertilizers leads to lower N losses compared with the treatments where N comes only from mineral fertilizers and to higher crop yields compared to the treatments where N comes from manure only.
- Nitrogen losses significantly increase with increasing fertilizer/manure rates.

Table 12.6 Nitrogen (N) losses in the lysimeter experiment for the four-year crop rotation

Treatment number and treatments	N input with fertilizer/manure, kg ha^{-1}		Average N concentration in the lysimeter water, mg l^{-1}			Average annual N losses with lysimeter water, kg ha^{-1}	$N\text{-}NO_3^-$ in the lysimeter water, %
	Total for 4 years	Annual	$N\text{–}NO_3^-$	$N\text{–}NH_4^+$	Total		
1.Zero inputs	0	0	6.8	0.7	7.5	15.3	91
2. FYM 40 t ha^{-1} + N60P60K60	352	88	13.5	1.4	14.9	30.4	91
3. FYM 80 t ha^{-1} + N60P60K60	584	146	15.3	1.5	16.8	34.3	91
4. FYM 320 t ha^{-1}	1856	464	52.2	2.5	54.7	111.7	95

Acknowledgements The study was funded by the All-Russian Scientific Research Institute of Organic Fertilizers and Peat—branch of the Federal Budget Scientific Institution "Verkhnevolzsky Federal Agrarian Scientific Center", Vladimir region, the Russian Federation.

References

Buchkina, N. P., Balashov, E. Y., Rizhiya, E. Y., & Smith, K. A. (2010). Nitrous oxide emission from light-textured arable soil of North-Western Russia: Effects of crops, fertilizers, manures and climate parameters. *Nutrient Cycling in Agroecosystems, 87*(3), 429–442.

Dokuchaev, V. V. (1949). *Izbrannye sochineniya* (Selected works). T.2. Moscow: Gosselhozizdat (In Russian).

GOST 26489-85. (1985). *Determination of exchangeable ammonium by CINAO method*. Gosstandart (GOST).

GOST 26951-86. (1986). *Determination of nitrates by ionometric method*. Gosstandart (GOST).

Kudeyarov, V. N. (1989). *Cycl azota i effektivnost udobrenij (Nitrogen cycle and efficiency of fertilizers)*. Moscow: Nauka. (In Russian).

Kutra, G., Gaigalis, K., & Šmitiene, A. (2006). Land use influence on nitrogen leaching and option for pollution mitigation. *Zemdirbyste-Agriculture, 93*(4), 119–129.

Lukin, S. M, Nikolski, K. S., Ryabkov, V. V., & Rusakova, I. V. (2014). Methods to reduce ammonia nitrogen losses during production and application of organic fertilizers. In K. W. Van der Hoek & N. P. Kozlova (Eds.), *Ammonia workshop 2012 Saint Petersburg. Abating ammonia emissions in the UNECE and EECCA region* (pp. 169–175). RIVM Report 680181001/SZNIIMESH Report. The Netherlands: Bilthoven.

Ministry of Agriculture of the Russian Federation. (2017). *Prikaz Ministerstva selskogo hozyaystva RF ot 6.07.2017 N 330* (Order of the Ministry of Agriculture of the Russian Federation from 6.07.2017 N 330). Available at: https://www.garant.ru/products/ipo/prime/doc/71634802/. Accessed September 2019 (In Russian).

Osipov, A. I., & Sokolov, A. I. (2001). *Ecologicheskaya bezopasnost i ustoichivoe razvitie. Kniga 4. Rol' azota v plodorodii pochv i pitanii rastenij* (Ecological safety and sustainable development. Book 4. Role of nitrogen to soil fertility and plant nutrition). St. Petersburg. (In Russian).

Romanenko, G. A., Komov, N. V., & Tyutyunnikov, A. I. (1996). *Zemelnye resursy Rossii, effektivnost ikh ispolzowaniya* (Land resources of Russia, efficiency of their use). Moscow: RAAS (In Russian).

ROSSTAT. (2014). *Vnesenie udobrenij pod urojay 2013 goda i provedenie rabot po khimicheskoj melioracii zemel* (Use of fertilizers and chemical land meliotration, 2013). http://www.gks.ru/wps/wcm/connect/rosstat_main/rosstat/ru/statistics/publications/catalog/doc_1265196018516. Accessed March 6, 2014 (In Russian).

Shilnikov, I. A., Sychev, V. G., Sheudghen, A Kh, Akanova, N. I., Bondareva, T. N., & Kizenek, S. V. (2012). *Poteri elementov pitaniya rastenij v agrobiogeo-khimicheskom krugovorote veshestv i sposoby ikh minimizatii (Loss of plant nutrients in agrobiogeochemical circulation and methods to minimize them)*. Moscow: VNIIA. (In Russian).

Shilova, E. I. (1955). Metod polucheniya pochvennogo rastvora v pripodnych usloviyach (The method of obtaining soil solution under natural conditions). *Pochvovedenie, 11*, 86–90. (In Russian).

Wild, A., & Cameron, K. C. (1980). Nitrate leaching through soils and environmental considerations with special reference to recent work in the United Kingdom. In *Soil Nitrogen as Fertilizer or Pollutant.* Proceedings and report of a Research Coordination Meeting organized by the joint FAO/IAEA division of atomic energy in food and agriculture in cooperation with the Gesellschaft fuer Strahlen-und Umweltforschung G.m.b.H and held in Piracicaba, Brazil, 3 July 1978 (pp. 289–306). Panel Proceedings Series 132. Vienna: International Atomic Energy Agency.

WRB. (2006). *World reference base for soil resources 2006. A framework for international classification, correlation and communication.* World Soil Resources Reports 103. Rome: FAO.

Chapter 13
Sorghum Response to Nitrogen in Organic Carbon-Categorized Ferralsol and Andosol in Uganda

Patrick Musinguzi, Peter Ebanyat, John S. Tenywa, and Mateete Bekunda

Abstract Grain sorghum is an important staple food crop for smallholder farmers in Uganda, but the yields remain low due to the decline in soil fertility, particularly nitrogen (N) and phosphorus (P). A study was conducted on an Acric Ferralsol and Calcic Andosol in contrasting agro-ecological zones (AEZs) of Uganda to evaluate the value of soil organic carbon (SOC) as a proxy of soil fertility status in influencing N fertilizer responses. Sorghum yield response trials to N fertilizer were installed in each AEZ during the growing seasons of 2010–2011. Phosphorus and potassium (K) fertilizers were applied to alleviate nutrient limitations. All experiments were laid in fields of variable fertility categorized into low, medium and high SOC. Substantial yield gains to applied N were observed in medium to high SOC fields. Grain yields were considerably more responsive to increase in SOC and N fertilizer in an Acric Ferralsol, averaging 297% yield gain, than in the Calcic Andosol with 165% yield gain. Application of N fertilizer to soils with relatively high SOC content and fertility bears better yield benefits than the low SOC counterparts. Acric Ferralsol with SOC > 1.7% and Calcic Andosols with SOC of 1.2–1.9% registered highest yield responses to N, suggesting existence of critical SOC ranges for increased N use efficiency in sorghum.

Keywords Soil fertility · Categories · Agronomic efficiency · Grain sorghum · Agro-ecological zones

P. Musinguzi (✉) · P. Ebanyat · J. S. Tenywa
Department of Agricultural Production, Makerere University, Kampala, Uganda
e-mail: musipato@yahoo.com

P. Ebanyat
e-mail: peter.ebanyat@gmail.com

J. S. Tenywa
e-mail: johntenywa@gmail.com

P. Ebanyat
International Institute of Tropical Agriculture (IITA), Kampala, Uganda

M. Bekunda
International Institute of Tropical Agriculture (IITA), Arusha, Tanzania
e-mail: M.Bekunda@cgiar.org

M. A. Sutton et al. (eds.), *Just Enough Nitrogen*,
https://doi.org/10.1007/978-3-030-58065-0_13

13.1 Introduction

Grain sorghum (*sorghum bicolar* L.) is a major food crop in sub-Saharan Africa (SSA) (Wortmann et al. 2009). However, its productivity per unit area remains the lowest in the region due to a range of biophysical constraints particularly the occurrence of too little nitrogen (N) (Bekunda et al. 1997). Current grain yields range between 0.41 and 1.07 t ha^{-1}, and can potentially triple under fertilization to as high as 2.7 t ha^{-1} (Kaizzi et al. 2012). The high N depletion has been attributed to increasing human population pressure on soils, and weak environment policies (Sanchez et al. 1997). Consequently, it is projected that the demand for cereals will increase and the region would have to import more metric tons to feed the population that is rapidly rising. Uganda is among the countries that have been affected by nutrient depletion, and mean levels of nutrient lost per year for N, P and K are estimated at 21, 8 and 43 kg ha^{-1} respectively (Nkonya et al. 2005). Despite such reported nutrient loses, the country registers only 1% of small-scale farmers using fertilizer, which is lowest in the region. Soil quality indicators such as soil organic carbon (SOC) are also rapidly declining due to poor agronomic conservational practices such as continuous cultivation (Musinguzi et al. 2014). Efforts to restore soil fertility using external inputs are limited by the high cost of fertilizers and the low nutrient recovery efficiencies. The cost of fertilizer in sub-Saharan Africa including Uganda is 2–6 times the cost in Europe, making it the most expensive in the world (Sanchez et al. 2009). Despite the challenge, the demand for fertilizers, especially mineral N has been projected to increase in the next decades (Cassman et al. 2002; Sanchez et al. 2009; African Fertilizer Summit 2006). Efforts that could substantially increase N use efficiency (presented as good indicator to better N management) and crop yield would reduce fertilizer requirements and reactive N load in environment, and promote food security (Sutton et al. 2013). Site specific nutrient management is one appropriate tool that contributes to such efforts since crop nutrient requirement is tailored to given environment conditions. However, limited effort has been invested in exploring locally feasible site specific soil fertility indicators. Soil organic carbon is one of the potential indicators of soil fertility heterogeneity in tropical soils (Musinguzi et al. 2013). It is a major source of indigenous N supply, and its variability with fertility levels in different soil types cannot be ignored amidst increased scientific uncertainties of net immobilization/mineralization rates (Cassman et al. 2002; Musinguzi et al. 2013). The causes of variations in responsiveness of soils to N fertilizers are often associated to soil fertility status (Musinguzi et al. 2013; Vanlauwe et al. 2011) and environmental conditions (Cassman et al. 2002). The value of soil parameters such as SOC that are driving factors in influencing nutrient response have often been overlooked. Knowledge about SOC and its influence on N fertilizer applications on sorghum yields would be a major breakthrough to improved N management for small-scale farmers in the low N use regions. The study explored the contribution of SOC as a potential guide to efficient N fertilizer use in sorghum production on two soil types in two contrasting agro-ecological zones (AEZs) of Uganda.

13.2 Materials and Methods

13.2.1 The Study Area

Nitrogen response experimental trials on sorghum were conducted on two soil types located in two AEZs of Uganda; that is South Western Highlands (SWH) at 0.007104° S, 29.808920° E and Western Mid-High Farmlands (WMHF) at 0.1867° N, 30.0881° E. The study sites are dominated by smallholder farming under low-input traditional farming systems and are characteristic of different daily rainfall patterns in a year and in the seasons of March to June and September to November (Fig. 13.1). The two AEZs are characterized by different farming cultures such as preferences for sorghum varieties, and planting time. The elevation ranges from 2128 to 2322 m above sea level (m.a.s.l) in SWH and 1087–1132 m.a.s.l in WMHF. In SWH, crop production is done on sloping terraces and on a fairly undulating landscape in WMHF. The soils in the two AEZs were classified on the basis of the World Soil Resources Report by the International Union of Soil Scientists (IUSS) Working Group for the World Reference Base for Soil Resources (WRB 2006).

13.2.2 Research Approach and Sampling Strategy

The study was conducted using a combination of both the participatory and scientific approaches. Farmers were initially identified in each AEZs, and they aided the identification of fields of different soil fertility levels using their indigenous knowledge. Fields perceived to have poor, medium and good soils in terms of productivity were subsequently initially characterized for quick check of SOC and texture to guide field categorization. Detailed sampling and analysis of soil was later conducted. Soil was sampled from 0 to 15 cm depth using an auger. The soil was thoroughly mixed, and through quarter sampling, composite samples were obtained and processed for analysis for pH, SOC, extractable phosphorus (P) and potassium (K), magnesium (Mg), calcium (Ca), sodium (Na) and texture (Anderson and Ingram 1993). Using soil chemical analytical results, a Soil Fertility Index was estimated (Andrews et al. 2004; Mukashema 2007). The Soil Fertility Index value expresses the relative amounts of crop available nutrients in the soil obtained from the soil tests. This index provides an estimate of the possible soil fertility status, which is positively correlated with SOC. As such, SOC concentrations guided the process of classifying the fields into three soil fertility categories in each soil type. The Calcic Andosol was rated into low fertility ($< 1.2\%$ SOC, $n = 8$); medium fertility (1.2–1.9% SOC, $n = 9$); and high fertility ($> 1.9\%$ SOC, $n = 9$). The Acric Ferralsol was rated into low fertility ($< 1.2\%$ SOC, $n = 11$); medium fertility (1.2–1.7% SOC, $n = 11$) and high fertility ($> 1.7\%$ SOC, $n = 11$). Different SOC categorization limits were used in each soil type in the AEZs. The limits varied as a result of differences on SOC concentrations

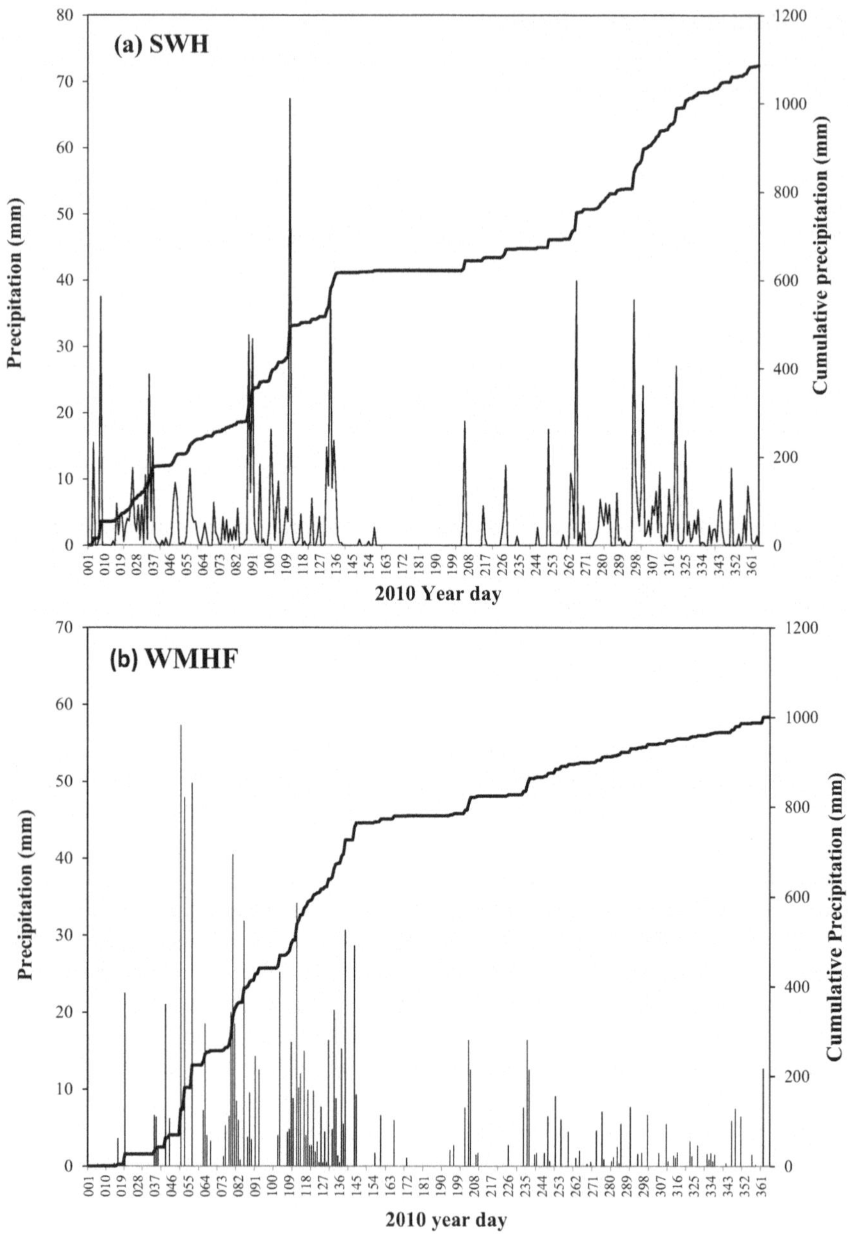

Fig. 13.1 Cumulative rainfall and daily precipitation for (**a**) South-Western Highlands (SWH)—Kabale, and (**b**) Western Mid High Farmland (WMHF)—Kasese, during the two cropping seasons in 2010

obtained, as influenced by the management history, climate, parent material, nature of landscape, and elevation in each AEZ.

A total of 26 experiments were conducted for the two seasons in WMHF, and 33 experiments for one season in SWH. The experimental plots in SWH were laid on the sloping terraces, that included the upper position (low SOC), medium position (medium SOC) and lower terrace position (high SOC). Sorghum (*Sekedo*) in WMHF and *Mabere* in SWH were planted at 5 cm depth, with a 60 cm and 10 cm inter and intra-row spacing. In WMHH, sowing was done on 15th March (first season) and 9th October 2010 (second season). In SWH, sowing was done on 17th January 2010, which is the normal planting period for the farmers. Plot sizes were 6 m × 5 m in WMHF, and 4 m × 3 m in SWH because of limited land due to fragmentation and nature of the landscape. The experimental sites were hand-ploughed twice to a depth of 0–15 cm, to a fine seedbed before planting. Seeding was by hand for all the experimental units, and depending on the weeds present, in-season weed control was done twice manually using a hoe for all crops as farmers' practice. No pesticides were applied since there was no evidence of disease and pest incidences in the fields. Nitrogen was applied at 0, 30, 60, 90 kg ha^{-1} as Urea. All experiments were laid-out in a randomized block design in a split plot type of arrangement with field types of low, medium and high fertility represented as main plots; differences in N fertilizer levels were treated as sub-plots. The farmers' sites represented blocks (replicates). Phosphorus and K were applied at blanket rates of 30 kg P ha^{-1} and 60 kg K ha^{-1} as Triple Super Phosphate and Muriate of Potash, respectively. Only 50% of N and K fertilizers were applied shortly after planting, and the remainder top-dressed after four weeks. Nitrogen and K nutrient based fertilizers were split applied by surface broadcasting at planting and after four weeks. The fertilizers were incorporated to 5–8 cm deep in soil with a hoe. Four weeks after planting, the second split of fertilizers was applied and placed about 10 cm to the side of the rows and incorporated. Only phosphorus fertilizer was basally applied at planting, in the planting holes (localized placement).

At physiological maturity, harvesting was done by cutting the plants at ground level considering the inner rows in each plot. In WMHF, harvesting was done on 29th June 2010 (first long rain season) and on 12th January 2011 (second short rain season). In SWH, harvesting was done on 20th August 2010. The harvested components were separated into grain and stover, and there were air-dried for about 7 days. The dried grain and stover were adjusted to water content of 140 g kg^{-1}. Total biomass and grain yield were converted to hectare equivalences. Agronomic efficiency was considered in this study; because of the capacity to define the yield production gains in applying fertilizers, and its uniqueness as an economically feasible tool that may require less laboratory costs thus most suitable for tropical farmers (Ladha et al. 2005).

Agronomic use efficiency (AE) in each soil type and crop was calculated as follows (Eq. 13.1):

$$AE = \frac{Y_F - Y_O}{N_F} \tag{13.1}$$

where Y_F = Yield obtained on applying a given N rate of fertilizer (kg N ha^{-1}); Y_O = Yield obtained without N fertilizer applied (control); N_F is the corresponding fertilizer rate (kg N ha^{-1}).

13.2.3 Data Analysis

Data were analyzed using the GenStat bio-statistical analysis software (13th edition). Simple descriptive statistics were generated from soil characteristics (Table 13.1). In order to cater for random effects of farmers' sites in a completely randomized design, a linear mixed model, using GenStat Restricted Maximum Likelihood Algorithms (RELM) directive was applied. Predicted means were generated for various treatments and these were separated using the Pooled Standard Error of Difference (SED) at $p \leq 0.05$. Tests for the effects of seasons, soil fertility levels, added N fertilizers, and their interactive effects on biomass and grain yield were conducted. Simple linear regression model fittings were constructed between SOC and yield in each soil type.

13.3 Results

An acidic Acric Ferralsols in SWH registered significantly different grain yield ($p < 0.05$) across the SOC categories and N fertilizer. Interactions between SOC categories and added N fertilizer were significant for yield ($p < 0.05$), (Fig. 13.2a). Soils with low SOC were responsive to added N, with the highest yield gain and AE registered with 90 kg N ha^{-1}. This ensued into triple yield gains to N from about 1000 to 3000 kg ha^{-1}. This, however, did not translate into desirable yields as observed in medium to high SOC fields. In medium SOC, double yield gains were obtained to N fertilization. Soils with high SOC (> 1.7%) were notably most responsive to N applications, except for rates higher than 60 kg N ha^{-1}. The highest AE gains were attained at 60 kg N ha^{-1} for high SOC and at 30 kg N ha^{-1} for medium SOC. Significant yield gains and high AE to N were generally obtained in soils with SOC > 1.2% (Fig. 13.2a). The local sorghum variety (*Mabere*) responded to fertilization irrespective of the SOC concentrations.

In a high pH Calcic Andosol in MWHF, all fields responded positively to applied N except for high fertility soils with SOC > 1.9% (Fig. 13.2b). High yield gains

Table 13.1 Mean values of soil properties (Standard Deviation in parentheses) for 0–15 cm topsoil for soil fertility categories derived using SOC in a Calcic Andosol and an Acric Ferralsol in Uganda

Experimental locations and soil classifications	Soil fertility categories based on SOC concentrations	pH	Sand	Silt	Clay	SOC	Total N	Available P (Bray 1)	Exchangeable cations			
									K	Ca	Na	Mg
			g/100 g soil					mg kg^{-1}	C mol (+) kg^{-1} soil			
Western Mid High Farmlands (WMHF)—Bukangara, Nyakiyumbu S/C (Calcic Andosol)	High fertility (> 1.9% SOC) (n = 8)	7.6 (0.59)	56 (4.5)	20 (3.7)	24 (2.5)	2.5 (0.27)	0.16 (0.03)	30.2 (14.2)	1.7 (1.3)	18.2 (6.2)	0.068 (0.01)	3.7 (0.7)
	Medium fertility (1.2–1.9% SOC) (n = 9)	7.2 (1.03)	64 (10.5)	19 (3.5)	17 (6.7)	1.604 (0.29)	0.14 (0.02)	25.6 (12.1)	1.0 (0.5)	16.1 (5.2)	0.062 (0.08)	3.8 (1.7)
	Low fertility (< 1.2% SOC) (n = 9)	7.2 (0.44)	60 (4.9.5)	23 (3.3)	17 (1.72)	1.05 (0.2)	0.11 (0.02)	24.8 (9.5)	1.2 (0.7)	12.6 (5.4)	0.068 (0.01)	3.2 (1.4)
South Western Highlands (SWH)—Kibuzigye, Bubaare S/C, Kabale district (Acric Ferralsol)	High fertility (> 1.7% SOC) (n = 11)	4.7 (0.43)	49 (11.2)	26 (4.3)	25 (9.2)	1.86 (0.21)	0.16 (0.01)	14.3 (9.1)	0.3 (0.2)	6.5 (1.7)	0.101 (0.02)	2.2 (0.4)
	Medium (1.2–1.7% SOC) (n = 11)	4.9 (0.33)	55 (8.1)	23 (4.4)	22 (5.6)	1.44 (0.16)	0.15 (0.03)	13.4 (6.9)	0.4 (0.2)	5.2 (2.4)	0.126 (0.02)	2.03 (0.5)
	Low (< 1.2% SOC) (n = 11)	4.4 (0.38)	59 (9.4)	20 (5.1)	21 (4.5)	0.99 (0.13)	0.09 (0.12)	9.4 (5.2)	0.5 (0.2)	5.8 (1.8)	0.105 (0.01)	1.89 (0.4)

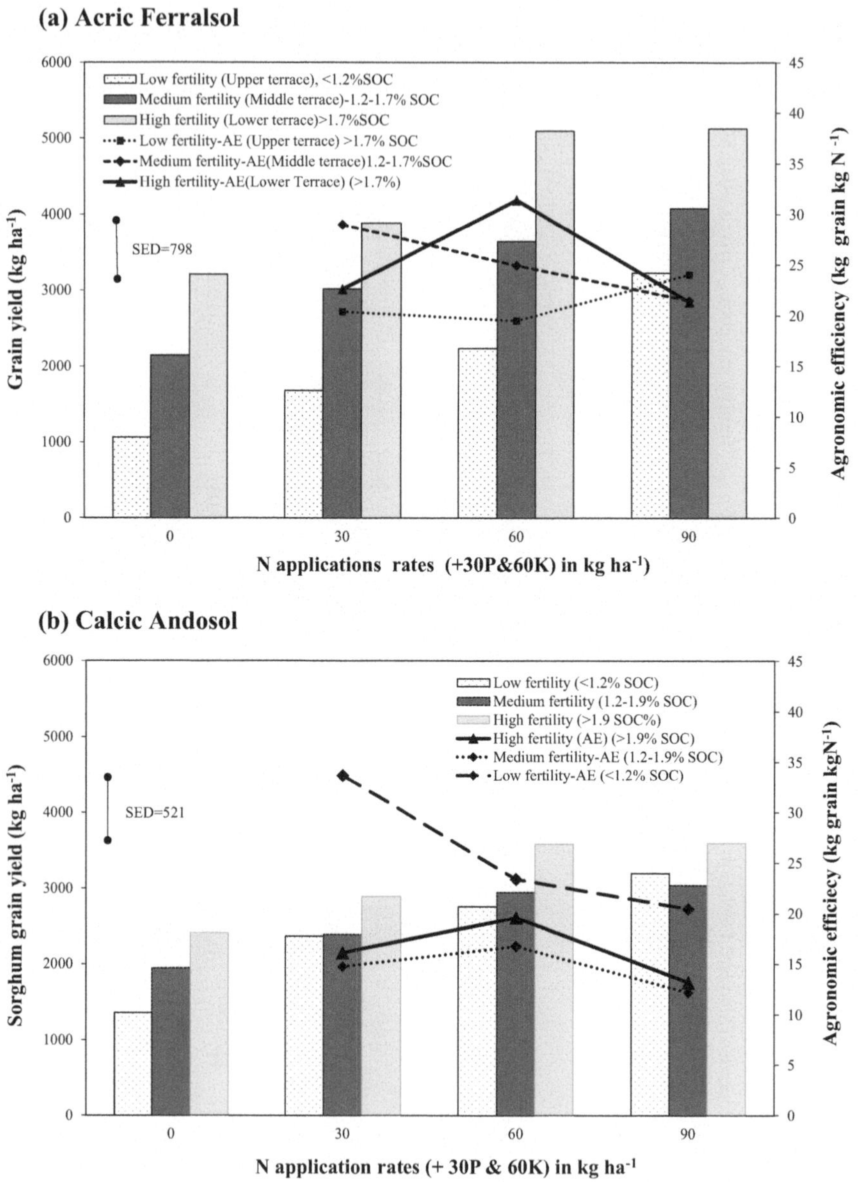

Fig. 13.2 Sorghum yield responses and agronomic efficiency to added N fertilizer in an Acric Ferralsol (**a**) and a Calcic Andosol (**b**)

were registered to added fertilizer in low SOC fields (< 1.2%). The yield was higher than what was observed in low SOC in an Acric Ferralsol (MWHF). No significant differences were observed between the two seasons, although the MWHF was characteristic of a uni-modal like precipitation pattern. No significant interactions at $p < 0.05$ between season and SOC categories were observed. In fields of low SOC, 30 kg N ha^{-1} resulted in the highest agronomic efficiency gain. In medium to high SOC (>1.2%), grain yields significantly increased with 60 kg N ha^{-1} (evident with high AE) as compared to the control ($p < 0.05$). Notably, there was a weak yield response to high N rate (i.e., 90 kg N ha^{-1}), for medium and high SOC fields.

In an Acric Ferralsol, an analysis of the relationship between yields and SOC concentrations under different N applications showed positive regression coefficients ranging from 0.59 to 0.79 (Fig. 13.3a, b). Yield increase per unit change in SOC concentrations was as high as 4055 and 4009 kg ha^{-1} under 60 and 90 kg N ha^{-1} respectively. Sorghum yield (top biomass and grain) was remarkably increased on changing SOC concentrations, particularly with added N fertilizer. In a calcic Andosol, a regression analysis between yield and SOC, under different N fertilizer rates showed significant positive regression coefficients ($R^2 > 0.6$). Grain yield increase of 582–1257 kg ha^{-1} per every unit change in SOC was registered for 0 and 30 kg N ha^{-1} (Fig. 13.4b). Biomass also increased with increase to N application rates. The yield gains was clearly different in the two soil types with higher yield gains per unit change in SOC observed in an Acric Ferralsols compared to a Calcic Andosol.

13.4 Discussion

Sorghum response to N fertilizer application was clearly different in the two AEZs with high yield gains per unit increase in SOC stronger in an acidic soil than in an alkaline (Calcic) soil. This is a contradicting observation since soils characteristic of moderately low pH exhibited high response as opposed to high pH soils. This can be attributed to the fact that when such a low fertility soil was alleviated with P and K fertilizers; there was a strong synergistic interaction with climate, texture, SOC quality, management and cultivar perhaps leading to high N use efficiency and yield. It also became apparent from this study that the optimum pH for sorghum may vary with attributes in an AEZ and type of cultivar. The local cultivar (*Mabere*) proved to be more tolerant to low pH soils than the *Sekedo* variety. This suggests existence of acid tolerant cultivars to pH that can be as low as 4.4–4.9, making *Mabere* cultivar a potential candidate for further research in breeding. Observed yields in an acidic soil were clearly influenced by SOC concentration and added mineral N interactions, as reflected by the significant increase in sorghum grain yield from 1 to 5 t ha^{-1}. Such strong grain yield responses confirmed N as a major limitation in crop production in Uganda and affirms SOC as good surrogate for N management (Kaizzi et al. 2006; Musinguzi et al. 2013).

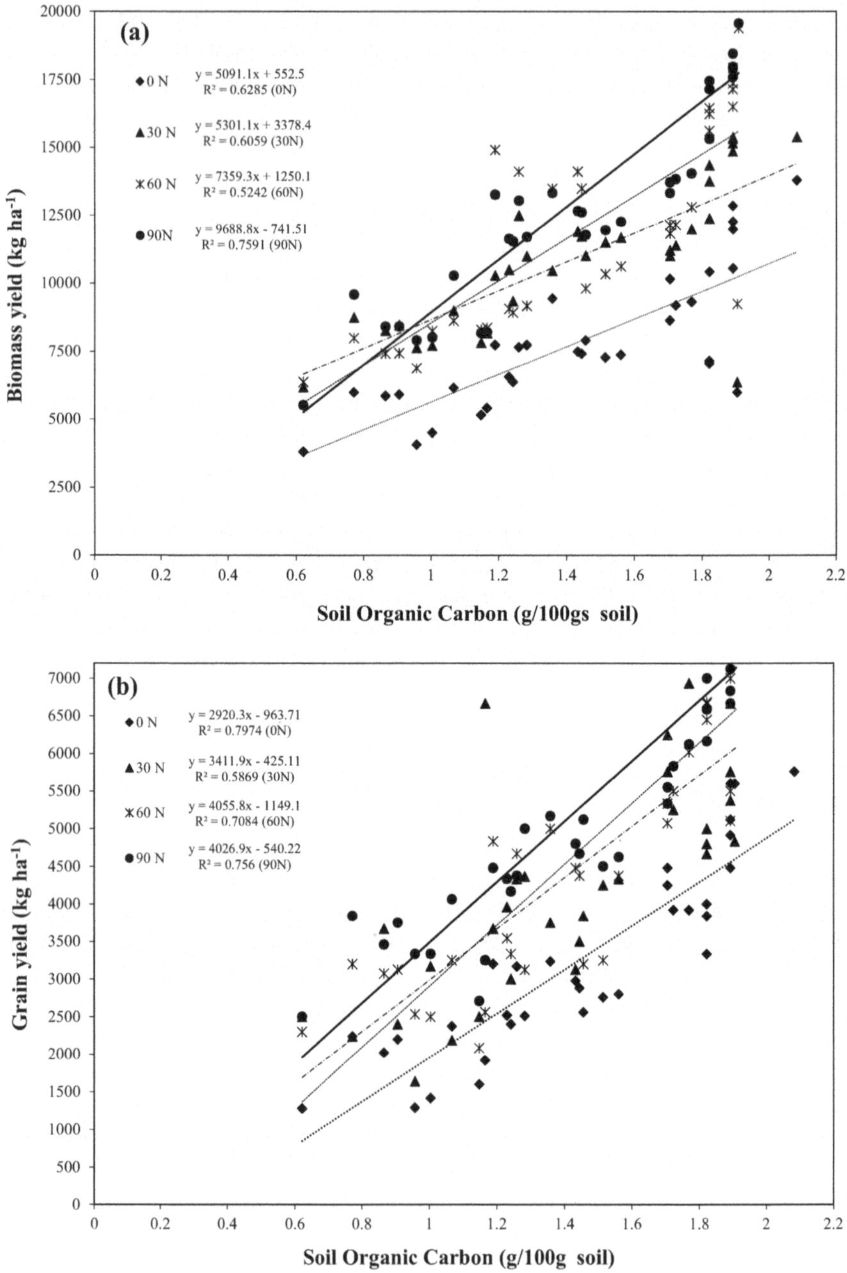

Fig. 13.3 Sorghum responses for biomass (**a**) and grain (**b**) to N fertilizers on soils of varying SOC levels with P + K applied to alleviate nutrient limiting conditions in an Acric Ferralsols

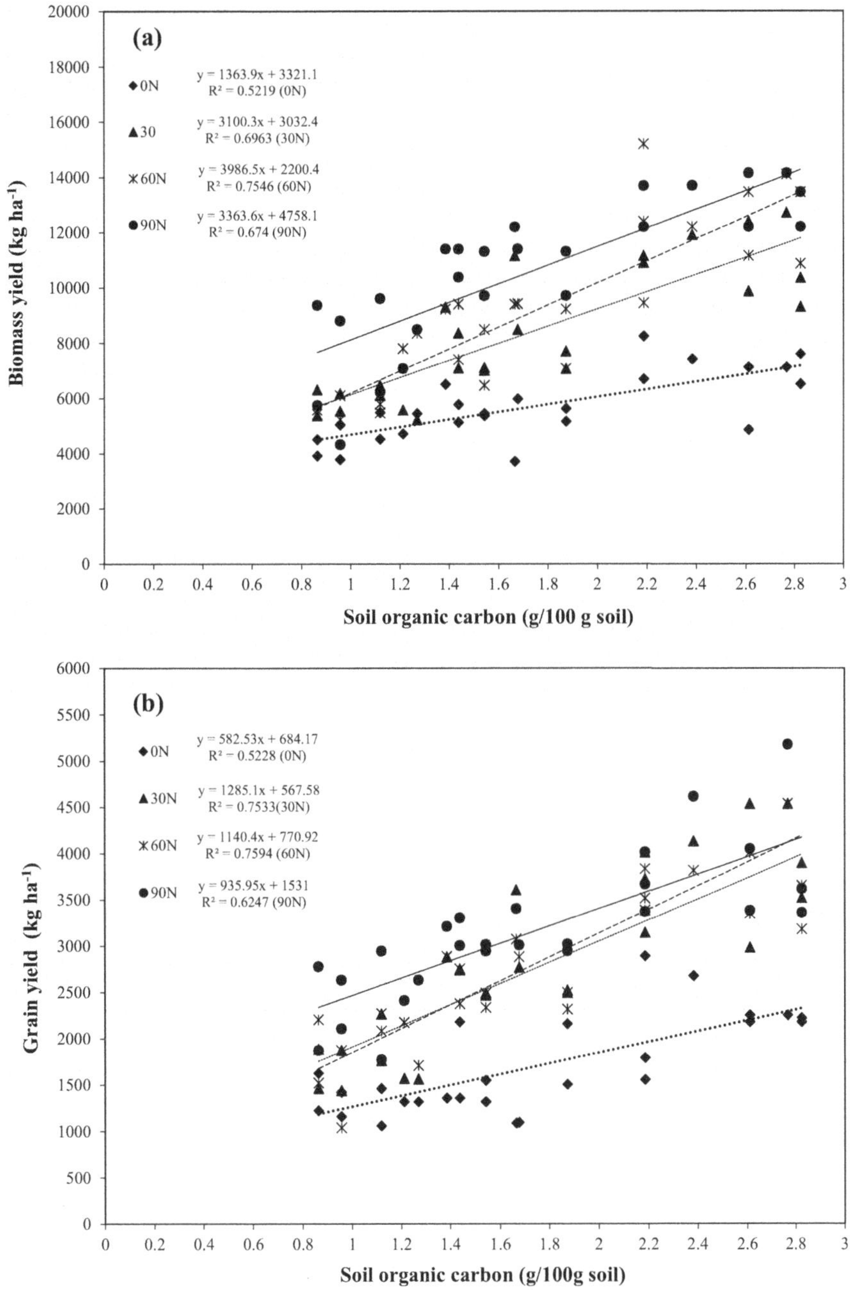

Fig. 13.4 Sorghum biomass (**a**) and grain (**b**) yield responses to N fertilizers on soils of varying SOC levels with P + K applied to alleviate limiting conditions in a calcic Andosol

High yield response to 30 and 60 kg N ha^{-1} in acidic soils with SOC > 1.2% demonstrated positive integrative benefits of SOC in enhancing soil quality and yields. In fields with SOC > 1.7%, significant yield gains and highest AE signaled high integral benefits of SOC to soil quality and in synchronizing positively with applied N. High N fertilizer rates for better yield was evident in low SOC, suggesting that low fertility fields are nutrient demanding and require more resources (monetary) to boost yield than high SOC fields. Despite N fertilization, low SOC fields could not provide the desired yield returns compared to high SOC fields. Such soils can be associated with poor physical, chemical and biological properties which can reduce N fertilizer effectiveness and its retention capacity, consequently influencing yield (Carter et al. 2003; Musinguzi et al. 2013).

Although SOC proved to be a valuable soil parameter in influencing yields, other soil attributes associated with the SWH agro-ecological zone cannot be underrated. Slope positions along a sloping bench terrace have been earlier reported as key factors that orchestrate soil erosion and high nutrient variability (Musinguzi et al. 2010). Upper terrace positions (on a sloping terrace) are known to have poor soil physical and chemical properties as compared to soils on the lower terrace position (Musinguzi et al. 2010). As such, soils in lower terrace position are associated with high SOC and improved soil properties such as silt + clay and low bulk density. Such soil environment enabled a positive interaction of N with SOC concentrations resulting in commendable yield responses in a low pH soil. However, it would be important for future studies to explore minimum threshold value of SOC below which yield response is not possible.

In a high pH Calcic Andosol, fields with SOC > 1.2% responded to N application resulting in highest yield. Soil organic carbon (> 1.2%) interacted positively with 60 kg N ha^{-1} showing the indisputable role of high SOC in enhancing crop productivity. High SOC and high soil pH are clearly important in tropical soil for improving nutrient availability and micro-biota activities. Calcic Andosols are also associated with soil attributes such as accumulated secondary carbonates, high phosphate retention capacity (due to active Al and Fe oxides), low bulk density, good root systems, good water storage properties, and high level of short-order range minerals (IUSS Working Group 2006). In case phosphorus is ameliorated, Andosols (as shown in this study) can potentially boost crop productivity within 1.2–1.9% SOC. The synergistic interaction between SOC and added N proved ideal to favour nutrient availability and nutrient uptake, as reflected by high AE. High fertility soil (> 1.9% SOC) were not very responsive to 90 kg N ha^{-1}, probably because the soil was biologically active resulting in high mineralization and N release, which influenced added N effectiveness. The soil possibly had reached a critical level for no response to added to N fertilizer. Such processes may not easily take place in acidic soils. The low response to applied N at high SOC (> 1.9%) suggests existence of maximum thresholds for effective N response in Andosols. Other AEZ attributes such as cumulative rainfall was different across the two seasons for MWHF zones (Fig. 13.1), but small yield differences suggested sufficient rainfall for sorghum. The two seasons might have had sufficient soil water content to support sorghum production, since it is also known to be a drought tolerant crop.

The two soils in different AEZs demonstrated differences in yield response to N fertilizer. Factors such as climate, type of cultivar, landscape features, texture, pH proved to be critical in understanding N response and yields in contrasting zones. In comparison with other studies in Uganda, grain yield obtained in these soils were higher than 2.27 t ha^{-1}, which was previously reported (Kaizzi et al. 2012). However, AE in these soils was lower than what was also reported by Kaizzi et al. (2012). Sorghum biomass (dry matter) was as high as 19 t ha^{-1} in an Acric Ferralsols, and 15 t ha^{-1} in a Calcic Andosol which were higher than reported yield in Uganda of about 9.84 t ha^{-1} on soils of low SOC and critically low P in North-Western and Eastern parts of Uganda (Kaizzi et al. 2012). The observations suggest the need to map soil responsiveness to applied N for different regions to optimize nutrient use, with much focus on SOC attributes of the different AEZs. The need for adopting site specific nutrient management approaches was apparent from this study. A specific nutrient management regime with emphasis on targeting right amounts of fertilizer to soils of similar attributes is inevitable as reported by Zingore et al. (2011). When comparing on-farm yields observed in this study to on-station sorghum yield at research stations, boosting soil to critical SOC concentrations under proper N application can potentially result in desired metric tons of grain, for cases where P and K limitations are alleviated.

13.5 Conclusion

An Acric Ferralsol was clearly more responsive to nitrogen (N) fertilizers compared to the Calcic Andosol. High soil fertility status characteristic of high soil organic carbon (SOC) concentrations played a substantial role in influencing high yield response and agronomic use efficiency (AE) to applied N in the different soil types. Fields with low SOC were responsive in the two soil types but did not translate in optimal yield gains obtained in high SOC. Soil organic carbon ranges above 1.2% for both soils were generally highly responsive to added N. Categorization of soil using SOC demonstrated its importance in understanding N responses in heterogeneous tropical farming systems. High yield gains obtained on adding N fertilizer in a relatively fertile soil suggested the need for building SOC to critical concentrations for higher returns in sorghum production. High SOC associated with very fertile soils such as Andosols can limit crop response to applied N fertilizer suggesting existence of maximum thresholds in some soil types.

Acknowledgements We are very grateful to World Phosphate Institute and Makerere University for funding this study. Special regards to the farmers, the research assistants and field-based facilitators for taking up the lead role in management of experimental trials.

References

African Fertilizer Summit. (2006). *Africa Fertilizer Summit proceedings.* Summit held in Abuja, Nigeria, 9–13 June 2006. USA, Muscle Shoals: IFDC.

Anderson, J., & Ingram, J. (1993). *Tropical soil biology and fertility: A handbook of methods.* Wallingford, UK: CAB International.

Andrews, S. S., Karlen, D. L., & Cambardella, C. A. (2004). The soil management assessment framework. *Soil Science Society of America Journal, 68*(6), 1945–1962.

Bekunda, M. A., Bationo, A., & Ssali, H. (1997). Soil fertility management in Africa: A review of selected research trials. *Soil Science Society of America Special Publication, 51,* 63–80.

Carter, M., Angers, D., Gregorich, E., & Bolinder, M. (2003). Characterizing organic matter retention for surface soils in eastern Canada using density and particle size fractions. *Canadian Journal of Soil Science, 83*(1), 11–23.

Cassman, K. G., Dobermann, A., & Walters, D. T. (2002). Agroecosystems, nitrogen-use efficiency, and nitrogen management. *Ambio, 31*(2), 132–140.

IUSS (International Union of Soil Scientists) Working Group. (2006). *World Reference Base for Soil Resources (WRB).* World Soil Resources Report, 103. Rome, Italy: FAO.

Kaizzi, C. K., Ssali, H., & Vlek, P. L. (2006). Differential use and benefits of Velvet bean and N fertilizers in maize production in contrasting agro-ecological zones of E. Uganda. *Agricultural Systems, 88*(1), 44–60.

Kaizzi, K. C., Byalebeka, J., Semalulu, O., Alou, I., Zimwanguyizza, W., Nansamba, A., et al. (2012). Sorghum response to fertilizer and nitrogen use efficiency in Uganda. *Agronomy Journal, 104*(1), 83–90.

Ladha, J. K., Pathak, H., Krupnik, T., Six, J., & van Kessel, C. (2005). Efficiency of fertilizer nitrogen in cereal production: Retrospects and prospects. *Advances in Agronomy, 87,* 85–156.

Mukashema, A. (2007). *Mapping and modelling landscape-based soil fertility change in relation to human induction. Case study: Gishwati Watershed of the Rwandan highlands.* The Netherlands: ITC.

Musinguzi, P., Tenywa, J. S., & Bekunda, M. (2010). Strategic nutrient management of field pea in southwestern Uganda. *African Journal for Food, Agriculture, Nutrition, and Development, 10,* 2695–2706.

Musinguzi, P., Tenywa, J. S., Ebanyat, P., Tenywa, M. M., Mubiru, N. D., Twaha, A. B., et al. (2013). Soil organic carbon thresholds and nitrogen management in tropical agroecosystems: Concepts and prospects. *Journal of Sustainable Development, 6*(12), 31–43.

Musinguzi, P., Ebanyat, P., Tenywa, J. S., Mwanjalolo, M., Basamba, T. A., Tenywa, M., et al. (2014). Using DSSAT-CENTURY Model to simulate soil organic carbon dynamics under a low-input maize cropping system. *Journal of Agricultural Sciences, 6*(6), 120–131.

Nkonya, E., Kaizzi, C., & Pender, J. (2005). Determinants of nutrient balances in a maize farming system in eastern Uganda. *Agricultural Systems, 85*(2), 155–182.

Sanchez, P. A., Shepherd, K. D., Soule, M. J., Place, F. M., Buresh, R. J., Izac, A.-M. N., et al. (1997). Soil fertility replenishment in Africa: An investment in natural resource capital. *Replenishing Soil Fertility in Africa, 6,* 1–46.

Sanchez, P. A., Denning, G. L., & Nziguheba, G. (2009). The African Green Revolution moves forward. *Food Security, 1*(1), 37–44.

Sutton, M. A., Bleeker, A., Howard, C., Bekunda, M., Grizzetti, B., Vries, W. D., et al. (2013). *Our Nutrient World: The challenge to produce more food and energy with less pollution.* Edinburgh: Centre for Ecology and Hydrology.

Vanlauwe, B., Kihara, J., Chivenge, P., Pypers, P., Coe, R., & Six, J. (2011). Agronomic use efficiency of N fertilizer in maize-based systems in sub-Saharan Africa within the context of integrated soil fertility management. *Plant and Soil, 339*(1–2), 35–50.

WRB. (2006). World reference base for soil resources 2006. *A framework for international classification, correlation and communication.* World Soil Resources Reports 103. Rome: FAO.

Wortmann, C. S., Mamo, M., Mburu, C., Letayo, E., Abebe, G., Kayuki, K. C., et al. (2009). *Atlas of Sorghum (Sorghum bicolor (L.) Moench): Production in Eastern and Southern Africa*. University of Nebraska-Lincoln, USA: INTSORMIL CRSP.

Zingore, S., Tittonell, P., Corbeels, M., Van Wijk, M., & Giller, K. (2011). Managing soil fertility diversity to enhance resource use efficiencies in smallholder farming systems: A case from Murewa District, Zimbabwe. *Nutrient Cycling in Agroecosystems, 90*(1), 87–103.

Chapter 14
Evaluating Resource Use Efficiency and Stock Balances of Nitrogen and Phosphorus Fertilizer Inputs: The Effect of Soil Supply Capacity in Tigray (Ethiopia)

Richard G. Kraaijvanger and Tom Veldkamp

Abstract In sub-Saharan Africa crop productivity is generally low, which affects food security and livelihoods. The application of mineral fertilizers in many cases is seen as a straightforward way to improve crop productivity. In Tigray, Northern Ethiopia, agricultural extension bureaus recommend the application of considerable amounts of fertilizers. Farmers, however, hesitate to adopt these recommendations and perceive that the use of fertilizers leads to "*addiction*". Different indicators are available to evaluate effectiveness of fertilizer application. We considered six different indicators: Agronomic Use Efficiency (AUE), Value-Cost-Ratio (VCR), Recovery Efficiency (RE), Capture Efficiency (CE), Soil Supply Capacity (SSC) and (partial) Nutrient Balances (NB). On-farm experiments were conducted for four years at 16 different locations. Crops involved were wheat, teff and hanfets. Experimental outcomes were evaluated using laboratory data on nitrogen, phosphorus and potassium (NPK) content of both soil and crops. Significant differences between the crops were found for CE, RE, NB and VCR. Wheat overall was found most extractive. Correlation between SSC and N-total and between RE and N-uptake was significant for all crops. For both nitrogen and phosphorus, NB correlated significantly with SSC for wheat and teff. Interaction between SSC, RE and NB demonstrated a significant trend for wheat: soils with higher SSC had lower NB and higher RE than soils with lower SSC. We concluded that achieving efficient use of mineral fertilizer goes at the cost of nutrient stock sustainability. The use of Integrated Soil Fertility Management-strategies is recommended to address these complex feedback

R. G. Kraaijvanger (✉)
Hogeschool van Hall Larenstein, University of Applied Sciences, P.O. Box 9001, 6880 GB Velp, The Netherlands
e-mail: richard.kraaijvanger@hvhl.nl

College of Dryland Agriculture, Mekelle University, P.O. Box 231, Mekelle, Ethiopia

T. Veldkamp
Faculty of Geo-Information Science and Earth Observation (ITC), Twente University, P.O. Box 217, 7500 AE Enschede, The Netherlands
e-mail: a.veldkamp@utwente.nl

M. A. Sutton et al. (eds.), *Just Enough Nitrogen*,
https://doi.org/10.1007/978-3-030-58065-0_14

and interaction mechanisms and to arrive at a sound balance between efficiency and sustainability of fertilizer use.

Keywords Sustainability · Resilience · Fertilizer use efficiency · Nutrient balances · Tigray

14.1 Introduction

In sub-Saharan Africa crop productivity is low and in many cases affecting food security and consequently livelihoods of rural communities. Many reasons are indicated for these low levels of crop yield, one of them being nutrient availability. The observation of systematic macro-level nutrient depletion (Stoorvogel et al. 1993; Sanchez and Swaminathan 2005) was the starting point for a series of initiatives to mitigate this depletion at site level (Jama and Pizarro 2008; Vanlauwe et al. 2010). A relatively simple and straightforward way to deal with depletion and to increase crop productivity is the application of mineral fertilizers. Sub-Saharan soils, however, are not always that responsive to fertilizer inputs. This lack of response often is attributed to depleted soils, probably also in combination with fixation of specific nutrients. Lack of response also might relate to a relatively high nutrient content of some soils; additional nutrients then are not required (Vanlauwe et al. 2011). In addition, factors like erratic rainfall or low levels of crop management interfere with successful response towards fertilizer application.

In Tigray, our study area, local extension bureaus recommend considerable quantities of mineral fertilizer to increase productivity. Farmers, however, frequently hesitate to adopt these recommendations due to various reasons, ranging from risk perception and high cost of fertilizer to an observed lack of response to fertilizer application (Kraaijvanger and Veldkamp 2015). Furthermore, a frequently heard statement of farmers in relation to the use of fertilizer is that it creates "*addiction*". In farmers' words: "*if mineral fertilizer was used one year, using it again in the next year will be needed to get acceptable produce*". Tittonell (2007) made a similar observation in Western Kenya. To these farmers it appears that applying fertilizers (as a curative action) results in a reduced capacity of the soil to supply nutrients. This observation strongly contrasts with the residual effect farmers normally observe in case of using traditional practices like applying manure or other organic fertilizers.

A wide range of indicators is available to evaluate the effectiveness of fertilizer application. Each indicator addresses a specific concern, mostly in relation to quantifying use efficiency or stock balance for specific nutrients. Agronomic Use Efficiency (AUE) deals with the increase in productivity as a result of the application of additional nutrients (Vanlauwe et al. 2011). It is an instrument to evaluate the agronomic efficiency of fertilizer use and to estimate optimal application rate. Value-cost ratio (VCR) is an indicator related to AUE and includes the costs (of fertilizer inputs) and revenues in order to estimate economic efficiency (Donovan et al. 1999).

Recovery Efficiency (RE) or Recovery Fraction relates to the fraction of applied nutrients that is returned in the harvested crop (Haefele et al. 2003; Chikowo et al. 2010) and is an example of a crop-based indicator. The amount of nutrients found in the harvested product compared to the total uptake is expressed in the so-called Capture Efficiency (CE), which is again a crop-based indicator addressing recovery (Chikowo et al. 2010). Partial and Full Nutrient Balance (NB) approaches (Stoorvogel et al. 1993; Haileslassie et al. 2005) go beyond fertilizer application and are used to estimate long-term changes in nutrient stocks.

Internal Nutrient Efficiency (INE) focuses on the quantity of grain that is produced compared to the uptake of nutrients in the above ground parts and is crop based (Haefele et al. 2003). Indigenous Nutrient Supply (INS) is defined as the uptake of a specific nutrient from (unfertilized) control plots, provided that other nutrients are not limiting (Haefele et al. 2003). Indigenous Nutrient Supply is an example of a soil-based indicator. Kurwakumire et al. (2014) used relationships based on soil properties like pH and Soil Organic Carbon (SOC) to estimate the value of INS for specific soils. Nutrient Uptake Efficiency is defined as the ratio of actual recovery and potential recovery in case all conditions are optimal, its scope essentially being agronomic, focusing on closing yield gaps (Janssen 1998).

Two related concepts are those of residual recovery and nutrient supply equivalents (Janssen 2011). Residual recovery quantifies the effect of application of fertilizer in the next growing season and is especially important in relation to phosphorus. Nutrient supply equivalents aim at a balanced supply of nutrients which is assumed to result in efficiency. Residual recovery is (soil) supply based, whereas supply equivalents stress the crop-physiological importance of more than one nutrient.

All above indicators underline that both resource use efficiency and sustainability are main concerns (van Noordwijk and Brussaard 2014). The attention given to various forms of efficiency is legitimate given the scarcity of resources and the ambition to feed the future world population. In the past, efficiency was merely explained in terms of maximum profit and productivity, resulting in high-input low-efficiency agriculture, which suited the motives of individual farmers. At present, simple demand-supply reasoning is no longer in all cases considered appropriate to address the challenges of the global community. In response to this, sustainability became more and more in vogue in the past 20 years. The nature of this sustainability is agronomic rather than environmental and is in relation to nutrients often connected to nutrient balances (De Jager et al. 2001). In the context of low-input farming the capacity of the soil to supply nutrients strongly relates to sustainability and is, for example, expressed in INS. In low-input high-efficiency agriculture (Koohafkan et al. 2012), soil nutrient buffers play a prominent role in supporting productivity and resilience of the involved agricultural systems.

Based on the outcomes of four years of on-farm experimentation in Tigray we explored the complex trade-offs and interactions between the concerns of efficiency, nutrient balances and soil supply capacity. To obtain a holistic picture we calculated six different indicators relating to resource use efficiency and nutrient stock for three main food crops in Tigray. Our focus was on nutrient supply capacity, which is a relatively new concept that links soil and crop and allows evaluations in time.

In addition, we commented on the "*addiction*" statement of the farmers and the remarkable sustainability of the traditional agricultural system in northern Ethiopia (Kraaijvanger and Veldkamp 2015).

14.2 Methods and Materials

14.2.1 Introduction

On-farm experiments were conducted in 16 different neighbourhoods for four consecutive years. In most cases, however, the actual location of an experimental site changed within these four years. In this chapter we included 37 experimental sites, considering only the first year of use as an experimentation site. The test crops were wheat, teff and hanfets. Hanfets is a traditional mixture of barley and wheat in variable ratios (Woldeamlak et al. 2001). In our research set up we challenged farmer groups to design their own experiments. In addition, we included replicated control treatments (unfertilized) and treatments with recommended fertilizer application of diammonium phosphate (DAP) and urea. Details are provided in Kraaijvanger and Veldkamp (2015). We used experimental outcomes in terms of yield, in combination with laboratory data on nitrogen, phosphorus and potassium (NPK) content of soil and crops to explore behaviour of soil and crops under fertilized and unfertilized conditions. In order to evaluate this behaviour we used six different indicators relating to uptake and source of nutrients used, with a focus on total uptake related to above ground biomass production (i.e., grain and straw). Nutrient uptake was calculated by multiplying crop produce with crop nutrient content.

All indicators considered were directly derived or slightly adapted from literature sources and defined as follows:

Agronomic Use Efficiency (AUE)
This indicator considers the effect of the applied fertilizer in terms of additional produce as compared to a control situation without nutrient inputs (Vanlauwe et al. 2011):

$$\mathrm{AUE} = (Y_\mathrm{t} - Y_\mathrm{c})/N_\mathrm{t} \tag{14.1}$$

Y_c=total dry matter produce control plots (kg/ha)
Y_t = total dry matter produce fertilized plots (kg/ha)
N_t = fertilizer nutrient input (kg/ha).

Soil Supply Capacity (SSC)
This indicator indicates the amount of nutrients that are supplied by the soil in case no fertilizer inputs are provided and represents the capacity of the soil to provide nutrients. SSC resembles Indigenous Nutrient Supply (INS), as was described by

Haefele et al. (2003). The difference is that INS presumes that the nutrient in question is limiting (and other nutrients are sufficient in supply). To estimate SSC, dry matter produce and crop nutrient content of the control plots were multiplied:

$$\mathrm{SSC} = N_{\mathrm{s}} = Y_{\mathrm{c}} * f_{\mathrm{c}} \quad (14.2)$$

N_{s} = soil nutrient supply (kg/ha)
f_{c} = nutrient content of produce control plots (fraction).

Recovery Efficiency (RE)
Resource Efficiency covers the efficiency of crops to use fertilizer resources. In this indicator uptake of nutrients of the above ground parts is compared to the input of fertilizer (Chikowo et al. 2010):

$$\mathrm{RE} = (Y_{\mathrm{t}} * f_{\mathrm{t}})/N_{\mathrm{t}} \quad (14.3)$$

f_{t} = nutrient content of produce fertilized plots (fraction).

Capture Efficiency (CE)
Capture Efficiency relates the uptake of nutrients to the total supply of nutrients by inputs (of fertilizer) and by the soil (Chikowo et al. 2010). In our case supply is estimated by considering SSC and direct fertilizer inputs:

$$\mathrm{CE} = Y_t * f_t / (N_t + N_s) \quad (14.4)$$

Partial Nutrient Balance (NB)
In a partial nutrient balance, output and input are compared to assess the possible risk for depletion (Haileslassie et al. 2005). In our case we considered nutrients contained in crop and straw:

$$\mathrm{OUTPUT} - \mathrm{INPUT} = \mathrm{NB} = (Y_t * f_t) - N_t \quad (14.5)$$

Value-Cost-Ratio (VCR)
To calculate a Value-Cost-Ratio the value of the additional produce resulting from fertilizer application is compared to the cost of this applied fertilizer (Donovan et al. 1999). In our case we took as a cost for the applied fertilizer 2500 ETB for the recommended 200 kg. The revenue from 1 kg was estimated 5 ETB kg^{-1} for wheat and hanfets and 8 ETB kg^{-1} for teff (data for 2013; ETB = Ethiopian *birr*; 25 ETB = 1 US$):

$$\mathrm{CR} = (Y_t - Y_c) * V_y / V_t \quad (14.6)$$

V_y = revenue produce (ETB/kg)
V_t = total cost fertilizer input (ETB).

14.2.2 Laboratory Analysis

For both fertilized and control plots, composite samples of the harvested parts (grains and straw) were analysed in the first experimentation year for total nitrogen (N), phosphorus (P) and potassium (K) content using wet destruction. In order to reduce costs, the number of laboratory analysis was restricted to three representative sites for wheat and hanfets and to two representative sites for teff. Averages for wheat, hanfets and teff were used to calculate the different indicators. Composite samples of the top soil (0–20 cm) of each experimental site were analysed for total N (Kjeldahl method), available P (Olsen method) and exchangeable K (ammonium-acetate extraction). Total N, available P and exchangeable K relate to medium term availability of respectively N, P and K.

14.2.3 Statistical Analysis

Means, standard deviations and coefficients of variation were calculated for both the yields observed in the specific sites and the outcomes of the crop analysis. Analysis of variance was used to evaluate differences between the crops for the specific indicators. In addition, correlations between specific variables were calculated. All statistics were conducted using MS-Excel.

14.3 Results and Discussion

Yields observed over the four experimentation years varied considerably and appeared to be site specific (Table 14.1). In general (grain) yields for wheat were highest. The application of recommended amounts of mineral fertilizers resulted in an increased nitrogen content for wheat, hanfets and teff (Table 14.2). With respect to phosphorus and potassium, differences between recommended application and controls were much less. In addition, straw contained a surprising high content of potassium in comparison to grains.

As expected, different crops responded differently with respect to nutrient uptake (Table 14.3). Wheat can be considered quite extractive; teff at the other hand is relatively mild in that respect. Differences between the crops were significant for the crop physiological indicators Recovery Efficiency (RE) and Capture Efficiency (CE) (for both N and P), for the environmental indicator Nutrient Balance (for both N and P) and for the economic indicator Value Cost Ratio (VCR). For the soil-based

Table 14.1 Average grain and straw production of fertilized and control plots for the sites involved (cv = coefficient of variation; n = number of plots)

			Control				Recommended fertilizer application (N and P)[a]			
Site	Year	Crop	Average grain (kg/ha)	Average straw (kg/ha)	cv grain (%)	n	Average grain (kg/ha)	Average straw (kg/ha)	cv grain (%)	n
Adigudat	2010	Wheat	1235	3127	19.8	6	1810	4663	2.2	3
Adigudat	2011	Wheat	1289	3988	14.6	2	2757	9360	13.5	4
Adigudat	2012	Wheat	975	1561	38.2	2	2588	3930		1
Adowro	2011	Wheat	1516	3571	18.9	3	3238	7245	16.8	3
Adowro	2013	Wheat	3439	4536	19	2	3549	5184		1
Awadu	2012	Wheat	874	1570	19.2	2	757	1587		1
Awadu	2013	Wheat	1712	2388	5.9	3	2478	2916		1
Biherawi	2011	Wheat	2137	4783	7.7	2	2139	6747	7.4	3
Biherawi	2013	Wheat	1381	1835	14.6	3	2709	3616		1
Dingelat	2010	Wheat	2077	2628	41.6	6	2348	2878	26.3	4
Dingelat	2013	Wheat	1053	1622	13.1	3	2322	3893	18.2	4
Endamariam	2010	Wheat	1105	1750	35.1	4	2010	3667	10.6	3
Gudowro	2010	Wheat	2397.5	3180	9.9	4	3720	5050	14.9	4
Gudowro	2012	Wheat	2307	3194	13.5	2	2906	3853	20.6	4
Machalawi	2010	Wheat	966	1766	15.2	5	1157	3090	17	6
Machalawi	2012	Wheat	1536	1973	13.2	3	1708	2651	16.1	2
Mayzagra	2011	Wheat	1956.2	4269	2.2	2	3061	6524	32.8	3
Mayzagra	2012	Wheat	2470.5	3658	45.2	2	2905	4179	13	2
Mayzagra	2013	Wheat	1157.3	1900	17.4	3	2243	3745	15.6	3
Mymisham	2011	Wheat	1350.9	2554	35.4	3	2936	7227	17.8	3

(continued)

Table 14.1 (continued)

			Control				Recommended fertilizer application (N and P)[a]			
Site	Year	Crop	Average grain (kg/ha)	Average straw (kg/ha)	cv grain (%)	n	Average grain (kg/ha)	Average straw (kg/ha)	cv grain (%)	n
Mymisham	2013	Wheat	1757.7	2235	24.9	3	3903	5063		1
Siluh	2010	Wheat	1676	2424	9.5	5	2260	4103	6	3
Siluh	2013	Wheat	935.7	1374	6.1	3	1759	3155	3.7	3
Tikuz	2013	Wheat	1111	2554	36.3	3	1554	5413		1
Zalaweni	2010	Wheat	1596.7	2127	33.2	3	1243	1908	5.4	4
Zalaweni	2013	Wheat	1046.3	1289	31.4	3	2135	2539		1
Adigudat	2013	Teff	1112	2883	2.1	3	405	2781	0	1
Munguda	2010	Teff	780	2843	14	4	1038	4588	9.3	3
Munguda	2011	Teff	805	2831	12.6	2	1079	6104	7.8	4
Adowro	2010	Hanfets	1314	2734	18.5	7	2040	4010	10.6	3
Awadu	2010	Hanfets	678	2193	10.5	4	910	3610	20.5	3
Biherawi	2010	Hanfets	1215	2423	17	4	1713	3973	2.8	3
Endamariam	2013	Hanfets	752	2421	31.2	3	949	4245		1
Machalawi	2013	Hanfets	1436	1854	16.5	3	2014	2553		1
Mymisham	2010	Hanfets	1150	3870	41.3	3	1730	5678	18.7	4
Zonghi	2010	Hanfets	1135	3518	32.8	4	1288	4510	12.8	4
Zonghi	2011	Hanfets	670	2168	36	2	1497	5250		1

[a]Recommended application for all sites was 61 kg N and 24 kg P except for the Awadu, Mayzagra, Siluh and Zalaweni sites (41 kg N and 24 kg P)

Table 14.2 Nutrient composition of grain and straw for different crops (cv = coefficient of variation)

		Control						Recommended fertilizer application (N and P)[a]					
Crop	n	Average N (g/kg)	cv N (%)	Average P (g/kg)	cv P (%)	Average K (g/kg)	cv K (%)	Average N (g/kg)	cv N (%)	Average P (g/kg)	cv P (%)	Average K (g/kg)	cv K (%)
Wheat (grain)	3	16.3	13	4.2	11	4.8	11	18.7	17	4	5	5	0
Wheat (straw)	3	3.1	20	1	37	16	6	3.6	17	1.1	42	19	9
Teff (grain)	2	15.5	7	4.1	3	5	0	16.4	3	4.2	9	5	0
Teff (straw)	2	5.2	3	1.2	22	8.3	14	4.9	12	1.1	47	8	35
Hanfets (grain)	3	17.9	15	4.3	8	5.7	10	19.3	15	4.6	6	5.7	10
Hanfets (straw)	3	4.6	29	1.6	39	15	18	5.1	42	1.7	48	14	14

[a]Recommended application for all sites was 61 kg N and 24 kg P except for the Awadu, Mayzagra, Siluh and Zalaweni sites (41 kg N and 24 kg P)

Table 14.3 Calculated average values for different indicators to evaluate the effect of fertilizer inputs

Indicator type	Indicator acronym (units)	Concern	Wheat	Teff	Hanfets
Agronomic use efficiency	AUE-N total (kg/kg)	Agronomic	45.6	23.7	33.9
	AUE-P total (kg/kg)	Agronomic	286.2	189.2	239.4
Soil supply capacity	SSC-N total (kg/ha)	Soil (properties)	33.8	23.0	30.3
	SSC-P total (kg/ha)	Soil (properties)	9.2	5.5	8.7
Recovery efficiency	RE- N total* (%)	Crop (physiology)	110.2	48.0	81.6
	RE-P total* (%)	Crop (physiology)	60.0	30.3	59.0
Capture efficiency	N-CE* (%)	Crop (physiology)	67.2	34.9	54.6
	P-CE* (%)	Crop (physiology)	43.0	23.3	43.2
(partial) Nutrient balance	NB-N* (kg/ha)	Environment	−4.5	33.3	11.0
	NB-P* (kg/ha)	Environment	9.6	16.7	9.8
Value to cost ratio	VCR*	Economic	2.1	0.6	1.0

*Significant difference between the crops ($p < 0.05$)

indicator Soil Supply Capacity (SSC) and for the agronomic indicator Agronomic Use Efficiency (AUE) differences between the crops were not significant. Differences in uptake efficiency are important in crop rotations. Continuous cultivation of wheat will result in much more depletion than rotations including the sequence wheat-hanfets-teff. In such rotations wheat is the fertilized component and able to capture the nutrients supplied. Fertilizing teff doesn't seem to make much sense as recoveries can be below 50% (Table 14.3). In addition, teff in many cases might start lodging when it is fertilized too much. In traditional (unfertilized) rotations, teff often is followed by legumes to obtain a soil enriched with N for the next (wheat) crop.

SSC-N significantly correlated with total N, but SSC-P was not significantly related to available P (Figs. 14.1 and 14.2). Although both total N and available P are related to medium term availability, soil supply of P was not all determined by available P, while supply of N indeed related to total N. In the context of our case study (low-input systems with traditional management), N-uptake (under non-fertilized conditions) primarily depended on mineralization of organic N, which is a main factor determining total N. P-supply likely interacted with adsorption and fixation by different soil components and with the low solubility of P in the soil solution. In addition, P-supply in the context of Tigray will be limited by the short growing period (about 100 days). This resulted in P-uptake being almost independent of available P.

For the fertilized plots, significant positive relationships for both N and P were found between supply capacity of the soil and recovery of mineral fertilizer by the crop (Figs. 14.3 and 14.4). This indicated that nutrient recovery increased and the fertilizer supplied was used more efficiently. In the case of wheat, N-recovery of above 100% required a Soil Supply Capacity of about half of the total input of N

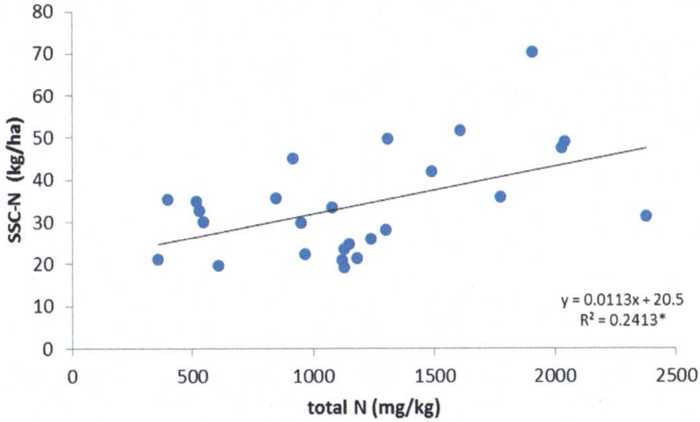

Fig. 14.1 Soil supply capacity (SSC) for N versus N-total for wheat (*indicates significant at $p =$ 0.05)

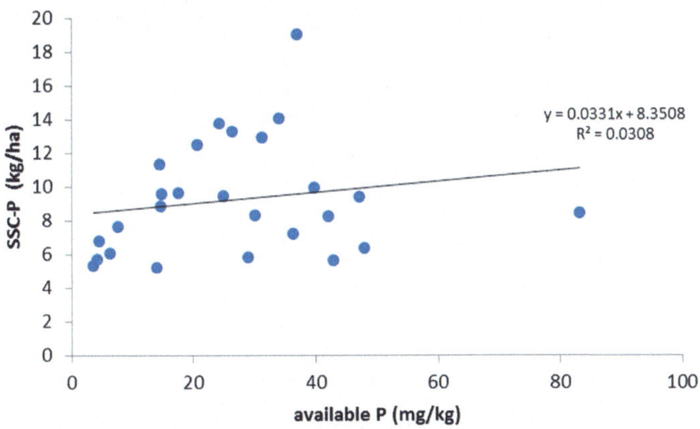

Fig. 14.2 Soil supply capacity (SSC) for P versus P-available for wheat

through fertilizer. The recovery for P increased and correlated with soil supply but never resulted in a recovery above 100%; P-applied consequently was not used fully. Low soil supply capacities apparently related to a higher probability for P-fixation; P being adsorbed rather than being used by the crop.

For wheat significant relationships were found between crop uptake and supply capacity of the soil (Fig. 14.5) and uptake above application level (41 and 64 kg/ha) in most cases was substantial. This indicated that the impact of soil supply on total uptake is important. Within our range of observations, contributions from the soil sometimes even exceeded fertilizer inputs.

For wheat and teff, partial nutrient balances (for N) demonstrated a significant negative relationship with Soil Supply Capacity (Fig. 14.6). For wheat, a higher soil supply resulted overall in negative balances: the presence of more easily available

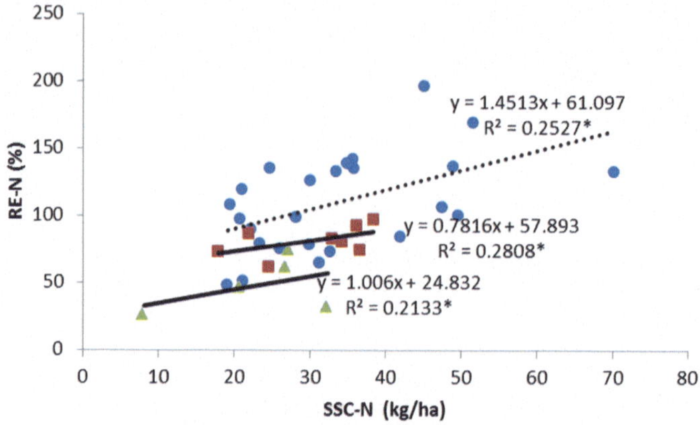

Fig. 14.3 Recovery efficiency (RE) versus soil supply capacity (SSC) for nitrogen for three grain crops (circles = wheat; squares = hanfets; triangles = teff; * = significant at $p = 0.05$)

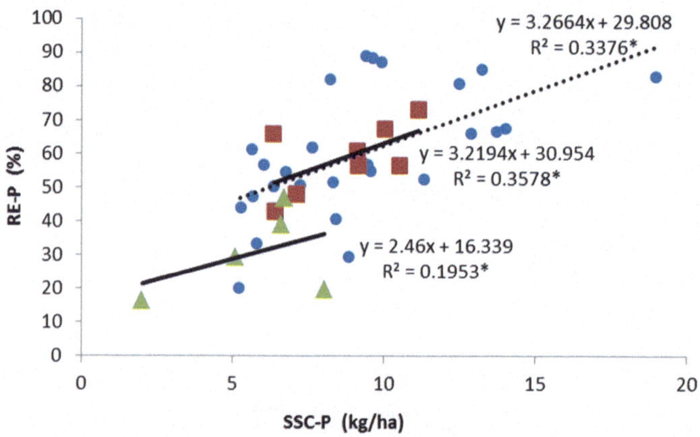

Fig. 14.4 Recovery efficiency (RE) versus soil supply capacity (SSC) for phosphorus for three grain crops (circles = wheat; squares = hanfets; triangles = teff; * = significant at $p = 0.05$)

nutrients in the soil, in combination with fertilizer input, apparently led to a higher level of extraction. This stronger extraction might be related with the promotion of root development and decomposition of organic matter by fertilizer inputs. In about half of the cases, nutrient balances were negative for wheat, despite the use of fertilizers. Wheat, as mentioned before, had a strong ability to extract nutrients. For both hanfets and teff, (partial) nutrient balances did not become negative within the range observed. These crops clearly were much less extractive.

Plotting partial nutrient balances and resource use efficiency (in terms of RE) for N and P resulted in two observations (Figs. 14.7 and 14.8):

(1) Higher Supply Capacities resulted in lower nutrient balances and higher resource use efficiencies. Different ranges of Soil Supply Capacity (for both N and P)

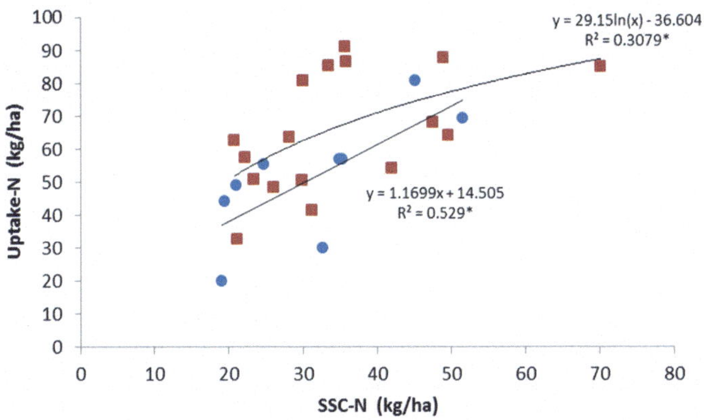

Fig. 14.5 Uptake of nitrogen versus soil supply capacity (SSC) for two different nitrogen input levels (circles = 41 kg N/ha; squares = 64 kg N/ha; * = significant at $p = 0.05$)

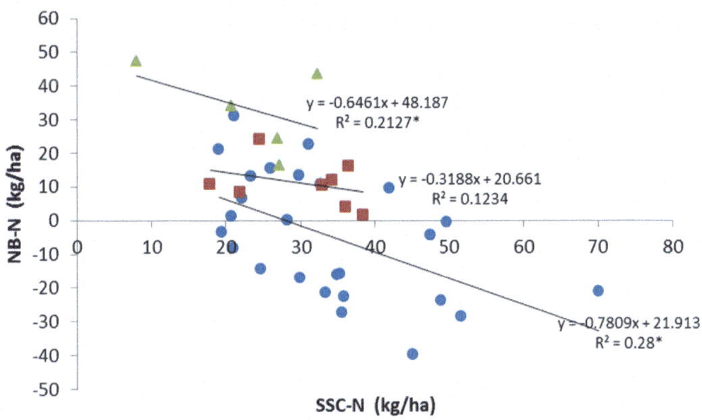

Fig. 14.6 Partial nutrient balances (NB) versus soil supply capacity (SSC) for nitrogen for different crops (circles = wheat; squares = hanfets; triangles = teff; * = significant at $p = 0.05$)

resulted in significantly different outcomes for nutrient balances and resource use efficiency. Consequently, soils with a high Soil Supply Capacity tended to deplete, even when recommended quantities of fertilizer were applied.

(2) Recovery Efficiency (RE) and (partial) Nitrogen Balance (NB) demonstrated a strong linear correlation. This correlation, however, related to the way RE and NB were calculated and the use of only two input levels for N and only one input level in the case of P. The strong intrinsic relation between both indicators is also demonstrated by the observation that in the case of N, a recovery of over 100% (automatically) resulted in a negative balance.

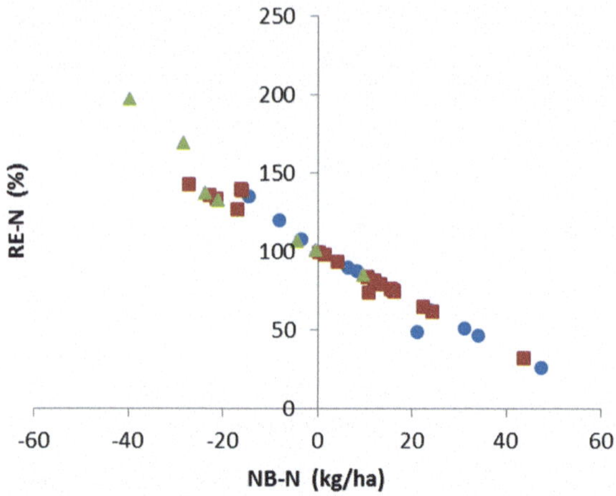

Fig. 14.7 Partial nutrient balances (NB) versus Recovery efficiency (RE) for nitrogen (wheat). ANOVA-difference between different ranges of SSC-N is significant at $p = 0.05$. (circles = SSC-N < 30 kg/ha; squares = SSC-N between 30 and 40 kg/ha; triangles = SSC-N > 40 kg/ha)

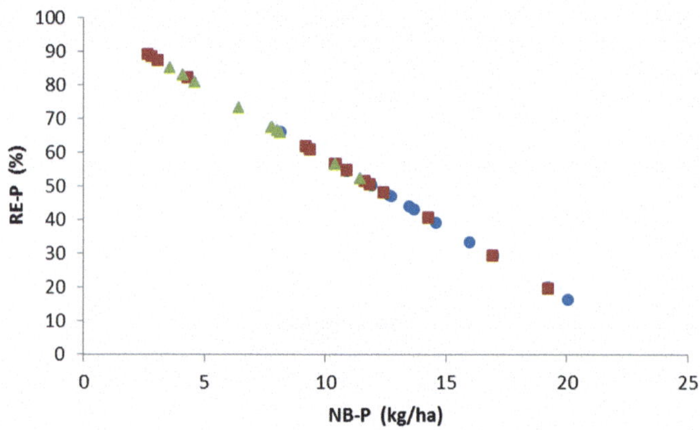

Fig. 14.8 Partial Nutrient balances (NB) versus recovery Efficiency (RE) for phosphorus. ANOVA-difference between different ranges of SSC-P is significant at $p = 0.05$. (circles = SSC-P < 7.5 kg/ha; squares = SSC-P between 7.5 and 10 kg/ha; triangles = SSC-P > 10 kg/ha)

14.4 Synthesis

Using different indicators to evaluate resource use efficiency and stock balances of fertilizer application in Tigray demonstrated that differences between the crops involved were significant and mainly related to different extractive capacities. As a consequence, the use of such indicators in a comparative way at scale levels above the field/crop scale level does not make much sense.

Farmers in Tigray make use of such crop specific differences. In their traditional rotations, wheat is the fertilized crop and is followed by crops like hanfets and teff. Our outcomes made clear that wheat was the most extractive crop followed by respectively hanfets and teff. As a consequence, the long-term sustainability of agricultural systems in Tigray can be explained by the traditional use of such rotations in combination with legumes, the use of crop residues and the practice of fallowing (Kraaijvanger and Veldkamp 2015).

Long-term sustainability of fertilizer application can be expressed by using nutrient balances; resource use efficiency of fertilizer application is a relatively short-term concern; relationships of Soil Supply Capacity (SSC) with both efficiency and nutrient-stock indicators were significant. Consequently, SSC appears a useful indicator in addition to the existing ones: (1) it combines short-term crop aspects (extraction) and long-term soil aspects (capacity); and (2) it allows the inclusion of a soil based temporal dimension for the evaluation of agricultural systems.

The observation of farmers in Tigray that the use of (mineral) fertilizers results in "*addiction*" was supported by our evaluation based on nutrient balances, resource use efficiency and soil supply capacity. The outcomes of the calculated (partial) nutrient balances demonstrated that, despite application of fertilizers, nutrient balances were in many cases negative, especially in the more fertile soils that were able to supply additional nutrients. As a consequence, SSC will be reduced and the system moves to a state with lower efficiency and less depletion. This (system) feedback will consequently reduce excessive extraction of nutrients. Apparently a trade-off existed between efficiency (aiming at crop supply) and stock supply. It is likely that these feedbacks were also responsible for supporting sustainable land use in Tigray for over 2500 years despite calculated negative nutrient balances at the higher scale levels (Kraaijvanger and Veldkamp 2015).

At first sight, applying fertilizers appears to reduce system losses, however, the common assumption that applying fertilizers in all cases will have beneficial effects does not hold. Soil stock supply, resource use efficiency and nutrient balances are clearly interconnected and cannot be separated. The only way to minimize the effect of these trade-offs is to re-use as much as possible crop residues and manure on top of the application of (mineral) fertilizers. Integrated Soil Fertility Management (ISFM) strategies (Vanlauwe et al. 2010) embraces such practices and is in this way able to address concerns of both resource use efficiency and nutrient stock sustainability. ISFM therefore entails a more feasible option to improve crop yield than the sole (and costly) input of mineral fertilizer. Still, development of systems fully sustainable with respect to nutrient balance remains difficult (Harris 1998). In addition to the reduction of NPK-stocks (Tittonell 2007), it is also possible that fertilizer inputs resulted in increased mining of trace elements. The absence of these essential nutrients then might lead to an additional loss of productivity and definitely requires further research.

14.5 Conclusion

Achieving efficient use of fertilizer will be at the cost of nutrient stock sustainability. In most literature the focus is merely on agronomic and crop performance, whereas soil continues to remain a *"black box"* that is to be filled with nutrients (i.e., mineral fertilizer inputs) in order to supply the nutrients required. However, the situation in reality is much more complex. Soils are not static *"black boxes"*, but interact with nutrient use efficiency and stock sustainability. At the same time, dynamic trade-offs exist between stock sustainability and fertilizer use efficiency. As a consequence, Soil Supply Capacity changes and crop production systems move back and forth from more to less fertile states. The presence of such complex feedbacks requires ISFM strategies to arrive at a sound balance between efficiency and sustainability of fertilizer use. Within the context of Tigray, these feedbacks were witnessed by the frequently heard statement that the use of fertilizers leads to "*addiction*", as well as by the long-term sustainability of traditional farming systems.

Acknowledgements We thank all farmers involved, staff from Mekelle University and BoARD (Bureau of Agriculture and Rural Development) for their support. In addition, we thank Zemen Mesfin, Mohammed Nur and Semere Gashaw for their field assistance and Sylvia de Jager for supporting laboratory analysis. Last but not least, we want to express our gratitude to many of the participants of N-2013 and particularly Dr. Mateete Bekunda for their thoughtful comments, which helped us to develop the present manuscript.

References

Chikowo, R., Corbeels, M., Mapfumo, P., Tittonell, P., Vanlauwe, B., & Giller, K. E. (2010). Nitrogen and phosphorus capture and recovery efficiencies, and crop responses to a range of soil fertility management strategies in sub-Saharan Africa. *Nutrient Cycling in Agroecosystems, 88*(1), 59–77.

De Jager, A., Onduru, D., Van Wijk, M. S., Vlaming, J., & Gachini, G. N. (2001). Assessing sustainability of low-external-input farm management systems with the nutrient monitoring approach: A case study in Kenya. *Agricultural Systems, 69*(1–2), 99–118.

Donovan, C., Wopereis, M., Guindo, D., & Nebie, B. (1999). Soil fertility management in irrigated rice systems in the Sahel and Savanna regions of West Africa: Part II. Profitability and risk analysis. *Field Crops Research, 61*(2), 147–162.

Haefele, S. M., Wopereis, M. C. S., Ndiaye, M. K., Barro, S. E., & Ould Isselmou, M. (2003). Internal nutrient efficiencies, fertilizer recovery rates and indigenous nutrient supply of irrigated lowland rice in Sahelian West Africa. *Field Crops Research, 80*(1), 19–32.

Haileslassi, A., Priess, J., Veldkamp, E., Teketay, D., & Lesschen, J. P. (2005). Assessment of soil nutrient depletion and its spatial variability on smallholders' mixed farming systems in Ethiopia using partial versus full nutrient balances. *Agriculture, Ecosystems and Environment, 108*(1), 1–16.

Harris, F. M. A. (1998). Farm-level assessment of the nutrient balance in northern Nigeria. *Agriculture, Ecosystems and Environment, 71*(1–3), 201–214.

Jama, B., & Pizarro, G. (2008). Agriculture in Africa: Strategies to improve and sustain smallholder production systems. *Annals of the New York Academy of Sciences, 1136*, 218–232.

Janssen, B. H. (1998). Efficient use of nutrients: An art of balancing. *Field Crops Research, 56*(1–2), 197–201.
Janssen, B. H. (2011). Simple models and concepts as tools for the study of sustained soil productivity in long-term experiments. II. Crop nutrient equivalents, balanced supplies of available nutrients, and NPK triangles. *Plant and Soil, 339*(1), 17–33.
Koohafkan, P., Altieri, M. A., & Holt Gimenez, E. (2012). Green Agriculture: Foundations for biodiverse, resilient and productive agricultural systems. *International Journal of Agricultural Sustainability, 10*(1), 61–75.
Kraaijvanger, R., & Veldkamp, T. (2015). Grain productivity, fertilizer response and nutrient balance of farming systems in Tigray, Ethiopia: a multi-perspective view in relation to soil fertility degradation. *Land Degradation and Development, 26*(7), 701–710.
Kurwakumire, N., Chikowo, R., Mtambanengwe, F., Mapfumo, P., Snapp, S., Johnston, A., et al. (2014). Maize productivity and nutrient and water use efficiencies across soil fertility domains on smallholder farms in Zimbabwe. *Field Crops Research, 164*(1), 136–147.
Sanchez, P. A., & Swaminathan, M. S. (2005). Hunger in Africa: the link between unhealthy people and unhealthy soils. *The Lancet, 365*(9457), 442–444.
Stoorvogel, J. J., Smaling, E. M. A., & Janssen, B. H. (1993). Calculating soil nutrient balances in Africa at different scales—I Supra-national scale. *Fertilizer Research, 35*(3), 227–235.
Tittonell P. A. (2007). *Msimu wa Kupanda—Targetting resources within diverse, heterogeneous and dynamic farming systems of East Africa*. Ph.D. Thesis, Wageningen University, Wageningen, The Netherlands.
Van Noordwijk, M., & Brussaard, L. (2014). Minimizing the ecological footprint of food: Closing yield and efficiency gaps simultaneously? *Current Opinion in Environmental Sustainability, 8*, 62–70.
Vanlauwe, B., Bationo, A., Chianu, J., Giller, K. E., Merckx, R., Mokwunye, U., et al. (2010). Integrated soil fertility management: Operational definition and consequences for implementation and dissemination. *Outlook on Agriculture, 39*(1), 17–24.
Vanlauwe, B., Kihara, J., Chivenge, P., Pypers, P., Coe, R., & Six, J. (2011). Agronomic use efficiency of N fertilizer in maize-based systems in sub-Saharan Africa within the context of integrated soil fertility management. *Plant and Soil, 339*(1), 35–50.
Woldeamlak, A., Bastiaans, L., & Struik, P. C. (2001). Competition and niche differentiation in barley (*Hordeum vulgare*) and wheat (*Triticum aestivum*) mixtures under rainfed conditions in the Central Highlands of Eritrea. *NJAS—Wageningen Journal of Life Sciences, 49*(1), 95–112.

Chapter 15
Rice Response to Nitrogen and Supplemental Irrigation Under Low Phosphorus and Potassium in Upland Production Systems in East Africa

Geoffrey Onaga, Joseph Kikafunda, George Bigirwa, Godfrey Asea, and Lizzy A. Mwamburi

Abstract Throughout upland rice ecologies, low soil fertility and moisture stress are the major factors limiting productivity and profitability. We conducted field experiments using 36 combinations of NPK fertilizer on a popular upland rice variety in East Africa (NERICA 4) to establish upland rice crop nutrient requirements under supplemental irrigation (SI) and rainfed (RF) conditions. NPK was applied in a factorial design by partially employing nutrient omission technique. The overall effect of NPK on the grain yield was more striking in SI, with 55% yield increase as compared to 40% in RF. Application of nitrogen (N), phosphorus (P) and potassium (K) fertilizers singly, in SI, increased the grain yield of NERICA 4 by 43%, 5% and 0.4%, respectively. In contrast, N increased grain yield by 20% in RF, and P and K had no significant effect on grain yield. Application of 120 kg N ha^{-1} alone, without P and K, however, led to a 44% decrease in agronomic efficiency (AE) in RF and a marginal increase in SI. Although maximum biomass was obtained with 120:40:40 kg NPK ha^{-1} in both SI and RF, the grain yield was not significantly different from 80:40:40 kg NPK ha^{-1}. Besides, the harvest index (HI) dropped by eight units in RF and increased only marginally in SI at 120:40:40 kg NPK ha^{-1}.

G. Onaga (✉) · J. Kikafunda · G. Asea
National Crop Resources Research Institute (NaCRRI), Namulonge, P.O. Box 7084, Kampala, Uganda
e-mail: geoffyonaga@gmail.com

J. Kikafunda
e-mail: josephkikafunda@yahoo.com

G. Asea
e-mail: grasea_99@yahoo.com

G. Bigirwa
Alliance for a Green Revolution in Africa, Eden Square, P.O. Box 66773, Westlands 00800, Nairobi, Kenya
e-mail: gbigirwa@agra-alliance.org

L. A. Mwamburi
Department of Biological Sciences, University of Eldoret, P.O. Box 1125, Eldoret 30100, Kenya
e-mail: lizzymwamburi@hotmail.com

M. A. Sutton et al. (eds.), *Just Enough Nitrogen*,
https://doi.org/10.1007/978-3-030-58065-0_15

The net profit, due to NPK, was 33% higher in SI than RF at 80:40:40 kg NPK ha^{-1}. Our data show that N is the most limiting nutrient, and applying N beyond 80 kg N ha^{-1} at the current P and K recommendation of 40 kg ha^{-1} for upland rice is less profitable.

Keywords Nitrogen · Upland rice · Rainfed · Agronomic efficiency · Grain yield

15.1 Introduction

Throughout upland rice ecologies, low rice yields are mainly attributed to moisture stress, low soil fertility and negative interactions between water and nutrient availability. Plant water stress brought about by either high atmospheric evaporation or decreased solution water potential causes reduced nutrient uptake by its effect on the rate of water flow through plants (Greenway and Klepper 1969; Haefele et al. 2010), or through active ion uptake mechanisms and passive efflux of ions (Erlandsson 1979). Upland rice production systems, where moisture levels are often below field capacity, are the most affected, and farmers rarely achieve yields higher than 2.5 tons ha^{-1}. This is exacerbated by limited or unbalanced plant nutrition. Moreover, negative interaction between water and nutrient availability has been reported in upland systems (Haefele et al. 2010). Among the essential plant nutrients, nitrogen (N), phosphorus (P) and potassium (K) are critical elements for plant growth and development (Dobermann and Fairhurst 2000; Miller and Cramer 2004). Low supply of N, P and K is sometimes highlighted as an even more important constraint than water availability (Fukai et al. 1998; Suriya-arunroj et al. 2000; Linquist and Sengxua 2001). Deficiency of N, P and K in upland rice limits tillering, panicle formation and grain filling, curtailing yield output per hectare (Wang et al. 2002; Yamah 2002).

In most African countries, where farmers have traditionally relied on fallow periods, fertilizer use is negligible and often rare. In fact, fertilizer use in Africa is estimated to have stagnated at 6–12 kg ha^{-1} $year^{-1}$ for the last 10 years (Sommer et al. 2013). Uganda, for instance, is among the lowest fertilizer users in the world, and soil nutrient depletion continues to be one of the major agricultural constraints (Musiime et al. 2005). According to the State of Uganda Population report 2010 (The Republic of Uganda/UNFPA 2010), it is estimated that between 1996 and 2000, nutrient fertilizer usage was 0.37 kg ha^{-1}, while nutrient mining was estimated to be 87 kg ha^{-1} $year^{-1}$ by 2008, with most upland areas at the risk of degradation.

The total area under upland rice in Uganda has increased from about 6000 ha in 2002 to about 60,000 ha in 2012, and upland rice dominates the new rice growing areas, suggesting the need for measures to sustain productivity. To our knowledge, the benefits of plant nutrition in upland rice production systems have remained obscure for decades, partly because of limited information on nutrient management, and limited empirical evidence of how fertilizer use rates compare to economically profitable levels. On the other hand, most smallholder farmers operate under varying conditions within the agricultural landscapes. This variability is explained largely by soil fertility gradients induced by either varying inherent soil fertility or management. Thus, soil analysis often falls short of accuracy due to the heterogeneous nature of the fields from which sampling is carried out. To overcome this, nutrient omission trials have been suggested as a tool to identify which of the macro-nutrients, N, P and K, are limiting crop growth and productivity. In this study, we investigated the effect of 36 combinations of NPK on rice yields in rainfed (RF) and supplemental irrigation (SI) upland rice systems. The 36 combinations were applied in a factorial design, in a nutrient omission technique aimed at establishing the crop needs under natural conditions. We used a popular upland rice variety (NERICA 4) that was released in 2003 for commercial production in most East African countries.

15.2 Materials and Methods

15.2.1 Site Information

Field experiments were conducted at the National Crops Resources Research Institute, Namulonge (0° 32′ N, 32° 37′ E, 1150 m above sea level), Uganda. Namulonge is located within the tropical wet and mild dry climate with slightly humid (65%) conditions. The area receives bimodal rainfall with two seasons having approximately the same length (3 months each). Rainfall amounts range from 800 to 1200 mm of annual precipitation and temperatures range from 16 to 28 °C. The soils represent a transition zone between red and yellow ferallitic soils derived from a basement complex. Before planting, soil analysis was done to provide an estimate of the nutrients that would be available to the crop. All soil analysis measurements were done according to the method described by Okalebo et al. (1993). The soil type at the experimental site was sandy-clay loam. Textural analysis values were 22% clay; 16% silt and 62% sand. The chemical properties at 0–20 cm soil depth were: pH 5.3; organic matter 31.5 g kg^{-1}; extractable phosphorus (P), 0.82 mg kg^{-1}; potassium (K), 88.6 mg kg^{-1}; calcium (Ca), 1.6 $cmol_c$ kg^{-1}; and magnesium (Mg), 1.13 $cmol_c$ kg^{-1}. Average annual rainfall amount was higher in 2008 than in 2009 (Fig. 15.1a) but did not cause a significant difference in crop performance in both years, most likely because of timely planting of the experiments. Temperature trends were relatively consistent with a minor increase in January for both years (Fig. 15.1b).

15.2.2 Treatments and Field Management

A factorial design in a split plot arrangement with three replications was used to determine the effect of NPK fertilizer and SI on NERICA 4 in 2008 and 2009. Trials were conducted in RF, and SI in which plants were irrigated with 20 mm of water using sprinklers every five days during windows of dry weather starting from panicle

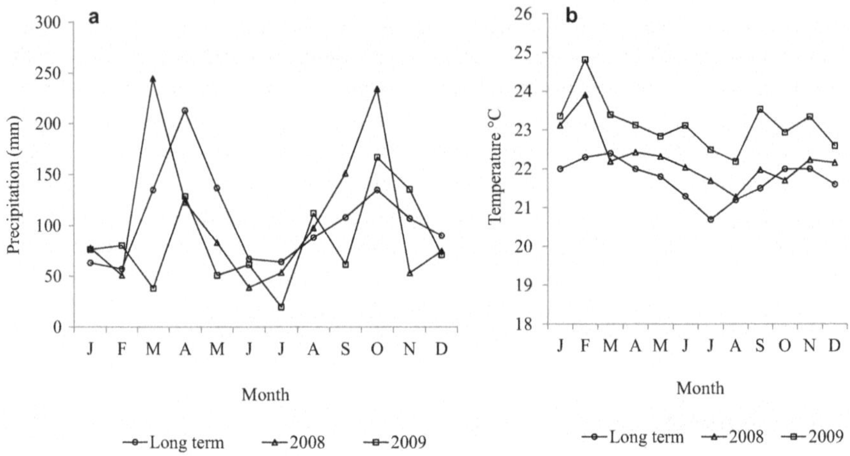

Fig. 15.1 Monthly average temperatures (**a**) and precipitation (**b**) of the two production years in comparison with the long term average

initiation stage. Thirty-six treatment combinations consisting of four levels of N (0, 40, 80 and 120 kg N ha^{-1}), and three levels of each of P and K (0, 20 and 40 kg ha^{-1}) were tested. The nutrients were supplied to the soil in the form of urea for N, triple super phosphate (TSP) for phosphorus and muriate of potash for potassium. Full quantity of P was applied and incorporated in the seedbed at planting. Potassium along with ½ fraction of N was applied at three weeks after planting. The remaining quantity of N was applied at panicle initiation stage. In each year, the experimental plot size was 10.5 m^2 consisting of eight rows, 5 m long with intra row spacing of 0.3 m. The plots received identical cultural treatments of ploughing, cultivation, seed rate, sowing method and pest control.

15.2.3 Sampling and Data Analysis

Four middle rows in each plot were selected and 1 m from either side had randomly selected plants tagged for recording plant height, tillers/m^2 and panicles/m^2. At maturity, panicles were harvested from the tagged plants and data on grain number per panicle was recorded. Grain yield, adjusted to moisture content of 14%, was also determined from the four middle rows using the formula in Fig. 15.2. Dry matter was determined after drying the straw to constant weight. Harvest index was calculated as a percentage of kernals over dry matter yield according to Fageria (2009). Agronomic efficiency (AE) was calculated based on the yield increase due to fertilizer application according to Haefele et al. (2010). All the data was subjected to analysis of variance following the split-plot model using SAS Statistical software (Version 9.3). The least significant difference was used to compare treatments within a factor and only

① $$\frac{W_i\,(Kg)}{DM_i\,(Kg)} = \frac{100\%}{100\% - Mc_i\,(\%)}$$

② $$\frac{W_t\,(Kg)}{DM_t\,(Kg)} = \frac{100\%}{100\% - MC_t\,(\%)}$$

③ $$DM_i\,(Kg) = \frac{W_i\,(Kg)*100\% - (100\% - Mc_i\,(\%))}{100\%}$$

④ $$DM_t\,(Kg) = \frac{W_t\,(Kg)*100\% - (100\% - Mc_t(\%))}{100\%}$$

⑤ $$DM_i\,(Kg) = DM_t\,(Kg)$$

⑥ $$\frac{W_i\,(Kg)*100\% - (100\% - Mc_i\,(\%))}{100\%} = \frac{W_t\,(Kg)*100\% - (100\% - Mc_t(\%))}{100\%}$$

Standard MC = 14%

⑦ $$W_t = W_i * \frac{100\% - Mc_i}{100\% - Mc_t}$$

⑧ $$W_t = W_i * \frac{100\% - Mc_i}{100\% - Mc_t}$$

⑨ $$\text{Yield (Kg/ha)} = \frac{W_i}{A} * \frac{100\% - MC_t}{100\% - 14} * 1000m^2$$

Fig. 15.2 Formula used to obtain grain yield per hectare standardized to 14% grain moisture. A four-row plot, with average row width of 30 cm was harvested. Where W_i is the initial weight of harvested grain with MC at 100%; MC_i is moisture content at harvest; MC_t is moisture content adjusted to 14%. W_t is the grain weight adjusted to 14% MC. DM is dry matter, which stays constant irrespective of MC. DM_i is dry matter at harvest; DM_t is dry matter at 14% moisture content (MC). A is the areas harvested

when the F-test of the variable was significant for that factor. The data on economic attributes was analyzed to assess the benefit of NPK at 80:40:40 with and without SI separately, in order to calculate the gross and net returns. Percentage grain yield difference between SI and RF due to NPK application was calculated using the formula:

$$YI_p = [(YI_{SI} - YI_{RF})/\ YI_{SI} * 100] \qquad (15.1)$$

where:

YI_p is percentage yield increase, and
YI_{SI} is yield increase in SI, and YI_{RF} is yield increase in RF.

15.3 Results

15.3.1 Grain Yield, Growth and Yield Component Attributes

NPK (120:40:40) increased grain yield by 55% and 40% in SI (Fig. 15.3a) and RF (Fig. 15.3b), respectively. However, the grain yield at 120 kg N ha^{-1} was not significantly different from 80 kg N ha^{-1} in RF even when P and K were applied at 40 kg ha^{-1}. Single application of N, P and K fertilizers in SI increased the grain yield of NERICA 4 from 2.46 to 4.31 tons ha^{-1} for N, 2.46 to 2.58 tons ha^{-1} for P and 2.46 to 2.47 for K; translating into 43%, 5%, and 0.4% yield increase due to N, P, and K, respectively. Interestingly, the grain yield gap was twice higher than the biomass gap when the two production systems (SI and RF) were compared (Figs. 15.3 and 15.4), and the biomass was more correlated to grain yield than the harvest index (HI) in SI compared to RF. At NPK rates of 120:40:40 kg ha^{-1}, HI dropped by eight units in RF (Fig. 15.6a). Moreover, the effect of lower NPK rates on HI was superior to higher rates in RF, with 40:40:40 NPK treatments producing the highest HI (0.45). Conversely, HI continued to increase in SI and was 11.5% and 20.5% higher in 40:40:40 kg ha^{-1} and 120:40:40 kg ha^{-1}, respectively. Despite the declining HI in RF at higher NPK rates, biomass was similar in RF at 80–120 kg N ha^{-1} (Fig. 15.4a), whereas that of SI continued to increase (Fig. 15.4b). At 0–40 kg ha^{-1} of P and K, biomass production was not significantly different from zero nitrogen addition (0 N)

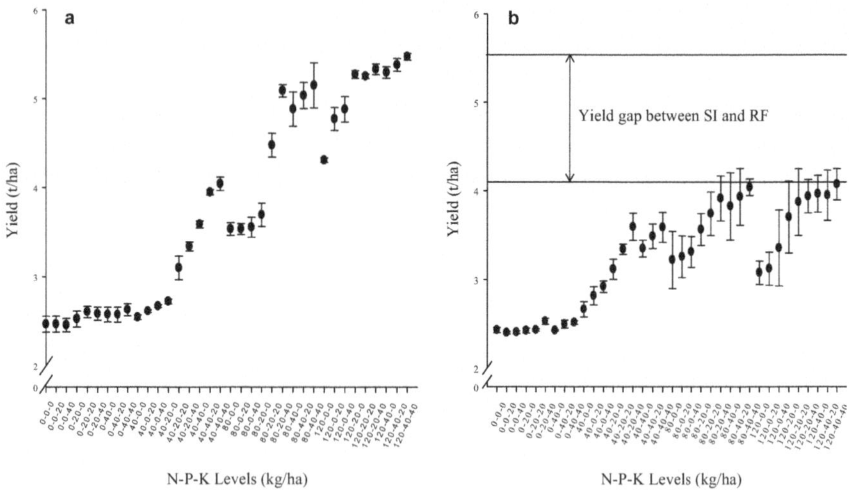

Fig. 15.3 Grain yield increase in relation to incremental rate of NPK fertilizer under supplemental irrigation (**a**) and rainfed (**b**) conditions. Yield gap is twice the biomass gap when the two production systems (SI and RF) are compared. The yield gap is the difference between the highest mean grain yields between SI and RF. Symbols indicate means and bars indicate standard errors. ($LSD_{(0.05)}$ = 0.78 and 0.17 for supplemental irrigation and rainfed conditions, respectively)

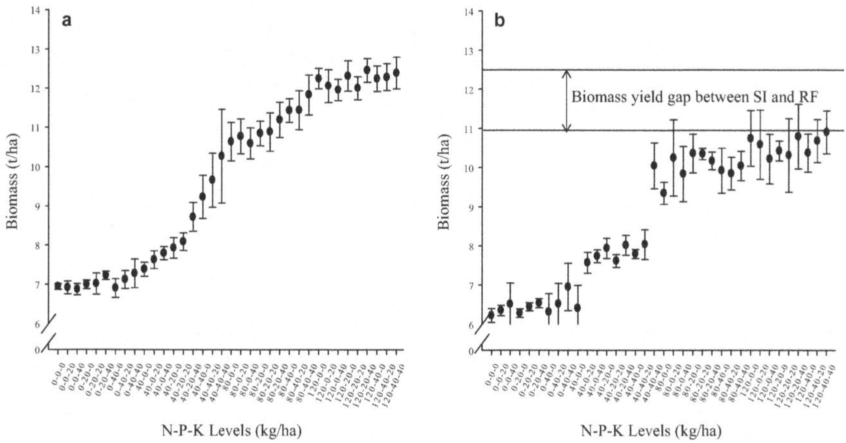

Fig. 15.4 Increase in biomass in relation to incremental rate of NPK fertilizer under supplemental irrigation (**a**) and rainfed (**b**) conditions. The biomass gap is the difference between the highest biomass yields between SI and RF. Symbols indicate means and bars indicate standard errors. ($LSD_{(0.05)} = 0.07$ and 0.10 for supplemental irrigation and rainfed conditions, respectively)

in both SI and RF; the same was true for 0–40 kg N ha^{-1} without P and K. Agronomic efficiency (AE) increased from 18 kg kg^{-1} to 24 kg kg^{-1} in SI (Fig. 15.5a). In contrast, average AE remained at suboptimal levels of 17 kg kg^{-1} and declined at N rates beyond 40 kg ha^{-1} in RF, which was similar to the pattern observed with HI. Productivity decline in terms of AE was also observed in SI when N fertilizer levels exceeded 80 kg ha^{-1}. Moreover, N application without P and K decreased AE by 30% and 47% in SI and RF, respectively (Fig. 15.5b). Panicle number (per m^2), tiller number, grain number (per panicle) and 1000 grain weight were also significantly influenced by both SI and NPK treatments. The panicle number (per m^2) ranged from 168 to 254 while tiller number ranged from 176 to 266 in SI across NPK application rates, which translated into 33% increase in tiller and panicle number (Fig. 15.6a). In RF, panicle number ranged from 167 to 223 while tiller number ranged from 176 to 239, which translated into 26% and 24% increase in tiller and panicle number, respectively. Grain number (per panicle) and 1000 grain weight also increased progressively with the increasing NPK levels. Interestingly, these yield parameters were also significantly influenced by PK in both SI and RF, except HI which dropped by 14% and 5% with increasing NK and NP, respectively (Fig. 15.6b). In general SI significantly augmented the effect of NPK on yield components than RF, and a combination of all the three nutrients was highly significant than when one was omitted.

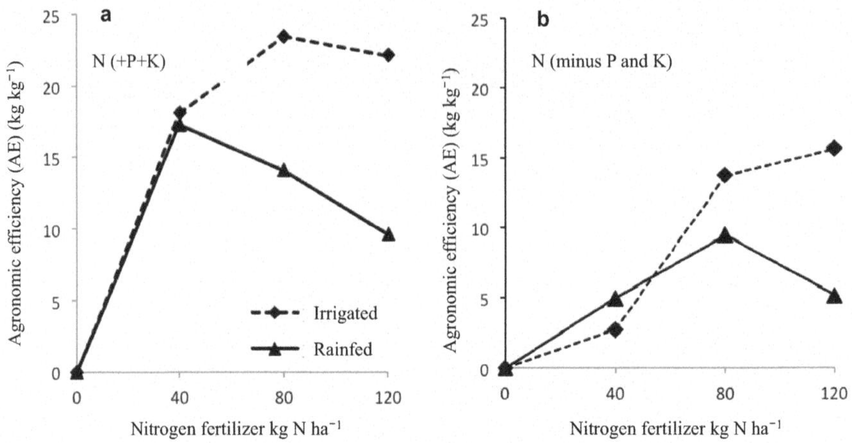

Fig. 15.5 Agronomic Efficiency (AE) of nitrogen fertilizer when applied with P and K, N (+ P + K) and without P and K, N (minus P and K) in supplemental irrigation (irrigated) and rainfed conditions. ($LSD_{(0.05)}$ = 2.26 and 3.67 for supplemental irrigation and rainfed conditions, respectively)

15.3.2 Economic Attributes

The values of the economic attributes (gross and net return) increased significantly with the rates of NPK applied in both SI and RF conditions (Table 15.1). However, the economic returns were more evident in SI compared to RF. The highest value of gross as well as net return was recorded at 120:40:40 NPK application rates in SI, whereas the RF crop had significantly higher returns at 80:20:20 NPK application. For comparison between the two systems, both gross and net returns were calculated based on NPK application rates of 80:40:40. Using this rate, gross and net returns of US$2127 and US$1323 were obtained under SI, respectively, whereas, RF conditions produced a gross and net return of US$1583 and US$1001, respectively. Average difference in net return between SI and RF at 80:40:40 NPK was US$322 ha^{-1}. This increase was 33% higher than the net returns obtained under RF conditions.

15.4 Discussion and Conclusions

Upland rice production in Africa suffers from low nutrient supply and moisture stress, which limit crop productivity and profitability. We examined the effects of supplementary irrigation (SI) and 36 NPK fertilizer application rates on the grain yield of the rice cultivar, 'NERICA 4', in Namulonge, Uganda over a period of two years. Several combinations were included to determine a suitable combination of N, P and K needed to improve upland rice productivity and profitability.

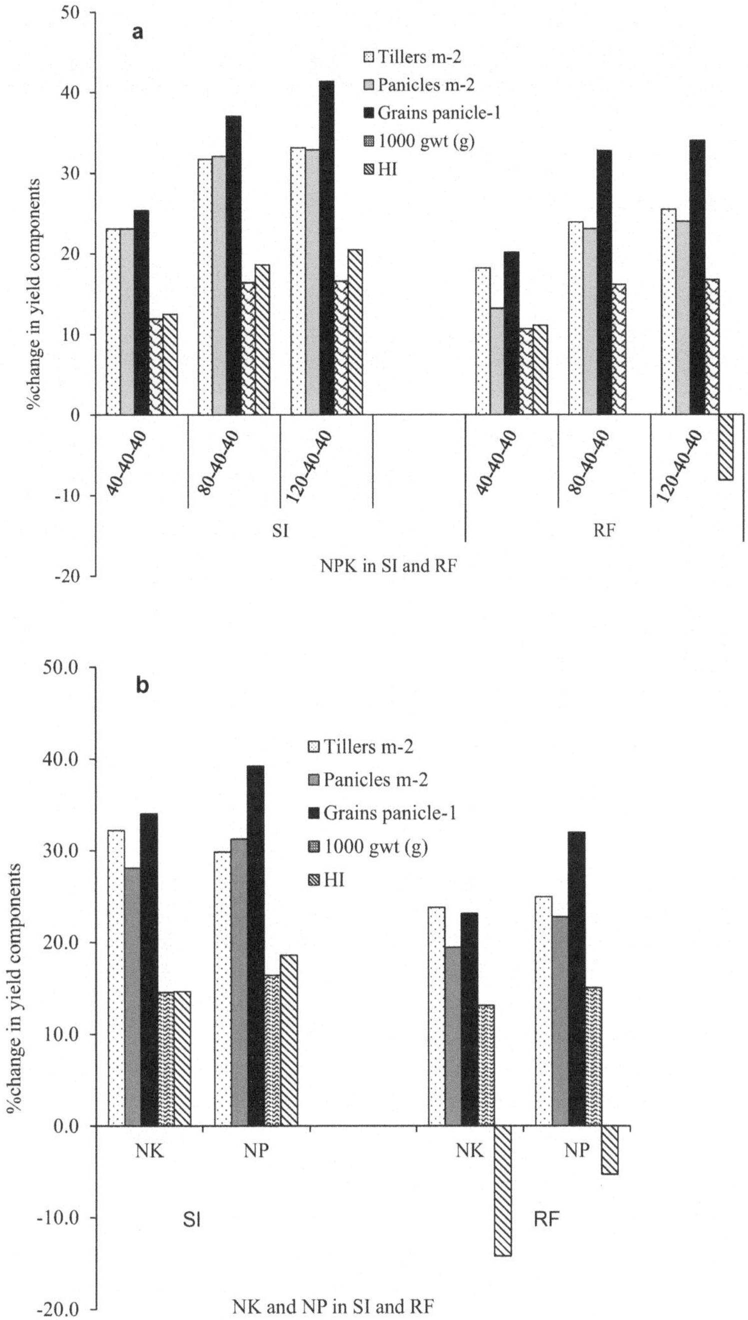

Fig. 15.6 Effect of NPK fertilizer on growth and yield components of NERICA 4 under supplemental irrigation (SI) and rainfed (RF) conditions. The percentage (%) change in each yield component is shown. **a** Compares the effect of all the three nutrients, N, P and K in varying amounts on yield components under SI and RF, whereas **b** compares the effect of NK and NP on yield components (legend)

Table 15.1 Effect of water treatments in respect with increasing NPK levels on economic returns due to grain yield of field grown NERICA 4

Fertilizer material	Price (US$)	Analysis (N-P-K)	N (%)	P_2O_5 (%)	K_2O (%)	Amount of fertilizer material needed (kg ha^{-1})	Costs ha^{-1} (US$)
Urea	38.9	46-0-0	46	0	0	174	135.3
Triple super phosphate	38.9	0-46-0	0	46	0	87	67.2
Muriate of potash	50	0-0-60	0	0	60	67	67
Other costs (US$)							
Seeds and labor							312.2
Additional cost, labor and fuel for SI							222.2
Total cost (SI)							**804**
Total cost (RF)							**581.8**
Selling price of rice kg^{-1}							0.39
Gross income (SI)							2127
Gross income (RF)							1583
Net profit (SI)							1323
Net profit (RF)							1001
Difference between SI and RF							**322**

Climatic measurements were collected in an attempt to explain the results of this study. However, the climatic influence on grain yield was not significant, and the data were negligibly different between the two years, and thus were not considered in the interpretation of the findings of this study.

The nutrient content of the soil at the experimental site was apparently insufficient for optimum crop yields, as reflected by the soil properties. In effect, the crop considerably responded to NPK in both SI and rainfed (RF) conditions, exhibiting significantly higher values of most of the crop-assessment attributes when compared to no NPK application.

Even though in RF the yield response was in a favorable range, the yield difference of 15% between SI and RF was substantial, and suggests that rice growers will need to match the crop nutrition with soil moisture to increase nutrient uptake at the time of crop nutrient need. Crop nutrient requirements change as plants develop. For example, rice has a greater N and P requirement in the early stage right through flowering, which decreases gradually until the dough stage; whereas the demand for K is lower at earlier growth of the plant, but increases from flowering until ripening

(Dobermann and Fairhurst 2000). Synchronizing these growth stages with nutrient and moisture supply will greatly improve rice productivity in uplands.

We found that increasing P and K levels at low N increased the grain yield only marginally, suggesting that adequate N supply is needed for productive utilization of P and K by the crop. Considering the percentage yield difference due to N (43%) as compared to 5% and 0.4% due to P and K, respectively, N is apparently the most limiting in Namulonge, and potentially has a synergistic effect on P and K uptake. A similar trend could be encountered across upland rice farming systems, considering the large significant difference in grain yield between zero and the other NPK rates used in this study.

Plots that were treated with a minimal difference between NPK ratios (e.g., 80:40:40) had significantly higher grain yields than plots with wide difference between the ratios (e.g., 120:0:20). This suggests that the practice of balanced nutrition is crucial for farmers to achieve optimum rice grain yield in uplands. Thus, the balanced application of NPK is likely to have a positive impact across upland rice production areas in East Africa.

We also found a reduction in grain yield of 24% when N was singly applied at 120 kg N ha^{-1}, even though there was an increase in the biomass. Because of this negative response, it is apparent that excessive addition of N fertilizer had a considerably negative effect on crop productivity when it is not balanced with P and K. Thus, it is not worth applying large amounts of N when soil is low in available P and K, as this may not only limit crop yields, but also cause financial losses to the grower. In fact, unbalanced supply of N, P and K is sometimes even highlighted as a more important constraint than water availability (Fukai et al. 1998; Suriya-arunroj et al. 2000; Linquist and Sengxua 2001). Besides this, some studies have shown that P and K increase total N uptake as well as grain yield (Horie et al. 1997; Inthapanya et al. 2001; Saito et al. 2006). Moreover there is a strong interaction between N and K in crop growth, thus crop response to applied N decreases when the exchangeable K content of the soil is below a critical target level (Belay et al. 2002; Cai and Qin 2006; Wang et al. 2007).

Our data show a significantly low response of grain yield to P and K when applied singly, which is consistent with the above findings. This is also consistent with the findings of George et al. (2001), who reported that application of only P had little effect on grain yield irrespective of increased P uptake. Both N and P are often associated with positive effect on tillers and panicles, and high yielding upland cultivars under high-input conditions are characterized by moderate panicle number in the Philippines ($\geq$ 300 panicles/m^2), tillering number in Brazil ($\geq$ 250 tillers/m^2), and by higher harvest index (HI) and intermediate height (Pinheiro and de Castro 2000; Wang et al. 2002; Saito et al. 2006).

Although it has not been previously reported in Uganda, it is not surprising to find such responses in cultivars adapted to conventional low-input systems, such as those existing in East Africa. In this study, NERICA 4 produced a maximum HI of 0.46, which was relatively higher than the HI reported by Saito et al. (2006) in Laos. However, the HI of NERICA 4 is still lower than 0.50, which is normally reported for improved semi-dwarf cultivars (Mae 1997). Moreover, HI dropped by eight units in RF and increased only marginally in SI at high NPK levels, which was also consistent with the low agronomic efficiency (AE). This might suggest that NERICA 4, despite its adaptation to low input systems, has comparatively a better yield performance. However, the low HI, despite having enhanced vegetative growth, may suggest limitation in the translocation of photosynthates to the grains at higher N, possibly explaining the double difference between the yield gap and the biomass gap. Thus, development of cultivars with a high correlation of nutrient and water use efficiency with HI and biomass in upland rice production systems remains to be explored.

Nevertheless, our data demonstrated that NPK significantly influenced the yield attributes of NERICA 4, including tiller and panicle number, 1000 grain weight and grains per panicle in both RF and SI. Moreover, AE of NERICA 4 was 30% higher in SI than in RF conditions, indicating that application of NPK considerably increased productivity in SI compared to RF. The low AE in RF is likely due to reduced photosynthetic rate as a consequence of limited moisture availability during short periods of dry weather, in which we applied water for the SI treatment. These findings point out the significance of SI in upland rice production and could be considered as a management strategy in semi-intensive upland rice systems. The contribution of SI and RF to gross and net return was consistent with NPK yield response trends (Table 15.1). Supplemental irrigation resulted in additional 33% economic returns at 80:40:40 NPK, which rationalizes the significance of moisture in rice mineral nutrition.

Overall, our data show that a dose of 80 kg N ha^{-1} has a profound influence on grain yield of upland rice in RF and SI. Although we found a slight grain yield increase by applying 120 kg N ha^{-1} in SI, it was insufficient to justify application of an additional 40 kg N ha^{-1} at the lower rates of P and K. Thus, applying nitrogen levels above 80 kg N ha^{-1} at the current P and K recommendation of 40 kg ha^{-1} for NERICA 4 could be counterproductive, and may contribute to excess nitrogen with negative consequences on the environment. The low AE and HI at higher NPK rates suggests the need for improvement of NERICA 4 or deployment of cultivars with AE values > 25 kg kg^{-1}.

Acknowledgements The authors would like to thank the Rockefeller Foundation for the financial support of this research. We extend our sincere thanks to Africa Rice for providing the germplasm and other capacity building efforts.

References

Belay, A., Claassens, A. S., & Wehner, F. C. (2002). Effect of direct nitrogen and potassium and residual phosphorus fertilizers on soil chemical properties, microbial components and maize yield under long-term crop rotation. *Biology and Fertility of Soils, 35,* 420–427.

Cai, Z. C., & Qin, S. W. (2006). Dynamics of crop yields and soil organic carbon in a long-term fertilization experiment in the Huang-Huai-Hai Plain of China. *Geoderma, 136,* 708–715.

Dobermann, A., & Fairhurst, T. (2000). *Rice: Nutrient disorders and nutrient management*. Potash & Phosphate Institute (PPI), Potash & Phosphate Institute of Canada (PPIC), and International Rice Research Institute (IRRI), p. 191.

Erlandsson, G. (1979). Efflux of potassium from wheat roots induced by changes in the water potential of the root medium. *Physiologia Plantarum, 47,* 1–6.

Fageria, N. K. (2009). *The use of nutrients in crop plants*. New York: CRC Press.

Fukai, S., Sittisuang, P., & Chanphengsay, M. (1998). Increasing production of rainfed lowland rice in drought prone environments—A case study in Thailand and Laos. *Plant Production Science, 1*(1), 75–82.

George, T., Magbanua, R., Roder, W., Van Keer, K., Trebuil, G., & Reoma, V. (2001). Upland rice response to phosphorus fertilization in Asia. *Agronomy Journal, 93,* 1362–1370.

Greenway, H., & Klepper, B. (1969). Relation between anion transport and water flow in tomato plants. *Physiologia Plantarum, 22,* 208–219.

Haefele, S. M., Sipaseuth, N., Phengsouvanna, V., Dounphady, K., & Vongsouthi, S. (2010). Agro-economic evaluation of fertilizer recommendations for rainfed lowland rice. *Field Crops Research, 119,* 215–224.

Horie, T., Ohnishi, M., Angus, J. F., Lewin, L. G., Tsukaguchi, T., & Matano, T. (1997). Physiological characteristics of high-yielding rice inferred from cross-location experiments. *Field Crops Research, 52,* 55–67.

Inthapanya, P., Sipaseuth, S. P., Sihathep, V., Chanphengsay, M., Fukai, S., & Basnayake, J. (2001). Genotypic performance of rainfed lowland rice under different fertilizer conditions in Laos. In S. Fukai & J. Basnayake (Eds.), *Increased lowland rice production in the mekong region. Proceedings of the International Workshop 101* (pp. 191–200). Laos: Vientiane.

Linquist, B., & Sengxua, P. (2001). *Nutrient management in rainfed lowland rice in the Lao PDR*. Los Banos (Philippines): International Rice Research Institute (IRRI).

Mae, T. (1997). Physiological nitrogen use efficiency in rice: nitrogen utilization, photosynthesis and yield potential. In T. Ando, K. Fugita, T. Mae, H. Matsumoto, S. Mori, & J. Sekiya (Eds.), *Plant nutrition for sustainable food production and environment* (pp. 51–60). The Netherlands, Dordrecht: Springer.

Miller, A. J., & Cramer, M. D. (2004). Root nitrogen acquisition and assimilation. *Plant and Soil, 274,* 1–36.

Musiime, O., Tenywa, M. M., Majaliwa, M. J. G., Lufafa, A., Nanfumba, D., Wasige, J., et al. (2005). Constraints to rice production in Bugiri district. *African Crop Science Journal, 7,* 1495–1499.

Okalebo, J. R., Gathua, K. W., & Woomer, P. L. (1993). *Laboratory methods of soil and plant analysis: A working manual*. Kenya, Nairobi: UNESCO.

Pinheiro, B., & de Castro, B. (2000). Upland rice in Brazil: Impact of altering plant type and grain appearance. In *Aerobic Rice Workshop proceedings*. 7–8 September. Los Banos (Philippines): International Rice Research Institute (IRRI).

Saito, K., Linquist, B., Atlin, G. N., Phanthaboon, K., Shiraiwa, T., & Horie, T. (2006). Response of traditional and improved upland rice cultivars to N and P fertilizer in northern Laos. *Field Crops Research, 96,* 216–223.

Sommer, R., Bossio, D., Desta, L., Dimes, J., Kihara, J., et al. (2013). *Profitable and sustainable nutrient management systems for East and Southern African smallholder farming systems—Challenges and opportunities*. http://repository.cimmyt.org/xmlui/bitstream/handle/10883/4035/99011.pdf. Accessed March 2020.

Suriya-Arunroj, D., Chaiyawat, P., Fukai, S., & Blamey, P. (2000). Identification of nutrients limiting rice growth in soils of Northeast Thailand under water-limiting and non-limiting conditions. *Plant Production Science, 3,* 417–421. https://doi.org/10.1626/pps.3.417.

The Republic of Uganda/UNFPA. (2010). *The State of Uganda population report 2010. Population and sustainable development: Emerging challenges, opportunities and prospects.*

Kampala, Uganda: Population Secretariat, Ministry of Finance, Planning and Economic Development, The Republic of Uganda, and United Nations Population Fund (UNFPA). Available at: http://npcsec.go.ug/wp-content/uploads/2013/06/The-State-of-Uganda-Population-Report-2010_opt.pdf. Accessed March 2020.

Wang, S. H., Cao, W., Jiang, D., Dai, T., & Zhu, X. (2002). Physiological characteristics and high yield techniques with SRI rice. In N. Uphoff, E. C. M. Fernandes, L. P. Yuan, J. Peng, S. Rafaralahy, & J. Rabenandrasana (Eds.), *Assessments of the System of Rice Intensification (SRI): Proceedings of an International Workshop, April 1–4, 2002, Sanya, China* (pp. 116–124). Ithaca, NY (USA): Cornell International Institute for Food, Agriculture and Development (CIIFAD).

Wang, X. B., Hoogmoed, W. B., Cai, D. X., Perdok, U. D., & Oenema, O. (2007). Crop residue, manure and fertilizer in dryland maize under reduced tillage in Northern China: II nutrient balances and soil fertility. *Nutrient Cycling in Agroecosystems, 79,* 17–34.

Yamah, A. (2002). The practice of the system of rice intensification in sierra leone. In N. Uphoff, E. C. M. Fernandes, L. P. Yuan, J. Peng, S. Rafaralahy, & J. Rabenandrasana (Eds.), *Assessments of the System of Rice Intensification (SRI): Proceedings of an International Workshop, April 1–4, 2002, Sanya, China* (pp. 23–25). Ithaca, NY (USA): Cornell International Institute for Food, Agriculture and Development (CIIFAD).

Chapter 16
Contribution of *Gliricidia sepium* Pruning and Fallow to Sweet Corn (*Zea mays* L. var. rugosa) Yield, Nitrogen Uptake, Release Pattern and Use Efficiency in a Humid Tropical Environment of Malaysia

P. Kathuli, A. R. Zaharah, and S. N. Nguluu

Abstract *Gliricidia sepium* is a fast growing legume shrub or tree with a wide range of environmental adaptation. Pruning of bushes at 1 m tall provides a source of crop nutrients. However, the value of pruning for increased agricultural productivity is not fully known. A study was carried out in a high rainfall humid tropical environment to investigate the contribution of *Gliricidia sepium* prunings (leaves and roots) and subsequent fallow with prunings to nitrogen (N) release pattern, sweet corn yield, nitrogen uptake and the N use efficiency (NUE) of the prunings. ^{15}N atom excess dilution and litter bag incubation techniques were used to partition N uptake and estimate N-release pattern of the leaves and roots. The prunings were applied as smooth stems and leaves at a rate of 120 kg N/ha in two split applications, 7–30 days after sweet corn germination. The crop was harvested after 75 days at the physiological maturity. The results showed that *Gliricidia Sepium* leaves and roots significantly ($p \leq 0.05$) increased sweet corn dry matter yield (3158 kg/ha) over the control (898 kg/ha). However, this was not significantly different from those treatments with leaf or root prunings in the presence of hedgerows. Use of root prunings as fallow showed significantly ($p \leq 0.05$) higher sweet corn dry matter (2175 kg/ha)

P. Kathuli (✉) · S. N. Nguluu
Kenya Agricultural and Livestock Research Organization, Katumani, P.O. Box 340-90100, Machakos, Kenya
e-mail: peterkathuli@yahoo.com

S. N. Nguluu
e-mail: nguluu@yahoo.com

P. Kathuli
National Dry-Land Research Centre, Katumani, Nairobi, Kenya

A. R. Zaharah
Department of Soil Science, Universiti Putra Malaysia (UPM), 43400 Serdang, Selangor, Malaysia
e-mail: zahaar@upm.my

S. N. Nguluu
South Eastern Kenya University (SEKU), P.O. Box 170-90100, Kitui, Kenya

M. A. Sutton et al. (eds.), *Just Enough Nitrogen*,
https://doi.org/10.1007/978-3-030-58065-0_16

yield than that (1082 kg/ha) obtained from leaf prunings application. Nitrogen partition using ^{15}N atom excess dilution showed that *Gliricidia sepium* fallow and leaf prunings contributed 25.9% and over 100% of total sweet corn N uptake and yield (as total dry matter), respectively. The incubation experiment showed that *Gliricidia sepium* leaves mineralize very fast, faster than roots, and should be re-applied within 20 days to maintain N-supply from the leaf prunings. It was concluded that *Gliricidia sepium* pruning and fallow contribute to the N uptake, growth and yield of the sweet corn with 5–9% NUE of the prunings. Further research is required to increase N use efficiency of *Gliricidia sepium* prunings and fallow in this study environment.

Keywords *Gliricidia sepium* pruning · Sweet corn yield · Nitrogen uptake · Nitrogen partition · Nitrogen release pattern · Humid tropical environment

16.1 Introduction

Gliricidia sepium is a nitrogen (N) fixing legume shrub and tree that grows well from an altitude of 0–1200 m above sea level and in regions with > 600 mm annual rainfall with annual temperatures of 20–27 °C (Elevitch and Francis 2006). The plant will tolerate a maximum daily temperature of 42 °C and will grow in acid to neutral soils with a pH range of 5–8.5 (Elevitch and Francis 2006). The plant will survive in dry seasons with less than 40 mm rainfall and fixes 13 kg N ha^{-1} $year^{-1}$. Grown as a bush, the plant has an ability to accumulate huge biomass particularly when pruned at least 2–4 times in a year (Barreto et al. 2012; Zaharah et al. 1996). The biomass has a high fertilizer equivalent content.

Gliricidia sepium can improve soil fertility from the second year after planting. The plant is pruned when it is 1 m in height to allow for sufficient root development and resilience. Its leaves and smooth stems have a C:N ratios of 10:1 and 13:1, respectively, while the roots have C:N ratios of 43:1. The C:N and polyphenol ratio controls the rate of decomposition of *Gliricidia sepium* pruning. The prunings have a polyphenol content of 2–4% and a nitrogen content of 3.6%, which favour decomposition on the soil. The leaves are expected to mineralize faster than the stems, while the roots are expected to mineralize slowly due to large C:N ratio and lignin + polyphenol: N ratio (Schwendener et al. 2005; Palm 1995).

The plant litter decomposes and accumulates into soil organic matter once it falls on the ground. The leaves and twigs mineralize very fast providing nutrients for in-situ crop or subsequent crop. Fallow with *Gliricidia sepium* similarly improves soil fertility from litter fall in the dry seasons and mineralization of both the accumulated organic matter and root decomposition.

Incorporation of the green leaves of *Gliricidia sepium* as manure has been shown to increase total top soil nitrogen, particulate soil organic matter, cation exchange capacity and water stable aggregate of a Kaolinitic soil in semi-arid north-eastern Brazil (Barreto et al. 2012). In comparison with other nitrogen fixing plants like *Albizia lebbeck* and *Leucaena leucocephala*, *Gliricidia sepium* is an intermediate N fixing plant, but with abundant nodules (Kadiata and Mulongoy 1995). These nodules contribute significantly to soil nitrogen in a *Gliricidia sepium* fallow (Kadiata and Mulongoy 1995). Furthermore, *Gliricidia sepium* does not depend much on its fixed nitrogen like other legume trees, and hence its suitability as a source of green manure and fallow soil quality improvement (Kadiata and Mulongoy 1995).

The main source of nitrogen for *Gliricidia sepium* is soil (Kadiata and Mulongoy 1995). *Gliricidia sepium* is one of the leguminous trees that can fix from 13–100 kg N ha^{-1} $year^{-1}$ with 45–100% fertilizer nitrogen equivalent (Danso et al. 1992; Elevitch and Francis 2006). Application of its prunings has been found to increase sweet corn (maize, *Zea mays* L. var. rugosa) grain yield significantly in a *Gliricidia sepium—Zea mays* intercropping in southern Malawi (Akinnifesi et al. 2006). Crop yield increase due to application of its prunings has been further reported in alley cropping trials by Amara et al. (1996).

Gliricidia sepium can influence both the supply and availability of nutrients in the soil through biological nitrogen fixation, subsoil exploitation and retrieval of nutrients from below the rooting zone and its green manure incorporation (Buresh and Tian 1997). The impact is realized when *Gliricidia sepium* prunings are incorporated alone or together with inorganic fertilizers. For example, Matilda (2009) reported a yield increase in cassava tubers due to use of *Gliricidia sepium* prunings combined with inorganic fertilizer. Sweet corn grain yield was found to be significantly increased by 55% in a *Gliricidia sepium* fallow compared to traditional fallow system (Hall et al. 2006) and three times (mean of 1–3.8 t/ha) in a trial over 10 years applying *Gliricidia* prunings in southern Malawi (Akinnifesi et al. 2006).

Gliricidia trees are either alley-cropped (i.e., as hedgerows) or planted in rotation with crops (i.e., short bushes) (Akinnifesi et al. 2006). The impact is greatest when the trees are pruned and the tree pruning biomass is applied and incorporated in the plough layer (Akinnifesi et al. 2006; Barrios et al. 1998). Legume trees and bushes in the field mainly build up soil fertility and particularly nitrogen through reduced leaching of nitrates and build-up of light fraction soil organic matter (Buresh and Tian 1997; Mekonnen et al. 1997). This leads to increased mineralizable nitrogen and hence available nitrogen for an in-situ or rotational crop. The influence of legume trees in the field further depends on the duration the trees have been in the field (Barrios et al. 1998), the lignin polyphenol to nitrogen ratio and the C:N ratio of the biomass (Palm 1995).

Three key questions are relevant when *Gliricidia sepium* prunings are used to supply nutrients for an in-situ crop:

(a) How do these prunings release the nutrients and what is their nitrogen use efficiency?
(b) How long can the application of prunings sustain nutrient supply to the crops in the field?
(c) What is the nutrient release pattern of the respective green manure source (leaves, stems, roots) of the *Gliricidia sepium* prunings?

Answers to these questions are vital in order to optimise nitrogen use efficiency (NUE) when *Gliricidia sepium* prunings are used as source of plant nutrients.

The estimation of NUE from *Gliricidia sepium* prunings is never precise, as the nitrogen uptake in study crops is never partitioned into sources from mineral fertilizer and applied *Gliricidia sepium* prunings. At present, there is insufficient understanding of nitrogen release pattern, NUE and subsequent crop N uptake from *Gliricidia sepium* prunings. Hence, there is a need for a study to gain the insights of *Gliricidia sepium* prunings on NUE, N release pattern, and subsequent crop N uptake in a humid tropical environment. This study was therefore conducted to investigate these factors, with the study focusing on the performance of sweet corn planted in a humid tropical environment in Malaysia.

16.2 Materials and Methods

16.2.1 Gliricidia Treatments

The site had been planted with young *Gliricidia sepium* trees, which were 18 months old. The trees were planted in such a way that the plants were cut off at age of 18 months to give way to 8 treatments replicated four times in a randomized complete block design (RCBD) where the sweet corn was planted. The treatments were:

1. P_1 = Control with *Gliricidia* tree hedgerows
2. P_2 = *Gliricidia* tree hedgerows with incorporation of *Gliricidia* leaves
3. P_3 = *Gliricidia* tree hedgerows and roots
4. P_4 = *Gliricidia* tree hedgerows + leaves + roots
5. P_5 = Control (no *Gliricidia* hedgerows, no leaves or roots)
6. P_6 = Application of *Gliricidia* leaves only
7. P_7 = Application of *Gliricidia* roots only (*Gliricidia sepium* fallow)
8. P_8 = Application of both leaves + roots of *Gliricidia*

Smooth *Gliricidia sepium* stems and leaves were applied as prunings in two split applications each at rate of 60 kg N/ha. This was equivalent to 12 kg of leaves per plot of $5 \times 3\ m^2$ dimensions. The prunings were applied 7–30 days after sweet corn germination and incorporated to provide a total of 120 kg N/ha. The leaves, young stems and roots were analysed prior to application to quantify the amount of nitrogen they contained. The smooth stems, leaves and roots contained 2%, 4% and 3% N, respectively. *Gliricidia sepium* fallow appeared when the shrub had been grown in the plot for 18 months, but was cleared for planting sweet corn.

16.2.2 Basal Fertilizers

All the plots were given a blanket fertilizer application of nitrogen (N), phosphorus (P) and potassium (K). Phosphorus and K were applied at 10 kg P/ha as triple superphosphate (TSP) and 75 kg K/ha as Muriate of potash (M.O.P) (elemental K and P, respectively). Phosphorus was applied as a single dose four days after sweet corn germination, while K was applied split in two doses, with a half of the amount being applied together with the TSP and the rest applied at the end of the second month. Nitrogen was applied at 40 kg N/ha as urea after germination in the big yield plot, leaving out the ^{15}N plot in the middle. In the absence of nitrogen from soil and pruning application, the nitrogen taken up by the sweet corn would be the labelled nitrogen which was estimated from the percentage atom excess in the sweet corn total dry matter using the isotope dilution technique.

16.2.3 ^{15}N Isotope Dilution Technique

Using an indirect ^{15}N labelling method it was possible to estimate nitrogen uptake from leaves, roots and to assess the influence of hedgerows on nitrogen derived from the prunings. The labelled fertilizer was applied in a micro-plot at 10% ^{15}N atom excess as ammonium sulphate at 40 kg N/ha before sweet corn planting in a plot size of 2×1 m^2, located at the centre of the big plot of 5×3 m^2. Using this labelled tracer it was possible to estimate the partitioning of N fertilizer sources in the sweet corn total dry matter.

16.2.4 Harvesting

The sweet corn crop was harvested at physiological maturity after 75 days. The ^{15}N micro-plot was harvested separately and the weight recorded. The big yield plot (yield from the experimental plot of 5×3 m^2, excluding the ^{15}N micro plot of 2×1 m^2 at the centre of the 5×3 m^2 plot) was harvested and weight recorded. The fresh yield from the ^{15}N micro-plot was added to the fresh weight of the big yield plot after sampling for moisture and laboratory analysis. The fresh sweet corn biomass from the ^{15}N plot and the big yield plot were chopped and sampled for moisture corrections and analysis of N-nutrient uptake from the fertilizer and the prunings and ^{15}N % atom dilution for partitioning N sources in the sweet corn in the ^{15}N micro-plot.

16.2.5 Incubation Experiment

An incubation experiment was conducted alongside the sweet corn field experimental plot to estimate the rate of turnover of both the *Gliricidia sepium* leaves and roots. Dry weight of leaves and roots were weighed separately into nylon bags in four replicates for each incubation time from day 0, 5, 10, 20, 30, 40, 50 and 70. At each time of sampling, the replicate samples for each incubation date were taken out from the soil, and dried until a constant weight was reached in the oven at 70 °C. The material was then ignited in a furnace at 450 °C for three hours, taken out of the furnace and allowed to cool in a desiccator and the weight taken. The weight of the material remaining was estimated from the percentage loss in weight after ignition. Nitrogen was analysed from a sample taken from the dried material before ignition.

16.2.6 Laboratory Analysis

The soil and plant samples were analysed according to methods given in ILCA (1991). Briefly, available P was extracted using the Bray II method, available Calcium, Magnesium and Potassium were extracted from the soil using buffered ammonium acetate at pH 7 while the trace elements were analysed using dilute double mineral acid (Kathuli et al. 2007). Total nitrogen in the soil and corn biomass was analysed using the Kjeldahl method (Sáez-Plaza et al. 2013; Bremner 1965). The ^{15}N atom excess was analysed using N^{-15} emission spectrometer (Rittenberg et al. 1939). Three samples per treatment were analysed for ^{15}N atom excess (%) and a mean analysis calculated for that treatment.

16.2.7 Data Analysis

The sweet corn yield and N uptake data from the big yield plot and the ^{15}N micro-plot was analysed for source of variations within the treatments and replicates by ANOVA (Analysis of Variance) using SAS (Statistical Analysis System) statistical package. Similarly, ANOVA was performed on the corn yield, N uptake and ^{15}N atom excess (^{15}N atom.exc.) of the sweet corn from ^{15}N plot. The % N in the sweet corn biomass was also analysed statistically using ANOVA. The amount of material (leaves or roots remaining after incubation and percentage nitrogen) remaining in the leaves and roots after 70 days of incubation was also analysed statistically using ANOVA to give the effect of time of incubation on the nitrogen release pattern of *Gliricidia sepium* leaves and roots.

Nitrogen derived from the pruning (NdfP) and nitrogen use efficiency of the prunings (NUE_p %) were calculated after analysis of ^{15}N atom excess in the sweet corn. The incubation experiment provided an indication whether the N release of

Gliricidia sepium pruning was in synchrony with sweet corn N uptake. In all the analyses, the means were ranked using Duncan's Multiple Range Test.

16.3 Results

16.3.1 Study Site

The study was conducted during 1995–1996 in a highly weathered tropical soil in a humid environment at the Puchong experimental site of the Universiti Putra Malaysia (UPM), which lies at 03′ 00′ 12.1 N and 101′ 39′ 33.1 E. The site is 30 km by road from Kuala Lumpur. The area has mean temperature of 30 °C and an annual rainfall of 2700 mm. The soils are classified as Ultisol (USDA 1975) and belong to the *Bungor* series under Malaysian soil classification. The soils are sandy clay loams and very acidic pH (H_2O): 4.7), low in CEC (6 cmol (+)/kg soil), low in exchangeable calcium and magnesium (< 0.1 cmol (+)/kg soil) and potassium below 0.02 cmol (+)/kg soil. Organic matter (and hence total nitrogen) was low with an average total N of 0.17%. Phosphorus was very low (4 mg/kg soil (Bray II). Trace element zinc was low (0.1 mg/kg). Exchangeable aluminium was in the range 40–111 mg/kg, copper was below 1 mg/kg, while manganese was in range of 1–3 mg/kg soil.

16.3.2 Field Experiments

The results of the study investigating the nitrogen release pattern and NUE of prunings from *Gliricidia sepium* and associated sweet corn yield at the Puchong experimental site are shown in Table 16.1.

The results of the parallel field incubation experiment showing the rate of decomposition of *Gliricidia sepium* leaves are shown in Fig. 16.1.

The nutrient release pattern of *Gliricidia sepium* leaves and roots in the same environment is shown in Figs. 16.2 and 16.3.

The partitioning of N uptake sources in the sweet corn is shown in Table 16.2.

The results showed that the combination of leaf and root prunings of *Gliricidia sepium* significantly ($p \leq 0.05$) increased sweet corn total dry matter yield and nitrogen uptake in the absence of *Gliricidia* hedgerows as compared with the control (rows P8 versus P5 in Table 16.1). This observation was the same whether it was made from the big yield plot or the ^{15}N labelled micro-plot. Although large mean differences were found, several of the other treatments were not significantly different (Table 16.1). The highest nitrogen use efficiency of prunings (NUE_p) was estimated for P2 (leaves application in plot with hedgerows, $NUE_p = 9\%$) followed by P6 and P8 ($NUE_p = 5\%$), each of which included treatment with leaf prunings.

Table 16.1 Contribution of *Gliricidia sepium* prunings to sweet corn yield, nitrogen uptake and nitrogen use efficiency at the Puchong experimental site of UPM Malaysia in 1994–1996

	Sweet corn total dry matter yield		Nitrogen partitioning and use					
Treatment	Big yield plot (kg/ha)	^{15}N plot (kg/ha)	% N	N uptake (kg/ha)	^{15}N % atom excess	% NdfP	NdfP (kg/ha)	NUE_p (%)
P1	1628^{bdc}	633^{abc}	1.749^{bc}	9.345^{bc}	3.784^{a}	0	0	0
P2	2686^{ab}	1225^{ab}	1.610^{bc}	19.455^{ab}	1.523^{c}	56.6	11.012	9.18
P3	2421^{abc}	1145^{ab}	1.463^{c}	17.873^{ab}	3.241^{ab}	7.6	1.358	1.13
P4	1836^{abcd}	600^{abc}	1.932^{b}	11.676^{abc}	2.657^{b}	24.3	2.837	2.36
P5	698^{d}	207^{c}	1.759^{bc}	3.300^{c}	3.509^{a}	0	0	0
P6	1082^{cd}	410^{c}	2.419^{a}	9.613^{bc}	1.298^{c}	63.0	6.056	5.05
P7	2175^{abcd}	827^{abc}	1.463^{c}	11.926^{abc}	3.579^{a}	0	0	0
P8	3158^{a}	1418^{a}	1.779^{bc}	24.330^{a}	2.600^{b}	25.9	6.301	5.25

Means with the same letter in the same column are not significantly different ($p \leq 0.05$)
P1 = control + Hedgerows, P2 = Hedgerows + leaf prunings, P3 = Hedgerows + root prunings, P4 = Hedgerows + leaf and root prunings, P5 = control, P6 = leaf prunings, P7 = Root prunings, P8 = Leaf and root prunings, NdfP = nitrogen derived from prunings, NUE_p = nitrogen use efficiency of prunings
^{a}highly significant, ^{ab}significant but not different from mean with letter a, ^{abc}less significant while ^{d}is least significant

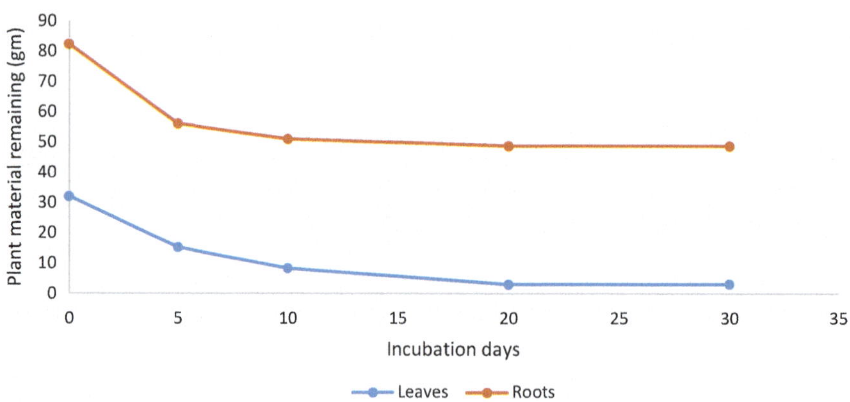

Fig. 16.1 Effect of incubation on decomposition of *Gliricidia sepium* leaves and roots at Puchong experimental site, UPM Malaysia

The results from the partition study of N uptake using the ^{15}N indirect labelling technique are shown in Table 16.2. The values indicate that sweet corn obtained nitrogen from each of applied fertilizer, soil and prunings in all the treatments that received *Gliricidia sepium* leaf prunings. However, there was more nitrogen derived

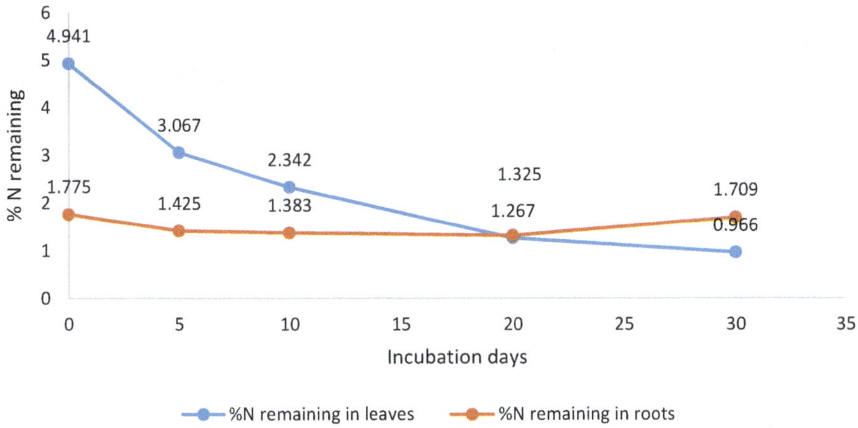

Fig. 16.2 Effect of incubation on N-release pattern of *Gliricidia sepium* prunings incorporated in soil in a humid tropical environment of Puchong, UPM Malaysia

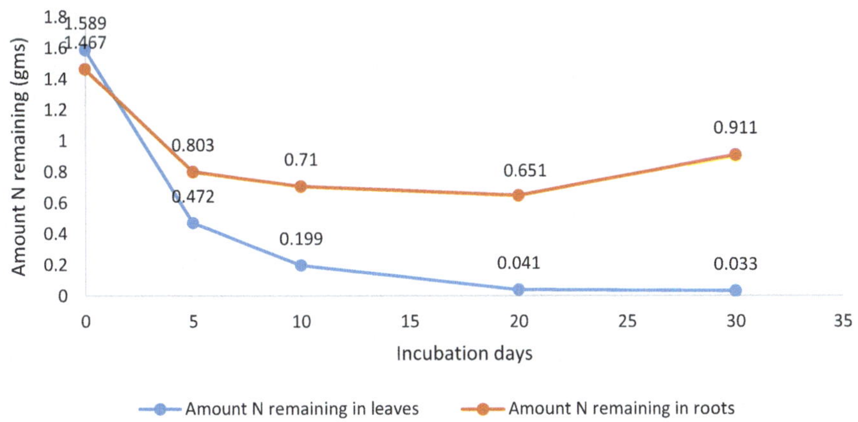

Fig. 16.3 Effect of time of incubation on amount of nitrogen release from *Gliricidia sepium* prunings incorporated in soil in a humid tropical environment of Puchong, UPM Malaysia

from the soil where there was *Gliricidia sepium* fallow with root prunings and from leaf prunings. Conversely, nitrogen derived from prunings was estimated to be largest when *Gliricidia* leaf prunings were applied. The ^{15}N percent atom excess analysis (Table 16.1) revealed that, there was significantly ($p \leq 0.05$) more ^{15}N atom excess taken up by sweet corn in those treatments that did not have *Gliricidia sepium* pruning (leaves) application.

The results of the parallel field incubation experiment on N release pattern and decomposition of *Gliricidia sepium* prunings (Figs. 16.1, 16.2 and 16.3) revealed that *Gliricidia sepium* leaves decompose significantly faster than its roots ($p \leq 0.05$). Approximately 50% of the nitrogen in leaves is released within 10–20 days after

Table 16.2 Partitioning of nitrogen sources in sweet corn in the ^{15}N plot using an indirect ^{15}N labelling technique at UPM experimental farm in Malaysia

Treatment	Nitrogen derived from fertilizer Ndff (kg/ha)	Nitrogen derived from soil Ndfs (kg/ha)	Nitrogen derived from prunings NdfP (kg/ha)
P1 = Control + Hedgerows	3.7	5.7	0.0
P2 = Hedgerows + leaf prunings	3.1	5.4	11.0
P3 = Hedgerows + root prunings	6.0	10.5	1.4
P4 = Hedgerows + leaf and root prunings	3.2	5.8	2.8
P5 = Control	1.2	2.1	0.0
P6 = Leaf prunings	1.3	2.2	6.1
P7 = Root prunings	4.4	27.5	0.0
P8 = Leaf and root prunings	6.6	11.4	6.3

Calculations based on Zapata (1990)

incorporation in the soil in the tropical humid environment. The *Gliricidia sepium* roots were decomposed much more slowly, with Fig. 16.3 showing a large recalcitrant fraction in the root nitrogen.

16.4 Discussion

Gliricidia Sepium prunings significantly increased the yield of sweet corn where the prunings were applied as leaves and smooth stems with retention of roots in the field with no hedgerows (treatment P8 as compared with control P5). By contrast, when hedgerows were present with application of leaves on a *Gliricidia sepium* fallow (treatment P4) the yield was not significantly increased compared with control P1. Several of the other treatments did not show significant differences, which reflects large scatter in the results, given the substantial differences between mean results. For example, the yields were not significantly different between treatments receiving pruning application with hedgerows (treatment P3; – roots retention + hedgerows, P2; – leaves pruning + hedgerows) and the treatment that retained roots (P7) without hedgerows.

16.4.1 Effect of Prunings Application and Fallow

Application of leaves pruning in *Gliricidia sepium* fallow (treatment P8) significantly increased sweet corn total dry matter yield from 698 kg/ha (treatment P5) to 3158 kg/ha. This represented 79% yield increase in comparison to control and 66% yield increase over the application of leaf prunings (P6). This was attributed to a combined supply of nitrogen from mineralization of *Gliricidia sepium* leaves and roots. The field had been planted with *Gliricidia sepium* for 18 months. This could have resulted in biological nitrogen fixation (BNF) of *Gliricidia sepium* and subsequent soil enrichment for the duration that *Gliricidia sepium* was in the field.

The *Gliricidia sepium* fallow (root prunings treatment P7) was associated with a higher sweet corn dry matter yield (from 698 to 2175 kg/ha, not significant, Table 16.1) and nitrogen uptake (from 3.3 to 11.93 kg N/ha, not significant, Table 16.1). Although not significant, this could point to N supply from decomposition of *Gliricidia sepium* roots and associated BNF of *Gliricidia sepium* over the duration the plant was in the field. This was corroborated by the fact that *Gliricidia sepium* fallow (P7) tended to increase sweet corn dry matter yield from 1082 kg/ha (treatment P6) to 2175 kg/ha. This represented 31% yield increase by 18 months old *Gliricidia sepium* fallow (Table 16.1).

16.4.2 Effect of Hedgerows and Pruning Applications

In general, the presence of hedgerows was associated with higher sweet corn yield and nitrogen uptake (Table 16.1). The sweet corn dry matter yields were increased by 14% (treatment P3) by planting in *Gliricidia sepium* fallow with hedgerows, 133% (P1) by having hedgerows and no pruning use and 148% in leaves with hedgerows plots (P2). Application of *Gliricidia sepium* prunings in the presence of hedgerows generally increased sweet corn dry matter yield in hedgerow combination with leaves, roots and control (treatments P6, P7 and P5 respectively) as shown in Table 16.1, except in plots receiving the application leaf prunings on *Gliricidia sepium* fallow. It was not clear why this happened and more data is required to explain this observation.

Similar results were found in the ^{15}N yield plot for both dry matter yield (DMY) and N uptake (Table 16.1). This supports the findings of Amara et al. (1996) which involved application of *Gliricidia sepium* prunings in alley cropping trials in Sierra Leone and the work of Jama et al. (1998) on improved fallows in western Kenya.

The treatment (P8), application of leaves with retention of roots, appears to have had the greatest yield increase over the other treatments because of the nitrogen released from the decomposing prunings and the initial soil fertility build-up from the *Gliricidia Sepium* fallow plots which were 18 months old. This observation is supported by earlier work of Jama et al. (1998); Barrios et al. (1998), (1996); Buresh and Tian (1997); Mekonnen et al. (1997) who reported soil fertility build up in soils under leguminous plants through biological nitrogen fixation. Incorporation of leaves added more nitrogen through mineralization. This has been found to occur in the top soils through an increase of light fraction organic matter, which leads to increase in mineralizable nitrogen and hence more nitrogen uptake (Barrios et al. 1998; Ikerra et al. 1999).

The N uptake was found to be significantly correlated to dry matter yield (DMY, kg/ha) of the sweet corn (Table 16.1; DMY = 111 (N uptake, kg/ha) + 494; $R^2 = 0.91$). Treatment P8 (leaves and roots pruning application) led to significant N uptake in the DMY of the sweet corn although it was not significantly different from those treatments (P2, P3, P4 and P7) (Table 16.1) which had pruning and hedgerows or roots retained in the plots (legume fallow rotation i.e., treatment P7). This can be attributed to the explanations given for corn dry matter yield of the effect of legume trees on the soil and the nutritional benefits of a subsequent crop (Buresh and Tian 1997; Jama et al. 1998). Notably, the absolute control treatment (P5) had the least N uptake (3.3 kg N/ha) (Table 16.1) which increased by three times with introduction of *Gliricidia sepium* hedgerows as expected in alley cropping (Amara et al. 1996). This was not significantly different from treatment (P6) which received leaf prunings alone. Sweet corn N uptake for treatment P2 (leaf prunings with *Gliricidia sepium* hedgerows) was double that of treatment P6 (leaf prunings without *Gliricidia* hedgerows) (Table 16.1, not significantly different).

16.4.3 Partitioning of Nitrogen Sources for the Sweet Corn

The nitrogen sources of the sweet corn were estimated from ^{15}N isotope dilution technique. The fundamental principle behind the ^{15}N isotope dilution technique is that if the plant has no other source of fertilizer, it will take up the labelled fertilizer in large amounts, i.e., if fertilizer applied had 10% atom excess ^{15}N, then there will be 10% atom excess of nitrogen taken up in the plant (Zapata 1990). The nitrogen derived from the fertilizer (Ndff) can be calculated as:

$$\%\text{Ndff} = \left[1 - (\text{atom excess in plant}/\text{atom excess in fertilizer applied})\right] \times 100 \quad (16.1)$$

If there is no dilution, then atom excess in plant = atom excess in fertilizer applied. Therefore the entire N taken by the plant can be considered has having come from the labelled fertilizer. The ^{15}N technique is a tag method to distinguish between fertilizer sources and can be used to identify sources of plant nitrogen. The technique can also be used to investigate the best method of fertilizer placement, effective depth of placement, best time of application and the effect of environment on fertilizer use. Nitrogen derived from pruning (NdfP) is given by the following equation:

$$\%\,\text{NdfP} = \left[1 - \left(\%^{15}\text{N atom excess in plant}/\%^{15}\text{N atom excess in control plot}\right)\right] * 100 \quad (16.2)$$

Nitrogen use efficiency from the pruning (NUE_p) is given by;

$$NUE_p\% = (NdfP\ /\text{amount of nitrogen applied as pruning}) * 100 \qquad (16.3)$$

These calculations are based on procedures of Zapata (1990) at the International Atomic Energy Agency (IAEA).

The results of partition of N uptake using the ^{15}N indirect isotope labelling technique (Table 16.2) and ^{15}N atom excess (Table 16.1) revealed that *Gliricidia sepium* prunings did significantly contribute to sweet corn dry matter yield and N uptake in the study environment. The significant effect of leaf prunings in lowering ^{15}N excess is clearly seen in Table 16.1.

The contribution of *Gliricidia* was greatest (11.0 kg N/ha (Nitrogen derived from pruning (NdfP = 11.0 kg N/ha) in the treatment P2 which had pruning application as leaves and had hedgerows. The interaction of leaves and roots (treatment P8) resulted in additional N uptake of 14.7 kg N/ha. However planting sweet corn in hedgerows decreased both N uptake and NdfP (Table 16.2) except where leaves were applied (treatment P2), which had 11 kg N/ha derived from pruning. The reason for this is not clear since there was significant increase in both the sweet corn dry matter yield and total N uptake. These results corroborated the findings of Ikerra et al. (1999) that *Gliricidia sepium* pruning increases top soil nitrogen and sweet corn yields and the findings of Srinivavasa Rao et al. (2011) that *Gliricidia sepium* pruning can decompose and enrich soil nitrogen from 5 to 150 days of incorporation in the soil.

Roots retention in plots previously planted with *Gliricidia sepium* contributed large amounts (from 10 to 27 kg N/ha) of nitrogen derived from the soil (Ndfs) and hence the high yield of the sweet corn in treatments P3, P4, P7 and P8 which had *Gliricidia sepium* roots (Table 16.2). This was attributed to nitrogen release from the *Gliricidia Sepium* litter and soil fertility improvement by the legume tree through symbiotically fixed nitrogen as reported by Buresh and Tian (1997); Jama et al. (1998); Barrios et al. (1998); Ikerra et al. (1999). These results further supported the findings of Martins et al. (2015) who reported that *Gliricidia sepium* can fix large amounts of nitrogen and contributes 40 kg N/ha from decomposition of its leaves in northern semi-arid Brazil. This nitrogen would benefit a subsequent crop as reported in this paper.

The interactions of leaves, roots and hedgerows showed that more nitrogen was derived from the soil in *Gliricidia sepium* fallows (treatments P8 and P7 with roots, Table 16.2) than in plots with leaves incorporation alone (treatment P6, Table 16.2). The plots with *Gliricidia sepium* fallow (treatment P7, Table 16.2) had more (27.56 kg N/ha) Ndfs than where *Gliricidia sepium* leaves were incorporated in a *Gliricidia sepium* fallow (Treatment P8, Table 16.2). This difference remains unexplained, as there is no apparent advantage of pruning application on a *Gliricidia sepium* fallow, although leaching of nitrate nitrogen could have differed. Nitrogen derived from pruning for leaves and roots appears to have been additive. Plots with leaves and roots had 6.3 kg N/ha NdfP while that with leaves only had 6.1 kg N/ha NdfP (Table 16.2).

Hedgerows were associated with a decrease in sweet corn nitrogen derived from pruning, except where leaves were incorporated (treatment P6, Table 16.2). Nitrogen from pruning can be maximized when leaves are applied in fields with hedgerows but not in *Gliricidia sepium* fallow and *Gliricidia sepium* fallow are productive when crops are planted with no hedgerows.

16.4.4 Decomposition and Nutrient Release of Gliricidia Sepium Pruning

The parallel field incubation results shown in Figs. 16.1, 16.2 and 16.3 revealed that the leaves of *Gliricidia sepium* mineralise and release nitrogen at a significantly faster rate than the roots. The leaves were found to decompose release half of their nitrogen within the first 20 days after application (Figs. 16.2 and 16.3). The roots were decomposed more slowly (Fig. 16.1), while Fig. 16.3 showed only a small reduction in total nitrogen. These results indicate that *Gliricidia Sepium* prunings (leaves and smooth stems) are a relevant source of nitrogen for sweet corn growing and will require subsequent application after every 20 days in order to maintain N supply from the decomposing prunings. It is here suggested that application of leaf prunings should start immediately after germination of the sweet corn at 60 kg N/ha fertilizer equivalent followed by another application 20 days after the first application, according to the observed pruning nitrogen release pattern (Fig. 16.3).

The present observations corroborate the findings of Hartemink and O'Sullivan (2001) who found that, *Gliricidia sepium* litter mineralized very fast and 50% of the leaf biomass had decomposed in 10 weeks in the humid lowlands of Papua New Guinea. The findings further corroborate the research of Srinivasa Rao et al. (2011) who reported that, *Gliricidia sepium* leaves can release 7 mg N/kg soil after 5 days of incubation and 121 mg N/kg soil after 150 days of incubation.

The roots are rather stable due to their high C: N ratio of 43 compared to that of leaves (C/N = 10.2) and did not release most of the nitrogen within 70 days of incubation. However, 18 months old *Gliricidia sepium* fallow can contribute over 25.4 kg N/ha of readily available nitrogen for a subsequent crop (Table 16.2). This supports the earlier observation by Buresh and Tian (1997) that leguminous trees can improve the nitrogen status of the soil through biological nitrogen fixation, retrieval of nutrients below the rooting zone of crops, recycled residues and reduction of nutrient losses from leaching and erosion.

16.4.5 Nitrogen Use Efficiency of Prunings

The nitrogen use efficiency of prunings (NUE_p) (Table 16.1) ranged from 5 to 9%. This indicates very low N recovery suggesting losses of nitrogen, such as through leaching or volatilization particularly from the application of leaf prunings. The area has high annual rainfall of 2700 mm and high annual mean temperature of 35 °C with sandy clay loam soils. Studies on litter decomposition and nitrogen release pattern indicated fast release of nitrogen in this environment. Nitrate form of nitrogen is released from mineralization of organic substrates like *Gliricidia sepium* pruning and nitrates are very soluble making them susceptible to leaching loss to the environment leading to pollution. It is suggested that, further research is required to optimize NUE_p. For example, NUE_p was found to be larger in sweet corn grown with *Gliricidia sepium* hedgerows (Table 16.1). For example, hedgerows might reduce leaching losses, as noted by Buresh and Tian (1997).

16.5 Conclusions

Gliricidia sepium prunings (especially young stems and leaves) mineralized to give nitrogen that contributed significantly to the nutrition of sweet corn and hence the yield. Mineralization of *Gliricidia sepium* leaf prunings is found to be faster than the roots. The observed time course of decomposition suggests that *Gliricidia sepium* leaf prunings should be applied within an interval of 20 days in order to maintain a steady supply of nitrogen to an in-situ crop. *Gliricidia sepium* fallow with incorporation of its prunings can provide available nitrogen in the soil leading to 3.5 times yield increase of a subsequent crop.

Nutrient release patterns of *Gliricidia sepium* pruning materials should be documented in their respective user environment to predict the required frequency of application. This may also help to optimise their nitrogen use efficiency because of the risk of losses, such as by leaching. Although not quantified in the present study, such losses may result from large amounts of rainfall and fast mineralization of prunings associated with high soil temperatures.

Acknowledgements The United Nations Development Programme (UNDP) and the International Atomic Energy Agency (IAEA) acknowledged for funding the project. Professor Zaharah of *Universiti Putra* Malaysia (UPM) is thanked for supervising the first author on use of isotopes in soil fertility studies. Thanks are also given to the Director Kenya Agricultural Research Institute (KARI) (now the Kenya Agricultural and Livestock Research Organization [KALRO]) for granting leave to study in Malaysia and to the technicians at UPM Soil Science Department for their assistance in soil analysis.

References

Akinnifesi, F. K., Makumba, W., & Kwesiga, F. R. (2006). Sustainable maize production using Gliricidia/maize intercropping in southern Malawi. *Experimental Agriculture, 42*(4), 441–457.

Amara, S. D., Sanginga, N. S., Danso, S. K. A., & Suale, D. S. (1996). Nitrogen contribution by multipurpose trees to rice and cowpea in alley cropping system in Sierra Leone. *Agroforestry Systems, 34*(2), 119–128.

Barreto, A. C., Chaer, G. M., & Fernandes, M. F. (2012). Hedgerow pruning frequency effects on soil quality and maize productivity in alley cropping with *Gliricidia Sepium* in Northeastern Brazil. *Soil and Tillage Research, 120,* 112–120.

Barrios, E., Buresh, R. J., & Sprent, J. I. (1996). Organic matter in soil particle size and density fractions from maize and legume cropping systems. *Soil Biology and Biochemistry Journal, 28*(2), 185–193.

Barrios, E., Kwesiga, F., Buresh, R. J., Coe, R., & Sprent, J. I. (1998). Relating preseason soil nitrogen to maize yield in tree legume-maize rotations. *Soil Science Society of America Journal, 62*(6), 1604–1609.

Bremner, J.M. (1965). Total nitrogen. In C.A. Black, D.D. Evans, L.E. Ensminger, L.E., White, J.L., Clark, F.E., & Dinauer, R.C. (Eds). *Methods of soil analysis Part 2.*

Buresh, R. J., & Tian, G. (1997). Soil improvement by trees in Sub-Sahara Africa. *Agroforestry Systems, 38,* 51–76.

Danso, S. K. A., Bowen, G. D., & Sanginga, N. (1992). Biological Nitrogen fixation in trees in agro-ecosystems. *Plant and Soil, 141,* 177–196.

Elevitch, C.R., & Francis, J.K. (2006). *Species profiles for Pacific Island agroforestry*. Ver. 2.1, 1–18. Available at: www.traditionaltree.org. Accessed August 2018.

Hall, N. M., Kaya, B., Dick, J., Skiba, U., Niang, A., & Tabo, R. (2006). Effect of improved fallow on crop productivity, soil fertility and climate-forcing gas emissions in semi-arid conditions. *Biology and Fertility of Soils, 42*(3), 224–230.

Hartemink, A. E., & O'Sullivan, J. N. (2001). Leaf litter decomposition of *Piper aduncum, Gliricidia sepium* and *Imperata cylindrica* in the humid lowlands of Papua New Guinea. *Plant and Soil, 230*(1), 115–124.

Ikerra, S. T., Maghembe, J. A., Smithson, P. C., & Buresh, R. J. (1999). Soil nitrogen dynamics and relationships with maize yields in a *Gliricidia*-maize intercrop in Malawi. *Plant and Soil, 211,* 155–164.

ILCA. (1991). *Methods of soil analysis Part 3. Chemical methods. Working document 13, ILCA, Addis Ababa, Ethiopia.* [NB: This working document was later published later in 1996 as SSSA Book series 5.3]. USA, WI: Soil Science Society of America, American Society of Agronomy.

Jama, B., Niang, A.I., Amadolo, B., Wolf, J., Rao, M.R., & Buresh, R.J. (1998). The potential of planted fallows to improve and conserve the fertility of nutrient-depleted soils of western Kenya. In *Proceedings of the 6th KARI Scientific Conference, Nairobi, November 9–13, 1998* (pp. 133–144). Nairobi: Kenya Agricultural Research Institute.

Kadiata, B. D., & Mulongoy, K. (1995). Early nitrogen fixation and utilization in Albizia Lebbeck, Leucaena Leucocephala and Gliricidia Sepium using nitrogen (^{15}N) labelling. *Communications in Soil Science and Plant Analysis, 26*(9/10), 1397–1409.

Kathuli, P., Gachimbi, L.N., Maingi, S.W., & Thuranira, E.G. (2007). Suitability of modified dilute double acid-Mehlich 1 (0.1NHCL + 0.025NH2SO4) and Olsen (0.5MNaHCO3 PH 8.5) methods for testing of available phosphorus for Kenyan soils. *Discovery and Innovation, Journal of the African Academy of Sciences and the Academy for the Developing World, 19*(3), 195–201.

Martins, J. C. R., Freitas, A. D. S., Menezes, R. S. C., & Sumpaio, E. V. S. (2015). Nitrogen symbiotically fixed by cowpea and *Gliricidia sepium* in traditional and agroforestry systems under semi-arid conditions. *Pesquisa Agropecuária Brasileira, 50*(2), 178–184.

Matilda, A. (2009). *Effect of Gliricidia Sepium and Senna Siamea pruning on the growth and root yield of cassava.* MPhil. Thesis submitted to the School of Postgraduate Studies, Kwame Nkrumah University of Science and Technology.

Mekonnen, K., Buresh, R. J., & Jama, B. (1997). Root and inorganic nitrogen distributions in Sesbania fallow, natural fallow and maize fields. *Plant and Soil, 188*(2), 319–327.

Palm, C. A. (1995). Contribution of agroforestry trees to nutrient requirements of inter-cropped plants. *Agroforestry Systems, 30,* 105–124.

Rittenberg, D., Keston, A. S., Rosebury, K. F., & Schoenheimer, B. (1939). Studies in protein metabolism. II. The determination of nitrogen isotopes in organic compounds. *Journal of Biological Chemistry, 127,* 291–299.

Sáez-Plaza, P., Navas, M.J., Wybraniec, S., Michalowski, T., & Asuero, A.G. (2013). An overview of the Kjeldahl method of nitrogen determination. Part II. Sample preparation, working scales, instrumental finish, and quality control. *Critical Reviews in Analytical Chemistry, 43,* 224–272. Taylor and Francis.

Schwendener, C. M., Lehmann, J., de Camargo, P. B., Luizão, R. C. C., & Fernandes, E. C. M. (2005). Nitrogen transfer between high- and low- quality leaves on a nutrient- poor Oxisol determined by ^{15}N enrichment. *Soil Biology & Biochemistry, 37*(4), 787–794.

Srinivasa Rao, Ch., Venkateswarlu, B., Dinesh Babu, M., Wani, S. P., Dixit, S., Sahrawat, K. L., & Sumanta K. (2011). *Soil health improvement with Gliricidia green leaf manuring in rainfed agriculture, on farm experiences.* Technical bulletin 2/2011. Central Research Institute for Dry land Agriculture (ICAR).

USDA. (1975). *Soil taxonomy. A basic system of soil classification for making and interpreting soil surveys.* Agriculture Handbook 436. Washington DC, USA: Soil Conservation Service, US Department of Agriculture (USDA).

Zaharah, A. R., Sharifuddin, H. A. H., & Anuar, R. (1996). Field measurements of N_2 fixation by *Gliricidia Sepium.* In O. Van Cleemput, G. Hofman, A. Vermoesen (Eds.). *Progress in nitrogen cycling studies. Proceedings of the 8th Nitrogen Workshop held at the University of Ghent, 5–8 September, 1994* (pp. 321–326). Developments in Plant and Soil Sciences, 68. Dordrecht: Springer Netherlands.

Zapata, F. (1990). Isotope techniques in soil fertility and plant nutrition studies. In G. Hardarson (Ed.) *Use of nuclear techniques in studies of soil-plant relationships* (pp. 61–127). Training course series No. 2. Vienna: International Atomic Energy Agency (IAEA).

Part IV
Nitrogen Impacts on Health, Ecosystems and Climate: Nitrogen Impacts on Health and Ecosystems

Chapter 17
Further Evidence of the Haber-Bosch—Harmful Algal Bloom (HB-HAB) Link and the Risk of Suggesting HAB Control Through Phosphorus Reductions Only

Patricia M. Glibert, Roxane Maranger, Daniel J. Sobota, and Lex Bouwman

The original "Haber-Bosch—Harmful Algal Bloom Link" paper was published in 2014 in *Environmental Research Letters* (Glibert et al. 2014a. (https://iopscience.iop.org/article/10.1088/1748-9326/9/10/105001.) The Haber Bosch-harmful algal bloom (HB-HAB) link, *Environmental Research Letters*, 9, 105001. https://doi.org/10.1088/1748-9326/9/10/105001). At the request of the editors of this volume, and with permission of IOP Publishing Ltd., under Creative Commons license CC BY 3.0 (https://creativecommons.org/licenses/by/3.0), much of that paper is reproduced herein. We have extended the discussion with new results that have come to light since that paper was published. We articulate why the continued calls for management efforts to focus on only phosphorus (P) at the expense of nitrogen (N) control (e.g., Schindler et al. 2008, 2016) are misguided at best. Overall, evidence continues to mount that our failure to aggressively control the use of N along with P, especially through fertilizer applications, will continue to hamper our ability to control the expanding problem of harmful algal blooms (HABs).

P. M. Glibert (✉)
University of Maryland Center for Environmental Science, Horn Point Laboratory, P.O. Box 775, Cambridge, MD 21613, USA
e-mail: glibert@umces.edu

R. Maranger
Groupe de Recherche Interuniversitaire en Limnologie et en Environnement Aquatique (GRIL), Département de Sciences Biologiques, Université de Montréal, Case Postale 6128, Succ. Centre-Ville, Montréal, Québec H3C 3J7, Canada
e-mail: r.maranger@umontreal.ca

D. J. Sobota
Environmental Solutions Division, Oregon Department of Environmental Quality, 811 SW 6th Ave., Portland, OR 97206, USA
e-mail: SOBOTA.Daniel@deq.state.or.us

L. Bouwman
Department of Earth Sciences—Geochemistry, Faculty of Geosciences, Utrecht University, P.O. Box 80021, 3508 TA Utrecht, The Netherlands
e-mail: lex.bouwman@pbl.n

PBL Netherlands Environmental Assessment Agency, P.O. Box 303, 3720 AH Bilthoven, The Netherlands

M. A. Sutton et al. (eds.), *Just Enough Nitrogen*,
https://doi.org/10.1007/978-3-030-58065-0_17

Abstract Large-scale commercialization of the Haber-Bosch (HB) process is resulting in intensification of nitrogen (N) fertilizer use worldwide. Globally N fertilizer use is far outpacing that of phosphorus (P) fertilizer. Much of the increase in N fertilizers is also now in the form of urea, a reduced form of N. Incorporation of these fertilizers into agricultural products is inefficient leading to significant environmental pollution and aquatic eutrophication. Of particular concern is the increased occurrence of harmful algal blooms (HABs) in waters receiving nutrient enriched runoff. Many phytoplankton causing HABs have physiological adaptive strategies that make them favoured under conditions of elevated N:P and supply of chemically reduced N (ammonium, urea). We propose that the HB-HAB link is a function of (1) the inefficiency of incorporation of N fertilizers in the food supply chain, the leakiness of the N cycle from crop to table, and the fate of lost N relative to P to the environment; and (2) adaptive physiology of many harmful algae to thrive in environments in which there is excess N relative to classic nutrient stoichiometric proportions and where chemically reduced forms of N dominate. The rate of HAB expansion is particularly pronounced in China where N fertilizer use has escalated very rapidly, where soil retention is declining, and where blooms have had large economic and ecological impacts. There, in addition to increased use of urea and high N:P based fertilizers overall, escalating aquaculture production adds to the availability of reduced forms of N, as does atmospheric deposition of ammonia. Harmful algal blooms in both freshwaters and marginal seas in China are highly related to these overall changing N loads and ratios. Without more aggressive N control the future outlook in terms of HABs is likely to include more events, more often, and they may also become more toxic.

Keywords Cyanobacteria · Dinoflagellates · Eutrophication · Harmful algal blooms · N:P ratio · Nitrogen · Phosphorus · Nitrogen fertilizer

17.1 Introduction: The Rate of Change in Fertilizer Use

Fertilizers, primarily those containing nitrogen (N) and phosphorus (P) that have played a major role in food production, have found their way into many water bodies across the globe. Major threats from nutrient enrichment are the creation of dead zones and blooms of toxic algae in inland and coastal waters. Although some nations limit or at least control the amount of fertilizers entering into waterways, the excessive use of fertilizers, particularly in the 1970 s and 1980s in developed countries, have left a nutrient legacy that continues to degrade aquatic environments. Furthermore, many regions around the world continue to apply significant amounts of fertilizers to their agricultural crops. Latecomers to big agriculture, such as India and China (countries that are applying fertilizers at phenomenal rates), will certainly face similar nutrient legacies in the future.

The great acceleration in food production was largely catalyzed by the development of industrial capability to convert N_2 gas in air into fertilizer. Such a capability was developed in the early part of the twentieth century (Smil 2001). The industrial fixation of N_2 to ammonia (NH_3), the Haber-Bosch (HB) process (Eq. 17.1) is considered to be one of the most important chemical reactions in the world (e.g. Smil 2001) and "the greatest single experiment in global geo-engineering ever made" (Sutton et al. 2013, p. 4).

$$N_2 + 3H_2 \rightleftarrows 2NH_3 \tag{17.1}$$

This reaction has produced the N fertilizers that have contributed to the 'green revolution', responsible for increased food production that has supported the expansion of human population from ~ 2 billion in the early twentieth century to > 7 billion people today (Smil 1999; Erisman et al. 2008).

Prior to World War II, the creation of reactive nitrogen (N_r) was largely due to natural processes, including biological N fixation and lightning, and population expansion kept pace with its creation (Galloway et al. 2002). After the mid-1940s and the commercialization and scaling up of the Haber-Bosch process, the manufacture and use of N expanded rapidly, from < 10 MT N $year^{-1}$ in 1950 to an expected > 220 MT N $year^{-1}$ by 2020 (MT = megatonnes; Fig. 17.1a, this chapter; Constant and Sheldrick 1992; FAO 2012; Heffer and Prud'homme 2016). In fact, 85% of all synthetic N fertilizers have been created since 1985 (Howarth 2008).

The rate of change in use of N fertilizers has eclipsed that of P fertilizers in large part due to this large-scale capacity for anthropogenic synthesis. Global use of N fertilizer has increased 9-fold, while that of P has increased 3-fold (Sutton et al. 2013; Fig. 17.1b, this chapter) and the upward trends in N:P in fertilizer application (Fig. 17.1c, this chapter) are apparent in virtually all regions of the globe (Fig. 17.1d–f, this chapter). In the United States (US), it has been estimated that there has been at least a 5-fold increase in N_r use on average compared to pre-industrial time (Houlton et al. 2013), but this increase is spatially variable ranging from negligible in unpopulated areas to > 35-fold in urban and agriculturally-intensive areas (Sobota et al. 2013).

Although the Haber-Bosch process is the conversion of N_2 to NH_3, nearly 60% of all N fertilizer now used throughout most of the world is in the form of urea ($CO(NH_2)_2$; Constant and Sheldrick 1992; Glibert et al. 2006; IFA 2014; Fig. 17.1a, this chapter). Urea production is an extension of the Haber-Bosch process, being produced by reacting CO_2 with anhydrous NH_3 under pressure at high temperatures. Multiple factors, including the less explosive nature of urea relative to ammonium-nitrate fertilizer (NH_4NO_3), make the transportation and storage of this synthetically produced N form much safer and the preferred choice for agricultural applications, particularly in developing countries. World use of urea as a fertilizer and feed additive has increased more than 100-fold in the past four decades (Glibert et al. 2006). From 2001–2010, global urea use grew on average at a rate of 3.8% $year^{-1}$, and from 2012 to 2017 approximately 55 new urea manufacturing plants were to be constructed worldwide, with half of these located in China (Heffer and Prud'homme 2013), contributing to another anticipated doubling by 2050 (Glibert et al. 2006). Eastern

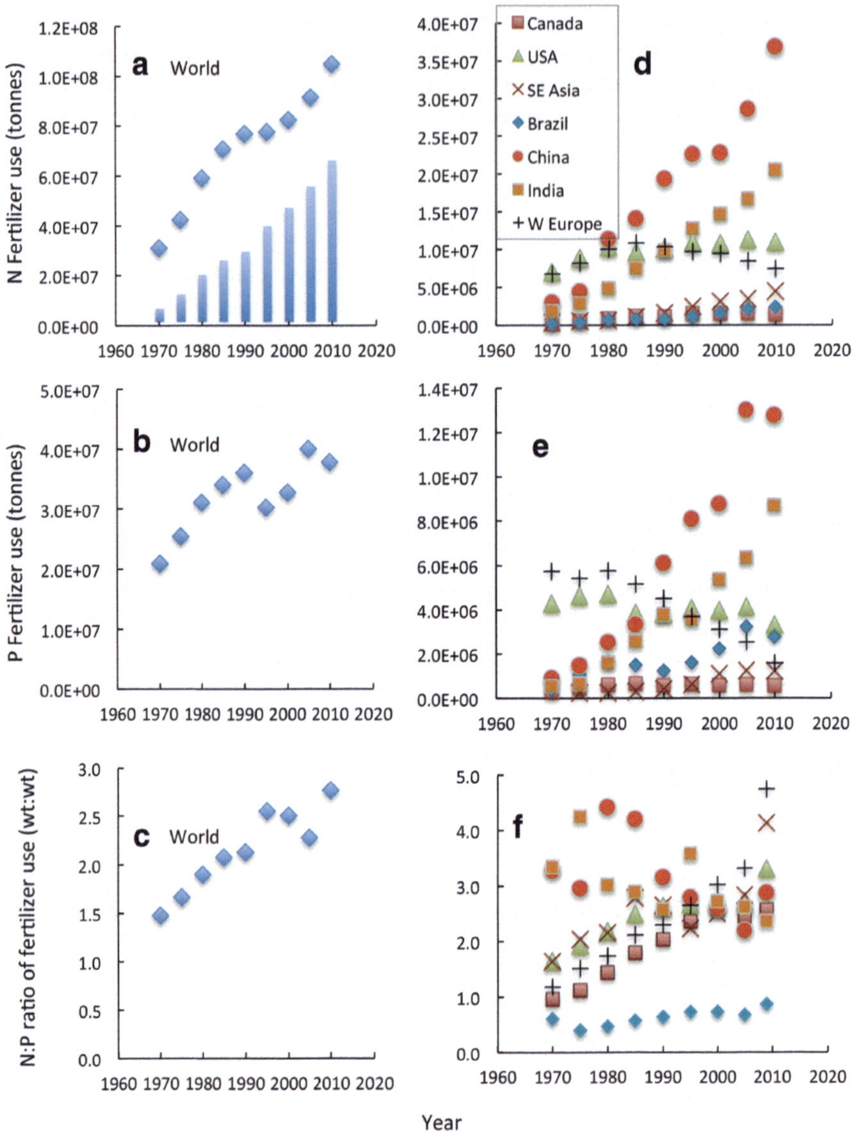

Fig. 17.1 N and P (as P_2O_5) fertilizer use and change in N:P ratio of fertilizer use by weight for the world (parts **a**, **b**, and **c** respectively) and for selected countries or regions (parts **b**, **d**, and **e** respectively). Superimposed on the world N use graph **a** is the fraction of N use as urea (bars). Total N and P data are from FAO (2012) and data are the average of the three preceding years for each five-year period; urea data are from Constant and Sheldrick (1992) through 1990 and estimated at 3.8% growth per year thereafter, comparable to urea data reported from IFA (2014). (Reproduced from Glibert et al. 2014a © IOP Publishing Ltd., under CC BY 3.0, https://creativecommons.org/licenses/by/3.0)

Europe and Central Asia, along with North America and Africa will account for 70% of overall capacity growth (Heffer and Prud'homme 2016).

These high use rates of N fertilizer do not equate to efficiency of agricultural plant incorporation, however. In fact, the incorporation of agricultural N into plant biomass is extremely inefficient. Although the efficiency of N use in experimental fields may be much higher than the global average of 50% (Balasubramanian et al. 2004), under practical conditions it is difficult to equate the N supply from fertilizer and from soil organic matter mineralization with the dynamics of crop N uptake demand (Dobermann and Cassman 2005). Considering the complete food chain, only ~ 10–30% of N applied actually reaches human consumers (Galloway et al. 2002; Houlton et al. 2013). The difficulty in improving N use efficiency in agriculture lies in the high mobility of N in the soil-plant system, and the variety of potential loss pathways ranging from NH_3 volatilization, denitrification, leaching and runoff, and other N transformation processes (Bouwman et al. 2009). Urea inputs are typically hydrolyzed to NH_4^+ in soil, but losses via volatilization and from runoff can be large and depend on the timing of application, weather, soil temperature and pH and other factors (Khakural and Alva 1995; Wali et al. 2003). In regions such as China where the rate of fertilization has risen rapidly, the rate of soil retention of the excess N is actually declining (Cui et al. 2013), leading to further environmental leakage.

The recovery of fertilizer P in crop products is also low (Syers et al. 2008) but its biogeochemistry leads to proportionately greater retention within soils and sediments than N (e.g., Rhue and Harris 1999; Smil 2000; Bouwman et al. 2009). The accumulation of residual soil P due to large fertilizer P surpluses over crop uptake during the 1960s, 1970s, and 1980s has led to an increased pool of plant available P in most industrialized countries; a similar development has been seen in later decades in India and China (Sattari et al. 2012, 2014).

Among the various fates of the "leaked N" are pathways that ultimately lead to N enrichment of lakes, rivers, and coastal waters. The major pathways for this leaked N include direct runoff, estimated to range up to 40% of inputs in large rivers (e.g., Howarth et al. 2006) and atmospheric volatilization of NH_4^+, and together these pathways can comprise more than half of the N input (e.g., Galloway et al. 2004). Phosphorus also runs off to receiving waters, but given the aforementioned change in patterns of fertilizer use and its biogeochemistry, the stoichiometry of the runoff has also changed in the last decades, leading to increasing N:P in receiving waters (e.g., Glibert et al. 2013). It has been estimated that the atmospheric deposition of nutrients in the ocean is now ~ 20 times the Redfield ratio for N:P (Jickells 2006; Peñuelas et al. 2012) and these changes are also having large consequences for N:P stoichiometry in lakes (Elser et al. 2009). This change in stoichiometry has been further compounded since the mid-1980s and 1990s when the major industrialized nations began curtailing P use by removing it from detergents, limiting its use in lawn fertilizers, and by upgrading sewage treatment processes which generally are more efficient in removing P than N (Litke et al. 1999; Van Drecht et al. 2009).

In general, urea concentrations in aquatic ecosystems are less than those of the inorganic N forms NO_3^- and NH_4^+, but depending on proximity of the land source to the water body, concentrations of urea in lacustrine, estuarine and coastal waters may be high, particularly when runoff occurs from heavily fertilized areas (Glibert et al. 2005a, 2006). Concentrations up to 25–50 μM-N have been reported in tributaries of the Chesapeake Bay (Lomas et al. 2002; Glibert et al. 2005b), and nearshore waters adjacent to the heavily fertilized Yaqui Valley, Mexico (Glibert et al. 2006), among other coastal areas (Kudela et al. 2008; Switzer 2008). Urea concentrations vary from undetectable to 150 μM–N in Lake Kinneret, Israel (Berman 1974), in Polish lakes (Siuda and Chrost 2006) and in lakes of central Canada where these concentrations also represented 10–50% of bioavailable N (Bogard et al. 2012). Urea is also part of the dissolved organic N (DON) pool that is typically a very large component of the available N in nutrient rich waters (Glibert et al. 2006; Solomon et al. 2010).

17.2 Eutrophication: An Aquatic Consequence of Fertilizer 'Over-Indulgence'

Eutrophication is the process by which waters are enriched in nutrients, leading to effects such as increased algal growth and development of high biomass blooms, changes in species diversity of both the primary and secondary producers, reductions in dissolved oxygen, fish kills, and the increased frequency of harmful algal blooms (HABs) (Nixon 1995; Cloern 2001). Harmful algal blooms are those proliferations of algae that can cause ecological harm to food webs when they accumulate in massive quantities. They can cause ecological, human health, and economic impacts when these cells produce toxic or other bioactive compounds and when the decay of high algal biomass results in hypoxia (Hallegraeff 1993; Glibert et al. 2005a; Backer and McGillicuddy 2006). The most common HABs are either dinoflagellates or cyanobacteria, although not all dinoflagellates or cyanobacteria are harmful, and not all HABs are made up of these species groups.

Cyanobacteria are the HAB functional group of proportionately greater concern in freshwater, while dinoflagellates are the HAB functional group of greater concern in estuarine and marine waters. Harmful algal blooms have been expanding globally, in spatial extent, in duration of blooms, and in intensity (Anderson et al. 2002, 2008; Glibert et al. 2005a) and a critical question has been the extent to which this change is associated with eutrophication and/or accelerated by climate or other factors (e.g., Paerl and Scott 2010; Wells et al. 2016). There are many reports of increases in HABs associated with eutrophication or nutrient loading (e.g., Anderson et al. 2002, 2008; Glibert et al. 2006, 2010; Heisler et al. 2008), but the complexity of the relationship is far from understood. Many nutrient reduction strategies are focused on reducing algal biomass (e.g., total chlorophyll), while appropriate efforts or best management practices that may specifically prevent toxic HABs have remained much more challenging (Jewett et al. 2008).

This chapter draws on a number of different resources for developing status and trends of fertilizer use and effects (Table 17.1). It is important to note that fertilizer statistics are variably reported as total capacity (permitted production by facility), actual production, amount purchased and usage. In a typical year, overproduction

Table 17.1 Table of data sources used herein

Data	Comment or description	Reference
Global and regional N consumption		FAO (2012)
Global and regional P consumption		FAO (2012)
Global urea consumption		Glibert et al. (2006)
Global region N:P consumption		FAO (2012)
Quebec N consumption (as amount purchased)		Statistics Canada
Quebec P consumption (as amount purchased)		Statistics Canada
Quebec urea/nitrogen solutions and ammonium nitrate consumption		Statistics Canada
Cyanobacteria l bloom incidence in Quebec	Based on observed blooms reported from over 450 lakes and rivers across the province	Ministère de l'environnement développement durable et lute contre les changements climatiques du Québec
China N consumption		FAO (2012)
China P consumption		FAO (2012)
China urea consumption		Zhang and Zhang (2008)
China retention capacity		Cui et al. (2013)
Duration of Microcystis blooms in Lake Tai (Taihu)		Duan et al. (2009)
Total number and aerial extent of HABs in China marginal seas	Values originally derived from SOA 2010	Wang et al. (2011)
Total number of HABs in Huanghai Sea region	Values originally derived from SOA 2010	Fu et al. (2012)
Average inorganic N:P in Huanghai Sea region		Fu et al. (2012)
Total number of HABs in northern South China Sea region		Wang et al. (2008)
Average inorganic N:P in northern South China Sea region	Average of water column concentrations in upper 200 m of the water column	Ning et al. (2009)

relative to demand amounts to a few percent (Heffer and Prud'homme 2013); actual usage depends on current climatic conditions as well as economic factors driving the price of the commodity. Additionally, most fertilizer data are reported as tonnes of product, not necessarily tonnes of N or P. Given the various formulations of the products, some of which include various additives or coatings, no attempt has been made to convert values to similar units; thus, where urea tonnage exceeds that of total N, values should not be taken to be percentages of each other. Our application here is only to compare trends and patterns, when all data are internally consistent.

China presents an interesting case study in terms of the relationships between increasing HAB frequency and the changes in fertilizer use and export. Fertilizer N use in China has escalated from about ~ 0.5 MT in the early 1960s to 42 MT around 2010, with the fraction of urea increasing nearly 5-fold over just the past 2 decades (Fig. 17.2a, this chapter; IFA 2014; FAO 2012; Zhang and Zhang, 2008). River export of N increased from 1980 to 2010 from ~ 500 to > 1200 kg N km^{-2} $year^{-1}$ in the Changjiang River (Yantgtze River), from ~ 100 to ~ 200 kg N km^{-2} $year^{-1}$ in the Huanghe River (Yellow River), and from ~ 400 to > 1200 kg N km^{-2} $year^{-1}$ in the Zhujiang River (Pearl River) basins, the latter having one of the largest N:P ratios in the world (Ti and Yan 2013). The annual N load from the Changjiang River to the coastal ocean is higher than loads from the Mississippi and the Amazon Rivers (Goolsby and Battaglin 2001; Duan et al. 2008). Atmospheric sources of NH_4^+, which is the dominant form of inorganic N on a regional basis, and which exceeds that of oxidized N forms in the Changjiang River basin, are due primarily to livestock excretion, fertilizer N use, and human waste (52%, 33% and 13% respectively, Xiao et al. 2010).

The number of HABs has increased in all waters of China in the past three decades and most inland and coastal waters are rated in the moderately to severely polluted range (Wang et al. 2011; Ti and Yan 2013). In Taihu, the largest lake in the Changjiang watershed, the duration of cyanobacterial *Microcystis* blooms has increased from ~ 1 month $year^{-1}$ to nearly 10 months $year^{-1}$ in the past 15 years (Duan et al. 2009; Fig. 17.2b, this chapter). The Changjiang watershed produces > 30% of China's agricultural output (Xing and Zu 2002) and in the Taihu region, N fertilizers from manures and mineral fertilizers have increased 8 to 10-fold since the 1950s (Chen et al. 2008) and fertilizer N use in Jiangsu Province (2.5 MT N $year^{-1}$) is 80% of total N inputs (PBL 2012). The change in HABs in Taihu is strongly related to the increase in urea and in the ratio of urea: P_2O_5 use scaled to the Changjiang watershed (Chen et al. 2003; Ye et al. 2007; Duan et al. 2009; Figs. 17.3b–d, this chapter; $r^2 = 0.85$ and 0.92, respectively, $p < 0.01$). These trends in availability of reduced forms of N and in increased N:P ratios in this region are further accelerated by aquaculture development and atmospheric deposition. In 2007, 13% of Chinese inland aquaculture production was in Jiangsu (China Fishery Yearbook, 2007), and large quantities of NH_4^+ and urea with high N:P ratio are also discharged from that source to Taihu. It should be noted that waste from aquaculture not only has a high N:P ratio (> 20) but it is also largely in chemically reduced N form, with significant amounts of urea (Bouwman et al. 2013). Also, 80% of the atmospheric deposition in this region has been associated with NH_4^+ mobilization from intense fertilization during the growing period (Chen et al. 2008). With intensive agriculture, N deposition onto the lake is a direct N input not accompanied by P.

In Chinese marginal seas there are similar trends of increased HABs over the past several decades (Wang et al. 2008; Wang et al. 2011; Fig. 17.3a, this chapter). When all China's marginal sea bloom occurrences are considered, there are strong

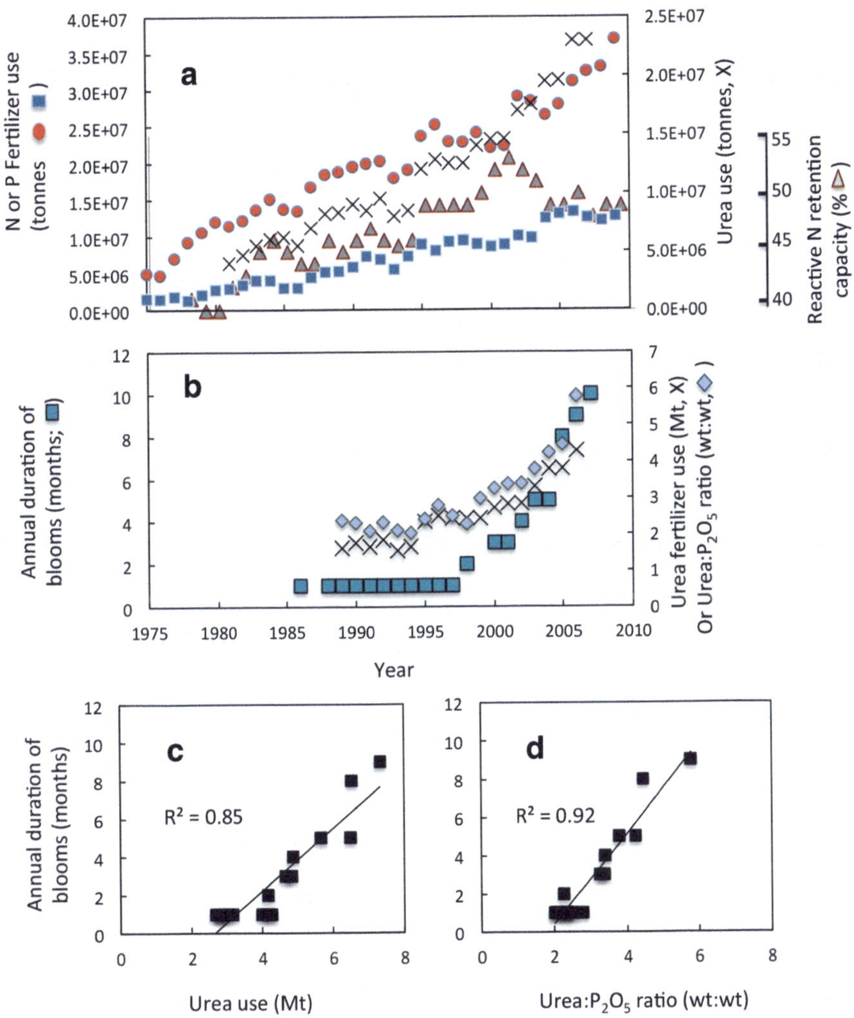

Fig. 17.2 **a**—Comparison of N (red circles) and P (blue squares) fertilizer use and the use of urea (black crosses) in China from 1975 to 2005. N and P data are from FAO (2012) while urea data are from Zhang and Zhang (2008) and represent the total of production and imports minus exports. Also shown (grey triangles) is the calculated reactive N retention capacity of the Chinese landscape as estimated by Cui et al. (2013). **b**—Change in annual duration of *Microcystis* blooms in Lake Tai (Taihu) in months, urea fertilizer use scaled to that in the Changjiang watershed and the ratio of use of urea:P_2O_5 fertilizer. *Microcystis* data are from Duan et al. (2009) reprinted with permission of the American Chemical Society, and total urea and P_2O_5 data for China are from Zhang and Zhang (2008). Parts **c** and **d**—Correlations between annual duration of blooms (months) and urea use in Changjiang watershed **c** and the urea:P_2O_5 ratio **d**. (Reproduced from Glibert et al. 2014a © IOP Publishing Ltd., under CC BY 3.0, https://creativecommons.org/licenses/by/3.0)

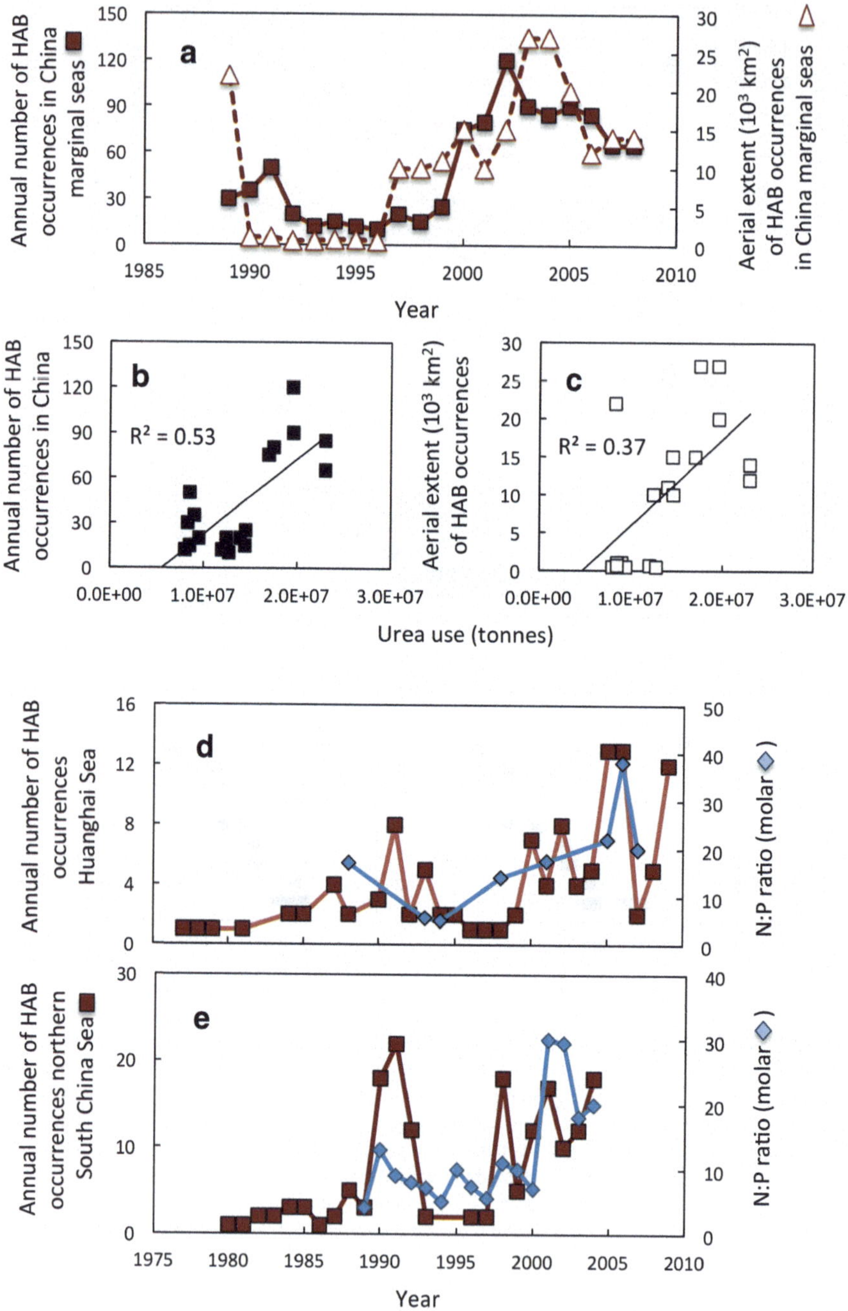
a
Annual number of HAB occurrences in China marginal seas
Aerial extent (10³ km²) of HAB occurrences in China marginal seas
Year
b
R² = 0.53
Annual number of HAB occurrences in China
c
R² = 0.37
Aerial extent (10³ km²) of HAB occurrences
Urea use (tonnes)
d
Annual number of HAB occurrences Huanghai Sea
N:P ratio (molar)
e
Annual number of HAB occurrences northern South China Sea
N:P ratio (molar)
Year

◂**Fig. 17.3** **a**—The total number of recorded harmful algal bloom (HAB) occurrences (red squares) along the coast of China and their aerial extent (open triangles). Data are from Wang et al. (2011), originally reported by SOA (2010; reproduced under Creative Commons license). Parts **b–c**—Correlations between annual number of HABs or areal extent of HABs in China marginal seas and urea use. **d**—The annual number of HAB occurrences in the Huanghai (Yellow) Sea region (red squares) and the molar N:P ratio of southern Huanghai Sea in spring (blue diamonds). Data are from Fu et al. (2012), with the HAB data originally reported by SOA (data reproduced with the permission of the Chinese Society of Oceanography and Springer-Verlag). **e**—The annual number of HAB occurrences in the northern South China Sea (red squares) and the molar N:P ratio in the water column. Data on HABs are from Wang et al. (2008) and N:P data are from Ning et al. (2009; reproduced under Creative Commons license) and represent the average values for the upper 200 m of the water column. (Reproduced from Glibert et al. 2014a © IOP Publishing Ltd., under CC BY 3.0, https://creativecommons.org/licenses/by/3.0)

relationships between both number of HAB reports and aerial HAB extent and N use as urea (Figs. 17.3b, c, this chapter; $r^2 = 0.53$ and 0.37, respectively, $p < 0.01$).

In both the Huanghai (Yellow) Sea region and in the northern South China Sea, the number of reported occurrences of HABs has increased in parallel with the long-term trend in N:P of the water column (Figs. 17.4d, e, this chapter; $r^2 = 0.46$ and 0.29, respectively, $p = 0.06$). As in the Taihu example, these relationships do not account for nutrient input from coastal aquaculture or from atmospheric NH_4^+ deposition that would enhance these relationships. Coastal N release from Chinese mariculture increased from 2000–2010 by 45% on average and up to 63% in some provinces while P increase was only 37% (Bouwman et al. 2013). In 2004, NH_4^+ and organic N comprised 75% of N in wet atmospheric deposition in southeast China (Chen et al. 2011).

At the mouth of the Changjiang River, molar NO_3^-:PO_4^{3-} values were ~ 30–40 in the 1960s but had risen to > 250 by the late 1990s (Shen and Liu 2009; Fig. 17.3d, this chapter). Occurrences of HABs in the East China Sea were rare in the 1970s, had increased to a few dozen over the 1980s and had increased to > 130 in just the five years from 1992–997 (Shen and Liu 2009) but the scale of the blooms soared in the years since; the spatial scale of the annual blooms increased from 1,000 s of km^2 in 2000 to > 15,000 km^2 by 2005 with many millions of dollars lost in high value aquaculture products due to associated fish kills (Li et al. 2009). The dinoflagellates *Prorocentrum donghaiense* and *Karenia mikmotoi* are among the common HABs now reported in East China Sea (e.g., Zhou et al. 2008; Li et al. 2009).

In the Huanghai Sea region, inorganic N:P ratios are now about twice Redfield proportions, and about 4-fold higher than in the 1990s (Fig. 17.3e, this chapter). There has also been nearly a 6-fold increase in HAB occurrences and a shift to proportionately more dinoflagellates compared to diatoms (Fu et al. 2012). In the South China Sea region, water column inorganic N:P ratios increased from ~ 2 in the mid-1980s to > 20 in the early 2000s (Ning et al. 2009). In addition to the increase in number of HABs, a change in species composition to increasing dominance of species

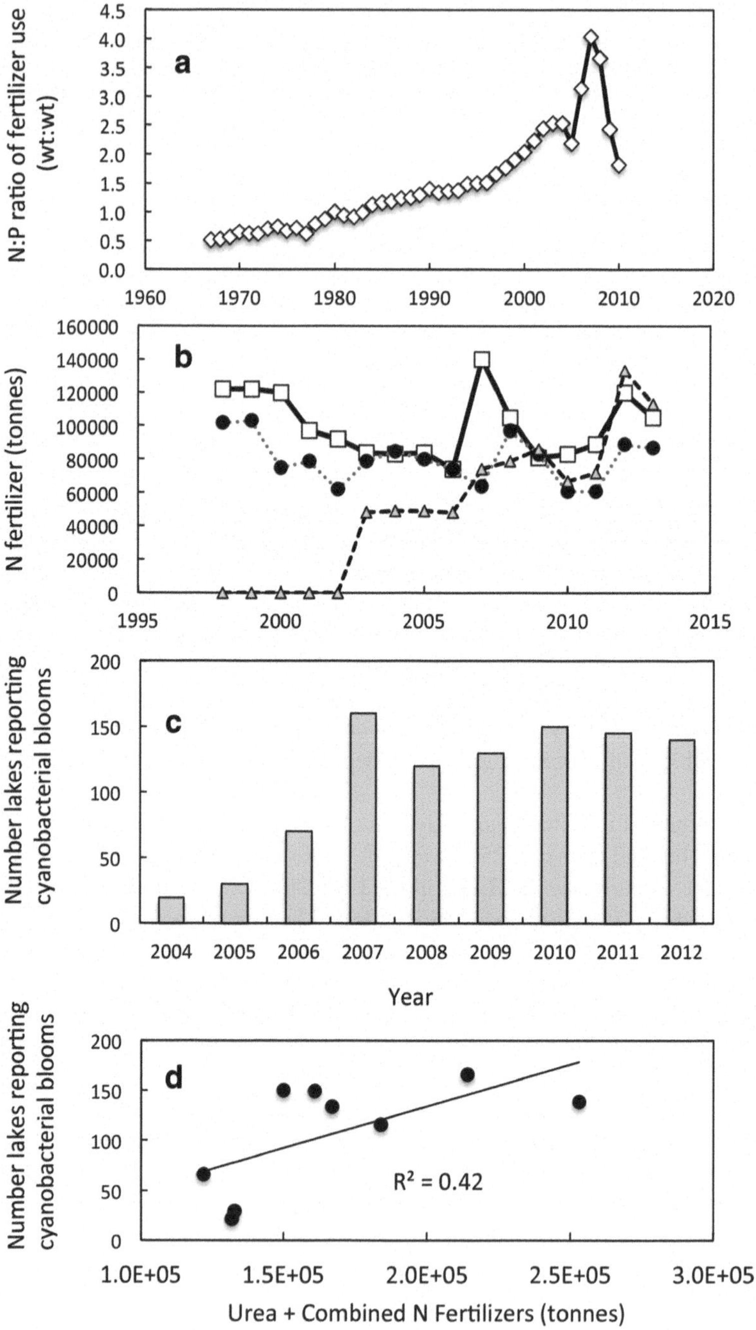
a
N:P ratio of fertilizer use (wt:wt)
4.5
4.0
3.5
3.0
2.5
2.0
1.5
1.0
0.5
0.0
1960
1970
1980
1990
2000
2010
2020
b
N fertilizer (tonnes)
160000
140000
120000
100000
80000
60000
40000
20000
0
1995
2000
2005
2010
2015
c
Number lakes reporting cyanobacterial blooms
200
150
100
50
0
2004
2005
2006
2007
2008
2009
2010
2011
2012
Year
d
Number lakes reporting cyanobacterial blooms
200
150
100
50
0
R² = 0.42
1.0E+05
1.5E+05
2.0E+05
2.5E+05
3.0E+05
Urea + Combined N Fertilizers (tonnes)

◂**Fig. 17.4** **a**—N:P ratio of fertilizer use (by weight) in Québec since 1967 illustrating year-to-year variability superimposed on long-term trends. **b**—Trends in use of different N fertilizer types since 1998 in Québec (urea-squares; NH_4NO_3-circles; urea-NH_4-NO_3 (UAN)- triangles); note the increasing use of UAN versus the more consistent use of urea and NH_4NO_3. **c**—Number of lakes with reported cyanobacteria blooms from over 450 systems surveyed in Québec since 2004. Note the time scale zooms into progressively recent years from Parts **a**–**c**; also note the peaks in all panels in the year 2007. **d**—Correlation between the number of lakes reporting cyanobacterial blooms from 2004–2012 and the use of urea plus combined N (UAN). (Reproduced from Glibert et al. 2014a © IOP Publishing Ltd., under CC BY 3.0, https://creativecommons.org/licenses/by/3.0)

such as *Chattonella, Gymnodinium breve,* and *Dinophysis* has occurred (Wang et al. 2008).

17.3 Nutrient Reduction Strategies Are Difficult and Controversial

It is well understood that reduction in eutrophication will require reductions in nutrients. The challenges are 1) where reductions should take place: source control (at the rate of application) or end-of-pipe (riparian buffers, wetlands, or within river/water body) and 2) which nutrient reduction strategies will be the most effective.

It is difficult to reduce nutrients in rivers. European and USA rivers currently feel the effects of excessive nutrient mobilization in the 1970s and 1980s, i.e., the *legacy* of historical nutrient use. This is because in the depletion phase, the landscape buffer or memory is now acting in a different way compared to the accumulation phase. With declining nutrient inputs, soils may be releasing nutrients by organic matter decomposition, and aquifers (particularly N) and sediments (particularly P) in lakes, reservoirs and rivers continue to deliver nutrients. The N concentration in the Mississippi has not decreased in recent decades, despite policies to reduce nutrient loading. The flow-normalized N export by the Mississippi has even increased since 1980, and the increased N concentration at low stream flows is a strong indication that NO_3^- delivery by groundwater has a strong effect on river concentrations (Sprague et al. 2011). Regulations in the European Union to reduce groundwater pollution by NO_3^- and nutrient discharges in wastewater (European Commission 1991a, b, 2000) have not led to a reduction in N concentrations in rivers (e.g., Meuse). While these regulations intend to reduce both nutrients, reduction of nutrient use in agriculture seems to have a direct effect on P, but not on N due to the legacy of historical N surpluses. Advances in wastewater treatment directly lead to a reduction of N and P discharge, but there has been a greater effort to reduce P than N from wastewater. Such a phenomenon of slow or no change in N and decreasing P is apparent not only in the Rhine and Meuse, but also in rivers draining into the English Channel, Atlantic, W. Mediterranean Sea and N. Adriatic Sea (Romero et al. 2013). Even the North Sea has witnessed increasing N:P ratios with total P inputs down 50–70%

but N inputs only down ~ 20% with nutrient reductions imposed in the late 1980s following years of large increases in both N and P (Lenhart et al. 2010; Passey et al. 2013; Burson et al. 2016). Consequently, Burson et al. (2016, p. 884) warn that, "further reductions of P loads without concomitant reduction of the N loads may be less effective in diminishing the risk of [HABs] by potentially toxic nano- and dinoflagellates."

One of the most central tenets of aquatic science is that algal biomass and production in lakes and other freshwaters is limited by the availability of P, while that in marine waters is more often limited by the availability of N (e.g., Ryther and Dunstan 1971; Schindler 1977). However, P limitation in lakes is not universal (Lewis and Wurtsbaugh 2008) as some regions are either naturally (Finlay et al. 2010) or culturally enriched (Bennett et al. 2001) in P relative to N, and excess N loading is changing the nutrient stoichiometry and limiting element in some coastal areas (e.g., Sylvan et al. 2006). It is commonly assumed that to control eutrophication the only focus should be on that nutrient which is classically considered "limiting". Such an argument is typically extended to promote enhanced P control over N control in freshwaters (e.g., Schindler et al. 2008; Wang and Wang 2009) for multiple reasons, among which is the notion that lakes are never limited in N since it is often assumed that N_2-fixing cyanobacteria will balance the deficit. Schindler et al. (2016) in a recent review cite numerous "success" stories of eutrophication reduction in lakes following P control, and they claim that no such successes have been observed with N control in any aquatic system. However, they fail to mention the numerous and increasing occurrences of non-N_2-fixing cyanobacterial blooms that are occurring in freshwaters, and for which P control alone has not been an effective control strategy. Contrary to the argument that P control effectively curtails cyanobacterial blooms there are many examples, in addition to the Taihu example given above (Fig. 17.2), that illustrate clearly that some of the most toxic cyanobacteria, namely *Microcystis* sp., are not N_2-fixing species and their abundances are increasingly predicted from increasing total N (TN) concentrations and from changes in TN to total P (TP) ratios (Smith 1983; Downing et al. 2001; Kosten et al. 2012; Dolman et al. 2012). It is also often erroneously stated that those who advocate N removal to counter freshwater eutrophication do not consider removing P. Dual reduction strategies are what is advocated here as well as elsewhere in the literature (Conley et al. 2009; Paerl et al. 2016). In fact, multiple other ecological and ecoservice benefits would be met by reducing N inputs (Vitousek et al. 1997) even according to those who advocate P only reductions (Schindler et al. 2016). Fragmenting sustainability arguments, i.e., focussing on one limiting nutrient to counter lake eutrophication only, as though the latter is disconnected from the landscape, undermines the need to act to protect multiple ecoservices at broader spatial scales. Indeed a more precautionary and transdisciplinary approach should be supported for greater overall sustainability needs (Brown 2010) rather than building narrow arguments around specific system types. Thus, the reduction of both nutrients should be the goal.

Further arguments for P relative to N control are that P does not have a gaseous form and therefore cannot be permanently removed from lakes, whereas N can be fixed from and lost to the atmosphere. Nitrogen cycling is relatively complex and to say that aquatic systems simply balance N needs accordingly, i.e., denitrify under conditions of excess N and fix in times of N deficit, is an oversimplification of reality. In the case of N deficits, it has been shown that rates of N_2-fixation do not offset N limitation (Scott and McCarthy 2010; Lewis et al. 2011). In the case of excess, there is growing evidence that the ability of a system to remove N is compromised when inputs are exceedingly high. Increased hypoxic and anoxic conditions, be they brought on by excess nutrients or climate warming, will uncouple nitrification from denitrification and/or favour dissimilatory nitrogen reduction to ammonium (Seitzinger 1988; Cornwell et al. 1999; Burgin et al. 2007, McCarthy et al. 2016) thus favouring internal recycling rather then net loss. In flowing systems, sediments, the primary sites for denitrification, simply become saturated when the concentrations in overlying waters are too high, which results in reduced removal efficiency (Mulholland et al 2008). Furthermore, changes in flow regimes as a function of land use change in watersheds accelerates N delivery through aquatic networks to coastal systems (Bettez et al. 2015). Eutrophication issues cannot be separated from hydrologic connectivity (Baulch 2013) as not all N, particularly in over enriched systems, can be expected to be lost via denitrification thus resulting in eutrophication related problems that may be spatially and temporally displaced from the original nutrient source (Conley et al. 2009; Paerl 2009; Glibert and Burkholder 2011a).

Most of the N inputs to aquatic ecosystems worldwide come from non-point sources, and as was mentioned in previous sections, the relative proportion of N to P in fertilizers applied to land is increasing. The P removal-only argument typically targets waste water treatment plants (Schindler et al. 2016). Thus, if we include point source inputs, the decision to remove P only further skews N: P ratios to receiving waters. Increasing N:P environments further favour HABs when the N form is disproportionately in chemically reduced form (i.e., urea, NH_4^+) relative to chemically oxidized form (i.e., NO_3^-). We thus propose that the HB-HAB link is a function of 1) the inefficiency of incorporation of N fertilizers in the food supply chain, the leakiness of the N cycle from crop to table, and the fate of the lost N to the environment as described above; and 2) adaptive physiology of many HABs to thrive in environments in which there is excess N relative to classic nutrient stoichiometric proportions and where chemically reduced forms of N are increasing.

There are a number of specific physiological strategies that allow certain types of algae to thrive under conditions of elevated N:P availability relative to classic Redfieldian proportions (Redfield 1934), but not all cells necessarily have all such adaptive strategies (Glibert and Burkholder 2011a). The first strategy is a low overall requirement for P. Very small cells, such as picocyanobacteria, have a lower requirement for P due to the smaller need for structural components in the cell (Finkel et al. 2010). The second strategy is the ability to "make do with less" which may be accomplished by physiological substitution of a P-containing lipid with a non-P-containing lipid (sulfolipid), and many cyanobacteria are able to do this (Van Mooy 2009). Thus, the cellular carbon (C) to P ratio of *Synechococcus* is about 100, whereas the C:P ratio in a typical diatom is about 50 (Finkel et al. 2010). The third strategy is the ability to acquire P in organic or particulate form, via alkaline phosphatase activity or mixotrophy, which may provide some cells a source of P not available to those cells dependent on inorganic P for their nutrition. Many dinoflagellates have a comparatively high cellular P requirement, and therefore the ability to consume particulate P may be an important reason why these types of cells can thrive when some others cannot. An added competitive benefit for these cells is that there may also be a growth advantage when feeding mixotrophically, compared to pure autotrophic growth (Jeong et al. 2004; Glibert et al. 2009; Flynn et al. 2013), thus mixotrophy is a major mode of nutrition by HABs in eutrophic waters (Burkholder et al. 2008) and may help to sustain blooms when dissolved nutrients are depleted.

Of particular concern is the association of increased toxin in many HABs under conditions of elevated N:P availability. Many cyanobacteria and dinoflagellate toxins are N-rich compounds and thus these cells require a supply of N in order to synthesize these metabolites. The most ubiquitous cyanotoxin in freshwater systems are microcystins (MCs), hepatotoxic compounds that can be lethal to mammals if ingested (Carmichael 1994; Chorus and Bartram 1999), some congeners of which are significantly more toxic than others (Sivonen and Jones 1999). *Microcystis* (Chroococcales), *Anabaena* (Nostocales), and *Planktothrix* (Oscillatoriales) are among the taxa that can synthesize MCs (Cronberg and Annadotter 2006). Total concentrations of MCs have been strongly related to N concentrations in several comparative studies (Giani et al. 2005; Rolland et al. 2005; Dolman et al. 2012; Monchamp et al. 2014). Although negative relationships between MC and N:P ratios have been reported for Lake Tai (Taihu), China (Otten et al. 2012), and in a cross system analysis of freshwaters (Orihel et al. 2012), the authors of the findings of negative correlations emphasize that such results are relevant only to hypereutrophic conditions in which total N concentrations are also high (> 100 μM-N). Furthermore, a reanalysis of the aforementioned comparative study reported highest MC concentration at intermediate N:P ratios (Scott et al. 2013). Excess N and high N:P ratios have also been related to MC production under controlled culture conditions (e.g., Lee et al. 2000; Oh et al. 2000; Vézie et al. 2002; Downing et al. 2005; Van de Waal et al. 2009). Some in situ evidence, albeit weak, suggests that increased N availability may influence the MC congener type to more toxic variants that have higher N content (e.g., Van de Waal et al. 2009) and that P limitation causes an increase in N-rich toxins of numerous HABs (Van de Waal et al. 2014).

Numerous marine HABs also show increased toxin production under conditions of elevated N:P ratios. As examples, under conditions of elevated N:P, haemolytic activity per cell increases by up to an order of magnitude in the prymnesiophytes *Prymensium parvum* and *Chrysochromulina* (now *Prymnesium*) *polylepis* (Johansson and Granéli 1999), and neurotoxin production increases in the diatom *Pseudo-nitzschia multiseries* and in the dinoflagellates *Karlodinium veneficum*, *Alexandrium* sp., and *Karenia brevis* (Granéli and Flynn 2006; Hardison et al. 2013).

Recent reviews of the physiological bases of N uptake as well as molecular and metatranscriptomic data lend considerable support to the emerging conclusion that diatoms are specialists in use of oxidized forms of N, while cyanobacteria and dinoflagellates are specialists in reduced forms of N (e.g., Glibert et al. 2016 and references therein). A considerable amount of experimental evidence supports the notion that freshwater cyanobacteria seem to favour reduced N forms (Blomqvist et al. 1994; Berman and Chava 1999). Extensive laboratory molecular evidence of the use of both urea and NH_4^+ by cyanobacteria (Flores and Herrero 2005), and gene expression data (Ginn et al. 2009) also lend support to this conclusion. There are a large number of mesocosm studies conducted in fresh and brackish systems that show that when enriched with oxidized vs reduced forms of N, even when the total N supply remains the same, proportionately more diatoms are produced under oxidized N conditions while more cyanobacteria and cryptophytes are produced under conditions of increasing reduced forms of N (e.g., Finlay et al. 2010; Donald et al. 2011, 2013; Glibert et al. 2014a, b; 2016 and references therein). These findings are consistent regardless of the original levels of nutrients: i.e., even when nutrients were at or near saturating levels prior to experimental enrichments, dichotomous communities emerged in response.

In situ evidence of the role of N forms in cyanobacterial community composition remain rare and influence on MC concentrations or congener composition even rarer. In Québec, Canada, the ratio of N:P in fertilizer acquisition increased steadily from 1977 to 1995, and accelerated from 1995 to peaks of > 3 (on a weight basis) from 2006–2008 (Fig. 17.4a, this chapter). There was a marked increase in the purchase of N fertilizer in the form of urea or products containing urea as of 2003 with a urea peak observed in 2007. That year was considered an "exceptionally favourable" year for agricultural commodities (Heffer and Prud'homme 2008), and a sustained higher overall use of urea products has occurred since that time (Fig. 17.4b, this chapter). Records of reported cyanobacterial incidences in lakes across the province suggest that the peak in events occurred in 2007 (Fig. 17.4c this chapter), when both the N:P ratio and urea use were high. Since then the incidences of blooms have remained high (Fig. 17.4c, this chapter). The correlation of incidences of blooms in these lakes with the increase in forms of N fertilizer containing urea (Fig. 17.4d, this chapter; $r^2 = 0.42$, $p = 0.058$) is suggestive of such an effect, although more years of data will be required to substantiate this relationship. While these patterns provide only indirect evidence, in a comparison over the course of the growing season of three lakes in Québec known to have toxic cyanobacteria, cyanobacterial community structure was primarily influenced by the availability of chemically reduced and organic N forms (DON and NH_4^+) and temperature (Monchamp et al. 2014). Nitrogen forms and concentration however did not influence congener composition or the toxicity of the dominant variant in that study, but cyanobacterial community structure did. This suggests complex interaction between the availability of N and other environmental variables in influencing community structure with an indirect effect on congener composition and overall bloom toxicity.

A recent study by Harris et al. (2016) extended the understanding of N form and toxins. They documented for midwestern USA reservoirs that elevated ratios of NH_4^+:NO_3^- provided conditions that favoured the production of secondary metabolites of cyanobacteria, among which were toxins from *Microcystis*. The metabolites microcystin, geosmin and 2-methylisoborneol (MIB) were all favoured when the ratio of NH_4^+:NO_3^- was elevated.

Several of the more potent marine HAB species show the same trend. For *Alexandrium tamarense*, the availability of urea has also been related to toxin content of the cells: the toxin content for urea-grown cells was found to be higher than that of NO_3^--grown cells, but not as high as cells grown on NH_4^+ (Leong et al. 2004). Furthermore, the biosynthesis of toxin when grown on urea appears to differ from that which occurs under NO_3^- or NH_4^+ growth conditions. For another dinoflagellate, *Karenia brevis*, up to 6-fold increases in toxin content have been observed during growth with elevated urea availability compared to controls without urea enrichment (Shimizu et al. 1993). For the toxic diatom, *Pseudo-nitzschia* sp., increases in toxicity in both laboratory cultures and natural field assemblages have also been found for cells growing on urea compared to those growing on NH_4^+ or NO_3^- (Cochlan et al. 2008; Kudela et al. 2008). On a global scale the simultaneous increase of total fertilizer N, particularly in the form of urea, and the frequency and extent of a number of HAB cyanobacteria and dinoflagellate species provides further, though indirect, evidence for a relationship between urea and HABs (Glibert et al. 2006, 2008).

17.4 Conclusions

The results herein lend support to the view that both N and P controls are necessary to reduce eutrophication in both freshwater and marine waters (e.g., Burkholder et al. 2006; Howarth and Paerl 2008; Conley et al. 2009; Paerl 2009; Glibert et al. 2011b, 2013). Loss of biodiversity, and effects on ecosystem and human health due to eutrophication are considered major challenges of our current day (e.g., Borja 2014). Given the known inefficiencies in all aspects of N use in both industrialized nations and throughout the world, the benefits of increased emphasis on N reduction, and improved N use efficiencies at all levels of the production side, would have far reaching benefits to ecosystems and especially to water quality (Houlton et al. 2013; Sobota et al. 2013, 2015. Calculated damage costs of loss of ecosystem services, eutrophication and human health due to loss of reactive N to the environment are large (Compton et al. 2011; Sobota et al. 2015). So too are the economic impacts to aquaculture, increased risks to human health and ecosystems, and losses to fisheries and ecosystem services due to HABs (Hoagland and Scatasta 2006). More work is needed to accurately quantify the sources and fluxes of N at all stages of the supply-to-loss pathways, and there is no question that control of N may be more challenging than control of P. While there have been, and will be, successes in reducing overall chlorophyll biomass in waters following reductions in P alone (i.e., no companion reductions in N), any wholesale claim that the only approach to reducing eutrophication is through P reductions (i.e., Schindler et al. 2016), is a product of those with blinders on the expanding HAB problem. Given our increasing knowledge of the physiological response of many HABs to increasing N:P and to increasing ratios of reduced:oxidized forms of N, we can conclude that without more aggressive N control the future outlook in terms of HABs is likely to include more events, more often, and such events may also be more toxic (O'Neil et al. 2012; Glibert et al. 2013, 2014b; Paerl and Otten 2013). Indeed, a recent modelling effort in which the physiological responses of HABs to altered N:P ratios and altered N form, even without any further increases in N loading, in conjunction with projected climate change effects, suggest an expansion in area and/or number of months annually conducive to development of several HAB genera along the NW European Shelf-Baltic Sea system and NE Asia by end of the century (Glibert et al. 2014c). Such projections alone should be cause for advancing our understanding of the relationships between HABs and nutrient loading, and together with the projected global expansion in N loading should be serve to sound the alarm that our existing approaches to nutrient management of sensitive coastal and freshwaters are not sufficient, particularly in the face of climate change and other stressors.

Acknowledgements This paper is based substantially on an earlier paper on this topic by Glibert et al. (2014a), The Haber Bosch-harmful algal bloom (HB-HAB) link, published in *Environmental Research Letters*, 9, 105001. https://doi.org/10.1088/1748-9326/9/10/105001, and has been revised and reproduced with the permission of IOP Publishing Ltd., under Creative Commons license CC BY 3.0, https://creativecommons.org/licenses/by/3.0), and with the encouragement of the editors of this volume, M.A. Sutton and K.E. Mason. Original discussions for this paper occurred at the 6th International Nitrogen Conference in Kampala, Uganda, 17–23 November 2013, based on presentations by Patricia M. Glibert, Roxane Maranger, and David J. Sobota. This is a contribution of the GEOHAB core research project on HABs and Eutrophication, and of SCOR Working Group 132 on Land Based Nutrient Pollution and Harmful Algae, funded through the Scientific Committee on Oceanic Research and the Land Ocean Interaction at the Coastal Zone Program, now Future Earth Coasts, and of the Groupe de Recherche Interuniversitaire en Limnologie et environment aquatique (GRIL). This is contribution number 4922 from the University of Maryland Center for Environmental Science.

References

Anderson, D., Glibert, P. M., & Burkholder, J. M. (2002). Harmful algal blooms and eutrophication: Nutrient sources, composition, and consequences. *Estuaries, 25*(4), 704–726.

Anderson, D. A., Burkholder, J. M., Cochlan, W. P., Glibert, P. M., Gobler, C. J., Heil, C. A., et al. (2008). Harmful algal blooms and eutrophication: Examining linkages from selected coastal regions of the United States. *Harmful Algae, 8*(1), 39–53.

Backer, L. C., & McGillicuddy, D. G., Jr. (2006). Harmful algal blooms at the interface between coastal oceanography and human health. *Oceanography, 19*(2), 94–106.

Balasubramanian, V., Alves, B., Aulakh, M., Bekunda, M., Cai, Z., Drinkwater, L., et al. (2004). Crop, environmental, and management factors affecting nitrogen use efficiency. In A. R. Mosier, J. K. Syers, & J. R. Freney (Eds.), *Agriculture and the nitrogen cycle* (pp. 19–33). Washington D.C.: Island Press.

Baulch, H. M. (2013). Asking the right questions about nutrient control in aquatic ecosystems. *Environmental Science and Technology, 47*(3), 1188–1189.

Bennett, E. M., Carpenter, S. R., & Caraco, N. F. (2001). Human impact on erodible phosphorus and eutrophication: A global perspective. *BioScience, 51*(3), 227–234.

Berman, T. (1974). Urea in waters of Lake Kinneret (Sea of Galilee). *Limnology and Oceanography, 19*(6), 977–980.

Berman, T., & Chava, S. (1999). Algal growth on organic compounds as nitrogen sources. *Journal of Plankton Research, 21*(8), 1423–1437.
Bettez, N. D., Duncan, J. M., Groffman, P. M., Band, L. E., O'Neil-Dunne, J., Kaushal, S. S., et al. (2015). Climate variation overwhelms efforts to reduce nitrogen delivery to coastal waters. *Ecosystems, 18*(8), 1319–1331.
Blomqvist, P., Pettersson, A., & Hyenstrand, P. (1994). Ammonium-nitrogen - a key regulatory factor causing dominance of non-nitrogen-fixing cyanobacteria in aquatic systems. *Archiv Für Hydrobiologie, 132*(2), 141–164.
Bogard, M. J., Donald, D. B., Finlay, K., & Leavitt, P. R. (2012). Distribution and regulation of urea in lakes of central North America. *Freshwater Biology, 57*(6), 1277–1292.
Borja, A. (2014). Grand challenges in marine ecosystems ecology. *Frontiers in Marine Sci*ence, 1(Article 1). https://doi.org/10.3389/fmars.2014.00001.
Bouwman, A. F., Beusen, A. H. W., & Billen, G. (2009). Human alteration of the global nitrogen and phosphorus soil balances for the period 1970–2050. *Global Biogeochemical Cycles, 23*(4), GBA04. https://doi.org/10.1029/2009gb003576.
Bouwman, L., Beusen, A., Glibert, P. M., Overbeek, C., Pawlowski, M., Herrera, J., et al. (2013). Mariculture: Significant and expanding cause of coastal nutrient enrichment. *Environmental Research Letters, 8*(4), 044026. https://doi.org/10.1088/1748-9326/8/4/044026.
Brown, V. A. (2010). Conducting an imaginative transdisciplinary inquiry. In V. A. Brown, J. A. Harris, & J. Y. Russell (Eds.), *Tackling wicked problems through the transdisciplinary imagination* (pp. 103–114). London: Earthscan, Routledge.
Burgin, A. J., & Hamilton, S. K. (2007). Have we overemphasized the role of denitrification in aquatic ecosystems? A review of nitrate removal pathways. *Frontiers in Ecology and the Environment, 5*(2), 89–96.
Burkholder, J. M., Dickey, D. A., Kinder, C., Reed, R. E., Mallin, M. A., McIver, M. R., et al. (2006). Comprehensive trend analysis of nutrients and related variables in a large eutrophic estuary: A decadal study of anthropogenic and climatic influences. *Limnology and Oceanography, 51*, 463–487.
Burkholder, J. M., Glibert, P. M., & Skelton, H. (2008). Mixotrophy, a major mode of nutrition for harmful algal species in eutrophic waters. *Harmful Algae, 8*(1), 77–93.
Burson, A., Stomp, M., Akil, L., Brussaard, C. P. D., & Huisman, J. (2016). Unbalanced reduction of nutrient loads has created an offshore gradient from phosphorus to nitrogen limitation in the North Sea. *Limnology and Oceanography, 61*, 869–888.
Carmichael, W. W. (1994). The toxins of cyanobacteria. *Scientific American, 270*(1), 78–86.
Chen, N., Hong, H., Zhang, L., & Cao, W. (2008). Nitrogen sources and exports in an agricultural watershed in Southeast China. *Biogeochemistry, 87*(2), 169–179.
Chen, N., Hong, H., Huang, Q., & Wu, J. (2011). Atmospheric nitrogen depositon and its long-term dynamics in a southeast China coastal area. *Journal of Environmental Management, 92*(6), 1663–1667.
Chen, Y., Fan, C., Teubner, K., & Dokulil, M. (2003). Changes in nutrients and phytoplankton chlorophyll-*a* in a large shallow lake, Taihu, China: an 8-year investigation. *Hydrobiologia, 506*(1–3), 273–279.
China Fishery Yearbook. (2007). *China Fishery Yearbook*. China Ministry of Agriculture Bureau of Fisheries. Beijing: Chinese Fishery Yearbook Publishing House.
Chorus, I., & Bartram, J. (Eds.). (1999). *Toxic cyanobacteria in water: A guide to their public health consequences, monitoring and management*. London: E & F.N. Spon.
Cloern, J. E. (2001). Our evolving conceptual model of the coastal eutrophication problem. *Marine Ecology Progress Series, 210*, 223–253.
Cochlan, W. P., Herndon, J., & Kudela, R. M. (2008). Inorganic and organic nitrogen uptake by the toxigenic diatom *Pseudo-nitzschia australis* (Bacillariophyceae). *Harmful Algae, 8*(1), 111–118.
Compton, J. E., Harrison, J. A., Dennis, R. L., Greaver, T. L., Hill, B. H., Jordan, S. J., et al. (2011). Ecosystem services altered by human changes in the nitrogen cycle: a new perspective for US decision making. *Ecology Letters, 14*(8), 804–815.

Conley, D. J., Paerl, H. W., Howarth, R. W., Boesch, D. F., Seitzinger, S. P., Havens, K. E., et al. (2009). Controlling eutrophication: Nitrogen and phosphorus. *Science, 323*(5917), 1014–1015.

Constant, K. M., & Sheldrick, W. F. (1992). *World Nitrogen Survey*. World Bank Technical paper 174, Asia Technical Department Series. Washington, D.C.: The World Bank.

Cornwell, J. C., Kemp, W. M., & Kana, T. M. (1999). Denitrification in coastal ecosystems: environmental controls and aspects of spatial and temporal scale. *Aquatic Ecology, 33*(1), 41–54.

Cronberg, G., & Annadotter, H. (2006). *Manual on aquatic cyanobacteria. A photo guide and a synopsis of their toxicity*. Copenhagen: International Society for the Study of Harmful Algae (ISSHA) and IOC of UNESCO.

Cui, S., Shi, Y., Groffman, P. M., Schlesinger, W. H., & Zhu, Y.-G. (2013). Centennial-scale analysis of the creation and fate of reactive nitrogen in China (1910–2010). *Proceedings of the National Academy of Sciences of the United States of America (PNAS), 110*(6), 2052–2057.

Dobermann, A., & Cassman, K. G. (2005). Cereal area and nitrogen use efficiency are drivers of future nitrogen fertilizer consumption. *Science in China, Series C: Life Sciences, 48*(Supplement 2), 745–758.

Dolman, A. M., Rücker, J., Pick, F. R., Fastner, J., Rohrlack, T., Mischke, U., et al. (2012). Cyanobacteria and cyanotoxins: The influence of nitrogen versus phosphorus. *PLoS ONE, 7*(6), e38757. https://doi.org/10.1371/journal.pone.0038757.

Donald, D. B., Bogard, M. J., Finlay, K., & Leavitt, P. R. (2011). Comparative effects of urea, ammonium, and nitrate on phytoplankton abundance, community composition, and toxicity in hypereutrophic freshwaters. *Limnology and Oceanography, 56*(6), 2161–2175.

Donald, D. B., Bogard, M. J., Finlay, K., Bunting, L., & Leavitt, P. R. (2013). Phytoplankton-specific response to enrichment of phosphorus-rich surface waters with ammonium, nitrate, and urea. *PLoS ONE, 8*(1), e53277. https://doi.org/10.1371/journal.pone.0053277.

Downing, J. A., Watson, S. B., & McCauley, E. (2001). Predicting cyanobacteria dominance in lakes. *Canadian Journal of Fisheries and Aquatic Sciences, 58*(10), 1905–1908.

Downing, T. G., Sember, C. S., Gehringer, M. M., & Leukes, W. (2005). Medium N: P ratios and specific growth rate comodulate microcystin and protein content in Microcystis aeruginosa PCC7806 and M. aeruginosa UV027. *Microbial Ecology, 49,* 468–473.

Duan, H., Ma, R., Xu, X., Kong, F., Zhang, S., Kong, W., et al. (2009). Two-decade reconstruction of algal blooms in China's Lake Taihu. *Environmental Science and Technology, 43,* 3522–3528.

Duan, S.-W., Liang, T., Zhang, S., Wang, L.-J., Zhang, X.-M., & Chen, X.-B. (2008). Seasonal changes in nitrogen and phosphorus transport in the lower Changjiang River before the construction of the Three Gorges Dam. *Estuarine Coastal Shelf Science, 79*(2), 239–250.

Elser, J. J., Andersen, T., Baron, J. S., Bergström, A. K., Jansson, M., Kyle, M., et al. (2009). Shifts in lake N: P stoichiometry and nutrient limitation driven by atmospheric nitrogen deposition. *Science, 326*(5954), 835–837.

Erisman, J. W., Sutton, M. A., Galloway, J. N., Klimont, Z., & Winiwarter, W. (2008). How a century of ammonia synthesis changed the world. *Nature Geoscience, 1,* 636–639.

European Commission. (1991a). *Council Directive 91/676/EEC concerning the protection of waters against pollution caused by nitrates from agricultural sources*. European Economic Community. Brussels: European Commission.

European Commission. (1991b). *Council Directive 91/271/EEC concerning urban waste water treatment*. European Economic Community. Brussels: European Commission.

European Commission. (2000). *Directive 2000/60/EC of the European Parliament and of the Council establishing a framework for the Community action in the field of water policy (EU Water Framework Directive)*. Brussels: European Commission.

FAO (Food and Agriculture Organization of the United Nations). (2012). FAOSTAT database collections. http://faostat.fao.org.

Finkel, Z. V., Beardall, J., Flynn, K. J., Quigg, A., Rees, T. A. V., & Raven, J. A. (2010). Phytoplankton in a changing world: Cells size and elemental stoichiometry. *Journal of Plankton Research, 32*(1), 119–137.

Finlay, K., Patoine, A., Donald, D. B., Bogard, M. J., & Leavitt, P. R. (2010). Experimental evidence that pollution with urea can degrade water quality in phosphorus-rich lakes of the northern great plains. *Limnology and Oceanography, 55*(3), 1213–1230.

Flores, E., & Herrero, A. (2005). Nitrogen assimilation and nitrogen control in cyanobacteria. *Biochemical Society Transactions, 33*(1), 164–167.

Flynn, K. J., Stoecker, D. K., Mitra, A., Raven, J. A., Glibert, P. M., Hansen, P. J., et al. (2013). Misuse of the phytoplankton-zooplankton dichotomy: the need to assign organisms as mixotrophs within plankton functional types. *Journal of Plankton Research, 35*(1), 3–11.

Fu, M., Wang, Z., Pu, X., Xu, Z., & Zhu, M. (2012). Changes in nutrient concentrations and N:P: Si ratios and their possible impacts on the Huanghai Sea ecosystem. *Acta Oceanologica Sinica, 31*(4), 101–112.

Galloway, J. N., Cowling, E. B., Seitzinger, S. P., & Socolow, R. H. (2002). Reactive nitrogen: Too much of a good thing? *Ambio, 31*(2), 60–63.

Galloway, J. N., Dentener, F. J., Capone, D. G., Boyer, E. W., Howarth, R. W., Seitzinger, S. P., et al. (2004). Nitrogen cycles: past, present, and future. *Biogeochemistry, 70*(2), 153–226.

Giani, A., Bird, D., Prairie, Y., & Lawrence, J. (2005). Empirical study of cyanobacterial toxicity along a trophic gradient of lakes. *Canadian Journal of Fisheries and Aquatic Sciences, 62*(9), 2100–2109.

Ginn, H., Pearson, L., & Neilan, B. (2009). Hepatotoxin biosynthesis and regulation in cyanobacteria—the putative involvement of nitrogen and iron homeostasis mechanisms. *Chiang Mai Journal of Science, 36*(2), 200–223.

Glibert, P. M., Anderson, D. M., Gentien, P., Granéli, E., & Sellner, K. G. (2005a). The global, complex phenomenon of harmful algal blooms. *Oceanography, 18*(2), 136–147.

Glibert, P. M., Trice, T. M., Michael, B., & Lane, L. (2005b). Urea in the tributaries of the Chesapeake and Coastal Bays of Maryland. *Water, Air, and Soil pollution, 160*(1–4), 229–243.

Glibert, P. M., Harrison, J., Heil, C., & Seitzinger, S. (2006). Escalating worldwide use of urea – a global change contributing to coastal eutrophication. *Biogeochemistry, 77*(3), 441–463.

Glibert, P. M., Mayorga, E., & Seitzinger, S. (2008). *Prorocentrum minimum* tracks anthropogenic nitrogen and phosphorus inputs on a global basis: application of spatially explicit nutrient export models. *Harmful Algae, 8*(1), 33–38.

Glibert, P. M., Burkholder, J. M., Kana, T. M., Alexander, J., Skelton, H., & Shilling, C. (2009). Grazing by *Karenia brevis* on *Synechococcus* enhances its growth rate and may help to sustain blooms. *Aquatic Microbial Ecology, 55,* 17–30.

Glibert, P. M., Allen, J. I., Bouwman, L., Brown, C., Flynn, K. J., Lewitus, A., et al. (2010). Modeling of HABs and eutrophication: status, advances, challenges. *Journal of Marine Systems, 83*(3–4), 262–275.

Glibert, P. M., & Burkholder, J. M. (2011). Harmful algal blooms and eutrophication: "strategies" for nutrient uptake and growth outside the Redfield comfort zone. *Chinese Journal of Oceanology and Limnology, 29*(4), 724–738.

Glibert, P. M., Fullerton, D., Burkholder, J. M., Cornwell, J., & Kana, T. M. (2011). Ecological stoichiometry, biogeochemical cycling, invasive species and aquatic food webs: San Francisco Estuary and comparative systems. *Reviews in Fisheries Science, 19*(4), 358–417.

Glibert, P. M., Kana, T. M., & Brown, K. (2013). From limitation to excess: the consequences of substrate excess and stoichiometry for phytoplankton physiology, trophodynamics and biogeochemistry, and implications for modeling. *Journal of Marine Systems, 125,* 14–28.

Glibert, P. M., Maranger, R., Sobota, D. J., & Bouwman, L. (2014a). The Haber Bosch-harmful algal bloom (HB-HAB) link. *Environmental Research Letters*, 9, 105001. https://doi.org/10.1088/1748-9326/9/10/105001. https://iopscience.iop.org/article/10.1088/1748-9326/9/10/105001.

Glibert, P. M., Wilkerson, F., Dugdale, R. C., Parker, A. E., Alexander, J., Blaser, S., & Murasko, S. (2014b). Phytoplankton communities from San Francisco Bay Delta respond differently to oxidized and reduced nitrogen substrates—even under conditions that would otherwise suggest nitrogen sufficiency. *Frontiers in Marine Science,* 1(article 17). https://doi.org/10.3389/fmars.2014.00017.

Glibert, P. M., Allen, J. I., Artioli, Y., Beusen, A., Bouwman, L., Harle, J., et al. (2014c). Vulnerability of coastal ecosystems to changes in harmful algal bloom distribution in response to climate change: projections based on model analysis. *Global Change Biology, 20*(12), 3845–3858. https://doi.org/10.1111/gcb.12662.

Glibert, P. M., Wilkerson, F., Dugdale, R. C., Raven, J. A., Dupont, C. L., Leavitt, P. R., et al. (2016). Pluses and minuses of ammonium and nitrate uptake and assimilation by phytoplankton and implications for productivity and community composition with emphasis on nitrogen-enriched conditions. *Limnology and Oceanography, 61,* 165–197.

Goolsby, D. A., & Battaglin, W. A. (2001). Long-term changes in concentrations and flux of nitrogen in the Mississippi River basin, USA. *Hydrological Processes, 15*(7), 1209–1226.

Granéli, E., & Flynn, K. (2006). Chemical and physical factors influencing toxin content. In E. Granéli & J. T. Turner (Eds.), *Ecology of Harmful Algae (Chapter 18* (pp. 229–241). Dordrecht: Springer.

Hallegraeff, G. M. (1993). A review of harmful algal blooms and their apparent global increase. *Phycologia, 32*(2), 79–99.

Hardison, D. R., Sunda, W. G., Shea, D., & Litaker, R. W. (2013). Increased toxicity of *Karenia brevis* during phosphate limited growth: Ecological and evolutionary implications. *PLoS ONE, 8*(3), e58545. https://doi.org/10.1371/journal.pone.0058545.

Harris, T. D., Smith, V. H., Graham, J. L., Van de Waal, D. B., Tedesco, L. P., & Clercin, N. (2016). Combined effects of nitrogen to phosphorus and nitrate to ammonia ratios on cyanobacterial metabolite concentrations in eutrophic Midwestern USA reservoirs. *Inland Waters, 6*(2), 199–210.

Heffer, P., & Prud'homme, M. (2008). Outlook for world fertilizer demand, supply, and supply/demand balance. *Turkish Journal of Agriculture and Forestry, 32,* 159–164.

Heffer, P., & Prud'homme, M. (2013). *Fertilizer outlook 2013-2017.* 81st IFA Annual Conference, Chicago (USA) 20–22 May 2013. Paris: International Fertilizer Industry Association (IFA). Available at: http://www.fertilizer.org. Accessed 22 June 2014.

Heffer, P., & Prud'homme, M. (2016). *Fertilizer outlook 2016–2020.* 84th IFA Annual Conference, Moscow (Russia) 30 May–1 June 2016. Paris: International Fertilizer Industry Association (IFA). Available at: http://www.fertilizer.org. Accessed 30 September 2016.

Heisler, J., Glibert, P. M., Burkholder, J. M., Anderson, D. M., Cochlan, W., Dennison, W. C., et al. (2008). Eutrophication and harmful algal blooms: A scientific consensus. *Harmful Algae, 8*(1), 3–13.

Hoagland, P., & Scatasta, S. (2006). The economic effects of harmful algal blooms. In E. Granéli and J.T. Turner (Eds.), *Ecology of Harmful Algae* (Chapter 30, pp. 391–402). Ecological Studies (Analysis and Synthesis), volume 189. Berlin, Heidelberg: Springer.

Houlton, B. Z., Boyer, E., Finzi, A., Galloway, J., Leach, A., Liptzin, D., et al. (2013). Intentional versus unintentional nitrogen use in the United States: trends, efficiency and implications. *Biogeochemistry, 114*(1–3), 11–23.

Howarth, R. W. (2008). Coastal nitrogen pollution: A review of sources and trends globally and regionally. *Harmful Algae, 8*(1), 14–20.

Howarth, R. W., & Paerl, H. (2008). Coastal marine eutrophication: Control of both nitrogen and phosphorus is necessary. *Proceedings of the National Academy of Sciences of the United States of America (PNAS), 105*(49), E103. https://doi.org/10.1073/pnas.0807266106.

Howarth, R. W., Swaney, D. P., Boyer, E. W., Marino, R., Jaworski, N., & Goodale, C. (2006). The influence of climate on average nitrogen export from large watersheds in the Northeastern United States. *Biogeochemistry, 79*(1–2), 163–86. https://doi.org/10.1007/s10533-006-9010-1.

IFA (International Fertilizer Association). (2014). *IFA database.* http://ifadata.fertilizer.org/ucSearch.aspx. Accessed December 2019.

Jeong, H. J., Yoo, Y. D., Kim, T. H., Kim, J. H., Kang, N. S., & Yih, W. H. (2004). Mixotrophy in the phototrophic harmful alga *Cochlodinium polykikoides* (Dinophycean): prey species, the effects of prey concentration and grazing impact. *The Journal of Eukaryotic Microbiology, 51*(5), 563–569.

Jewett, E.B., Lopez, C.B., Dortch, Q., Etheridge, S.M., Backer, L.C. (2008). Harmful algal bloom management and response: Assessment and plan. Interagency Working Group on Harmful Algal Blooms, Hypoxia, and Human Health of the Joint Subcommittee on Ocean Science and Technology. Washington DC.

Jickells, T. (2006). The role of air-sea exchange in the marine nitrogen cycle. *Biogeosciences, 3,* 271–280.

Johansson, N., & Granéli, E. (1999). Influence of different nutrient conditions on cell density, chemical composition and toxicity of *Prymnesium parvum* (Haptophyta) in semi-continuous cultures. *Journal of Experimental Marine Biology and Ecology, 239*(2), 243–258.

Khakural, B. R., & Alva, A. K. (1995). Hydrolysis of urea in two sandy soils under citrus production as influenced by rate and depth of placement. *Communications in Soil Science Plant Analysis, 26*(13–14), 2143–2456.

Kosten, S., Huszar, V. L. M., Bécares, E., Costa, L. S., Van Donk, E., Hansson, L.-A., et al. (2012). Warmer climates boost cyanobacterial dominance in shallow lakes. *Global Change Biology, 18*(1), 118–126.

Kudela, R. M., Lane, J. Q., & Cochlan, W. P. (2008). The potential role of anthropogenically derived nitrogen in the growth of harmful algae in California, USA. *Harmful Algae, 8*(1), 103–110.

Lee, S. J., Jang, M. H., Kim, H. S., Yoon, B. D., & Oh, H.-M. (2000). Variation of microcystin content of Microcystis aeruginosa relative to medium N: P ratio and growth stage. *Journal of Applied Microbiology, 89*(2), 323–329.

Lenhart, H.-J., Mills, D. K., Baretta-Bekker, H., van Leeuwen, S. M., van der Molen, J., Baretta, J. W., et al. (2010). Predicting the consequences of nutrient reduction on the eutrophication status of the North Sea. *Journal of Marine Systems, 81*(1–2), 148–170.

Leong, S. C. Y., Murata, A., Nagashima, Y., & Taguchi, S. (2004). Variability in toxicity of the dinoflagellate *Alexandrium tamarense* in response to different nitrogen sources and concentration. *Toxicon, 43*(4), 407–415.

Lewis, W. M., Jr., & Wurtsbaugh, W. A. (2008). Control of lacustrine phytoplankton by nutrients: Erosion of the phosphorus paradigm. *International Review of Hydrobiology, 93*(4–5), 446–465.

Lewis, W. M., Jr., Wurtsbaugh, W. A., & Paerl, H. W. (2011). Rationale for control of anthropogenic nitrogen and phosphorus to reduce eutrophication of inland waters. *Environmental Science and Technology, 45*(24), 10300–10305.

Li, J., Glibert, P. M., Zhou, M., Lu, S., & Lu, D. (2009). Relationships between nitrogen and phosphorus forms and ratios and the development of dinoflagellate blooms in the East China Sea. *Marine Ecology Progress Series, 383,* 11–26.

Litke, D.W. (1999). Review of phosphorus control measures in the United States and their effects on water quality. U.S. Geological Survey, Water–Resources Investigations Report 99-4007. Denver, Colorado: National Water-Quality Assessment Program.

Lomas, M. W., Trice, T. M., Glibert, P. M., Bronk, D. A., & McCarthy, J. J. (2002). Temporal and spatial dynamics of urea uptake and regeneration rates and concentrations in Chesapeake Bay. *Estuaries, 25*(3), 469–482.

McCarthy, M. J., Gardner, W. S., Lehmann, M. F., & Bird, D. F. (2016). Benthic nitrogen regeneration, fixation, and denitrification in a temperate, eutrophic lake: Effects on the nitrogen budget and cyanobacteria blooms. *Limnology and Oceanography, 61*(4), 1406–1423.

Monchamp, M. E., Pick, F. R., Beisner, B. E., & Maranger, R. (2014). Variation in microcystin concentration and composition in relation to cyanobacterial community structure. *PLoS ONE, 9*(1), e85573. https://doi.org/10.1371/journal.pone0085573.

Mulholland, P. J., Helton, A. M., Poole, G. C., Hall, R. O., Hamilton, S. K., Peterson, B. J., et al. (2008). Stream denitrification across biomes and its response to anthropogenic nitrate loading. *Nature, 452,* 202–205.

Ning, X., Lin, C., Hao, Q., Liu, C., Le, F., & Shi, J. (2009). Long term changes in the ecosystem in the northern South China Sea during 1976-2004. *Biogeosciences, 6,* 2227–2243.

Nixon, S. W. (1995). Coastal marine eutrophication: a definition, social causes, and future concerns. *Ophelia, 41,* 199–219.

O'Neil, J. M., Davis, T. W., Burford, M. A., & Gobler, C. J. (2012). The rise of harmful cyanobacteria blooms: the potential roles of eutrophication and climate change. *Harmful Algae, 14,* 313–334.

Oh, H.-M., Lee, S. J., Jang, M.-H., & Yoon, B.-D. (2000). Microcystin production by *Microcystis aeruginosa* in a phosphorus-limited chemostat. *Applied Environmental Microbiology, 66,* 176–179.

Orihel, D. M., Bird, D. F., Brylinsky, M., Chen, H., Donald, D. B., Huang, D. Y., et al. (2012). High microcystin concentrations occur only at low nitrogen-to-phosphorus ratios in nutrient-rich Canadian lakes. *Canadian Journal of Fisheries and Aquatic Sciences, 69*(9), 1457–1462.

Otten, T. G., Xu, H., Qin, B., Zhu, G., & Paerl, H. W. (2012). Spatiotemporal patterns and ecophysiology of toxigenic *Microcystis* blooms in Lake Taihu, China: Implications for water quality management. *Environmental Science and Technology, 46*(6), 3480–3488.

Paerl, H. W. (2009). Controlling eutrophication along the freshwater–marine continuum: dual nutrient (N and P) reductions are essential. *Estuaries and Coasts, 32*(4), 593–601.

Paerl, H. W., & Otten, T. G. (2013). Blooms bite the hand that feeds them. *Science, 342*(6157), 433–434.

Paerl, H. W., & Scott, J. T. (2010). Throwing fuel on the fire: Synergistic effects of excessive nitrogen inputs and global warming on harmful algal blooms. *Environmental Science and Technology, 44*(20), 7756–7758.

Paerl, H. W., Scott, J. T., McCarthy, M. J., Newell, S. E., Gardner, W. S., Havens, K. E., et al. (2016). It takes two to tango: when and where dual nutrient (N & P) reductions are needed to protect lakes and downstream ecosystems. *Environmental Science and Technology, 50*(20), 10805–10813.

Passey, P., Gypens, N., Billen, G., Garnier, J., Thieu, V., Rousseau, V., et al. (2013). A model reconstruction of riverine nutrient fluxes and eutrophication in the Belgian coastal zone since 1984. *Journal of Marine Systems, 128,* 106–122.

PBL Netherlands Environmental Assessment Agency. (2012). *Roads from RIO + 20. Pathways to achieve global sustainability goals by 2050.* The Hague, The Netherlands: PBL Netherlands Environmental Assessment Agency. Available online at: http://www.pbl.nl/en/publications/2012/roads-from-rio20. Accessed December 2019.

Peñuelas, J., Sardans, J., Rivas-Ubach, A., & Janssens, I. A. (2012). The human-induced imbalance between C, N and P in Earth's life system. *Global Change Biology, 18*(1), 3–6.

Redfield, A. C. (1934). On the proportions of organic derivatives in sea water and their relation to the composition of plankton. In: *James Johnstone Memorial Volume* (pp. 176–192). Liverpool: University of Liverpool Press.

Rhue, R. D., & Harris, W. G. (1999). Phosphorus sorption/desorption reactions in soil and sediments. In K. R. Reddy, G. A. O'Connor, & C. L. Schelske (Eds.), *Phosphorus biogeochemistry in subtropical systems* (pp. 187–206). Boca Raton, FL, USA: Lewis Publishers (imprint of CRC Press LLC).

Rolland, A., Bird, D. F., & Giani, A. (2005). Seasonal changes in composition of the cyanobacterial community and the occurrence of hepatotoxic blooms in the eastern townships, Québec. *Canada. Journal of Plankton Research, 27*(7), 683–694.

Romero, E., Garnier, J., Lassaletta, L., Billen, G., Le Gendre, R., Riou, P., et al. (2013). Large-scale patterns of river inputs in southwestern Europe: Seasonal and interannual variations and potential eutrophication effects at the coastal zone. *Biogeochemistry, 113*(1–3), 481–505.

Ryther, J. H., & Dunstan, W. M. (1971). Nitrogen, phosphorus and eutrophication in the coastal marine environment. *Science, 171*(3975), 1008–1013.

Sattari, S. Z., Bouwman, A. F., Giller, K. E., & van Ittersum, M. K. (2012). Residual soil phosphorus as the missing piece in the global phosphorus crisis puzzle. *Proceedings of the National Academy of Sciences of the United States of America (PNAS USA), 109*(16), 6348–6353.

Sattari, S. Z., van Ittersum, M. K., Giller, K. E., Zhang, F., & Bouwman, A. F. (2014). Key role of China and its agriculture in global sustainable phosphorus management. *Environmental Research Letters, 9*(5), 054003. https://doi.org/10.1088/1748-9326/9/5/054003.

Schindler, D. W. (1977). Evolution of phosphorus limitation in lakes. *Science, 195*(4275), 260–262.

Schindler, D. W., Hecky, R. E., Findlay, D. L., Stainton, M. P., Parker, B. R., Paterson, M. J., et al. (2008). Eutrophication of lakes cannot be controlled by reducing nitrogen input: results of a 37-year whole-ecosystem experiment. *Proceedings of the National Academy of Sciences of the United States of America (PNAS USA), 105*(32), 11254–11258.

Schindler, D. W., Carpenter, S. R., Chapra, S. C., Hecky, R. E., & Orihel, D. M. (2016). Reducing phosphorus to curb lake eutrophication is a success. *Environmental Science and Technology, 50*(17), 8923–8929.

Scott, J. T., & McCarthy, M. J. (2010). Nitrogen fixation may not balance the nitrogen pool in lakes over timescales relevant to eutrophication management. *Limnology and Oceanography, 55*(3), 1265–1270.

Scott, J. T., McCarthy, M. J., Otten, T. G., Steffen, M. M., Baker, B. C., Grantz, E. M., et al. (2013). Comment: An alternative interpretation of the relationship between TN:TP and microcystins in Canadian lakes. *Canadian Journal of Fisheries and Aquatic Sciences, 70*(8), 1265–1268.

Seitzinger, S. P. (1988). Denitrification in freshwater and coastal marine ecosystems: ecological and geochemical significance. *Limnology and Oceanography, 33*(4), 702–724.

Shen, Z.-L., & Liu, Q. (2009). Nutrients in the Changjiang River. *Environmental Monitoring and Assessment, 153*(1–4), 27–44.

Shimizu, Y., Watanabe, N., & Wrensford, G. (1993). Biosynthesis of brevetoxins and heterotrophic metabolism in *Gymnodinium breve*. In P. Lassus, G. Arzul, E. Erard, P. Gentien, & C. Marcaillou (Eds.), *Harmful marine algal blooms* (pp. 351–357). Paris: Lavoisier Publishing.

Siuda, W., & Chrost, R. J. (2006). Urea and ureolytic activity in lakes of different trophic status. *Polish Journal of Microbiology, 55*(3), 211–225.

Sivonen, K., & Jones, G. J. (1999). Cyanobacterial toxins. In I. Chorus & J. Bartram (Eds.), *Toxic cyanobacteria in water: A guide to their public health consequences, monitoring and management* (pp. 41–111). London: E & F.N. Spon.

Smil, V. (1999). Detonator of the population explosion. *Nature, 400,* 415.

Smil, V. (2000). Phosphorous in the environment: natural flows and human interferences. *Annual Review of Energy and the Environment, 25,* 53–88.

Smil, V. (2001). *Enriching the earth: Fritz Haber, Carl Bosch, and the transformation of world food*. Cambridge, United Kingdom: The MIT Press.

Smith, V. H. (1983). Low nitrogen to phosphorus ratios favor dominance by blue-green-algae in lake phytoplankton. *Science, 221*(4611), 669–671.

SOA (State Oceanographic Administration of China). (2010). *Marine disaster communiqué of China (1989–2009).* State Oceanographic Administration of China.

Sobota, D. J., Compton, J. E., & Harrison, J. A. (2013). Reactive nitrogen inputs to US lands and waterways: how certain are we about sources and fluxes? *Frontiers in Ecology and the Environment, 11*(2), 82–90.

Sobota, D. J., Compton, J. E., McCrackin, M. L., & Singh, S. (2015). Cost of reactive nitrogen release from human activities to the environment in the United States. *Environmental Research Letters, 10,* 025006. https://doi.org/10.10088/1748-9326/10/2/025006.

Solomon, C. M., Collier, J. L., Berg, G. M., & Glibert, P. M. (2010). Role of urea in microbes in aquatic systems: a biochemical and molecular review. *Aquatic Microbial Ecology, 59,* 67–88.

Sprague, L. A., Hirsch, R. M., & Aulenbach, B. T. (2011). Nitrate in the Mississippi River and its tributaries, 1980 to 2008: are we making progress? *Environmental Science and Technology, 45*(17), 7209–7216.

Sutton, M. A., Bleeker, A., Howard, C. M., Bekunda, M., Grizzetti, B., De Vries, W., Van Grinsven, H. J. M., Abrol, Y. P., Adhya, T. K., Billen, G., Davidson, E. A., Datta, A., Diaz, R., Erisman, J. W., Liu, X. J., Oenema, O., Palm, C., Raghuram, N., Reis, S., Scholz, R. W., Sims, T., Westhoek, H., & Zhang, F. S. (2013). *Our Nutrient World: The challenge to produce more food and energy with less pollution.* Global Overview of Nutrient Management. Centre for Ecology and Hydrology, Edinburgh on behalf of the Global Partnership on Nutrient Management and the International Nitrogen Initiative.

Switzer, T. (2008). Urea loading from a spring storm-Knysna estuary. *South Africa. Harmful Algae, 8*(1), 66–69.

Syers, J. K., Johnston, A. A., & Curtin, D. (2008). *Efficiency of soil and fertilizer phosphorus use. Reconciling changing concepts of soil phosphorous behaviour with agronomic information.* Rome: Food and Agriculture Organization of the United Nations (FAO).

Sylvan, J. B., Dortch, Q., Nelson, D. M., Maier Brown, A. F., Morrison, W., & Ammerman, J. W. (2006). Phosphorus limits phytoplankton growth on the Louisiana shelf during the period of hypoxia formation. *Environmenal Science & Technology, 40*(24), 7548–7553.

Ti, C., & Yan, X. (2013). Spatial and temporal variations of river nitrogen exports from major basins in China. *Environmental Science and Pollution Research, 20*(9), 6509–6520.

Van de Waal, D. B., Verspagen, J. M. H., Lürling, M., Van Donk, E., Visser, P. M., & Huisman, J. (2009). The ecological stoichiometry of toxins produced by harmful cyanobacteria: An experimental test of the carbon-nutrient balance hypothesis. *Ecology Letters, 12*(12), 1326–1335.

Van de Waal, D. B., Smith, V. H., Declerck, S. A., Stam, E. C. M., & Elser, J. J. (2014). Stoichiometric regulaitn of phytoplankton toxins. *Ecology Letters, 17*(6), 736–742.

Van Drecht, G., Bouwman, A. F., Harrison, J., & Knoop, J. M. (2009). Global nitrogen and phosphate in urban waste water for the period 1970–2050. *Global Biogeochemical Cycles*, 23(4), GB0A03. https://doi.org/10.1029/2009gb003458.

Van Mooy, B. A. S., Fredricks, H. F., Pedler, B. E., Dyhrman, S. T., Karl, D. M., Koblížek, M., et al. (2009). Phytoplankton in the ocean use non-phosphorus lipids in response to phosphorus scarcity. *Nature, 458,* 69–72.

Vézie, C., Rapala, J., Vaitomaa, J., Seitsonen, J., & Sivonen, K. (2002). Effect of nitrogen and phosphorus on growth of toxic and nontoxic *Microcystis* strains and on intracellular microcystin concentrations. *Microbial Ecology, 43,* 443–454.

Vitousek, P. M., Aber, J. D., Howarth, R. W., Likens, G. E., Matson, P. A., Schindler, D. W., et al. (1997). Human alteration of the global nitrogen cycle: sources and consequences. *Ecological Applications, 7*(3), 737–750.

Wali, P., Kumar, V., & Singh, J. P. (2003). Effect of soil type, exchangeable sodium percentage, water content, and organic amendments on urea hydrolysis in some tropical Indian soils. *Australian Journal of Soil Research, 41*(6), 1171–1176.

Wang, B., Xie, L., & Sun, X. (2011). Water quality and marginal seas off China in the last two decades. International Journal of Oceanography, article ID 731828. https://doi.org/10.1155/2011/731828.

Wang, H., & Wang, H. (2009). Mitigation of lake eutrophication: loosen nitrogen control and focus on phosphorus abatement. *Progress in Natural Science, 19*(10), 1445–1451.

Wang, S. F., Tang, D. L., He, F. L., Fukuyo, Y., & Azanza, R. (2008). Occurrences of harmful algal blooms (HABs) associated with ocean environments in the South China Sea. *Hydrobiologia, 596*(1), 79–93.

Wells, M. L., Trainer, V. L., Smayda, T. J., Karlson, B. S. O., Trick, C. G., Kudela, R. M., et al. (2016). Harmful algal blooms and climate change: learning from the past and present to forecast the future. *Harmful Algae, 49,* 68–93.

Xiao, H.-Y., Tang, C.-G., Xiao, H.-W., Liu, X.-Y., & Liu, C.-Q. (2010). Mosses indicating atmospheric nitrogen deposition in the Yangtze River drainage basin China. *Journal of Geophysical Research, 115*(D14), D14301. https://doi.org/10.1029/2009JD012900.

Xing, G. X., & Zu, Z. L. (2002). Regional nitrogen budgets for China and its major watersheds. *Biogeochemistry, 57*(1), 405–427.

Ye, C., Xu, Q., Kong, H., Shen, Z., & Yan, C. (2007). Eutrophication conditions and ecological status in typical bays of Lake Taihu in China. *Environmental Monitoring and Assessment, 135*(1–3), 217–225.

Zhang, W., & Zhang, F. (2008). *The fertilizer situation and outlook in China.* PowerPoint presentation. China Agriculture University. Available online at: http://www.firt.org/sites/default/files/Zhang_FertilizerMarketinChina_presentation.pdf. Accessed 22 June 2014.

Zhou, M.-J., Shen, Z.-L., & Yu, R.-C. (2008). Responses of a coastal phytoplankton community to increased nutrient input from the Changjiang (Yangtze) River. *Continental Shelf Research, 28*(12), 1483–1489.

Chapter 18
Human Health Effects of Exposure to Nitrate, Nitrite, and Nitrogen Dioxide

Jean D. Brender

Abstract Human exposure to nitrogen (N) pollution via nitrate in food and water and nitrogen oxides in the ambient air is reviewed. Increased risks of myocardial infarction, respiratory problems, and asthma have been linked to higher exposures to nitrogen oxides in ambient air. Excess nitrate/nitrite exposure in food and water may be harmful to human health by: (1) its contribution to the endogenous formation of *N*-nitroso compounds, demonstrated carcinogens and teratogens in animal models; (2) its potential role in methemoglobinemia; and (3) in high doses, its ability to competitively inhibit iodine uptake and induce changes in the thyroid. High nitrate levels in drinking water have been linked to methemoglobinemia in infants ("blue baby syndrome") and children. In 2006, an International Agency for Research on Cancer expert panel concluded that ingested nitrate/nitrite under conditions that result in endogenous nitrosation is probably carcinogenic to humans; subsequent studies have continued to support this conclusion. Maternal exposures to higher nitrate levels in drinking water have been associated with a variety of adverse pregnancy outcomes. In contrast, maternal dietary intake of nitrate/nitrite has not been associated with birth defects in offspring unless coupled with prenatal nitrosatable drug exposure. Further research is indicated on the beneficial and harmful effects of nitrate on human health. Given what is currently known about the harmful effects of nitrogen oxides and ingested nitrate/nitrite, it appears prudent to support measures to reduce nitrogen emissions in the atmosphere and to continue to maintain current limits and guidelines on nitrate concentration in drinking water.

Keywords Birth defects · Cancer · Methemoglobinemia · Nitrate · Nitrogen dioxide

---

J. D. Brender (✉)
Epidemiology and Biostatistics, Texas A&M University, School of Public Health, 77843-1266, College Station, TX, USA
e-mail: jdbrender@tamu.edu

M. A. Sutton et al. (eds.), *Just Enough Nitrogen*,
https://doi.org/10.1007/978-3-030-58065-0_18

18.1 Introduction

Nitrogen (N) is ubiquitous in the environment and an essential constituent of amino acids, the building blocks of protein (Smil 2002). During the past 5–10 years, the literature has mushroomed on the potential beneficial effects of nitrate ingestion. Ingested nitrate and nitrite contribute to the in vivo generation of nitric oxide. Nitric oxide has been observed to enhance endothelial function through reduction of arterial stiffness, inflammation, and intimal thickness (Lidder and Webb 2013). Furthermore, oral administration of nitrate has been noted in human volunteers to reduce blood pressure, inhibit platelet aggregation, and reduce oxygen needs during exercise (Lundberg et al. 2011). Nitrogen emissions from agriculture and from fossil fuel combustion can contaminate air, food, and water. For example, nitrogen emitted to the atmosphere can contribute to adverse health outcomes and environmental effects via its role in fine particulate matter ($PM_{2.5}$) and tropospheric ozone formation (Sutton et al. 2011). This chapter will review and summarize the potential harmful effects of "too much nitrogen" in the form of nitrogen dioxide, nitrate, and nitrite on human health.

18.2 Human Exposure Pathways to Nitrogen Dioxide, Nitrate and Nitrite

Humans are exposed to nitrogen and nitrogen compounds through inhalation and ingestion. Various types of nitrogen oxides are inhaled from both indoor and outdoor sources, with nitrogen dioxide being one of the oxides generating the most concern. Sources of nitrogen dioxide in outdoor air include motor vehicle emissions, industrial processes, and natural sources such as lightening and forest fires (Hesterberg et al. 2009; Wolfe and Patz 2002). Indoor exposures can be a significant source of exposure, especially with poorly vented gas stoves and gas heaters (Wolfe and Patz 2002). Because nitrogen dioxide occurs with other air pollutants, it is often difficult to distinguish the adverse health effects associated with nitrogen dioxide from those of other co-pollutants, such as $PM_{2.5}$, sulfur dioxide and ozone.

Along with nitrogen in proteins that are considered essential for life, nitrogen is ingested in the form of nitrate or nitrite which might have adverse health effects if ingested in excessive amounts. Vegetables are the major source of nitrate in human diets, with celery, lettuce, radishes, and spinach contributing the most nitrate (> 250 mg nitrate/100 g fresh weight) (Hord et al. 2009; Santamaria 2006). Generally, drinking water contributes a small percentage of daily nitrate intake, unless nitrate concentration exceeds 10 mg/L as nitrate-nitrogen or 45 mg/L as nitrate (National Academy of Sciences 1981), the maximum contaminant level (MCL) set by the U.S. Environmental Protection Agency for public drinking water supplies. Cured meats contain some of the highest levels of nitrite (National Academy of Sciences 1981), although endogenous conversion of nitrate to nitrite is also a significant source of nitrite with approximately 5% of ingested nitrate in food and water being converted to nitrite in the saliva (Choi 1985).

18.3 Nitrogen Dioxide and Human Health

In healthy human volunteers, the reported acute effects of inhalation of nitrogen oxides are not uniformly adverse, although nitrogen dioxide is a respiratory irritant when inhaled at higher concentrations (Hesterberg et al. 2009). Furthermore, a recent study of healthy young volunteers conducted by Huang et al. (2012) indicated that nitrogen dioxide enhanced the acute cardiovascular effects of ambient particulate matter; nitrogen dioxide and particulate matter are common co-pollutants.

Results of recently published cohort studies and meta-analyses have indicated that exposure to nitrogen dioxide, especially in conjunction with higher levels of particulate matter, are associated with reduced lung function, increased symptoms in children with asthma, myocardial infarction, and less favorable outcomes for patients with heart failure. In a cohort study of 1185 children conducted in Manchester (UK) that took into account exposures to air pollutants encountered both indoors and outdoors, long-term particulate matter (PM_{10}) and nitrogen dioxide exposures were associated with small, but statistically significant reductions in lung growth volume (Mölter et al. 2013). Higher concentrations of nitrogen dioxide indoors were also associated with increased respiratory symptoms in 150 asthmatic children in Baltimore (USA), with each 20-ppb increase in nitrogen exposure significantly associated with an increase in number of days with limited speech due to wheezing, coughing without a cold, nocturnal awakenings due to respiratory symptoms, and respiratory difficulties while exercising (Breysse et al. 2010).

With respect to adult health, the findings of a recent meta-analysis (21 studies) of the relation between air pollutants and myocardial infarction indicated that for every 10 micrograms per cubic meter increase in nitrogen dioxide, risk of a myocardial infarction increased about 1.1% (Mustafic et al. 2012). Another meta-analysis examined the effects of air pollution on persons with heart failure, with 28 studies available for review on nitrogen dioxide and decompensated heart failure (Shah et al. 2013). For every 10 ppb increase in nitrogen dioxide, hospitalization due to worsening heart failure or death increased by 1.7% (c.f. 2.5% for $PM_{2.5}$); both hospitalization and mortality significantly increased with every 10 ppb increase in ambient nitrogen dioxide.

18.4 Nitrate and Nitrite Ingestion and Human Health

Three primary mechanisms have been proposed regarding the potential harmful effects of nitrate/nitrite ingestion, including: (1) the formation of methemoglobin; (2) contribution to the endogenous formation of *N*-nitroso compounds, some of which are carcinogens and teratogens, and (3) in high doses, competitive inhibition of iodine uptake in the thyroid gland (Ward 2009). In some study populations, excess and/or higher concentrations of nitrate/nitrite in food and/or water have been associated with methemoglobinemia, some types of cancers, thyroid disorders, and adverse pregnancy outcomes.

18.4.1 Methemoglobinemia

Nitrite ions oxidize ferrous iron in hemoglobin to ferric iron, thereby forming methemoglobin, a substance that, unlike hemoglobin, does not carry oxygen. Excess formation of methemoglobin in the blood can lead to a condition called methemoglobinemia in which the body tissues lack adequate oxygen. Infants are more vulnerable to this condition than adults and can develop the "blue baby syndrome" from too much nitrate/nitrite ingestion. In the U.S., the drinking water standard for nitrate is based on preventing this syndrome in infants, although the standard has been disputed as being too stringent for several reasons (Avery 1999), including the observation that well water associated with cases of methemoglobinemia was also frequently contaminated with human/animal feces and that some affected infants had diarrhea, another potential cause of methemoglobinemia in young children.

On the other hand, higher methemoglobin levels in infants and children have been observed with higher nitrate intake from food and water in studies conducted in Africa, the Middle East, and Eastern Europe (Abu Naser et al. 2007; Sadeq et al. 2008; Super et al. 1981; Zeman et al. 2002). In a study conducted in Namibia, Super et al. (1981) observed methemoglobin levels greater than 3% among 33% of infants in a high water nitrate region (> 20 mg/L nitrate-N) compared with 13% of infants in a low water nitrate region ($\leq$ 20 mg/L nitrate-N). Moroccan children who drank well water with nitrate concentrations greater than 50 mg/L as nitrate were 1.8 times more likely (95% confidence interval [CI]) 1.2, 2.6) than children on well water with nitrate concentrations less than or equal to 50 mg/L to have methemoglobin levels greater than 2% (relative risk and respective confidence interval calculated from data presented in Table 2 in Sadeq et al. 2008). Infantile methemoglobinemia was most strongly associated with nitrate and nitrite exposure through diet via feeding of formula and tea made with water containing high levels of nitrate in a study of children in Transylvania, Romania (Zeman et al. 2002). In a study conducted among infants in three municipalities located in the Gaza Strip (Palestine), the highest proportion of infants with methemoglobin levels greater than 8% resided with families living in the area with the highest nitrate concentrations in wells used for drinking water (mean of 195 ppm as nitrate) (Abu Nasser et al. 2007).

18.4.2 Nitrate/Nitrite Ingestion and Cancer

Nitrite (from dietary sources and nitrate reduced to nitrite by oral bacteria) can react with amines or amides in an acidic environment, such as that found in the human stomach, to form nitrosamines and nitrosamides. Some of these compounds have been found to be potent carcinogens in animal studies. Endogenous formation of *N*-nitroso compounds contributes 40–75% of exposure to these compounds in humans (Tricker 1997). Numerous studies have been conducted on the relation between nitrate/nitrite exposure in food/drinking water and various types of cancer, with mixed findings. In 2006, the International Agency for Research on Cancer (IARC) convened a panel of experts to consider the carcinogenicity of ingested nitrate and nitrite. Based on the available literature at that time, the panel concluded that there was sufficient evidence that nitrite in conjunction with amines and amides was a factor in causing cancer in humans and experimental animals, but that there was inadequate evidence regarding whether nitrate in food and drinking water was associated with cancer in humans or experimental animals (IARC Working Group on the Evaluation of Carcinogenic Risks to Humans 2010). The panel classified ingested nitrate or nitrite under conditions that result in endogenous nitrosation as *probably carcinogenic to humans (Group 2A).*

Since the IARC Working Group meetings in 2006, over 30 additional studies have been published on the relation between ingested nitrate/nitrite and various types of cancers in humans. Findings from these studies continue to support the overall conclusions of the IARC panel. Although studies published since 2006 have reported findings on associations of nitrate/nitrite with a wide variety of cancers, one-third of the studies focused on gastrointestinal cancers including those of the esophagus, stomach, and colon/rectum. In a multi-center U.S. case-control study, esophageal and gastric cancers were associated with high meat/nitrite intake (Navarro Silvera et al. 2011), although nitrate and nitrite were not associated with esophageal or gastric cancer in a large prospective cohort study (Cross et al. 2011). In a study that took into account nitrite from dietary sources and nitrate from both drinking water and dietary sources, distal stomach cancer was associated with higher water nitrate ingestion coupled with higher intake of processed meat, but these joint exposures were not associated with esophageal cancer (Ward et al. 2008). On the other hand, the investigators observed elevated odds ratios for stomach and esophageal cancers with increasing intake of nitrate and nitrite from animal sources. In a study population in Mexico City, odds ratios ranged from 1.3 to 2.0 for gastric cancer associated with the highest intakes of total nitrite as well as nitrate or nitrite from animal sources (Hernandez-Ramirez et al. 2009).

In a large prospective cohort study conducted in an eight-state area of the U.S., increasing intake of nitrate from processed meats was significantly associated with increasing risk of colorectal cancer (p-value for trend 0.001), while less of a trend was noted with nitrite intake from processed meats (p-value for trend 0.055) (Cross et al. 2010). In this study population, both red and processed meat consumption were significantly associated with colorectal cancer, and the investigators hypothesized that increased nitrate and nitrite consumption was one of the underlying mechanisms of the meat associations, in addition to heme iron and heterocyclic amines. Investigators observed increased risks of colorectal cancer with higher nitrate intake only among study participants who also had a low vitamin C intake (< 83.9 mg per day) in a cohort of women in Shanghai, China (DellaValle et al. 2014). Vitamin C has been demonstrated to inhibit the formation of *N*-nitroso compounds through a rapid reduction of nitrous acid to nitric oxide and the production of dehydroascorbic acid (Bartsch et al. 1988). DellaValle et al. (2014) suggested that the increased risk of colorectal cancer among the Shanghai women with low vitamin C intake supported the hypothesis of endogenous formation of *N*-nitroso compounds as a mechanism for colorectal cancer risk. Ward et al. (2007) also observed increased risk of renal cell carcinoma with higher nitrate intake from drinking water among subgroups with above the median red meat intake or below the median vitamin C intake.

18.4.3 Nitrate/Nitrite Ingestion and the Thyroid Gland

Nitrate has been observed to be a competitive inhibitor of iodide uptake by the human sodium iodide symporter in *in vitro* test systems (Tonacchera et al. 2004) which may cause reduced production of thyroid hormones. Slovakian children living in three villages with high nitrate concentrations in water wells were found to have significantly increased thyroid volume and increased frequency of signs of subclinical thyroid disorders in contrast to a comparison group of children residing in four villages with relatively low concentrations of nitrate (Tajtakova et al. 2006). In several Amish communities in Pennsylvania (USA), higher drinking water nitrate (> 6.5 mg/L nitrate-N) was significantly associated with subclinical hypothyroidism in women but not men (Aschebrook-Kilfoy et al. 2012). In another cohort of women residing in Iowa (USA), higher dietary intake of nitrate was associated with an increased prevalence of hypothyroidism, but higher drinking water nitrate (> 5 mg/L nitrate-N) was not (Ward et al. 2010).

18.4.4 Maternal Nitrate/Nitrite Ingestion and Adverse Pregnancy Outcomes

Findings from studies in human populations have suggested that elevated levels of nitrate in drinking water, in some instances below the U.S. Environmental Protection Agency (USEPA) MCL, might increase risk for adverse pregnancy outcomes, including prematurity (delivery before 37 weeks gestation) (Bukowski et al. 2001; Jakucionyte et al. 2001; Joyce et al. 2008), intrauterine growth retardation (less than 10% predicted fetal weight for gestational age) (Bukowski et al. 2001), congenital malformations, and neonatal mortality (Aschengrau et al. 1993). Other studies found no association between higher maternal intake of nitrate from drinking water and congenital malformations (Aschengrau et al. 1993), including congenital heart defects (Zierler et al. 1988), or prematurity and low birth weight birth (Super et al. 1981).

As of 2014, five published studies have reported a relationship between higher drinking water nitrate concentrations and risk for central nervous system malformations in offspring, four specifically with neural tube defects (Arbuckle et al. 1988; Brender et al. 2004; Brender et al. 2013; Croen et al. 2001; Dorsch et al. 1984). Other congenital malformations associated with higher maternal exposures to drinking water nitrates have included musculoskeletal defects (Dorsch et al. 1984), specifically limb deficiencies (Brender et al. 2013); oral cleft defects (Brender et al. 2013; Dorsch et al. 1984); and congenital heart defects (Brender et al. 2013; Cedergren et al. 2002). These drinking water studies are limited by the focus on associations of outcomes with single contaminants, i.e., nitrate, instead of exposures to mixtures of contaminants. In a recently published study on chemical mixtures in U.S. groundwater including 383 public wells in 35 states, nitrate was noted to frequently occur with other contaminants, especially with pesticides, arsenic and other trace metals, and water disinfection by-products (Toccalino et al. 2012). Other co-pollutants might account for the adverse effects noted with relatively low concentrations of nitrate or perhaps some type of synergistic effect might occur with these mixtures.

In contrast to significant associations found between drinking water nitrate and birth defects, results from recent studies have not indicated a positive association between maternal nitrate/nitrite intake from food and birth defects in offspring, including neural tube defects, oral clefts, or limb deficiencies (Croen et al. 2001; Brender et al. 2004; Huber et al. 2013). On the other hand, higher dietary nitrate/nitrite intake has been observed to strengthen associations between prenatal nitrosatable drug exposures and adverse pregnancy outcomes including birth defects (Brender et al. 2004, 2011, 2012) and prematurity (Vuong et al. 2016). Nitrosatable drugs are capable of reacting with nitrite in an acidic environment, such as in the stomach, to form *N*-nitroso compounds (Brambilla and Martelli 2007).

With data from the U.S. National Birth Defects Prevention Study (NBDPS), Brender et al. (2011) observed the strongest associations between tertiary amine drugs (nitrosatable) and anencephaly by total nitrite intake in the upper two tertiles; the odds ratios associated with tertiary amines from the lowest to highest tertiles of total nitrite intake (sum of nitrite and 5% nitrate intake) were 1.16 (95% CI 0.59, 2.29), 2.19 (95% CI 1.25, 3.86), and 2.51 (95% CI 1.45, 4.37). In the NBDPS, higher intakes of nitrate and nitrite also strengthened associations between prenatal nitrosatable drugs and cleft lip, cleft palate, conotruncal heart defects, single ventricle, and atrioventricular septal defects (Brender et al. 2012, 2013). Findings from animal studies have indicated that nitrite and some nitrosatable compounds are teratogenic only in combination with one another, and not when administered separately (Teramoto et al. 1980). Results from recently published studies conducted by Brender et al. (2011, 2012, 2013) support the hypothesis that endogenous formation of *N*-nitroso compounds might be teratogenic in humans.

Authors of previous publications on nitrate in drinking water and birth defects have suggested that observed associations might be due to the endogenous formation of *N*-nitroso compounds. To test this assumption, Brender et al. (2013) assessed the effects of higher nitrate intake from drinking water on associations between prenatal nitrosatable drug exposure and birth defects in offspring to NBDPS participants residing in Iowa and Texas (USA). Stronger associations between nitrosatable drug exposure and birth defects were not observed with higher daily intake of nitrate from drinking water in this study population. The authors suggested that this lack of effect modification by water nitrate intake might be explained by the relative low contribution (6%) of drinking water to daily nitrate intake in the study population (Brender et al. 2013).

18.5 Conclusions—Human Health Effects of Too Much Nitrogen

In summary, recent data indicate that too much inhaled nitrogen dioxide may reduce lung function in children, increase risk of myocardial infarction, increase morbidity and mortality in persons with heart failure, and exacerbate symptoms in children with asthma. Higher intake of nitrate/nitrite has been associated with methemoglobinemia in infants and young children and altered thyroid function in children and adults. On the other hand, harmful health effects of nitrogen exposure may be contingent on the presence of other substances, such as ingested nitrosatable compounds, like heme iron (cancer), prenatal intake of certain types of drugs (preterm birth and birth defects), and higher particulate matter in the ambient air (cardiovascular diseases) . Because various forms of nitrogen occur in conjunction with other air and water contaminants, adverse health effects might be due to other co-pollutants instead of nitrogen. Finally, harmful effects might occur in persons with deficient intake of vitamins and other nutrients, such as vitamin C, a demonstrated inhibitor of nitrosation. Further research is indicated on the impact of too much nitrogen on human health, recognizing that nitrogen is essential for life and, in the right amounts, beneficial to cardiovascular health.

Meanwhile, it appears prudent to support measures to reduce nitrogen emissions in the atmosphere and to continue to maintain current limits and guidelines on nitrate concentrations in drinking water. Since private wells in agricultural areas have some of the highest concentrations of nitrate, persons who obtain their drinking water from these sources should consider having their wells tested and/or obtain their drinking water from another source, e.g., bottled water, if testing is not possible or water nitrate concentrations are above 10 mg/L nitrate-N. The public should be encouraged to eat a healthy diet with fruits and vegetables rich in vitamin C and to avoid heavy intake of red meat. Women, who are considering having children or who are pregnant, should check with their health care providers before taking medications and should maintain a healthy diet with adequate vitamin C intake along with prenatal supplements as recommended by their health care providers.

Acknowledgements The author's research on nitrate, nitrite, and nitrosatable drugs and birth defects was supported by Award Numbers 5R01ES015634 and 3R01ES015634-03S1 from the National Institute of Environmental Health Sciences (NIEHS); Centers for Disease Control and Prevention, Birth Defects Branch Cooperative Agreement U01DD000494; and Texas Department of State Health Services Contract 2012-039849. The content is solely the responsibility of the author and does not necessarily represent the official views of NIEHS, the National Institutes of Health, or the Centers for Disease Control and Prevention. The author gratefully acknowledges the following collaborators in this research: Dr. Martha Werler, Dr. Peter Weyer, Dr. Paul Romitti, Dr. John C. Huber Jr., Dr. Joseph Sharkey, Dr. Qi Zheng, Dr. Binayak Mohanty, Dr. Peter Langlois, Dr. Lucina Suarez, Dr. Mark Canfield, Dr. Sadia Malik, Dr. Mayura Shinde, Dr. Ann Vuong, Dr. John S. Griesenbeck, Dr. Dipankar Dwivedi, Ms. Katherine Kelley, Mr. Scott A. Horel, Ms. Jiji Kantamneni, Dr. Shiv Patel, Ms. Michelle Steck, Ms. Amy Hanson, and Dr. Benjamin Zhan. The content in this chapter was first presented at the 6th International Nitrogen Conference on November 19, 2013 in Kampala, Uganda. The author would like to express her appreciation to SCOPE (Scientific Committee on Problems in the Environment) for travel support to the conference.

References

Abu Naser, A. A., Ghbn, N., & Khoudary, R. (2007). Relation of nitrate contamination of groundwater with methaemoglobin level among infants in Gaza. *Eastern Mediterranean Health Journal, 13*(5), 994–1004.

Arbuckle, T. E., Sherman, G. J., Corey, P. N., Walters, D., & Lo, B. (1988). Water nitrates and CNS birth defects: a population-based case-control study. *Archives of Environmental Health, 43*(2), 162–167.

Aschebrook-Kilfoy, B., Heltshe, S. L., Nuckols, J. R., Sabra, M. M., Shuldiner, A. R., Mitchell, B. D., Airola, M., Holford, T. R., Zhang, Y., & Ward, M. H. (2012). Modeled nitrate levels in well water supplies and prevalence of abnormal thyroid conditions among the Old Order Amish in Pennsylvania. *Environmental Health,* 11(6), PMC3305600. Doi 1186/1476-069X-11-6.

Aschengrau, A., Zierler, S., & Cohen, A. (1993). Quality of community drinking water and the occurrence of late adverse pregnancy outcomes. *Archives of Environmental Health, 48*(2), 105–113.

Avery, A. A. (1999). Infantile methemoglobinemia: Reexamining the role of drinking water nitrates. *Environmental Health Perspectives, 107*(7), 583–586.

Bartsch, H., Ohshima, H., & Pignatellli, B. (1988). Inhibitors of endogenous nitrosation. Mechanisms and implications in human cancer prevention. *Mutation Research, 202*(2), 307–324.

Brambilla, G., & Martelli, A. (2007). Genotoxic and carcinogenic risk to humans of drug-nitrite interaction products. *Mutation Research, 635*(1), 17–52.

Brender, J. D., Olive, J. M., Felkner, M., Suarez, L., Marckwardt, W., & Hendricks, K. A. (2004). Dietary nitrites and nitrates, nitrosatable drugs, and neural tube defects. *Epidemiology, 15*(3), 330–336.

Brender, J. D., Werler, M. M., Kelley, K. E., Vuong, A. M., Shinde, M. U., Zheng, Q., Huber, J. C. Jr., Sharkey, J. R., Griesenbeck, J. S., Romitti, P. A., Langlois, P. H., Suarez, L., Canfield, M. A., & National Birth Defects Prevention Study. (2011). Nitrosatable drug exposure during early pregnancy and neural tube defects in offspring: National Birth Defects Prevention Study. *American Journal of Epidemiology, 174*(11), 1286–1295.

Brender, J. D., Werler, M. M., Shinde, M. U., Vuong, A. M., Kelley, K. E., Huber, J. C. Jr., Sharkey, J. R., Griesenbeck, J. S., Romitti, P. A., Malik, S., Suarez, L., Langlois, P. H., Canfield, M. A., & National Birth Defects Prevention Study. (2012). Nitrosatable drug exposures during the first trimester of pregnancy and selected congenital malformations. *Birth Defects Research Part A Clinical and Molecular Teratology, 94*(9), 701–713.

Brender, J. D., Weyer, P. J., Romitti, P. A., Mohanty, B. P., Shinde, M. U., Vuong, A. M., Sharkey, J. R., Dwivedi, D., Horel, S. A., Kantamneni, J., Huber, J. C. Jr., Zheng, Q., Werler, M. M., Kelley, K. E., Griesenbeck, J. S., Zhan, F. B., Langlois, P. H., Suarez, L., Canfield, M. A., & National Birth Defects Prevention Study. (2013). Prenatal nitrate intake from drinking water and selected birth defects in offspring of participants in the National Birth Defects Prevention Study. *Environmental Health Perspectives, 121*(9), 1083–1089.

Breysse, P. N., Diette, G. B., Matsui, E. C., Butz, A. M., Hansel, N. N., & McCormack, M. C. (2010). Indoor air pollution and asthma in children. *Proceedings of the American Thoracic Society, 7*(2), 102–106.

Bukowski, J., Somers, G., & Bryanton, J. (2001). Agricultural contamination of groundwater as a possible risk factor for growth restriction or prematurity. *Journal of Occupational and Environmental Medicine, 43*(4), 377–383.

Cedergren, M. I., Selbing, A. J., Lofman, O., & Kallen, B. A. (2002). Chlorination byproducts and nitrate in drinking water and risk for congenital cardiac defects. *Environmental Research, 89*(2), 124–130.

Choi, B. C. (1985). N-Nitroso compounds and human cancer. A molecular epidemiologic approach. *American Journal of Epidemiology, 121*(5), 737–743.

Croen, L. A., Todoroff, K., & Shaw, G. M. (2001). Maternal exposure to nitrate from drinking water and diet and risk for neural tube defects. *American Journal of Epidemiology, 153*(4), 325–331.

Cross, A. J., Ferrucci, L. M., Risch, A., Graubard, B. I., Ward, M. H., Park, Y., et al. (2010). A large prospective study of meat consumption and colorectal cancer risk: an investigation of potential mechanisms underlying this association. *Cancer Research, 70*(6), 2406–2414.

Cross, A. J., Freedman, N. D., Ren, J., Ward, M. H., Hollenbeck, A. R., Schatzkin, A., et al. (2011). Meat consumption and risk of esophageal and gastric cancer in a large prospective study. *American Journal of Gastroenterology, 106*(3), 432–442.

Dellavalle, C. T., Xiao, Q., Yang, G., Shu, X. O., Aschebrook-Kilfoy, B., Zheng, W., et al. (2014). Dietary nitrate and nitrite intake and risk of colorectal cancer in the Shanghai Women's Health Study. *International Journal of Cancer, 134*(12), 2917–2926. https://doi.org/10.1002/ijc.28612.

Dorsch, M. M., Scragg, R. K., McMichael, A. J., Baghurst, P. A., & Dyer, K. F. (1984). Congenital malformations and maternal drinking water supply in rural South Australia: a case-control study. *American Journal of Epidemiology, 119*(4), 473–486.

Hernández-Ramirez, R. U., Galván-Portillo, M. V., Ward, M. H., Agudo, A., González, C. A., Oñate-Ocaña, L. F., et al. (2009). Dietary intake of polyphenols, nitrate and nitrite and gastric cancer risk in Mexico City. *International Journal of Cancer, 125*(6), 1424–1430.

Hesterberg, T. W., Bunn, W. B., McClellan, R. O., Hamade, A. K., Long, C. M., & Valberg, P. A. (2009). Critical review of the human data on short-term nitrogen dioxide (NO_2) exposures: evidence for NO_2 no-effect levels. *Critical Review of Toxicology, 39*(9), 743–781.

Hord, N. G., Tang, Y., & Bryan, N. S. (2009). Food sources of nitrates and nitrites: the physiologic context for potential health benefits. *American Journal of Clinical Nutrition, 90*(1), 1–10.

Huang, Y. C., Rappold, A. G., Graff, D. W., Ghio, A. J., & Devlin, R. B. (2012). Synergistic effects of exposure to concentrated ambient fine pollution particles and nitrogen dioxide in humans. *Inhalation Toxicology, 24*(12), 707–790.

Huber, J. C. Jr., Brender, J. D., Zheng, Q., Sharkey, J. R., Vuong, A. M., Shinde, M. U., Griesenbeck, L. S., Langlois, P. H., Canfield, M. A., Romitti, P. A., Weyer, P. J., & National Birth Defects Prevention Study. (2013). Maternal dietary intake of nitrates, nitrites and nitrosamines and selected birth defects in offspring: a case-control study. *Nutrition Journal, 12*(34). https://doi.org/10.1186/1475-2891-12-34.

IARC Working Group on the Evaluation of Carcinogenic Risks in Humans. (2010). IARC monographs on the evaluation of carcinogenic risks to humans. Ingested nitrate and nitrite, and cyanobacterial peptide toxins. *IARC Monographs on the Evaluation of Carcinogenic Risks in Humans, 34,* 1–412.

Jakucionyte, L., Grazuleviciene, R., & Rutkoviene, V. (2001). Drinking water contamination by nitrate and low birth weight risk in rural population. In I. Linkow & J. Palma-Oliveira (Eds.), *Assessment & Management of Environmental Risks* (pp. 357–362). Dordrecht, The Netherlands: Kluwer Academic Publishers.

Joyce, S. J., Cook, A., Newnham, J., Brenters, M., Ferguson, C., & Weinstein, P. (2008). Water disinfection by-products and pre-labor rupture of membranes. *American Journal of Epidemiology, 168*(5), 514–521.

Lidder, S., & Webb, A. J. (2013). Vascular effects of dietary nitrate (as found in green leafy vegetables and beetroot) via the nitrate-nitrite-nitric oxide pathway. *British Journal of Clinical Pharmacology, 75*(3), 677–696.

Lundberg, J. O., Carlstrom, M., Larsen, F. J., & Weitzberg, E. (2011). Roles of dietary inorganic nitrate in cardiovascular health and disease. *Cardiovascular Research, 89*(3), 525–532.

Mölter, A., Agius, R. M., de Vocht, F., Lindley, S., Gerrard, W., Lowe, L., et al. (2013). Long-term exposure to PM_{10} and NO_2 in association with lung volume and airway resistance in the MAAS Birth Cohort. *Environmental Health Perspectives, 121*(10), 1232–1238.

Mustafic, H., Jabre, P., Caussin, C., Murad, M. H., Escolano, S., Tafflet, M., et al. (2012). Main air pollutants and myocardial infarction. A systematic review and meta-analysis. *JAMA, 307*(7), 713–721.

National Academy of Sciences. (1981). *The health effects of nitrate, nitrite, and N-nitroso compounds*. Washington, D.C.: National Academy of Sciences.

Navarro Silvera, S. A., Mayne, S. T., Rishch, H. A., Gammon, M. D., Vaughan, T., Chow, W. H., et al. (2011). Principal component analysis of dietary and lifestyle patterns in relation to risk of subtypes of esophageal and gastric cancer. *Annals of Epidemiology, 21*(7), 543–550.

Sadeq, M., Moe, C. L., Attarassi, B., Cherkaoui, I., Elaouad, R., & Idrissi, L. (2008). Drinking water nitrate and prevalence of methemoglobinemia among infants and children aged 1–7 years in Moroccan areas. *International Journal of Hygiene and Environmental Health, 211*(5–6), 546–554.

Santamaria, P. (2006). Nitrate in vegetables: toxicity, content, intake and EC regulation. *Journal of Science of Food and Agriculture, 86*(1), 10–17. https://doi.org/10.1002/jsfa.2351.

Shah, A. S., Langrish, J. P., Nair, H., McAllister, D. A., Hunter, A. L., Donaldson, K., et al. (2013). Global association of air pollution and heart failure: a systematic review and meta-analysis. *The Lancet, 382*(9897), 1039–1948.

Smil, V. (2002). Nitrogen and food production: proteins for human diets. *Ambio, 31*(2), 126–131.

Super, M., Heese, H. D., MacKenzie, D., Dempster, W. S., du Plessis, J., & Ferreira, J. J. (1981). An epidemiological study of well-water nitrates in a group of South West African/Namibian infants. *Water Research, 15*(11), 1265–1270.

Sutton, M. A., Howard, C. M., Erisman, J. W., Billen, G., Bleeker, A., Grennfelt, P., et al. (Eds.). (2011). *The European Nitrogen Assessment—sources, effects and policy perspectives*. Cambridge: Cambridge University Press.

Tajtáková, M., Semanová, Z., Tomková, Z., Szökeová, E., Majoros, J., Rádiková, Z., et al. (2006). Increased thyroid volume and frequency of thyroid disorders signs in schoolchildren from nitrate polluted area. *Chemosphere, 62*(4), 559–564.

Teramoto, S., Saito, R., & Shirasu, Y. (1980). Teratogenic effects of combined administration of ethylenethiourea and nitrite in mice. *Teratology, 21*(1), 71–78.

Toccalino, P. L., Norman, J. E., & Scott, J. C. (2012). Chemical mixtures in untreated water from public-supply wells in the U.S. – occurrence, composition, and potential toxicity. *Science of the Total Environment, 431,* 262–270. https://doi.org/10.1016/j.scitotenv.2012.05.044.

Tonacchera, M., Pinchera, A., Dimida, A., Ferrarini, E., Agretti, P., Vitti, P., et al. (2004). Relative potencies and additivity of perchlorate, thiocyanate, nitrate, and iodide on the inhibition of radioactive iodide uptake by the human sodium iodide symporter. *Thyroid, 14*(12), 1012–1019.

Tricker, A. R. (1997). *N*-nitroso compounds and man: sources of exposure, endogenous formation and occurrence in body fluids. *European Journal of Cancer Prevention, 6*(3), 226–268.
Vuong, A. M., Shinde, M. U., Brender, J. D., Shipp, E. M., Huber, J. C. Jr., Sharkey, J. R., McDonald, T. J., Werler, M. M., Kelley, K. E., Griesenbeck, J. S., Langlois, P. H., Canfield, M. A., & National Birth Defects Prevention Study Investigators. (2016). Prenatal exposure to nitrosatable drugs, dietary intake of nitrites, and preterm birth. *American Journal of Epidemiology, 183*(7), 634–642.
Ward, M. H., Rusiecki, J. A., Lynch, C. F., & Cantor, K. P. (2007). Nitrate in public water supplies and the risk of renal cell carcinoma. *Cancer Causes and Control, 18*(10), 1141–1151.
Ward, M. H., Heinemann, E. F., Markin, R. S., & Weisenburger, D. D. (2008). Adenocarcinoma of the stomach and esophagus and drinking water and dietary sources of nitrate and nitrite. *International Journal of Occupational and Environmental Health, 14*(3), 193–197.
Ward, M. H. (2009). Too much of a good thing? Nitrate from nitrogen fertilizers and cancer. *Reviews on Environmental Health, 24*(4), 357–363.
Ward, M. H., Kilfoy, B. A., Weyer, P. J., Anderson, K. E., Folsom, A. R., & Cerhan, J. R. (2010). Nitrate intake and risk of thyroid cancer and thyroid disease. *Epidemiology, 21*(3), 389–395.
Wolfe, A. H., & Patz, J. A. (2002). Reactive nitrogen and human health: acute and long-term implications. *Ambio, 31*(2), 120–125.
Zeman, C. L., Kross, B., & Vlad, M. (2002). A nested case-control study of methemoglobinemia risk factors in children of Transylvania Romania. *Environmental Health Perspectives, 110*(8), 817–822.
Zierler, S., Theodore, M., Cohen, A., & Rothman, K. J. (1988). Chemical quality of maternal drinking water and congenital heart disease. *International Journal of Epidemiology, 17*(3), 589–594.

Chapter 19
Nitrogen Deposition to China's Coastal Seas: Status and Ecological Impacts

Enzai Du

Abstract Rapid increase in nitrogen (N) emissions in mainland China may have enhanced N deposition to adjacent coastal seas. Assessing the status and characteristics of N deposition in coastal regions is of significant importance to understand how it affects the adjacent marine ecosystems. This chapter reviews the status of N deposition to China's coastal seas and its ecological impacts by synthesizing recent literature. Wet N deposition at China's coastline ranged from 0.76 to 6.13 g N m^{-2} $year^{-1}$ and averaged 1.75 g N m^{-2} $year^{-1}$ in the 2000s. Estimated dry N deposition was lower, whereas large uncertainty remained. Overall, the average of total N deposition to China's coastline was around 2.01 g N m^{-2} $year^{-1}$, leading to an estimate of total atmospheric inputs of 4.56 Tg N $year^{-1}$ to China's coastal seas. The effects of N deposition on primary production in China's seas differ by regions depending on background nutrient status. Nitrogen deposition can also affect phytoplankton composition via regulating competitive interactions among species. This chapter recommends that a long-term network be established to monitor wet and dry N deposition to coastal areas and that modelling tools are needed to combine site-observed data at regional scale. Novel experiments are needed to assess the integrative impacts of N deposition and explore their mechanisms. Moreover, critical loads should be determined for each coastal region to improve the marine ecosystem management.

Keywords Nitrogen deposition · Wet deposition · Dry deposition · Net primary production · Phytoplankton composition · Coastal region · China

19.1 Introduction

Rapid industrialization and urbanization along with intensified agricultural activities in mainland China have increased reactive nitrogen (N_r) emissions to the atmosphere

E. Du (✉)
State Key Laboratory of Earth Surface Processes and Resource Ecology, School of Natural Resources, Faculty of Geographical Science, Beijing Normal University, Beijing 100875, China
e-mail: enzaidu@bnu.edu.cn

M. A. Sutton et al. (eds.), *Just Enough Nitrogen*,
https://doi.org/10.1007/978-3-030-58065-0_19

(Ohara et al. 2007; Wang et al. 2009; Liu et al. 2013). Consequently, elevated atmospheric concentrations of NO_x and NH_3 (Richter et al. 2005; van der A et al. 2008; Clarisse et al. 2009) can be transported and deposited to the Pacific Ocean (Cooper et al. 2010). The long- or short-range transportation of N_r to the coastal and the open seas can induce positive (e.g., increasing nutrient source for primary production) or negative (e.g., biodiversity loss, harmful alga blooms, ocean acidification) ecological impacts (Jickells 2005; Doney et al. 2007; Duce et al. 2008).

Assessing the magnitudes and patterns of N deposition is relevant to understand how nitrogen (N) deposition affects the coastal and adjacent marine ecosystems. This study synthesizes observed data from various studies to assess N deposition to China's coastal seas and reviews their ecological impacts. Although organic N has been found to be directly taken up by phytoplankton (Peierls and Paerl 1997) to fuel primary productivity in coastal and marine ecosystems (Duce et al. 2008), and it can contribute more than one fourth to the total N deposition (Neff et al. 2002; Cornell et al. 2003; Du and Liu 2014), the limited measurements of organic N deposition and its large uncertainty hold back the assessment of organic N deposition in this study.

19.2 Status of Nitrogen Deposition to Coastal Regions of China

Direct measurements and indirect estimation from $^{15}N/^{14}N$ isotope signal in coral cores both suggest an increase in N deposition in coastline and nearby seas of China. For instance, wet N deposition at the site of Yuen Ng Fan (World Meteorological Organization (WMO)/Global Atmospheric Watch (GAW) site, Hong Kong) increased from 0.77 g N m^{-2} $year^{-1}$ in 1989 to 1.05 g N m^{-2} $year^{-1}$ in 1992 (Ayers 1996). The wet N deposition at two sites in Zhuhai (Acid Deposition Monitoring Network in East Asia (EANET) site, close to Hong Kong) averaged 1.56 g N m^{-2} $year^{-1}$ between 2000 and 2004 (EANET, 2006) and were much higher than the level of 0.85 g N m^{-2} $year^{-1}$ at Hong Kong between 1989 and 1992 (Ayers 1996). At Qianliyan Island (Shandong Province), wet N deposition increased from 0.77 g N m^{-2} $year^{-1}$ in 1998 to 1.27 g N m^{-2} $year^{-1}$ between 2004 and 2005 (Zhang et al. 2000; Bi 2006). In addition, a recent assessment of $^{15}N/^{14}N$ in coral core signals a significant increase in N deposition to South China Sea during the past three decades (Ren et al. 2017).

To estimate wet N deposition to China's coastal regions, I synthesized measured data at 13 sites located from north to south, including Dalian Jiaotong University (Liaoning Province; Jiang and Shi 2007), Qianliyan Island (Shandong Province; Zhang et al. 2007), Fulongshan (Shandong Province; Bi 2006), Shanghai (Shanghai; Mei and Zhang 2007), Shengsi (Zhejiang Province; Zhang et al. 2007), Zhoushan (Zhejiang Province; Bi 2006), Fenghua (Zhejiang Province; Song 2008), Xiaoping (Fujian Province; EANET 2006), Hongwen (Fujian Province; EANET 2006), South China Botanical Garden (Guangdong Province; Huang et al. 2010), Shenzhen PKU graduate school (Guangdong Province; Niu et al. 2008), Zhuaxian Dong (Guangdong Province; EANET 2006) and Xiangzhou (Guangdong Province; EANET 2006). The results indicate that wet N deposition to China's coastline (including shoreside sites and island sites) ranged from 0.76 to 6.13 g N m^{-2} $year^{-1}$ with a geometric mean of 1.75 (median = 1.61) g N m^{-2} $year^{-1}$ in the 2000s. Metropolitan regions of Shanghai and Guangzhou were two hotspots of wet N disposition with average rates of 6.13 and 4.05 g N m^{-2} $year^{-1}$, respectively. Wet ammonium deposition ranged from 0.46 to 2.79 g N m^{-2} $year^{-1}$ with a geometric mean of 0.94 (median = 0.83) g N m^{-2} $year^{-1}$. Wet nitrate deposition ranged from 0.30 to 3.34 g N m^{-2} $year^{-1}$, showing a geometric mean 0.77 (median = 0.57) g N m^{-2} $year^{-1}$. The ratio of NH_4^+–N/NO_3^-–N ranged from 0.36 to 2.45 and averaged 1.34 ± 0.54 (median = 1.39), indicating a higher contribution of ammonium than nitrate.

Nitrogen deposition is an essential nutrient input to the marine ecosystems (Zhang and Liu 1994; Wan et al. 2002). Assuming that wet N deposition declines to a background low level with increasing distance to the coastline, the inorganic N fluxes of wet deposition to the four Chinese seas (Bohai Sea, Yellow Sea, East China Sea and South China Sea) can be roughly estimated according to a simple equation Eq. (19.1):

$$F_N = 0.5 \times N_{dep} \times A \times 10^{-6} \tag{19.1}$$

where F_N (Tg N $year^{-1}$) is the atmospheric N inputs from wet deposition, N_{dep}(g N m^{-2} $year^{-1}$) is the mean N deposition at the coastline for each sea region, and A (km^2) is the surface area for each sea region. Table 19.1 summarizes the estimated N fluxes from wet deposition to the four Chinese sea regions. The total N inputs from wet deposition to China's coastal seas were estimated to be 3.34 Tg N $year^{-1}$. Previous studies have indicated that N deposition generally shows a power-law decrease with increasing distance from the N emission hotspots (Du et al. 2014, 2015). Therefore, this simplified method may overestimate the total N inputs via wet deposition.

Dry N deposition has been estimated at several coastal sites on the Yellow Sea and East China Sea (Table 19.2). Generally, dry deposition was much higher at shoreside sites (2.11 g N m^{-2} $year^{-1}$, Fulongshan, Shandong Province) than at island sites (0.20 ~ 0.31 g N m^{-2} $year^{-1}$). The limited observations in coastal regions showed lower dry deposition than wet deposition, a finding confirmed by a recent assessment based on measurements at six coastal sites (Luo et al. 2014). The ratio of NH_x-N/NO_x-N (0.4 ~

Table 19.1 Estimated annual nitrogen inputs (Tg N $year^{-1}$) via wet deposition to China's coastal seas

Region	Area(km^2)	NH_4^+–N	NO_3^-–N	Inorganic N
Bohai sea	7.7×10^4	0.04	0.05	0.09
Yellow sea	3.8×10^5	0.12	0.11	0.23
East China sea	7.7×10^5	0.27	0.29	0.56
South China Sea	3.5×10^6	1.49	0.98	2.46
Total	4.7×10^6	1.92	1.42	3.34

Table 19.2 Dry nitrogen deposition (g N m^{-2} year^{-1}) at coastal sites on the Yellow Sea and East China Sea. Data are limited to four sites, including Qianliyan Island (QLY, Shandong Province), Fulongshan (FLS, Shandong Province), Shengsi (SS, Zhejiang Province), Zhoushan (ZS, Zhejiang Province)

Site	Period	NH_x–N	NO_x–N	Inorganic-N	References
QLY	1999–2003	0.10	0.11	0.21	Zhang et al. (2007)
	2004–2005	0.09	0.22	0.31	Bi (2006)
FLS	2004–2005	0.85	1.26	2.11	Bi (2006)
ZS	2003–2004	0.10	0.17	0.27	Bi (2006)
SS	1999–2003	0.10	0.10	0.20	Zhang et al. (2007)

1.0) in dry deposition was lower than those of wet deposition, implying a dominance of oxidized N. Unavailability of sufficient observed data limited a precise estimate on the N inputs to China's coastal seas via dry deposition. Nevertheless, assuming background dry deposition averaged as 0.26 g N m^{-2} year^{-1} for China's seas, the N inputs from dry deposition was estimated to be 1.22 Tg N year^{-1}.

Riverine fluxes were estimated to contribute only < 15% of the N requirement by phytoplankton growth and are confined to the estuaries (Liu et al. 2009). Therefore, other sources such as atmospheric deposition may play a major role for primary production in China's coastal seas. The average total N deposition to China's coastline was around 2.01 g N m^{-2} year^{-1}, leading to an estimate of total atmospheric inputs of 4.56 Tg N year^{-1} to China's coastalseas, with wet deposition contributing to two thirds of the total. Moreover, N deposition to China's coastal seas accounted for one third of atmospheric N inputs to the continental China (~ 12–16 Tg N year^{-1}) (Lü and Tian 2007; Zhao et al. 2017) and ~ 23% of the mainland anthropogenic N emissions (~ 20 Tg N year^{-1}) (Liu et al. 2011, 2013).

19.3 Ecological Impacts of Nitrogen Deposition on China's Coastal Seas

Nitrogen is the most common macronutrient that limits phytoplankton growth in most marine ecosystems (Vitousek and Howarth 1991). This review shows that external N inputs via atmospheric deposition are important N sources to China's coastal and oceanic ecosystems (e.g., Yang et al. 2014). However, the effect of enhanced N deposition on new production differs by regions due to different background nutrient status. For instance, wet nitrate deposition in the Yellow Sea has been estimated to account for about 4.3–9.2% of the nitrate requirement for the new production and a contribution three times higher would be expected if dry nitrate deposition, and wet and dry ammonium deposition were included (Chung et al. 1998). In the South China Sea, N addition experiments have shown that phytoplankton growth is N limited and nitrate additions dramatically stimulate phytoplankton growth (Chen et al. 2004). Analysis of data sets for atmospheric N deposition, satellite chlorophyll-a, and air mass back trajectories reveals that N deposition contributes approximately 20% of the annual new production in the South China Sea (Kim et al. 2014). This finding is latterly supported by an assessment of $^{15}N/^{14}N$ in coral cores, suggesting that atmospheric N deposition represented about 20% of annual total N input to the surface of the northern South China Sea (Ren et al. 2017). In the Chinese estuaries and nearby waters, primary production is likely limited by phosphorus rather than N (Huang et al. 1989; Zhang 1994; Zou et al. 2001). Therefore, N deposition may exert very limited fertilization effects on primary production in regions with high levels of N inputs (Zou et al. 2000). For instance, N deposition appears to support about 0.1–9% of the new production in the East China Sea (Nakamura et al. 2005).

Nitrogen deposition also affects species composition of phytoplankton community via regulating competitive interactions. Experimental results have shown that inorganic N additions have different effects on phytoplankton species and can lead to changes in phytoplankton composition (Wang and Jiao 2002). Based on a field incubation experiment with nutrient addition in the South China Sea, microphytoplankton exhibits higher sensitivity response to nutrient uptake than picophytoplankton, shifting the size-fractionation proportion in favor of the microphytoplankton (Cui et al. 2016). Artificial inputs of N as well as an increase of the N:P ratio has been found to reduce the Shannon's index of planktonic species (Qu et al. 2000). Furthermore, N deposition may stimulate harmful blooms and result in serious loss of species (Zhang 1994; Zhang et al. 2003). Anthropogenic N deposition may also lead to other effects such as water acidification and increasing N_2O emissions (Doney et al. 2007; Duce et al. 2008) but unfortunately these effects have rarely been assessed in China's seas.

19.4 Research Needs and Policy Implications

China's coastal regions and nearby seas currently receive high levels of N deposition but large uncertainty still exists in the measurement of dry N deposition. Wet N deposition to China's coastline ranged from 0.76 to 6.13 g N m^{-2} $year^{-1}$ and averaged 1.75 g N m^{-2} $year^{-1}$ in the 2000s, whereas the limited observations showed that dry N deposition in coastal regions was likely to be much lower than wet deposition. Overall, the average of total N deposition at China's coastline was around 2.01 g N m^{-2} $year^{-1}$, leading to an estimate of total atmospheric inputs of 4.56 Tg N $year^{-1}$ to China's coastal seas, with two-thirds contributed by wet deposition. However, the rates and composition of N deposition may change over time. For instance, a recent study based on measurements at six sites along China's coastline shows that wet N deposition ranged from 1.42 to 2.52 g N m^{-2} $year^{-1}$ in early 2010s, with nitrate contributing slightly higher than ammonium (Luo et al. 2014). More precise assessment of the spatial-temporal patterns of N deposition needs integrative efforts of a long-term network to monitor N deposition and modelling tools to combine the observed data to a regional scale.

Nitrogen deposition exerts significant effects on primary production and phytoplankton composition in China's coastal seas. Due to limited research efforts, the overall impact of N deposition to these ecosystems is far less clear. Novel experiments and modelling tools are needed to assess the integrative impacts of N deposition on ecosystem processes and biogeochemical cycles as well as their interactions with acidification, eutrophication and climate change. Nitrogen deposition to China has been found to stabilize at a high level during the recent decade (2005–2015) (Yu et al. 2019) and it is expected to decrease in the near future. It is of great importance to assess how the new trend in anthropogenic N deposition will affect the health and function of China's marine ecosystems. Moreover, critical loads should be determined for each sea region and used in the marine ecosystem management.

Acknowledgements The author would like to express special thanks to the organizers and participants of the 6th International Nitrogen Conference (18–22, November 2013, Kampala, Uganda). This work was partially supported by Fok Ying-Tong Education Foundation (161015).

References

Ayers, G. P. (1996). Acid deposition in Hong Kong. *Atmospheric Environment, 30,* 1581–1587.

Bi, Y. F. (2006) *Atmospheric nutrient deposition at the East China coast and its impact on marine primary production.* Master's thesis. Ocean University of China.

Chen, Y. L. L., Chen, H. Y., Karl, D. M., & Takahashi, M. (2004). Nitrogen modulates phytoplankton growth in spring in the South China Sea. *Continental Shelf Research, 24*(4), 527–541.

Chung, C. S., Hong, G. H., Kim, S. H., Lim, J. H., Park, J. K., & Yang, D. B. (1998). Shore based observation on wet deposition of inorganic nutrients in the Korean Yellow Sea coast. *The Yellow Sea, 4,* 30–39.

Clarisse, L., Clerbaux, C., Dentener, F., Hurtmans, D., & Coheur, P. F. (2009). Global ammonia distribution derived from infrared satellite observations. *Nature Geoscience, 2*(7), 479–483.

Cooper, O. R., Parrish, D. D., Stohl, A., Trainer, M., Nédélec, P., Thouret, V., et al. (2010). Increasing spring time ozone mixing ratios in the free troposphere over western North America. *Nature, 463*(21), 344–348.

Cornell, S. E., Jickells, T. D., Cape, J. N., Rowland, A. P., & Duce, R. A. (2003). Organic nitrogen deposition on land and coastal environments: A review of methods and data. *Atmospheric Environment, 37,* 2173–2191.

Cui, D. Y., Wang, J. T., Tan, L. J., & Dong, Z. Y. (2016). Impact of atmospheric wet deposition on phytoplankton community structure in the South China Sea. *Estuarine Coastal Shelf Science, 173,* 1–8.

Doney, S. C., Mahowald, N., Lima, I., Feely, R. A., Mackenzie, F. T., Lamarque, J. F., et al. (2007). Impact of anthropogenic atmospheric nitrogen and sulfur deposition on ocean acidification and the inorganic carbon system. *Proceedings of the National Academy of Sciences, 104*(37), 14580–14585.

Du, E., & Liu, X. (2014). High rates of wet nitrogen deposition in China: A synthesis. In M. A. Sutton, K. E. Mason, L. J. Sheppard, H. Sverdrup, R. Haeuber, & W. K. Hicks (Eds.), *Nitrogen Deposition, Critical Loads And Biodiversity – Proceedings of the International Nitrogen Initiative Workshop, Linking Experts of the Convention on Long-Range Transboundary Air Pollution and the Convention On Biological Diversity* (Chapter 6, pp. 49–56). Dordrecht, The Netherlands: Springer.

Du, E., Jiang, Y., Fang, J., & de Vries, W. (2014). Inorganic nitrogen deposition in China's forests: Status and characteristics. *Atmospheric Environment, 98,* 474–482.

Du, E., de Vries, W., Liu, X., Fang, J., Galloway, J. N., & Jiang, Y. (2015). Spatial boundary of urban 'acid islands' in southern China. *Scientific Reports, 5,* 12625. https://doi.org/10.1038/srep12625.

Duce, R. A., LaRoche, J., Altieri, K., Arrigo, K. R., Baker, A. R., Capone, D. G., et al. (2008). Impacts of atmospheric anthropogenic nitrogen on the open ocean. *Science, 320,* 893–897.

EANET. (2006). *Periodic report on the state of acid deposition in East Asia. Part I: Regional assessment.* Acid Deposition Monitoring Network in East Asia (EANET). PDF available at: http://www.eanet.asia/product. Accessed Sept 2019.

Huang, D. Y., Xu, Y. C., Zhou, B., Zhang, H. H., & Lan, J. B. (2010). Wet deposition of nitrogen and sulfur in Guangzhou, a subtropical area in South China. *Environmental Monitoring and Assessment, 171,* 429–439.

Huang, L. M., Chen, Q. C., & Yuan, W. B. (1989). Characteristics of chlorophyll distribution and estimation of primary productivity in Daya Bay. *Asian Marine Biology, 6,* 115–128.

Jickells, T. (2005). External inputs as a contributor to eutrophication problems. *Journal of Sea Research, 54*(1), 58–69.

Jiang, X. L., & Shi, K. (2007). An initial study on wet deposition of Nitrogen in Dalian. *Urban Environment & Urban Ecology, 20*(6), 25–35 (in Chinese, with English abstract).

Kim, T.-W., Lee, K., Duce, R., & Liss, P. (2014). Impact of atmospheric nitrogen deposition on phytoplankton productivity in the South China Sea. *Geophysical Research Letters, 41*(9), 3156–3162.

Liu, S. M., Hong, G. H., Zhang, J., Ye, X. W., & Jiang, X. L. (2009). Nutrient budgets for large Chinese estuaries. *Biogeosciences, 6*(10), 2245–2263.

Liu, X., Duan, L., Mo, J., Du, E., Shen, J., Lu, X., et al. (2011). Nitrogen deposition and its ecological impact in China: An overview. *Environmental Pollution, 159*(10), 2251–2264.

Liu, X., Zhang, Y., Han, W., Tang, A., Shen, J., Cui, Z., et al. (2013). Enhanced nitrogen deposition over China. *Nature, 494,* 459–462.

Lü, C. Q., & Tian, H. Q. (2007) Spatial and temporal patterns of nitrogen deposition in China: synthesis of observational data. *Journal of Geophysical Research*, 112(D22), D22S05.

Luo, X. S., Tang, A. H., Shi, K., Wu, L. H., Li, W. Q., Shi, W. Q., et al. (2014). Chinese coastal seas are facing heavy atmospheric nitrogen deposition. *Environmental Research Letters, 9*(9), 095007.

Mei, X. Y., & Zhang, X. F. (2007) Nitrogen in wet deposition in Shanghai area and its effects on agriculture ecosystem. *Chinese Journal of Eco-Agriculture, 15*(1), 16–18 (in Chinese, with English abstract).

Nakamura, T., Matsumoto, K., & Uematsu, M. (2005). Chemical characteristics of aerosols transported from Asia to the East China Sea: An evaluation of anthropogenic combined nitrogen deposition in autumn. *Atmospheric Environment, 39,* 1749–1758.

Neff, J. C., Holland, E. A., Dentener, F. J., McDowell, W. H., & Russell, K. M. (2002). The origin, composition and rates of organic nitrogen deposition: A missing piece of the nitrogen cycle? *Biogeochemistry, 57*(58), 99–136.

Niu, Y. W., He, L. Y., & Hu, M. (2008). Chemical characteristics of atmospheric precipitation in Shenzhen. *Environmental Science, 29*(4), 1014–1019 (in Chinese, with English abstract).

Ohara, T., Akimoto, H., Kurokawa, J., Horii, N., Yamaji, K., Yan, X., et al. (2007). An Asian emission inventory of anthropogenic emission sources for the period 1980–2020. *Atmospheric Chemistry and Physics, 7*(16), 4419–4444.

Peierls, B. L., & Paerl, H. W. (1997). Bioavailability of atmospheric organic nitrogen deposition to coastal phytoplankton. *Limnology and Oceanography, 42,* 1819–1823.

Qu, K. M., Chen, B. J., Yuan, Y. X., & Xin, F. (2000). A preliminary study on influence of N and P on population of planktonic diatoms in seawater. *Chinese Journal of Applied Ecology, 11*(13), 445–448 (in Chinese, with English abstract).

Ren, H., Chen, Y. C., Wang, X. T., Wong, G. T., Cohen, A. L., DeCarlo, T. M., et al. (2017). 21st-century rise in anthropogenic nitrogen deposition on a remote coral reef. *Science, 356*(6339), 749–752.

Richter, A., Burrows, J. P., Nüß, H., Granier, C., & Niemeier, U. (2005). Increase in tropospheric nitrogen dioxide over China observed from space. *Nature, 437,* 129–132.

Song, L. (2008). *An initial study on atmospheric nitrogen deposition in typical ecosystems in China.* Master's Dissertation, 5–14. China Agriculture University.

Van der A, R. J., Eskes, H. J., Boersma, K. F., van Noije, T. P. C., Van Roozendael, M., De Smedt, I., Peters, D. H. M. U., & Meijer, E. W. (2008). Trends, seasonal variability and dominant NO_x source derived from a ten year record of NO_2 measured from space. *Journal of Geophysical Research, 113,* D04302.

Vitousek, P. M., & Howarth, R. W. (1991). Nitrogen limitation on land and in the sea—how can it occur? *Biogeochemistry, 13,* 11–87.

Wan, X. F., Wu, Z. M., Chang, Z. Q., & Zhang, X. L. (2002). Reanalysis of atmospheric flux of nutrients to the South Yellow Sea and the East China Sea. *Marine Environmental Science, 21*(4), 14–18 (in Chinese, with English abstract).

Wang, S. W., Liao, Q. J. H., Hu, Y. T., & Yan, X. Y. (2009). A preliminary inventory of NH_3 emission and its temporal and spatial distribution of China. *Journal of Agro-Environment Science, 28,* 619–626 (in Chinese, with English abstract).

Wang, Y., & Jiao, N. Z. (2002) Response of phytoplankton to nutrient addition in Jiaozhou Bay. *Marine Sciences, 26*(4), 8–13 (in Chinese, with English abstract).

Yang, J. Y., Hsu, S. C., Dai, M. H., Hsiao, S. Y., & Kao, S. J. (2014). Isotopic composition of water-soluble nitrate in bulk atmospheric deposition at Dongsha Island: Sources and implications of external N supply to the northern South China Sea. *Biogeosciences, 11*(7), 1833–1846.

Yu, G., Jia, Y., He, N., Zhu, J., Chen, Z., Wang, Q., et al. (2019). Stabilization of atmospheric nitrogen deposition in China over the past decade. *Nature Geoscience, 12*(6), 424–429.

Zhang, G. S., Chen, H. T., Zhang, J., & Liu, S. (2003). Nutrient elements in the atmospheric wet deposition in Changjiang River estuary. *Chinese Journal of Applied Ecology, 14*(7), 1107–1111 (in Chinese, with English abstract).

Zhang, G. S., Zhang, J., & Liu, S. M. (2007). Characterization of nutrients in the atmospheric wet and dry deposition observed at the two monitoring sites over Yellow Sea and East China Sea. *Journal of Atmospheric Chemistry, 57,* 41–57.

Zhang, J. (1994). Atmospheric wet deposition of nutrient elements: correlation with harmful biological blooms in northwest Pacific coastal zones. *Ambio, 58,* 464–468.

Zhang, J., & Liu, M. G. (1994). Observations on nutrient elements and sulphate in atmospheric wet deposition over the northwest Pacific coastal oceans-Yellow Sea. *Marine Chemistry, 47,* 173–189.

Zhang, J. L., Yu, Z. G., Zhang, J., & Chen, N. (2000). Wet Deposition (precipitation) of major elements at two sites of northwestern Yellow Sea. *Environmental Chemistry, 19*(4), 352–356 (in Chinese, with English abstract).

Zhao, Y., Zhang, L., Chen, Y., Liu, X., Xu, W., Pan, Y., et al. (2017). Atmospheric nitrogen deposition to China: A model analysis on nitrogen budget and critical load exceedance. *Atmospheric Environment, 153,* 32–40.

Zou, L., Chen, H. T., & Zhang, J. (2000). Experimental examination of the effects of atmospheric wet deposition on primary production in the Yellow Sea. *Journal of Experimental Marine Biology and Ecology, 249,* 111–121.

Zou, L., Zhang, J., Pan, W. X., & Zhan, Y. P. (2001). In situ nutrient enrichment experiment in the Bohai and Yellow Sea. *Journal of Plankton Research, 23*(10), 1111–1119.

Chapter 20
Anthropogenic Nitrogen Loads to Freshwater: A High-Resolution Global Study

Mesfin M. Mekonnen and Arjen Y. Hoekstra

Abstract At a spatial resolution level of 5 by 5 arc minute, we estimate anthropogenic nitrogen (N) loads to freshwater, globally. The global anthropogenic N load to freshwater systems from diffuse and point sources in the period 2002–2010 was 32.6 million tonnes per year. China contributed about 45% to this global anthropogenic N load. The USA was the second largest contributor (7%), followed by Russia (6%) and India (5%). Three quarters of the N loads came from diffuse sources (agriculture), 23% from domestic point sources and 2% from industrial point sources. Among the crops, production of cereals had the largest contribution to the N loads (18%, of which 7% wheat and 6% maize), followed by vegetables (15%) and oil crops (11%). It is estimated that, globally, 18% of the total N input on crop fields in the form of artificial fertilizer and manure leaches to freshwater systems.

Keywords Global · Nitrogen loads · Diffuse pollution · Nitrogen leaching · Point sources · Crops

20.1 Introduction

Over the last few decades, global population, crop production, fertilizer application rates, sewage emissions and fossil fuel combustion have markedly increased (Galloway et al. 2014). These changes have significantly altered the global biogeochemical cycle of nitrogen (N). Altogether, human activities have more than doubled the rate at which biologically available nitrogen enters the terrestrial biosphere compared to preindustrial levels (Galloway et al. 2004). A large fraction of the anthropogenically mobilized N enters ground and surface water and is transported

M. M. Mekonnen (✉)
Robert B. Daugherty Water for Food Global Institute, University of Nebraska, Lincoln, NE, USA
e-mail: mmmekonnen@ua.edu

Department of Civil, Construction and Environmental Engineering, University of Alabama, Tuscaloosa, AL 35487, USA

A. Y. Hoekstra (Deceased 2019)
Twente Water Centre, University of Twente, P.O. Box 217, 7500 AE Enschede, The Netherlands

M. A. Sutton et al. (eds.), *Just Enough Nitrogen*,
https://doi.org/10.1007/978-3-030-58065-0_20

by rivers to coastal seas (Bouwman et al. 2009; Galloway et al. 2004; Kanakidou et al. 2012; Seitzinger et al. 2010). Nitrogen leaching or running off from agricultural soils can cause groundwater pollution, eutrophication of lakes, rivers and coastal zones, loss of biodiversity, hypoxia and fish kills (Carpenter et al. 1998; Diaz and Rosenberg 2008; Seitzinger et al. 2010; Tilman 1999; Vitousek et al. 1997).

There have been various attempts at estimating N flows at different spatial scales, from country scale (Howarth et al. 2002; Lesschen et al. 2007; Van Breemen et al. 2002) and regional scale (Howarth et al. 1996; Thieu et al. 2010; Van der Struijk and Kroeze 2010; Yasin et al. 2010) to global scale (Bouwman et al. 2009; Bouwman et al. 2005a, b; Bouwman et al. 2011; Galloway et al. 2004; Harrison et al. 2005; Kroeze and Seitzinger 1998; Liu et al. 2010; Mayorga et al. 2010; Morée et al. 2013; Seitzinger et al. 2005; Seitzinger and Kroeze 1998; Seitzinger et al. 2010; Smil 1999; Turner et al. 2003; Van Drecht et al. 2009). Some of these studies focus on N loads from diffuse sources, while others focus on urban waste. Some of the studies estimate N loads at the level of countries or regions as a whole, without paying attention to the spatial variation within a country or region. The studies that apply a high spatial resolution do not explicitly distinguish between different crops and different nitrogen application rates per crop. Nitrogen loads are thus specified by region but not attributed to specific crops. In this study we carry out a global assessment of anthropogenic N loads to freshwater from diffuse and point sources at a spatial resolution of 5 by 5 arc-minute (~10 × 10 km near the equator) for the period 2002–2010 and calculate the anthropogenic N load from agriculture per type of crop. This is useful input for assessing nitrogen-related grey water footprints and water pollution levels (Mekonnen and Hoekstra 2015).

20.2 Method and Data

Diffuse N loads to freshwater from agriculture were estimated for 126 crops separately. We took spatial crop distributions from Monfreda et al. (2008). The application rate of artificial fertilizer per crop per country was calculated using three sources of artificial fertilizer data. Primarily, we used the dataset of IFA et al. (2002), which provides artificial fertilizer application rates per crop for 88 countries. We used FAO (2012b) and Heffer (2009) to complement this dataset. Since the application rates provided in these data sources refer to different years, these were adjusted to fit FAO (2012a) country average artificial fertilizer consumption per year for the period 2002–2010. The manure input was calculated at grid cell level by multiplying livestock density by the animal-specific excretion rates. The volume of manure actually applied on cropland was estimated by accounting for the collection rate and the allocation of collected manure over croplands versus pasture. In this study, we considered manure inputs on croplands (including managed grasslands), but did not further study manure inputs on grazing lands. We grouped the crops into leguminous, irrigated and non-leguminous in order to estimate the N input through bio-fixation. Atmospheric nitrogen deposition rates for the year 2000 were taken from Dentener et al. (2006).

The nutrient input through irrigation water was calculated for irrigated croplands by multiplying the nitrogen content of irrigation water (in kg of N per cubic metre) by the irrigation application rate (in m^3/ha per year). The nitrogen removal with harvested crops was estimated by multiplying the crop yield by the crop-specific nitrogen content. The nitrogen removal with crop residues was calculated by multiplying the yield of crop residue by the nutrient content of the crop residue and a residual removal factor. We adopted the approach of Liu et al. (2010) to calculate the nutrient loss through erosion. We employed the empirical model of Bouwman et al. (2002a) to calculate ammonia volatilization and Bouwman et al. (2002b) to estimate N loss through N_2O and NO from the application of animal manure and artificial fertilizers. Denitrification (emission of N_2) in the soil was calculated as a fraction of the nitrogen surplus after accounting for ammonia volatilization and nitrogen removal with the harvest of crop and crop residue (Van Drecht et al. 2003). Leaching of N—the movement of N from the soil to the groundwater—was estimated by assuming balance of N in the soil in the long term. A small fraction of what is calculated here as N leaching may actually leave the field through horizontal runoff, but this flow is generally considered minor compared to the actual leaching flow. Finally, we estimated, still at grid cell level, the anthropogenic nitrogen load to freshwater (i.e., the load due to artificial fertilizer and manure application) by multiplying the total leached volume by the fraction of N input from artificial fertilizer and manure to the total N input (which also includes the amounts of N added through biofixation, deposition and irrigation water).

To estimate the N loads from diffuse sources, we followed two approaches: the Bouwman et al. (2011) approach and the De Willigen (2000) approach. While the former uses a full soil balance approach, the latter estimates nitrogen leaching using a regression model. Both models take into account precipitation and soil properties: in the Bouwman et al. (2011) model for estimating gaseous losses from the soil and in the De Willigen (2000) to estimate the leaching fraction. The soil parameters were obtained from (Batjes 2012). The precipitation data for the period 2002–2010 were obtained from the Climate Research Unit of the University of East Anglia (Mitchell and Jones 2005). The rooting depths for individual crops were obtained from Allen et al. (1998).

Nitrogen loads from point sources were estimated based on dietary per capita protein consumption per country over the period from 2002 to 2010, using data from FAOSTAT (FAO 2012a), following the approach of Van Drecht et al. (2009). The N intake through food is estimated by assuming an average of 16% N content in the protein consumed (Block and Bolling 1946; FAO 2003). About 97% of the N intake is assumed to be excreted in the form of urine and faeces and the remainder 3% is lost via sweat, skin, hair, blood and miscellaneous (Calloway et al. 1971; FAO et al. 1985; Kimura et al. 2004; Morée et al. 2013). Data on connection to public sewerage system and the distribution of the different treatment types was collected from different sources (European Commission 2014; OECD 2014; UNSD 2014; Van Drecht et al. 2009). Since there is lack of data on industrial emissions, we have estimated the N load from the industrial sector as a function of the urban domestic load by assuming a ratio of industrial to urban households N load of 0.10 (Billen et al. 1999; Brion et al. 2008; Liu 2005; Luu et al. 2012; Quynh et al. 2005).

20.3 Global Loads of Nitrogen to Freshwater

20.3.1 Nitrogen Loads from Diffuse Sources

The global anthropogenic nitrogen load to freshwater per crop category is presented in the last row of Table 20.1. The table shows all terms in the nitrogen balance of the soil: both nitrogen inputs (artificial fertilizer, manure, bio-fixation, atmospheric deposition and supply through irrigation water) and nitrogen outputs (N removal with harvested crops and crop residues, erosion, gaseous losses and leaching). The largest share of the nitrogen input in croplands comes from artificial fertilizers, which account for about 50% of the total input. Nitrogen input from manure accounts for 20% of the total input, and biofixation for 18%. Cereal crops account for about 62% of the N input from artificial fertilizer and 32% of the manure. Oil crops contribute 11% of the N input from artificial fertilizer, followed by 'other crops', which contribute 9%. The 'other crops' category (mainly fodder crops) also accounts for a large contribution (25%) to the N input in the form of manure. About three-quarters of N input from bio-fixation came from oil crops. Relative to total N input, bio-fixation was most important in the case of oil crops (e.g., soybean, groundnuts) and pulses. Cereal crops account for the largest nitrogen removal with harvested crops (42%) and crop residues (60%). Oil crops are second in this respect, responsible for 24% of the N removed with harvested crops and 25% of the N removed with crop residues.

The total N leaching from croplands was 35 million tonne N/year, of which 70% (24.4 million tonne N/year) originated from anthropogenic sources (fertilizers, manure). Looking at the contribution of the different crop categories to the anthropogenic N load to freshwater, we see that the largest share (23%) came from cereal crops, followed by vegetables (19%), oil crops (15%) and fruits (12%). Our result shows that globally about 18% of the total N input in the form of artificial fertilizer and manure leaches to freshwater systems.

Table 20.2 presents the estimated N leached from croplands using the method of De Willigen (2000). The global total happens to be the same as what we obtained following the approach of Bouwman et al. (2011), but the distribution over different crop categories differs. In particular, the two methods differ in their estimate of N leaching from cereal production. The regression model of De Willigen (2000) is based on experimental data, and could therefore give unreasonably high leaching values when used for input parameter values (for precipitation, clay content, and layer thickness) outside the ranges used in deriving the regression equation. De Willigen (2000) stresses that the equation is only valid when used for interpolation within the ranges of the data used for the regression. Thus, from our point of view, the estimates

Table 20.1 Global nitrogen (N) inputs to and outputs from croplands per crop category, based on the Bouwman et al. (2011) approach (million tonne per year). The leaching from anthropogenic sources is calculated as a fraction of total N leaching. Period: 2002–2010

Balance term	Cereals	Vegetables	Oil crops	Fruits	Roots & tubers	Sugar crops	Pulses	Nuts	Other crops[a]	Total
Artificial fertilizer	60	5.8	11	4.3	3.0	2.6	0.8	0.5	8.7	96
Manure	12	4.2	5.5	3.1	1.9	0.5	0.6	0.3	9.4	38
Bio-fixation	5.3	0.2	25	0.2	0.2	0.1	2.2	0.03	0.7	34
Atmospheric N deposition	6.7	0.6	2.3	0.5	0.5	0.2	0.6	0.1	1.3	13
N supply in irrigation water	5.3	0.4	1.3	0.6	0.2	0.4	0.2	0.1	1.6	10
Total N inputs	**90**	**11**	**45**	**8.7**	**5.7**	**3.9**	**4.4**	**1.0**	**22**	**191**
N removed with harvested crops	31	0.8	15	0.9	1.0	0.5	1.6	0.02	12	63
N removed with crop residue	12	0.2	4.7	0.3	0.1	0.1	0.8	0.01	1.0	20
Total N removed with crop and crop residues	**43**	**1.0**	**20**	**1.2**	**1.1**	**0.5**	**2.4**	**0.0**	**12.5**	**82**
N budget (available for gaseous & leaching loss)	46	10	24	7.5	4.6	3.4	2.0	1.0	9.2	108
Erosion	8.8	0.7	2.9	0.9	0.8	0.4	0.9	0.1	2.2	18

(continued)

Table 20.1 (continued)

Balance term	Cereals	Vegetables	Oil crops	Fruits	Roots & tubers	Sugar crops	Pulses	Nuts	Other crops[a]	Total
NH_3 volatilization	5.1	0.6	1.1	0.4	0.3	0.2	0.1	0.04	0.5	8.3
Denitrification (N_2)	24	3.5	9.9	2.5	1.4	1.0	0.7	0.3	1.6	45
N_2O emission	0.4	0.04	0.2	0.1	0.1	0.03	0.1	0.01	0.1	1.0
NO	0.6	0.04	0.2	0.05	0.05	0.02	0.1	0.01	0.2	1.2
N leaching	**7.1**	**5.3**	**10**	**3.6**	**2.0**	**1.7**	**0.1**	**0.5**	**4.6**	**35**
Total N outputs	**90**	**11**	**45**	**8.7**	**5.7**	**3.9**	**4.4**	**1.0**	**21.7**	**191**
Leaching from anthropogenic sources	**5.7**	**4.8**	**3.7**	**3.0**	**1.7**	**1.3**	**0.04**	**0.4**	**3.9**	**24**

[a]Including fodder crops, coffee, tea, cocoa, spices and fibre crops

Table 20.2 Global nitrogen (N) input, removal and leaching to freshwater systems per crop category following the De Willigen (2000) approach (million tonne per year). Period: 2002–2010

Balance term	Cereals	Vegetables	Oil crops	Fruits	Roots & tubers	Sugar crops	Pulses	Nuts	Other crops	Total
Artificial fertilizer	60	5.8	11	4.3	3.0	2.6	0.8	0.5	8.7	96
Manure	12	4.2	5.5	3.1	1.9	0.5	0.6	0.3	9.4	38
N from decomposition of soil organic matter	115	4	37	5.4	5.1	5.3	6.7	0.7	16	195
Total N inputs	**187**	**14**	**54**	**13**	**10**	**8.4**	**8.1**	**1.5**	**34**	**330**
N removed with harvested crops	31	0.8	15	0.9	1.0	0.5	1.6	0.02	11.5	63
N removed with crop residue	12	0.2	4.7	0.3	0.1	0.1	0.8	0.01	1.0	20
Total N removed	**43**	**1.0**	**20**	**1.2**	**1.1**	**0.5**	**2.4**	**0.03**	**13**	**82**
N leaching	**19**	**2.6**	**4.5**	**1.5**	**1.8**	**1.0**	**1.0**	**0.2**	**3.2**	**35**

based on the balance approach of Bouwman et al. (2011) are to be considered more accurate than the estimates based on the empirical approach of De Willigen (2000), so that in the remaining analysis we will use the figures from Table 20.1.

20.3.2 Nitrogen Loads from Point Sources

The global N load to the freshwater system from point sources was about 8.2 million tonne of N per year (91% domestic and 9% industry). China contributed most (about 30%) to this global N load from point sources, followed by India and the USA, which each contributed 7%. Considering the per capita N load to freshwater systems from point sources, we find the highest values in countries with low treatment percentages. In Japan, with an urban wastewater treatment coverage of 67% and N removal rate of 10%, the load from point sources is 2.5 kg N per year per capita, while in Germany, with an urban wastewater treatment coverage of 100% and N removal rate of 76%, this is 0.82 kg N per year per capita.

20.3.3 Total Anthropogenic Nitrogen Load to Freshwater

The global anthropogenic N load to freshwater systems from both diffuse and point sources in the period 2002–2010 was 32.6 million tonnes per year, of which 24.4 million tonnes were from agriculture. China contributed about 45% to this global anthropogenic N load. The USA was the second largest contributor (7%), followed by Russia (6%) and India (5%). The contribution of the agricultural sector is largest in nearly 40% of the countries of the world, which also hosts three quarters of the world population. In India, agriculture contributed 68% to national anthropogenic N loads to freshwater. This was 72% in the USA and 84% in both China and Russia. In Japan, the domestic sector gives the largest contribution to the national anthropogenic N load (68%), as well as in Turkey (50%).

The spatial variation of the anthropogenic nitrogen loads was quite significant (Fig. 20.1). The largest loads per hectare were found in South-eastern China, Northern India, Western Europe, Mid-western USA, the Nile delta in Egypt, South East Brazil and the Central Valley in Chile. The high nitrogen loads closely correlate to the places with high artificial fertilizer and manure application rates and population densities.

Figure 20.2 shows the contribution of different product categories and regions to the global anthropogenic nitrogen load to freshwater. The largest contribution (75%) comes from diffuse sources, i.e., from artificial nitrogen fertilizers and manure applied on croplands. Relatively large percentages come from cereal production (18%, of which 7% wheat and 6% maize), production of vegetables (15%, of which 1.1% tomatoes and smaller percentages for a large variety of other vegetables) and oil crops (11%, of which 3.1% soybean, another 3.1% rapeseed and 2.4% cotton). Nitrogen loads from the domestic sector account for 23% of the total and the industrial

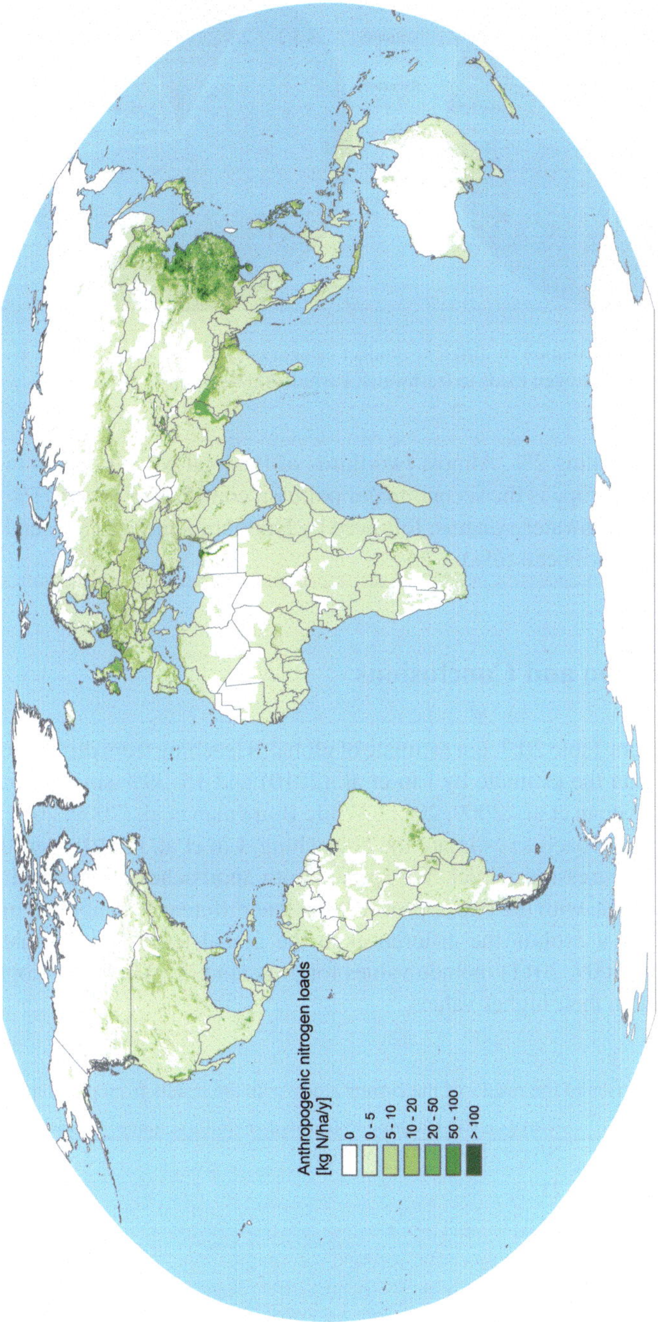

Fig. 20.1 Map of global anthropogenic nitrogen loads to freshwater systems from diffuse and point sources. Period: 2002–2010

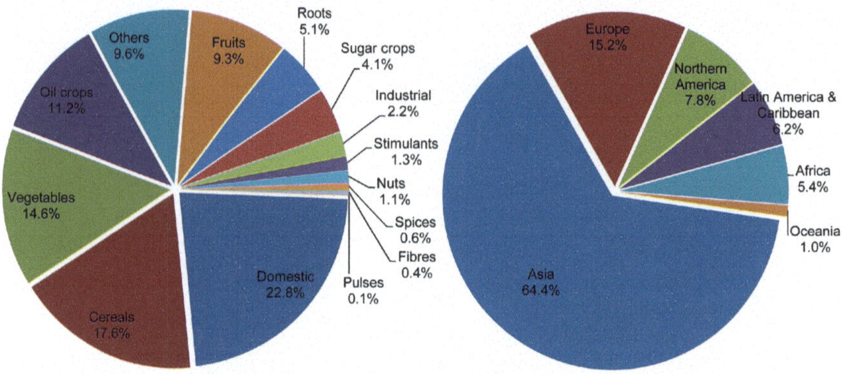

Fig. 20.2 Relative contribution of different product categories (left) and different regions (right) to total anthropogenic nitrogen loads to freshwater. Period: 2002–2010

sector for the remaining 2%. Almost two-thirds of the total N load occurs in Asia, mainly in China. Europe is the second major polluter, contributing about 15% to the global N load to freshwater systems, followed by Northern America (8%) and Latin America and the Caribbean (6%).

20.4 Discussion and Conclusions

As we can see from Table 20.3, our estimate of global N leaching from diffuse sources is 52% larger than the estimate by Liu et al. (2010) and 15–39% smaller than the estimate by Bouwman et al. (2009, 2011). While Bouwman et al. (2009, 2011) used the soil balance approach to estimate the N leaching, Liu et al. (2010) used the De Willigen (2000) regression model. We applied both approaches and arrived at the same global estimate with the two approaches, so that differences in applied method cannot immediately explain the differences at the global level. The estimates by Bouwman et al. (2009, 2011) include values for grassland in addition to croplands, which may explain their higher values.

Table 20.3 Comparison of the results of the current study with the results from previous studies

Study	Nitrogen (N) leaching to freshwater (million tonne N/year) from diffuse sources	Study period
Liu et al. (2010)	23	2000
Bouwman et al. (2011)	57	2000
Bouwman et al. (2009)	41	2000
Current study	35	2002–2010

The estimates of N loads to freshwater from both diffuse and point sources are based on a number of assumptions and global datasets, leading to significant uncertainties. First, due to a lack of spatially distributed data, a number of assumptions had to be made regarding, for example, artificial fertilizer application rates per crop and per country, nutrient removal by crop harvest and removed crop residues, and manure production and application rates. Second, to estimate leaching of nitrogen, the study assumed a long-term steady state condition in the soil regarding N content, which might not hold true in all places. Third, emissions from domestic sources were based on protein consumption, wastewater treatment coverage and nutrient removal in the wastewater treatment plants, while other point sources such as household solid waste, urban livestock and other domestic animal wastes were not included. Fourth, due to a lack of data, the emission from the industrial sector was estimated as a certain fraction of that from the domestic sector. Nevertheless, the study provides a good indication of the general magnitude and the spatial distribution of the total N loads to freshwater.

Our results show that globally about 18% of the N inputs onto croplands in the form of artificial fertilizer and manure were lost to freshwater through leaching. Nitrogen leaching intensities per hectare were highest in Western Europe, South-eastern China and Northern India, where the artificial fertilizer and manure application rates were largest. The N loads per hectare from point sources were largest in highly populated areas with low wastewater treatment coverage, like in the Nile delta in Egypt, Japan, central Mexico and Bangladesh. The large N loads to freshwater put pressure on many freshwater systems in the world, leading to groundwater pollution, eutrophication of lakes, rivers and coastal zones, degradation of ecosystems and loss of biodiversity (Carpenter et al. 1998; Seitzinger et al. 2010; Tilman 1999; Tilman et al. 2001; Vitousek et al. 1997; Vitousek et al. 2009). There is a need to balance the benefit of N application onto crop fields and the associated environmental problems. In some developing countries, raising agricultural productivities requires additional nutrients while in other places, in both developing and industrial countries, crops receive excessive amounts of nutrients. In many places, N application rates can be reduced without affecting agricultural productivities (Vitousek et al. 2009).

Acknowledgements This research was partly supported by the European Community Framework Programme 7, Compiling and Refining Environmental and Economic Accounts (CREEA), grant agreement no.: 265134.

References

Allen, R. G., Pereira, L. S., Raes, D., & Smith, M. (1998). *Crop evapotranspiration: Guideline for computing crop water requirement*. FAO Irrigation and Drainage Paper No. 56. Rome: FAO.

Batjes, N. H. (2012). *ISRIC-WISE derived soil properties on a 5 by 5 arc-minutes global grid (version 1.2b)*. Wageningen: ISRIC – World Soil Information.

Billen, G., Garnier, J., Deligne, C., & Billen, C. (1999). Estimates of early-industrial inputs of nutrients to river systems: implication for coastal eutrophication. *Science of the Total Environment, 243–244*, 43–52.
Block, R. J., & Bolling, D. (1946). The amino acid composition of proteins and foods. *Science, 103*(2675), 431–432.
Bouwman, A. F., Boumans, L. J. M., & Batjes, N. H. (2002a). Estimation of global NH_3 volatilization loss from synthetic fertilizers and animal manure applied to arable lands and grasslands. *Global Biogeochemical Cycles, 16*(2), 1024. https://doi.org/10.1029/2000gb001389.
Bouwman, A. F., Boumans, L. J. M., & Batjes, N. H. (2002b). Modelling global annual N_2O and NO emissions from fertilized fields. *Global Biogeochemical Cycles, 16*(4). https://doi.org/10.1029/2001gb001812.
Bouwman, A. F., Van Drecht, G., Knoop, J. M., Beusen, A. H. W., & Meinardi, C. R. (2005a). Exploring changes in river nitrogen export to the world's oceans. *Global Biogeochemical Cycles*, 19(1), GB1002. https://doi.org/10.1029/2004gb002314.
Bouwman, A. F., Van Drecht, G., & Van der Hoek, K. W. (2005b). Global and regional surface nitrogen balances in intensive agricultural production systems for the period 1970–2030. *Pedosphere, 15*(2), 137–155.
Bouwman, A. F., Beusen, A. H. W., & Billen, G. (2009). Human alteration of the global nitrogen and phosphorus soil balances for the period 1970–2050 Human alteration of the global nitrogen and phosphorus soil balances for the period 1970–2050. *Global Biogeochemical Cycles*, 23(4), GB0A04. https://doi.org/10.1029/2009gb003576.
Bouwman, L., Goldewijk, K. K., Van der Hoek, K. W., Beusen, A. H. W., Van Vuuren, D. P., Willems, J., et al. (2011). Exploring global changes in nitrogen and phosphorus cycles in agriculture induced by livestock production over the 1900–2050 period. *Proceedings of the National Academy of Sciences, 110*(52), 21195–21196.
Brion, N., Jans, S., Chou, L., & Rousseau, V. (2008). Nutrient loads to the Belgian coastal zone. In V. Rousseau, C. Lancelot, & D. Cox (Eds.), *Current Status of Eutrophication in the Belgian Coastal Zone* (pp. 17–43). Brussels: Presses Universitaires de Bruxelles.
Calloway, D. H., Odell, A. C. F., & Margen, S. (1971). Sweat and miscellaneous nitrogen losses in human balance studies. *The Journal of Nutrition, 101*(6), 775–786.
Carpenter, S. R., Caraco, N. F., Correll, D. L., Howarth, R. W., Sharpley, A. N., & Smith, V. H. (1998). Nonpoint pollution of surface waters with phosphorus and nitrogen. *Ecological Applications, 8*(3), 559–568.
De Willigen, P. (2000). *An analysis of the calculation of leaching and denitrification losses as practised in the NUTMON approach*. Wageningen: Plant Research International.
Dentener, F., Stevenson, D., Ellingsen, K., van Noije, T., Schultz, M., Amann, M., et al. (2006). The global atmospheric environment for the next generation. *Environmental Science and Technology, 40*(11), 3586–3594.
Diaz, R. J., & Rosenberg, R. (2008). Spreading dead zones and consequences for marine ecosystems. *Science, 321*(5891), 926–929.
European Commission. (2014). *Eurostat - Wastewater collection and treatment*. European Commission. http://epp.eurostat.ec.europa.eu/portal/page/portal/statistics/search_database. Accessed 19 Feb 2014.
FAO. (2003). *Food energy - methods of analysis and conversion factors*. Food and Nutrition Paper. Rome: Food and Agricultural Organization (FAO).
FAO. (2012a). *FAOSTAT on-line database*. Rome: Food and Agriculture Organization (FAO). http://faostat.fao.org. Accessed 10 Jan 2012.
FAO. (2012b). *Fertistat on-line database*. Rome: Food and Agriculture Organization (FAO). www.fao.org/ag/agl/fertistat. Accessed 10 Jan 2012.
FAO., WHO., & UNU. (1985). *Energy and protein requirements*. Report of a Joint FAO/WHO/UNU Expert Consultation. World Health Organization Technical Report Series. Geneva: World Health Organization.

Galloway, J. N., Dentener, F. J., Capone, D. G., Boyer, E. W., Howarth, R. W., Seitzinger, S. P., et al. (2004). Nitrogen cycles: Past, present, and future. *Biogeochemistry, 70*(2), 153–226.
Galloway, J. N., Winiwarter, W., Leip, A., Leach, A. M., Bleeker, A., & Erisman, J. W. (2014). Nitrogen footprints: Past, present and future. *Environmental Research Letters, 9,* 115003. https://doi.org/10.1088/1748-9326/9/11/115003.
Harrison, J. A., Caraco, N., & Seitzinger, S. P. (2005). Global patterns and sources of dissolved organic matter export to the coastal zone: Results from a spatially explicit, global model. *Global Biogeochemical Cycles, 19*(4), GB4S04. https://doi.org/10.1029/2005gb002480.
Heffer, P. (2009). *Assessment of fertilizer use by crop at the global level 2006/07-2007/08.* Paris: International Fertilizer Industry Association.
Howarth, R., Billen, G., Swaney, D., Townsend, A., Jaworski, N., Lajtha, K., et al. (1996). Regional nitrogen budgets and riverine N & P fluxes for the drainages to the North Atlantic Ocean: Natural and human influences. *Biogeochemistry, 35*(1), 75–139.
Howarth, R., Sharpley, A., & Walker, D. (2002). Sources of nutrient pollution to coastal waters in the United States: Implications for achieving coastal water quality goals. *Estuaries and Coasts, 25*(4), 656–676.
IFA., IFDC., IPI., PPI., & FAO. (2002). *Fertilizer use by crops.* 5th Edition. Rome: Food and Agriculture Organization
Kanakidou, M., Duce, R. A., Prospero, J. M., Baker, A. R., Benitez-Nelson, C., Dentener, F. J., Hunter, K. A., Liss, P. S., Mahowald, N., Okin, G. S., Sarin, M., Tsigaridis, K., Uematsu, M., Zamora, L. M., & Zhu, T. (2012). Atmospheric fluxes of organic N and P to the global ocean. *Global Biogeochemical Cycles*, 26(3), GB3026. https://doi.org/10.1029/2011gb004277.
Kimura, S. D., Liang, L., & Hatano, R. (2004). Influence of long-term changes in nitrogen flows on the environment: A case study of a city in Hokkaido Japan. *Nutrient Cycling in Agroecosystems, 70*(3), 271–282.
Kroeze, C., & Seitzinger, S. P. (1998). Nitrogen inputs to rivers, estuaries and continental shelves and related nitrous oxide emissions in 1990 and 2050: A global model. *Nutrient Cycling in Agroecosystems, 52*(2), 195–212.
Lesschen, J., Stoorvogel, J., Smaling, E., Heuvelink, G., & Veldkamp, A. (2007). A spatially explicit methodology to quantify soil nutrient balances and their uncertainties at the national level. *Nutrient Cycling in Agroecosystems, 78*(2), 111–131.
Liu, Y. (2005). *Phosphorus flows in China: Physical profiles and environmental regulation.* PhD Thesis. Wageningen: Wageningen University.
Liu, J., You, L., Amini, M., Obersteiner, M., Herrero, M., Zehnder, A. J. B., et al. (2010). A high-resolution assessment on global nitrogen flows in cropland. *Proceedings of the National Academy of Sciences, 107*(17), 8035–8040.
Luu, T., Garnier, J., Billen, G., Le, T., Nemery, J., Orange, D., et al. (2012). N, P, Si budgets for the Red River Delta (northern Vietnam): How the delta affects river nutrient delivery to the sea. *Biogeochemistry, 107*(1–3), 241–259.
Mayorga, E., Seitzinger, S. P., Harrison, J. A., Dumont, E., Beusen, A. H. W., Bouwman, A. F., et al. (2010). Global nutrient export from WaterSheds 2 (NEWS 2): Model development and implementation. *Environmental Modelling and Software, 25*(7), 837–853.
Mekonnen, M. M., & Hoekstra, A. Y. (2015). Global gray water footprint and water pollution levels related to anthropogenic nitrogen loads to fresh water. *Environmental Science and Technology, 49*(21), 12860–12868.
Mitchell, T. D., & Jones, P. D. (2005). An improved method of constructing a database of monthly climate observations and associated high-resolution grids. *International Journal of Climatology, 25*(6), 693–712.
Monfreda, C., Ramankutty, N., & Foley, J. A. (2008). Farming the planet: 2. Geographic distribution of crop areas, yields, physiological types, and net primary production in the year 2000. *Global Biogeochemical Cycles, 22*(1), GB1022. https://doi.org/10.1029/2007gb002947.

Morée, A. L., Beusen, A. H. W., Bouwman, A. F., & Willems, W. J. (2013). Exploring global nitrogen and phosphorus flows in urban wastes during the twentieth century. *Global Biogeochemical Cycles, 27*(3), 836–846. https://doi.org/10.1002/gbc.20072.

OECD. (2014). *OECD Environmental Data, Compendium 2006–2008: Inland waters*. Paris: OECD. http://www.oecd.org/statistics/. Accessed 19 Feb 2014.

Quynh, L. T. P., Billen, G., Garnier, J., Théry, S., Fézard, C., & Minh, C. V. (2005). Nutrient (N, P) budgets for the Red River basin (Vietnam and China). *Global Biogeochemical Cycles, 19*(2), GB2022. https://doi.org/10.1029/2004gb002405.

Seitzinger, S. P., & Kroeze, C. (1998). Global distribution of nitrous oxide production and N inputs in freshwater and coastal marine ecosystems. *Global Biogeochemical Cycles, 12*(1), 93–113. https://doi.org/10.1029/97gb03657.

Seitzinger, S. P., Harrison, J. A., Dumont, E., Beusen, A. H. W., & Bouwman, A. F. (2005). Sources and delivery of carbon, nitrogen, and phosphorus to the coastal zone: An overview of Global Nutrient Export from Watersheds (NEWS) models and their application. *Global Biogeochemical Cycles*, 19(4), GB4S01. https://doi.org/10.1029/2005gb002606.

Seitzinger, S. P., Mayorga, E., Bouwman, A. F., Kroeze, C., Beusen, A. H. W., Billen, G., Van Drecht, G., Dumont, E., Fekete, B. M., Garnier, J., & Harrison, J. A. (2010). Global river nutrient export: A scenario analysis of past and future trends. *Global Biogeochemical Cycles, 24*(4), GB0A08. https://doi.org/10.1029/2009gb003587.

Smil, V. (1999). Nitrogen in crop production: An account of global flows. *Global Biogeochemical Cycles, 13*(2), 647–662. https://doi.org/10.1029/1999gb900015.

Thieu, V., Mayorga, E., Billen, G., & Garnier, J. (2010). Subregional and downscaled global scenarios of nutrient transfer in river basins: Seine-Somme-Scheldt case study. *Global Biogeochemical Cycles, 24*(4), GB0A10. https://doi.org/10.1029/2009gb003561.

Tilman, D. (1999). Global environmental impacts of agricultural expansion: The need for sustainable and efficient practices. *Proceedings of the National Academy of Sciences, 96*(11), 5995–6000.

Tilman, D., Fargione, J., Wolff, B., D'Antonio, C., Dobson, A., Howarth, R., et al. (2001). Forecasting agriculturally driven global environmental change. *Science, 292*(5515), 281–284.

Turner, R. E., Rabalais, N. N., Justic, D., & Dortch, Q. (2003). Global patterns of dissolved N, P and Si in large rivers. *Biogeochemistry, 64*(3), 297–317.

UNSD. (2014). *Environmental indicators and selected time series - Population connected to wastewater collecting and treatment system*. United Nations Statistics Division. http://unstats.un.org/unsd/environment/Time%20series.htm#InlandWaterResources. Accessed 19 Feb 2014.

Van Breemen, N., Boyer, E. W., Goodale, C. L., Jaworski, N. A., Paustian, K., Seitzinger, S. P., Lajtha, K., Mayer, B., van Dam, D., Howarth, R. W., Nadelhoffer, K. J., Eve, M., & Billen, G. (2002). Where did all the nitrogen go? Fate of nitrogen inputs to large watersheds in the northeastern U.S.A. *Biogeochemistry, 57–58*(1), 267–293.

Van der Struijk, L. F., & Kroeze, C. (2010). Future trends in nutrient export to the coastal waters of South America: Implications for occurrence of eutrophication. *Global Biogeochemical Cycles, 24*(4), GB0A09. https://doi.org/10.1029/2009gb003572.

Van Drecht, G., Bouwman, A. F., Knoop, J. M., Beusen, A. H. W., & Meinardi, C. R. (2003). Global modeling of the fate of nitrogen from point and nonpoint sources in soils, groundwater, and surface water. *Global Biogeochemical Cycles, 17*(4), 1115. https://doi.org/10.1029/2003gb002060.

Van Drecht, G., Bouwman, A. F., Harrison, J., & Knoop, J. M. (2009). Global nitrogen and phosphate in urban wastewater for the period 1970 to 2050. *Global Biogeochemical Cycles, 23*(4), GB0A03. https://doi.org/10.1029/2009gb003458.

Vitousek, P. M., Aber, J. D., Howarth, R. W., Likens, G. E., Matson, P. A., Schindler, D. W., et al. (1997). Technical report: Human alteration of the global nitrogen cycle: Sources and consequences. *Ecological Applications, 7*(3), 737–750.

Vitousek, P. M., Naylor, R., Crews, T., David, M. B., Drinkwater, L. E., Holland, E., et al. (2009). Nutrient imbalances in agricultural development. *Science, 324*(5934), 1519–1520.

Yasin, J. A., Kroeze, C., & Mayorga, E. (2010). Nutrients export by rivers to the coastal waters of Africa: Past and future trends. *Global Biogeochemical Cycles, 24*(4), GB0A07. https://doi.org/10.1029/2009gb003568.

Chapter 21
Atmospheric Nitrogen Deposition in Spain: Emission and Deposition Trends, Critical Load Exceedances and Effects on Terrestrial Ecosystems

Héctor García-Gómez, Héctor Calvete-Sogo, Ignacio González-Fernández, Isaura Rábago, Victoria Bermejo, Fernando Valiño, Javier Sanz, Susana Elvira, and Rocío Alonso

Abstract The Mediterranean Basin presents an extraordinary biological richness, but very little information is available on the threat posed by air pollution, in particular reactive nitrogen (N_r). Atmospheric N deposition in Spain (up to 30 kg N ha^{-1} year^{-1}) is lower than loads recorded in central Europe, but some evidences indicate that N enrichment is already occurring in some ecosystems. Total N deposition exceeds empirical N critical loads in some protected areas of the Spanish Natura 2000 network. The habitats showing the highest risk are the Pyrenean grasslands, mountain forests of *Pinus uncinata* or *Abies pinsapo*, Mediterranean sclerophyllous forests in Catalonia and the oro-Mediterranean heathlands of the Cantabrian Range.

H. García-Gómez · H. Calvete-Sogo · I. González-Fernández · I. Rábago · V. Bermejo · F. Valiño · J. Sanz · S. Elvira · R. Alonso (✉)
Ecotoxicology of Air Pollution, CIEMAT, Avenida Complutense 40, Edif.70, Madrid 28040, Spain
e-mail: rocio.alonso@ciemat.es

H. García-Gómez
e-mail: hector.garcia@ciemat.es

H. Calvete-Sogo
e-mail: hector.calvete@ciemat.es

I. González-Fernández
e-mail: ignacio.gonzalez@ciemat.es

I. Rábago
e-mail: isaura.rabago@ciemat.es

V. Bermejo
e-mail: victoria.bermejo@ciemat.es

F. Valiño
e-mail: fernando.valino@ciemat.es

J. Sanz
e-mail: j.sanz@ciemat.es

S. Elvira
e-mail: susana.elvira@ciemat.es

M. A. Sutton et al. (eds.), *Just Enough Nitrogen*,
https://doi.org/10.1007/978-3-030-58065-0_21

Nitrogen deposition effects have already been reported in some of these ecosystems. Further research is needed to define Mediterranean-specific empirical N critical loads and to understand the possible interactions between N and other air pollutants and climate.

Keywords Nitrogen deposition · Risk assessment · Natura 2000 · Ozone interactions

21.1 Introduction

Nitrogen (N) deposition is considered the third most important driver affecting biodiversity after changes in land use and climate change (Sala et al. 2000). The Mediterranean Basin presents an extraordinary biological richness and it is recognized as one of the 25 Global Biodiversity Hotspots for conservation priorities (Myers et al. 2000). However, very little information is available on the threat that air pollution, and in particular reactive N and its interaction with other important air pollutants in the area (i.e., tropospheric ozone), can pose to biodiversity in the Mediterranean region (Ochoa-Hueso et al. 2011).

European policies have resulted in substantial reductions of the atmospheric emissions of N compounds in the last two decades in Europe (EEA 2011). In Spain, in contrast, emissions of NH_3 increased 14% for the period 1990–2011, while NO_x emissions increased until 2007, and decreased afterwards resulting in 20% lower emissions in 2011 than in 1990, the base year for the National Emission Ceilings Directive (MAGRAMA 2013). Both gases in 2011 were above the national emissions ceilings: 7% above the ceiling of 353 Gg of NH_3 and 2% above the ceiling of 847 Gg of NO_2. Emissions of N_2O have fluctuated around 15% above or below the emissions in 1990 being 8% lower in 2011 than in 1990 (MAGRAMA 2013).

21.2 Atmospheric Nitrogen Deposition in Spain

Wet deposition of total inorganic N measured in different monitoring networks in Spain for the period 2005–2008 ranged 0.7–13.3 kg ha^{-1} year^{-1} (García-Gómez et al. 2014a). Average wet deposition of oxidized N was 12% higher than of reduced N. Air quality models, such as EMEP and CHIMERE, provide acceptable estimations of total N wet deposition in Spain, although estimates should be applied with caution in regions with complex topography and with the influence of local emissions (García-Gómez et al. 2014a).

Dry deposition in Mediterranean ecosystems can represent the main input of atmospheric N deposition with values up to 65–95% of the total deposition (Bytnerowicz and Fenn 1996). Previous studies performed in Spain estimated that dry deposition represents 50–80% of total N deposition in *Quercus ilex* and *Pinus halepensis* forests

(Àvila and Rodà 2012; García-Gómez et al. 2014b; Sanz et al. 2002). Further research is needed to estimate dry deposition, particularly in seasonally dry regions such as the Mediterranean area, to determine the effectiveness of emissions control strategies.

Modelled total N deposition in Spain showed a decreasing distribution along a NE-SW axis, with higher deposition (maxima values of 22.9 kg N ha^{-1} $year^{-1}$) in the northern and eastern coastal regions than inland and southern areas (Fig. 21.1). This distribution pattern responds to the spatial distribution of the expected three main drivers: regional emissions, precipitation distribution and transboundary contribution (García-Gómez et al. 2014a).

21.3 Evidences of Nitrogen Enrichment in Spanish Natural Ecosystems

Little information is available on N deposition effects that could be occurring in Spanish natural ecosystems but some evidences of N enrichment have been reported. A continuous increase of nitrophilous species (plants, lichens, mosses) has been detected in the Iberian Peninsula for the period 1900–2008 using the Global Biodiversity Information Facility (GBIF) database (Ariño et al. 2011). Moreover, the areas showing an increase in the percentage of nitrophilous species in general agree with the spatial distribution of high atmospheric N deposition. Also an increase in the N content in bryophytes, but not in vascular plants, has been observed in herbarium specimens collected in Spain throughout the last century (Peñuelas and Filella 2001).

On the other hand, the concentration of NO_3^- in headwater streams of relatively undisturbed catchments of North-East Spain have increased since the 1980s indicating a N exportation from terrestrial to aquatic ecosystems (Àvila and Rodà 2012; Camarero and Aniz 2010). Although other factors such as climate change could be playing a role in streamwater nitrate concentrations, this trend has been related to atmospheric deposition and it could be considered a sign of the onset of eutrophication. (Àvila and Rodà, 2012). In bigger catchments, the rises of nitrate concentration have been related to the influence of agricultural cover (Lassaletta et al. 2009).

Furthermore, in Catalonia in North-East Spain, a region with high atmospheric N deposition, the leaf N content and stoichiometry of forest trees has been related mainly with climatic variables but also with chemical properties of soil and anthropogenic N emissions (Sardans et al. 2011). Similarly, a nutritional imbalance has been reported in *Abies pinsapo* forests, an endemic Mediterranean fir growing in southern Spain (close to the Strait of Gibraltar) caused by chronic atmospheric N inputs (Blanes et al. 2013).

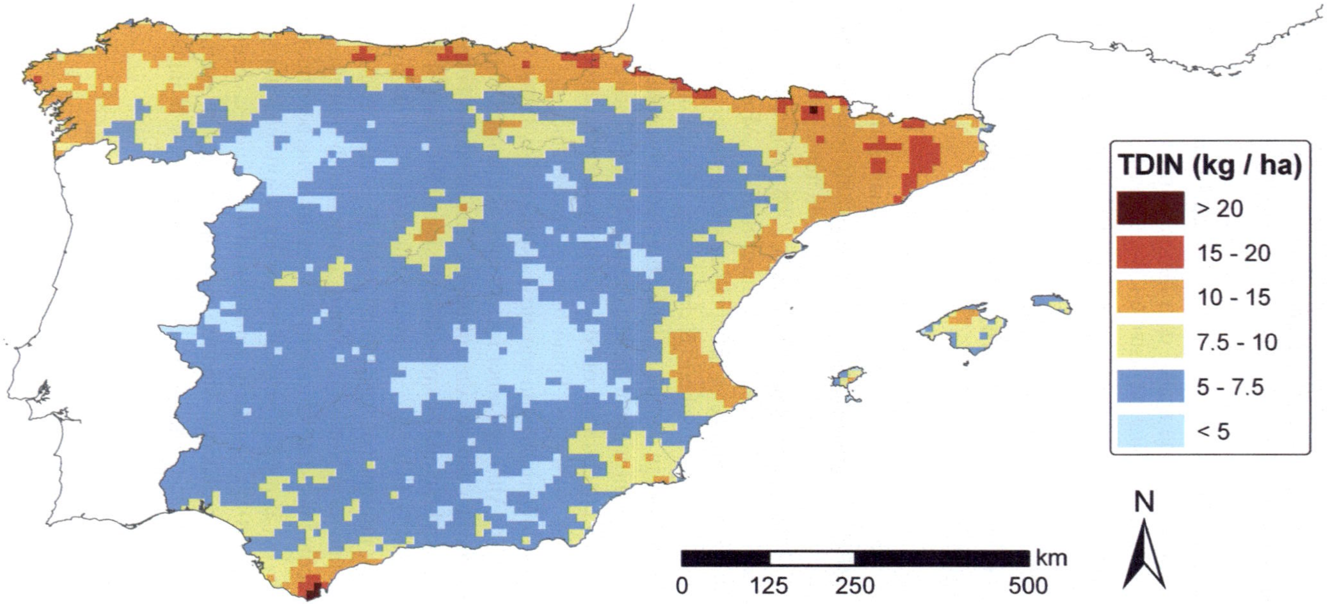

Fig. 21.1 Total deposition of inorganic nitrogen in Spain (excluding Canary Islands), estimated with the CHIMERE model for the year 2008 (modified from García-Gómez et al. 2014a)

21.4 Exceedance of Critical Loads

Critical loads (CL) are deposition thresholds defined under the UNECE's Convention on Long-range Transboundary Air Pollution (CLRTAP) for the protection of ecosystem function and structure. Different approaches have been adopted to define N critical loads involving either modelling or empirically based on field evidence. Considering the particularities of the Mediterranean region, with co-occurrence of other pressures and high seasonality, it is especially recommended to determine empirical CL (De Vries et al. 2007; Fenn et al. 2011).

Empirical CL ranging from 3 to 25.9 kg N ha^{-1} $year^{-1}$ have been proposed for the different vegetation types in Mediterranean California (Fenn et al. 2011). However, scarce proposals exist for European Mediterranean ecosystems, characterized by a longer history of human management which can modify the CL value. An empirical CL of 26 kg N ha^{-1} $year^{-1}$ has been proposed for broadleaf evergreen semi-natural woodlands (Pinho et al. 2012), while 2.4 kg N ha^{-1} $year^{-1}$ has been proposed for European forests including seven sites in Spain (Giordani et al. 2014).

Recently revised values of the empirical CL (Bobbink and Hetteling 2011) were used to estimate CL exceedances in the Spanish Natura 2000 network (García-Gómez et al. 2014a; Fig. 21.2). The highest occurrence of habitats threatened by N deposition was estimated for North-East Spain and in mountain ranges close to emission sources. Natural grasslands located in the north (Pyrenees, Cantabrian Range) were the habitats with the highest risk of N effects. Other high-altitude vegetation types like *Pinus uncinata* or *Abies pinsapo* forests, oro-Mediterranean heathlands and *Cytisus purgans* formations seemed to be highly threatened by N deposition. Also the Mediterranean sclerophyllous forests (holm oak) close to Barcelona City (North-East Spain) showed important CL exceedances. The abundance of mountain areas threatened should be taken into account in the design of monitoring networks.

21.5 Nitrogen Effects in Spanish Terrestrial Ecosystems

Natural grasslands were estimated to be the most threaten habitat based on empirical CL exceedances (García-Gómez et al. 2014a). However, little information is available on the possible effects of atmospheric N deposition. Fertilization experiments have been focused on maximizing biomass production for forage use. However, N fertilization reduces the proportion of clover biomass in perennial pastures (3% of Spanish surface), changing pasture structure and fodder quality (Calvete-Sogo et al. 2011). Nitrogen effects on annual grasslands (17% of Spanish surface and understory of broadleaf evergreen forests) have shown a significant interaction with ozone (O_3) in open-top chamber experiments in Spain. O_3 exposure reduced the fertilization effect of enhanced N availability, while N could counteract pernicious O_3 effects on plant and flower biomass production, but only at moderate O_3 levels (Sanz et al. 2011, 2013; Calvete-Sogo et al. 2014; Fig. 21.3).

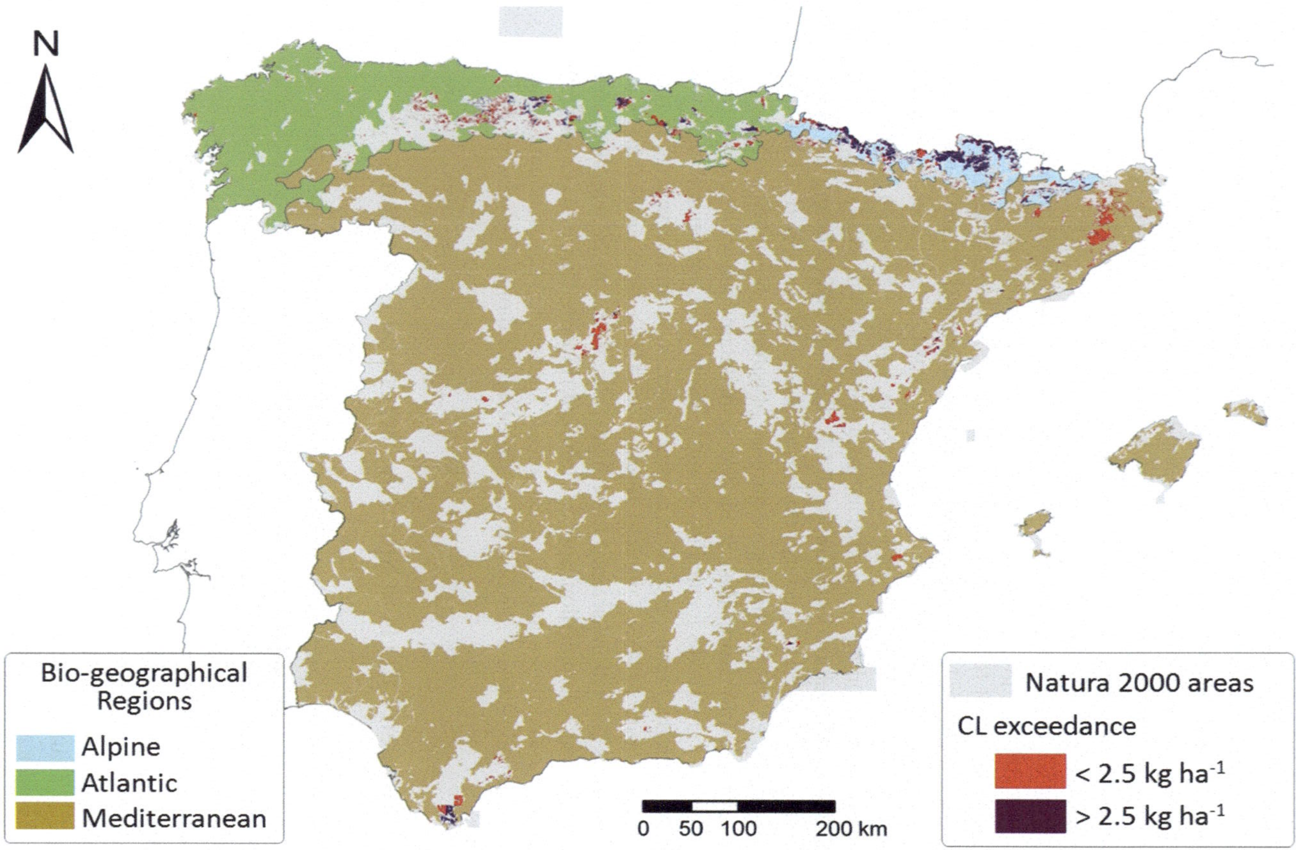

Fig. 21.2 Exceedances of empirical critical loads (CLexc) within the Natura 2000 network in Spain (excluding Canary Islands), according to CL from Bobbink and Hetteling (2011) and atmospheric N deposition estimated with CHIMERE for 2008 (modified from García-Gómez et al. 2014a)

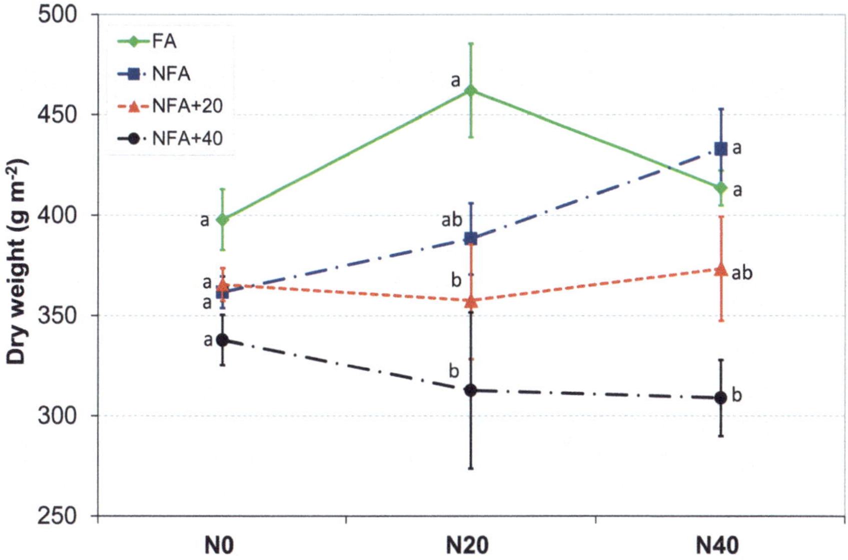

Fig. 21.3 Green biomass harvested in annual pastures exposed to different O_3 and N treatments (mean ± SE). FA: charcoal filtered air; NFA: non filtered air; NFA + 20: NFA supplemented with 20 nl l^{-1} of O_3; NFA + 40: NFA supplemented with 40 nl l^{-1} of O_3; N0: soil N background; N20: addition of 20 kg N ha^{-1}; N40: addition of 40 kg N ha^{-1}. Different letters indicate significant differences among O_3 treatments (modified from Calvete-Sogo et al. 2014)

Broadleaf evergreen forests in North-East Spain, with holm oak as the main species, have shown signs of N enrichment with seasonal increases of NO_3^- concentration in stream waters during peak runoff periods outside the growing season. However, these ecosystems are considered still far from N saturation, since most of the deposited N is retained within the ecosystem (Àvila and Rodà 2012). Moreover, aboveground net primary production was enhanced by N fertilization in these forests (Rodà et al. 1999). Nonetheless, other early effects such as changes in species composition of sensitive communities, like lichens, have not been assessed.

Nitrogen effects have been described in *Abies pinsapo* mountain forests along a gradient of atmospheric N deposition in the south of Spain close to the Strait of Gibraltar. Chronic N deposition reduces fine root growth and shifts forests from N limitation to P limitation. This induced nutritional imbalance has been related to a decrease in photosynthetic nutrient use efficiency (Blanes et al. 2013).

Fertilization experiments in mountain heathlands of the Cantabrian Range have shown an increase in the abundance of arthropod herbivores in the short-term (Cuesta et al. 2008) and a significant increase in total plant richness in the long term, due to an increase in the number of perennial herbs without displacing the dominant woody species (Calvo et al. 2007). In semi-arid ecosystems, N fertilization affects soil nutrient cycling and fertility, and alters the functioning of biological soil crusts (Ochoa-Hueso et al. 2013).

21.6 Conclusions

Atmospheric N deposition in Spain could be affecting the biodiversity and health of natural ecosystems in Spain. More information is needed to quantify N deposition, particularly dry deposition. Atmospheric deposition networks in Spain should include monitoring stations in mountain areas where most of the empirical CL exceedances have been detected. Further research is needed to define specific CL for the protection of Spanish ecosystems. Moreover, in a changing world, reactive N interactions with other factors such as ozone or drought need to be considered.

Acknowledgements This research was funded by the Spanish projects Consolider Montes CSD2008-00040, EDEN project CGL2009-13188-C03-02, Comunidad de Madrid-Agrisost S2009AGR-1630, and the EU-FP7-ENV-2011 ECLAIRE project. This study was also supported by the Spanish Ministry of Agriculture, Food and Environment.

References

Ariño, A. H., Gimeno, B. S., Pérez de Zabalza, A., Ibáñez, R., Ederra, A., & Santamaría, J. M. (2011). Influence of nitrogen deposition on plant biodiversity at Natura 2000 sites in Spain. In W. K. Hicks, C. P. Whitfield, W. J. Bealey, & M. A. Sutton, (Eds.), *Nitrogen deposition and Natura 2000. Science & practice in determining environmental impacts.* (pp. 140–146). Brussels: COST office.

Àvila, A., & Rodà, F. (2012). Changes in atmospheric deposition and streamwater chemistry over 25 years in undisturbed catchments in a Mediterranean mountain environment. *Science of the Total Environment, 434,* 18–27.

Blanes, M. C., Viñegla, B., Merino, J., & Carreira, J. A. (2013). Nutritional status of *Abies pinsapo* forests along a nitrogen deposition gradient: Do C/N/P stoichiometric shifts modify photosynthetic nutrient use efficiency? *Oecologia, 171,* 797–808.

Bobbink, R., & Hettelingh, J. P. (2011). *Review and revision of empirical critical loads and dose-response relationships.* Bilthoven: Coordination Centre for Effects (RIVM).

Bytnerowicz, A., & Fenn, M. E. (1996). Nitrogen deposition in California forests: A review. *Environmental Pollution, 92,* 127–146.

Calvete-Sogo, H., Pérez Corona, E., Bermejo, V., Sanz, J., García, H., Elvira, S., & Alonso, R. (2011). *Nitrogen effects on Spanish grasslands: literature review.* Poster presentation at 24th Task Force Meeting of the UNECE ICP Vegetation, Rapperswil, Switzerland, 31st Jan–2nd Feb 2011.

Calvete-Sogo, H., Elvira, S., Sanz, J., González-Fernández, I., Sánchez-Martín, L., Alonso, R., et al. (2014). Current ozone levels threaten gross primary production and yield of Mediterranean annual pastures and nitrogen modulates the response. *Atmospheric Environment, 95,* 197–206.

Calvo, L., Alonso, I., Marcos, E., & de Luis, E. (2007). Effects of cutting and nitrogen deposition on biodiversity in Cantabrian heathlands. *Applied Vegetation Science, 10,* 43–52.

Camarero, L., & Aniz, M. (2010). Surface waters monitoring in the LTER-Aigüestortes node: Trends and indicators of the impacts of N and S atmospheric deposition. *Ecosistemas, 19,* 24–41. (in Spanish).

Cuesta, D., Taboada, A., Calvo, L., & Salgado, J. M. (2008). Short- and medium-term effects of experimental nitrogen fertilization on arthropods associated with Calluna vulgaris heathlands in north-west Spain. *Environmental Pollution, 152,* 394–402.
De Vries, W., Kros, H., Reinds, G., Wamelink, W., Mol, J., van Dobben, H., Bobbink, R., Emmett, B., Smart, S., Evans, C., Schlutow, A., Kraft, P., Belyazid, S., Sverdrup, H., van Hinsberg, A., Posch, M., Hettelingh, J. -P. (2007). *Developments in deriving critical limits and modeling critical loads of nitrogen for terrestrial ecosystems in Europe.* Alterra-rapport 1382. Wageningen and CCE, Bilthoven: Alterra.
EEA (European Environment Agency). (2011). *Air quality in Europe—2011 report.* European Environment Agency Technical Report 12/2011. Luxembourg: Publications Office of the European Union.
Fenn, M. E., Allen, E. B., & Geiser, L. H. (2011). Mediterranean California. In L. H. Pardo, M. J. Robbin-Abbot & C. T Driscoll (Eds.), *Assessment of nitrogen deposition effects and empirical critical loads of nitrogen for ecoregions of the United States* (Chapter 13). Delaware: Forest Service Publications (USDA).
García-Gómez, H., Garrido, J. L., Vivanco, M. G., Lassaletta, L., Rabago, I., Avila, A., Tsyro, S., Sánchez, G., González Ortiz, A., González-Fernández, I., & Alonso, R. (2014a). Nitrogen deposition in Spain: Modeled patterns and threatened habitats within the Natura 2000 network. *Science of the Total Environment, 485–486,* 450–460.
García-Gómez, H., Valiño, F., Izquieta, S., Aguillaume, L., González-Fernández, I., Calvete-Sogo, H., Elustondo, D., Santamaría, J. M., Àvila, A., & Alonso, R. (2014b). Atmospheric nitrogen inputs in the Mediterranean evergreen (*Quercus ilex*) forests: Results from EDEN Project. In *8th Symposium on Ecosystem Behaviour (BIOGEOMON 2014).* Bayreuth, Germany.
Giordani, P., Calatayud, V., Stofer, S., Seidling, W., Granke, O., & Fischer, R. (2014). Detecting the nitrogen critical loads on European forests by means of epiphytic lichens. A signal-to-noise evaluation. *Forest Ecology and Management, 311,* 29–40.
Lassaletta, L., García-Gómez, H., Sánchez-Gimeno, B., & Rovira, J. V. (2009). Agriculture-induced increase in nitrate concentrations in stream waters of a large Mediterranean catchment over 25 years (1981–2005). *Science of the Total Environment, 407,* 6034–6043.
MAGRAMA (Spanish Ministry of Agriculture, Food and Environment). (2013). *Informe de la calidad del aire en España—2012.* Madrid: Subdirección Gral. de Calidad del Aire y Ambiente Industrial.
Myers, N., Mittermeier, R. A., Mittermeier, C. G., Da Fonseca, G. A., & Kent, J. (2000). Biodiversity hotspots for conservation priorities. *Nature, 403,* 853–858.
Ochoa-Hueso, R., Allen, E. B., Branquinho, C., Cruz, C., Dias, T., Fenn, M. E., et al. (2011). Nitrogen deposition effects on Mediterranean-type ecosystems: An ecological assessment. *Environmental Pollution, 159,* 2265–2279.
Ochoa-Hueso, R., Maestre, F. T., de los Ríos, A., Valea, S., Theobald, M. R., Vivanco, M. G., Manrique, E., & Bowker, M. A. (2013). Nitrogen deposition alters nitrogen cycling and reduces soil carbon content in low-productivity semiarid Mediterranean ecosystems. *Environmental Pollution, 179,* 185–193.
Peñuelas, J., & Filella, I. (2001). Herbaria century record of increasing eutrophication in Spanish terrestrial ecosystems. *Global Change Biology, 7,* 427–433.
Pinho, P., Theobald, M. R., Dias, T., Tang, Y. S., Cruz, C., Martins-Loução, M. A., et al. (2012). Critical loads of nitrogen deposition and critical levels of atmospheric ammonia for semi-natural Mediterranean evergreen woodlands. *Biogeosciences, 9,* 1205–1215.
Rodà, F., Mayor, X., Sabaté, S., & Diego, V. (1999). Water and nutrient limitations to primary production. In F. Rodà, J. Retana, C. A. Gracia, & J. Bellot (Eds.), *Ecology of Mediterranean evergreen oak forests. Ecological Studies* (Vol. 137, pp. 253–269). Berlin: Springer.
Sala, O. E., Chapin, F. S., Armesto, J. J., Berlow, E., Bloomfield, J., Dirzo, R., et al. (2000). Global biodiversity scenarios for the year 2100. *Science, 287,* 1770–1774.

Sanz, M. J., Carratalá, A., Gimeno, C., & Millán, M. M. (2002). Atmospheric nitrogen deposition on the east coast of Spain: Relevance of dry deposition in semi-arid Mediterranean regions. *Environmental Pollution, 118,* 259–272.

Sanz, J., Bermejo, V., Muntifering, R., González-Fernández, I., Gimeno, B. S., Elvira, S., et al. (2011). Plant phenology, growth and nutritive quality of *Briza maxima*: Responses induced by enhanced ozone atmospheric levels and nitrogen enrichment. *Environmental Pollution, 159,* 423–430.

Sanz, J., Calvete, H., González-Fernández, I., Lin, J., García-Gómez, H., Muntifering, Alonso, R., & Bermejo-Bermejo, V. (2013). Foliar senescence is the most sensitive response to ozone in *Bromus hordeaceus* and is modulated by nitrogen input. *Grass and Forage Science, 94,* 765–772.

Sardans, J., Rivas-Ubach, A., & Peñuelas, J. (2011). Factors affecting nutrient concentration and stoichiometry of forest trees in Catalonia (NE Spain). *Forest Ecology and Management, 262,* 2024–2034.

Part V
Nitrogen Impacts on Health, Ecosystems and Climate: Nitrogen, Climate Change and Trace Gas Enrichment

Chapter 22
Nitrogen Aspects of the Free-Air CO_2 Enrichment (FACE) Study for Paddy Rice Ecosystems

Kentaro Hayashi, Keisuke Ono, Takeshi Tokida, Miwa Y. Matsushima, Midori Yano, Sakae Toyoda, Genki Katata, Nobuko Katayanagi, Tamon Fumoto, Hirofumi Nakamura, and Toshihiro Hasegawa

Abstract Anticipated future increase in atmospheric carbon dioxide (CO_2) is environmentally problematic leading to climate change owing to its radiative forcing, whereas it enhances primary production of terrestrial ecosystems, including rice plants in paddy fields. However, responses of rice plants to elevated CO_2 levels ([eCO_2]) are also subject to nitrogen (N) availability, as well as climate change. A free-air CO_2 enrichment (FACE) facility enables to simulate [eCO_2] in an actual

---

K. Hayashi (✉) · K. Ono · T. Tokida · M. Yano · N. Katayanagi · T. Fumoto
Institute for Agro-Environmental Sciences, National Agriculture and Food Research Organization (NARO), 3-1-3, Kannondai, Tsukuba, Ibaraki 305-8604, Japan
e-mail: kentaroh@affrc.go.jp

K. Ono
e-mail: onok@affrc.go.jp

T. Tokida
e-mail: tokida@affrc.go.jp

M. Yano
e-mail: yanom230@affrc.go.jp

N. Katayanagi
e-mail: katayan@affrc.go.jp

T. Fumoto
e-mail: tamon@affrc.go.jp

M. Y. Matsushima
Chiba University, 648, Matsudo, Chiba 271-8510, Japan
e-mail: matsushima@faculty.chiba-u.jp

S. Toyoda
Tokyo Institute of Technology, 4259, Nagatsuta, Yokohama, Kanagawa 226-8502, Japan
e-mail: toyoda.s.aa@m.titech.ac.jp

G. Katata
Ibaraki University, 2-1-1, Bunkyo, Mito, Ibaraki 310-8512, Japan
e-mail: genki.katata.mirai@vc.ibaraki.ac.jp

H. Nakamura
Taiyo Keiki Co. Ltd, 1-12-3, Nakajujo, Tokyo, Kita 114-0032, Japan
e-mail: 1623nakamura@gmail.com

M. A. Sutton et al. (eds.), *Just Enough Nitrogen*,
https://doi.org/10.1007/978-3-030-58065-0_22

paddy field, where we can investigate how rice plants and paddy rice ecosystems respond to [eCO_2] in conjunction with other factors, such as N availability and temperature rising. We implemented a three-year project focusing on N aspects of the FACE study (abbreviated as FACE-N) during 2010–2013 using the Tsukuba FACE for single cropping of paddy rice in central Japan. In this chapter, the concept, components, and major achievements of FACE-N are summarized, and prospective future research with respect to responses of N-related processes in paddy rice ecosystems to [eCO_2] is suggested.

Keywords Elevated CO_2 · FACE · Fertilization · Nitrogen cycle · Paddy field

22.1 Introduction

Atmospheric carbon dioxide (CO_2) levels have steadily been increasing by 40% since pre-industrial times. At the same time, climate change resulting from the radiative forcing of greenhouse gases including CO_2 is of concern (IPCC 2013). Elevated CO_2 levels, (here indicated by [eCO_2]) and climate change might affect rice cropping and paddy rice ecosystems. Free-air CO_2 enrichment (FACE) study in an actual paddy field is advantageous to elucidate how rice plants and paddy rice ecosystems respond to [eCO_2].

[eCO_2] enhances photosynthesis, called the CO_2 fertilization effect, and is essentially advantageous to rice growth (Sakai et al. 2006). [eCO_2] also induces partial stomatal closure, which reduces transpiration and thus enhances water use efficiency (Yoshimoto et al. 2005). However, responses of yield and quality of paddy rice to [eCO_2] vary among years (weather) and rice varieties (Hasegawa et al. 2013; Usui et al. 2014). Mechanisms causing these differences in [eCO_2] response are the primary interest of the current FACE study of paddy rice cropping.

Climate change (e.g., temperature rising) and [eCO_2] affect physiology and phenology of rice plants, which in turn affect the carbon (C) cycle in paddy rice ecosystems. In this regard, behavior of methane (CH_4), a potent greenhouse gas stronger than CO_2 on a weight basis, is particularly important, since paddy fields are a major anthropogenic source of atmospheric CH_4. The annual global CH_4 emission from paddy fields during 2000–2009 is estimated to be in a range of 33–40 Tg CH_4 year^{-1}, which accounts for 9–13% of the anthropogenic emissions (IPCC 2013). Elevated CO_2 concentrations have been found to increase paddy field CH_4 emissions through the increase in labile C supply from rice roots to paddy soil (Tokida et al. 2010, 2011).

On the other hand, effects of [eCO_2] and climate change on the nitrogen (N) cycle in paddy rice ecosystems have also been under discussion. Nitrogen availability

T. Hasegawa
Tohoku Agricultural Research Center, National Agriculture and Food Research Organization (NARO), 4, Akahira, Shimokuriyagawa, Morioka, Iwate 020-0198, Japan
e-mail: thase@affrc.go.jp

(strongly affected by fertilization) was an important factor even in early studies of rice cropping under [eCO_2] and climate change (e.g., Kim et al. 2003). However, few studies have been conducted to elucidate responses of the N cycle in paddy rice ecosystems to [eCO_2] and climate change, except for some studies of a specific N process like biological N fixation (Cheng et al. 2001; Hoque et al. 2001, 2002).

Given the circumstances, we planned and implemented a three-year project to investigate comprehensively responses of relevant processes in the N cycle of paddy rice ecosystems to [eCO_2] and temperature rising using a FACE facility for a rice paddy field in central Japan (abbreviated as FACE-N). In this chapter, we summarize the concept, components, and major achievements of FACE-N, and suggest prospective future research on the N aspects of FACE study.

22.2 Study Site and FACE Design

Tsukuba FACE, a FACE facility for single cropping of paddy rice, is located in a paddy area in central Japan (35°58'27"N, 139°59'32"E, 10 m.a.s.l.), where the Institute for Agro-Environmental Sciences, NARO, Japan has performed relevant studies since 2010. Four rectangular bays were used for the FACE experiments. A CO_2-enrichment plot was set in each bay, accompanying an ambient plot without a CO_2 treatment, to ensure a pairwise comparison of experimental treatments. The target level of [eCO_2] was +200 ppm on average to ambient levels. Nitrogen fertilizer application rates (e.g., zero, normal, and rich) and water temperature (control and +2 °C) were also embedded as sub treatments in the CO_2 treatment. Details of the experimental design of Tsukuba FACE are described in Nakamura et al. (2012). The main soil properties of the study site are shown in Hasegawa et al. (2013). The website of Tsukuba FACE also provides relevant information including a publication list (http://www.niaes.affrc.go.jp/outline/face/english/index.html).

All experimental plots received equal amounts of phosphorus (P) and potassium (K) in early April, before ploughing, when PK compound fertilizer was applied at a rate of 100 kg ha^{-1} as P_2O_5 and 100 kg ha^{-1} as K_2O. The fields were kept submerged after late April. For experimental plots with standard N fertilization (80 kg N ha^{-1} in total), three kinds of N fertilizer were applied in early May just prior to puddling: 20 kg N ha^{-1} as urea, 40 kg N ha^{-1} as one type of controlled-release fertilizer (polymer-coated urea; type LP100, JCAM Agri. Co. Ltd, Tokyo, Japan), and 20 kg ha^{-1} as another type of controlled-release fertilizer (type LP140, JCAM Agri. Co. Ltd), and were then incorporated into the plowed layer by puddling. Concentrations of atmospheric NH_3 shown in this chapter were measured on a weekly basis using a filter-pack method (Hayashi et al. 2013a). Exchange fluxes were calculated by the aerodynamic gradient method (Hayashi et al. 2013a).

22.3 Concept and Components of FACE-N Project

Figure 22.1 illustrates a schematic of N aspects of FACE study for paddy rice ecosystems, in which FACE-N mainly targeted Focus 2 (disturbance in N cycle). Focus 1 (growth, yield and quality of rice) was the primary and mainstream target in Tsukuba FACE by other projects. Focus 3 (interaction with C cycle) is a challenging target of a subsequent project of FACE-N (Sect. 22.5, this chapter).

FACE-N consisted of three themes as follows: (1) atmosphere–rice paddy exchanges of reactive nitrogen (N_r), (2) rice–soil N related processes, and (3) modelling of atmosphere–rice–soil N processes. Each theme had specific subthemes.

Theme (1) was composed of subthemes such as monitoring of wet N deposition of N_r compounds (nitrogen compounds other than molecular N, abbreviated as N_r) mainly focused on ammonium and nitrate, paddy field–atmosphere exchanges of gaseous and particulate N_r, and rice plant emission of ammonia (NH_3). These results are summarized in Sect. 22.4.1, this chapter.

Theme (2) was composed of a number of field studies, e.g., isotopic signatures (Sect. 22.4.2, this chapter, for details), N mineralization, soil ammonium dynamics (dissolved versus adsorbed), N supply rates of controlled-release urea, biological N fixation rates evaluated by acetylene reduction activity, and net biological N fixation evaluated by changes in total N in the surface soil (Hayashi et al. 2014).

Theme (3) encompassed modification and validation of two mechanistic models, i.e., DNDC-Rice (Fumoto et al. 2008) and SOLVEG (Katata et al. 2013), and

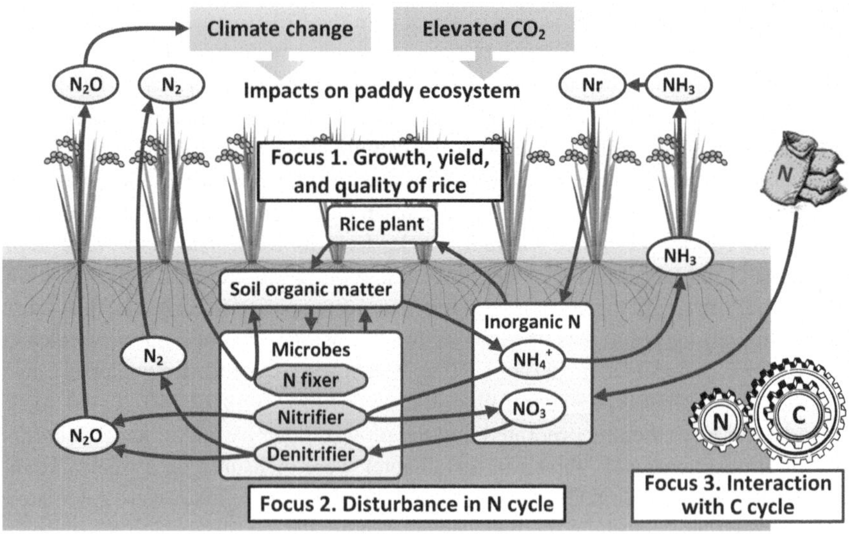

Fig. 22.1 Research interests of FACE study for paddy rice ecosystems with respect to nitrogen. In this figure the label N_r indicates to other reactive nitrogen compounds in the atmosphere contributing to atmospheric deposition, including mainly ammonia, ammonium, nitric acid gas, and nitrate

modelling relationships between stomatal conductance and photosynthesis based on field-scale flux measurements (Ono et al. 2013). DNDC-Rice excels at reproducing soil processes, and SOLVEG includes complicated processes at the land-atmosphere interface, although both models explicitly describe the atmosphere–plant–soil system. Katayanagi et al. (2013) verified DNDC-Rice for its application to evaluate the N budget in a paddy rice ecosystem. Section 22.4.3, this chapter, shows the modification and validation results of SOLVEG applied to a paddy rice ecosystem.

22.4 Excerption of Major Achievements

22.4.1 Nitrogen Deposition and NH_3 Emission

Wet and dry deposition can supply N_r to paddy rice ecosystems, whereas gas and particle emissions lose N from paddy fields. We measured wet N deposition and exchange fluxes of gases and particles of N_r as a current baseline on a weekly mean basis from September 2010 to September 2012, in which day-night separation was employed in the weekly mean measurements to reduce the errors in flux calculation originating from the long averaging time (Hayashi et al. 2013a). On average, annual wet N deposition (as ammonium and nitrate) was 9.0 kg N ha^{-1} year^{-1}, and annual N exchange (as NH_3, nitric acid gas, nitrous acid gas, particulate ammonium, and particulate nitrate) was 3.4 kg N ha^{-1} year^{-1} of deposition, i.e., 12.4 kg N ha^{-1} year^{-1} in total as a net input from the atmosphere. It is noted that annual N dry deposition of 8.2 kg N ha^{-1} year^{-1} was countered by annual N emission of 4.8 kg N ha^{-1} year^{-1} to result in the annual N exchange of 3.4 kg N ha^{-1} year^{-1}. Here, NH_3 showed the largest flux both for the dry deposition (3.5 kg N ha^{-1} year^{-1}) and emission (3.2 kg N ha^{-1} year^{-1}).

Ammonia is an important N_r gas because of its relatively high air concentrations and large deposition and emission fluxes, with substantial potential emissions from fertilized rice crops. Figure 22.2 shows the air concentrations and exchange fluxes of NH_3 at Tsukuba FACE. NH_3 was a resident component in the atmosphere with an average concentration of 3.5 μg N m^{-3} (20 °C, 1013 hPa) at a height of 6 m above the ground surface. The exchange fluxes of NH_3 showed emission in many cases in the daytime of cropping seasons. By contrast deposition at night deposition mainly occurred throughout the measurements (Fig. 22.2). Overall, the mean net flux of NH_3 was 2.4 kg N ha^{-1} year^{-1} of emission for day and 2.7 kg N ha^{-1} year^{-1} of deposition for night, giving a total net deposition of 0.3 kg N ha^{-1} year^{-1}.

The current understanding is that a paddy field is usually a sink of atmospheric NH_3 (Hayashi et al. 2013b), but N fertilizer application (Hayashi et al. 2006, 2008a) and postharvest field burning (Hayashi et al. 2013b) change it into a temporary source of atmospheric NH_3. Rice plants themselves can be a source of NH_3 emissions (Hayashi et al. 2008a, b, 2011), as well as the paddy surface (i.e., soil or floodwater). In this

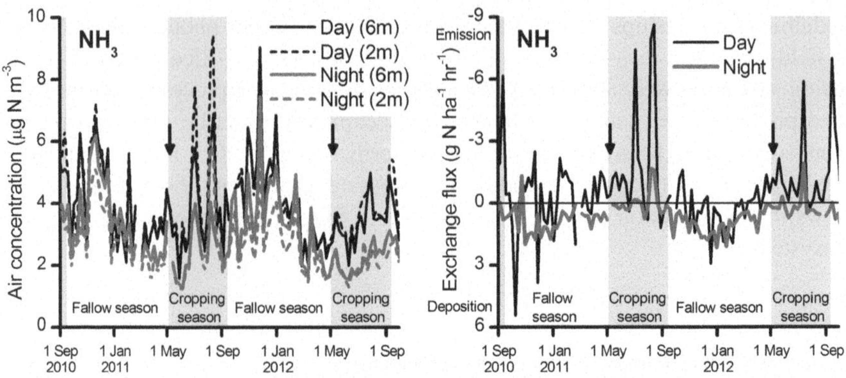

Fig. 22.2 Measured air concentrations and exchange fluxes of NH_3 at Tsukuba FACE. Vertical arrows show the timing of one-shot fertilizer application which was applied and incorporated into the plowed layer by puddling. Paddy fields were flooded roughly from late April to late August

regard, Miyazawa et al. (2014) elucidated that [eCO_2] decreases the photorespiratory NH_3 production, but does not decrease the NH_3 compensation point of rice leaves. These results suggest that suppression of RuBisCO oxygenation by [eCO_2] does not decrease potential leaf NH_3 emission in rice plants.

It was expected that NH_3 emission would be small at Tsukuba FACE in the cropping seasons, because the N application method in the area, i.e., one-shot application and incorporation of controlled-release urea based fertilizer, was effective to limit NH_3 emission. Contrary to the expectation, however, large NH_3 emissions were observed occasionally in the daytime in the cropping seasons (Fig. 22.2). Rice plants might contribute to the large emissions (up to 8.6 g N ha^{-1} $year^{-1}$) as a channel transporting ammoniacal N from the paddy soil to the atmosphere. Future research is needed to clarify the causes, including whether this represents a plant-based or soil-based emission source.

22.4.2 *Isotopic Signatures*

Early studies suggested that [eCO_2] stimulates denitrification and nitrous oxide (N_2O) emissions from upland fields particularly with a high N application rate (van Groenigen et al. 2011; Dijkstra et al. 2012), through increases in carbon availability in soil by the CO_2 fertilization effect, and enhancement of anoxic conditions in soil by the reduced transpiration resulting from partial stomatal closure (Ineson et al. 1998; Robinson and Conroy 1999; Kammann et al. 2008; Niboyet et al. 2011). By contrast, only a few reports describe the [eCO_2] effect on N_2O emission from paddy fields (Cheng et al. 2006; Pereira et al. 2013). Thus, elucidation of N_2O production mechanisms in submerged soil is required for better understanding of N_2O dynamics under [eCO_2] and even at ambient CO_2 levels.

In order to elucidate the N_2O dynamics in paddy fields, we analyzed microbial metabolic processes using intramolecular distribution of N isotopes (the relative abundance of ^{15}N in the central (α) and terminal (β) N atoms in the asymmetric N_2O molecule, $\delta^{15}N^{\alpha}-\delta^{15}N^{\beta} = {}^{15}N$-site preference, abbreviated as SP) (Toyoda and Yoshida 1999). The SP enables pathways of microbial metabolism such as nitrification and denitrification to be distinguished (Toyoda et al. 2002; Sutka et al. 2006). We collected soil gases during the initial flood irrigation in early spring when episodic N_2O emissions were observed, and measured the SP and conventional elemental (bulk) isotope ratios of N_2O as well as isotope ratios of soil inorganic N compounds. Our results (Yano et al. 2014) indicated that surface-emitted N_2O was mainly produced at shallow depths above the rising groundwater table, and it resulted mainly from N_2O production by bacterial denitrification. After the submergence of soil surface in late April, N_2O temporarily accumulated in the soil, and then most N_2O was reduced to N_2 in the soil. Another isotopic analysis for the rice growing season under ambient and elevated CO_2 levels is underway.

22.4.3 *Modelling of NH_3 Exchange Using SOLVEG*

NH_3 exchange at a paddy field is bidirectional as shown in Sect. 22.4.1, this chapter. Depending on relative abundances of ammoniacal N in paddy fields and in the atmosphere, rice plants and/or the ground surface (floodwater or soil) can absorb atmospheric NH_3 (NH_3 deposition) or emit NH_3 to the atmosphere (NH_3 emission) . Accurate reproduction of NH_3 exchange at a paddy field using a mechanistic model is still a key issue under development. Accordingly, we developed new modules of SOLVEG to calculate the floodwater temperature and the NH_3 concentrations inside rice leaves to estimate compensation points and in floodwater or on the soil surface. SOLVEG is a one-dimensional multi-layer model that consists of four modules for the atmosphere near the surface, soil, vegetation, and radiation within the vegetation canopy. The basic equations related to gas exchange processes are described in Katata et al. (2011, 2013).

Katata et al. (2013) reported that the modified SOLVEG successfully reproduced the NH_3 exchange fluxes observed at a paddy field in central Japan (Hayashi et al. 2012). We also simulated how NH_3 exchange transitioned with rice growth, and revealed that NH_3 recapture (capture of NH_3 emitted from the floodwater under the rice canopy) increased with rice growth. At the maturing stage of rice, most of the NH_3 emitted from floodwater was recaptured by the overlying rice canopy, and therefore was not further transferred to the atmosphere (Katata et al. 2013). This finding is useful as it points to a mechanism for supplemental fertilization of the plants that can improve the N use efficiency for paddy rice cropping, while at the same time reducing NH_3 volatilization loss to the atmosphere. It also opens up considerations of how fertilizer application and canopy structure interactions could be used to maximize this canopy recapture of nitrogen.

22.5 Conclusions

FACE-N has incorporated a variety of N-related studies in paddy field ecosystems that form a basis for assessing their responses under [eCO_2]. The major components of atmospheric N deposition and bi-directional NH_3 exchange have been quantified, which require further elucidation in relation to their response under [eCO_2]. Results to date already show that [eCO_2] increased photorespiration and associated intracellular NH_3 production, however this did not significantly alter NH_3 compensation points of rice leaves.

Further development in N aspects of FACE study is highly desirable, e.g., modification of mechanistic models to reproduce the N cycle in paddy rice ecosystems under [eCO_2] and climate change. A multidisciplinary study of microbial ecology with environmental science is also a good example (e.g., Okubo et al. 2014, effects of [eCO_2] on flora of root-associated bacteria). Since 2014, a subsequent three-year project of FACE-N has been working to elucidate relationships between N and CH_4 behaviors in paddy rice ecosystems under [eCO_2] (FACE-C $\times$ N).

Acknowledgements This study was supported by a Grant-in-Aid for Scientific Research, No. 22248026 and 2625206 provided by the Japan Society for the Promotion of Science. Tsukuba FACE was established and maintained by "Development of Technologies for Mitigation and Adaptation to Climate Change in Agriculture, Forestry and Fisheries", a project provided by the Ministry of Agriculture, Forestry and Fisheries, Japan.

References

Cheng, W., Inubushi, K., Yagi, K., Sakai, H., & Kobayashi, K. (2001). Effects of elevated carbon dioxide concentration on biological nitrogen fixation, nitrogen mineralization and carbon decomposition in submerged rice soil. *Biology and Fertility of Soils, 34,* 7–13.

Cheng, W., Yagi, K., Sakai, H., & Kobayashi, K. (2006). Effects of elevated atmospheric CO_2 concentrations on CH_4 and N_2O emission from rice soil: An experiment in controlled-environment chambers. *Biogeochemistry, 77,* 351–373.

Dijkstra, F. A., Prior, S. A., Runion, G. B., Torbert, H. A., Tian, H., Lu, C., et al. (2012). Effects of elevated carbon dioxide and increased temperature on methane and nitrous oxide fluxes: Evidence from field experiments. *Frontiers in Ecology and the Environment, 10,* 520–527.

Fumoto, T., Kobayashi, K., Li, C., Yagi, K., & Hasegawa, T. (2008). Revising a process-based biogeochemistry model (DNDC) to simulate methane emission from rice paddy fields under various residue management and fertilizer regimes. *Global Change Biology, 14,* 382–402.

Hasegawa, T., Sakai, H., Tokida, T., Nakamura, H., Zhu, C., Usui, Y., et al. (2013). Rice cultivar responses to elevated CO_2 at two free-air CO_2 enrichment (FACE) sites in Japan. *Functional Plant Biology, 40,* 148–159.

Hayashi, K., Nishimura, S., & Yagi, K. (2006). Ammonia volatilization from the surface of a Japanese paddy field during rice cultivation. *Soil Science and Plant Nutrition, 52,* 545–555.

Hayashi, K., Nishimura, S., & Yagi, K. (2008a). Ammonia volatilization from a paddy field following applications of urea: Rice plants are both an absorber and an emitter for atmospheric ammonia. *Science of the Total Environment, 390,* 485–494.

Hayashi, K., Hiradate, S., Ishikawa, S., & Nouchi, I. (2008b). Ammonia exchange between rice leaf blades and the atmosphere: Effect of broadcast urea and changes of xylem sap and leaf apoplastic ammonium concentrations. *Soil Science and Plant Nutrition, 54*, 807–818.

Hayashi, K., Tokida, T., & Hasegawa, T. (2011). Potential ammonia emission from flag leaves of paddy rice (Oryza sativa L. cv. Koshihikari). *Agriculture, Ecosystems & Environment, 144*, 117–123.

Hayashi, K., Ono, K., Tokida, T., Mano, M., Takimoto, T., Miyata, A., et al. (2012). Atmosphere-rice paddy exchanges of inorganic aerosols and relevant gases during a winter and a summer weeks. *Journal of Agricultural Meteorology, 68*, 55–68.

Hayashi, K., Matsuda, K., Ono, K., Tokida, T., & Hasegawa, T. (2013a). Amelioration of the reactive nitrogen flux calculation by a day/night separation in weekly mean air concentration measurements. *Atmospheric Environment, 79*, 462–471.

Hayashi, K., Mano, M., Ono, K., Takimoto, T., & Miyata, A. (2013b). Four-year monitoring of atmospheric ammonia using passive samplers at a single-crop rice paddy field in central Japan. *Journal of Agricultural Meteorology, 69*, 229–241.

Hayashi, K., Tokida, T., Matsushima, M. Y., Ono, K., Nakamura, H., & Hasegawa, T. (2014). Free-air CO_2 enrichment (FACE) experiment on net nitrogen fixation at the paddy soil surface under submerged conditions. *Nutrient Cycling in Agroecosystems, 98*, 57–69.

Hoque, M. M., Inubushi, K., Miura, S., Kobayashi, K., Kim, H.-Y., Okada, M., et al. (2001). Biological dinitrogen fixation and soil microbial biomass carbon as influenced by free-air carbon dioxide enrichment (FACE) at three levels of nitrogen fertilization in a paddy field. *Biology and Fertility of Soils, 34*, 453–459.

Hoque, M. M., Inubushi, K., Miura, S., Kobayashi, K., Kim, H.-Y., Okada, M., et al. (2002). Nitrogen dynamics in paddy field as influenced by free-air CO_2 enrichment (FACE) at three levels of nitrogen fertilization. *Nutrient Cycling in Agroecosystems, 63*, 301–308.

Ineson, P., Coward, P. A., & Hartwig, U. A. (1998). Soil gas fluxes of N_2O, CH_4 and CO_2 beneath Lolium perenne under elevated CO_2: The Swiss free air carbon dioxide enrichment experiment. *Plant and Soil, 198*, 89–95.

IPCC (2013) *Climate Change 2013: The Physical Science Basis*. Contribution of Working Group I to the Fifth Assessment Report of the Intergovernmental Panel on Climate Change. In T.F. Stocker, D. Qin, G.-K. Plattner, M. Tignor, S.K. Allen, J. Boschung, A. Nauels, Y. Xia, V. Bex, P.M. Midgley, (Eds.). Cambridge, UK and New York, US: Cambridge University Press.

Kammann, C., Muller, C., Grunhage, L., & Jager, H. J. (2008). Elevated CO_2 stimulates N_2O emissions in permanent grassland. *Soil Biology & Biochemistry, 40*, 2194–2205.

Katata, G., Nagai, H., Zhang, L., Held, A., Serca, D., & Klemm, O. (2011). Development of an atmosphere–soil–vegetation model for investigation of radioactive materials transport in the terrestrial biosphere. *Progress in Nuclear Science and Technology, 2*, 530–537.

Katata, G., Hayashi, K., Ono, K., Nagai, H., Miyata, A., & Mano, M. (2013). Coupling atmospheric ammonia exchange process over a rice paddy field with a multi-layer atmosphere-soil-vegetation model. *Agricultural and Forest Meteorology, 180*, 1–21.

Katayanagi, N., Ono, K., Fumoto, T., Mano, M., Miyata, A., & Hayashi, K. (2013). Validation of the DNDC-Rice model to discover problems in evaluating the nitrogen balance at a paddy field for single-cropping of rice. *Nutrient Cycling in Agroecosystems, 95*, 255–268.

Kim, H.-Y., Lieffering, M., Kobayashi, K., Okada, M., & Miura, S. (2003). Seasonal changes in the effects of elevated CO_2 on rice at three levels of nitrogen supply: A free air CO_2 enrichment (FACE) experiment. *Global Change Biology, 9*, 826–837.

Miyazawa, S., Hayashi, K., Nakamura, H., Hasegawa, T., & Miyao, M. (2014). Elevated CO_2 decreases the photorespiratory NH_3 production but does not decrease the NH_3 compensation point in rice leaves. *Plant and Cell Physiology, 55*, 1582–1591.

Nakamura, H., Tokida, T., Yoshimoto, M., Sakai, H., Fukuoka, M., & Hasegawa, T. (2012). Performance of the enlarged rice-FACE system using pure-CO_2 installed in Tsukuba, Japan. *Journal of Agricultural Meteorology, 68*, 15–23.

Niboyet, A., Brown, J. R., Dijkstra, P., Blankinship, J. C., Leadley, P. W., Le Roux, X., et al. (2011). Global change could amplify fire effects on soil greenhouse gas emissions. *PLoS ONE, 6,* e20105. https://doi.org/10.1371/journal.pone.0020105.

Okubo, T., Tokida, T., Ikeda, S., Bao, Z., Tago, K., Hayatsu, M., et al. (2014). Effects of elevated carbon dioxide, elevated temperature, and rice growth stage on the community structure of rice root-associated bacteria. *Microbes and Environments, 29,* 184–190.

Ono, K., Maruyama, A., Kuwagata, T., Mano, M., Takimoto, T., Hayashi, K., et al. (2013). Canopy-scale relationships between stomatal conductance and photosynthesis in irrigated rice. *Global Change Biology, 19,* 2209–2220.

Pereira, J., Figueiredo, N., Goufo, P., Carneiro, P., Morais, R., Carranca, C., et al. (2013). Effects of elevated temperature and atmospheric carbon dioxide concentration on the emissions of methane and nitrous oxide from Portuguese flooded rice fields. *Atmospheric Environment, 80,* 464–471.

Robinson, D., & Conroy, J. P. (1999). A possible plant-mediated feedback between elevated CO_2, denitrification and the enhanced greenhouse effect. *Soil Biology & Biochemistry, 31,* 43–53.

Sakai, H., Hasegawa, T., & Kobayashi, K. (2006). Enhancement of rice canopy carbon gain by elevated CO_2 is sensitive to growth stage and leaf nitrogen concentration. *New Phytologist, 170,* 321–332.

Sutka, R. L., Ostrom, N. E., Ostrom, P. H., Breznak, J. A., Gandhi, H., Pitt, A. J., et al. (2006). Distinguishing nitrous oxide production from nitrification and denitrification on the basis of isotopomer abundances. *Applied and Environmental Microbiology, 72,* 638–644.

Tokida, T., Fumoto, T., Cheng, W., Matsunami, T., Adachi, M., Katayanagi, N., et al. (2010). Effects of free-air CO_2 enrichment (FACE) and soil warming on CH_4 emission from a rice paddy field: Impact assessment and stoichiometric evaluation. *Biogeosciences, 7,* 2639–2653.

Tokida, T., Adachi, M., Cheng, W. G., Nakajima, Y., Fumoto, T., Matsushima, M., et al. (2011). Methane and soil CO_2 production from current-season photosynthates in a rice paddy exposed to elevated CO_2 concentration and soil temperature. *Global Change Biology, 17,* 3327–3337.

Toyoda, S., & Yoshida, N. (1999). Determination of nitrogen isotopomers of nitrous oxide on a modified isotope ratio mass spectrometer. *Analytical Chemistry, 71,* 4711–4718.

Toyoda, S., Yoshida, N., Miwa, T., Matsui, Y., Yamagishi, H., Tsunogai, U., et al. (2002). Production mechanism and global budget of N_2O inferred from its isotopomers in the western North Pacific. *Geophysical Research Letters, 29,* 1037.

Usui, Y., Sakai, H., Tokida, T., Nakamura, H., Nakagawa, H., & Hasegawa, T. (2014). Heat-tolerant rice cultivars retain grain appearance quality under free-air CO_2 enrichment. *Rice, 7,* 1–9.

Van Groenigen, K. J., Osenberg, C. W., & Hungate, B. A. (2011). Increased soil emissions of potent greenhouse gases under increased atmospheric CO_2. *Nature, 475,* 214–216.

Yano, M., Toyoda, S., Tokida, T., Hayashi, K., Hasegawa, T., Koba, K., et al. (2014). Isotopomer analysis of production, consumption and soil-to-atmosphere emission processes of N_2O at the beginning of flood irrigation. *Soil Biology & Biochemistry, 70,* 66–78.

Yoshimoto, M., Oue, H., & Kobayashi, K. (2005). Energy balance and water use efficiency of rice canopies under free-air CO_2 enrichment. *Agricultural and Forest Meteorology, 133,* 226–246.

Chapter 23
Nitrous Oxide (N_2O) Emissions from Forests, Grasslands and Agricultural Soils in Northern Spain

María Eréndira Calleja-Cervantes, Ximena Huerfano, Iskander Barrena, José María Estavillo, Pedro M. Aparicio-Tejo, Carmen Gonzalez-Murua, and Sergio Menéndez

Abstract This chapter presents a study of nitrous oxide (N_2O) emissions from (a) forest soils (*Pinus radiata* and *Fagus sylvatica*) at different growth stages and from (b) grassland and agricultural soils in the edapho-climatic conditions of northern Spain. A mixed clover–ryegrass sward, wheat and oats crops were fertilized with ammonium sulphate nitrate, cattle slurry and/or sewage sludge. The mitigation effectiveness of the nitrification inhibitor (NI) 3,4-dimethylpyrazole phosphate (DMPP) was also assayed. Nitrous oxide losses from forests in northern Spain are limited by nitrogen (N) deposition, resulting in lower losses than temperate forest in the rest of Europe. The NI DMPP seems to be an advisable strategy to mitigate N_2O emissions, both in grasslands and in crops. Sewage sludge application should be further studied but significant reduction in N_2O emission is possible with DMPP. The results presented suggest that an emission factor around 0.1–0.3%, depending on crop type, time of

M. E. Calleja-Cervantes · P. M. Aparicio-Tejo
Institute for Multidisciplinary Research in Applied Biology (IMAB), Edificio Jerónimo de Ayanz, Campus Arrosadia, 31006 Pamplona, Navarra, Spain
e-mail: tlallihuacan@gmail.com

P. M. Aparicio-Tejo
e-mail: pmapariciotejo@unavarra.es

X. Huerfano · I. Barrena · J. M. Estavillo · C. Gonzalez-Murua · S. Menéndez (✉)
Department of Plant Biology and Ecology, University of the Basque Country UPV/EHU, Apdo. 644, E-48080 Bilbao, Spain
e-mail: sergio.menendez@ehu.eus

X. Huerfano
e-mail: enithximena.huerfano@ehu.eus

I. Barrena
e-mail: iskander.barrena@ehu.eus

J. M. Estavillo
e-mail: jm.estavillo@ehu.eus

C. Gonzalez-Murua
e-mail: carmen.gmurua@ehu.eus

M. A. Sutton et al. (eds.), *Just Enough Nitrogen*,
https://doi.org/10.1007/978-3-030-58065-0_23

year and fertilizer type, should be applied for this area instead of the default value of 1% suggested by the Intergovernmental Panel on Climate Change (IPCC).

Keywords Beech · 3,4-dimethylpyrazole phosphate (DMPP) · Nitrous oxide · Oats · Radiata pine · Sewage sludge · Slurry · Wheat · Northern Spain

23.1 Introduction

Soils, whether for forestry or agriculture, are usually regarded as net sources of nitrous oxide (N_2O), but also as temporary sinks (Chapuis-Lardy et al. 2007). In northern Spain, forests play an important role in this regard because of the large areas they occupy. Soils on agricultural land receive high inputs of mineral fertilizers that lead to deleterious effects on the environment (Peñuelas et al. 2012). In addition, the environment is also under pressure due to large quantities of organic residues produced by rapid urban development and the intensification of livestock farms. One solution to these issues is the use of organic residues as soil amendments (Evanylo et al. 2008). Notwithstanding the beneficial effects of using organic sources in reducing mineral fertilizer application, soil gaseous emissions still represent an important challenge that has to be explored. An important strategy that targets nutrient loss reduction is the use of nitrification inhibitors (NIs).

In the rural areas of northern Spain, forestry and agriculture are major land use types. Agricultural soils represent 25% of the total surface area, mainly covered by crops and grasslands (Asner et al. 2004). Forestry is one of the most important activities of the primary sector, being a traditional activity in rural areas it has a major influence on the local environment and economy. Forests cover around 68% of the total territory, with plantations for wood production of radiata pine (*Pinus radiata* D. Don) and native beech forest (*Fagus sylvatica* L.) being the most representative (Inventario Forestal CAE 2005). Douglas fir (*Pseudotsuga menziesii* (Mirb.) Franco) is another valuable plantation tree.

Agriculture and forestry both have a major impact on the environment via nitrous oxide emissions. Nitrous oxide is a powerful greenhouse gas (GHG) that accounts for 5% of the annual increase in the global atmospheric radiative forcing (IPCC 2007). Soils are the dominant source of N_2O worldwide (around 65% of global N_2O emissions). In order to mitigate N_2O emissions various fertilizer-related strategies are being tested to find the most suitable option for each targeted situation (e.g., forests and agriculture). We examined two main strategies: the use of NIs and alternative organic-residue derived fertilizers. Nitrification inhibitors delay the bacterial oxidation of ammonia to nitrite in the soil for a certain period of time. One of these NIs is 3,4-dimethylpyrazol phosphate (DMPP) which is applied at very low rates compared to the recommended doses for other NIs. The application of DMPP has shown to reduce nitrogen (N) losses by N_2O emissions (Weiske et al. 2001; Macadam et al. 2003; Menéndez et al. 2006).

Treated sewage sludge (SS) is an inevitable by-product of municipal wastewater treatment. In order to obtain the cleanest water possible the fact that SS is rich in N, phosphorus (P) and organic matter needs to be taken into consideration. This condition makes SS a good candidate as an alternative fertilizer and reliable treatments and management practices have been developed to ensure its application is innocuous (Fytili and Zabaniotou 2008). Therefore, SS has been studied lately to determine its mitigation potential.

All countries that approved the Kyoto Protocol have the obligation to report their national GHG inventories. The Intergovernmental Panel on Climate Change (IPCC 2006) has established a default emission factor (EF) of 1.0% of the N applied for the calculation of national emissions of N_2O from agricultural soils. The default EF is irrespective of the type of fertilizer or manure. Neither application time, nor application method and application level are considered. Nevertheless, individual countries can use a different EF that corresponds to country-specific soil type or farm management. With the aim of providing information for national inventories and to estimate an adjusted EF for the edapho-climatic conditions of northern Spain, we present this brief review regarding N_2O emissions from soils.

23.2 Materials and Methods

23.2.1 Field Experiment 1

The experiment was carried out in three different forest types: radiata pine (*Pinus radiata* D. Don), Douglas fir (*Pseudotsuga mensiezii* Mirb.) and beech (*Fagus sylvatica* L.). In addition, two stages of beech forest (young and mature) and three stages of radiata pine (new plantation, young, and mature) were included. Both radiata pine and Douglas fir stands were located at Artzentales (43°13′38″ N, 3°11′54″ W, 350 m), while beech stands were located at the Natural Park of Gorbea (43°6′27″ N, 2°48′16″ W, 400 m).

23.2.2 Field Experiment 2

This work was conducted on typical cut grassland under Atlantic conditions. A randomized complete block factorial design with four replicates was established, each experimental plot covered an area of 12 m^2 (4 × 3 m). Two fertilizers were applied: ammonium sulphate nitrate (26%) (ASN treatment) or cattle slurry (S treatment). The slurry had a total N content of 2.04% (w/w); NH_4^+–N = 0.23% (w/w) and C:N ratio = 11.22. DMPP was applied or omitted with both fertilizers. In ASN, DMPP was applied as the marketed product *ENTEC® 26,* containing 0.8% of NH_4^+-N (ENTEC treatment). A treatment with no fertilizer was included as a control.

23.2.3 Field Experiment 3

This work was performed in a wheat crop during two years under humid Mediterranean conditions. A randomized complete block factorial design with four replicates was established, with an individual plot size of 40 m^2. Three main treatments were applied: a control treatment without fertilizer, a treatment with ammonium sulphate nitrate (ASN 26%) and a treatment consisting in *ENTEC® 26*. Fertilization was split in two: the first at tillering, when 60 kg N ha^{-1} were applied; the second one at stem elongation, when 120 kg N ha^{-1} were applied. A fourth treatment with *ENTEC® 26* applied at a rate of 180 kg N ha^{-1} at tillering was also included.

23.2.4 Field Experiment 4

This experiment utilized a long-term field experiment which was established 20 years before GHG measurements. The trial was conducted at the cereal (oats) experimental station of Arazuri, which is a humid- template- Mediterranean area. A randomized complete block factorial design with four replicates was established, with an individual plot size of 35 m^2. Five treatments were applied annually: an unfertilized control treatment, a mineral fertilizer (ammonium sulphate nitrate), two treatments consisting in different doses of treated SS (40 t ha^{-1} $year^{-1}$, 80 t ha^{-1} $year^{-1}$) and a treatment with SS applied every 3 years (40 t ha^{-1} 3 $year^{-1}$). The mineral treatment was split in two applications: 60 kg N ha^{-1} at tillering and 40 kg N ha^{-1} at stem elongation. The SS was spread to soil surface at seeding and incorporated into the soil at 0–30 cm depth.

23.2.5 Nitrous Oxide Measurement

In experiments 1, 3 and 4, N_2O emissions were measured using the closed chamber technique (Menéndez et al. 2008) and gas chromatography (GC) (Agilent, 7890A) with an electron capture detector (ECD) for sample analysis. In experiment 2, N_2O emissions were measured using a closed air circulation technique in conjunction with a photoacoustic infrared gas analyzer (Brüel and Kjaer 1302 Multi-Gas Monitor) during 40 min after insertion of the chamber (Menéndez et al. 2006). In chamber experiments fluxes were calculated from the linear concentration increase in the chamber headspace over time. Cumulative N_2O emissions for each experiment were estimated by averaging the rate of emission between two successive measurements, multiplying that average rate by the length of the period between the measurements, and adding that amount to the previous cumulative total.

23.3 Results and Discussion

Table 23.1 shows N_2O losses from different forest soils in northern Spain. Daily fluxes (data not shown) were lower than those from temperate forests in other areas of Europe (Barrena et al. 2013). The low atmospheric nitrogen deposition in this area is responsible for a nitrogen deficit which seems to trigger these relatively low gas fluxes. In addition, the ground vegetation cover may play an important role in enhancing the soil nitrogen turnover rate (Ambus and Zechmeister-Boltenstern 2007) that led to different N_2O emission rates in both beech and radiata pine forest types. Also, the vegetation appears to be a contributing factor of the differences between growth stages for each forest type. In the case of Douglas fir, the litter horizon is probably responsible for higher N_2O emissions as a consequence of an aerobic nitrification process enhanced by a lower soil water content achieved along the year and by a simultaneous anaerobic denitrification process.

Table 23.2 presents N_2O losses from grasslands in experiment 2. The cattle slurry significantly increased N_2O emissions due to its intrinsic organic matter while mineral fertilizer showed no effect. The highest losses took place in spring. As discussed by Menéndez et al. (2009) and Merino et al. (2005), the soil water content and temperature determine N_2O emissions. DMPP can reduce N_2O emissions by up to a 69%, when it is applied with cattle slurry. Nevertheless, its capacity to reduce emissions is not always significant or required, as when environmental conditions favour denitrification to N_2 (Menéndez et al. 2012).

Nitrous oxide emissions from wheat were measured during the whole period of the crop (Table 23.3). The application of DMPP reduced N_2O emissions significantly to control levels. In spite of the increase induced by fertilizer addition, losses were lower than the 1% of applied N suggested by IPCC (2006). In fact, the emission factor of the different treatments did not exceed 0.2% during the two assayed years.

The effect of SS application in the oat crop is presented in Table 23.4. The application of SS significantly increased the N_2O losses with respect to the control and ASN treatments. However, N_2O losses were similar between SS treatments, regardless of the application rate. So, despite receiving double the amount of N, the EF for the highest dose (80 t ha^{-1} $year^{-1}$) was 0.35%. In addition, the EF was reduced compared to that of the 40 t ha^{-1} $year^{-1}$ which was the highest in this assay. This result is understandable as the EF is conditioned by the emission and the application rate.

23.4 Conclusions

Nitrous oxide losses from forests in northern Spain are limited by N deposition, resulting in lower losses than temperate forest in the rest of Europe. The NI DMPP seems to be an advisable strategy to mitigate N_2O emissions, both in grasslands and in crops. Sewage sludge application should be further studied but significant reduction in N_2O emission is possible with DMPP. The results presented suggest

Table 23.1 Cumulative N_2O emissions (g N_2O–N ha^{-1} $year^{-1}$) from different forests at different growth stages

	Young Beech		Mature Beech		New pine		Young pine		Mature pine		Douglas fir	
2010	103 ± 60	n.s.	16 ± 37	n.s.	97 ± 34	n.s.	17 ± 30	n.s.	93 ± 30	n.s.	222 ± 84	n.s.
2011	167 ± 58		15 ± 20		49 ± 32		28 ± 39		91 ± 104		308 ± 74	
Mean	135 ± 54	ab	15 ± 22	b	73 ± 30	b	22 ± 33	b	92 ± 64	b	265 ± 76	a

Differences between years for each stand were analyzed using a T-student test ($p < 0.05$; n = 6). Different letters indicate significantly different mean annual fluxes between forest stands using Duncan Test ($p < 0.05$; n = 6)

Table 23.2 Cumulative N_2O emissions from grasslands during different seasons. The length of each experimental period and the application rate of fertilizer are in brackets. Percentages of reduction induced by DMPP are shown

	Spring						Autumn		
	Kg N_2O–N ha^{-1} (59 days; 97 kg N ha^{-1} applied)		Emission factor (%)	Kg N_2O–N ha^{-1} (29 days; 70 kg N ha^{-1} applied)		Emission factor	Kg N_2O-N ha^{-1} (22 days; 135 kg N ha^{-1} applied)		Emission factor (%)
Control	2.68 c		–	1.55 a		–	3.2 b		–
ASN	4.44 c	9%	1.81	–		–	–		–
ENTEC® 26	4.05 c		1.41	–		–	–		–
S	15.49 a	30%	13.21	2.28 a	8%	1.04%	11.5 a	69%	6.15
S + DMPP	10.98 b		8.56	2.10 a		0.78	3.6 b		0.30

Different letters within a column indicate significantly different rates ($p < 0.05$; n = 4)

Table 23.3 Cumulative N_2O emissions from wheat crop. The length of each experimental period is in brackets. Percentages of reduction induced by DMPP and emission factors are also shown

	2011			2012		
	Kg N_2O–N ha^{-1} (228 days)		Emission factor (%)	Kg N_2O–N ha^{-1} (238 days)		Emission factor (%)
Control	1.31 b		–	0.25 c		–
60 + 120 ASN	1.62 a		0.17	0.59 a		0.19
60 + 120 ENTEC® 26	1.39 b	14%	0.04	0.49 ab	17%	0.13
180 ENTEC® 26	1.35 b	17%	0.02	0.43 b	27%	0.10

Different letters within a column indicate significantly different rates ($p < 0.05$; n = 4)

Table 23.4 Cumulative N_2O emissions and emission factors (EF) from oats crop for a period of 261 days

	Kg $N_2O^{-}N$ ha^{-1}	Emission factor (%)
Control	0.58 b	–
60 + 40 ASN	0.91 b	0.33
40 t ha^{-1} $year^{-1}$ SS	4.39 a	0.91
80 t ha^{-1} $year^{-1}$ SS	3.49 a	0.35
40 t ha^{-1} 3 $year^{-1}$ SS	1.92 a	0.32

Different letters within a column indicate significantly different rates ($p < 0.05$; n = 4)

that an emission factor around 0.1–0.3%, depending on crop type, time of year and fertilizer type, should be applied for this area instead of the default value of 1% suggested by IPCC (2006).

Acknowledgements This work was funded by the Spanish Government (AGL2012-37815-C05-02 and 05), and by the Basque Government (GV-IT-526-10). M.E. Calleja-Cervantes, X. Huerfano and I. Barrena are the recipients of a predoctoral fellowship from the Public University of Navarre, the Ministry of Economy and Competitiveness and the Department of Education, Universities and Research of the Basque Government, respectively.

References

Ambus, P., & Zechmeister-Boltenstern, S. (2007). Denitrification and N-cycling in forest ecosystems. In H. Bothe, S. J. Ferguson, & W. E. Newton (Eds.), *Biology of the nitrogen cycle* (pp. 343–358). Amsterdam: Elsevier Science BV.

Asner, G. P., Elmore, A. J., Olyer, L. P., Martin, R. E., & Harris, A. T. (2004). Grazing systems, ecosystem responses, and global change. *Annual Review of Environment and Resources, 29,* 261–299.

Barrena, I., Menéndez, S., Duñabeitia, M., Merino, P., Stange, C. F., Spott, O., et al. (2013). Greenhouse gas fluxes (CO_2, N_2O and CH_4) from forest soils in the Basque Country: Comparison of different tree species and growth stages. *Forest Ecology and Management, 310,* 600–611.

Chapuis-Lardy, L., Wrage, N., Metay, A., Chotte, J. L., & Bernoux, M. (2007). Soils, a sink for N_2O? A review. *Global Change Biology, 13,* 1–17.

Evanylo, G., Sherony, C., Spargo, J., Starner, D., Brosius, M., & Haering, K. (2008). Soil and water environmental effects of fertilizer-, manure-, and compost-based fertility practices in an organic vegetable cropping system. *Agriculture, Ecosystems & Environment, 127,* 50–58.

Fytili, D., & Zabaniotou, A. (2008). Utilization of sewage sludge in EU application of old and new methods - A review. *Renewable and Sustainable Energy Reviews, 12,* 116–140.

Inventario Forestal CAE. (2005). *Inventario Forestal CAE 2005.* Departamento de Agricultura. Pesca y Alimentación, Gobierno Vasco.

IPCC. (2006). *IPCC Guidelines for National Greenhouse Gas Inventories.* Prepared by the National Greenhouse Gas Inventories Programme. Japan: IGES.

IPCC. (2007). *Climate change 2007: The physical science basis. Contribution of Working Group I to the Fourth Assessment Report of the Intergovernmental Panel on Climate Change.* Cambridge: Cambridge University Press.

Macadam, X. M. B., Prado, A., Merino, P., Estavillo, J. M., Pinto, M., & González-Murua, C. (2003). Dicyandiamide and 3,4-dimethyl pyrazole phosphate decrease N_2O emissions from grassland but dicyandiamide produces deleterious effects in clover. *Journal of Plant Physiology, 160,* 1517–1523.

Menéndez, S., Merino, P., Pinto, M., González-Mura, C., & Estavillo, J. M. (2006). 3,4-Dimethylpyrazol phosphate effect on nitrous oxide, nitric oxide, ammonia, and carbon dioxide emissions from grasslands. *Journal of Environmental Quality, 35,* 973–981.

Menéndez, S., López-Bellido, R. J., Benítez-Vega, J., González-Murua, C., López-Bellido, L., & Estavillo, J. M. (2008). Long-term effect of tillage, crop rotation and N fertilization to wheat on gaseous emissions under rainfed Mediterranean conditions. *European Journal of Agronomy, 28,* 559–569.

Menéndez, S., Merino, P., Pinto, M., González-Murua, C., & Estavillo, J. M. (2009). Effect of N-(n-butyl) thiophosphoric triamide and 3,4 dimethylpyrazole phosphate on gaseous emissions from grasslands under different soil water contents. *Journal of Environmental Quality, 38,* 27–35.

Menéndez, S., Barrena, I., Setien, I., González-Murua, C., & Estavillo, J. M. (2012). Efficiency of nitrification inhibitor DMPP to reduce nitrous oxide emissions under different temperature and moisture conditions. *Soil Biology & Biochemistry, 53,* 82–89.

Merino, P., Menéndez, S., Pinto, M., González-Murua, C., & Estavillo, J. M. (2005). DMPP reduces nitrous oxide emissions from grassland after slurry application. *Soil Use and Management, 21,* 53–57.

Peñuelas, J., Sardans, J., Rivas-Ubach, A., & Janssens, I. A. (2012). The human-induced imbalance between C, N and P in earth's life system. *Global Change Biology, 18,* 3–6.

Weiske, A., Benckiser, G., Herbert, T., & Ottow, J. C. G. (2001). Influence of the nitrification inhibitor 3,4-dimethylpyrazole phosphate (DMPP) in comparison to dicyandiamide (DCD) on nitrous oxide emissions, carbon dioxide fluxes and methane oxidation during 3 years or repeated application in field experiments. *Biology and Fertility of Soils, 34*(2), 109–117.

Chapter 24
Effect of Climate Change and Crop-Year on the Yield and Nitrogen Fertilizer Efficiency in Winter Wheat (*Triticum aestivum* L.) Production

Peter Pepó

Abstract Winter wheat has a decisive role in Hungarian crop production. The sowing area of wheat varies between 1.0 and 1.2 million ha. The national average yield of wheat was 5.0–5.5 t ha^{-1} in the 1980s but nowadays the country average one varies between 3.0 and 5.0 t ha^{-1} depending on the climatic factors of crop-year. The total Hungarian Nitrogen (N), Phosphorus (P) and Potassium (K) fertilizer usage was 250–300 kg ha^{-1} in the 1980s, but it has decreased to 80–100 kg ha^{-1} presently. Results of our long-term experiments on chernozem soil proved that the crop-year and the climatic factors (mainly the water supply: quantity of rainfall XE "rainfall" and its distribution) have strong effects on the natural nutrient utilization, yield surpluses of N-fertilization, the maximum yield and the optimum N(+PK) fertilizer doses of different winter wheat genotypes. The nutrient (mainly nitrogen) utilization of winter wheat was modified by abiotic (climatic factors) and biotic (leaf-, stem- and spike-diseases) stresses. In the optimum crop-year and agro-technical models, the maximum yields of winter wheat varied between 7 and 9 t ha^{-1}. In unfavourable climatic and agronomic conditions, the yields of winter wheat dropped to 3–6 t ha^{-1}.

Keywords Nitrogen · Wheat · Climate · Fertilization efficiency

24.1 Introduction

Winter wheat has a determinative role in Hungarian crop production. The sowing area of wheat varies between 1.0 and 1.2 million ha. The Hungarian country average yield of wheat was 5.0–5.5 t ha^{-1} in the 1980s, but nowadays the average yield varies between 3.0 and 5.0 t ha^{-1} depending on the climatic factors of crop-year.

Many foreign and Hungarian experimental results of scientific papers proved that the climatic conditions of crop-years strongly modified the yield of wheat (Pepó 2004), the yield-stability (Balla et al. 2006; Pepó and Győri 2005) and the baking

P. Pepó (✉)
Faculty of Agricultural and Food and Environmental Sciences, Institute of Crop Sciences, University of Debrecen, H-4032 Böszörményi str 138, Debrecen, Hungary
e-mail: pepopeter@agr.unideb.hu

M. A. Sutton et al. (eds.), *Just Enough Nitrogen*,
https://doi.org/10.1007/978-3-030-58065-0_24

quality (Pepó et al. 2005; Szentpétery et al. 2005a, b), even in favourable, intensive agro-technical circumstances. The nitrogen (N) and other nutrient supply has a determinative role in sustainable wheat production (Pepó 2004). It is possible to reduce the unfavourable, negative agro-ecological, weather effects by using optimum nutrient supply, fertilization and appropriate variety-selection.

The frequency of dry crop-years increased from 22.5 to 52.6% in Hungary in the last 150 years. The results of (Balogh and Pepó 2008; Birkás et al. 2006; Olesen and Bindi 2002) showed that as a result of global climate change the yield of crops have dropped and yield fluctuation has increased. To a certain extent, the unfavourable abiotic stress effects (weather) can be reduced by appropriate variety selection and optimum use of agro-technical factors. Among agro-technical elements the optimum nutrient- and water supply (Fowler 2003; Pepó 2007) and crop rotation (Hornok 2008) have outstandingly high importance.

24.2 Material and Methods

Our long-term experiment was set up in 1983 on a chernozem soil in Hajdúság (East-Hungary). The experimental location is found on Látókép Research Farm, 15 km of Debrecen (latitude: 47° 30′, longitude: 21° 30′, elevation above the Adriatic Sea: 118 m).

Analytical data for initial soil conditions showed that, as regards its soil physics, the area can be classified as having loam soil with nearly neutral pH value (pH_{KCl} 6.46). It has medium humus content (2.76% in the 0–0.2 m upper soil layer) and a humus layer of about 0.8 m. The phosphorus (P) and potassium (K) supplies of the soil can be regarded as medium (AL–P_2O_5 133 mg kg^{-1}) and good (AL K_2O 240 mg kg^{-1}), respectively. The long-term experiment had a split-plot arrangement with four repetitions. Increasing NPK doses were applied (the basic dose was N = 30 kg ha^{-1}, P_2O_5 = 22.5 kg ha^{-1}, K_2O = 26.5 kg ha^{-1}), together with inputs at 2×, 3×, 4×, and 5× of the basic dose.

A second long-term experiment was set up in 1983 on chernozem soil on the Látókép Research Station of the University of Debrecen in the Hajdúság region (Eastern Hungary). The following factors were examined in the second long-term experiment:

- Crop rotation: bi-culture (maize, wheat), tri-culture (pea-wheat-maize).
- Fertilization: control, N = 50 kg ha^{-1}, P_2O_5 = 35 kg ha^{-1}, K_2O = 40 kg ha^{-1}, and inputs at 2×, 3×, and 4× these rates.
- Irrigation: irrigated and non-irrigated.

24.3 Results and Discussion

The basic element of sustainable winter wheat production is to match a suitable, adaptable genotype to local agro-ecological conditions. If we know the site and variety circumstances well (possibilities and barriers), then we could build up the agro-technical management. Every agro-technical factor has an important role in the manifestation of yield-ability, yield-stability and baking quality of wheat genotypes. The role and weight of every agro-technical element could be modified by the intensity of crop management. According to the results of our long-term experiments over two decades, we can state that the roles of agro-ecological factors (weather, soil) were decisive (together 60%) in the wheat crop-model for extensive management, and we found that the nutrient supply, fertilization and genotype had the most important effects in the wheat crop-model for intensive, sustainable management (Fig. 24.1).

Nutrient supply and fertilization have the key-role in sustainable wheat production because, on the one hand, fertilization directly and indirectly modifies every other agro-technical factor (planting-technology, crop protection etc.). On the other hand, non-optimum fertilization causes some harmful environmental effects (e.g., NO_3–N accumulation in different soil layers etc.).

Our long-term experimental results proved that the weather of crop-years and genotypes strongly modified the yields of wheat varieties even on the chernozem soil characterized by excellent water- and nutrient husbandry (Table 24.1). Of the different varieties the average maximum yield was 6.5 t ha^{-1} (in 2002 it was 6369 kg ha^{-1}; in 2006 it was 6981 kg ha^{-1}; respectively). In favourable, good crop-years we obtained much bigger maximum yields, but there were differences between the crop-years characterized by favourable agro-meteorological parameters (in 2004 it was 8241 kg ha^{-1}, in 2005 it was 8230 kg ha^{-1} for the mean of varieties). The maximum yield differences of varieties were significant in average and good crop-years. In a very dry crop-year (2003) the maximum yields in optimum N + PK treatments were moderated (4321 kg ha^{-1} in the mean of varieties) because of limited water supply and nutrient uptake of wheat.

The crop-year can modify, not only the maximum yields of wheat varieties, but the optimum N + PK doses too. In the crop-years characterized by excellent water supply (the rainy crop-years: 2004 and 2005) the optimum N-doses were moderated

Extensive model				Intensive model
32 %	←	Crop-year	→	16 %
28 %	←	Soil	→	9 %
8 %	←	Variety	→	18 %
17 %	←	Tillage	→	9 %
10 %	←	Fertilization	→	29 %
5 %	←	Crop protection	→	19 %
100 %	←	**Total**	→	**100 %**

Fig. 24.1 Effect of management factors on the yields in different wheat models. *Source* Pepó (2006)

Table 24.1 Maximum yields of winter wheat varieties and their fertilizer doses (yield kg ha^{-1}, N_{opt} kg ha^{-1}, chernozem soil)

Variety	2002	2003	2004	2005	2006	Average
GK Öthalom	6247$_{(120)}$	4343$_{(90)}$	8442$_{(60)}$	8126$_{(60)}$	6761$_{(150)}$	6784
Fatima	6903$_{(90)}$	4916$_{(90)}$	8317$_{(60)}$	8853$_{(60)}$	–	7247
GK Ati	6540$_{(90)}$	4129$_{(90)}$	8425$_{(60)}$	–	7532$_{(150)}$	6657
GK Kapos	–	–	8709$_{(60)}$	8596$_{(60)}$	6617$_{(120)}$	7974
GK Petur	6629$_{(120)}$	4366$_{(60)}$	8342$_{(60)}$	–	7353$_{(150)}$	6673
Lupus	5915$_{(90)}$	4523$_{(60)}$	7583$_{(60)}$	7725$_{(30)}$	7439$_{(120)}$	6637
Mv Csárdás	5980$_{(90)}$	3648$_{(90)}$	–	–	6237$_{(120)}$	5288
Saturnus	–	–	7870$_{(60)}$	7850$_{(60)}$	6930$_{(150)}$	7550
Average	**6369**	**4321**	**8241**	**8230**	**6981**	**6851**
Yield interval, t ha^{-1}	5.9–6.9	3.6–4.9	7.6–8.7	7.7–8.9	6.2–7.4	5.3–8.0
Min–Max, %	93–108	84–114	92–106	94–108	89–108	77–116
Interval of yield fluctuation, %	15	30	14	14	19	39
Interval of N_{opt} kg ha^{-1}	90–120	60–90	60	30–60	120–150	72–100
$LSD_{5\%}$	381	272	435	385	220	–

$LSD_{5\%}$ = Least Significant Difference at probability of $p = 0.05$
Numbers in brackets = the optimum N-fertilizer doses in each variety

to be low (30–60 kg ha^{-1} + PK) on chernozem soil. In rainy crop-years the wheat genotypes could better utilize both the natural nutrients of soil and the nutrients of fertilizers. In a dry crop-year (2003) the optimum N-doses were closer to the average (60–90 kg ha^{-1} + PK). We obtained the largest variability of optimum N doses for wheat varieties (90–150 kg ha^{-1} + PK) in the crop-years characterized by average agro-meteorological parameters.

Winter wheat is one of the most fertilizer-responsive field crops. Our scientific long-term experimental data proved that the fertilization of wheat resulted in excellent yield-surpluses on chernozem soil characterized by excellent nutrient husbandry. The efficiency of fertilizers was determined by the water supply of the crop-years. Yield-surpluses by fertilization proved that the efficiency of fertilization was better in favourable (good water supply) crop-years (in 2000, 2001, 2004 and 2005 yield surpluses varied between 3559–4250 kg ha^{-1}). The yield-surpluses by fertilization compared with the control (in the average of varieties) were slightly moderated in the average water-supplied crop-years (in 1999, 2002 and 2006 yield surpluses varied between 2091 and 3067 kg ha^{-1}). The crop-year 2003 was hot and dry, with a huge water deficit in the vegetation period, which moderated the efficiency of fertilizer use (the yield-surplus was 940 kg ha^{-1} in the mean of varieties) (Table 24.2).

The yields of winter wheat in 2007 and 2008 well-reflected the weather stress effects and the interrelation of those agro-technical factors that decrease or increase yields (Table 24.3). The dry crop-year of 2007 had an unfavourable effect on the vegetative and generative growth and yield formation of winter wheat. The stress

Table 24.2 Effect of crop-year on the control and maximum yield of winter wheat (Debrecen, 1999–2006) (average of varieties)

Crop-year	Control yield (kg ha^{-1})	Maximum yield (kg ha^{-1})	Yield-surplus (kg ha^{-1})	Rainfall in veg. period (mm)	Rainfall deviation from 30 year average (mm)
1998/1999	4042	6598	2556	470.4	+69.5
1999/2000	4041	8296	4250	312.9	−88.0
2000/2001	3193	7226	4033	430.2	+29.3
2001/2002	4466	6555	2091	184.6	−216.3
2002/2003	3447	4387	940	279.3	−121.6
2003/2004	4713	8573	3860	376.5	−24.4
2004/2005	4539	8098	3559	410.4	+9.5
2005/2006	3949	7016	3067	476.5	+37.6

Table 24.3 The effects of crop-years and agro-technical elements on the yields of winter wheat (Debrecen, 2007–2008, chernozem soil)

Treatments	2007		2008	
	Non-irrigated	Irrigated	Non-irrigated	Irrigated
Bi-culture				
Control	1892	2330	3015	2892
N_{50} + PK	3420	4002	5043	4870
N_{100} + PK	5048	5932	6260	6517
N_{150} + PK	5590	6926	7065	6882
N_{200} + PK	5205	7835	6772	6585
Tri-culture				
Control	4426	5328	7228	7350
N_{50} + PK	6273	7012	8112	7874
N_{100} + PK	6913	8492	6346	6108
N_{150} + PK	7279	8016	6036	6242
N_{200} + PK	6842	7582	5440	5149
$LSD_{5\%}$	872			

$LSD_{5\%}$ = Least Significant Difference at probability of $p = 0.05$

caused by the unfavourable dry crop-year could significantly be decreased by optimal use of agro-technical factors. In the dry crop-year of 2007, among agro-technical factors, irrigation, fertilization and crop rotation had significant influence on the yields of wheat. In 2007, the yields of wheat ranged between 1892 and 5590 kg ha^{-1} (non-irrigated) and between 2330 and 7835 kg ha^{-1} (irrigated) in bi-culture, and between 4426–7279 kg ha^{-1} (non-irrigated) and 5328–8492 kg ha^{-1} (irrigated) in tri-culture, respectively. Our results proved that in a dry crop-year (2007) the yield increasing effect of irrigation itself was extremely limited in the absence of sufficient nutrient supply. In 2007, the yield increase caused by irrigation ranged between 438 kg ha^{-1} (bi-culture) and 902 kg ha^{-1} (tri-culture) in the control treatment (abiotic stress caused by nutrient deficiency). Contrary to this, the yield increase by irrigation was significantly higher in the optimal NPK treatment. In this case, the yield increase caused by irrigation was 2630 kg ha^{-1} in bi-culture (after maize forecrop which had higher water uptake) and 1579 kg ha^{-1} in tri-culture (after pea forecrop that has lower demand for water) (Fig. 24.2).

The yield-increasing effect of fertilization and the optimal fertilizer dose were equally influenced by crop-year, crop rotation and irrigation (Fig. 24.3). Compared with the control treatment, the yield surpluses of the optimal NPK dose treatment in

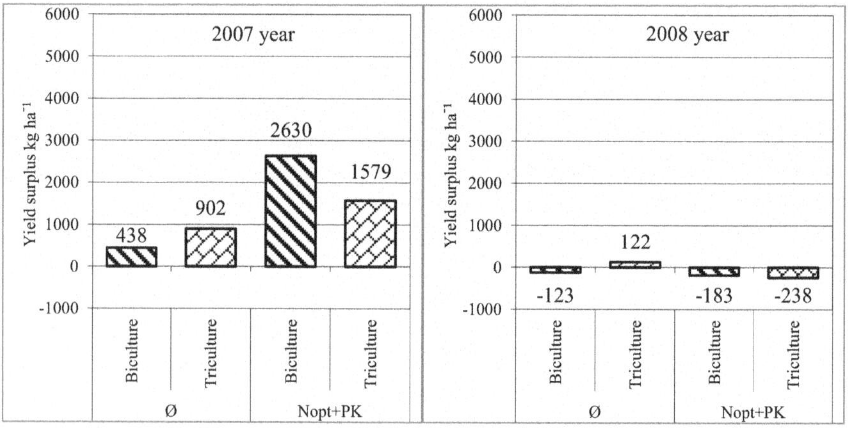

Fig. 24.2 The effect of irrigation on the yield surpluses of winter wheat (Debrecen, 2007–2008)

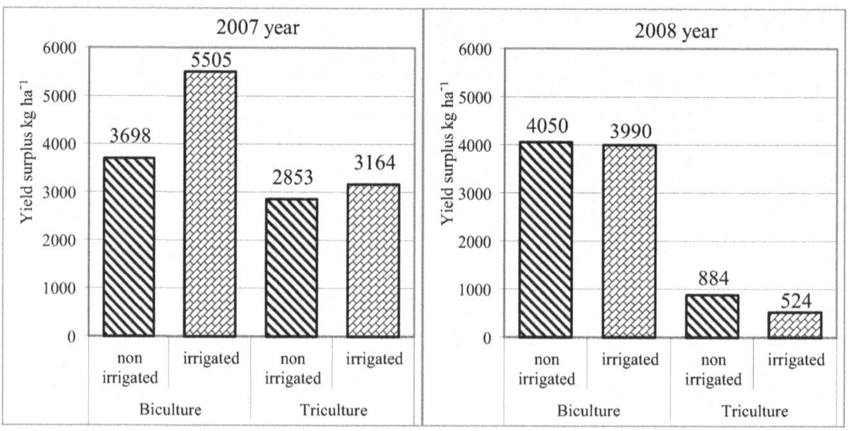

Fig. 24.3 The effect of fertilization on the yield surpluses of winter wheat (Debrecen, 2007–2008)

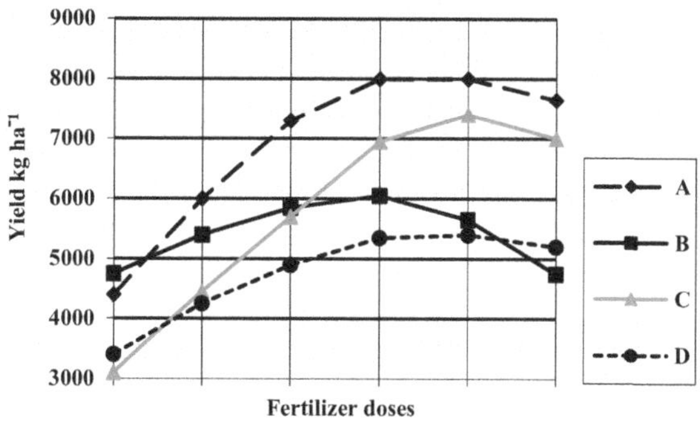

Fig. 24.4 Fertilizer response types of wheat (chernozem soil, continental climatic conditions). A = modern genotype- excellent natural and fertilizer utilization. B = extensive genotype- excellent natural and bad fertilizer utilization. C = intensive genotype- bad natural and excellent fertilizer utilization. D = out of date genotype- bad natural and fertilizer utilization

2007 were 3698 kg ha^{-1} (non-irrigated) and 5505 kg ha^{-1} (irrigated) in bi-culture, while the yield surpluses were significantly lower in tri-culture, 2853 kg ha^{-1} in non-irrigated treatment and 3164 kg ha^{-1} in irrigated treatment, respectively. In 2008, irrigation was not applied, so the crop rotation determined the yield increase caused by fertilization. For 2008, the yield increase caused by fertilization was 4050 kg ha^{-1} (non-irrigated) and 3990 kg ha^{-1} (irrigated) in bi-culture and 884 kg ha^{-1} (non-irrigated) and 524 kg ha^{-1} (irrigated) in tri-culture, respectively.

In sustainable winter wheat production, it is a very important issue to use variety-specific fertilization. It is important to take into consideration both the fertilizer-response of wheat varieties (variety-specific) and the ecological conditions (site-specific) to create sustainable fertilization of wheat. The variety-specific fertilization of wheat involves determination of the natural nutrient utilization of genotypes (yields in control treatment), the efficiency of fertilizer use (grain yield surplus per one unit NPK) XE "grain" , the utilization of genetic yield ability (maximum yield), and the fertilizer demand of genotypes (optimum NPK doses). On the basis of our long-term experiment spanning more than 20 years, the genotypes of wheat could be classified into four groups: modern genotype, extensive genotype, intensive genotype, and out-of-date genotype (Fig. 24.4).

24.4 Conclusions

The effect of crop-year as an indicator of abiotic stress on winter wheat was studied on chernozem soil in a long-term experiment in 2007 and 2008. In a dry year (2007), in non-irrigated circumstances, the maximum yields of winter wheat were 5590 kg ha^{-1} in bi-culture, 7279 kg ha^{-1} in tri-culture, while in the crop-year with favourable water supply (2008), the maximum yields were 900–1500 kg ha^{-1} higher (7065 kg ha^{-1}

in bi-culture, 8112 kg ha^{-1} in tri-culture). In the dry crop-year (2007), in bi-culture the $N_{150-200}$ + PK treatments and in tri-culture the $N_{100-150}$ + PK treatments proved to be optimal, while in the crop-year with favourable water supply, the N_{150} + PK (bi-culture) and N_{50} + PK fertilization treatments proved to be optimal, respectively.

Our long-term experiments proved that, by harmonizing the optimal agro-technical factors (irrigation, crop rotation, fertilization), even in unfavourable, dry crop-years (abiotic stress effect), similar yields can be obtained (8500 kg ha^{-1} in tri-culture, with irrigation in the N_{100} + PK treatment) as in the crop-year with favourable water supply (8100 kg ha^{-1} in tri-culture, with no irrigation in the N_{50} + PK treatment).

The negative effects of abiotic stress in unfavourable crop-years can be reduced and even eliminated, however this requires extremely intensive agro-technical management and high inputs.

Our long-term experiments carried out on chernozem soil proved that the efficiency of fertilization was strongly modified by crop-year. In the crop-years characterized by good water-supply the yield-surpluses of fertilization varied between 3.6 and 4.3 t ha^{-1}. In the crop-years characterized by average water supply, the yield surpluses varied between 2.1 and 3.1 t ha^{-1}. In the crop-years characterized by drought the average yield surplus of the varieties was 0.9 t ha^{-1}. The wheat genotypes tested differed strongly in their natural nutrient utilization, fertilizer-responses, maximum yields and optimum NPK doses. This means that it is very important to use variety-specific fertilization in sustainable wheat production.

Acknowledgements This research was realized in the frame of TÁMOP 4.2.4. A/2-11-1-2012-0001 "National Excellence Program—Elaborating and operating an inland student and researcher personal support system". The project was subsidized by the European Union and co-financed by the European Social Fund.

References

Balla, K., Bedő, Z., & Veisz, O. (2006). Effect of heat and drought stress on the photosynthetic processes of wheat. *Cereal Research Communications, 34*(1), 381–385.

Balogh, Á., & Pepó, P. (2008). Crop-year effects on the fertilizer responses of winter wheat (*Triticum aestivum* L.) genotypes. *Cereal Research Communications, 36*(3), 731–734.

Birkás, M., Dexter, A. R., Kalmár, T., & Bottlik, L. (2006). Soil quality—soil condition—production stability. *Cereal Research Communications, 34*(1), 135–138.

Fowler, D. B. (2003). Crop nitrogen demand and grain protein concentration of spring and winter wheat. *Agronomy Journal, 95*(2), 260–265.

Hornok, M. (2008). Effects of the most important agrotechnical elements on the yield of winter wheat. *Cereal Research Communications, 36*(3), 1243–1246.

Olesen, J. E., & Bindi, M. (2002). Consequences of climate change for European agricultural productivity, land use and policy. *European Journal of Agronomy, 16*(4), 239–262.

Pepó, P. (2004). Effect of crop-year on the yield of winter wheat varieties in long-term experiment. *Növénytermelés, 53*(4), 339–350.

Pepó, P. (2006). Az őszi búza fajtaspecifikus trágyázása. *Gyakorlati Agrofórum, 17*(10), 4–9.

Pepó, P. (2007). The role of fertilization and genotype in sustainable winter wheat (*Triticum aestivum* L.) production. *Cereal Research Communications, 35*(2), 917–920.

Pepó, P., & Győri, Z. (2005). A study of the yield stability of winter wheat varieties. *Cereal Research Communications, 33*(4), 769–776.

Pepó, P., Sipos, P., & Győri, Z. (2005). Effects of fertilizer application on the baking quality of winter wheat varieties in a long term experiment under continental climatic conditions in Hungary. *Cereal Research Communications, 33*(4), 825–832.

Szentpétery, Zs., Jolánkai, M., Kleinheincs, Cs., & Szöllősi, G. (2005a). Effect of nitrogen top-dressing on winter wheat. *Cereal Research Communications, 33*(2/3), 619–626.

Szentpétery, Zs., Hegedűs, Z., & Jolánkai, M. (2005b). Impact of agrochemicals on yield quality and pesticide residues of winter wheat varieties. *Cereal Research Communications, 33*(2/3), 635–640.

Part VI
Management Tools and Assessment: Implementing Nitrogen Management Policies

Chapter 25
DNMARK: Danish Nitrogen Mitigation Assessment: Research and Know-how for a Sustainable, Low-Nitrogen Food Production

Tommy Dalgaard, Steen Brock, Birgitte Hansen, Berit Hasler, Ole Hertel, Nicholas J. Hutchings, Brian H. Jacobsen, Lars Stoumann Jensen, Chris Kjeldsen, Brian Kronvang, Jørgen E. Olesen, Jan K. Schjørring, Torben Sigsgaard, Morten Graversgaard, Fatemeh Hashemi, Katrine Turner, Henrik Vejre, Wim de Vries, and Irene A. Wiborg

Abstract The aim of this chapter is to present the Danish Nitrogen Mitigation Assessment (www.DNMARK.org), an ongoing five-year multidisciplinary research alliance, focusing on the quantification of nitrogen (N) flows and solutions scenarios for a more sustainable N use in Denmark. As one of the world's most agriculturally intensive countries, with a long N regulation history, and state-of-the-art monitoring of developments in key indicators for nitrogen losses, use and efficiency, Denmark is a case of special interest. Based on the results and recommendations from the European Nitrogen Assessment, DNMARK addresses all parts of the N cascade,

T. Dalgaard (✉) · N. J. Hutchings · C. Kjeldsen · J. E. Olesen · M. Graversgaard · F. Hashemi · K. Turner
Department of Agroecology, Aarhus University, Blichers Allé 20, 8830 Tjele, Denmark
e-mail: tommy.dalgaard@agro.au.dk

N. J. Hutchings
e-mail: nick.hutchings@agro.au.dk

C. Kjeldsen
e-mail: chris.kjeldsen@agro.au.dk

J. E. Olesen
e-mail: jeo@agro.au.dk

M. Graversgaard
e-mail: morten.graversgaard@agro.au.dk

F. Hashemi
e-mail: fh@agro.au.dk

K. Turner
e-mail: katrine.grace.turner@gmail.com

S. Brock
Department of Culture and Society, Aarhus University, Jens Chr. Skous Vej 7, 8000 Aarhus C, Denmark
e-mail: filsb@cas.au.dk

M. A. Sutton et al. (eds.), *Just Enough Nitrogen*,
https://doi.org/10.1007/978-3-030-58065-0_25

and demonstrates results both at the landscape scale and the national scale. The national N flows and N balance are accounted for 1990–2010, and methods for the downscaling of these results to regional pilot study areas are developed, together with approaches for the integrated assessment and modelling of the three main types of solution scenarios defined: (i) New production chains with a more efficient use and recycling of N, (ii) Geographically differentiated N-measures implemented by cost-effective instruments with localized planning and management of agricultural landscapes, and (iii) Changed consumption patterns driving land use change and reducing N use.

Keywords Denmark · Nitrogen assessment · Nitrogen flows · Solution scenarios · Sustainable nitrogen management

25.1 Aims and Objective

Allying ten Danish research groups, more than 20 private and public stakeholder partners and key international partners, the overall objective of the DNMARK research alliance is to identify new pathways to significantly reduce nitrogen (N) pollution and increase N efficiency, thus making Denmark a leader in both resource efficient agriculture and mitigation of N-derived impacts from agricultural production on the environment, climate, public health and the economy.

Specifically the aims of the six Research Components (RC1–RC6) are to:

B. Hansen
Geological Survey of Denmark and Greenland (GEUS), Lyseng Allé 1, 8270 Højbjerg, Denmark
e-mail: bgh@geus.dk

B. Hasler · O. Hertel
Department of Environmental Sciences, Aarhus University, Frederiksborgvej 399, 4000 Roskilde, Denmark
e-mail: bh@envs.au.dk

O. Hertel
e-mail: oh@envs.au.dk

B. H. Jacobsen
Department of Food and Resource Economics, University of Copenhagen, Rolighedsvej 25, 1870 Frederiksberg C, Denmark
e-mail: brian@ifro.ku.dk

L. S. Jensen · J. K. Schjørring
Department of Plant and Environmental Sciences, University of Copenhagen, Thorvaldsensvej 40, 1871 Frederiksberg C, Denmark
e-mail: lsj@plen.ku.dk

J. K. Schjørring
e-mail: jks@plen.ku.dk

B. Kronvang
Department of Bioscience, Aarhus University, Vejlsøvej 25, 8600 Silkeborg, Denmark
e-mail: bkr@bios.au.dk

- Develop new methods to analyse time series of national N flows, and the effect of innovative mitigation scenarios on future agricultural production and food/biomass consumption (RC1).
- Assess the potential for locally-targeted N mitigation measures at landscape-scale by analysing and modelling a number of pilot study areas with extensive spatial data coverage (RC2). Provide policy-relevant knowledge about catchment scale policy implementation, and hereby extend the research on cost-effectiveness and implementation of N measures (RC3).
- Enhance the collaboration between the individual Danish N research environments via Ph.D. and post-doc research education on high-impact topics affecting N mitigation (RC4).
- Synthesize results, and communicate with farmers, consumers and the wider public how the detrimental effects of N can be reduced via changes in the management of N in the whole chain from production to consumption of food and bioenergy (RC5, RC6).

25.2 Background and Hypothesis

The availability of industrial N fertilizers led to a large expansion of agricultural production (including livestock), while a reliable reactive N supply remains essential for the yield and stability of agricultural production (Jensen et al. 2011). However, large losses of reactive N species, primarily from agricultural systems, have considerable adverse effects on the environment and human health. The European Nitrogen Assessment (Sutton et al. 2011a, b) estimated the cost of reactive N emissions in Europe to be €70–320 billion per year, which outweighs the direct economic benefits of reactive N in agriculture. The highest costs were associated with reductions in air and water quality, and related health and nature effects, though these estimates are still associated with large uncertainties. The benefits of reducing N loading to

T. Sigsgaard
Department of Public Health, Aarhus University, Bartholin Allé 2, 8000 Aarhus C, Denmark
e-mail: ts@ph.au.dk

H. Vejre
Department of Geosciences and Natural Resource Management, University of Copenhagen, Rolighedsvej 23, 1858 Frederiksberg C, Denmark
e-mail: hv@ign.ku.dk

W. de Vries
Wageningen University and Research (WUR), Droevendaalsesteeg 3, 6708PB Wageningen, The Netherlands
e-mail: wim.devries@wur.nl

I. A. Wiborg
Knowledge Centre for Agriculture, Agro Food Park 20, 8200 Aarhus N, Denmark
e-mail: iaw@seges.dk

improve water quality to comply with the Water Framework Directive have been estimated in Danish case studies (Hasler et al. 2009), indicating a positive benefit-cost ratio for many but not all catchments. Achieving cost-efficient mitigation of N losses is therefore highly sensitive to spatial targeting, the choice of policy instruments and farmers' responses to these (Beharry-Borg et al. 2012).

There are many different forms of reactive nitrogen (e.g., NH_3, NO_3, NO, N_2O and NO_x) that move through biogeochemical processes. This implies that one atom of N may take part in many environmental processes before it is immobilised or finally converted back to N_2. Referred to as the nitrogen cascade (Galloway et al. 2003), this explains why policy measures targeting one N species (e.g., nitrate or ammonia) may have large positive or negative effects on other species (e.g., nitrous oxide). These strong inter-linkages require a holistic approach to solve problems related to excess reactive N (Rygnestad et al. 2002).

Denmark has already implemented a number of measures to reduce losses of reactive N to the environment (Dalgaard et al. 2011a, 2014; Kronvang et al. 2008), but there is still a need to make substantial further reductions. If major further reductions are to be achieved cost-effectively, it is essential to include measures that control the flows of N between agriculture and society (including urban waste) and between agriculture and other ecosystems (including the harvesting of biomass for bioenergy use). Since the flows of reactive N display large spatial variations, such measures will also need a stakeholder engagement process and landscape or catchment approach in order to maximize their efficiency.

Our central hypothesis is that the cycling of reactive N can be significantly improved through targeted measures at national, landscape and farm management scales, and that the design of policies to promote such measures requires a comprehensive understanding of the cycling of reactive N and its holistic impact on ecosystems and socio-economy at national and landscape scales. A clear understanding of the effectiveness of the incentive structures used to implement measures is crucial, as is the need to understand the influence of the social and institutional context of regulation. The development and use of this knowledge will help prioritise new, innovative measures and technologies for dealing with the N problem, thereby minimizing the costs and maximizing the benefits. *We thereby hypothesise significantly higher cost-effectiveness of targeted regulation that takes into account the local areas' vulnerability to the specific N pollution.*

In the Danish Nitrogen Mitigation Assessment (DNMARK) alliance, we believe that interdisciplinary research and the integration of the whole range of public and private stakeholders into the research chain are key to the identification of solution pathways for a more sustainable N management and utilisation. This must include two-way communication of research results, to be tested in the real world. In this way DNMARK can move the frontier of interdisciplinary research, and facilitate the improvements needed in N management for a sustainable bioeconomy development.

Policy targets and solution scenarios are used in the DNMARK project to structure discussion with stakeholders and promote collaboration between the constituent Research Components. The three policy targets are to reduce losses of: (1) NO_3^- to comply with the EU Water Framework Directive, (2) NH_3 to comply with the Clean

Air for Europe nature and health targets, (3) N_2O by 50%, to contribute to national and international commitments for greenhouse gas (GHG) emissions. In addition, we focus on *three contrasting solution scenarios*:

1. New production chains with more efficient use and recycling of N,
2. Geographically differentiated N-measures based on intelligent planning and management of agricultural landscapes, and
3. Changed consumption patterns driving land use change and reducing N use.

25.3 Methodology and Results

As illustrated in Fig. 25.1, the development of methodologies for sustainable N management are divided into the five Research Components (RC1–RC5), and a coordination and synthesis component (RC6).

25.3.1 RC1 National Nitrogen Model

In this component a national Danish nitrogen model (inputs, outputs and losses) for the period 1990–2010 is constructed. The model aims to quantify all major terrestrial and aquatic flows (Leip et al. 2011) using the *top-down* method developed during the European Nitrogen Assessment. The model will subsequently be refined in response to stakeholder comments and the consequences of solution scenarios. Subsequently, an alternative national scale model for the agricultural sector will be developed using a *bottom-up* approach based on farm-scale data and an existing farm N model (Happe et al. 2011) to allow a robust comparison with the top-down approach. Revised national N budgets for 1990–2010 and future national N budgets

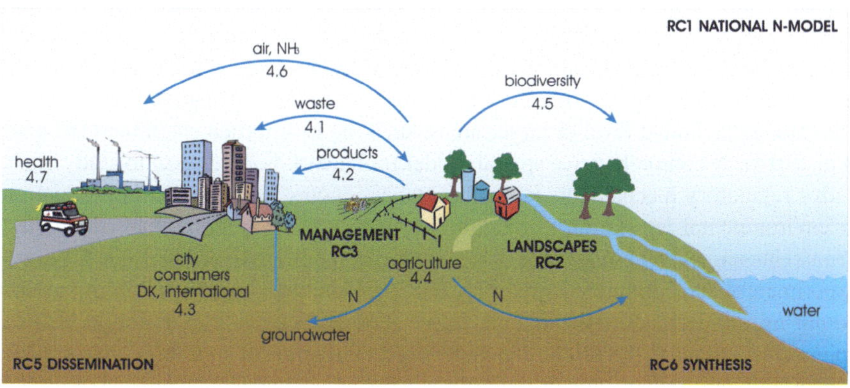

Fig. 25.1 Focus areas for the DNMARK research components RC1–RC6

reflecting the consequences of solution scenarios will be included in the Danish Nitrogen Assessment (RC6).

25.3.2 RC2 DNMARK Landscapes

In collaboration with local municipalities and farmers' unions, case landscapes/watersheds with both low and high NO_3^- retention are inventoried with farm and landscape data in GIS (Dalgaard et al. 2011b). A detailed N budget is constructed using inputs from RC1 and more detailed local data.

Local scenarios for the case landscapes are formulated where N reduction effects of various changes in landscape, agricultural practice and technical installations are modelled. Forecasting scenarios allow the prediction of effects of changes on open trajectories, whereas back-casting scenarios allow optimization of landscape and farm management tools with fixed aims in terms of N loss reduction (Bende-Michl et al. 2011). The scenarios are formulated with stakeholders and in an iterative process with RC3, RC4.1–4.7 and RC5, ensured by a sequence of scenario-building followed by local workshops. In the last part of the RC we evaluate the extent of other externalities brought about by the various N mitigation efforts. Selected ecosystem services (wildlife habitats, flood control, cultural heritage, recreation etc.) are mapped in a baseline scenario, and the development in the provision of services assessed for additional scenarios (Andersen et al. 2013). The assessment requires development of new methodologies, in close collaboration with municipal stakeholders and includes assessment of the global effect of local land use change.

25.3.3 RC3 Management Strategies

Scenarios for cost-effective N reductions to allowable loads for freshwater and marine waters are formulated with input data from RC1 and RC2. Furthermore, cost-effective spatial distribution of measures to achieve the N load reductions are modelled by applying existing cost minimization models at national level and at pilot study catchment level or larger areas, such the Limfjorden catchment (Konrad et al. 2014). A comprehensive spatial model framework is developed to study alternative regulatory mechanisms (subsidies, taxes, quotas, spatial zoning, production or environmental legislation), modelling farm behaviour as individual optimizing firms (Hansen and Hansen 2012). The model outcomes are tested and refined using experimental data (Beharry-Borg et al. 2012) to test the extent to which the model output mimics behavioural outcome using alternative economic assumptions and approaches. Finally, the studies include implementation of measures that require cooperation between agents, as their effectiveness is dependent on scale and spatial adjacency. Spatially defined subsidy schemes and agglomeration bonus schemes are investigated in a Payment for Ecosystem Services (PES) modelling framework. This includes a special focus on construction of wetlands, buffer strips and watercourse maintenance, which needs collaboration between farmers at sub-catchment level. We use a heuristic optimization approach building on agent-based modelling frameworks (Touza et al. 2012). Promising PES schemes are evaluated using experiments in workshops (Beharry-Borg et al. 2012) where farmers evaluate the potential for implementing such schemes in the agri-environmental policy mix. The result are compared to results from real cases where such payment schemes have been implemented, e.g., the Swiss example where farmers receive bonus payments when their fields are part of habitat networks (Wätzold and Drechsler 2011).

25.3.4 RC4 Critical Nitrogen Impact Issues

This RC focuses on gaps in our current N knowledge, and comprises in-depth studies of critical N issues in relation to a sustainable agriculture and food production. These issues have been identified by the alliance partners as of key importance to the quantification of N flows or to the mitigation of N losses in Denmark, and were also prioritized in the European Nitrogen Assessment report as research needs. Moreover, each of these research education projects adds to the core competences of the research alliance members supervising the projects:

RC4.1 *Urban-rural N recycling from waste*: There is scope for increased recycling of N in urban waste residuals (WR) from new emerging technologies for municipal solid waste and waste water treatment, e.g., biosolids, composts, struvite precipitates, digestate (Svirejeva-Hopkins et al. 2011). We are screening a range of WR together with the industry and for a subset quantify fertilizer value, improvement options, medium to long-term effects on soil quality and emissions (laboratory and field tests in long-term trials).

RC4.2 *Cost-benefit assessment of N measures to improve surface water quality*: The aim is to bridge costs analyses and benefit analyses in order to answers questions like: What is the optimal water quality in a given catchment from a cost-benefit viewpoint? How to proceed to cover the whole country, where local benefit analyses are not possible? How much can the costs of implementation be reduced if on site specific N-reduction potential can be obtained (Jensen et al. 2013)?

RC4.3 *Sustainable, low-N food consumption*: It is investigated whether the dual aims of reducing environmental N loss and reducing the protein share of healthy diets are congruent or conflicting, and on which scale. The work will estimate food demand component of the N map nationally and selected local areas, determine trends and drivers for consumption of Danish food products–nationally and internationally and analyse alternative interventions to change food demand behaviour at the local level.

RC4.4a *Watershed N Management*: Managing N in more sustainable ways is an important and challenging task that must be synthesized by science and society in forms that are useful for policymakers, farmers and society in general. This involves bringing different disciplines and stakeholders together at different geographical scales, from national, watershed, landscape to local farm scale and with different stakeholder involvement processes in sustainable nitrogen management (Graversgaard et al. 2014, 2016). Going from farm to watershed based N management shows significant potentials for increased productivity combined with lower N-losses (Dalgaard et al. 2011b). Areas vulnerable to N-losses are selected from RC2, and new watershed management concepts are developed together with local stakeholders and farm advisory services. This local embedded type of knowledge when combined with biophysical knowledge of nitrogen and watershed processes is important to identifying new and more geographical targeted solutions to environmental problems (Voinov and Gaddis 2008). The aim of RC4.4a is to involve multiple stakeholders at multiple scales in the management of N. For this, new sustainable N management models and participatory concepts are being developed. Through a series of scenario workshops, in different test watersheds, an N-bio-physical model is being integrated with inputs from stakeholders to further enhance the resource effectiveness of N usage. The work is identifying the potentials for developing a fully integrated socio-biophysical model.

RC4.4b *Assessing Spatially Differentiated Nitrogen Mitigation in Agriculture:* Developing and implementing a new, targeted and differentiated regulation of agricultural use of nitrogen (Natur- og Landbrugskommissionen 2013), and improved management of N in agriculture is seen as necessary to achieve a sustainable balance between the production of food and other biomass, and the unwanted effects of N on water pollution (Dalgaard et al. 2012). According to Article 16 of EC Regulation No.746/96, all EU member states are obliged to monitor and evaluate the environmental, agricultural, and socio-economic impacts of their agri-environment programs. Therefore in order to provide policy makers with the necessary information for responsible political action, research on the possible environmental and economic impacts of different N-mitigation strategies and regulation in catchment scale is essential. Thus the aim is to develop methodologies for spatial estimation of N-leaching for different mitigation options in Denmark and the Baltic Sea region. This is based on a review study of existing N mitigation options, land use and land management scenarios. Assessment of the spatially differentiated N measures is being done by comparison of different methods to estimate N leaching. Ecological and economic efficiency of spatially differentiated measures are being assessed by establishing spatially differentiated scenarios of N management for selected catchments.

RC4.5 *Nitrogen mitigation, Ecosystems Services mapping and biodiversity management*: Ecosystem services (ES) are in this context a way of understanding the ecological resources in the landscape and clarify the important processes and products we depend on from a functional nature (e.g., wastewater treatment, recreation opportunities and food production) (Turner et al. 2016). The intensive land use, combined with high nitrogen emissions have a heavy impact on nature, and affect the functionality of ecosystem services (Tilman et al. 2002). We focus on how composition and distribution of ecosystem services (ES) correlates with, among others, N mitigation options, N vulnerable areas, and agricultural production, to quantify synergy effects between N mitigation and biodiversity protection (Dise et al. 2011). Spatial distributions of the proposed ES and N management are analysed, and based (among others) on the method of Turner et al. (2014) compared with the effects in different scenarios.

RC4.6 *Agricultural airborne N-pollution, particle pollution and public health effects*: The aim is to assess the contribution from agricultural N-emissions to negative health effects from ambient air particle exposure of the Danish population. This is based on state-of-the-art source apportionments and exposure assessment as a basis for epidemiological study with health register data, using a GIS approach. The integrated system approach, based on impact-pathway is adjusted to assess the health-related economic externalities of agricultural air pollution (Brandt et al. 2011) based on the refined DNMARK dataset.

RC4.7 *Groundwater N-pollution and public health effects*: The aim is to assess the contribution from N polluted groundwater to negative health effects on the Danish population. This is based on an epidemiological study of people exposed to nitrate containing drinking water and assess the incidence of cancer (e.g., colon cancers; van Grinsven et al. 2010) by combining drinking water quality data (Hansen et al. 2011) with health register data using a GIS approach.

25.3.5 RC5 Stakeholder Involvement and Dissemination

Local dissemination of the DNMARK results and solution scenarios is being developed and tested in the local landscapes including meetings with local farmer groups, where the agenda is to identify measures at farm level that might improve resource efficiency and climate change adaptation, including carbon (C) and N footprint. Through the utilisation of cognitive mapping, and the facilitation of learning processes between the multiple stakeholders at local development workshops, action plans for sustainable N-management strategies at local and regional level are being produced, allowing results to be disseminated through the national advisory services.

25.3.6 RC6 Management and Synthesis

Annual solution scenarios workshops are used to facilitate stakeholder integration and crosscutting research and dissemination activities. The general solution scenario pathways were defined during the inception phase (Year 0) and form the basis for the work in RC1–RC5. The national scale baseline and preliminary scenario results (Year 1) feed into more specific landscape scenarios (Year 2), management mitigation options to be quantified and discussed in Year 3, and the effects of the RC4 specific key N-topics to be synthesized in Year 4. These workshops and the specified RC deliveries ensure that results feedback to the final Year 5 synthesis scenarios to be developed both at landscape and national scales in the final DNMARK assessment, and the continuous dissemination of results (Fig. 25.2).

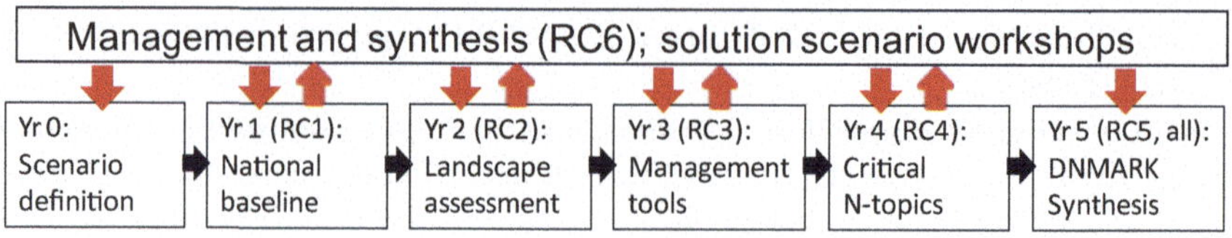

Fig. 25.2 Development of solution scenarios during the series of annual project meetings organized by RC6 as a cross cutting activity for stakeholder involvement, research activity coordination, and dissemination of results (co-organised by the Research Components (RCs) mentioned in brackets). Yr = year

25.4 Examples of Results and Relevance to Stakeholders

Major stakeholder-relevant results of the DNMARK programme are summarized in Table 25.1.

25.5 Conclusions

Agricultural food and biomass production are the main sources of reactive nitrogen pollution, causing N concentrations in air and water exceeding critical levels for eutrophication, significant green-house gas emissions, landscape and biodiversity deterioration, and severe human health impairments. In parallel, N is the main limiting factor to increased agricultural productivity. Many research-based N mitigation measures have already been implemented in Danish agriculture, yet N pollution and the related costs for society and the food sector remain unacceptably high. Future societal development is likely to require N pollution to be significantly reduced while it is likely that global food and biomass production will need to increase. Thus, innovative, cross-sectoral solutions to reduce N losses and ally public and private stakeholders are crucial for the development of a sustainable biobased economy.

The DNMARK cross-disciplinary research alliance is identifying barriers and developing research-based solutions to meet these N challenges, emphasizing both costs and benefits of different development pathways. In this integrated project with core private and public partners we are focusing on three main solution scenarios: (i) New production chains with a more efficient use and recycling of N, (ii) Geographically differentiated N-measures based on intelligent planning and manage-ment of agricultural landscapes, and (iii) Changed consumption patterns driving land use change and reducing N use.

For the first time, a consistent Danish framework for N flows is being established, along with landscape study sites and economic evaluation models. Ph.D. research studies are focusing on critical N issues of relevance to the participating private and public stakeholders. Project management and dissemination activities will ensure that results are synthesized and disseminated nationally and internationally.

Table 25.1 Main DNMARK results and relevance to selected private and public stakeholders

Output	Relevance to stakeholders
Solution scenarios to increased N efficiency, and significantly reduced N-footprint	It is recognized by Danish agriculture and major agribusinesses that significant improvements in N utilisation is crucial for the further development of the sector The Danish government needs to comply with EU directives and international treaties, and development of a targeted strategy to reduce N-pollution is demanded
The first full, dynamic model for N flows in Denmark	It is attractive for international research institutions to test new N-models in Denmark, and develop world leading agro-environmental databases. This is also important for documenting effects of national N-mitigation initiatives
Landscape platforms to test effects of local N mitigation initiatives	Locally adapted N-mitigation solutions are important to municipalities' mandatory spatial planning and sustainable development strategies. New planning tools are needed, to integrate the full range of ecosystems services and the comparative, often disparate, advantages for local agribusinesses and rural development
Models to assess cost–effectiveness of targeted and voluntary N measures	The societal costs of Danish N pollution amounts to billions of Danish Krona (DKK) year^{-1} (van Grinsven et al. 2013), and cost-effectiveness analyses and modelling of N measures are of high priority to the Danish Government. The development of models for more efficient, location-specific instruments, including voluntary actions and improved collaboration between agents, is required to develop the bio-economy
New insights from post-docs and Ph.D. studies in high-impact N topics	High-priority areas for further N research include the development of new urban-rural N recycling technologies, water policy cost-benefit analyses, sustainable food consumption and healthy lifestyles, agricultural watershed management and optimal N use, Ecosystem Services Assessment, and public health effects of air and water N pollution
Dissemination and synthesis of knowledge	Via significant investments the Danish farm advisory services and agroindustry intend to integrate project results into new, more holistic local advisory services for the sustainable development of farming and food production in Denmark

With the Danish Nitrogen Mitigation Assessment (DNMARK), we aim to develop Denmark's international position in this area, and bring together Danish research environments, dealing with the N problem in the production and consumption chains of food and bioenergy. More than 20 public and private stakeholder partners and key international partners are actively involved, and contributing to a fruitful process of building the alliance and creating innovative research. Further information can be found at www.dnmark.org.

Acknowledgements We thank The Danish Council for Strategic Research/Innovationsfonden (Ref. 12-132421) and the Aarhus University Research Foundation for funding the www.dnmark.org Strategic Research Alliance, presented in this present paper.

References

Andersen, P. S., Vejre, H., Dalgaard, T., & Brandt, J. (2013). An indicator-based method for quantifying farm multifunctionality. *Ecological Indicators, 25,* 166–179.

Beharry-Borg, N., Smart, J. C. R., Termansen, M., & Hubaek, K. (2012). Evaluating farmers' likely participation in a payment program for water quality protection in the UK uplands. *Regional Environmental Change, 13*(3), 633–647.

Bende-Michl, U., Volk, M., Newham, L., Harmel, D., & Dalgaard, T. (2011). Monitoring strategies and scale appropriate hydrologic and biogeochemical modelling for natural resource management. *Environmental Modelling and Software, 26,* 538–542.

Brandt, J., Silver, J. D., Christensen, J. H., Andersen, M. S., Bønløkke, J. H., et al. (2011). *Assessment of Health-Cost Externalities of Air Pollution at the National Level using the EVA modelling system.* CEEH Scientific Report No. 3. Centre for Energy, Environment & Health Report series. Roskilde, Denmark: CEEH.

Dalgaard, T., Bienkowski, J. F., Bleeker, A., Dragosits, U., Drouet, J. L., Durand, P., et al. (2012). Farm nitrogen balances in six European landscapes as an indicator for nitrogen losses and basis for improved management. *Biogeosciences, 9,* 5303–5321.

Dalgaard, T., Hansen, B., Hasler, B., Hertel, O., Hutchings, N. J., Jacobsen, B. H., et al. (2014). Policies for agricultural nitrogen management—Trends, challenges and prospects for improved efficiency in Denmark. *Environmental Research Letters, 9,* 115002.

Dalgaard, T., Hutchings, N., Dragosits, U., Olesen, J. E., Kjeldsen, C., Drouet, J. L., et al. (2011a). Effects of farm heterogeneity and methods for upscaling on modelled nitrogen losses in agricultural landscapes. *Environmental Pollution, 159,* 3183–3192.

Dalgaard, T., Olesen, J. E., Petersen, S. O., Petersen, B. M., Jørgensen, U., Kristensen, T., et al. (2011b). Developments in greenhouse gas emissions and net energy use in Danish agriculture–How to achieve substantial CO_2 reductions? *Environmental Pollution, 159,* 3193–3203.

Dise, N. B., Ashmore, M., Belyazid, S., Bleeker, A., Bobbink, R., de Vries, W., et al. (2011). Nitrogen as a threat to European terrestrial biodiversity. In M. A. Sutton, C. M. Howard, J. W. Erisman, G. Billen, A. Bleeker, P. Grennfelt, H. van Grinsven, & B. Grizzetti (Eds.), *The European Nitrogen Assessment: Sources, effects and policy perspectives (Chap. 20)* (pp. 463–494). Cambridge: Cambridge University Press.

Galloway, J. N., Aber, J. D., Erisman, J. W., Seitzinger, S. P., Howarth, R. W., Cowling, E. B., et al. (2003). The nitrogen cascade. *BioScience, 53,* 341–356.

Graversgaard, M., Dalgaard, T., Kjeldsen, C., Wiborg, I. A., & Olesen, J. E. (2014). *New ways of organizing sustainable management of nitrogen in agriculture—An adaptive framework for stakeholder involvement*. IARU International Scientific Congress on sustainability, Copenhagen, October 22–24, 2014.

Graversgaard, M., Thorsøe, M. H., Kjeldsen, C., & Dalgaard, T. (2016). Evaluating public participation in Denmark's water councils: How policy design and boundary judgements affect water governance. *Outlook on Agriculture, 45*(4), 225–230.

Hansen, L. B., & Hansen, L. G. (2012). Can non-point phosphorus emissions from agriculture be regulated efficiently using input-output taxes? IFRO Working Paper, No. 2012/4. Denmark: University of Copenhagen, Department of Food and Resource Economics (IFRO).

Hansen, B., Thorling, L., Dalgaard, T., & Erlandsen, M. (2011). Trend reversal of nitrate in Danish groundwater. *Environmental Science and Technology, 45*, 228–234.

Happe, K., Hutchings, N., Dalgaard, T., & Kellermann, K. (2011). Modelling the interactions between regional farming structure, nitrogen losses and environmental regulation. *Agricultural Systems, 104*, 281–291.

Hasler, B., Brodersen, S. L., Christensen, L. P., Christensen, T., Dubgaard, A., et al. (2009). *Denmark: Assessing economic benefits of good ecological status under the EU water Framework Directive. Testing practical guidelines in Odense River Basin. Case study report.* Denmark: Aquamoney.

Jensen, C. L., Dubgaard, A., Jacobsen, B. H., Olsen, S. B., & Hasler, B. (2013). A practical CBA-based screening procedure for identification of river basins where the costs of fulfilling the WFD requirements may be disproportionate—Applied to the case of Denmark. *Journal of Environmental Economics and Policy, 2*(2), 164–200.

Jensen, L. S., Schjoerring, J. K., van der Hoek, K. W., Poulsen, H. D., Zevenbergen, J. F., Pallière, C., et al. (2011). Benefits of nitrogen for food, fibre and industrial production. In M. A. Sutton, C. M. Howard, J. W. Erisman, G. Billen, A. Bleeker, P. Grennfelt, H. van Grinsven, & B. Grizzetti (Eds.), *The European Nitrogen Assessment: Sources, effects and policy perspectives (Chap. 3)* (pp. 32–61). Cambridge: Cambridge University Press.

Konrad, M. T., Andersen, H. E., Thodsen, H., Termansen, M., & Hasler, B. (2014). Cost-efficient reductions in nutrient loads; identifying optimal spatially specific policy measures. *Water Resources and Economics, 7*, 39–54.

Kronvang, B., Andersen, H. E., Børgesen, C. D., Dalgaard, T., Larsen, S. E., Bøgestrand, J., et al. (2008). Effects of policy measures implemented in Denmark on nitrogen pollution of the aquatic environment. *Environmental Science and Policy, 11*, 144–152.

Leip, A., Achermann, B., Billen, G., Bleeker, A., Bouwman, L., et al. (2011). Integrating nitrogen fluxes at the European scale. In M. A. Sutton, C. M. Howard, J. W. Erisman, G. Billen, A. Bleeker, et al. (Eds.), *The European nitrogen assessment* (Chap. 16, pp. 345–376). Cambridge: Cambridge University Press.

Natur- og Landbrugskommissionen. (2013). *Natur- og Landbrugskommissionen Final Report.* Anbefalinger fra Natur-og Landbrugs Kommissionen. Danish Government Commission on Nature and Agriculture. http://www.naturoglandbrug.dk/slutrapport_2013.aspx?ID=52071. Accessed January 19, 2016.

Rygnestad, H., Jensen, J. D., Dalgaard, T., & Schou, J. S. (2002). Cross-achievements between policies for drinking water protection. *Journal of Environmental Management, 64*, 77–83.

Sutton, M. A., Howard, C., Erisman, J. W., Billen, G., Bleeker, A., Grennfelt, P., et al. (Eds.). (2011a). *The European nitrogen assessment*. Cambridge: Cambridge University Press.

Sutton, M. A., Oenema, O., Erisman, J. W., Leip, A., van Grinsven, H., & Winiwarter, W. (2011b). Too much of a good thing. *Nature, 472*, 159–162.

Svirejeva-Hopkins, A., Reis, S., Magid, J., Nardoto, G., Barles, S., Bouwman, A. F., et al. (2011). Nitrogen flows and fate in urban landscapes. In M. A. Sutton, C. M. Howard, J. W. Erisman, G. Billen, A. Bleeker, P. Grennfelt, H. van Grinsven, & B. Grizzetti (Eds.), *The European nitrogen assessment (Chap. 12* (pp. 249–270). Cambridge: Cambridge University Press.

Tilman, D., Cassman, K. G., Matson, P. A., Naylor, R., & Polasky, S. (2002). Agricultural sustainability and intensive production practices. *Nature, 418*, 671–677.

Touza, J., Drechsler, M., Smart, J. C. R., & Termansen, M. (2012). Emergence of cooperative behaviours in the management of mobile ecological resources. *Environmental Modelling and Software, 45,* 52–63.

Turner, K. G., Anderson, S., Gonzales-Chang, M., Costanza, R., Courville, S., Dalgaard, T., et al. (2016). Towards an integrated assessment of land degradation and restoration: methods, data, and models. *Ecological Modelling, 319,* 190–207.

Turner, K., Odgaard, M. V., Bøcher, P., Dalgaard, T., & Svenning, J. C. (2014). Bundling ecosystem services: Trade-offs and synergies in a cultural landscape. *Landscape and Urban Planning, 125,* 89–104.

Van Grinsven, H. J. M., Holland, M., Jacobsen, B. H., Klimont, Z., Sutton, M. A., & Willems, W. J. (2013). Costs and benefits of nitrogen for Europe and implications for mitigation. *Environmental Science and Technology, 47,* 3571–3579.

Van Grinsven, H. J. M., Rabl, A., & Kok, T. M. (2010). Estimation of incidence and social cost of colon cancer due to nitrate in drinking water in the EU: a tentative cost-benefit assessment. *Environmental Health, 9,* 58. https://doi.org/10.1186/1476-069X-9-58.

Voinov, A., & Gaddis, E. J. B. (2008). Lessons for successful participatory watershed modeling: A perspective from modeling practitioners. *Ecological Modelling, 216,* 197–207.

Wätzold, F., & Drechsler, M. (2011). *Agglomeration payment, agglomeration bonus or homogeneous payment?* Conference presentation, "PES and their institutional dimension" conference, Berlin, November 10th, 2011.

Part VII
Management Tools and Assessment: Manure Management

Chapter 26
Farm Level Assessment of Nitrogen Use Efficiency as Part of Environmental Management

Aleksandr Briukhanov, Eduard Vasilev, Natalia Kozlova, Dmitry Maksimov, Ekaterina Shalavina, and Igor Subbotin

Abstract Analysis of statistical data and field studies has revealed the problems associated with agricultural manure utilization in the North-West Federal District of Russia. Large-scale livestock farms face a difficult task as the majority of them do not have sufficient agricultural land to utilize the amount of animal and poultry manure produced. This surplus results in accumulation of animal and poultry manure near the farms, and increases the risk of nutrient discharge into the water sources and ammonia emissions into the atmosphere. Nitrogen (N) flux control is considered a part of environmental management at farm level. It includes environmental assessment of agricultural enterprises based on the nutrient use efficiency (nitrogen budgets) and assessment of manure handling technologies on the basis of Best Available Technique (BAT) approaches. Results of the environmental assessment of large-scale livestock enterprises in the Leningrad Region are presented and show that the actual distance of organic fertilizer transportation (R_{ac}) exceeds the cost-effective distance (R_{ce}) for all types of farms (cattle farms, pig farms, poultry factories). The results suggest that animal/poultry manure processing and its use as organic fertilizer is unprofitable unless financial support systems for the introduction of agricultural technologies with minimal nutrient (nitrogen, in the first place) loss are in place. According to

A. Briukhanov (✉) · E. Vasilev · N. Kozlova · D. Maksimov · E. Shalavina · I. Subbotin
Federal State Budgetary Scientific Institution "Federal Scientific Agroengineering Center VIM", Branch "Institute for Engineering and Environmental Problems in Agricultural Production (IEEP)", Saint Petersburg, Russian Federation
e-mail: sznii@yandex.ru

E. Vasilev
e-mail: sznii6@yandex.ru

N. Kozlova
e-mail: Natalia.kozlova@sznii.ru

D. Maksimov
e-mail: maximov@sznii.ru

E. Shalavina
e-mail: shalavinaev@mail.ru

I. Subbotin
e-mail: itmo1652@mail.ru

M. A. Sutton et al. (eds.), *Just Enough Nitrogen*,
https://doi.org/10.1007/978-3-030-58065-0_26

preliminary estimates, if such financial support were to be put in place, the input of nitrogen into the environment may be reduced by up to 30%.

Keywords Manure management · Ecology · Nitrogen efficiency · Economic incentives

26.1 Introduction

According to the priority directions and programmes adopted in Russia, the step-by-step development of agriculture is based on intensification of production. Intensive development involves construction of large-scale livestock and poultry complexes and, consequently, greater risk of negative environmental impact. In livestock and poultry farming the major source of hazard is the system of animal/poultry manure processing and application. At the same time, the crop growing subsector in Russia requires high-quality organic fertilizers, to the extent that even if animal and poultry stock were to increase substantially, all the manure produced may not be sufficient to meet the demand (Briukhanov and Volkov 2009).

The experience from intensively developing regions, such as the Leningrad Region, for example, demonstrates the fact that even under the positive balance between livestock and crop production there might arise serious environmental problems with animal/poultry manure utilization. It results, in the first place, from low agricultural production profitability in general and the inability of farms to make independent decisions on the introduction of state-of-the-art and reliable technologies for manure handling (Briukhanov et al. 2013; Afanasiev et al. 2012). The survey shows that important factors, which have a negative effect on environmentally sound utilization of animal/poultry manure, are as follows:

- Insufficient knowledge of agricultural producers on the most reliable, economically and ecologically substantiated technologies for animal/poultry manure handling.
- Non-observance of the technological regulations for animal/poultry manure processing and organic fertilizers application.
- Inefficient coordination by the executive authorities of the location of newly built livestock complexes and reconstruction of operating complexes in terms of their environmental impact (processing technologies and logistics of animal/poultry manure use).
- Lack of economic incentives to introduce the practices of environmentally safe processing of animal/poultry manure; usually such practices are cash consuming and have a material effect on the self-cost of products, especially those of pig and poultry farms, which have no agricultural land of their own.

Therefore, to ensure environmentally safe agricultural farming in the North-West of Russia, in the Leningrad Region in particular, the attention should be directed to addressing the above problems.

The main trend in animal/poultry manure use today, and also in the foreseeable future, is its application as a high-quality organic fertilizer to improve the soil fertility. Technologies, such as biogas generation via anaerobic digestion, for the generation of heat and electrical energy are considered only in exceptional circumstances due to the specific features of the Russian legislation and economy.

Of great importance are the sanitary and hygienic properties of the fertilizers produced (absence of pathogens and odour) as they directly affect the air, surface and ground water pollution, as well as agrochemical indices such as nutrient content, absence of weed seeds, and homogeneous structure.

The basic technologies for animal/poultry manure processing and use, which are recommended and applied in the North-West of Russia, the Leningrad Region in particular, are as follows (Afanasiev et al. 2012; Popov et al. 2015; Maximov et al. 2014):

- For bedding (solid) animal/poultry manure: preparation of solid organic fertilizers (composts) using moisture absorbing materials (peat, straw, wood waste, etc.).
- For semi-liquid and liquid manure and manure effluents: long-term storage or separation into the solid and liquid fractions with the subsequent composting of the solid fraction and the long-term storage of the liquid fraction.
- As an exception, when an agricultural enterprise has absolutely no land for manure application, a possible option is an integrated treatment of the liquid fraction to reach the discharge standard onto the filtration fields.
- To produce the organic fertilizers with enhanced properties from the poultry manure a bio-fermentation technology in chamber or drum-type bioreactors may be applied.

It should be noted, however, that currently the Russian market offers a wide range of manure handling technologies, machines and equipment. This makes it difficult for farmers to choose the technologies and machines most appropriate for the conditions on a particular farm. To partially address this concern, the institute has designed an online database of technologies, machines and equipment for animal/poultry manure processing. The database was designed with the support of several international projects.[1] It includes both the above technologies and a number of other practices [biogas generation, combustion, integrated cleaning, etc. (Foged 2010)].

The database supports the final decision-making and is created on the basis of expert knowledge in the form of a data model and algorithms for choosing the technology options. Using a number of previously entered farm operation factors, such as type and number of livestock, availability and amount of land for organic fertilizer application, etc., the system suggests the techniques most suitable for each case and calculates the consolidated economic indices of each offered technique. This system is available online to all interested parties at http://eco.sznii.ru.

[1]Listed in the **Acknowledgments** at the end of this chapter.

26.2 Methods

Increased efficiency of organic fertilizer production from animal and poultry manure and lower nitrogen losses can be achieved by using the scientifically proven methods and good machine-based practices for organic fertilizer production and soil application. These tasks may be completed with the use of optimization methods, which enable the production of organic fertilizers with minimal costs under specific farm conditions (Briukhanov and Volkov 2009). The ecological safety factors for the technology applied and the mechanical setup are the critical criteria in technology selection. Currently, it is suggested that the ratio between the volume of nitrogen supplied to plants and the volume of manure nitrogen (N) at the farm exit be used to calculate an ecological safety factor (Briukhanov and Volkov 2009; Popov et al. 2015):

$$\kappa ecs = \frac{Q_N^1}{Q_N} \tag{26.1}$$

where

κecs = ecological safety factor;
Q_N^1 = amount of nitrogen supplied to plants (%);
Q_N = amount of nitrogen in fresh manure (%).

In order to determine the amount of nitrogen supplied to plants, the nitrogen losses from the technological operations were estimated and the critical spots in the

Table 26.1 Nitrogen losses from technological operations of manure processing

Technological operation index number	Description of a technological operation	Nitrogen losses (%)
1	Transportation	1–2
2	Intermediate storage	2–5
3	Rough separation	2
4	Fine separation in settlement tanks	2–5
5	Biological cleaning	15
6	Mixing with additives	2
7	Composting	10–25
8	Drying	5–20
9	Storage	5–20
10	Application	5–15

technological chain of nitrogen losses were identified (Table 26.1) (Bryukhanov and Volkov 2009).

The results show that the main focus of the subsequent work to reduce the nitrogen losses should be on enhancing the technologies for manure composting, storage and application of produced organic fertilizers.

Investigations carried out by IEEP (Popov et al. 2015) revealed that nutrient flow control at regional and farm level is an important tool to reduce the nitrogen and phosphorus (P) inputs to the environment. With this aim in mind, the institute has elaborated the decision-making guidelines for the local executive agencies, responsible for agricultural development, on the siting of new livestock complexes and modernization of existing ones in terms of their environmental impact (processing technologies and logistics of animal/poultry manure use) (Fig. 26.1). This methodological approach guides the initiation of the coordination of activities aimed at ensuring the ecological safety of agricultural production at regional level.

The offered scheme describes the interaction between the two stakeholders: an agricultural enterprise (investor) and the Regional Agriculture Committee or Agriculture Ministry, responsible for the development of agriculture. Planning for the construction of a new enterprises or the capacity increase of existing agricultural enterprises is carried out in the following sequence:

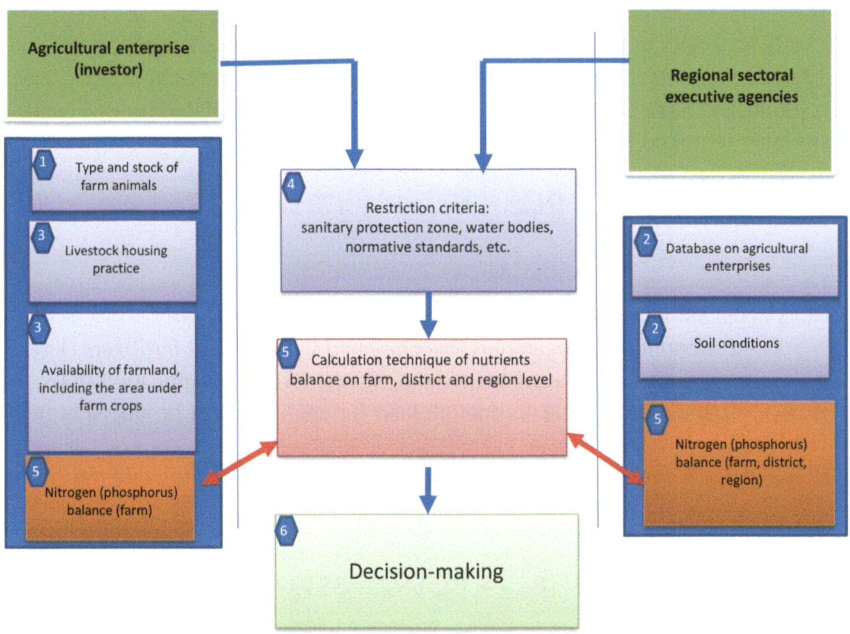

Fig. 26.1 The block scheme of decision-making on environmentally sound location and operation of animal/poultry farms

1. Specification of farm profile and production capacity (the initial data source is the agricultural enterprise).
2. Choosing the site for construction (the initial data source is the agricultural enterprise and the Regional Agriculture Committee or Agriculture Ministry).
3. Specification of space-and-layout design and technical and technological options (the initial data source is the agricultural enterprise).
4. Checking the compliance with the restriction criteria, such as sanitary protection zone, water bodies, normative standards, etc. (the initial data source is the agricultural enterprise and the Regional Agriculture Committee or Agriculture Ministry).
5. Comparison of nutrients (N, P) balance in the designed enterprise and N and P balance in the district selected for the farm location (the initial data source is the agricultural enterprise and the Regional Agriculture Committee or Agriculture Ministry).
6. Decision-making on the farm location or production expansion. In the case of N and/or P surplus in the designed enterprise and it not being economically viable at the district(s) level, it is necessary to revise the results of actions 1, 2, and 3.

The complex index of environmental impact of a livestock farm (N_{bal}, P_{bal}) is an estimated value of N and P balance with due account of available land for their application. The Baltic Marine Environment Protection Commission (also known as the Helsinki Commission—HELCOM) requires the application limit to be below 170 kg N ha^{-1} and 25 kg P ha^{-1}. These values were used in the calculations.

To determine this index the amount of nitrogen and phosphorus in organic fertilizers produced on the considered agricultural enterprise is calculated per one hectare of available farmland for this enterprise.

$$N_{bal} = \left(N_{prod} + N_{pur}\right)/S_{av} - 170\ \text{kg ha}^{-1} \quad (26.2)$$

where

N_{prod} is the amount of nitrogen in the organic fertilizer (kg);
N_{pur} is the amount of nitrogen purchased/applied with mineral fertilizers (kg);
S_{av} is the cultivated area, available for the organic fertilizer application (ha).
If $N_{bal} > 0$, the enterprise lacks enough land for application of produced organic fertilizers.
If $N_{bal} < 0$, the enterprise has enough land to use the produced organic fertilizers.

$$P_{bal} = \left(P_{prod} + P_{pur}\right)/S_{av} - 25\ \text{kg ha}^{-1} \quad (26.3)$$

where

P_{prod} is the amount of phosphorus in the organic fertilizer (kg);
P_{pur} is the amount of phosphorus purchased/applied with mineral fertilizers (kg).

Table 26.2 Nitrogen and phosphorus content in organic fertilizers

Organic fertilizer source	N (kg/t)	P (kg/t)
Cattle manure	4.8	0.8
Pig manure	2.4	0.7
Poultry manure	10	2.1

If $P_{bal} > 0$, the enterprise lacks enough land for produced organic fertilizers application.
If $P_{bal} < 0$, the enterprise has enough land to use the produced organic fertilizers.

To calculate the tentative nitrogen and phosphorus content in organic fertilizers produced by the basic technologies it is suggested to use the following values from Table 26.2 (based on our own measurements).

It should be noted that along with the ecological indices (nutrients balance) an economic component of manure production and organic fertilizer application should be taken into consideration.

The basis for ecological and economic substantiation of processing and use of animal/poultry manure as an organic fertilizer can be assessed using indicators such as specific capital expenditures, specific operating expense, ecological effect of organic fertilizer use and the cost-effective transportation distance of organic fertilizers (R_{ce}).

Table 26.3 presents some tentative values for specific operating expenses of processing and transportation of organic fertilizers within the cost-effective transportation distance (R_{ce}) for the basic technologies for animal/poultry manure processing. The cost-effective transportation distance (R_{ce}) meets the condition, when the received additional (net) profit (ecological effect from the use of organic fertilizers) exceeds the expenditures on processing and transportation of fertilizers. The net profit from the use of organic fertilizers is determined as the extra yield cost minus related harvesting costs.

26.3 Results

Specific operating expenses on processing and transportation of organic fertilizers were calculated accounting for: depreciation deductions on renovation of machines and equipment; labour costs with social welfare deductions; the costs associated with maintenance, repairs of machines and equipment, buildings, and structures; fuel and electricity costs.

Transportation distances longer than the cost-effective ones result in higher operational costs, negative economic effects on organic fertilizers use, and ultimately, in higher self-cost of the end products. Analysis of the agricultural enterprises in the Leningrad Region shows that the actual distance of organic fertilizers transportation (R_{ac}) exceeds the cost-effective distance (R_{ce}) for all the different types of

Table 26.3 Some tentative indicators of technologies for animal/poultry manure processing

Processing technologies	Cattle			Pigs		Poultry	
	Specific operational expenses on processing and transportation of manure, (RUR/t)		Cost-effective transportation distance (km)	Specific operational expenses on processing and transportation of manure (RUR/t)	Cost-effective transportation distance (km)	Specific operational expenses on processing and transportation of manure (RUR/t)	Cost-effective transportation distance (km)
	Liquid	Solid					
Passive composting	–	375	7–8	–	–	650	32–33
Composting with regular aeration	–	370	7–8	–	–	645	32–33
Disinfection through the long-term storing	320	–	7–8	320	4–5	–	–
Separation into solid and liquid fractions with subsequent composting of the solid fraction and the long-term storing of the liquid fraction	390	–	10–11	290	7–8	–	–
Biofermentation in chamber bioreactors	–	–	–	–	–	1215	43–44
Biofermentation in drum-type bioreactors	–	–	–	–	–	1500	43–44

farm (cattle farms, pig farms, poultry factories). Therefore, animal/poultry manure processing and its use as an organic fertilizer becomes non-profitable.

Table 26.4 presents the initial data for the formulation of economic incentives for manure processing and the application of the resulting organic fertilizers, which are based on reimbursement of transportation costs when the actual transportation distance exceeds the cost-effective one.

The data from Table 26.4 were used to calculate the expenses, which have the negative effect on the economic efficiency of the use of organic fertilizers. Calculation results are shown in Table 26.5.

Table 26.5 shows that expenses per hectare of fertilized area vary from 2400 to 8400 rubles, which is not profitable due to the long transportation distances.

26.4 Discussion and Conclusions

The above approaches are relevant to Russia as currently significant reforms in national environmental legislation are in progress. On 21st July 2014, the Federal Law of the Russian Federation No. 219-FZ "Concerning the Introduction of Amendments to the Federal Law on Environmental Protection and Certain Legislative Acts of the Russian Federation" was adopted and came into force on 1st January 2015. This Law provides for the introduction of Best Available Techniques (BATs) and also reduces the payment for the negative environmental impact by the amount of actual expenses on environmental measures introduced.

The regulatory acts in force (Management Directive for Agro-Industrial Complex 2008) include the requirements on compliance of availability of agricultural land to the amount of manure produced on a farm and on compliance of fertilizer application rates to the types of farm crops grown and soil nutrient content. In practice, these requirements are not always observed (Maximov et al. 2014).

The manure management issues were considered here through the example of the Leningrad Region. Calculation of nutrient balance for typical farms in the Leningrad Region (Afanasiev and Kozlova 2011) has shown the nitrogen surplus to be below 60 kg ha^{-1} $year^{-1}$, indicating the relatively low potential environmental risk from the agricultural activity on the farms under consideration. The farms have sufficient land to utilize the amount of manure produced. The values of nitrogen use efficiency (NUE) are found to be at the lower limit of the range for this class of mixed productions, $NUE \leq 0.3$ (Bittman et al. 2014). The basic potential for more efficient nutrient use in such farms is in plant growing through the use of all the nutrients from manure.

One of the main problems in the region under consideration is the high concentration of production facilities in one area and, consequently, the large amounts of manure that require transportation before they can be used as fertilizer.

On average in the Leningrad Region, with its highly developed agriculture, the animal density is less than three farm animal units per 1 km^2, i.e., less than one animal unit per ha of arable land. At the same time, the average farm size in the Leningrad Region is 1300 head for cattle farms and 1,500,000 head for poultry factories. There

Table 26.4 Initial data to substantiate the need for economic motivation of manure processing and use of organic fertilizers

Organic fertilizer source	Specific operating expenses (RUR/t)	Recommended application rate (t/ha)	Cost-effective transportation distance (R_{ce}) (km)	Actual transportation distance (R_{ac}) (km)	Cost of ton-km of organic fertilizer, (RUR/t/km)
Cattle manure	350	35	9	12	30
Pig manure	300	35	7	15	30
Poultry manure	800	12	40	45	40

Table 26.5 Substantiation of economic motivation of manure processing and organic fertilizers use

Organic fertilizer source:	R_{ac}–R_{ce} (km)	Transportation costs (RUR/t)	Per hectare expenses, with the account for application rates (RUR/ha)
Cattle manure	3	90	3150
Pig manure	8	240	8400
Poultry manure	5	200	2400

are large-scale pig complexes with the capacity of over 25,000 head and the medium-scale farms for 6000 pigs in the region. So in a large-scale farm the distance from the manure storing facility to the field may be up to 50 km.

It is well known that in order to stipulate the practical introduction of environmental standards at farm level a number of countries (Denmark, Switzerland) (Oenema et al. 2011; Menzi et al. 2014) use certain economic instruments. Based on the technical and economic analysis results shown in Tables 26.3, 26.4 and 26.5 the proposals for financial support of manure handling operations were drafted for the specific conditions of the region under consideration (Briukhanov 2012).

Major ecological and economic benefits of the organic fertilizers produced from animal/poultry manure are the higher quality characteristics of farmland soils and higher crop yields. That is why it seems reasonable to consider the proposed subsidies in the framework of the Russian Federation Government Resolution No. 869 dated 4th October 2013 "On approval of subsidy allocation rules from the federal budget to the budgets of the Russian Federation subjects on rendering decoupled income support to agricultural producers in the field of crop production".

According to this Resolution, the subsidies for decoupled income support for agricultural producers in the sphere of crop production are financed from the budgets of the Russian Federation (local budgets), including the funds from the Federal budget by way of co-financing. The co-financing is for compensating a portion of expenditures on agro-technological work packages, raising the environmental safety of farming, increasing the soil fertility and quality per hectare of farm crop area. Thus, financial support systems for the introduction of agricultural technologies with the minimal nutrient (nitrogen, in the first place) loss will substantially reduce the negative environmental impact of farming. According to preliminary estimates the input of nitrogen into the environment may be reduced by up to 30%; in the Leningrad Region it may amount to 5000 tons of nitrogen per year.

Acknowledgements Part of the above research was funded from EU-HELCOM projects "Baltic Hazardous Waste and Agricultural Releases Reduction—BaltHazAR" (2009—2013) and "Implementation of the HELCOM Baltic Sea Action Plan—BASE project" (2013–2014) and the international project "Sustainable animal and poultry manure handling in farms of Leningrad Oblast, Russia", realized by Nordic Environment Finance Corporation (NEFCO) as Implementing Agency and Leningrad Region Government, with Northern Dimension Environmental Partnership (NDEP), the Ministry of the Environment of Finland (FMoE), the Ministry of Agriculture and Forestry of Finland (FMoAF) as Financiers (2011–2014).

References

Afanasiev, A. V., & Kozlova, N. P. (2011). Farm N balance as a tool of environmental assessment. *Bulletin of All-Russian Research Institute of Mechanization of Livestock Production, 22*(3), 197–202. [In Russian].

Afanasiev, A. V., Maximov, D. A., Afanasiev, V. N., & Briukhanov, A Yu. (2012). Substantiation of environmentally sound technologies of animal and poultry manure processing on the farms in the North-Western Federal District of the Russian Federation. *Agroecological Journal, 2,* 32–35. [In Russian].

Bittman, S., Dedina, M., Howard, C. M., Oenema, O., & Sutton, M. A. (Eds.). (2014). *Options for ammonia mitigation: Guidance from the UNECE task force on reactive nitrogen.* Edinburgh, UK: Centre for Ecology and Hydrology.

Briukhanov, A. Yu. (Ed.). (2012). *Recommendations on arrangement of full-scale ecological monitoring and control of animal and poultry manure processing and application systems. The order of development of Technological Regulations.* Saint Petersburg: SZNIIMESH [In Russian].

Briukhanov, A. Yu., & Volkov, A. N. (2009). Ecological and economic assessment of technologies for organic fertilizer production from animal/poultry manure. *Technologies, machines and equipment for mechanized plant and livestock production. Collection of Research Papers* (pp. 166–170). Saint Petersburg: SZNIIMESH [In Russian].

Briukhanov, A Yu., Afanasiev, V. N., & Kozlova, N. P. (2013). Substantiation of environmentally friendly location and operation of animal and poultry farms in Leningrad Region. *Bulletin of All-Russian Research Institute of Mechanization of Livestock Production, 4*(12), 93–101. [In Russian].

Foged, H. L. (2010). *Best available technologies for manure treatment—For intensive rearing of pigs in baltic sea region EU Member States.* Stockholm: Baltic Sea 2020.

Management Directive for Agro-Industrial Complex (RD-APK 1.10.15.02-08). (2008). *Recommended Practice for Engineering Designing of Animal and Poultry Manure Removal Systems and the Systems of Animal and Poultry Manure Preparation for Application.* Moscow: Ministry of Agriculture of the Russian Federation [In Russian].

Maximov, D. A., Bryukhanov, A. Yu., Subbotin, I. A., & Vasilyev, E. V. (2014). Selecting manure management technologies to reduce ammonia emissions from big livestock farms in the North-western Federal District of Russia. In K. W. Van der Hoek & N. P. Kozlova (Eds.), *Ammonia workshop 2012 Saint Petersburg. Abating ammonia emissions in the UNECE and EECCA region* (pp. 161–168). RIVM Report 680181001/SZNIIMESH Report. Bilthoven, The Netherlands: RIVM.

Menzi, H., Kupper, T., & Achermann, B. (2014). Ammonia emissions in Switzerland. In K. W. Van der Hoek & N. P. Kozlova (Eds.), *Ammonia workshop 2012 Saint Petersburg. Abating ammonia emissions in the UNECE and EECCA region* (pp. 99–107). RIVM Report 680181001/SZNIIMESH Report. Bilthoven, The Netherlands: RIVM.

Oenema, O., Bleeker, A., Braathen, N. A., Budňáková, M., Bull, K., et al. (2011). Nitrogen in current European policies. In M. A. Sutton, C. M. Howard, J. W. Erisman, G. Billen, A. Bleeker, et al. (Eds.), *The European nitrogen assessment* (pp. 62–81). Cambridge University Press.

Popov, V., Maksimov, D., & Briukhanov, A. (2015). Approaches of agro-engineering ecology to reduction of adverse environmental impact of agricultural practices. In V. D. Popov (Ed.), *Environmentally Friendly Agriculture and Forestry for Future Generations: Proceedings of International Scientific XXXVI CIOSTA & CIGR SECTION V Conference, 26–28 May, 2015* (pp. 288–290). Saint Petersburg: Saint-Petersburg State Agrarian University.

Part VIII
Management Tools and Assessment: Multi-species and Agroforestry Systems

Chapter 27
Agroforestry and Opportunities for Improved Nitrogen Management

Gerry Lawson, William J. Bealey, Christian Dupraz, and Ute M. Skiba

Abstract Multi-species systems include herbaceous mixtures (cover cropping, living mulches, intercropping), woody mixtures and herbaceous-woody mixtures, i.e., "agroforestry". Agroforestry systems have particular potential to improve nitrogen (N) availability to both the tree and herbaceous components, and to increase N-conservation and recycling. This is achieved through the following: (a) deeper and more extensive root distributions, (b) greater root and shoot turnover, (c) N-fixation in either tree or crop roots, (d) facultative mycorrhizal associations, (e) greater interception of light and water resources, (f) improved soil structure and organic matter content, and (g) control of erosion and leaching. Ammonia recapture by tree-foliage from livestock emissions reduces N deposition on nearby sensitive ecosystems, and contributes to reductions in net N_2O emissions. Trees, especially if they are not N-fixers, are also likely to reduce soil N_2O emissions by drying the soil, increasing soil-aeration, and removing excess nitrate concentrations throughout most of the year. Agroforestry is identified as one of the best options for climate change mitigation and adaptation, but more measurements are required on fluxes of N_2O and CH_4 from agroforestry systems of different type and age.

Keywords Herbaceous mixtures · Woody mixtures · Agroforestry · Root safety-net · Ammonia recapture · Climate-change mitigation · Climate-smart agriculture

---

G. Lawson (✉) · W. J. Bealey (✉) · U. M. Skiba
UK Centre for Ecology and Hydrology, Edinburgh Research Station, Bush Estate, Penicuik, Midlothian, EH26 0QB, UK
e-mail: gerrylawson2@gmail.com

W. J. Bealey
e-mail: bib@ceh.ac.uk

U. M. Skiba
e-mail: ums@ceh.ac.uk

C. Dupraz
INRAE, UMR, ABSys, University of Montpellier, 2 Place Viala, 34060 Montpellier, France
e-mail: Christian.dupraz@inra.fr

M. A. Sutton et al. (eds.), *Just Enough Nitrogen*,
https://doi.org/10.1007/978-3-030-58065-0_27

27.1 Introduction

The terrestrial nitrogen (N) cycle involves many biological and non-biological processes such as mineralisation, nitrification, denitrification and nitrogen fixation. These processes are mediated by the physical and chemical properties of soils, meteorological conditions, ecosystem type and management methods and results in an altered distribution of nitrogen to plant and microbial biomass, and changes in gaseous losses to the air, leaching and erosion.

Trees can affect the nitrogen cycle in many ways, for example, they improve the physical, chemical and biological properties of the soil, and they recycle nitrogen which would otherwise be lost from the system. Land use systems in which trees are grown in combination with agriculture on the same land are termed "agroforestry" (European Commission 2013). A recent comparison of the 20 "greening" options available to EU Member States within the direct payments regime of the Common Agricultural Policy, gave agroforestry the highest rating (Warner et al. 2016). A recent study demonstrated that around 45% of the world's agricultural land has a tree crown-cover greater than 10% (Zomer et al. 2016). Yet the term "agroforestry" is little known amongst farmers outside the tropics, mainly due to policy disincentives (Lawson et al. 2016). Perhaps a greater understanding of its potential role in supplying nitrogen to crops, and in buffering nitrogen pollution from farmland, will encourage the wider uptake in temperate regions?

This chapter is in two halves, focusing on the twin facets of nitrogen cycling—firstly as a vital nutrient, and secondly as a damaging pollutant.

Section 27.2 considers the role of nitrogen as a nutrient in three types of multi-species system:

- herbaceous mixtures—where one of the species is a nitrogen fixer, normally a legume, and the species are mixed in time and/or space;
- woody mixtures—where two or more components in the mixture are trees or shrubs, such as in forestry mixtures or energy coppice;
- herbaceous-woody mixtures (agroforestry)—where one or more elements in the mixture are trees, and others are either a crop or pasture species, and where the system is purposefully managed to ensure more efficient utilisation of light, water and nutrient resources than in comparable monocultures.

Section 27.3 considers the potential abatement of nitrogen pollution in agroforestry systems, in terms of erosion control, retention of aqueous nitrogen and reduction of atmospheric N emissions.

27.2 Increased Nitrogen Supply in Mixtures

27.2.1 Herbaceous Mixtures

Cover cropping with grasses or legumes is a long established method for weed control and enhancement of fertility in both temperate and tropical areas (Erenstein 2003; Hartwig and Ammon 2002; Malezieux et al. 2009). Cover crops can reduce erosion, increase soil organic matter, control weeds, reduce soil compaction, fix nitrogen, and scavenge for nitrogen. In areas where excess nitrogen is already a problem they can provide a sink for nitrogen and hold it for other crops to use during the next growing season. Even legumes tend to use soil nitrogen rather than fixing their own, if it is available (Hooda et al. 1998).

Not all cover crop species provide the same benefits, and herbaceous species are often mixed together for different purposes. For example, forage radish (*Raphanus sativus* var. *longipinnatus*) can suppress weeds and reduce compaction in autumn, but, because it is killed by frost, it does not provide a living root system or cover in the spring. Red clover (*Trifolium pratense*) captures atmospheric nitrogen, but may not suppress weeds when seeded in the heat of summer without a companion species. Cereal rye (*Secale cereale*) can stop nitrogen from leaching, but may deprive the following cash crop of nitrogen. Some species do not perform well in mixtures because they are too competitive, and complementary species need to be selected based on knowledge of their phenology, growth forms, nitrogen acquisition strategies, weed suppression and attractiveness to beneficial insects and pollinators. All farmers have different needs, and increasing use is being made of computerised databases of species information to enable them to select appropriate mixes (OSCAR Project 2016; White et al. 2016).

Living mulches are another type of cover crop, but they grow alongside the main crop. They are planted either before the main crop or with it and are maintained as a living ground cover throughout the growing season. Examples of nitrogen-fixing species are birdsfoot trefoil (*Lotus corniculatus* L.) or crownvetch (*Coronilla varia* L.). If the living mulch is a perennial, it may be possible to maintain it from year to year without the need for reseeding (Mohammadi 2012). Living mulches are commonly used in vineyards and orchards, and have been used with a variety of crops, where they can bring yield and biodiversity benefits (Pelosi et al. 2009; Villenave et al. 2009).

Intercropping is the growing of two or more crop species simultaneously. Andrews and Kassam (1976) identified four types: (a) mixed intercropping, with no distinct row arrangement; (b) row intercropping; (c) strip intercropping and (d) relay intercropping, where the growth cycles of the two species overlap to some degree. Intercropping is a regular practice in the tropics and provides many benefits for smallholder farmers. Yields can be expressed as Land Equivalent Ratios (LERs), i.e., the ratio of intercrop-yield to the combined yield of the crops grown separately (Mead and Willey 1980). Land Equivalent Ratios of maize and bean mixtures in the tropics of up to 2 have been reported, although alley cropping or relay intercropping is much less successful if growth is limited by water supply rather than nitrogen (Sanchez 1999). In temperate conditions, intercrops are seldom beneficial if grain production is the main requirement (Crookston and Hill 1979), but can make economic sense when high quality silage or forage is required. Benefits include: enhanced supplies of nitrogen and phosphorus, ground cover, water use efficiency and resistance to pests and diseases (Anil et al. 1998). Temperate legumes are often grown together and Mikić et al. (2012) reported LERs of up to 1.4. They suggest four main principles for these legume-intercrops: (a) same time of sowing, (b) similar growing habit, (c) similar cutting time, and (d) one component has good standing ability (supporting crop) and another is susceptible to lodging (supported crop).

Carbon and nitrogen mineralisation kinetics are intimately linked in herbaceous mixtures, and they often lead to higher soil quality, greater microbiological activity and enhanced nutrient availability compared with conventional farming. Rotations are versatile and additional organic fertilizers can be added (Hansen et al. 2001; Marinari et al. 2010; Shannon et al. 2002). The increase in microbial biomass and activity under organic management leads to increased mineralisation and nutrient availability for plants (Tu et al. 2006; Wang et al. 2003).

The rhizosphere is a favourable environment for microbial processes, including nitrification and denitrification which lead to NO/N_2O emissions and thereby losses of N that could otherwise be used by the plants. Legumes are a reliable source of mineral N for plants but also for nitrifiers/denitrifiers. This can lead to larger NO/N_2O emissions than from a non-legume system, but emissions will be less if nitrate is actively absorbed by living roots. The benefits of legumes and agroforest systems (i.e., less synthetic N requirement, higher organic matter possibly leading to higher yield) should be balanced against the potential increase in NO/N_2O emissions.

In summary, herbaceous mixed cropping systems, based on careful choice of species, can have: (a) higher overall productivity, (b) better control of pests and diseases, (c) enhanced ecological services, and (d) greater economic profitability. However, increased emissions of the greenhouse gas N_2O are a concern, and research is needed on the relationship of yield to NO/N_2O emissions, and on the potential role of agroforestry tree-roots, together with cover-crops in reducing overall greenhouse gas (GHG) emissions.

27.2.2 Woody Mixtures

The great majority of timber plantations are monocultures, but there are advantages to be gained using species mixtures. These can include enhanced biodiversity, reduced risks of pest attack, greater resistance to physical damage (e.g., wind), adaptability to adverse conditions and climates, and enhanced overall productivity. Yield gains are most apparent when the canopies and root systems have different phenologies and geometries (Cannell et al. 1992).

One of the most common scenarios is to grow a valuable non-nitrogen-fixing species alongside a N-fixing species. This technique is primarily used in low fertility tropical or subtropical plantations (Bouillet et al. 2013), partially because of the wider range of N-fixing leguminous trees available (Wormald 1992), but it also has a role also in temperate forestry. Some examples are as follows:

- Trials and reviews have confirmed the positive effect of mixtures with N-fixing species on the diameter growth of non N-fixing tree species but the yield advantage depends on soils, silviculture and species (Debell et al. 1997; Erskine et al. 2006; Kelty 2006; Piotto 2008; Redondo-Brenes and Montagnini 2006). Most of the benefit is gained through enhanced levels of ammonium and nitrate in the soils, but there is evidence of direct transfer of nitrogen between the roots of N-fixing trees and those non-fixers (Paula et al. 2015).
- A review of 18 studies of eucalyptus mixed with N-fixing tree species F (Forrester et al. 2012) suggested that mixtures were only advisable if: (a) eucalyptus overtopped the more shade tolerant N-fixing species, (b) rates of cycling of N and P were significantly higher than in monocultures, and (c) the leaf litter of the N-fixers was able to decompose rapidly.
- Binkley et al. (2003) found that at 10-year old *Eucalyptus saligna* had greater total biomass in monocultures than in mixtures with the nitrogen-fixing tree *Falcataria moluccana*, but that individual eucalyptus trees in the mixtures had larger diameters. By year 20, 65% of the eucalyptus biomass in the mixed plots was in trees over 30 cm dbh (diameter at breast height), whereas it was only 35% in the monoculture.
- Mixtures of poplar and alder (N-fixing) for short rotation coppice have been investigated in Europe (Teissier du Cros et al. 1984; Van der Meiden 1961) and North America (Heilman and Stettler 1985; Radwan and DeBell 1988) as potential bio-energy crops on farmland and have showed significant yield advantages on poorer soils.
- Herbaceous nitrogen-fixing ground cover species, like perennial lupins (*Lupinus* spp.) are regularly used to assist the establishment of tree plantations on sandy soils, moraines, and land restoration sites (Marrs et al. 1982; Turvey and Smethurst 1983).
- In temperate regions there are fewer nitrogen-fixing leguminous trees to use in forestry, with false acacia (*Robinia pseudoacacia*) (Rédei et al. 2008) and honey locust (*Gleditsia triacanthos*) as the main options. The latter forms N-fixing "bacteroids" in root-hairs rather than root-nodules (Bryan et al. 1996). There are also 24 species of trees or shrubs which form nodular associations with the *Frankia* genus of actinomycetes—of which *Alnus* and *Casuarina* are the main genera (Baker 1990; Russo 2005).
- Binkley (2005) reviewed data on mixtures of timber species and N-fixing species of *Alnus and Ceanothus* in temperate locations and Leucaena, Falcataria, and Casuarina in the tropics. Rates of N accretion averaged 7.3 g N m^{-2} $year^{-1}$ across all studies, with a range from 0.8 to 15.3 g N m^{-2} $year^{-1}$. Overall, these soils accumulated about 12–15 g of carbon (C) for every g of N accumulated— this

increase in soil-carbon was not seen with an equivalent addition of inorganic nitrogen fertilizer.

Coppice-with-standards is a traditional silvicultural method in many parts of Europe, combining timber production from widely spaced standard trees with fuel-wood or other products from coppice-stools (Dupouey et al. 2002). Nitrogen-fixing alder coppice is an option for the coppice species. When the standard trees are in regular rows the system is sometimes called "alley coppice" (Tosi et al. 2014). These spatial- and species-mixtures of woody species will have greater biodiversity than monocultures of short-rotation coppice, and greater resilience in the fact of disease attack (Morhart et al. 2014).

27.2.3 Herbaceous-Woody Mixtures (Agroforestry)

The growth in diameter of trees is often increased in agroforestry, particularly on good agricultural soils and when the trees are in close proximity to crops rather than left on uncultivated fallow (Cardinael et al. 2015a; Chifflot et al. 2005; Thevathasan et al. 2016). Crop growth can also be enhanced in these systems (Lorenz and Lal 2014; Sanchez 1995), and this is usually linked to an increase in the quantity and quality of soil organic matter (SOM) and increased uptake of nitrate and ammonium from the soil, and reduced losses of N from the system (Fig. 27.1 and Sect. 27.3). So, for example, pines are frequently grown amongst tree-lupins (*Lotus uliginosus*) in New Zealand (Gadgil et al. 1986; Knowles 1991; West 1997) to increase nitrogen supply, and common walnut (*Juglans regia*) is traditionally intercropped with N-fixing forage legumes in France (Dupraz et al. 1998). Intercropping increases the nitrogen content and growth of the walnut trees, particularly in years with average or above average rainfall. Conversely, the beneficial impact of agroforestry systems on the yield of some nitrogen fixing grain crops such as chickpea has been observed (Mahieu et al. 2015). Whether a nitrogen fixing tree or crop component will increase the overall yield will depend on whether the system is predominantly limited by nitrogen or water supply (Kass et al. 1997). Two linked factors affect the retention and uptake of N in agroforestry systems: SOM and the availability of soil nitrate and ammonium.

27.2.3.1 Increased Soil Organic Matter

Silvoarable systems, where the trees are planted in rows with annual or perennial crops in alleys, have been shown to significantly increase SOM compared with crop-monocultures (Kumar and Nair 2011; Lorenz and Lal 2014), particularly if cover-crops are also used in the alleys (Canet and Schreiber 2015). The combined input of leaf and root litter from silvoarable systems into SOM pools is larger in silvoarable systems than in monocultures since the total input of leaf and root litter is greater,

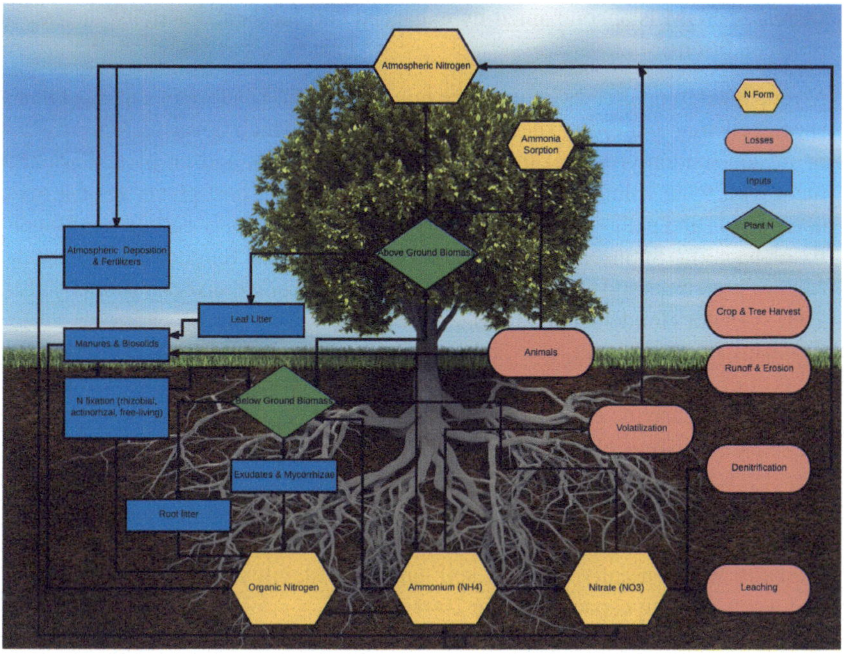

Fig. 27.1 Simplified nitrogen cycle in an agroforestry system showing the linkages between plant N, N form, inputs and losses. Crop growth can be enhanced in these systems, there can be an increase in the quantity and quality of soil organic matter (SOM). There can also be sorption of gaseous ammonia by the tree canopy, and reduced losses of N from the system via increased uptake of nitrate and ammonium from the soil by the tree root "safety-net (see chapter text for more details)

more evenly distributed through the year, and occurs to a greater depth (Chenu et al. 2016). While agroforestry plots may have lower total tree-root biomass than forest plots, the roots of individual trees penetrate very widely. For example Dupraz et al. (1995), in the south of France, found that a plot of 100 trees ha^{-1} of 10 m height and 6 m crown diameter fully colonised the surface soil with roots. The projected area of tree roots at the soil surface was four times that of the projected area of the crown.

Silvopastoral systems, where trees are planted into permanent pasture, also tend to increase SOM but the increase only becomes apparent once the trees are well established, and requires measurements to be made under the rooting layer of grasses (Haile et al. 2008; Howlett et al. 2011). Soil-carbon content in the upper horizons requires pasture production to be maintained by regular grazing or mowing (Delmer 2015; Ferreiro-Dominguez et al. 2016; Upson et al. 2016). In addition to being deposited in deeper soil layers, the carbon from tree roots is held in organic fractions which decompose more slowly than those from dead grass-roots. Hence the total potential for long-term soil-carbon sequestration of silvopastoral systems is likely to be higher than in either permanent grasslands or in forests (Baah-Acheamfour et al. 2015; Beckert et al. 2016; Sharrow and Ismail 2004).

Increased SOM is accompanied by enhanced soil biological activity, at least in aerobic conditions:

- Mycorrhizal fungi greatly improve the mineral nutrition and water supply of many species of tree and crop. They develop mycelium which cover (ectomyrrhizae) or invade (endomyrrhizae) the fine roots of both trees and crops and connect them with wide volumes of soil. They can also protect host plants against toxic compounds and pathogens (Cameron et al. 2013), and supply carbon to nitrogen-fixing bacteria (Sponseller et al. 2016). Endomycorrhizal fungi often colonise the roots of both trees and crops, especially species of *Glomeromycetes* (Ingleby et al. 2007; van der Heijden et al. 2015).
- Exudates from trees can have both positive and negative effects on crops. Thus roots of many tree species will provide a habitat from which *Glomus* endomycorrhizae can escape the effects of a fungicide excreted by rape roots, and recolonise subsequent sowings of wheat or barley (Dodd 2000). Whereas the leaves of some tree species (e.g., poplar or walnut) produce soluble phenolic compounds which inhibit the development of a number of arbuscular mycorrhizal fungi species (Piotrowski et al. 2008). Endomycorrhizal fungi synthesise glycoproteins like glomalin which greatly improve soil aggregate stability, and therefore fertility and water-holding capacity (Rillig 2004).
- Nitrogen-fixing bacterial symbionts, whether rhizobial or actinorhizal species, are important in agroforestry, and there is much evidence that the fixed atmospheric nitrogen is used by the non-fixing components of both silvopastoral (Daudin and Sierra 2008; Goh et al. 1996; Plath et al. 2011) and silvoarable (Mahieu et al. 2015; Nygren et al. 2012) systems. Intercropping of olive and perennial legumes is common in Mediterranean areas (Anglade et al. 2015) and helps to both lower N fertilizer needs and reduce soil erosion (Vallebona et al. 2016). Alfalfa (*Medicago sativa* L.) is an important perennial forage species which gives high yields even under rainfed conditions (Mantino et al. 2016), and which is traditionally used in agroforestry systems (Querné et al. 2017).

27.2.3.2 Increased Ammonium and Nitrate Uptake

The accumulation of ammonium in the soil is affected by the comparative rates of the following: (a) mineralisation of organic nitrogen in the soil, or application of ammonium containing fertilizers and manures, (b) uptake of ammonium by microbes as a source of nitrogen for growth, (c) uptake of ammonium by plants as a source of nitrogen for growth, (d) volatilisation of ammonia, (e) nitrification (biological conversion of ammonium to nitrate), (f) loss of nitrate by leaching (which increases the rate of nitrification), and (g) plant uptake of nitrate as a source of nitrogen for growth (which increases the rate of nitrification) (Fig. 27.1). Therefore, a low concentration of ammonium in the soil does not indicate low mineralisation rates but may indicate high rates of nitrification, volatilisation or microbial and plant uptake. Vegetation with a high C:N ratio will have a lower rate of nitrogen mineralisation.

Nitrification is the oxidation of ammonium nitrogen to nitrites and nitrates, and is carried out predominately by a range of chemoautotrophic (or chemolithotrophic) microorganisms. The two primary nitrifying bacteria genera are *Nitrosomonas* and *Nitrobacter. Nitrosomonas* carries out the oxidation of ammonium to nitrite to obtain energy and *Nitrobacter* oxidizes nitrite to nitrate for the same purpose. Both genera use CO_2 as their sole carbon source for growth as they are obligate autotrophs and strict aerobes. In addition heterotrophic nitrification by fungi can be of importance in low-N soils, such as forests (Kuroiwa et al. 2011).

The effectiveness of agroforestry trees in absorbing soil ammonium and nitrate depends on many factors whose relative importance is difficult to determine:

- Uptake is greatest when trees and crops are actively growing—Andrianarisoa et al. (2015) showed that, in the growing season, soil mineral nitrogen in the top metre of soil of an agroforestry plot was less than half the level observed in a crop control;
- Uptake is increased by a tree root "safety net" in alleys between tree rows, with fine roots in place throughout the year—noting that there is evidence that trees can produce fine roots at depth (e.g., > 2.5 m) even in winter when leaves are absent, and that the fine roots are longer living than at the surface (Germon et al. 2015);
- Uptake is increased if the root architecture of trees and crops is as different as possible, for example if tree roots have been encouraged to grow more deeply using a succession of winter cereals to dry the surface layers (Cardinael et al. 2015b; Mulia and Dupraz 2006), and by deep ploughing before planting trees to break up any soil-pans, or by regular ploughing of alleys to induce deeper rooting (Newaj et al. 2013);
- Turnover of tree fine-roots is greater in agroforestry than in forest monocultures (Lehmann and Zech 1998), because of greater disturbance and above ground pruning, and this leads to greater accumulation SOM on a per tree basis;
- Mineralisation of nitrogen in organic matter to ammonium is stimulated by temperature, moisture, pH and is greatly affected by the "quality" of the SOM itself—most deciduous tree leaves and fine roots decompose rapidly (Prieto et al. 2016), but branches and coarse roots provide slower-release nutrients;
- Microclimate extremes are reduced in agroforestry systems—this can have positive or negative effects on mineralisation of SOM, but it is common to observe higher carbon and nitrogen under trees in wood pastures, tropical grasslands (Bernardi et al. 2016), savannas (Wilson and Wild 1991), Mediterranean dehesas (Uribe et al. 2015) and in temperate conditions (Williams et al. 1999);
- Accumulations of nitrate in warmer conditions are often followed by a rapid decrease due to a combination of plant uptake, denitrification, immobilization and leaching (Warren et al. 1997);
- Tree roots with access to groundwater will take advantage of the often high nitrate concentration (Grimaldi et al. 2012; Wang et al. 2012) providing that the trees are actively growing.

27.3 Decreased Nitrogen Pollution from Mixtures

The annual cost to the EU of the environmental impacts of nitrogen pollution has been estimated at between €70 billion and €320 billion (UWE 2013). The majority of reactive nitrogen (N_r) used in agriculture is not taken up by crops and a range of methods to improve losses have been suggested. Oenema et al. (2009) considered that the three most promising of these are: (i) balanced fertilization combined with improved crop and manure management; (ii) low-protein animal feeding and improved herd management; and (iii) ammonia (NH_3) emissions abatement measures.

The following sections consider the use of trees in agriculture to reduce nitrogen pollution in terms of reduced soil-erosion and losses in drainage and atmospheric emissions. We have not considered the use of tree prunings or tree-fodder-banks as a means of reducing the emission of methane, but these are long established animal feeds, particularly in dry periods when pasture has stopped growing (Papanastasis et al. 1999), and new species and systems are currently being evaluated (Novak et al. 2016).

27.3.1 *Reduced Nitrogen Pollution in Erosion, Leaching and Floods*

Soil erosion has a major effect on ecosystem services such as agricultural productivity, drinking water quality and carbon stocks. A recent report on water erosion from edited European countries indicated an average loss of 2.32 t ha^{-1} $year^{-1}$, with the highest rate in Italy (6.60 t ha^{-1} $year^{-1}$) and lowest in the Netherlands (0.25 t ha^{-1} $year^{-1}$) (Panagos et al. 2015b). Four million hectares of croplands in Europe have an unsustainable water erosion rate of more than 5 t ha^{-1} $year^{-1}$ (Panagos et al. 2015b), and an additional 42 million ha is estimated to be affected by significant wind erosion (EEA 2014).

Agroforestry and other soil conservation measures, like low or reduced tillage, have a particular role in combatting erosion. Tree roots on their own do not prevent much erosion, and some of the highest erosion losses observed in Europe are from olive plantations where the soil is continually ploughed and left bare: e.g., 61–184 t ha^{-1} $year^{-1}$ for olive plantations in southern Spain (Vanwalleghem et al. 2011). Furthermore, leaves in the canopy can increase the kinetic energy of throughfall from rain, and sometimes increase erosion (Goebes et al. 2015). However the herbaceous or shrub vegetation in the tree strips will have a significant effect in reducing erosion, particularly when these follow the slope contours (Panagos et al. 2015a). Soil aggregate stability is greatly improved in the tree rows (Fig. 27.2) because of root exudates, mycorrhizal activity and higher organic matter inputs above and below ground.

In drier areas, nitrogen mineralisation is reduced in summer, and a combination of lower rainfall and increased soil water storage under trees limits the risk of nitrogen

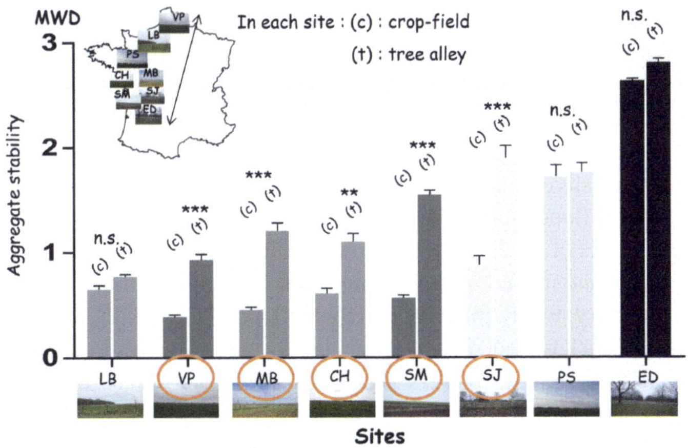

Fig. 27.2 Comparison of the aggregate stability in crop-field (c) and tree alley (t), within eight silvoarable trials in France. Asterisks indicate significant statistical differences between treatments (Mean comparison: *, **, *** indicate respectively $p < 0.05$, 0.01, $p < 0.001$. 'n.s.' indicates non-significant difference (Monnier et al. 2016)

leaching during autumn and winter. An increased uptake of nitrogen during the extended periods of crop and tree growth contributes to the reduction in N-leaching, and has been observed both in agroforestry field-trials, and process-based models of nitrogen cycling (Andrianarisoa et al. 2015). Some consequences of tree-intercrops, like increased mineralisation and reduced denitrification, will increase the production of nitrates, while others will enhance their uptake. Vertical movement of soluble nutrients to crop roots can occur via hydraulic lift in tree roots (Bayala et al. 2014; Sun et al. 2013) and actively transpiring trees will absorb water and nutrients from shallow water tables to which they have access, while trees as linear buffers can play an important role in absorbing nitrates from the horizontal flow of saturated groundwater towards watercourses (Grandgirard et al. 2014), particularly if sited some distance away from the stream itself (Komor and Magner 1996).

Dupraz et al. (2011) show that with deep and well-drained soils in France, which are sensitive to leaching, the effect of a stand of 50 mature trees per ha, with a crown cover of 30%, may effectively prevent any leaching, providing that rainfall is spread across a number of rain events.

Tree canopies can also recapture and recycle some of the ammonia lost in storage of manure and in its spreading on farmland (see Sect. 27.3.2.1). 30% of the nitrogen excreted in European animal housing systems is lost during storage and 18% immediately after application to land (Oenema et al. 2009). Thus 48% of the nitrogen excreted in animal housing is lost during storage and immediately after application, and its recapture will help reduce the need for synthetic fertilizers (Sutton et al. 2011). Efficient use of manure and slurries is also favoured by integrated crop-livestock systems (Moraine et al. 2014), and agroforestry can form an important part of these systems.

The water balance is also modified under trees: with implications for the nitrogen cycle. Watershed-scale studies on the hydrological impact networks of hedges (as in the bocage landscape) confirm that these networks modify the water balance, sustain the water flow of rivers during drought events, and reduce the amount of soil entering the rivers (Durand et al. 2002). The floodplains of some Mediterranean rivers have been planted with silvoarable tree rows in a fishbone pattern to help collect flood detritus and divert floodwater away from (Servaire 2007) the riverbanks, reducing the dangerous peaks of the flash floods downstream (Dupraz et al. 2011). In regions liable to flooding, and indeed on any sloping land liable to erosion, there is scope for "swale" agroforestry, where raised banks (berms) are planted with trees, and ditches can temporarily stock excess runoff, allowing it to infiltrate (Watte 2014).

Trees planted in riparian buffers and hedge-networks reduce nitrogen loading in watercourses through plant uptake, microbial denitrification, and soil retention (Beaujouan et al. 2001): their effectiveness is linked to width and length if the tree-strips, and also factors like soil type, subsurface flows, soil saturation, organic carbon supply, nitrogen inputs etc. (Benhamou et al. 2013; Mayer et al. 2007). In nitrate vulnerable zones these benefits can be of considerable economic value to society and efforts are underway in the USA to develop models, and linkages to GIS systems, which quantify the "nitrogen credits" payable to farmers who establish different types of riparian buffers (Boleman and Jacobson 2016; Delgado et al. 2010; Wiseman et al. 2014).

Forested areas with high nitrogen loads can become nitrogen saturated when the soil resources of ammonium and nitrate exceed the requirements of trees, understory vegetation and microorganisms (Aber et al. 1989). Once a riparian buffer is saturated its capacity for nitrogen uptake drops off causing leaching and one solution to this is that timber, coppice and crops in these areas should be more intensively harvested to maintain the buffering capacity of the system (Laudon et al. 2011).

27.3.2 Reduced Gaseous Nitrogen Pollution

Sensitive habitats in the middle of intensive agricultural areas or next to livestock housing are vulnerable to the effects of excess nitrogen inputs, leading to losses of the most sensitive species (e.g., lichens and bryophytes), and an increase in more invasive type species that prefer high rates of nitrogen (e.g., coarse grasses). This results in a gradual decrease in biodiversity—i.e., a net loss in species (Stevens et al. 2010).

There are important options available using trees to abate gaseous nitrogen pollution from agricultural sources. Approaches for the recapture of ammonia and the mitigation of nitrous oxide emissions are explored in the following sections.

27.3.2.1 Ammonia Recapture by Trees

Trees are a particularly effective scavenger of air pollutants (including ammonia, particulate matter, NO_x) due to their effect on turbulence (Beckett et al. 2000; Nowak et al. 2006). Having a higher roughness length (and lower aerodynamic resistance) aids mechanical turbulence and promotes dry deposition to the leaf surface. Dry deposition rates to trees exceed those to grassland by typically a factor of 3–20 (Fowler et al. 2004; Gallagher et al. 2002). This implies that the conversion of grassland and arable land to trees or targeted management of existing wooded areas can be used to promote the removal of ammonia from the atmosphere.

Trees can be used to mitigate NH_3 air pollution by the following methods: (a) reducing emissions from slurry lagoons by reducing the wind speed over the water surface; (b) recapture emissions by the trees themselves through increased turbulence and deposition velocities; (c) increase the dispersion above the canopy through increased mixing, thereby reducing deposition to nearby sensitive habitats (Fig. 27.3). At low wind speed the capture of NH_3 is greater, whereas at higher wind speeds the residence time of the plume in the tree belt is shorter, so that the amount dry deposited becomes less (Asman 2008).

Reduced nitrogen comprises mainly gaseous NH_3, aerosol NH_4^+ and wet deposited NH_4^+. Dry deposition of reduced N shows strong local-scale variability and is closely correlated to emissions from agricultural livestock emissions. For land close to sources (e.g., < 1 km) dry deposition is driven by the gaseous form (NH_3) as conversion to NH_4^+ has not had time to occur. Furthermore, the dry deposition velocity of NH_3 is about five times higher than for particulate NH_4^+ (Ferm 1998).

The capture of ammonia by surrounding vegetation has been studied by Patterson et al. (2008), who observed lower NH_3 concentrations when potted trees were present

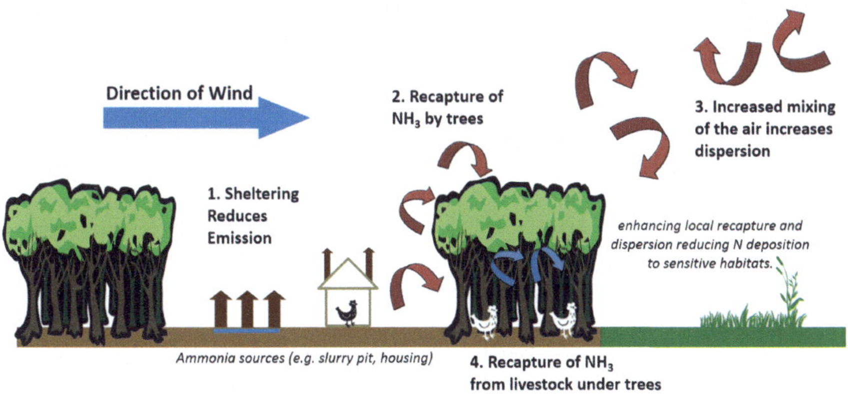

Fig. 27.3 Effect of trees on capturing and dispersing ammonia emissions by sheltering of storage pits, and recapture downwind of animal housing. Shelter can reduce emissions (Step 1), but as the airflow enters a tree belt wind speed is reduced and NH_3 capture on foliage occurs (Step 2), while part of the air-plume is pushed upwards and flows over the top of the trees, where greater turbulence leads to further deposition and local recapture (Step 3). (Bealey and Sutton 2011)

downwind of the poultry house fans compared with when the trees were removed (16.4 vs. 19.3 ppm). Wind tunnel experiments by Famulari et al. (2015) showed that significant NH_3 was recaptured using 2 m tall *Picea abies* (Norway spruce) placed in five rows in a wind tunnel. This ranged 30% for low concentration releases (180 ppbV) to 15% for high concentration releases (750 ppbV), with more NH_3 being recaptured under wetter conditions (up to 43%).

Bealey et al. (2014) used MODDAS-THETIS, a coupled turbulence and deposition turbulence model, to examine the relationships between tree canopy structure and ammonia capture for three source types—animal housing, slurry lagoon, and livestock under a tree canopy. The estimated capture efficiencies varied substantially depending on the assumptions made for canopy length, leaf area index, leaf area density, and canopy height. A maximum of 27% of the emitted ammonia was captured by a tree strip downwind of an animal housing source, for the slurry lagoon the maximum was 19%, while livestock in a silvopastoral system under trees attained a maximum of 60% recapture.

Theobald et al. (2004) looked at species choice for maximum ammonia recapture in the UK, particularly in urban environments. Criteria examined included canopy structure, suitability for site conditions, relative growth rates and suitability for management. Preferred species had a high nitrogen requirement or a tolerance of nitrogen, with a high growth rate and leaf area index (LAI). A mix of evergreen and deciduous species was preferred, together with the ability to withstand coppicing. Less suitable species were cherry (*Prunus avium*) (low LAI), sweet chestnut (*Castanea sativa*) (large leaves) and oak (*Quercus robur* or *Q. petraea*) (slow growth rate). More suitable species for use in mixtures included beech (*Fagus sylvatica*) (shade tolerant), field maple (*Acer campestre*) (suitable for most soils and can be coppiced), birch (*Betula* sp.) (suitable on most soils, can coppice), Scots pine (*Pinus sylvestris*) (high LAI) and Sitka spruce (*Picea sitchensis*) (high LAI and fast growing).

Atmospheric deposition of nitrogen is becoming an increasingly important input on agricultural and forest soils. Worrall et al. (2016), for example, estimate for 2012 that 17% (421 kton N year^{-1}) of total N inputs in the UK were provided by atmospheric deposition compared to 13% (298 kton N year^{-1}) in 1990. Enhanced interception of this atmospheric fertilizer input is a seldom mentioned benefit of agroforestry.

27.3.2.2 Nitrous Oxide Emission Abatement

Agriculture in the EU produced 464.3 Mt CO_2 equivalent of greenhouse gases in 2012, which was 10.1% of total EU emissions (Domingo et al. 2014). Emissions from the agricultural sector of non-CO_2 gases have declined by 22% since 1990 (23% for N_2O and 22% for CH_4), but much of this reduction was due to a decrease in livestock numbers across Europe (26% cattle, 33% sheep). This decrease in animal numbers has halted and it will be difficult for agriculture in Europe, and in most temperate regions, to make reductions anywhere near those in other sectors of the economy.

The emissions of N_2O that result from anthropogenic N inputs or N mineralisation occur through both a direct pathway (i.e., directly from the soils to which the N is added/released), and through two indirect pathways: (i) following volatilisation of NH_3 and NO_x from managed soils and from fossil fuel combustion and biomass burning, and the subsequent redeposition of these gases and their products, and (ii) after leaching and runoff of N, mainly as NO_3. Microbial nitrification and denitrification processes are responsible for the production of N_2O and other nitrogen oxides (NO_x). Nitrous oxide emissions are directly the result of denitrification, and are faster in heavy soils from high-rainfall areas, which are rich in nutrients. It is strongly linked to the use of nitrogen fertilizers at times of year when there is insufficient uptake by plant roots (Butterbach-Bahl et al. 2013). Methane (CH_4) is produced by eructation in ruminants, and anaerobic fermentation of manure and other organic soils, such as in mires. In contrast, CH_4 can be oxidised to CO_2 in well aerated soils. The impact of N_2O and CH_4 emissions in the GHG agricultural balance is related to their 100-year global warming potentials (GWP), which are much higher than that of CO_2 ($CO_2 = 1$, $CH_4 = 25$, $N_2O = 298$) (IPCC 2013).

Sorption of ammonia on tree foliage also reduces atmospheric nitrogen pollution, including eutrophication and emission of N_2O). According to the IPCC (2006), at least 1% of any reactive nitrogen (N_r) which is prevented from entering the atmosphere will be reduced to N_2O, and more recently a figure of 2% has been suggested (Butterbach-Bahl et al. 2011). The fate of nitrogen taken up by vegetation is a key factor.

The best way of controlling N_2O is to limit the conditions needed for its emission from the soil. There is evidence in forests that the N_2O flux from the soil is correlated to nitrate concentration in wet periods and to the soil-water filled pore space in drier periods (Skiba et al. 1993), with a negative correlation between N_2O emissions and the presence of fine-roots of fast-growing trees (Weintraub et al. 2014). Agroforestry systems, where the fine-roots of trees or crops are growing rapidly for much of the year, can lower N_r and humidity levels in the soil, and limit the conditions which favour denitrification and N_2O production (Kim et al. 2016). However, the use of nitrogen-fixing trees in agroforestry systems on low-fertility soils trees can increase reactive N and N_2O emissions (Rosenstock et al. 2014).

27.4 Discussion and Conclusions

Species mixtures generally, and agroforestry in particular, offer added value to farmers and society because of more effective use and recycling of nitrogen in agriculture or timber production, and because of reduced erosion and pollution from aqueous or atmospheric sources. In some circumstances the yields of agroforestry can exceed those of monocultures, and their environmental services include: (a) carbon sequestration, (b) enhanced biological and landscape diversity, (c) capture of ammonia from livestock emissions on tree foliage, (d) diversified economic products, and (e) adaptation options to climate change. A well distributed and actively-growing network of tree roots, particularly when this is distributed under the layer of crop-roots provides a "safety-net" which prevents N-leaching for much of the year, and which will dry and aerate the soil in a way that helps reduce denitrification and nitrous oxide production. Nitrous oxide emissions may not be reduced in agroforestry systems containing N-fixing trees or crops, but are likely to be lower than if similar quantities of inorganic N fertilizer were applied. More measurements are needed in mature agroforestry systems to confirm or deny the suggestions from N-flux models.

This chapter has emphasised that there is significant potential for carbon sequestration in agroforestry and also for the reduction in net N_2O, NH_3 and CH_4 emissions from soils containing actively growing crop and tree roots. Agroforestry has been identified as one of the best options for climate change mitigation and adaptation: however more measurements are required on fluxes of N_2O and CH_4 in different types and ages of agroforestry system, especially in temperate systems.

Incentive policies are being investigated in the US and Europe to establish "nitrogen credits" for farmers in the same way as "carbon credits" were developed in the Kyoto Clean Development Mechanism. This chapter shows that there may be an opportunity for researchers, farmers and policy makers to work together to develop and parameterise whole-farm GHG models which calculate the effects of "climate smart" agroforestry activities on emissions, and which could be used as a tool to determine payments based on predicted impacts.

Acknowledgements We are grateful to Yogan Monnier of INRA, UMR AMAP, Montpellier, for permission to include Fig. 27.2 and to Kevin Hicks for valuable comments.

References

Aber, J. D., Nadelhoffer, K. J., Steudler, P., & Melillo, J. M. (1989). Nitrogen saturation in northern forest ecosystems. *BioScience, 39*(6), 378–386.

Andrews, D. J., & Kassam, A. H. (1976). *The importance of multiple cropping in increasing world food supplies*. Madison: American Society of Agronomy.

Andrianarisoa, K. S., Dufour, L., Bienaimé, S., Zeller, B., & Dupraz, C. (2015). The introduction of hybrid walnut trees (*Juglans nigra* × *regia* cv. NG23) into cropland reduces soil mineral N content in autumn in southern France. *Agroforestry Systems, 90*(2), 193–205.

Anglade, J., Billen, G., & Garnier, J. (2015). Relationships for estimating N_2 fixation in legumes: incidence for N balance of legume-based cropping systems in Europe. *Ecosphere*, 6(3), article 37, 1–24.

Anil, L., Park, J., Phipps, R. H., & Miller, F. A. (1998). Temperate intercropping of cereals for forage: A review of the potential for growth and utilization with particular reference to the UK. *Grass and Forage Science, 53*(4), 301–317.

Asman, W. A. H. (2008). *Entrapment of ammonia, odour compounds, pesticide sprays and pathogens by shelterbelts*. DJF Plant Science Report 135. Aarhus University: Faculty of Agricultural Sciences.

Baah-Acheamfour, M., Chang, S. X., Carlyle, C. N., & Bork, E. W. (2015). Carbon pool size and stability are affected by trees and grassland cover types within agroforestry systems of western Canada. *Agriculture, Ecosystems and Environment, 213,* 105–113.

Baker, D. D. (1990). Actinorhizal plants: Underexploited trees and shrubs for forestry and agroforestry. *Nitrogen Fixing Tree Research Reports, 8,* 3–7.

Bayala, J., Sanou, J., Teklehaimanot, Z., Kalinganire, A., & Ouédraogo, S. J. (2014). Parklands for buffering climate risk and sustaining agricultural production in the Sahel of West Africa. *Current Opinion in Environmental Sustainability, 6,* 28–34.

Bealey, W. J., & Sutton, M. A. (2011). *Agroforestry Systems for Ammonia Abatement (SAMBA).* Unpublished presentation given at the Farm Woodland Forum Annual General Meeting, 30th June–1st July 2011, Holmewood Hall, Holme, Peterborough, UK. Available online at: https://www.agroforestry.ac.uk/sites/www.agroforestry.ac.uk/files/downloads/2011_meeting/bealey_pp.pdf. Accessed July 2019.

Bealey, W. J., Loubet, B., Braban, C. F., Famulari, D., Theobald, M. R., Reis, S., et al. (2014). Modelling agro-forestry scenarios for ammonia abatement in the landscape. *Environmental Research Letters, 9*(12), 125001. https://doi.org/10.1088/1748-9326/9/12/125001.

Beaujouan, V., Durand, P., & Ruiz, L. (2001). Modelling the effect of the spatial distribution of agricultural practices on nitrogen fluxes in rural catchments. *Ecological Modelling, 137*(1), 93–105.

Beckert, M. R., Smith, P., Lilly, A., & Chapman, S. J. (2016). Soil and tree biomass carbon sequestration potential of silvopastoral and woodland-pasture systems in North East Scotland. *Agroforestry Systems, 90*(3), 371–383.

Beckett, K. P., Freer Smith, P. H., & Taylor, G. (2000). Effective tree species for local air quality management. *Journal of Arboriculture, 26*(1), 12–19.

Benhamou, C., Salmon-Monviola, J., Durand, P., Grimaldi, C., & Merot, Ph. (2013). Modeling the interaction between fields and a surrounding hedgerow network and its impact on water and nitrogen flows of a small watershed. *Agricultural Water Management, 121,* 62–72.

Bernardi, R. E., de Jonge, I. K., & Holmgren, M. (2016). Trees improve forage quality and abundance in South American subtropical grasslands. *Agriculture, Ecosystems and Environment, 232,* 227–231.

Binkley, D. (2005). How nitrogen-fixing trees change soil carbon. In D. Binkley & O. Menyailo (Eds.), *Tree species effects on soils: implications for global change. Proceedings of the NATO Advanced Research Workshop on Trees and Soil Interactions, Implications to Global Climate Change August 2004 Krasnoyarsk, Russia* (pp. 155–164). The Netherlands: Springer.

Binkley, D., Senock, R., Bird, S., & Cole, T. G. (2003). Twenty years of stand development in pure and mixed stands of *Eucalyptus saligna* and nitrogen-fixing *Facaltaria moluccana. Forest Ecology and Management, 182*(1–3), 93–102.

Boleman, P. J., & Jacobson, M. G. (2016). Nitrogen credit trading as agroforestry payment for ecosystem service for riparian buffer establishment on Pennsylvania farmland. In M. Gosme (Editor in Chief), *3rd European Agroforestry Conference—Montpellier, France, 23–25 May 2016. Book of abstracts* (pp. 159–162). EURAF (European Agroforestry Federation).

Bouillet, J.-P., Laclau, J.-P., de Moraes Gonçalves, J. L., Voigtlaender, M., Gava, J. L., Leite, F. P., et al. (2013). *Eucalyptus* and *Acacia* tree growth over entire rotation in single- and mixed-species plantations across five sites in Brazil and Congo. *Forest Ecology and Management, 301,* 89–101.

Bryan, J. A., Berlyn, G. P., & Gordon, J. C. (1996). Toward a new concept of the evolution of symbiotic nitrogen fixation in the Leguminosae. *Plant and Soil, 186*(1), 151–159.

Butterbach-Bahl, K., Baggs, E. M., Dannenmann, M., Kiese, R., & Zechmeister-Boltenstern, S. (2013). Nitrous oxide emissions from soils: How well do we understand the processes and their controls? *Philosophical Transactions of the Royal Society of London B: Biological Sciences, 368*(1621), 20130122. https://doi.org/10.1098/rstb.2013.0122.

Butterbach-Bahl, K., Nemitz, E., Zaehle, S., Billen, G., Boeckx, P., Erisman, J. W., Garnier, J., Upstill-Goddard, R., Kreuzer, M., Oenema, O., Reis, S., Schaap, M., Simpson, D., de Vries, W., Winiwarter, W., Sutton, M. A. (2011). *Nitrogen as a threat to the European greenhouse gas balance*. In M. A. Sutton, C. M. Howard, J. W. Erisman, G. Billen, A. Bleeker, et al. (Eds.), *The European Nitrogen Assessment. Sources, effects and policy perspectives* (Chap. 19, pp. 434–462). Cambridge: Cambridge University Press.

Cameron, D. D., Neal, A. L., van Wees, S. C. M., & Ton, J. (2013). Mycorrhiza-induced resistance: More than the sum of its parts? *Trends in Plant Science, 18*(10), 539–545.

Canet, A., & Schreiber, K. (2015). *La couverture végétale des sols et les pratiques agroforestières, au service de territoires productifs et durables*. AFAF (Association Française d'Agroforesterie).

Cannell, M. G. R., Malcolm, D. C., & Robertson, P. A. (Eds.)(1992). *The ecology of mixed-species stands of trees*. Special publication of the British Ecological Society, 11. Blackwell Scientific Publications.

Cardinael, R., Chevallier, T., Barthès, B. G., Saby, N. P. A., Parent, T., Dupraz, C., et al. (2015a). Impact of alley cropping agroforestry on stocks, forms and spatial distribution of soil organic carbon—A case study in a Mediterranean context. *Geoderma, 259–260*, 288–299.

Cardinael, R., Mao, Z., Prieto, I., Stokes, A., Dupraz, C., Kim, J. H., et al. (2015b). Competition with winter crops induces deeper rooting of walnut trees in a Mediterranean alley cropping agroforestry system. *Plant and Soil, 391*(1–2), 219–235.

Chenu, C., Cardinael, R., Autret, B., Chevallier, T., Girardin, C., & Mary, B. (2016). Agricultural practices that store organic carbon in soils: is it only a matter of inputs? In *Conference Abstracts of the European Geosciences Union General Assembly, 17–22 April 2016, Vienna, Austria. Geophysical Research Abstracts, 18, EGU2016-14094*. EGU (European Geosciences Union).

Chifflot, V., Bertoni, G., Cabanettes, A., & Gavaland, A. (2005). Beneficial effects of intercropping on the growth and nitrogen status of young wild cherry and hybrid walnut trees. *Agroforestry Systems, 66*(1), 13–21.

Crookston, R. K., & Hill, D. S. (1979). Grain yields and land equivalent ratios from intercropping corn and soybeans in Minnesota. *Agronomy Journal, 71*(1), 41–44.

Daudin, D., & Sierra, J. (2008). Spatial and temporal variation of below-ground N transfer from a leguminous tree to an associated grass in an agroforestry system. *Agriculture, Ecosystems and Environment, 126*(3–4), 275–280.

Debell, D. S., Cole, T. G., & Whitesell, C. D. (1997). Growth, development, and yield in pure and mixed stands of Eucalyptus and Albizia. *Forest Science, 43*(2), 286–298.

Delgado, J. A., Gross, C. M., Lal, H., Cover, H., Gagliardi, P., McKinney, S. P., et al. (2010). A new GIS nitrogen trading tool concept for conservation and reduction of reactive nitrogen losses to the environment. *Advances in Agronomy, 105*, 117–171.

Delmer, A. (2015). *Tree planting in pasture and soil organic carbon* (M.Sc. thesis). School of Energy, Environment and Agrifood Land Reclamation and Restoration, Cranfield University, UK.

Dodd, J. C. (2000). The role of arbuscular mycorrhizal fungi in agro- and natural ecosystems. *Outlook on Agriculture, 29*(1), 55–62.

Domingo, J., De Miguel, E., Hurtado, B., Metayer, N., Bochu, J.-L, & Pointereau, P. (2014). *Measures at farm level to reduce greenhouse gas emissions from EU agriculture* (No. IP/B/AGRI/IC/2013_154). European Union: Directorate-General for Internal Policies; Policy Department B: Structural and Cohesion Policies, Agriculture and Rural Development.

Dupouey, J. L., Dambrine, E., Laffite, J. D., & Moares, C. (2002). Irreversible impact of past land use on forest soils and biodiversity. *Ecology, 83*(11), 2978–2984.

Dupraz, C., Dauzat, M., Suard, B., Girardin, N., Olivier, A. (1995). Root extension of young wide-spaced Prunus avium trees in an agroforest as deduced from the water budget. In J. H. Ehrenreich, D. L. Ehrenreich, & H. W. Lee (Eds.), *Proceedings of the 4th North-American Agroforestry Conference, Growing a Sustainable Future* (pp. 46–50). Boise, Idaho: Association for Temperate Agroforestry.

Dupraz, C., Liagre, F., Querne, A., Andrianarisoa, S., & Talbot, G. (2011). *L'agroforesterie peut-elle permettre de réduire les pollutions diffuses azotées d'origine agricole?* Rapport de contrat de recherche no. 2009–0009 avec l'Agence de l'Eau Rhône-Méditerranée-Corse (no. codique INRA: 24000236). Montpellier: INRA.

Dupraz, C., Simorte, V., Dauzat, M., Bertoni, G., Bernadac, A., & Masson, P. (1998). Growth and nitrogen status of young walnuts as affected by intercropped legumes in a Mediterranean climate. *Agroforestry Systems, 43*(1–3), 71–80.

Durand, P., Gascuel-Odoux, C., & Cordier, M. O. (2002). Parameterisation of hydrological models: A review and lessons learned from studies of an agricultural catchment (Naizin, France). *Agronomie, 22*(2), 217–228.

EEA. (2014). *The European environment: state and outlook 2015: An integrated assessment of the European environment*. Copenhagen: European Environment Agency.

Erenstein, O. (2003). Smallholder conservation farming in the tropics and sub-tropics: a guide to the development and dissemination of mulching with crop residues and cover crops. *Agriculture, Ecosystems and Environment, 100*(1), 17–37.

Erskine, P. D., Lamb, D., & Bristow, M. (2006). Tree species diversity and ecosystem function: Can tropical multi-species plantations generate greater productivity? *Forest Ecology and Management, 233*(2–3), 205–210.

European Commission. (2013, December 17). *Regulation (EU) No 1305/2013 of the European Parliament and of the Council on support for rural development by the European Agricultural Fund for Rural Development (EAFRD) and repealing Council Regulation (EC) No. 1698/2005.*

Famulari, D., Twigg, M. M., Robertson, A., Grigorova, E., Braban, C. F., et al. (2015). Quantifying NH_3 depletion by a woodland: a wind tunnel experiment. In W. J. Bealey, C. F. Braban, M. R. Theobald, D. Famulari, Y. S. Tang, et al. (Eds.), *Agroforestry systems for ammonia abatement*. Defra Report AC0201 (pp. 13–14). UK: Centre for Ecology & Hydrology.

Ferm, M. (1998). Atmospheric ammonia and ammonium transport in Europe and critical loads: A review. *Nutrient Cycling in Agroecosystems, 51*(1), 5–17.

Ferreiro-Domínguez, N., Rigueiro-Rodríguez, A., Rial-Lovera, K. E., Romero-Franco, R., & Mosquera-Losada, M. R. (2016). Effect of grazing on carbon sequestration and tree growth that is developed in a silvopastoral system under wild cherry (*Prunus avium* L.). *CATENA, 142*, 11–20.

Forrester, D. I., Pares, A., O'Hara, C., Khanna, P. K., & Bauhus, J. (2012). Soil organic carbon is increased in mixed-species plantations of Eucalyptus and nitrogen-fixing Acacia. *Ecosystems, 16*(1), 123–132.

Fowler, D., Skiba, U., Nemitz, E., Choubedar, F., Branford, D., Donovan, R., et al. (2004). Measuring aerosol and heavy metal deposition on urban woodland and grass using inventories of ^{210}Pb and metal concentrations in soil. *Water, Air, and Soil Pollution: Focus, 4*(2–3), 483–499.

Gadgil, R. L., Charlton, J. F. L., Sandberg, A. M., & Allen, P. J. (1986). Relative growth and persistence of planted legumes in a mid-rotation radiata pine plantation. *Forest Ecology and Management, 14*(2), 113–124.

Gallagher, M. W., Nemitz, E., Dorsey, J. R., Fowler, D., Sutton, M. A., et al. (2002). Measurements and parameterizations of small aerosol deposition velocities to grassland, arable crops, and forest: Influence of surface roughness length on deposition. *Journal of Geophysical Research, D: Atmospheres, 107*(D12), AAC 8-1–AAC 8-10. https://doi.org/10.1029/2001jd000817.

Germon, A., Cardinael, R., Prieto, I., Mao, Z., Kim, J. H., Stokes, A., et al. (2015). Unexpected phenology and lifespan of shallow and deep fine roots of walnut trees grown in a Mediterranean agroforestry system. *Plant and Soil, 401*(1–2), 409–426.

Goebes, P., Bruelheide, H., Härdtle, W., Kröber, W., Kühn, P., Li, Y., et al. (2015). Species-specific effects on throughfall kinetic energy in subtropical forest plantations are related to leaf traits and tree architecture. *PLoS ONE, 10*(6), e0128084. https://doi.org/10.1371/journal.pone.0128084.

Goh, K. M., Mansur, I., Mead, D. J., & Sweet, G. B. (1996). Biological nitrogen fixing capacity and biomass production of different understorey pastures in a *Pinus radiate*-pasture agroforestry system in New Zealand. *Agroforestry Systems, 34*(1), 33–49.

Grandgirard, D., Combaud, A., Mercadal, L., Liagre, A. M., Bachevillier, Y., & Marin, A. (2014). *Agroforestry systems design at parcel scale within the territory (FR): Towards a spatial decision support system for catched water quality improvement*. EURAF (European Agroforestry Federation).

Grimaldi, C., Fossey, M., Thomas, Z., Fauvel, Y., & Merot, P. (2012). Nitrate attenuation in soil and shallow groundwater under a bottomland hedgerow in a European farming landscape. *Hydrological Processes, 26*(23), 3570–3578.

Haile, S. G., Nair, P. K. R., & Nair, V. D. (2008). Carbon storage of different soil-size fractions in Florida silvopastoral systems. *Journal of Environmental Quality, 37*(5), 1789–1797.

Hansen, B., Alrøe, H. F., & Kristensen, E. S. (2001). Approaches to assess the environmental impact of organic farming with particular regard to Denmark. *Agriculture, Ecosystems and Environment, 83*(1–2), 11–26.

Hartwig, N. L., & Ammon, H. U. (2002). Cover crops and living mulches. *Weed Science, 50*(6), 688–699.

Heilman, P., & Stettler, R. F. (1985). Mixed, short-rotation culture of red alder and black cottonwood: Growth, coppicing, nitrogen fixation, and allelopathy. *Forest Science, 31*(3), 607–616.

Hooda, P. S., Moynagh, M., Svoboda, I. F., & Anderson, H. A. (1998). A comparative study of nitrate leaching from intensively managed monoculture grass and grass–clover pastures. *The Journal of Agricultural Science, 131*(3), 267–275.

Howlett, D. S., Mosquera-Losada, M. R., Nair, P. K. R., Nair, V. D., & Rigueiro-Rodríguez, A. (2011). Soil carbon storage in silvopastoral systems and a treeless pasture in northwestern Spain. *Journal of Environmental Quality, 40*(3), 825–832.

Ingleby, K., Wilson, J., Munro, R. C., & Cavers, S. (2007). Mycorrhizas in agroforestry: spread and sharing of arbuscular mycorrhizal fungi between trees and crops: Complementary use of molecular and microscopic approaches. *Plant and Soil, 294*(1–2), 125–136.

IPCC. (2006). *2006 IPCC guidelines for national greenhouse gas inventories*. Intergovernmental Panel on Climate Change (IPCC). Prepared by the National Greenhouse Gas Inventories Programme. In H. S.Eggleston, L. Buendia, K. Miwa, T. Ngara, & K. Tanabe (Eds.). Japan: IGES.

IPCC. (2013). *IPCC fifth assessment report (AR5)*. Intergovernmental Panel on Climate Change.

Kass, D. C. L., Sylvester-Bradley, R., & Nygren, P. (1997). The role of nitrogen fixation and nutrient supply in some agroforestry systems of the Americas. *Soil Biology and Biochemistry, 29*(5–6), 775–785.

Kelty, M. J. (2006). The role of species mixtures in plantation forestry. *Forest Ecology and Management, 233*(2–3), 195–204.

Kim, D.-G., Kirschbaum, M. U. F., & Beedy, T. L. (2016). Carbon sequestration and net emissions of CH4 and N2O under agroforestry: synthesizing available data and suggestions for future studies. *Agriculture, Ecosystems and Environment, 226*, 65–78.

Knowles, R. L. (1991). New Zealand experience with silvopastoral systems: A review. *Forest Ecology and Management, 45*(1–4), 251–267.

Komor, S. C., & Magner, J. A. (1996). Nitrate in groundwater and water sources used by riparian trees in an agricultural watershed: a chemical and isotopic investigation in Southern Minnesota. *Water Resources Research, 32*(4), 1039–1050.

Kumar, B. M., & Nair, P. K. R. (2011). *Carbon sequestration potential of agroforestry systems: Opportunities and challenges* (Advances in Agroforestry series Vol. 8). Netherlands: Springer.

Kuroiwa, M., Koba, K., Isobe, K., Tateno, R., Nakanishi, A., Inagaki, Y., et al. (2011). Gross nitrification rates in four Japanese forest soils: Heterotrophic versus autotrophic and the regulation factors for the nitrification. *Journal of Forest Research, 16*(5), 363–373.

Laudon, H., Sponseller, R. A., Lucas, R. W., Futter, M. N., Egnell, G., Bishop, K., et al. (2011). Consequences of more intensive forestry for the sustainable management of forest soils and waters. *Forests, 2*, 243–260.

Lawson, G. J., Balaguer, F., Palma, J., & Papanastasis, V. (2016). Options for Agroforestry in the CAP 2014-2020. In M. Gosme (Editor in Chief), *3rd European Agroforestry Conference—Montpellier, France, 23–25 May 2016. Book of abstracts* (pp. 425–428). EURAF (European Agroforestry Federation).

Lehmann, J., & Zech, W. (1998). Fine root turnover of irrigated hedgerow intercropping in Northern Kenya. *Plant and Soil, 198*(1), 19–31.

Lorenz, K., & Lal, R. (2014). Soil organic carbon sequestration in agroforestry systems: A review. *Agronomy for Sustainable Development, 34*(2), 443–454.

Mahieu, S., Metay, A., Brunel, B., & Dufour, L. (2015). Nitrogen fluxes in chickpea grown in Mediterranean agroforestry systems. *Agroforestry Systems, 90*(2), 313–324.

Malézieux, E., Crozat, Y., Dupraz, C., Laurans, M., Makowski, D., Ozier-Lafontaine, H., et al. (2009). Mixing plant species in cropping systems: Concepts, tools and models: A review. *Agronomy for Sustainable Development, 29*(1), 43–62.

Mantino, A., Ragaglini, G., Tozzini, C., Cappucci, A., Mele, M., & Bonari, E. (2016). Yield and nutritive value of alfalfa (*Medicago sativa* l.) in an olive (*Olea europaea* L.) alley-cropping practice. In M. Gosme (Editor in Chief), *3rd European Agroforestry Conference—Montpellier, France, 23–25 May 2016. Book of abstracts* (pp. 403–406). EURAF (European Agroforestry Federation).

Marinari, S., Lagomarsino, A., Moscatelli, M. C., Di Tizio, A., & Campiglia, E. (2010). Soil carbon and nitrogen mineralization kinetics in organic and conventional three-year cropping systems. *Soil and Tillage Research, 109*(2), 161–168.

Marrs, R. H., Owen, L. D. C., Roberts, R. D., & Bradshaw, A. D. (1982). Tree lupin (*Lupinus arboreus* SIMS) an ideal nurse crop for land restoration and amenity plantings. *Arboricultural Journal, 6*(3), 161–174.

Mayer, P. M., Reynolds, S. K., Jr., McCutchen, M. D., & Canfield, T. J. (2007). Meta-analysis of nitrogen removal in riparian buffers. *Journal of Environmental Quality, 36*(4), 1172–1180.

Mead, R., & Willey, R. W. (1980). The concept of a 'land equivalent ratio' and advantages in yields from intercropping. *Experimental Agriculture, 16*(03), 217–228.

Mikić, A., Ćupina, B., Mihailović, V., Krstić, Đ., Đorđević, V., Perić, V., et al. (2012). Forage legume intercropping in temperate regions: Models and ideotypes. In E. Lichtfouse (Ed.), *Sustainable agriculture reviews book series, SARV* (Vol. 11, pp. 161–182). Netherlands: Springer.

Mohammadi, G. R. (2012). Living mulch as a tool to control weeds in agroecosystems: A Review. In A. J. Price (Ed.) *Weed control* (Chap. 4, pp. 75–100). InTech. Available online open access at: https://www.intechopen.com/books/weed-control/living-mulch-as-a-tool-to-control-weeds-in-agroecosystems.

Monnier, Y., Le Bissonnais, Y., Maeght, J.-L., Erktan, A., & Stokes, A. (2016). Effects of tree alleys on soil structure and fertility: is there a general pattern? In M. Gosme (Editor in Chief), *3rd European Agroforestry Conference—Montpellier, France, 23–25 May 2016. Book of abstracts* (pp. 85-87). EURAF (European Agroforestry Federation).

Moraine, M., Duru, M., Nicholas, P., Leterme, P., & Therond, O. (2014). Farming system design for innovative crop-livestock integration in Europe. *Animal, 8*(8), 1204–1217.

Morhart, C. D., Douglas, G. C., Dupraz, C., Graves, A. R., Nahm, M., Paris, P., et al. (2014). Alley coppice—A new system with ancient roots. *Annals of Forest Science, 71*(5), 527–542.

Mulia, R., & Dupraz, C. (2006). Unusual fine root distributions of two deciduous tree species in southern France: What consequences for modelling of tree root dynamics? *Plant and Soil, 281*(1–2), 71–85.

Newaj, R., Dhyani, S. K., Alam, B., Ajit, Prasad, R., Handa, A. K., et al. (2013). Long term effect of root management practices on rooting pattern in *Dalbergia sissoo* and grain yield of mustard under agrisilviculture system. *Range Management and Agroforestry, 34*(1), 47–50.

Novak, S., Liagre, F., & Emile, J. C. (2016). Integrating agroforestry into an innovative mixed crop-dairy system. In M. Gosme (Editor in Chief), *3rd European Agroforestry Conference—Montpellier, France, 23–25 May 2016. Book of abstracts* (pp. 397–399). EURAF (European Agroforestry Federation).

Nowak, D. J., Crane, D. E., & Stevens, J. C. (2006). Air pollution removal by urban trees and shrubs in the United States. *Urban Forestry and Urban Greening, 4*(3–4), 115–123.

Nygren, P., Fernandez, M. P., Harmand, J. M., & Leblanc, H. A. (2012). Symbiotic dinitrogen fixation by trees: An underestimated resource in agroforestry systems? *Nutrient Cycling in Agroecosystems, 94*(2–3), 123–160.

Oenema, O., Witzke, H. P., Klimont, Z., Lesschen, J. P., & Velthof, G. L. (2009). Integrated assessment of promising measures to decrease nitrogen losses from agriculture in EU-27. *Agriculture, Ecosystems and Environment, 133*(3–4), 280–288.

OSCAR Project. (2016). The cover crop and living mulch toolbox. *European Conservation Agriculture Federation.* Available at: https://web5.wzw.tum.de/oscar/toolbox/database/index.html Accessed November 16, 2016.

Panagos, P., Borrelli, P., Meusburger, K., Alewell, C., Lugato, E., & Montanarella, L. (2015a). Estimating the soil erosion cover-management factor at the European scale. *Land Use Policy, 48,* 38–50.

Panagos, P., Borrelli, P., Poesen, J., Ballabio, C., Lugato, E., Meusburger, K., et al. (2015b). The new assessment of soil loss by water erosion in Europe. *Environmental Science and Policy, 54,* 438–447.

Papanastasis, V. P., Tsiouvaras, C. N., Dini, O., Vaitsis, T., Stringi, L., et al. (1999). *Selection and utilization of cultivated fodder trees and shrubs in the Mediterranean region.* Options Méditerranéennes, Série B: Etudes et Recherches (Vol. 23). Montpellier: CIHEAM/EC-DGVI.

Patterson, P. H., Adrizal, Hulet, R. M., Bates, R. M., Despot, D. A., et al. (2008). The potential for plants to trap emissions from farms with laying hens. 1. Ammonia. *The Journal of Applied Poultry Research, 17*(1), 54–63.

Paula, R. R., Bouillet, J. P., Trivelin, P., Zeller, B., Goncalves, J., Nouvellon, Y., et al. (2015). Evidence of short-term belowground transfer of nitrogen from *Acacia mangium* to *Eucalyptus grandis* trees in a tropical planted forest. *Soil Biology and Biochemistry, 91,* 99–108.

Pelosi, C., Bertrand, M., & Roger-Estrade, J. (2009). Earthworm community in conventional, organic and direct seeding with living mulch cropping systems. *Agronomy for Sustainable Development, 29*(2), 287–295.

Piotrowski, J. S., Morford, S. L., & Rillig, M. C. (2008). Inhibition of colonization by a native arbuscular mycorrhizal fungal community via *Populus trichocarpa* litter, litter extract, and soluble phenolic compounds. *Soil Biology and Biochemistry, 40*(3), 709–717.

Piotto, D. (2008). A meta-analysis comparing tree growth in monocultures and mixed plantations. *Forest Ecology and Management, 255*(3–4), 781–786.

Plath, M., Mody, K., Potvin, C., & Dorn, S. (2011). Do multipurpose companion trees affect high value timber trees in a silvopastoral plantation system? *Agroforestry Systems, 81*(1), 79–92.

Prieto, I., Stokes, A., & Roumet, C. (2016). Root functional parameters predict fine root decomposability at the community level. *The Journal of Ecology, 104*(3), 725–733.

Querné, A., Battie-laclau, P., Dufour, L., Wery, J., & Dupraz, C. (2017). Effects of walnut trees on biological nitrogen fixation and yield of intercropped alfalfa in a Mediterranean agroforestry system. *European Journal of Agronomy, 84,* 35–46.

Radwan, M. A., & DeBell, D. S. (1988). Nutrient relations in coppiced black cottonwood and red alder. *Plant and Soil, 106*(2), 171–177.

Rédei, K., Osváth-Bujtás, Z., & Veperdi, I. (2008). Black locust (*Robinia pseudoacacia* L.) improvement in Hungary: a review. *Acta Silvatica et Lignaria Hungarica, 4,* 127–132.

Redondo-Brenes, A., & Montagnini, F. (2006). Growth, productivity, aboveground biomass, and carbon sequestration of pure and mixed native tree plantations in the Caribbean lowlands of Costa Rica. *Forest Ecology and Management, 232*(1–3), 168–178.

Rillig, M. C. (2004). Arbuscular mycorrhizae, glomalin, and soil aggregation. *Canadian Journal of Soil Science, 84*(4), 355–363.

Rosenstock, T. S., Tully, K. L., Arias-Navarro, C., Neufeldt, H., Butterbach-Bahl, K., & Verchot, L. V. (2014). Agroforestry with N_2-fixing trees: Sustainable development's friend or foe? *Current Opinion in Environmental Sustainability, 6,* 15–21.

Russo, R. O. (2005). Nitrogen-fixing trees with actinorhiza in forestry and agroforestry. In D. Werner & W. E. Newton (Eds.), *Nitrogen fixation in agriculture, forestry, ecology, and the environment* (pp. 143–171). Nitrogen fixation: origins, applications, and research progress book series (NITR, Vol. 4). Netherlands: Springer.

Sanchez, P. A. (1995). Science in agroforestry. *Agroforestry Systems, 30*(1–2), 5–55.

Sanchez, P. A. (1999). Improved fallows come of age in the tropics. *Agroforestry Systems, 47*(1–3), 3–12.

Servaire, M. (2007). *Etude de la faisabilité de la mise en place d'agroforesterie sur la plaine du Vistre*. ENESAD. Available online at: http://www.agroof.net/agroof_projets/documents/vistre/RAPPORT_Vistre.pdf.

Shannon, D., Sen, A. M., & Johnson, D. B. (2002). A comparative study of the microbiology of soils managed under organic and conventional regimes. *Soil Use and Management, 18*(s1), 274–283.

Sharrow, S. H., & Ismail, S. (2004). Carbon and nitrogen storage in agroforests, tree plantations, and pastures in western Oregon, USA. *Agroforestry Systems, 60*(2), 123–130.

Skiba, U., Smith, K. A., & Fowler, D. (1993). Nitrification and denitrification as sources of nitric oxide and nitrous oxide in a sandy loam soil. *Soil Biology and Biochemistry, 25*(11), 1527–1536.

Sponseller, R. A., Gundale, M. J., Futter, M., Ring, E., Nordin, A., Näsholm, T., et al. (2016). Nitrogen dynamics in managed boreal forests: Recent advances and future research directions. *Ambio, 45*(Supplement 2), 175–187.

Stevens, C. J., Thompson, K., Grime, J. P., Long, C. J., & Gowing, D. J. G. (2010). Contribution of acidification and eutrophication to declines in species richness of calcifuge grasslands along a gradient of atmospheric nitrogen deposition. *Functional Ecology, 24*(2), 478–484.

Sun, S.-J., Meng, P., Zhang, J.-S., & Wan, X. (2013). Hydraulic lift by *Juglans regia* relates to nutrient status in the intercropped shallow-root crop plant. *Plant and Soil, 374*(1–2), 629–641.

Sutton, M., Billen, G., Bleeker, A., Erisman, J. W., Grennfelt, P., et al. (2011). Technical summary. In M. A. Sutton, C. M. Howard, J. W. Erisman, G. Billen, A. Bleeker, et al. (Eds.), *The European nitrogen assessment. Sources, effects and policy perspectives* (pp. xxxv–li). Cambridge: Cambridge University Press.

Teissier du Cros, E., Jung, G., & Bariteau, M. (1984). Alder—Frankia interaction and alder—poplar association for biomass production. *Plant and Soil, 78*(1–2), 235–243 (Thematic issue: A. D. L. Akkermans, D. Baker, K. Huss-Danell, J. D. & Tjepkema (Eds.), *Proceedings of the Workshop on FRANKIA SYMBOSES*).

Theobald, M. R., Loubet, B., Cellier, P., Hargreaves, K. J., Nemitz, E., & Sutton, M. A. (2004). *AMBER: Ammonia mitigation by enhanced recapture. Impact of vegetation and/or other on-farm features on net ammonia emissions from livestock farms*. Final report on project WA0719 to Defra (Land Management Improvement Division). Edinburgh: Centre for Ecology & Hydrology.

Thevathasan, N. V., Gordon, A. M., Wotherspoon, A., Graungaard, K., Dunfield, K., et al. (2016). Tree-based intercropping systems: A potential land-use solution for climate change mitigation in Canadian agricultural landscapes. In M. Gosme (Editor in Chief), *3rd European Agroforestry Conference–Montpellier, France, 23–25 May 2016. Book of abstracts* (pp. 204-207). EURAF (European Agroforestry Federation).

Tosi, L., Nahm, M., Paris, P., Pisanelli, A., Lunny, R., et al. (2014). *Social and economical evaluation of innovative alley coppice systems mixing timber trees with bioenergy wood crops in agroforestry systems*. EURAF posters. Available at: http://hdl.handle.net/10400.5/7246.

Tu, C., Ristaino, J. B., & Hu, S. (2006). Soil microbial biomass and activity in organic tomato farming systems: Effects of organic inputs and straw mulching. *Soil Biology & Biochemistry, 38*(2), 247–255.

Turvey, N. D., & Smethurst, P. J. (1983). Nitrogen fixing plants in forest plantation management. In J. C. Gordon & C. T. Wheeler (Eds.), *Biological nitrogen fixation in forest ecosystems: foundations and applications* (pp. 233–259). Forestry Sciences book series (FOSC, Vol. 9). Dordrecht: Springer.

Upson, M. A., Burgess, P. J., & Morison, J. (2016). Soil carbon changes after establishing woodland and agroforestry trees in a grazed pasture. *Geoderma, 283*, 10–20.

Uribe, C., Inclan, R., Hernando, L., Roman, M., Clavero, M. A., Roig, S., et al. (2015). Grazing, tilling and canopy effects on carbon dioxide fluxes in a Spanish dehesa. *Agroforestry Systems, 89*(2), 305–318.

UWE. (2013). *Nitrogen pollution and the European environment implications for air quality policy.* Science for Environment Policy In-Depth Report for the European Commission. Bristol: University of the West of England (UWE), Science Communication Unit.

Vallebona, C., Mantino, A., & Bonari, E. (2016). Exploring the potential of perennial crops in reducing soil erosion: A GIS-based scenario analysis in southern Tuscany, Italy. *Applied Geography, 66,* 119–131.

Van der Heijden, M. G. A., Martin, F. M., Selosse, M.-A., & Sanders, I. R. (2015). Mycorrhizal ecology and evolution: The past, the present, and the future. *The New Phytologist, 205*(4), 1406–1423.

Van der Meiden, H. A. (1961). De els in populierenbeplantingen [Alder in mixture with poplar]. *Nederlands bosbouw tijdschrift, 33,* 168–171.

Vanwalleghem, T., Amate, J. I., de Molina, M. G., Fernández, D. S., & Gómez, J. A. (2011). Quantifying the effect of historical soil management on soil erosion rates in Mediterranean olive orchards. *Agriculture, Ecosystems and Environment, 142*(3–4), 341–351.

Villenave, C., Rabary, B., Chotte, J.-L., Blanchart, E., & Djigal, D. (2009). Impact of direct seeding mulch-based cropping systems on soil nematodes in a long-term experiment in Madagascar. *Pesquisa Agropecuária Brasileira, 44*(8), 949–953.

Wang, W. J., Dalal, R. C., Moody, P. W., & Smith, C. J. (2003). Relationships of soil respiration to microbial biomass, substrate availability and clay content. *Soil Biology & Biochemistry, 35*(2), 273–284.

Wang, L., Duggin, J. A., & Nie, D. (2012). Nitrate-nitrogen reduction by established tree and pasture buffer strips associated with a cattle feedlot effluent disposal area near Armidale, NSW Australia. *Journal of Environmental Management, 99,* 1–9.

Warner, D. J., Tzilivakis, J., Green, A., & Lewis, K. A. (2016). A guidance tool to support farmers with ecological focus areas–the benefits of agroforestry for ecosystem services and biodiversity. In M. Gosme (Editor in Chief), *3rd European Agroforestry Conference—Montpellier, France, 23–25 May 2016. Book of abstracts* (pp. 373–375). EURAF (European Agroforestry Federation).

Warren, G. P., Atwal, S. S., & Irungu, J. W. (1997). Soil nitrate variations under grass, sorghum and bare fallow in semi-arid Kenya. *Experimental Agriculture, 33*(3), 321–333.

Watte, J. (2014). Featured farm: Agroforestry on contour swales in Flanders: venturing into aquaculture. *EURAF Newsletter*, 6/2014 (November), 7–9. Available online at: https://euraf.isa.utl.pt/sites/default/files/pub/docs/euraf_newsletter_6_2014.pdf.

Weintraub, S. R., Russell, A. E., & Townsend, A. R. (2014). Native tree species regulate nitrous oxide fluxes in tropical plantations. *Ecological Applications, 24*(4), 750–758.

West, G. G. (1997). *The effect of Maku Lotus on Radiata pine growth in the Kaiangaroa Forest. Master of Forestry Science degree thesis.* New Zealand: Canterbury University.

White, C., Barbercheck, M., DuPont, T., Finney, D., Hamilton, A., et al. (2016). *Making the most of mixtures: Considerations for winter cover crops in temperate climates.* Penn State Extension–Agroecology in Practice (EE0166). The Pennsylvania State University. Available online at: http://articles.extension.org/pages/72973/making-the-most-of-mixtures:-considerations-for-winter-cover-crops-in-temperate-climates.

Williams, D. G., Wallace, P., McKeon, G. M., Hall, W., Katjiua, M., & Abel, N. (1999). *Effects of trees on native pasture production on the southern tablelands.* A report for the RIRDC/LWRRDC/FWPRDC Joint Venture Agroforestry Program, publication no. 99/165, project no. UCA-3A. Rural Industries Research and Development Corporation (RIRDC), Joint Venture Agroforestry Program.

Wilson, J. R., & Wild, D. (1991). Improvement of nitrogen nutrition and grass growth under shading. In H. M. Shelton & W. W. Stür (Eds.), *Forages for plantation crops. Proceedings of a workshop, Sanur Beach, Indonesia, 27–29 June 1990* (pp. 77–82). ACIAR Proceedings. Australian Centre for International Research (ACIAR).

Wiseman, J. D., Burchell, M. R., Grabow, G. L., Osmond, D. L., & Messer, T. L. (2014). Groundwater nitrate concentration reductions in a riparian buffer enrolled in the NC Conservation Reserve Enhancement Program. *Journal of the American Water Resources Association, (JAWRA), 50*(3), 653–664.

Wormald, T. J. (1992). *Mixed and pure forest plantations in the tropics and subtropics*. Based on the work of T. J. Wormald. FAO Forestry Paper 103. Rome: Food and Agriculture Organization of the United Nations (FAO).

Worrall, F., Burt, T. P., Howden, N., & Whelan, M. J. (2016). The UK's total nitrogen budget from 1990 to 2020: A transition from source to sink? *Biogeochemistry, 129*(3), 325–340.

Zomer, R. J., Neufeldt, H., Xu, J., Ahrends, A., Bossio, D., Trabucco, A., et al. (2016). Global tree cover and biomass carbon on agricultural land: The contribution of agroforestry to global and national carbon budgets. *Scientific Reports, 6,* 29987. https://doi.org/10.1038/srep29987.

Part IX
Management Tools and Assessment: Regional and Global Nitrogen Assessment

Chapter 28
Global Nitrogen and Phosphorus Pollution

Bruna Grizzetti, Gilles Billen, Eric A. Davidson, Wilfried Winiwarter, Wim de Vries, David Fowler, Clare M. Howard, Albert Bleeker, Mark A. Sutton, Luis Lassaletta, and Josette Garnier

Abstract Humans have altered the natural nitrogen and phosphorus biogeochemical cycles by the massive input of fertilizers to the agricultural system to boost production. As a result a large amount of nitrogen and phosphorus have been mobilised and delivered to the environment, creating threats to aquatic and terrestrial ecosystem functioning and human health. Based on recent studies published in the literature,

---

B. Grizzetti (✉)
European Commission, Joint Research Centre (JRC), Ispra, Italy
e-mail: Bruna.GRIZZETTI@ec.europa.eu

B. Grizzetti · G. Billen · L. Lassaletta · J. Garnier
Sorbonne Université, CNRS, EPHE, UMR, 7619 Metis, 4 Place Jussieu, 75005 Paris, France
e-mail: gilles.billen@upmc.fr

L. Lassaletta
e-mail: luis.lassaletta@upm.es

J. Garnier
e-mail: josette.garnier@upmc.fr

E. A. Davidson
University of Maryland Center for Environmental Science, Appalachian Laboratory, 301 Braddock Road, Frostburg, MD 21532, USA
e-mail: edavidson@umces.edu

W. Winiwarter
International Institute for Applied Systems Analysis (IIASA), Laxenburg, Austria
e-mail: winiwarter@iiasa.ac.at

W. de Vries
Wageningen University and Research (WUR), Droevendaalsesteeg 3, 6708 PB Wageningen, The Netherlands
e-mail: wim.devries@wur.nl

D. Fowler · C. M. Howard · M. A. Sutton
UK Centre for Ecology & Hydrology, Edinburgh Research Station, Bush Estate, Penicuik, Midlothian, EH26 0QB, UK
e-mail: dfo@ceh.ac.uk

C. M. Howard
e-mail: cbritt@ceh.ac.uk

M. A. Sutton et al. (eds.), *Just Enough Nitrogen*,
https://doi.org/10.1007/978-3-030-58065-0_28

this research describes and quantifies the main fluxes of the current global nitrogen and phosphorus cycles (year 2000–2010).

Keywords Global biogeochemical cycle · Nitrogen · Phosphorus · Aquatic · Terrestrial

28.1 Introduction

In the last 50 years humans have significantly increased the input of nitrogen (N) and phosphorus (P) fertilizers to sustain agricultural production. This has altered the natural N and P biogeochemical cycles, and the large additional amount of N and P delivered to the environment is posing threats to human health and the aquatic and terrestrial ecosystems (Sutton et al. 2011, 2013a). The effects being related both to the amount of new nutrients in the ecosystem and the alteration of the nutrient relative ratio (Shantz et al. 2016; Beusen et al. 2016; Grizzetti et al. 2012). The understanding and quantification of nutrient fluxes in global biogeochemical cycles is fundamental to assessing the impact of human activities and proposing effective remediation measures. As the impacts and processes involved are global, analysing the fluxes at this scale is important. Based on recent studies published in the literature, this research describes the current global N and P biogeochemical cycles, quantifying the main fluxes between environmental compartments and highlighting the impact of anthropogenic activities on the alteration of the fluxes. This work was developed as part of the global assessment on nutrient cycles (Sutton et al. 2013a).

28.2 Methodology

We describe graphically the N and P global cycles around the year 2000–2010 combining the fluxes reported in recent studies published in the literature. Four main components were considered: terrestrial ecosystems, fresh waters, oceans, and the atmosphere, and the main pools within each component. To highlight the human impact on the nutrient cycle, we distinguished between natural fluxes (in green), human altered fluxes mainly related to food production (in blue) and natural fluxes

M. A. Sutton
e-mail: ms@ceh.ac.uk

A. Bleeker
National Institute of Public Health and Environment (RIVM), P.O. Box 1, 3720 BA Bilthoven, The Netherlands
e-mail: albert.bleeker@rivm.nl

L. Lassaletta
CEIGRAM/Department of Agricultural Production, Universidad Politécnica de Madrid, Madrid 28040, Spain

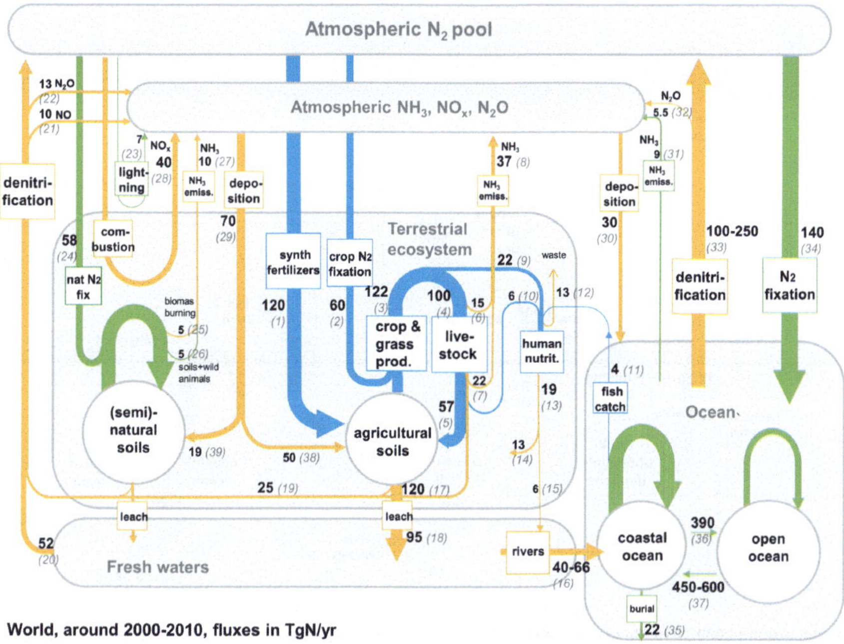

Fig. 28.1 Global nitrogen cycle around years 2000–2010. The arrows show the nitrogen fluxes across environmental pools and compartments (green: natural fluxes, blue: intended human altered fluxes, orange: natural fluxes affected by anthropogenic activities). The figures in black indicate the nitrogen fluxes in Tg N/yr (yr = year) and the figures within brackets are the legend and refer to the accompanying table including the references for each flux (see Table 28.1). Note that not all figures add exactly, due to the use of different data sources (Sutton et al. 2013a). Figure reproduced from Sutton et al. (2013a) *Our Nutrient World: The challenge to produce more food and energy with less pollution* (with permission of the UK Centre for Ecology & Hydrology, © Centre for Ecology and Hydrology)

affected by anthropogenic activities (in orange) (Figs. 28.1 and 28.2), with arrow sizes proportional to the relevance of the fluxes, using a graphical approach similar to the one presented in the European Nitrogen Assessment (Sutton et al. 2011).

28.3 Results

The global nitrogen cycle is presented in Fig. 28.1 and the relative references are reported in Table 28.1. The global phosphorus cycle is shown in Fig. 28.2 and the relative references are reported in Table 28.2.

Overall, the efficiency of nutrient use is very low: considering the full chain, on average over 80% of nitrogen and 25–75% of phosphorus that are consumed (where not temporarily stored in agricultural soils) end up lost to the environment. This

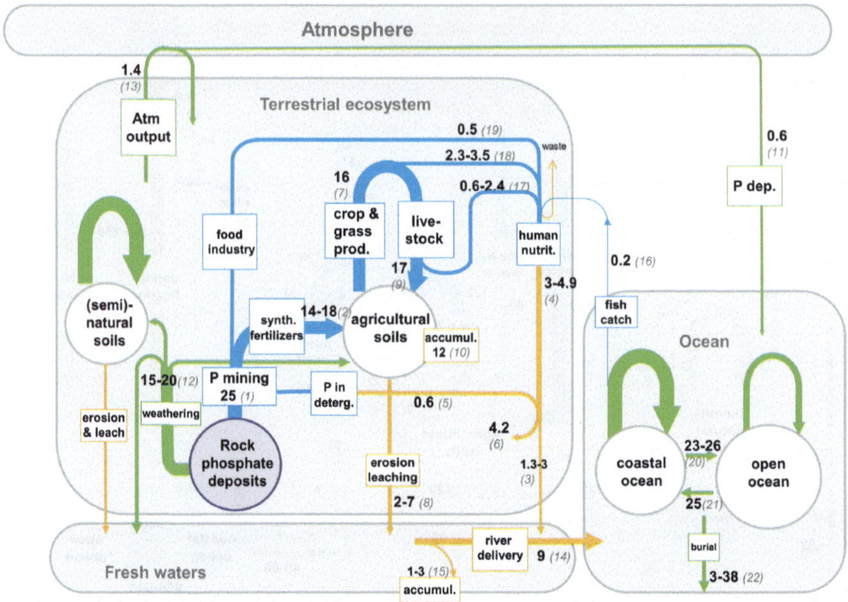

Fig. 28.2 Global phosphorus cycle around the years 2000-2010. The arrows show the phosphorus fluxes across environmental pools and compartments (green: natural fluxes, blue: intended human altered fluxes, orange: natural fluxes affected by anthropogenic activities). The figures in black indicate the phosphorus fluxes in Tg P/yr (yr = year) and the figures within brackets are the legend and refer to the accompanying table including the references for each flux (see Table 28.2). Note that not all figures add exactly, due to the use of different data sources (Sutton et al. 2013a). Figure reproduced from Sutton et al. (2013a) *Our Nutrient World: The challenge to produce more food and energy with less pollution* (with permission of the UK Centre for Ecology & Hydrology, © Centre for Ecology and Hydrology)

causes pollution through emissions of the greenhouse gas nitrous oxide (N_2O) and ammonia (NH_3) to the atmosphere, and losses of nitrates (NO_{3-}), phosphate and organic N and P compounds to water.

Further substantial increase of synthetic N and P fertilizer use of around 40–50% is expected over the next 50 years in order to feed the growing world population and in response to current trends in dietary lifestyle increasing consumption of animal products (Sutton et al. 2013a). These changes are expected to increase N and P emissions to the environment. In addition, according to current projections of climate and land use changes, biological and anthropogenic fixation will increase during the twenty-first century, as well as NH_3 emission to the atmosphere (Fowler et al. 2015).

The consequences of not taking action include: further warming effects, from increasing atmospheric N_2O; continuing deterioration via eutrophication of fresh waters and coastal seas; declines in soil quality, threatening ecosystem services and

Table 28.1 Global nitrogen fluxes around year 2000 reported in the literature. *Legend* refers to the values reported in Fig. 28.1

Legend	Global Nitrogen Fluxes	Tg N/year	Reference (Cited in the reference and/or additional references)
1	Fertilizer consumption	120	Fowler et al. (2013); Galloway et al. (2008); Bouwman et al. (2013)
2	N_2 crop fixation	50–70	Fowler et al. (2013); Herridge et al. (2008)
3	Crops & grass production	122	Billen et al. (2013)
4	Crops & grass for livestock production	100	Billen et al. (2013)
5	N back to agricultural soils	57	Based on Billen et al. (2013); and Sutton et al. (2013a)
6	NH_3 emissions—agricultural system—from crops & grass	15	Sutton et al. (2013b)
7	NH_3 emissions—agricultural system—from livestock	22	Sutton et al. (2013b)
8	NH_3 emissions—agricultural system (total)	37	Sutton et al. (2013b)
9	Crops for human nutrition	22	Billen et al. (2013)
10	Livestock for human nutrition	6	Billen et al. (2013)
11	Fish landing	3.7	Voss et al. (2013); and Maranger et al. (2008)
12	Food waste	13	Billen et al. (2013)
13	Human excretion	19	Billen et al. (2013)
14	Waste water treatment	13	Billen et al. (2013)
15	Sewage	6–8	Billen et al. (2013); Beusen et al. (2016)
16	Riverine input to oceans	40–66	Voss et al. (2013); Voss et al. (2011); Seitzinger et al. (2005); and Beusen et al. (2016)
17	Surplus in agricultural soils	120	Billen et al. (2013)
18	Input from agricultural soils to aquifers and rivers	95	Billen et al. (2013)

(continued)

Table 28.1 (continued)

Legend	Global Nitrogen Fluxes	Tg N/year	Reference (Cited in the reference and/or additional references)
19	Soil denitrification	25	Billen et al. (2013)
20	Denitrification in aquatic systems	52	Billen et al. (2013)
21	NO emissions from soils	10	Fowler et al (2013)
22	N_2O emissions from soils	10–13	Fowler et al. (2013); and Davidson and Kanter (2014)
23	Lightening	2–10	Fowler et al. 2013; (Levy et al. 1996; Tie et al. 2002)
24	N_2 natural fixation in terrestrial ecosystem s	58	Fowler et al. (2013); Vitousek et al. (2013)
25	NH_3 emissions—biomass burning	5.5	Sutton et al. (2013b)
26	NH_3 emissions—natural soils	4.9	Sutton et al. (2013b)
27	NH_3 emissions—natural ecosystems	10.4	Sutton et al. (2013b)
28	Combustion	30–40	Fowler et al. (2013); Van Vuuren et al. (2011)
29	Wet and dry deposition on soils	70	Fowler et al. (2013); Dentener et al. (2006); and Duce et al. (2008)
30	Wet and dry deposition on oceans	30	Fowler et al. (2013); Dentener et al. (2006); and Duce et al. (2008)
31	NH_3 emissions—oceans (& volcanoes)	8.6	Sutton et al. (2013b)
32	N_2O emissions from the ocean	3.8–5.5	Voss et al. (2013); IPCC (2007); Duce et al. (2008)
33	Denitrification in oceans	100–250	Voss et al. (2011); and Voss et al. (2013)

(continued)

Table 28.1 (continued)

Legend	Global Nitrogen Fluxes	Tg N/year	Reference (Cited in the reference and/or additional references)
34	N_2 Fixation by oceans	140	Voss et al. (2013); Deutsch et al. (2007); Duce et al. (2008)
35	Burial in oceans	22	Voss et al. (2013)
36	Flux from coastal ocean to open ocean	390	Voss et al. (2013)
37	Flux from open ocean to coastal ocean	450–600	Voss et al. (2013)
38	Wet and dry deposition of NH_x and NO_y on agricultural soils	50	Based on Dentener et al. (2006); and Duce et al. (2008)
39	Wet and dry deposition of NH_x and NO_y on natural soils	19	Based on Dentener et al. (2006); and Duce et al. (2008)

Table 28.2 Global phosphorus fluxes around year 2000 reported in the literature. The *Legend* refers to the values reported in Fig. 28.2. Some fluxes within the ocean refer to pre-anthropogenic values

Legend	Global Phosphorus Fluxes	Tg P/year	References
1	Rock phosphate mining	25	Calculated from Scholz and Wellmer (2013)
2	Fertilizer consumption	14–18	Heffer and Prud'Homme (2012); Bouwman et al. (2013); and Bouwman et al. (2009)
3	Sewage	1–3	Van Vuuren et al. (2010); Van Drecht et al. (2009); Mackenzie et al. (2002); and Beusen et al. (2016)
4	Human emissions	3–4.9	Calculated from (Van Drecht et al. 2009; Cordell et al. 2009)
5	Detergents	0.6	Calculated from Van Drecht et al. (2009)
6	Waste water treatment	4.2	Calculated from Van Drecht et al. (2009)
7	Harvest and grazing	16	Bouwman et al. (2009)
8	Input to fresh waters	2–7	Bouwman et al. (2009); (2013); Van Vuuren et al. (2010)

(continued)

Table 28.2 (continued)

Legend	Global Phosphorus Fluxes	Tg P/year	References
9	Input from manure	17	Bouwman et al. (2009, 2013)
10	Accumulation in soils	12	Bouwman et al. (2013); Bouwman et al. 2009; (see also Bennett et al. 2001; and references herein)
11	Deposition on ocean	0.6	Mahowald et al. (2008); (see also Kanakidou et al. 2012)
12	Weathering	15–20	Bennett et al. (2001); and references herein
13	Source of atmospheric P	1.39	Mahowald et al. (2008); (see also Benitez-Nelson 2000; and references herein)
14	Riverine input to oceans	4–9	Beusen et al. 2016; Seitzinger et al. 2010; Beusen et al. 2005); (see also Mackenzie et al. 2002; Bennett et al. 2001; Benitez-Nelson 2000; and references herein)
15	Accumulation in fresh waters	1–3.1	Bennett et al. (2001); and references herein
16	P in food supply—Fish, Seafood	0.2	Estimated using FAOSTAT data
17	P in food supply—Meat, Milk and Eggs	0.6–2.4	Estimated using FAOSTAT data; Van Vuuren et al. (2010); Cordell et al. (2009)
18	P in food supply—Cereals, Fruits and Vegetables	2.3–3.5	Estimate using FAOSTAT data; Cordell et al. (2009); (see also Van Vuuren et al. 2010)
19	Added Food Phosphates	0.5	Estimated
20	Flux from coastal ocean to open ocean	23–26	Calculated from Slomp and Van Cappellen 2007; Mackenzie et al. 2002
21	Flux from open ocean to coastal ocean	25	Calculated from Slomp and Van Cappellen 2007
22	Burial in ocean sediments	3–38	Calculated from Slomp and Van Cappellen 2007; Mackenzie et al. 2002; Benitez-Nelson 2000; and references herein

biodiversity; and deterioration of air quality shortening and reducing the quality of human life.

Acknowledgements The authors would like to thank Kevin Hicks for editing the manuscript.

References

Benitez-Nelson, C. R. (2000). The biogeochemical cycling of phosphorus in marine systems. *Earth-Science Reviews, 51*(1–4), 109–135.

Bennett, E. M., Carpenter, S. R., & Caraco, N. F. (2001). Human impact on erodable phosphorus and eutrophication: A global perspective. *BioScience, 51*(3), 227–234.

Beusen, A. H. W., Dekkers, A. L. M., Bouwman, A. F., Ludwig, W., & Harrison, J. (2005). Estimation of global river transport of sediments and associated particulate C, N, and P. *Global Biogeochemical Cycles, 19*(4), GB4S05. https://doi.org/10.1029/2005gb002453.

Beusen, A. H. W., Bouwman, A. F., Van Beek, L. P. H., Mogollón, J. M., & Middelburg, J. J. (2016). Global riverine N and P transport to ocean increased during the 20th century despite increased retention along the aquatic continuum. *Biogeosciences, 13,* 2441–2451.

Billen, G., Garnier, J., & Lassaletta, L. (2013). The nitrogen cascade from agricultural soils to the sea: Modelling nitrogen transfers at regional watershed and global scales. *Philosophical Transactions of the Royal Society B: Biological Sciences, 368*(1621), 20130123. https://doi.org/10.1098/rstb.2013.0123.

Bouwman, A. F., Beusen, A. H. W., & Billen, G. (2009). Human alteration of the global nitrogen and phosphorus soil balances for the period 1970–2050. *Global Biogeochemical Cycles, 23,* GB0A04. https://doi.org/10.1029/2009gb003576.

Bouwman, L., Goldewijk, K. K., Van Der Hoek, K. W., Beusen, A. H. W., Van Vuuren, D. P., Willems, J., et al. (2013). Exploring global changes in nitrogen and phosphorus cycles in agriculture induced by livestock production over the 1900–2050 period. *Proceedings of the National Academy of Sciences of the United States of America (PNAS), 110,* 20882–20887.

Cordell, D., Drangert, J. O., & White, S. (2009). The story of phosphorus: Global food security and food for thought. [Review]. *Global Environmental Change, 19*(2), 292–305.

Davidson, E. A., & Kanter, D. (2014). Inventories and scenarios of nitrous oxide emissions. *Environmental Research Letters, 9*(10), 105012. https://doi.org/10.1088/1748-9326/9/10/105012.

Dentener, F., Drevet, J., Lamarque, J. F., Bey, I., Eickhout, B., Fiore, A. M., Hauglustaine, D., Horowitz, L. W., Krol, M., Kulrestha, U. C., Lawrence, M., Galy-Lacaux, C., Rast, S., Shindall, D., Stevenson, D., Van Noije, T., Atherton, C., Bell, N., Bergman, D., Butler, T., Cofala, J., Collins, B., Doherty, R., Ellingsen, K., Galloway, J., Gauss, M., Montanaro, V., Müller, J. F., Pitari, G., Rodriguez, J., Sanderson, M., Solmon, F., Strahan, S., Schultz, M., Sudo, K., Szopa, S., & Wild, O. (2006). Nitrogen and sulfur deposition on regional and global scales: A multimodel evaluation. *Global Biogeochemical Cycles, 20*(4), GB4003. https://doi.org/10.1029/2005gb002672.

Deutsch, C., Sarmiento, J. L., Sigman, D. M., Gruber, N., & Dunne, J. P. (2007). Spatial coupling of nitrogen inputs and losses in the ocean. *Nature, 445*(7124), 163–167.

Duce, R. A., LaRoche, J., Altieri, K., Arrigo, K. R., Baker, A. R., Capone, D. G., et al. (2008). Impacts of atmospheric anthropogenic nitrogen on the open ocean. *Science, 320*(5878), 893–897.

Fowler, D., Coyle, M., Skiba, U., Sutton, M. A., Cape, J. N., Reis, S., et al. (2013). The global nitrogen cycle in the Twenty-first century. *Philosophical Transactions of the Royal Society B: Biological Sciences, 368*(1621), 20130164. https://doi.org/10.1098/rstb.2013.0164.

Fowler, D., Steadman, C. E., Stevenson, D., Coyle, M., Rees, R. M., Skiba, U. M., et al. (2015). Effects of global change during the 21st century on the nitrogen cycle. *Atmospheric Chemistry and Physics, 15,* 13849–13893.

Galloway, J. N., Townsend, A. R., Erisman, J. W., Bekunda, M., Cai, Z. C., Freney, J. R., et al. (2008). Transformation of the nitrogen cycle: Recent trends, questions, and potential solutions. *Science, 320*(5878), 889–892.

Grizzetti, B., Bouraoui, F., & Aloe, A. (2012). Changes of nitrogen and phosphorus loads to European seas. *Global Change Biology, 18,* 769–782.

Heffer, P., & Prud'homme, M. (2012). *Fertilizer Outlook 2012–2016*. Paper presented at the 80th IFA Annual Conference, Doha (Qatar), 21–23 May 2012.

Herridge, D. F., Peoples, M. B., & Boddey, R. M. (2008). Global inputs of biological nitrogen fixation in agricultural systems. *Plant and Soil, 311*(1–2), 1–18.

IPCC. (2007). *Climate Change 2007: The Physical Science Basis. Contribution of Working Group I to the Fourth Assessment Report of the Intergovernmental Panel on Climate Change*. Cambridge University Press.

Kanakidou, M., Duce, R.A., Prospero, J. M., Baker, A. R., Benitez-Nelson, C., Dentener, F. J., Hunter, K. A., Liss, P. S., Mahowald, N., Okin, G. S., Sarin, M., Tsigaridis, K., Uematsu, M., Zamora, L. M., & Zhu, T. (2012). Atmospheric fluxes of organic N and P to the global ocean. *Global Biogeochemical Cycles*, 26, GB3026. https://doi.org/10.1029/2011gb004277.

Levy, H., Moxim, W. J., & Kasibhatla, P. S. (1996). A global three-dimensional time-dependent lightning source of tropospheric NO_X. *Journal of Geophysical Research-Atmospheres, 101*(D17), 22911–22922.

Mackenzie, F. T., Vera, L. M., & Lerman, A. (2002). Century-scale nitrogen and phosphorus controls of the carbon cycle. *Chemical Geology, 190*(1–4), 13–32.

Mahowald, N., Jickells, T. D., Baker, A. R., Artaxo, P., Benitez-Nelson, C.R., Bergametti, G., Bond, T. C., Chen, Y., Cohen, D. D., Herut, B., Kubilay, N., Losno, R., Luo, C., Maenhaut, W., McGee, K. A., Okin, G. S., Siefert, R. L., & Tsukuda, S. (2008). Global distribution of atmospheric phosphorus sources, concentrations and deposition rates, and anthropogenic impacts. *Global Biogeochemical Cycles, 22*(4), GB4026. https://doi.org/10.1029/2008gb003240.

Maranger, R., Caraco, N., Duhamel, J., & Amyot, M. (2008). Nitrogen transfer from sea to land via commercial fisheries. *Nature Geoscience, 1*(2), 111–113.

Scholz, R. W., & Wellmer, F.-W. (2013). Approaching a dynamic view on the availability of mineral resources: what we may learn from the case of phosphorus? *Global Environmental Change, 23*(1), 11–27.

Seitzinger, S. P., Harrison, J. A., Dumont, E., Beusen, A. H. W., & Bouwman, A. F. (2005). Sources and delivery of carbon, nitrogen, and phosphorus to the coastal zone: An overview of Global Nutrient Export from Watersheds (NEWS) models and their application. *Global Biogeochemical Cycles, 19*(4), GB4S01. https://doi.org/10.1029/2005gb002606.

Seitzinger, S. P., Mayorga, E., Bouwman, A. F., Kroeze, C., Beusen, A. H. W., Billen, G., Van Drecht, G., Dumont, E., Fekete, B. M., & Harrison, J. A. (2010). Global river nutrient export: A scenario analysis of past and future trends. *Global Biogeochemical Cycles, 24*. https://doi.org/10.1029/2009gb003587.

Shantz, A. A., Lemoine, N. P., & Burkepile, D. E. (2016). Nutrient loading alters the performance of key nutrient exchange mutualisms. *Ecology Letters, 19*(1), 20–28.

Slomp, C. P., & Van Cappellen, P. (2007). The global marine phosphorus cycle: Sensitivity to oceanic circulation. *Biogeosciences, 4*(2), 155–171.

Sutton, M., Howard, C. M., Erisman, J. W., Billen, G., Bleeker, A., Grennfelt, P., et al. (2011). *The European Nitrogen Assessment: Sources, effects and policy perspectives*. Cambridge: Cambridge University Press.

Sutton, M. A., Bleeker, A., Howard, C. M., Bekunda, M., Grizzetti, B., de Vries, W., van Grinsven, H. J. M., Abrol, Y. P., Adhya, T. K., Billen, G., Davidson, E. A., Datta, A., Diaz, R., Erisman, J. W., Liu, X.J., Oenema, O., Palm, C., Raghuram, N., Reis, S., Scholz, R.W., Sims, T., Westhoek, H., & Zhang, F. S. (2013a). *Our Nutrient World: The challenge to produce more food and energy with less pollution*. Global Overview of Nutrient Management. Edinburgh, UK: Centre for Ecology and Hydrology, on behalf of the Global Partnership on Nutrient Management and the International Nitrogen Initiative.

Sutton, M. A., Reis, S., Riddick, S. N., Dragosits, U., Nemitz, E., Theobald, M. R., Tang, Y. S., Braban, C. F., Vieno, M., Dore, A. J., Mitchell, R. F., Wanless, S., Daunt, F., Fowler, D., Blackall, T. D., Milford, C., Flechard, C. R., Loubet, B., Massad, R., Cellier, P., Personne, E., Coheur, P. F., Clarisse, L., Van Damme, M., Ngadi, Y., Clerbaux, C., Skjoth, C. A., Geels, C., Hertel, O., Kruit, R. J. W., Pinder, R. W., Bash, J. O., Walker, J.., Simpson, D., Horvath, L., Misselbrook, T. H., Bleeker, A., Dentener, F., & de Vries, W. (2013b). Towards a climate-dependent paradigm of ammonia emission and deposition. *Philosophical Transactions of the Royal Society B: Biological Sciences, 368*(1621), 20130166. https://doi.org/10.1098/rstb.2013.0166.

Tie, X. X., Zhang, R. Y., Brasseur, G., & Lei, W. F. (2002). Global NO_x production by lightning. *Journal of Atmospheric Chemistry, 43*(1), 61–74. https://doi.org/10.1023/a:1016145719608.

Van Drecht, G., Bouwman, A. F., Harrison, J., & Knoop, J. M. (2009). Global nitrogen and phosphate in urban wastewater for the period 1970 to 2050. *Global Biogeochemical Cycles, 23*, GB0A03. https://doi.org/10.1029/2009gb003458.

Van Vuuren, D. P., Bouwman, A. F., & Beusen, A. H. W. (2010). Phosphorus demand for the 1970-2100 period: A scenario analysis of resource depletion. *Global Environmental Change, 20*(3), 428–439.

Van Vuuren, D. P., Bouwman, L. F., Smith, S. J., & Dentener, F. (2011). Global projections for anthropogenic reactive nitrogen emissions to the atmosphere: an assessment of scenarios in the scientific literature. *Current Opinion in Environmental Sustainability, 3*(5), 359–369.

Vitousek, P. M., Menge, D. N. L., Reed, S. C., & Cleveland, C. C. (2013). Biological nitrogen fixataion: Rates, patterns, and ecological controls in terrestrial ecosystems. *Philosophical Transactions of the Royal Society B, 368*(1621), 20130119. https://doi.org/10.1098/rstb.2013.0119.

Voss, M., Baker, A., Bange, H. W., Conley, D., Cornell, S., Deutsch, B., et al. (2011). Nitrogen processes in coastal and marine systems. In M. A. Sutton, C. M. Howard, J. W. Erisman, G. Billen, A. Bleeker, P. Grennfelt, H. van Grinsven, & B. Grizzetti (Eds.), *The European nitrogen assessment: Sources, effects and policy perspectives (Chapter 8* (pp. 147–176). Cambridge: Cambridge University Press.

Voss, M., Bange, H. W., Dippner, J. W., Middelburg, J. J., Montoya, J. P., & Ward, B. (2013). The marine nitrogen cycle: recent discoveries, uncertainties and the potential relevance of climate change. *Philosophical Transactions of the Royal Society of London. Series B, Biological sciences, 368*(1621), 20130121. https://doi.org/10.1098/rstb.2013.0121.

Chapter 29
A First Approach to the Calculation of Nitrogen Footprint in Lisbon, Portugal

Cláudia M. d. S. Cordovil, Vitória Gonçalves, Amarilis de Varennes, Allison M. Leach, and James N. Galloway

Abstract Food and energy production have negative impacts due to the buildup of reactive nitrogen in the environment and subsequent nitrogen (N) impacts. To quantify the nitrogen footprint left in the environment by an individual, scientists from the University of Virginia USA developed a model to calculate nitrogen footprints, the N-Calculator, which transforms data on food and energy consumption into kg N per year lost into the environment. To calculate the average N footprint for an individual in the area of Lisbon, one thousand Lisbon inhabitants were surveyed at random using the N-Footprint questionnaire. The average N footprint estimated for Portugal was 25.5 kg N *per capita* per year and the average personal N footprint for the Lisbon area was 24.7 kg N *per capita* per year. The average N footprint for men and women was similar and, despite a tendency for people over the age of 65 to have a lower footprint, there was no statistical difference across the age groups studied (e.g., the N-Footprint for > 65 and 21–30 years of age was 23.5 and 25.8 kg N *per capita* per year, respectively). Losses of N to the environment were always higher for food consumption than for energy. Generally, the food types consumed do not comply with the recommendations of the Portuguese Association of Nutritionists for a healthy diet.

C. M. d. S. Cordovil (✉) · V. Gonçalves · A. de Varennes
Universidade de Lisboa, Instituto Superior de Agronomia, CEF, Tapada da Ajuda, 1349-017 Lisboa, Portugal
e-mail: cms@isa.ulisboa.pt

V. Gonçalves
e-mail: vitoriampgnascimento@gmail.com

A. de Varennes
e-mail: adevarennes@isa.ulisboa.pt

A. M. Leach
University of New Hampshire, The Sustainability Institute at UNH, 107 Nesmith Hall, 131 Main St., Durham, NH 03824, USA
e-mail: Allison.Leach@unh.edu

J. N. Galloway
Environmental Sciences Department, University of Virginia, 291 McCornik Road, P.O. Box 400123, Charlottesville, VA 22904, USA
e-mail: jng@virginia.edu

M. A. Sutton et al. (eds.), *Just Enough Nitrogen*,
https://doi.org/10.1007/978-3-030-58065-0_29

Keywords Food consumption · Nitrogen losses · Nitrogen footprint

29.1 Introduction

Nitrogen (N) is one of the most important elements on earth and constitutes 78% of the atmosphere in the form of di-nitrogen (N_2), which cannot be directly used by the majority of plant species and is not readily available for living organisms (Erisman et al. 2008; Galloway et al. 2003). All the N compounds on Earth can be divided in two groups: non-reactive N_2, and reactive N (N_r) which includes all the other N compounds present in the Earth's atmosphere and biosphere (Galloway et al. 2003). When N_r is released into the environment it can affect air, water, and soil quality, change biodiversity and ecosystems and alter greenhouse gases balance (Galloway et al. 2008).

Human activities are the main driver for N loss into the environment, and food and energy production are the main causes for N_r accumulation. In 2008 Galloway et al. reported that ~77% of anthropogenic N_r is generated from food production, ~16% from energy production, and only ~9% from industrial uses. To raise awareness about the excess of N_r in the environment, Leach et al. (2012) developed the first personal nitrogen footprint tool (the N-Calculator) to calculate the amount of N_r lost to the environment from personal consumption of food and energy, including that associated with production practices (Leach et al. 2012). The N-Calculator tool allows the user to calculate and track and reduce their N footprint.

The objective of this work was to calculate the average *per capita* N Footprint for Portugal and for a specific region in Portugal, Lisbon, by adapting the N-Calculator proposed by Leach et al (2012) to typical consumption habits in Portugal. Parameters adjusted for the Portuguese diet differ in some cases from the existing parameters for the United States (US), United Kingdom (UK), Germany (DE) and the Netherlands (NL) (www.n-print.org).

29.2 Materials and Methods

To calculate the average *per capita* N footprint in Lisbon we used the model N-Calculator developed by Leach et al. (2012), by conducting the survey available at www.n-print.org.

At the time of the last census in Portugal (INE 2014) there were 2,821,876 residents in the Lisbon area. The survey (Table 29.1) was performed in different parts of the city and surrounding area, including the oldest neighborhoods of Lisbon, the new residential parts of the city, as well as the commercial and industrial surrounding areas, in order to cover the full range of characteristic areas of the Lisbon region. People invited to take part in the questionnaire survey were all Portuguese and residents in Portugal. They were chosen at random on the street, and their age range identified. Interviews

Table 29.1 Questions in the survey, divided by the main areas of consumption

	Questions
Identification	Gender and the age of the person taking the survey
Food	How many weekly doses consumed: poultry meat, pig meat, bovine meat, milk, butter and yogurts, cheese, fish and shellfish, animal fat, mutton and goat meat, eggs, cereals (wheat), rice, other cereals (including breakfast flakes), pasta, fruit, beans and grain, starchy roots, dry fruits, olive oil and olives, cooked vegetables, fresh vegetables, sugar and sweeteners, oil crops and spices)
	How many eggs eaten per week?
Drinks	How many glasses of wine drunk per week?
	How many glasses of alcoholic white spirits drunk per week?
	How many cups of coffee and/or tea drunk per week?
	How many glasses of beer drunk per week?
	How many glasses of soft drinks drunk per week?
Others	Do you smoke? If yes, how many cigarettes do you smoke per day?
	Is your house linked to a sewerage system with tertiary treatment of sewage?
Transport	How many hours of plane travelled this year?
	How many km travelled in public transport per week?
	How many km travelled travel by car per week?
	How many km travelled by motorcycle per week?
	Do you have your own vehicle? If yes, what kind of fuel do you use?

Some modifications were made to the original model survey. The most relevant changes are the inclusion of rice, olives and olive oil, and of gas cylinder as an energy source

included a brief presentation of the N problem and the importance of assessing N footprints, to motivate and inform the interviewees. The survey consisted of questions about food consumption and energy consumption habits, including household appliances and types of transportation used.

Survey data obtained in one thousand interviews was used as input data to calculate the participant's N footprint by using the N-calculator formulas proposed by Leach et al. (2012) adapted to the food consumption habits in Portugal (INE 2014). Food categories were changed accordingly and the following food types were added to the calculations: butter and yogurts, animal fats, mutton and goat meat, other cereals (including flakes), oil and olive oil, cooked vegetables, fresh vegetables, oil crops, wine, coffee and tea, beer, and soft drinks. Portions used to calculate the N-footprint correspond to Portuguese data and are described in Table 29.2 (APN 2011; INE 2014). We compared the daily recommended intake with actual daily intake and then adjusted the input into the N-calculator, because the actual intake is different to that recommended in the food wheel defined by the Portuguese Association of Nutritionists (APN 2011; INE 2014) (Table 29.3).

The FAO database was used to obtain the values for food supply, expressed in kg per person per year, and protein supply in g per person per day (FAO 2011), in order

Table 29.2 Comparison of food types: recommended and actual consumption (statistics) expressed as percentage of the total diet (APN 2011; INE 2014)

Food type	Recommended (%)	Consumed (%)
Fruit	20	14
Oils and fats	2	6
Dairy products	18	20
Meat, fish and eggs	5	16.3
Pulses	4	0.7
Cereals and root crops	28	30
Vegetables	23	13

Table 29.3 Recommended food portion sizes for Portugal (APN 2011)

Food Type	Portion Sizes (g)
Animal Products	
Poultry meat	120
Pork meat	120
Beef meat	120
Milk	275
Butter & yogurts	220
Cheese	44
Fish and seafood	120
Animal fats	30
Mutton und goat meat	120
Eggs	50
Vegetal products	
Cereals (wheat)	50
Rice	110
Other cereals (inc. flakes)	35
Mass	110
Fruits	150
Beans & grains	10
Starchy roots	125
Oil and olives oil	30
Vegetables cooked	160
Fresh vegetables	160
Oil crops	30
Drinks	
Wine	240
Alcoholic beverages	240
Coffee or tea	8
Beer	350
Soft drinks	330

Table 29.4 Data for average percentage of food waste for Europe (FAO 2011)

Food type	Food waste (%)
Cereals	27
Roots and tubers	23
Oilseeds and pulses	5
Fruit and vegetables	27
Meat	15
Fish and seafood	19
Milk	7

to calculate N-footprint from food consumption. Protein data was converted to kg per person per year, and multiplied by a factor of 0.16 to calculate N supply from proteins (Leach et al. 2012). Nitrogen loss data related to food waste was taken from the FAO database (2011) (Table 29.4). When a food waste factor was not available for a specific food category, we used the factor for the most similar food category. We also used the assumption that roughly one-third of food produced for human consumption is lost or wasted globally (FAO 2011).

In order to relate the units of N_r released to the environment per unit of N_r consumed, Leach et al. (2012) created the concept of a virtual N factor to estimate $N_{r\ released}/N_{r\ consumed}$. The virtual N factors used were derived from Stevens et al. (2014) (Table 29.5). We used factors from Europe, because to date there are no specific virtual N factors available for Portugal. For other food categories, we used the N virtual factor provided for similar production processes (Leach et al. 2012).

The calculation of the energy N Footprint was based on consumption data for three sectors: housing (electricity, natural gas and gas cylinder), transportation (public transport, car and motorcycle, plane) and goods and services. Average housing energy consumption by type was derived from data obtained in Portuguese official statistics (Pordata 2011). Emission factors not available in Portugal, were obtained from Dutch data (Leach et al. 2012). The national emission of N_r/N by sector was obtained by multiplying the emission factor with the average figure for housing energy consumption and the number of households in Portugal (4,043,726 according to INE 2014). The size of the Portuguese population is approximately 10,562,178 based on the most recent census for 2014 (INE 2014). In the transportation sector, the average energy consumption and the emission factors were obtained from Dutch data (Leach et al. 2012), except for motorcycles. For those, Italian factors (Leach and Galloway 2011) were considered to be more representative of Portuguese consumption standards (Table 29.6). The total N_r/N emission in Portugal in 2009 was 103.3 kt (APA 2013), with only 52.9% of people in Portuguese travelling by car, 21.9% using public transport modes, and 1.1% travelling by motorcycle (IMTT 2010). Emission of N_r/N from aircraft was estimated at 0.128 kg hour^{-1} (APA 2013). As the goods and services sector shows great complexity in quantification of specific emission factors, the Dutch values considered by Leach et al. (2012) were used in this study.

Table 29.5 Virtual nitrogen (N) factors from Europe (Stevens et al. 2014)

Food type	Virtual N factor
Poultry meat	3.2
Pig meat	4.4
Beef	7.9
Milk (except butter)	3.9
Cheese	3.9
Eggs	4.4
Fish and seafood	2.9
Animal fats	5.2
Mutton and goat meat	5.2
Stimulants	8.2
Cereals	1.3
Rice	1.3
Fruits	8.2
Pulses	0.5
Starchy roots	1.1
Vegetables	8.2
Nuts	0.5
Alcoholic beverages	1.3
Oil crops	0.5
Spices	8.2
Sugar and sweeteners	8.2
Vegetable oils	8.2

Table 29.6 Emission factors and average energy consumption by sector and type

Residential homes	Emission factor	Unit	AVG per capita energy consumption	Unit
Electricity	0.000102	kg N/kwh	387.68	kwh/month
Natural gas	0.000495	kg N/m^3	5.76	m^3/month
Gas cylinder	0.000495	kg N/m^3	4.8	m^3/month
Transportation	Emission factor	Unit	AVG distance/time traveled per capita	Unit
Airplane	0.128000	kg N/hour	0.31	hours/year
Public transport	0.000043	kg N/km	26.1	km/week
Private car	0.000101	kg N/km	171	km/week
Motorcycles	0.000001	kg N/km	10.72	km/week

The average *per capita* N energy footprint was calculated according to Leach et al. (2012) by adding all categories.

29.3 Results

For Portugal the average *per capita* N Footprint was calculated as 25.45 kg N/capita/year. The average N footprint from food consumption was 5.68 kg N/capita/year and the average N footprint from food production was 17.70 kg N/capita/year. The average N footprint from energy was 2.07 kg N/capita/year, consisting of 0.57 kg from housing and 1 kg from transportation. The average N-footprint from goods and services was 0.5 kg N/capita/year. In Fig. 29.1, there is a comparison between N-footprints in Portugal, and the national N footprints in the US, the Netherlands, UK and Germany. The N-footprint for the US is 41.20 kg N/capita/year, for the Netherlands 24.80 kg N/capita/year, for the UK 27.1 kg N/capita/year and for Germany 24 kg N/capita/year. Globally, the US has the largest per capita N-footprint, and Portugal and the Netherlands have similar N-footprint values.

Figure 29.2 shows the comparison between the average N-footprint in Portugal and the Lisbon area. The average N-footprint for an individual in Lisbon was calculated as 24.73 kg N/capita/year, lower than that for Portugal, 25.45. We observed that the average N-footprint for men and women was similar, 24.22 and 25.10 kg N/capita/year for males and females respectively. The N-footprint for energy

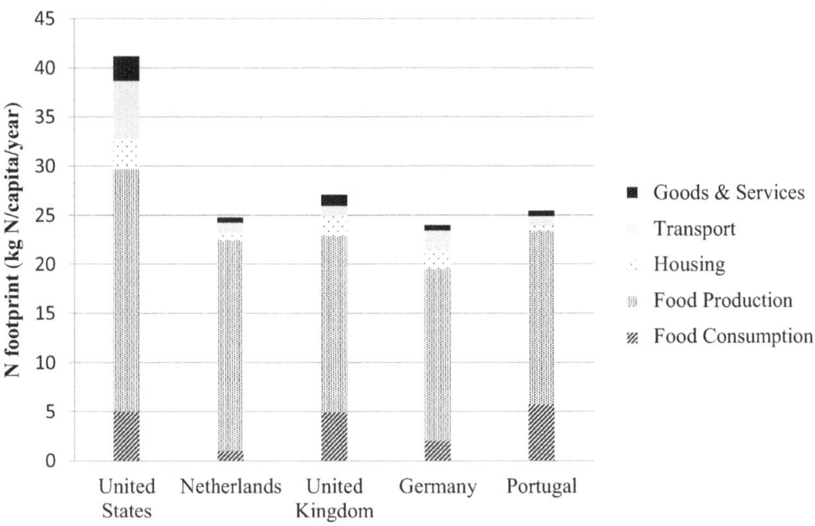

Fig. 29.1 Comparison of current Nitrogen Footprint *per capita* in the US, the Netherlands (Leach et al. 2012), UK, Germany (Stevens et al. 2014), and Portugal

Fig. 29.2 Comparison of Nitrogen Footprint *per capita* in Lisbon (present sampling), and Portugal

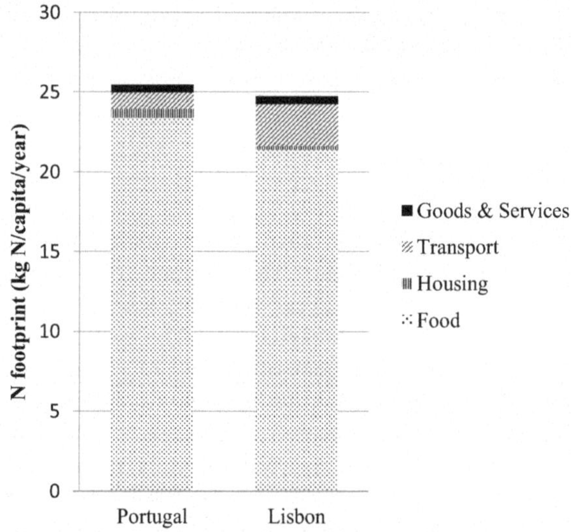

for men was 3.35 kg N/capita/year and for women was 3.33 kg N/capita/year. In the food sector the N-footprint was 21.77 kg N/capita/year for women and 20.87 kg N/capita/year for men. The differences were not statistically different ($p <$ 0.05).

There was a slight, but not statistically significant variation, between N-footprints for individuals within the different age classes surveyed (Fig. 29.3). For example, the results show that the N-footprint for individuals tends to be lower for people older than 65 years old (23.51 kg N/capita/year) and is higher for people between 21–30 years old (25.83 kg N/capita/year).

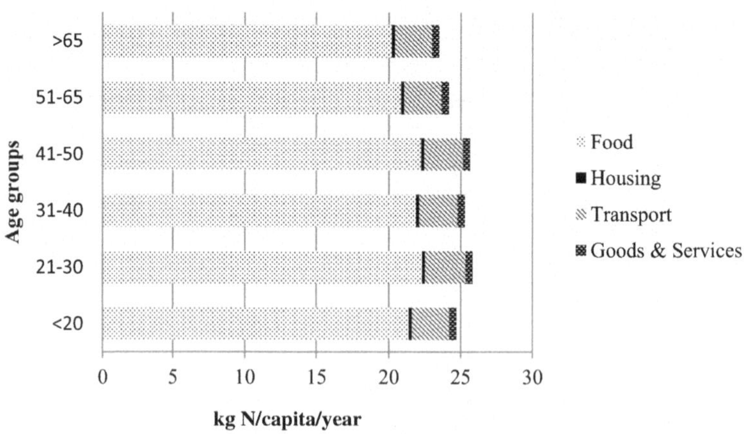

Fig. 29.3 Nitrogen Footprint *per capita* for different age groups (kg N/capita/year)

29.4 Discussion

Food is by far the most significant component of the N-footprint in all countries studied, and food production has a higher contribution than food consumption. There is a substantial difference between Portugal and the US in terms of energy and food consumption, as food culture and consumption habits are different. The average *per capita* N-footprint for Portugal is more similar to that of others European countries as shown in Fig. 29.1.

The average N-footprint of an individual for Lisbon, for both genders and different ages was slightly lower than the average *per capita* N-footprint for Portugal. Although Lisbon is the largest city and the capital of Portugal, in the North of the country the income *per capita* is higher which leads to higher beef consumption than in the centre and South. There are also lower prices for commodities in the North (INE 2014) and food portions are also bigger in the North of the country than in Lisbon due to the food culture The people that were surveyed may not be representative of Portugal overall as there are differences in consumption patterns throughout the country, but they are a representative sample of the Lisbon population. Moreover, since all of the surveys were made in the Lisbon area, it is not likely that they representative of the rural population from the interior of the country. On the other hand, almost all the fast food brands have their locations in cities along the coast of the country where most of the Portuguese population lives. There are also other factors that may have led to the differences found in the numbers reported in this survey. For example, surveyed people may not have accurately reported their food consumption because of recall bias. Portugal is also one of the countries in Europe with less fast food consumption (Vogli et al. 2014), which may also contribute to the N-footprint for the country as a whole being slightly larger (although not significantly) than that for the capital Lisbon.

A significant proportion of the population is aged over 65 and eats smaller portions of food, so their consumption and food waste are reduced. They also do not tend to use a car or travel by plane, and do not need to use transport modes for a daily commute to work, because most of them are retired. Therefore, the average N-footprint of an individual has a tendency to be lower for people over 65 years of age. However, people between 21 and 30 years old are more active and tend to eat larger portions of food.

29.5 Conclusions

The Portuguese population has an N-footprint similar to that of the Netherlands. The average *per capita* N-footprint for Lisbon is very similar to the average N-footprint for the whole of Portugal. Despite a tendency for different N-footprints according to age group, no statistically significant difference was found. In order to increase the representativeness of the results, more surveys are needed in the North and in the

South of Portugal. More information related to all emission factors for the specific Portuguese situation are required, as well as average values of energy and transport consumption in Portugal. To better estimate food consumption N-footprints, it is also necessary to collect national data relating to food waste and generate specific virtual N-factors for Portugal.

There are several ways of reducing the N-footprint of the food and energy production sectors, especially related to dietary choice. The N-Footprint approach has the potential to help Portuguese consumers understand the ways in which their habits and choices influence their N-footprint. The results need to be communicated to the general public as a first step towards reducing the impacts of N loss on human health and the environment.

Acknowledgements This work was partially funded by national funds from Fundação para a Ciência e a Tecnologia through the research unit **UID/AGR/04129/2013 (LEAF)**, ADISA, IEFP, Caixa Geral de Depósitos bank, Valorfito and EPCA consulting.

References

APA. (2013). *Portuguese Environment Agency (APA) data*. Online access at http://www.apambiente.pt/. Accessed Mar 2014.

APN. (2011). *Portuguese Association of Nutritionists (APN) data*. Online access at http://www.apn.org.pt/. Accessed Mar 2013.

Erisman, J. W., Sutton, M. A., Galloway, J. N., Klimont, Z., & Winiwarter, W. (2008). How a century of ammonia synthesis changed the world. *Nature Geoscience, 1*, 636–639.

FAO. (2011). *Food and Agriculture Organization of the United Nations (FAO) database*. Online access at http://www.fao.org/home/en/. Accessed in Mar 2016.

Galloway, J. N., Aber, J. D., Erisman, J. W., Seitzinger, S. P., Howarth, R. W., Cowling, E. B., et al. (2003). The nitrogen cascade. *BioScience, 53*(4), 341–356.

Galloway, J. N., Raghuram, N., & Abrol, Y. P. (2008). A perspective on reactive nitrogen in a global. Asian and Indian context. *Current Science, 94*(11), 1375–1381.

IMTT. (2010). *Institute for Mobility and Transports (IMTT) data*. Online access at http://www.imtt.pt. Accessed in Mar 2014.

INE. (2014). *Portuguese national statistics institute (INE) statistics*. Online access at www.ine.pt. Accessed in Mar 2014.

Leach, A. M., & Galloway, J. N. (2011). *N-PRINT Workshop Report*. The Royal Institution of Great Britain in London, UK, 3–4 December 2011. Available online at www.n-print.org.

Leach, A. M., Galloway, J. N., Bleeker, A., Erisman, J. W., Kohn, R., & Kitzes, J. (2012). A nitrogen footprint model to help consumers understand their role in nitrogen losses to the environment. *Environmental Development, 1*, 40–66.

Pordata. (2011). *Portuguese national statistics database*. Organized by Foundation Francisco Manuel dos Santos (FFMS). http://www.pordata.pt/. Accessed in Mar 2014.

Stevens, C. J., Leach, A. M., Dale, S., & Galloway, J. N. (2014). Personal nitrogen footprint tool for the United Kingdom. *Environmental Science: Processes & Impacts, 16*(7), 1563–1569.

Vogli, R., Kouvonen, A., & Gimeno, D. (2014). The influence of market deregulation on fast food consumption and body mass index: A cross-national time series analysis. *Bulletin of the World Health Organization, 92*(2), 99–107A.

Part X
Management Tools and Assessment: International Nitrogen Initiative Regional Centres

Chapter 30
The INI European Regional Nitrogen Centre: Concepts and Vision

Wilfried Winiwarter, Wim de Vries, Bruna Grizzetti, W. Kevin Hicks, Hans J. M. van Grinsven, and Maren Voss

Abstract In the global setting of the International Nitrogen Initiative (INI), the European Centre facilitates enhanced cooperation and integration among European researchers, policy makers and practitioners on environmental issues related to reactive nitrogen. INI-Europe represents a region that is characterized by agronomic challenges posed by high population density and the associated large food demand, but a declining economic share of agriculture. It is largely an area of excess nitrogen, a fact that is increasingly being recognized by stakeholders and environmental policy. INI-Europe aims to promote awareness building and to provide scientific information to stakeholders and the policy process, in order to facilitate implementation of measures to reduce environmental nitrogen loads and associated impacts.

W. Winiwarter (✉)
International Institute for Applied Systems Analysis (IIASA), Laxenburg, Austria
e-mail: winiwarter@iiasa.ac.at

W. de Vries
Wageningen University and Research (WUR), Droevendaalsesteeg 3, 6708 PB Wageningen, The Netherlands
e-mail: wim.devries@wur.nl

B. Grizzetti
European Commission, Joint Research Centre (JRC), Ispra, Italy
e-mail: Bruna.GRIZZETTI@ec.europa.eu

Sorbonne Université, CNRS, EPHE, UMR 7619 Metis, 4 Place Jussieu, 75005 Paris, France

W. K. Hicks
Stockholm Environment Institute (SEI), Department of Environment and Geography, University of York, York, YO10 5DD, UK
e-mail: kevin.hicks@york.ac.uk

H. J. M. van Grinsven
PBL Netherlands Environmental Assessment Agency, Bezuidenhoutseweg 30, 2594 AV The Hague, The Netherlands
e-mail: Hans.vanGrinsven@pbl.nl

M. Voss
Leibniz Institute for Baltic Sea Research, Warnemuende, Germany
e-mail: maren.voss@io-warnemuende.de

M. A. Sutton et al. (eds.), *Just Enough Nitrogen*,
https://doi.org/10.1007/978-3-030-58065-0_30

Keywords International Nitrogen Initiative · Europe · Nitrogen sources · Nitrogen effects · Nitrogen management

30.1 Introduction—Scope of INI-Europe

As much as Carbon represents energy use in a very general sense, the element Nitrogen (N) is associated with supporting life on earth. Reactive nitrogen (N_r), the sum of all chemically active forms of N available to organisms, drives all organic growth processes. As such it is indispensable for life, while the additional input of N is a major requirement for supporting food production for an ever increasing global population. This increase triggers an interference with N's natural cycle, which also strongly affects the structure and function of ecosystems (Sutton et al. 2013).

Anthropogenic activities have roughly doubled the amount of N cycled globally compared to pre-industrial levels (Galloway et al. 2004, 2008). As the unintended consequence of food production, various global and regional environmental problems derive from this increased availability of N, including: (i) human health impacts related to ozone formation induced by nitrogen oxides (NO_x) as well as ammonia (NH_3) and NO_x induced formation of particulate matter, (ii) global warming due to emissions of nitrous oxides (N_2O) and NO_x-induced formation of tropospheric ozone, and (iii) eutrophication of terrestrial and aquatic ecosystems (Erisman et al. 2013). Addressing these problem areas individually is often not effective, due to possible "pollution swapping", e.g., when measures preventing NH_3 emissions induce the respective N to reappear as increased N_2O emissions and N leaching. Nevertheless, policy approaches to tackle nitrogen problems are still mainly oriented according to problem areas and environmental media and are not sufficiently integrated.

The overall goal of the International Nitrogen Initiative (INI) is to optimize N's beneficial role in sustainable food production and to minimize N's negative effects on human health and the environment. Therefore INI-Europe (the European Centre of INI) supports scientific and policy approaches that address reactive nitrogen as a cross-cutting topic and look for integrated ways to handle N-related pollution. INI-Europe sees it as a benefit to both agricultural and manufacturing industry when regulation provides one consistent set of limiting measures, rather than industry having to deal with quite different demands and legislation from individual policy angles.

INI-Europe has been active in providing appropriate information on the scientific background as well as in devising appropriate mitigation measures of N-related pollution. It also looks into ways of implementing scientific, engineering and policy tools in practice to solve nitrogen related problems.

30.2 Achievements and Activities

INI-Europe was established with the start of INI in 2004 as one of its regional centres at the Energy Research Centre of the Netherlands (ECN). In this first phase, it initiated and coordinated the ESF-programme "Nitrogen in Europe" and the COST Action 729 on nitrogen fluxes in the atmosphere-biosphere system (see Bleeker and Erisman 2011), which included European wide workshops on issues such as nitrogen deposition and biodiversity conservation (Hicks et al. 2011).

From 2007 to 2012, INI-Europe was situated at the Centre for Ecology & Hydrology (CEH), Edinburgh. The European Nitrogen Assessment (ENA) (Sutton et al. 2011a), published in this period, represents a comprehensive overview of the fate of N_r in the environment, the threats posed by excess N_r, and the challenges to resolve these issues. The monetary valuation of N_r damage (Sutton et al. 2011b) is merely one aspect of the full picture. Further assessment-type studies that INI-Europe was heavily involved in included an overview on the global nutrient requirements for food production (Sutton et al. 2013) and a UNEP study on the greenhouse gas N_2O (Alcamo et al. 2013).

During the period of concern for this paper, 2013–2016, INI-Europe was based at the Austrian Academy of Sciences in Vienna, before continuing to its next site at the Stockholm Environment Institute (SEI), based at York University, UK. The Centre keeps close interaction with its community by way of a constantly updated website (http://initrogen.org/europe) and via an e-mail newsletter ("nitrogen alerts", see: https://bit.ly/2lSnplt) that is sent to a distribution list of 200+ subscribers. It links to policy processes by way of organizing specific sessions at events related to the science-policy interface (e.g., INI Europe chaired a working group on "Nitrogen" at the "Saltsjöbaden V" workshop in 2013) as well as meetings of agro-scientists (e.g., INI-Europe co-sponsored the 18th N-workshop in Lisbon in 2014 and moderated a working group "Building a blueprint for food with less N").

30.3 Current Challenges

INI-Europe facilitates enhanced cooperation and integration among European researchers, policy makers and practitioners to cover environmental issues related to N_r. INI-Europe's task includes to inventory, review and synthesize work from existing related activities like those of EUROSTAT and OECD, the Task Force on Reactive Nitrogen (TFRN) and its expert panels operating under the Convention on Long-Range Transboundary Air Pollution (LRTAP) . It also includes contributions to work in research projects under the EU's 7th Framework or Horizon 2020 programmes or similar activities. Below, seven key focal areas are briefly described that summarize the aims, topics, and rationale behind the approaches that are taken to better understand and resolve issues related to N_r.

30.3.1 Regional Nitrogen Budgets and Nitrogen Pathways/Emissions

Understanding the quantities of N_r flows between environmental pools allows key flows to be differentiated from less important ones. Budget approaches (farm, soil, land and total N budgets) enable the use and comparison of very different data sets and thus can help identify possible intervention points in N_r cycles. Spatially disaggregated soil and land N budgets have been derived for the EU (De Vries et al. 2011a; Leip et al. 2011), partly as contributions to the ENA. Based on these activities, the Expert Panel on Nitrogen Budgets operating under the TFRN prepared a guidance document on national total N budgets (UN-ECE 2013). This guidance is now being applied to individual countries. One study was published on N budgets in Denmark (Hutchings et al. 2014), which provided evidence of the significant reductions of N flows in Denmark over a 20 year period.

30.3.2 Nitrogen Compounds in Air and Water and Societal Concerns About Their Effects on Ecosystems and Human Health

Environmental policy is more strongly driven by impacts based on real threats than by mere quantities of N_r flows. The cost of N pollution for society (and vice versa the benefits of N policies) therefore can be expressed by the increase or decrease of adverse impacts for human health, ecosystems health and climate stability. The economic value of N pollution can be estimated by surveying the willingness-to-pay (WTP) of citizens to prevent these impacts. INI-Europe has successfully applied such a cost-benefit analysis based on WTP for the 27 Member States of the European Union (EU27) for N emissions in 2000 in the ENA (Sutton et al. 2011b). Results have been extended and refined for the year 2008 (Van Grinsven et al. 2013). One conclusion was that, for most N emissions and at the current level of mitigation, the (monetary) benefit of reduced impacts exceeds the cost of N abatement policies. Another striking result was that the social costs of N pollution by agriculture exceeded the economic benefits from applying N to increase crop yield for the primary agricultural sector, but not for the agro-food complex. This conclusion provides justification for sustainable extensification as an alternative for sustainable intensification, which currently appears to be viewed as the solution for producing sufficient food with minimal pollution (Van Grinsven et al. 2015). However, using the WTP concept to estimate the societal benefit of decreased N pollution is still considered less than ideal. The WTP data used for the ENA are somewhat outdated, and data on WTP to prevent ecosystem damage rely almost solely on the Baltic Sea and exclude terrestrial ecosystems. Another approach was successfully tested to comparatively valuate different ecosystems service endpoints (Bateman et al. 2013), but certain assets were not accessible to be monetized. There is, therefore, some danger that such aspects are not valuated and get ignored. Thus, a useful option may be to find non-monetary means of comparison, for example, by using approaches comparing relative risks.

30.3.3 Effects of Nitrogen Inputs on Terrestrial Ecosystems, in Interaction with Air Quality and Climate Change

The proper assessment of N_r effects on European ecosystems needs evaluation of the intertwined relationships with other drivers (e.g., the effects of other pollutants such as tropospheric ozone, O_3, of climate change or of an increase in CO_2 concentrations) and with other nutrients including phosphorus (P). Effects become apparent on functional aspects of ecosystems such as productivity and carbon sequestration on the one hand, and on structural aspects such as biodiversity (specifically plant species diversity) on the other hand. The effects of nitrogen deposition on vegetation and biodiversity have recently been investigated at the global level (Sutton et al. 2014). Interactions of N_r with the carbon cycle lead to elevated carbon sequestration by forests and in soils, while parallel interactions with climate change and other air pollutants occur (De Vries 2014; De Vries et al. 2014a, b; De Vries and Posch 2011). Soil nutrient availability has been identified as a dominant driver in carbon sequestration (De Vries 2014), and N deposition may account for approximately 10% of the global carbon sequestration (De Vries et al. 2014a). While N deposition was the main driver for enhanced forest growth in Europe in the past, climate change and CO_2 rise will be dominant in the future due to the expected reductions in N deposition and increased CO_2 concentrations and temperature (De Vries and Posch 2011). Furthermore, long-term elevated N deposition may cause growth decline due to limited availability of P and base cations (Kint et al. 2012; Braun et al. 2010; De Vries et al. 2014a). More research is needed on the long-term level of carbon sequestration due to N deposition induced by ammonia emission from agriculture, which currently largely offsets the warming effects of nitrous oxide emissions from agriculture De Vries et al. 2011b).

30.3.4 Nitrogen Inputs and Effects on Aquatic/Marine Ecosystems

Quantifying the N emission to the water system is complex. Before reaching the sea, the anthropogenic N input in the river basin passes through the continuum formed by soils, ground waters, riparian zones, floodplains, rivers, lakes and estuaries where processes of elimination, immobilisation, transformation and transport take place (Billen et al. 1991; Bouwman et al. 2013). Part of N is removed, and a fraction is emitted to the atmosphere as N_2O. In Europe, N export from land to coastal waters is substantial (Billen et al. 2011), and the N:P ratio in the nutrient load to the European sea has steadily increased over the last 20 years (Grizzetti et al. 2012). In aquatic and marine ecosystems, the availability of N relative to P and silica affects productivity and carbon sequestration as well as eutrophication (Voss et al. 2013). These elements and their interaction should get more attention in view of the eutrophication of European surface waters.

Policies and management strategies have the potential to reduce the excess of nitrogen. The relationship between source areas and marine pollution needs to be further explored (see for example Stålnacke et al. 2014, and references therein on the nitrogen losses from agriculture in the Baltic Sea Region), but the development of management plans to reduce diffuse nitrogen water pollution and the assessment of their effectiveness are challenging, due to the complexity of processes and the spatial and temporal scales involved (Bouraoui and Grizzetti 2014). Understanding the effectiveness of policies remains a priority. Recent estimates show that the implementation of the current EU legislation and recommendations could reduce nitrogen and phosphorus pressures on European surface waters (Bouraoui et al. 2014). However, the decrease of nutrient loads alone will not guarantee the successful achievement of the good ecological status for aquatic ecosystems. There is a need to link the nutrient reduction targets to ecological effects. Recent advancements were reported for lakes at the European scale (Poikane et al. 2014).

30.3.5 Link Nitrogen and Phosphorus Use with Food Productivity and Assess Regional Transfers in Relation to Food Productivity and Environmental Impacts

A connection between food production and the application of N_r and P can be drawn via meaningful indicators, which need to be transparent and easily applicable. The nitrogen use efficiency (NUE) and phosphorus use efficiency (PUE), defined as the N or P contained in a useful product or consumed by humans versus the N or P fixed or applied, fulfill these conditions. These indicators may address the field level, where NUE/PUE equals the N or P in harvested products versus N or P applied, or they may refer to the whole food chain (from farm to fork), where NUE/PUE equals the N or P consumed versus the N or P fixed or mined. Various studies have shown that NUEs and PUEs from farm to fork at global scale are below 10–20% (e.g., Bouwman et al. 2013). Significant changes of NUE, often decreases, at the country level have been observed over the last 50 years (Lassaletta et al. 2014). Improving efficiency in the use of N and P (as reflected in the indicators) is a key factor to increase productivity at the same or even lower N and P inputs, thus decreasing environmental loads. The indicators help identify intervention points including the reduction of food waste, increased recycling of household, animal and crop waste and increased efficiency in N and P application (e.g., Bodirsky et al. 2014).

An approach to extend NUE into an indicator covering the full life cycle of human consumption is the N_r footprint that has been calculated for a number of countries now, including Austria (Pierer et al. 2014). This analysis demonstrates the dominance of food production (three quarters of which are due to animal products) to the overall footprint. A further extension of indicator applications (Galloway et al. 2014) covers institutional, national and consumer-oriented labelling indicators to assess and compare the N_r loads to be attributed to a respective entity.

30.3.6 Assess Regional Boundaries for Nitrogen and Phosphorus Use in View of Food Production and Environmental Impacts

The concept of Planetary Boundaries (Rockström et al. 2009) identifies levels of anthropogenic perturbations below which the risk of destabilization of the earth system is likely to remain low. Extending from the original concept, an improved planetary N boundary needs to account for both human needs for N in view of food security and risk indicators in view of environmental impacts, while accounting for spatial variability (De Vries et al. 2013). Focusing on eutrophication of aquatic ecosystems as the most relevant environmental concern, these authors calculated an N boundary of 62–82 Tg N $year^{-1}$, depending on the critical N concentration used. More recently, the link between the P and N boundaries was included based on the coupling of these elements by the N:P ratio in growing plant tissue and in aquatic organisms the range in N boundaries thus derived varied from 73–132 Tg N $year^{-1}$ (Steffen et al. 2015), all being considerably higher than the original assessment of 35 Tg N $year^{-1}$ (Rockström et al. 2009). For a future world population of 9 billion people, a minimum global N fixation need of ~50–80 Tg N $year^{-1}$ was estimated (De Vries et al. 2013) with the lower number assuming a 25% increase in NUE.

More work will be needed to find a balance between arguments for the sustainable needs of a global world population, and the boundaries related to the multiple threats posed by excess N_r (and P) application. This requires a regional (at least country based) assessment of N and P loads needed in view of optimal food production, critical loads of N and P, and of other adverse impacts, caused by elevated NH_3 and N_2O emissions to air and N and P runoff to surface water.

30.3.7 Improving Nitrogen Management Across Europe and Current Best Practice in Europe

Agricultural scientists have been able to develop a range of options that reduce the release of N_r into the environment. Considerable discrepancy is seen, however, when comparing these options with the ambition to which they have been included in legal requirements, or the level of practical implementation of such knowledge. The scientific community has often failed to communicate its results effectively. At the same time, in many European countries the agricultural industry does not recognize its responsibility to control the release of effluents, especially while striving for economic competitiveness in a global market in a sector with decreasing economic weight. Often, the most dedicated and influential stakeholders choose a producers' perspective, which leads to a "science-policy gap", where regulation remains distant from what is technically feasible. Attempts have started to involve producer groups or farming extension services into the scientific discussions to recommend appropriate emission abatement measures, with the idea to involve stakeholders at an earlier stage into activities to improve N management (Amon et al. 2014). This allows the identification of those mitigation options that will increase agricultural productivity while reducing environmental effects (win-win options), which will foster their realization in practice.

30.4 Future Activities

Policy advancements in terms of reducing N_r release to the environment differ strongly amongst countries in Europe. Currently, on the EU level, a proposal for improving air quality, which would also affect N_r, is still under discussion. The German Advisory Council on the Environment recently published a report calling for a national nitrogen strategy (SRU 2015), and the Dutch government made available a draft Dutch Nitrogen Programme for public consultation (see http://pas.natura2000.nl). These are all steps in the right direction, but given the current science-policy gap, considerably more effort is needed to further reduce the release of N_r to levels in agreement with sustainability goals. INI-Europe serves as a platform to harmonize and exchange information available in individual European countries, to support scientists and to allow policy makers to implement the best available integrated measures to combat N_r-related effects, working to ensure that advice is in line with scientific progress made in other parts of the world.

Acknowledgements The work of INI Europe described here was hosted by the Austrian Academy of Sciences, Commission of Climate and Air Quality. Part of the activities described here were performed under the ECLAIRE project (EU 7th framework programme).

References

Alcamo, J., Leonard, S. A., Ravishankara, A. R., & Sutton, M. A. (Eds.). (2013). *Drawing down N_2O to protect climate and the ozone layer*. A UNEP Synthesis Report. Nairobi: United Nations Environment Programme (UNEP).

Amon, B., Winiwarter, W., Anderl, M., Baumgarten, A., Dersch, G., Guggenberger, T., et al. (2014). Farming for a better climate (FarmClim): Design of an Inter- and Transdisciplinary Research Project aiming to address the "Science-Policy Gap". *GAIA, 23*(2), 118–124.

Bateman, I. J., Harwood, A. R., Mace, G. M., Watson, R. T., Abson, D. J., Andrews, B., et al. (2013). Bringing ecosystem services into economic decision-making: Land use in the United Kingdom. *Science, 341,* 45–50.

Billen, G., Lancelot, C., Meybeck, M. (1991). N, P, and Si retention along the aquatic continuum from land to ocean. In R. F. C. Mantoura, J.-M. Martin, & R. Wollast (Eds.), *Ocean margin processes in global change: Report of the Dahlem Workshop on Ocean Margin Processes in Global Change, Berlin, 1990, March 18–23* (pp. 19–44). Physical Chemical and Earth Science Research Reports, Pc9. Chichester: Wiley.

Billen, G., Silvestre, M., Grizzetti, B., Leip, A., Garnier, J., Voss, M., et al. (2011). Nitrogen flows from European regional watersheds to coastal marine waters. In M. A. Sutton, C. M. Howard, J. W. Erisman, G. Billen, A. Bleeker, P. Grennfelt, H. van Grinsven, & B. Grizzetti (Eds.), *The European Nitrogen Assessment: Sources, effects and policy perspectives* (Chapter 13, pp. 271–297). Cambridge: Cambridge University Press.

Bleeker, A., & Erisman, J. W. (Eds.). (2011). *Final Report COST Action 729: Assessing and Managing Nitrogen Fluxes in the Atmosphere-Biosphere System in Europe*. Wageningen: Wageningen Academic Publishers.

Bodirsky, B. L., Popp, A., Lotze-Campen, H., Dietrich, J. P., Rolinski, S., Weindl, I., et al. (2014). Reactive nitrogen requirements to feed the world in 2050 and potential to mitigate nitrogen pollution. *Nature Communications, 5,* 3858.

Bouraoui, F., & Grizzetti, B. (2014). Modelling mitigation options to reduce diffuse nitrogen water pollution from agriculture. *Science of the Total Environment, 468–469,* 1267–1277.

Bouraoui, F., Thieu, V., Grizzetti, B., Britz, W., & Bidoglio, G. (2014). Scenario analysis for nutrient emission reduction in the European inland waters. *Environmental Research Letters, 9,* 125007.

Bouwman, A. F., Bierkens, M. F. P., Griffioen, J., Hefting, M. M., Middelburg, J. J., Middelkoop, H., et al. (2013). Nutrient dynamics, transfer and retention along the aquatic continuum from land to ocean: Towards integration of ecological and biogeochemical models. *Biogeosciences, 10*(1), 1–23.

Braun, S., Thomas, V. F. D., Quiring, R., & Flückiger, W. (2010). Does nitrogen deposition increase forest production? The role of phosphorus. *Environmental Pollution, 158,* 2043–2052.

De Vries, W. (2014). Forest ecology: Nutrients trigger carbon storage. *Nature Climate Change, 4,* 425–426.

De Vries, W., & Posch, M. (2011). Modelling the impact of nitrogen deposition, climate change and nutrient limitations on tree carbon sequestration in Europe for the period 1900–2050. *Environmental Pollution, 159,* 2289–2299.

De Vries, W., Leip, A., Reinds, G. J., Kros, J., Lesschen, J. P., & Bouwman, A. F. (2011a). Comparison of land nitrogen budgets for European agriculture by various modeling approaches. *Environmental Pollution, 159,* 3253–3267.

De Vries, W., Kros, J., Reinds, G. J., & Butterbach-Bahl, K. (2011b). Quantifying impacts of nitrogen use in European agriculture on global warming potential. *Current Opinion in Environmental Sustainability, 3,* 291–302.

De Vries, W., Kros, J., Kroeze, C., & Seitzinger, S. P. (2013). Assessing planetary and regional nitrogen boundaries related to food security and adverse environmental impacts. *Current Opinion in Environmental Sustainability, 5,* 392–402.

De Vries, W., Du, E., & Butterbach-Bahl, K. (2014a). Short and long-term impacts of nitrogen deposition on carbon sequestration by forest ecosystems. *Current Opinion in Environmental Sustainability, 9–10,* 90–104.

De Vries, W., Dobbertin, M. H., Solberg, S., van Dobben, H. F., & Schaub, M. (2014b). Impacts of acid deposition, ozone exposure and weather conditions on forest ecosystems in Europe: An overview. *Plant and Soil, 380,* 1–45.

Erisman, J. W., Galloway, J. N., Seitzinger, S., Bleeker, A., Dise, N. B., Petrescu, R., et al. (2013). Consequences of human modification of the global nitrogen cycle. *Philosophical Transactions of the Royal Society, 368*(1621), 20130116. https://doi.org/10.1098/rstb.2013.0116.

Galloway, J. N., Dentener, F. J., Capone, D. G., Boyer, E. W., Howarth, R. W., Seitzinger, S. P., et al. (2004). Nitrogen cycles: Past, present, and future. *Biogeochemistry, 70*(2), 153–226.

Galloway, J. N., Townsend, A. R., Erisman, J. W., Bekunda, M., Cai, Z. C., Freney, J. R., et al. (2008). Transformation of the nitrogen cycle: Recent trends, questions, and potential solutions. *Science, 320*(5878), 889–892.

Galloway, J. N., Winiwarter, W., Leip, A., Leach, A. M., Bleeker, A., & Erisman, J. W. (2014). Nitrogen footprints: Past, present and future. *Environmental Research Letters, 9,* 115003.

Grizzetti, B., Bouraoui, F., & Aloe, A. (2012). Changes of nitrogen and phosphorus loads to European seas. *Global Change Biology, 18,* 769–782.

Hicks, W. K., Whitfield, C. P., Bealey W. J., & Sutton, M. A. (Eds.). (2011). *Nitrogen deposition and Natura 2000: Science & practice in determining environmental impacts.* COST729/Nine/ESF/CCW/JNCC/SEI Workshop Proceedings. Brussels: COST office—European Cooperation in Science and Technology.

Hutchings, N. J., Nielsen, O.-K., Dalgaard, T., Mikkelsen, M. H., Børgesen, C. D., Thomsen, M., et al. (2014). A nitrogen budget for Denmark; developments between 1990 and 2010, and prospects for the future. *Environmental Research Letters, 9,* 115012.

Kint, V., Aertsen, W., Campioli, M., Vansteenkiste, D., Delcloo, A., & Muys, B. (2012). Radial growth change of temperate tree species in response to altered regional climate and air quality in the period 1901–2008. *Climatic Change, 115,* 343–363.

Lassaletta, L., Billen, G., Grizzetti, B., Anglade, J., & Garnier, J. (2014). 50 year trends in nitrogen use efficiency of world cropping systems: The relationship between yield and nitrogen input to cropland. *Environmental Research Letters, 9,* 105011.

Leip, A., Achermann, B., Billen, G., Bleeker, A., Bouwman, L., de Vries, W., et al. (2011). Integrating nitrogen fluxes at the European scale. In M. A. Sutton, C. M. Howard, J. W. Erisman, G. Billen, A. Bleeker, P. Grennfelt, H. van Grinsven, & B. Grizzetti (Eds.), *The European Nitrogen Assessment: Sources, effects and policy perspectives* (Chapter 16, pp. 345–376), Cambridge: Cambridge University Press.

Pierer, M., Winiwarter, W., Leach, A. M., & Galloway, J. N. (2014). The nitrogen footprint of food products and general consumption patterns in Austria. *Food Policy, 49,* 128–136.

Poikane, S., Portielje, R., van den Berg, M., Phillips, G., Brucet, S., Carvalho, L., et al. (2014). Defining ecologically relevant water quality targets for lakes in Europe. *Journal of Applied Ecology, 51,* 592–602.

Rockström, J., Steffen, W., Noone, K., Persson, Å., Chapin, F. S. I. I. I., Lambin, E., et al. (2009). A safe operating space for humanity. *Nature, 461,* 472–475.

SRU—Sachverständigenrat für Umweltfragen. (2015). *Stickstofff – Lösungsstrategien für ein drängendes Umweltproblem.* Berlin: SRU. http://www.umweltrat.de.

Stålnacke, P., Bechmann, M., & Iital, A. (2014). Introduction: Nitrogen losses from agriculture in the Baltic Sea region. *Agriculture, Ecosystems & Environment, 198* (Special issue: Nitrogen losses from agriculture in the Baltic Sea region), 1–3.

Steffen, W., Richardson, K., Rockström, J., Cornell, S. E., Fetzer, I., Bennett, E. M., et al. (2015). Planetary boundaries: Guiding human development on a changing planet. *Science, 347,* 6223. https://doi.org/10.1126/science.1259855.

Sutton, M., Howard, C., Erisman, J. W., Billen, G., Bleeker, A., Greenfelt, P., et al. (Eds.). (2011a). *The European Nitrogen Assessment: Sources, effects and policy perspectives.* Cambridge: Cambridge University Press.

Sutton, M. A., Oenema, O., Erisman, J. W., Leip, A., van Grinsven, H., & Winiwarter, W. (2011b). Too much of a good thing—Curbing nitrogen emissions is a central environmental challenge for the twenty-first century. *Nature, 472,* 159–161.

Sutton, M. A., Bleeker, A., Howard, C. M., Bekunda, M., Grizzetti, B., de Vries, W., et al. (2013). *Our Nutrient World: The challenge to produce more food and energy with less pollution.* Global

Overview of Nutrient Management. Edinburgh: Centre for Ecology and Hydrology on behalf of the Global Partnership on Nutrient Management and the International Nitrogen Initiative.

Sutton, M. A., Mason, K. E., Sheppard, L. J., Sverdrup, H., Haeuber, R., & Hicks, W. K. (Eds.). (2014). Nitrogen deposition, critical loads and biodiversity. In *Proceedings of the International Nitrogen Initiative Workshop, linking experts of the Convention on Long-range Transboundary Air Pollution and the Convention on Biological Diversity*. Dordrecht: Springer.

UN-ECE. (2013). *Guidance document on national nitrogen budgets. United Nations Economic Commission for Europe*. Executive Body for the Convention on Long-range Transboundary Air Pollution. ECE/EB.AIR/119. Geneva: UN-ECE.

Van Grinsven, H. J., Erisman, J. W., de Vries, W., & Westhoek, H. (2015). Potential of extensification of European agriculture for a more sustainable food system, focusing on nitrogen. *Environmental Research Letters, 10*(2), 025002.

Van Grinsven, H. J. M., Holland, M., Jacobsen, B. H., Klimont, Z., Sutton, M. A., & Willems, W. J. (2013). Costs and benefits of nitrogen for Europe and implications for mitigation. *Environmental Science and Technology, 47,* 3571–3579.

Voss, M., Bange, H. W., Dippner, J. W., Middelburg, J. J., Montoya, J. P., & Ward, B. (2013). The marine nitrogen cycle: Recent discoveries, uncertainties and the potential relevance of climate change. *Philosophical Transactions of the Royal Society of London. Series B, Biological Sciences, 368*, 20130121. https://doi.org/10.1098/rstb.2013.0121.

Chapter 31
The INI African Regional Nitrogen Centre: Challenges and Opportunities in Africa

Cargele Masso, Mateete Bekunda, and Fredrick Mhina Mngube

Abstract The International Nitrogen Initiative (INI) aims at optimizing nitrogen (N) use in food and energy production, while minimizing environmental risks. In Africa, nitrogen management must address the 'too little' and 'too much' paradox. Too little nitrogen is used in food production, which has led to chronic food insecurity and malnutrition. Conversely, too much nitrogen load in water bodies due mainly to excessive soil erosion, leaching, limited nitrogen recovery from wastewater, and from atmospheric deposition, still contributes to eutrophication in some areas. Significant research has been conducted to improve N use for production, whereas little has been done to be effective in addressing the 'too much' issue. The current research gaps must be addressed, and supportive policies operationalized, to maximize nitrogen benefits, while reducing the negative impacts of nitrogen on the environment. Innovation platforms involving key stakeholders in Africa are required to address the full chain of nitrogen use efficiency.

Keywords Africa · International Nitrogen Initiative · Nitrogen management · Nitrogen use efficiency · Eutrophication

31.1 Introduction

Africa's agricultural lands continue being degraded with an annual estimated economic cost of up to 18% of the gross domestic product (Nkonya et al. 2011)

C. Masso (✉)
International Institute of Tropical Agriculture (IITA), 1st Main IRAD Road, Nkolbisson, P.O. Box 2008 (Messa), Yaoundé, Cameroon
e-mail: C.Masso@cgiar.org

M. Bekunda
International Institute of Tropical Agriculture (IITA), Arusha, Tanzania
e-mail: M.Bekunda@cgiar.org

F. M. Mngube
Lake Victoria Basin Commission, Kisumu, Kenya
e-mail: mngube@lvbcom.org

M. A. Sutton et al. (eds.), *Just Enough Nitrogen*,
https://doi.org/10.1007/978-3-030-58065-0_31

because of poor agronomic practices and nutrient depletion (Tittonell and Giller 2013; Sutton et al. 2013). Over 80% of agricultural lands are for instance nitrogen (N) deficient (Liu et al. 2010). Barriers such as availability and costs of inputs, poor economic returns of input use due to limited market opportunities or quality of the inputs, limited financial capacity and access to credits, and insufficient extension services among others have drastically affected adoption of N inputs (Akpan et al. 2012a, 2012b; Akudugu et al. 2012; IFDC 2006).

Limited research capacity in most regions of Africa, particularly for long term trials, has also added to the difficulty of improving nitrogen agronomic use efficiency. Soil acidification, low organic matter content, deficiencies of various nutrients and reduced microbial activities are among factors affecting crop responses to applied nitrogen (Fairhurst 2012; Nezomba et al. 2015). Adequate diagnosis of the actual limiting factors to inform the application of integrated soil fertility management is therefore required to optimize N agronomic use efficiency (Giller et al. 2011) and increase the sustainability of agricultural intensification (Vanlauwe et al. 2015).

Continuous nitrogen depletion has contributed not only to the persistent food insecurity for both quantity and quality (Marler and Wallin 2006), but also excessive erosion due to insufficient land cover, deforestation, and encroachment to marginal lands, and consequently N load into water bodies (Leip et al. 2014). In highly populated regions of Africa such as the Lake Victoria catchment, inadequate systems for municipal wastewater treatment have also resulted in excessive N load into water bodies and contributed to eutrophication of certain sections of the Lake (LVBC 2012). Atmospheric deposition also contributes to N load into the African environment including surface water (Galy-Lacaux and Delon 2014). Other sources of N load into water bodies include erosion in infrastructure construction, runoff of feed and food waste from both municipal and industrial areas, N runoff and leaching mainly from commercial farms, and insufficient treatment of wastewater from industry (e.g., fisheries) among others.

Nitrogen management in Africa must take into consideration the ‘too little’ and ‘too much’ paradox. This would require focused research programs and supportive policies. Recent efforts have been mainly limited to improving food security and overlooked environmental challenges related to the full N cycle and various N sources. Existing policies lack focus on nitrogen; in most of the cases they should be improved, strengthened, and importantly operationalized. This overview of challenges and opportunities of nitrogen management in Africa therefore underscores issues that must be addressed to optimize nitrogen use efficiency, recommends solution pathways, and highlights the potential contribution of the African Centre of the International Nitrogen Initiative (INI) to operationalize retained solutions.

31.2 Background of the INI African Centre

The African Centre (hereafter the Centre) is one of the six Centres of the International Nitrogen Initiative (i.e., Africa, East Asia, Europe, Latin America, North America, and South Asia). Since 2003, the Centre has been involved in various INI activities including, but not limited to, (i) the formation of an executive committee to coordinate the Centre's activities, (ii) the contribution of book chapter in 'Advances in Integrated Soil Fertility Management in sub-Saharan Africa: Challenges and Opportunities' (Bekunda et al. 2007), (iii) the contribution to INI's policy and various conferences including hosting the 2013 INI conference in Kampala, Uganda (N2013). During N2013 the Kampala Statement-for-Action on Reactive Nitrogen [i.e., all nitrogen compounds, except N_2 (Shibata et al. 2014)] in Africa and globally was promulgated (see Sutton et al. 2020, Chap. 38, this volume). To improve the efficiency of the Centre, formalization of a network of committed stakeholders interested in N management in Africa and institutionalization of the Centre including host-organization and financing mechanisms have been recommended.

31.3 Current Challenges

31.3.1 Low Nitrogen Use for Production

Nutrient depletion including N is a critical issue in Africa. In certain countries, less than 1% of farmers are using fertilizers (Nkonya et al. 2011). Most of the countries, particularly in sub-Saharan Africa, have not been able to meet the target of 50 kg nutrients ha^{-1} set in the 2006 Abuja Declaration following the Africa Fertilizer Summit (IFDC 2006; Wanzala 2011). On average, the current nutrient rate is less than 10 kg ha^{-1} (Dittoh et al. 2012), although Sheahan and Barrett (2014) reported 26 kg nutrients ha^{-1} based on data from six countries i.e. Ethiopia, Malawi, Niger, Nigeria, Tanzania, and Uganda. This has led not only to soil nitrogen depletion varying from 9–60 kg N ha^{-1} $year^{-1}$ (Stoorvogel and Smaling 1990; Zhou et al. 2014), but also to the current yield gaps that could reach over 300% for crops such as cereals and legumes (Mutegi and Zingore 2014). Low nitrogen use has often been associated with high costs of N fertilizers and poor economic returns (Guo et al. 2009), poor quality of N fertilizers (Bold et al. 2015) or rhizobia (Jefwa et al. 2014), and importantly the limited financial capacity of resource-disadvantaged smallholder farmers (Akudugu et al. 2012). The issue has been exacerbated by poor extension services to promote good agronomic practices including integrated soil fertility management and lack of effective operationalization of supportive policies (Akpan et al. 2012a, 2012b; Dittoh et al. 2012; Kiptot et al. 2016).

31.3.2 High Nitrogen Loss to the Environment

Despite the low N use for production, significant N losses still occur in the African context, which exacerbate the nutrient depletion on agricultural lands. For instance, the N atmospheric deposition in Africa is equivalent to the current rate of fertilizer use i.e. 4–15 kg N ha^{-1} $year^{-1}$ (Galy-Lacaux and Delon 2014; Vet et al. 2014). The portion of this N that falls on agricultural lands represents a significant N input, but it becomes a significant risk to the environment when it ends up in water bodies or other areas where it cannot be used for plant growth; the beneficial portion still has to be quantified.

Part of the N in the atmospheric deposition is a result of N volatilization and emission. Greenhouse gases (GHGs) estimates in Africa are mainly based on publications of the Emissions Database for Global Atmospheric Research (EDGAR) (Hickman et al. 2011), which reported an African contribution of up to 20% of the global nitrous oxide (N_2O) emissions. Studies in the Nyungwe forest in Rwanda estimated NO-N and N_2O-N emissions at 0.8–5.1 and 2.8–5.5 kg ha^{-1} $year^{-1}$ respectively (Gharahi Ghehi et al. 2014). Previously, Werner et al. (2007a) had reported N_2O-N emissions in the range of 1.4–3.8 kg ha^{-1} $year^{-1}$ in the Kakamega forest in Kenya. Werner et al. (2007b) estimated N_2O-N emissions from tropical and subtropical forests in Africa at approximately 4×10^{12} g $year^{-1}$. In addition to greenhouse gases, N leaching in Nyungwe forest has been estimated at 20.8 kg inorganic N ha^{-1} $year^{-1}$ mainly as nitrates ($\approx$ 95%) (Cizungu Ntaboba 2015). However, the paucity of information has not allowed for a comprehensive quantification of N losses at the continental scale.

31.3.3 Limited Research Capacity

The paucity of information on the full chain of N use efficiency has been related to the limited research capacity and the research priorities of most national and international research organizations. In general, the human choices in terms of food consumption, as well as energy and transport, drive nitrogen use efficiency pathways. Food consumption is mainly linked to nitrogen use efficiency through food, feed, and animal production, and food supply, whereas energy and transport are related to unintended N fixation through combustion (Sutton et al. 2013). In the African context, while selected investigations have been made to improve the nitrogen agronomic use efficiency in food and feed production, quantification of N use efficiency in the pools of food supply, manure, sewage, and combustion has been too scarce to be representative, which has resulted in excessive uncertainties in N budgets (Rufino et al. 2014; Zhou et al. 2014). There is a need to allocate resources along the full chain of N use efficiency so that comprehensive data are collected for accurate N budget determination at the continental level. This would also include quantification of N exported out, and imported in Africa in various commodities containing significant N such as feed and food products among others.

31.4 Current Opportunities

31.4.1 Nitrogen Agronomic Use Efficiency

Nitrogen agronomic use efficiency is defined as the yield gain per unit amount of N, when plots with and without applied N are compared (Dobermann 2005). In Africa, it is more preferred than nitrogen use efficiency as the data could be comparable to data obtained worldwide when good agronomy is practiced (Vanlauwe et al. 2011). Conversely, N use efficiency may exceed 100% in Africa due to insufficient use of N inputs (Edmonds et al. 2009), while it is generally lower than 70% in countries with sufficient N use for production (Sutton et al. 2013).

In Africa, N agronomic use efficiency is still low because of poor agronomic practices including blanket fertilizer recommendations (CAB International 2012), too low fertilizer application rates to result in significant effect, unbalanced fertilization where the focus is put for instance on the macronutrients nitrogen, phosphorus and potassium (NPK) without secondary and micro-nutrients (NAAIAP 2014). Recent interventions including integrated soil fertility management (i.e., improved seeds, use of balanced fertilization, organic inputs, liming materials, water management, and appropriate tillage practices among others) showed that N agronomic use efficiency could be doubled with adoption of good agronomic practices (Vanlauwe et al. 2011, 2015). The dilemma is that in Africa farming is mainly practiced by resource-disadvantaged smallholder farmers who cannot afford most of the inputs at the actual market prices (Alobo Loison 2015). Supportive policies would be required to improve crop productivity and N agronomic use efficiency (Dittoh et al. 2012), while minimizing negative impacts on the environment. For instance, Denning et al. (2009) reported improved maize production in Malawi as a result of the fertilizer subsidy program, that included N, but timely supply of the fertilizers and other inputs was recommended to improve the efficiency and effectiveness of the program (Chirwa and Dorward 2013). Similarly, as recommended in the sustainable land management strategy of the Lake Victoria Basin Commission in East Africa, other cost-effective options of N use for production like nutrient recycling through utilization of feed and food wastes and sewage, as well as biological nitrogen fixation, should be exploited.

31.4.2 Innovation Platforms to Improve Nitrogen Agronomic Use Efficiency

Studies have been conducted to improve N agronomic use efficiency. However, most of the data are not found in the public domain. There is a need to improve the collaboration between national and international research organizations in Africa to facilitate data sharing and identification of the knowledge gaps and research areas. Networks like the Tropical Soil Biology and Fertility (TSBF), the African Soil Science Society (ASSS), the Africa Soil Information Service (AfSIS), and the

International Nitrogen Management System (INMS) among others represent good avenues for collaborative research. Given the current context, the additional research areas must address not only agronomic challenges, but also environmental, socio-economic, and policy issues. The ultimate goal would be improving access to high quality inputs, promote agronomic practices that take into consideration local agro-climatic conditions and crop requirements while minimizing environmental risks, and increase farming profitability through enhanced market opportunities.

Improved collaboration between scientists and relevant stakeholders would, for instance, enhance access or acquisition of comprehensive data for more accurate N budget at the continental scale through better identification of all the types of data required for the whole N cycle. Importantly, the innovation platforms would be useful to identify and engage experts and scientists who have the necessary capacity to assess N use efficiency of the various pools linked to (i) food consumption and diet choices, as well as (ii) energy consumption and transport choices so as to address the current knowledge gaps. This would be important to generate scientific evidence that can be used to inform policy decisions intended to promote good agronomic practices to optimize N agronomic use efficiency, improve farmer access to innovative technologies and profitability of the farming systems, and protect the environment against any harm associated with poor nitrogen management.

In addition to above, the study conducted by Lake Victoria Basin Commission (LVBC 2012) noted other key opportunities to ensure sufficient and proper handling and application of nitrogen fertilizers. These include improvement of land tenure systems to enhance ownership and equity and consequently encourage investments and harmonization of national and regional policies and laws that governing investments and management of nitrogen. The innovation platforms would therefore play an advocacy role to facilitate cooperation and coordination of regional and continental efforts to ensure effectiveness and efficiency of interventions, and promote investments for sustainable N management.

31.5 Near Future Perspectives

Based on current challenges and opportunities related to N management in Africa, the African Centre of INI has identified the following issues as the focus areas in the near future particularly in the context of INMS:

- Create a Nitrogen Innovation Platform including multidisciplinary stakeholders to address the various pools of the full chain of N use efficiency;
- Assess the various pools of reactive N including N in food and feed production and consumption systems, energy production and transport systems, volatilization, emissions, and leaching among others, to ensure sufficient data collection for a more accurate N budget at the continental scale;
- Identify the sources and environmental fate of N in atmospheric deposition;

- Promote good agronomic practices intended to optimize N agronomic use efficiency, while minimizing environmental risks related to insufficient or excessive use of N;
- Advocate for policies conducive to increased profitability of nitrogen use for food and feed production;
- Educate consumers for adoption of good dietary practices that maximize N (i.e., protein) use efficiency, while minimizing N losses in food and feed wastes and sewage;
- Advocate for policies intended to promote energy production and transport practices that minimize unintended N fixation or maximize the recovery of such N for use in feed and food production.

31.6 Conclusion

The success of the African Centre of the International Nitrogen Initiative will depend on how effectively and efficiency the current issue of 'too little' and 'too much' nitrogen is addressed. The perceived satisfaction of affected and interested stakeholders will certainly be based on improved food and feed production as a result of proper and sufficient use of N for production (i.e., increased N agronomic use efficiency), effective reduction of N depletion, and reduced environmental pollution related to N at the continental scale. This will undoubtedly require strengthening the research capacity of the Centre including involvement of national and international experts and scientists, the institutionalization of the Centre for increased visibility, and development of financing mechanisms of the priority N research areas of the Centre. To improve stakeholders' uptake of practices that enhance N use efficiency across the full chain of N use, supportive policies or empowerment would be required to facilitate access to (i) innovative or alternative technologies, (ii) relevant knowledge and the know-how, (iii) profitable farming systems (i.e., market opportunities), and (iv) credit (financing opportunities).

Acknowledgements The authors would like to thank all the stakeholders who participated in the African Centre workshop organized on 4–5 November 2014 and acknowledge the financial support of the International Nitrogen Initiative management during the workshop. The facilitation of the International Institute of Tropical Agriculture (IITA) and the Lake Victoria Basin Commission (LVBC) is also highly appreciated. The authors gratefully acknowledge preparatory funding toward the establishment of the International Nitrogen Management System (INMS) East Africa Demonstration Activity.

References

Akpan, S. B., Nkanta, V. S., & Essien, U. A. (2012a). A double-durdle model for fertilizer adoption and optimum use among farmers in southern Nigeria. *Tropicultura, 30,* 249–253.

Akpan, S. B., Udoh, E. J., & Nkanta, V. S. (2012b). Factors influencing fertilizer use intensity among smallholder crop farmers in Abak agricultural zone in Akwa Ibom State, Nigeria. *Journal of Biology, Agriculture and Healthcare, 2,* 54–65.

Akudugu, M. A., Guo, E., & Dadzie, S. K. (2012). Adoption of modern agricultural production technologies by farm households in Ghana: what factors influence their decisions? *Journal of Biology, Agriculture and Healthcare, 2,* 1–13.

Alobo Loison, S. (2015). Rural livelihood diversification in sub-Saharan Africa: A literature review. *Journal of Development Studies, 51,* 1125–1138.

Bekunda, M., Galloway, J., Syers, K., & Scholes, M. (2007). Background, current status and the African context of the International Nitrogen Initiative. In A. Bationo, B. Waswa, J. Kihara, & J. Kimetu (Eds.), *Advances in integrated soil fertility management in sub-Saharan Africa: challenges and opportunities* (pp. 115–119). Dordrecht: Springer.

Bold, T., Kaizzi, K. C., Svensson, J., Yanagizawa-Drott, D. (2015). *Low quality, low returns, low adoption: Evidence from the market for fertilizers and hybrid seed in Uganda.* International Growth Centre. http://www.amitsa.org. Accessed 20 Dec 2015.

CAB International. (2012). *Enhancing national agriculture research capacities in developing and fine-tuning fertilizer recommendations within an integrated soil fertility management framework.* Proposal submitted to the Alliance for Green Revolution in Africa (AGRA). Nairobi: CAB International.

Chirwa, E., & Dorward, A. (2013). *Agricultural input subsidies—The recent Malawi experience.* Oxford: Oxford University Press.

Cizungu Ntaboba, L. (2015). Input, dynamics and loss of reactive nitrogen in a central African tropical mountain forest and *eucalyptus* plantation. Ph.D. thesis. Ghent: Ghent University.

Denning, G., Kabambe, P., Sanchez, P., Malik, A., Flor, R., Harawa, R., Nkhoma, P., et al. (2009). Input subsidies to improve smallholder maize productivity in Malawi: Towards an African green revolution. *PLoS Biology, 7*(1), 2–10, e1000023. https://doi.org/10.1371/journal.pbio.1000023.

Dittoh, S., Omotosho, O., Belemvire, A., Akuriba, M., Haider, K. (2012). *Improving the effectiveness, efficiency and sustainability of fertilizer use in sub-Saharan Africa.* Briefing Paper Number 3. New Delhi: Global Development Network.

Dobermann, A. R. (2005). *Nitrogen use efficiency—State of the art.* Paper presented at the IFA International Workshop on Enhanced-Efficiency Fertilizers, Frankfurt, Germany, 28–30 June 2005. University of Nebraska, Lincoln, US: Agronomy and Horticulture—Faculty Publications. Available online at: http://digitalcommons.unl.edu/agronomyfacpub/316. Accessed 6 March 2016.

Edmonds, D. E., Abreu, S. L., West, A., Caasi, D. R., Conley, T. O., Daft, M. C., et al. (2009). Cereal nitrogen use efficiency in sub-Saharan Africa. *Journal of Plant Nutrition, 32*(12), 2107–2122.

Fairhurst, T. (2012). *Handbook for integrated soil fertility management.* Africa Soil Health Consortium. Nairobi: CAB International.

Galy-Lacaux, C., & Delon, C. (2014). Nitrogen emission and deposition budget in West and Central Africa. *Environmental Research Letters, 9*(12), 125002. https://doi.org/10.1088/1748-9326/9/12/125002.

Gharahi Ghehi, N., Werner, C., Hufkens, K., Kiese, R., Van Ranst, E., Nsabimana, D., et al. (2014). N_2O and NO emission from the Nyungwe tropical highland rainforest in Rwanda. *Geoderma Regional, 2*(3), 31–49.

Giller, K. E., Tittonell, P., Rufino, M. C., van Wijk, M. T., Zongore, S., Mapfumo, P., et al. (2011). Communicating complexity: Integrated assessment of trade-offs concerning soil fertility management within farming systems to support innovation and development. *Agricultural Systems, 104,* 191–203.

Guo, Z., Koo, J., Wood, S. (2009). *Fertilizer profitability in East Africa: A spatially explicit policy analysis.* Contributed paper prepared for presentation at the International Association of Agricultural Economics Conference, Beijing, China, August 16–22, 2009.

Hickman, J. E., Havlikova, M., Kroeze, C., & Palm, C. A. (2011). Current and future nitrous oxide emissions from African agriculture. *Current Opinion in Environmental Sustainability, 3,* 370–378.

IFDC. (2006). *Africa Fertilizer Summit Proceedings, Abuja, Nigeria June 9–13, 2006.* Florence (Alabama, United States of America): International Fertilizer Development Center (IFDC).

Jefwa, J. M., Pypers, P., Jemo, M., Thuita, M., Mutegi, E., Laditi, M. A., Faye, A., et al. (2014) Do commercial biological and chemical products increase crop yields and economic returns under smallholder farmer conditions?. In B. Vanlauwe, P. Asten, G. Blomme (Eds.), *Challenges and opportunities for agricultural intensification of the humid highland systems of sub-Saharan Africa* (Chapter 7, pp. 81–96). Cham (ZG), Switzerland: Springer International Publishing Switzerland.

Kiptot, E., Karuhanga, M., Franzel, S., & Nzigamasabo, P. B. (2016). Volunteer farmer-trainer motivations in East Africa: Practical implications for enhancing farmer-to-farmer extension. *International Journal of Agricultural Sustainability, 14*(3), 339–356.

Leip, A., Leach, A., Musinguzi, P., Tumwesigye, T., Olupot, G., Tenywa, J. S., et al. (2014). Nitrogen-neutrality: A step towards sustainability. *Environmental Research Letters, 9,* 115001. https://doi.org/10.1088/1748-9326/9/11/115001.

Liu, J., You, L., Amini, M., Obersteiner, M., Herreco, M., Zehnder, B., et al. (2010). A high-resolution assessment on global nitrogen flow in cropland. *Proceedings of the National Academy of Sciences of the United States of America (PNAS), 107*(17), 8035–8040.

LVBC. (2012). *A basin-wide strategy for sustainable land management in the Lake Victoria Basin.* Kisumu: Lake Victoria Basin Commission (LVBC).

Marler, J. B., & Wallin, J. R. (2006). *Human health, the nutritional quality of harvested food and sustainable farming systems.* Bellevue, WA: Nutrition Security Institute. Available online at: http://www.nutritionsecurity.org. Accessed 21 March 2015.

Mutegi, J., & Zingore, S. (2014). *Closing yield gaps in sub-Saharan Africa through integrated soil fertility management.* ISFM Policy Highlights No. 1. Nairobi: International Plant Nutrition Institute (IPNI).

NAAIAP. (2014). *Soil suitability evaluation for maize production in Kenya.* Nairobi: National Accelerated Agricultural Input Access Programme (NAAIAP).

Nezomba, H., Mtambanengwe, F., Tittonell, P., & Mapfumo, P. (2015). Point of no return? Rehabilitating degraded soils for increased crop productivity on smallholder farms in eastern Zimbabwe. *Geoderma, 239–240,* 143–145.

Nkonya, E., Place, F., Pender, J., Mwanjololo, M., Okhimamhe, A., Kato, E., Crespo, S., et al. (2011). *Climate risk management through sustainable land management in sub-Saharan Africa.* International Food Policy Research Institute (IFPRI) Discussion Paper 01126. Washington, DC: International Food Policy Research Institute (IFPRI).

Rufino, M. C., Brandt, P., Herrero, M., & Butterbach-Bahl, K. (2014). Reducing uncertainty in nitrogen budgets for African livestock systems. *Environmental Research Letters, 9*(10), 105008. https://doi.org/10.1088/1748-9326/9/10/105008.

Sheahan, M., & Barrett, C. B. (2014). *Understanding the agricultural input landscape in sub-Saharan Africa—Recent plot, household, and community-level evidence.* Washington, DC: World Bank Group, Africa region, Office of the Chief Economist.

Shibata, H., Cattaneo, L. R., Leach, A. M., & Galloway, J. (2014). First approach to the Japanese nitrogen footprint model to predict the loss of nitrogen to the environment. *Environmental Research Letters, 9*(11), 115013. https://doi.org/10.1088/1748-9326/9/11/115013.

Stoorvogel, J. J., & Smaling, E. M. A. (1990). *Assessment of soil nutrient depletion in sub-Saharan Africa: 1983–2000. Volume II: Nutrient balances per crop and per land use system.* Report 28. Wageningen, The Netherlands: The Winard Staring Centre.

Sutton, M. A., Bleeker, A., Howard, C. M., Bekunda, M., Grizzetti, B., de Vries, W., van Grinsven, H. J. M., et al. (2013). *Our Nutrient World: The challenge to produce more food and energy with less pollution.* Global Overview of Nutrient Management. Edinburgh: Centre for Ecology and Hydrology on behalf of the Global Partnership on Nutrient Management and the International Nitrogen Initiative.

Sutton, M. A., Ebanyat, P., Raghuram, N., Bekunda, M., Tenya, J., Winiwarter, W., Bleeker, A., et al. (2020). The Kampala Statement-for-Action on Reactive Nitrogen in Africa and Globally. In M. A. Sutton, K. E. Mason, A. Bleeker, W. K. Hicks, C. Masso, N. Raghuram, S. Reis, et al. (Eds.), *Just enough nitrogen: Perspectives on how to get there for regions with too much and too little nitrogen* (Chapter 38, this volume). Springer.

Tittonell, P., & Giller, K. E. (2013). When yield gaps are poverty traps: The paradigm of ecological intensification in African smallholder agriculture. *Field Crops Research, 143,* 76–90.

Vanlauwe, B., Descheemaeker, K., Giller, K. E., Huising, J., Merckx, R., Nziguheba, G., et al. (2015). Integrated soil fertility management in sub-Saharan Africa. *Soil, 1,* 491–508.

Vanlauwe, B., Kihara, J., Chivenge, P., Pypers, P., Coe, R., & Six, J. (2011). Agronomic use efficiency of N fertilizer in maize-based systems in sub-Saharan Africa within the context of integrated soil fertility management. *Plant and Soil, 339,* 35–50.

Vet, R., Richard, S., Carou, A. S., Shaw, M., Ro, C.-U., Aas, W., et al. (2014). A global assessment of precipitation chemistry and deposition of sulfur, nitrogen, sea salt, base cations, organic acids, acidity and pH, and phosphorus. *Atmospheric Environment, 93,* 3–100.

Wanzala, M. (2011). *The Abuja declaration on fertilizers for an African green revolution—status of implementation at regional and national levels.* Johannesburg: Policy Alignment and Program Development Directorate, The New Partnership for Africa's Development (NEPAD).

Werner, C., Butterbach-Bahl, K., Haas, E., Hickler, T., Kiese, R. (2007b). A global inventory of N_2O emissions from tropical rainforest soils using a detailed biogeochemical model. *Global Biogeochemical Cycles, 21,* GB3010. https://doi.org/10.1029/2006gb.

Werner, C., Kiese, R., & Butterbach-Bahl, K. (2007a). N_2O, CH_4 and CO_2 flux measurements from tropical rainforest soils in western Kenya. *Journal of Geophysical Research-Atmospheres, 112,* D03308. https://doi.org/10.1029/2006JD007388.

Zhou, M., Brandt, P., Pelster, D., Rufino, M. C., Robinson, T., & Butterbach-Bahl, K. (2014). Regional nitrogen budget of the Lake Victoria Basin, East Africa: Syntheses, uncertainties, and perspectives. *Environmental Research Letters, 9,* 105009. https://doi.org/10.1088/1748-9326/9/10/105009.

Chapter 32
The INI South Asian Regional Nitrogen Centre: Capacity Building for Regional Nitrogen Assessment and Management

N. Raghuram, Y. P. Abrol, H. Pathak, T. K. Adhya, and M. K. Tiwari

Abstract The South Asian Nitrogen Centre (SANC) was born out of the initiative of the Indian Nitrogen Group (ING) under the Society for Conservation of Nature (SCON), formalized in the 4th International Nitrogen Conference in Brazil in 2007. It hosted the 5th International Nitrogen Initiative (INI) Conference in New Delhi in 2010 and has since been instrumental in mainstreaming the importance of reactive nitrogen (N_r) in India and South Asia through advocacy, workshops and publications. The scoping study on nutrient management for the South Asia Co-operative Environment Programme (SACEP) has led to the intergovernmental recognition of sustainable management of nitrogen (N) and other nutrients at the ministerial level. The trio of ING-SCON-SANC has ensured a strong representation for the South Asian region in the UNEP Global Partnership on Nutrient Management (GPNM) and developed regional cooperation with the East Asian centre of the INI in China. It steered the first GPNM Global Environment Facility (GEF) project involving Chilika Lake in India and Manila Bay in the Philippines, and has highlighted the importance of N and other nutrients at an INI side event in COP-11 in Hyderabad. Over the

N. Raghuram (✉)
University School of Biotechnology, Guru Gobind Singh Indraprastha University, Dwarka Sec. 16C, New Delhi 110078, India
e-mail: raghuram@ipu.ac.in

N. Raghuram · Y. P. Abrol · H. Pathak · T. K. Adhya · M. K. Tiwari
South Asian Nitrogen Centre, Indian Nitrogen Group, Sustainable India Trust and Society for Conservation of Nature, F-4, A-Block, NASc Complex, DPS Marg, Pusa, New Delhi, India 110 012
e-mail: hpathak.iari@gmail.com

T. K. Adhya
e-mail: adhyas@yahoo.com

M. K. Tiwari
e-mail: mk.tiwari2005@gmail.com

H. Pathak
ICAR-National Institute of Abiotic Stress Management, Baramati, Maharashtra, India

T. K. Adhya
School of Biotechnology, Kalinga Institute of Industrial Technology (KIIT), Campus XI, Patia, Bhubaneswar 751024, Odisha, India

M. A. Sutton et al. (eds.), *Just Enough Nitrogen*,
https://doi.org/10.1007/978-3-030-58065-0_32

last few years, SANC has been the main driver of the need for a regional N assessment as a part of the development of an international N management system. These initiatives led to the inclusion of the South Asian Nitrogen Assessment as a regional demonstration under the GEF project on the International Nitrogen Management System (INMS), and the Indo-UK Virtual Nitrogen Centre on Nitrogen Efficiency in Whole-cropping Systems (NEWS India-UK). Thus, the South Asian Nitrogen Centre, through its local partners (ING-SCON) has played a pivotal role in setting the agenda for research and policy on N_r at the Indian and South Asian levels. Its immediate task is to implement the comprehensive assessment of N_r in South Asia and showcase the local and regional capacity for science-led policies for sustainable N management in the region.

Keywords Nitrogen management · Nitrogen assessment · Nitrogen policy · Mitigation strategies · Nitrogen use efficiency

32.1 Introduction: Origins of INI in South Asia

By the time the International Nitrogen Initiative (INI) formed in 2003, there were independent concerns among some Indian scientists that the importance of reactive nitrogen (N_r) in all its dimensions was not being fully recognized in South Asia. There were several scientists working on various aspects of the nitrogen (N) cycle in the region, but they were largely confined to their narrow disciplinary boundaries. The need for an integrative approach to research and policy regarding N_r was first addressed by the authors in 2004 under the Society for Conservation of Nature (SCON), a voluntary body of scientists registered in New Delhi since 1998. They brought together some concerned Indian experts from diverse backgrounds to discuss the issue of N_r in agriculture, industry and environment. In the next couple of years, SCON held a series of nationwide consultations with the Indian National Academy of Agricultural Sciences (NAAS), Indian Government's Department of Biotechnology (DBT) and Indian National Science Academy (INSA), with active support from other related agencies such as the Ministry of Environment and Forests (MoEF) and Council of Scientific and Industrial Research (CSIR). These consultations culminated in the formation of the Indian Nitrogen Group (ING) under SCON as an outcome of the INSA workshop in 2006. Soon, ING came in contact with the INI formed in 2003, and persuaded it to set up a South Asian Nitrogen Centre for better regional and international coordination. This proposal was formalized during the 4th International Nitrogen Conference in Brazil in 2007. ING-SCON has since been instrumental in establishing the South Asian Nitrogen Centre for INI and in hosting the 5th International Nitrogen Conference in New Delhi in 2010, with Dr. M. S. Sachdev as the first Director of the South Asian Nitrogen Centre (SANC), as well as the Conference Coordinator.

32.2 Early Indian Activities

In the meantime, ING began documenting what is known about N_r in India and what was not. The early discussions at NAAS on N_r and N use efficiency in Indian agriculture led to the adoption of a policy paper (NAAS 2005). A compilation of position papers and reviews from the nationwide consultations on agriculture, health and environment in the Indian context was published as a book (Abrol et al. 2007). A more comprehensive rigorous compilation of reviews was published as a special section in one India's oldest and best known broad-scope scientific journals, *Current Science* (Abrol et al. 2008). Simultaneously, an extensive mapping exercise was undertaken to identify the various aspects of N management in the Indian context, the various universities, national and local institutions involved in them, as well as the various national funding agencies that supported them.

A map of the district-wise use of inorganic N fertilizer in Indian agriculture was made for the first time, and analyses of long term trends highlighted the declining partial factor productivity of N with increasing fertilizer input (Fig. 32.1). Soon, ING began to identify gap areas and advocating with the funding agencies and scientific institutions to catalyze research in those areas, leading to the elevation of reactive nitrogen as a priority area for research by some of the agencies. They include the Indian Council of Agricultural Research (ICAR), and the Ministries of Earth Sciences and Science and Technology. The outcomes include research projects funded on N-use efficiency in agriculture under the National Initiative for Climate Resilient Agriculture, among others.

32.3 Nitrogen Budget for Indian and World Agriculture

During 2010, it is estimated that 16.6 Mt of N fertilizer was used in Indian agriculture, while animal manure and wastes, crop nitrogen fixation and deposition (including rain, irrigation water and grazing of animal) contributed 1.4, 1.4 and 2.7 Mt, respectively (Fig. 32.2). In 2010, removal of N by agricultural crops in the country was 9.8 Mt. Besides, the losses of N included leaching and run-off (3.1 Mt), ammonia volatilization (4.1 Mt), and denitrification (3.1 Mt). This resulted in an accumulation of 1.9 Mt N in the agricultural soil-water system. In comparison to the annual N budget for Indian agriculture, about 120 Mt of fertilizer N was used for global food production during 2000–2010. The other sources, namely, animal manure and wastes, crop nitrogen fixation and deposition (including rain, irrigation water and grazing of animal) contributed 57, 60 and 70 Mt, respectively. An approximately equal amount of fertilizer N was removed by the harvested crop (122 Mt). The losses of N through leaching and run-off, ammonia volatilization and denitrification were

37, 95 and 25 Mt, respectively, resulting in a positive balance (soil accumulation) of 28 Mt of N annually.

During 1970 to 2010, Indian agriculture increased its fertilizer N consumption by about 11 times, whereas the uptake by crop increased by three times and loss of N increased by four times. India's share in N loss mechanisms may increase as more

a

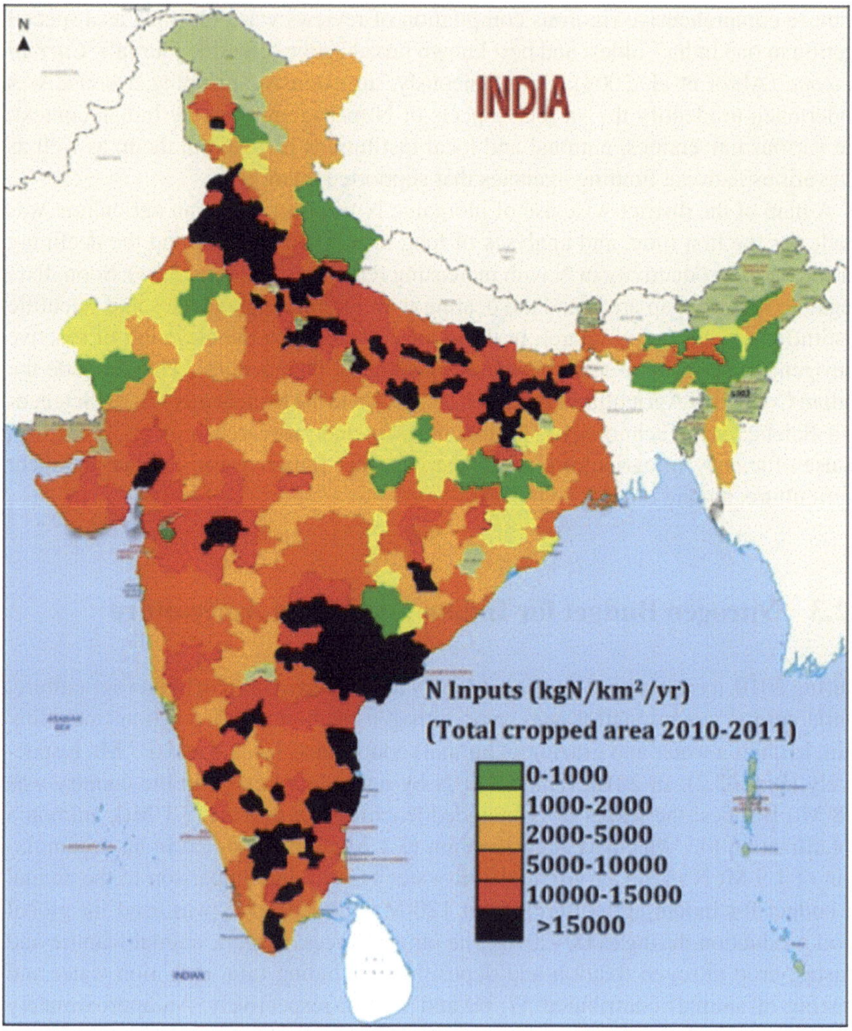

Fig. 32.1 **a** District-wise variation the annual input of inorganic nitrogen fertilizer in India in 2010–11, compiled by SCON using data from the Directorate of Economics and Statistics, Government of India and Fertilizer Association of India (Sutton et al. 2013)

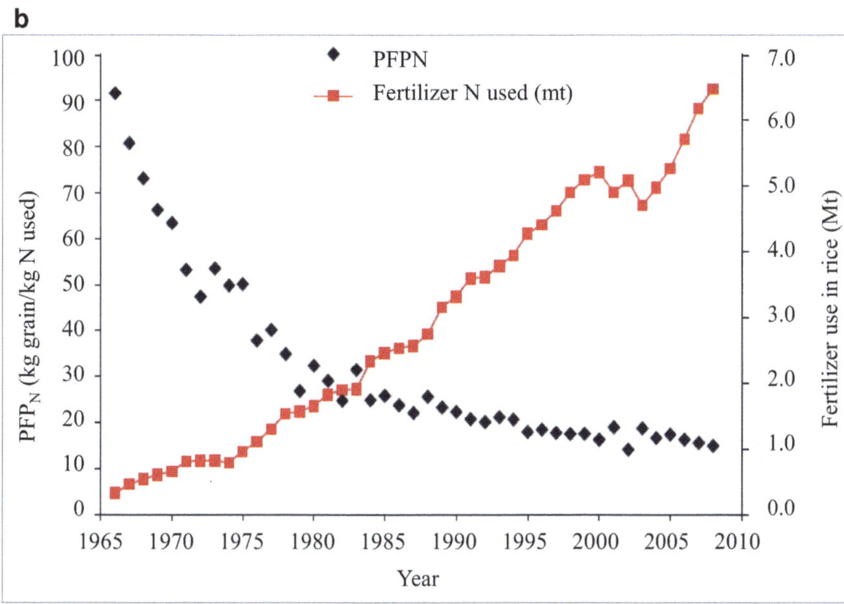

Fig. 32.1 (continued) **b** Long-term trends in inorganic N fertilizer use and rice partial factor productivity (kg grain/kg N applied) in India (Adhya et al. 2010b)

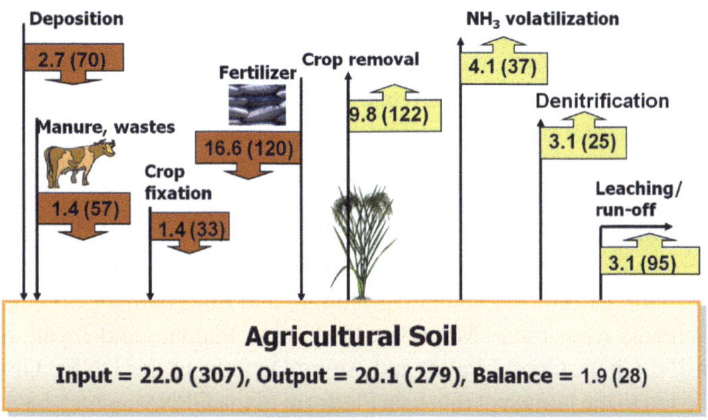

Fig. 32.2 Annual budget of nitrogen (Mt) in agricultural soils of India and World (in parenthesis) during the year 2010. Original graphic, from Pathak (unpublished), World values from Sutton et al. (2013)

fertilizer N is expected to be applied in future to produce more food for the growing population. On the other hand North American and European countries are gradually reducing their N accumulation in soil. Therefore, in India research on the ways to increase the efficiency of applied fertilizer N such as use of nitrification inhibitor and slow release fertilizer, deep placement of N, conjunctive use of organic manure, and use of better agronomic practices needs to be intensified. Attempts also need to be made for including more and more legumes in cropping systems to reduce chemical fertilizer N application in soil.

32.4 Regional and International Forays

Even as the SANC was being set up, the partnership between ING-SCON and INI catered to the increasing international interest in N_r in the Indian and South Asian context. Some early examples of this cooperation include an IGBP-WCRP-SCOPE Report on N_r in Indian agriculture, industry and environment (Singh et al. 2010) and an Annual Review on the environmental reach of Asia (Galloway et al. 2008). Scientists from ING-SCON were the only representatives from South Asia in the first meeting of the newly formed Global Partnership on Nutrient Management (GPNM) by the United Nations Environment Programme (UNEP) in The Hague in October 2009. The meeting drew up an action programme on nutrient assessment and management for the global partnership and formed a steering committee for its implementation. The representation of ING-SCON and SANC in the steering committee as well as in the action programme led to the inclusion of Lake Chilika from India in the first collaborative GPNM project funded by the Global Environment Facility (GEF). The project was titled "Global foundations for reducing nutrient enrichment and oxygen depletion from land based pollution, in support of the Global Nutrient Cycle". The South Asian component of this project assessed the nutrient scenario of Lake Chilika and developed its first ever ecosystem health report card (Chilika lake 2012), which was updated further (Chilika lake 2016, available at http://www.chilika.com/publication.php). This model of the ecosystem health report card has been subsequently adopted by the Philippines government for Manila bay, which was also a part of the GPNM project.

In June 2010, SANC hosted a joint consultation of the South Asia and East Asia Regional centres of the INI in New Delhi, on 'Reactive Nitrogen Flows and Budget in the South and East Asia Region in Agriculture and Environment'. It was attended by 30 participants from India, Malaysia, Sri Lanka, Vietnam and Japan, as well as by the then INI Chair, Cheryl Palm, and Anjan Datta from the UNEP GPNM. The meeting also led to the launch of the Asia Platform of the GPNM, which has expanded further since then, through subsequent meetings held in Beijing, China (2011) and Danang, Vietnam (2015). Their main focus was to facilitate scientific exchanges and sharing of data on the nutrient scenario in Asia and the technologies and best practices for nutrient use efficiency management in the region. These events were very important to foster regional collaborations, modelling/forecasting, demonstrations of technological and managerial interventions and their wider dissemination.

The most important event on reactive N in South Asia was SANC's hosting of the 5th International Nitrogen Conference on 'Nitrogen Management for Sustainable Development—Science, Technology and Policy (N2010)' of INI in New Delhi, India, 3–7 December 2010. It was attended by 345 scientists from 36 countries and had

175 oral and 130 poster presentations spanning a whole range of subject disciplines related to food security, energy security and industry, human health and degradation, ecosystem health and biodiversity, climate change and integration. The Delhi Declaration was adopted at the conclusion of the conference, which called upon "*the UN bodies such as UNEP, FAO, UN-Habitat, WHO, UNDP, UNFCCC, CBD, CLRTAP and other regional organizations, national governments, scientific communities including CGIAR, industries, policy makers, International Nitrogen Initiative and the civil society to address nutrient deficiencies, move towards increased efficiencies in each segment of nitrogen cycle management, in order to reduce the adverse effects. Approaches should consider the use of incentives, make full use of re-cycling and ensure the treatment of discharges….Identification, communication, and promotion of best practices require collaboration among many stakeholders including governments, scientists, practitioners, and policy makers at global, regional and national levels. The formation of the Global Partnership on Nutrient Management (GPNM) facilitated by UNEP is a welcome development in this regard…*"

An important contribution of the N2010 conference in India was to trigger extensive consultations towards a national N assessment in India. Two national brainstorming workshops were held by ING-SCON in New Delhi, during February and March 2012. The first workshop was on 'Simulating Nitrogen Dynamics in Major River Basins of India' and discussed the methodological aspects of reactive N assessment and modelling in India. This was followed by another workshop on "Backward Integration of Reactive N in Major Indian River Basins". Both these workshops were supported by the Ministry of Earth Sciences of the Government of India. This newly formed ministry had accumulated long-term data on coastal nutrient pollution from its predecessor, the Department of Ocean Development, but the nutrient loading into the sea from land-based sources was not traced systematically. Therefore, ING-SCON brought together scientists representing different ministries of the Government of India, national institutes of CSIR and ICAR and several central and state universities including agricultural universities. They agreed on the urgent need for a national assessment of N_r, including the role of land-ocean connections and options for interventions to reduce nutrient pollution. Using a common methodological framework for the whole country, project proposals were developed for major river basins of India and submitted to the Ministry of Earth Sciences. They have been received very well, but are yet to be funded.

During the Conference of Parties (COP11) to the Convention of Biological Diversity (CBD) that took place in India in October 2012, SANC organized a special side event (ID: 2776) at the venue of CBD-COP11 in Hyderabad on the 18th October 2012, in association with GPNM, ING-SCON, UNEP, GPA and GEF. The event was titled "The challenge to produce more food & energy with less pollution: Towards a Global Nitrogen Assessment". It was chaired by David Coates from the CBD Secretariat, Montreal, and attended by over 20 participants, including Tom Hammond from UNEP/GEF and several other key players from the National Biodiversity Authority of India, fertilizer industry, leading Indian/South-Asian NGOs, scientists and others. It was highlighted that nutrient issues have not received adequate attention, despite the fact that these are at the heart of all three Rio Conventions. The main presentation was on the natural N cycle and its anthropogenic distortions, its implications for environment, biodiversity and climate change and the need for global action to address it. It highlighted the efforts of INI at the global level and ING at the Indian level to quantitatively assess the anthropogenic perturbations to the N cycle and the role of UNEP in catalysing global engagement, especially through GPNM. The main findings of the various documents produced by INI, UNEP, GPNM and ING and other scientific literature were summarized regarding the scale of the nitrogen problem and its impacts on air, soil, water quality, health, biodiversity, ecosystem services and climate change. The need for a global assessment of reactive nitrogen cycle to support policies and actions was highlighted.

The SANC brought in 10 contributors to the Global Overview on Nutrient Management (Sutton et al. 2013), commissioned by UNEP for GPNM, which received extensive media coverage and scholarly citations internationally.

32.5 The South Asian Situation

South Asia is one of the most populous and the fastest growing regions in the world, mainly comprising of the sub-Himalayan countries, Afghanistan, Bangladesh, Bhutan, India, Maldives, Nepal, Pakistan, and Sri Lanka. Together, they cover an area of about 4.5 million km^2 (over 1.7 million square miles), with less than 5% of the world's land mass, 14% of the global arable land, 2.73% of the world forest area and 4% of the world's coastline and yet support over 25% of the world's population or over 45% of Asia's population.

South Asia has a diversity of ecosystems from lush tropical forest to harsh, dry desert, a huge diversity of languages and religions, but shared history and culture, which sets its people apart from the rest of the world. The countries of the region also share natural resource concerns such as depletion of water quality and quantity, dwindling forests and coastal resources, and soil degradation resulting from nutrient depletion and salinization.

Several South Asian countries are parties to all three Rio Conventions (that are relevant to N_r) and the Manila Declaration (agreed at the 3rd Intergovernmental Review of the Global Programme of Action for the Protection of the Marine Environment from Land-based Activities), and are also actively involved in the discussion of sustainable development goals at the global level. At the regional level, there are several agreements/programmes/mechanisms such as the South Asian Seas Programme (SASP), South Asia Environment and Natural Resources Information centre (SENRIC), South Asia Coral Reef Task Force (SACRTF), the Malé Declaration on control and prevention of air pollution and its likely transboundary effects, and the South Asia Biodiversity Clearing House Mechanism. The South Asia Co-operative Environment Programme (SACEP) was established in 1982 by the governments of South Asia to promote and support protection, management and enhancement of the environment in the region. The member countries SACEP are Afghanistan, Bangladesh, Bhutan, India, Maldives, Nepal, Pakistan and Sri Lanka.

The SANC and other partners of ING-SCON took up a scoping study for the SACEP on the nutrient pollution of the coastal and marine systems in South Asia, as an input document for a sub-regional workshop on nutrient management and the Bay of Bengal Large Marine Ecosystem (BOBLME) project. This study of 2013 reviewed the available information from India, Bangladesh, Sri Lanka, Maldives, and Pakistan, identified constraints and recommended technological, managerial and policy measures in a coordinated, sub-regional approach. It covered N use efficiency in cropping, livestock, aquaculture etc.; it outlined the N losses to freshwater, coastal and marine environments; and it identified the critical marine habitats affected. The SACEP ministerial level acceptance of this scoping study titled "*Nutrient loading and Eutrophication of coastal waters of the South Asian seas*" is an important milestone in the intergovernmental recognition of reactive N as an important aspect of nutrient loading at the South Asian level. The SACEP study has also revealed the gaps in our documented knowledge on the quantitative aspects of reactive N and lays the foundation for a South Asian N assessment.

32.6 The Case for a South Asian Nitrogen Assessment

Uneven development within and between countries of South Asia, as well as the huge diversity of soil types, water availability, climate, socio-economic and governance factors contribute to the problem of too much N in some areas/sectors and too little in others. For example, the intensive N fertilizer use is common in the irrigated cropping areas, including flood-prone areas, contributing to run-off losses and volatilization of N_r, while there is too little N use in the vast rainfed cropping areas, leading to soil degradation. Similarly, intensive livestock farming is common in peri-urban areas compared to its relatively thin spread in the rural areas, with little or no regulation on N-losses in both cases. Usable N is lost to ground water and surface water bodies through agricultural run-offs, sewage, animal and human excreta, and also into the air due to emission of reactive N compounds from agricultural soils, livestock, sewage dumps, residue burning, vehicular and industrial emissions (e.g., Food/Beverage Manufacturing, Slaughter Houses, Textile, Paper and Pulp, Agro-Based etc.).

There are pockets of coastal eutrophication around the Indian peninsula, and N-loading has also been observed in several lakes and other inland water bodies. In addition high levels of nitrates have been reported in the ground water in some places. There has been a slow, but growing awareness in the region regarding the leakage of N_r from production systems and its environmental consequences, partly due to the work of the ING and the SANC. But the lack of their systematic geographical or chronological documentation and assessment has hampered credible trend analyses and thus, prevented informed decisions on sustainable N management. National, regional and global investments in the quantification of the N_r scenario in the South Asian region is essential for a more accurate understanding of the global N-cycle, as well as for the development of a realistic International Nitrogen Management System. Given that South Asia is the most populous and the fastest growing region of the world with a distinct socio-economic, cultural and climatic profile, this region also offers a tropical testing ground for the validation of assumptions made on the basis of Western experience. This could in turn enable the adoption of more informed means of estimating the region's N-budget as well as its contribution to the global N-budget.

32.7 Policies and Practices

The prevailing priority for food production input use efficiency in South Asia has encouraged ignorance about the leakage of reactive N from fertilizers, dairy/livestock, fisheries, sewage, fossil fuel burning etc. Government policies or regulations against leakages of N_r have either been lacking or ineffective or unimplemented in most South Asian countries. There are policies for fertilizer N-dosage recommendations and N pollution limits for potable water and air, which can be considered as N-specific. Other related policies are meant to improve soil health and agricultural N-management, such as the Indian Government's nutrient-based fertilizer subsidy, soil health laboratories and soil-health cards. Policies such as banning of burning agricultural and municipal residues, dung-cakes, etc., exist in many places, but their enforcement has been poor. There are no emission/effluent standards of reactive N for specific sectors and point sources such as cropping, livestock, poultry, aquaculture, sewage, solid waste etc. There is also a lack of policies and incentives for recovery/recycling N_r. The need for specific standards/norms for specific ecosystems and ecosystem services (e.g., potable waters or recreational waters) is not yet recognized and therefore not yet included in the existing policies.

The adoption of N-use efficient fertilizer formulations such as neem-coated urea, and practices such as deep placement, leaf colour charts, integrated nutrient management practices for better demand-supply of N in agriculture are growing slowly, especially in India, Bangladesh and to some extent elsewhere in South Asia. But there are also challenges associated with changing existing practices. Farmers tend to exceed the recommended doses of N-fertilizers in irrigated crops, partly due to their low cost and government subsidy. Another area is the habitual residue burning by farmers, municipal workers despite the practice being banned in many places. The widespread use of dung-cakes, firewood and other inefficient fuels may partly be due to the lack of affordable access to better fuels as well as the lack of better monetary or use value for dung (especially if cattle owners are not growing their own crops). Another problem could be the lack of a well-developed recycling industry for recycling human and animal wastes as manures to offset inorganic fertilizers.

32.8 Conclusions and Prospects

The emerging indications on the N_r scenario in South Asia underscore the need for its comprehensive assessment in the best interest of the countries of the region, as well as that of the world. A sustainable N-management system has to be built on a strong local scientific capacity for quantitative assessment of the N-cycle, and a credible governance or management system that takes informed decisions and monitors their implementation. Fortunately, significant scientific capacities already exist in some of the countries of this region. The activities of the SANC and its partners have already generated sufficient interdisciplinary interest in some individual countries as well as at the intergovernmental level through SACEP. What is required are adequate investments at the national, South Asian and global level in developing integrated scientific understanding of the N cycle in South Asia. The resulting outcomes should be used to take informed decisions on how to tackle the too-much and too-little N use areas/sectors for sustainable N management.

The governments of South Asia can do a lot by acknowledging the problems of managing N_r in their governance programmes and incorporating them in their priorities for research and policy, making timely investments, taking informed decisions, and monitoring implementation throughout the command chain. For example, the governments can fund national N assessments to galvanize and institutionalise domestic researchers, and identify priority sectors and areas for action. Enabling policies and incentives are also needed for recovery/recycling reactive N and other nutrients from intensive animal husbandry, poultry, aquaculture, sewage, solid waste etc. The private sector can also help in the process by developing affordable technologies, products and services for recovery and recycling of nutrients from all available sources.

Acknowledgements The authors gratefully acknowledge preparatory support from UN Environment and the Global Environment Facility (GEF) for the South Asian demonstration activity under the International Nitrogen Management System (INMS) and from the Indian Department of Biotechnology (DBT) and the UK Biotechnology and Biological Sciences Research Council (BBSRC) and Natural Environment Research Council (NERC) for the NEWS India-UK virtual joint centre.

References

Abrol, Y. P., Raghuram, N., & Sachdev, M. S. (Eds.). (2007). *Agricultural nitrogen use and its environmental implications*. New Delhi: IK International.

Abrol, Y. P., Raghuram, N., & Chanakya, H. N. (2008). Reactive nitrogen in Indian agriculture, environment and health. *Current Science*, *94*(11), 1375–1477.

Adhya, T. K., Adhikari, A., Muralidhar, M., & Ayyappan, S. (2010a). *Nitrogen dynamics in the coastal regions of India*. ING Bulletins on Regional Assessment of Reactive Nitrogen, No. 16 (Ed. H. Pathak). New Delhi: SCON-ING.

Adhya, T. K., Shukla, A. K., & Panda, D. (2010b). *Nitrogen losses, N-use efficiency and N-management in rice and rice- based cropping system*. ING Bulletins on Regional Assessment of Reactive Nitrogen, No. 2 (Ed. Bijay-Singh). New Delhi: SCON-ING.
Galloway, J., Dentener, F., Marmer, E., Cai, Z., Abrol, Y. P., Dadhwal, V. K., et al. (2008). The environmental reach of Asia. *Annual Review of Environment and Resources, 33*, 461–481.
NAAS. (2005). *Policy options for efficient nitrogen use*. Policy Paper no. 33, (Eds. Y.P. Abrol, B.N. Johri). New Delhi: National Academy of Agricultural Sciences.
Singh, Y., Bijay-Singh, & Thind, H. S. (2010). *Green manure approaches to crop production and sustainable agriculture*. ING Bulletins on Regional Assessment of Reactive Nitrogen, No. 14 (Ed. H. Pathak). New Delhi: SCON-ING.
Sutton, M. A., Bleeker, A., Howard, C. M., Bekunda, M., Grizzetti, B., de Vries, W., van Grinsven, H. J. M., et al. (2013). *Our nutrient world: The challenge to produce more food and energy with less pollution*. Global Overview of Nutrient Management. Centre for Ecology & Hydrology, Edinburgh on behalf of the Global Partnership on Nutrient Management and the International Nitrogen Initiative.

Further Reading

Abrol, Y. P., Pandey, R., Raghuram, N., & Ahmad, A. (2012). Nitrogen cycle sustainability and sustainable technologies for nitrogen fertilizer and energy management. *Journal of the Indian Institute of Science, 92*, 37–62.
Aneja, V. P., Schlesinger, W. H., Erisman, J. W., Sharma, M., Behera, S. N., & Battye, W. (2010). *Reactive nitrogen emissions from crop and livestock farming in India*. ING Bulletins on Regional Assessment of Reactive Nitrogen, No. 3, (Ed. H. Pathak). New Delhi: SCON-ING.
Balasubramanian, V., Raghuram, N., Abrol, Y. P., Sachdev, M. S., Pathak, H., & Bijay-Singh (2010). *Reactive nitrogen: Good, bad and ugly. A comprehensive status report* (Ed. Bijay-Singh). New Delhi: SCON-ING.
Benbi, D. K., & Bijay-Singh. (2010). *Role of soil nitrogen in efficient management of fertilizer nitrogen in India*. ING Bulletins on Regional Assessment of Reactive Nitrogen. No. 6 (Ed. H. Pathak). New Delhi: SCON-ING.
Bijay-Singh, Tiwari, M.K., & Abrol, Y. P. (2008). *Reactive nitrogen in agriculture, industry and environment in India*. IGBP-WCRP.SCOPE Report Series. 3. New Delhi: Indian National Science Academy.
Chanakya, H. N., & Sharatchandra, H. C. (2010). *Nitrogen and sustainability issues of waste management in urban ecosystems*. ING Bulletins on Regional Assessment of Reactive Nitrogen. No. 5 (Ed. N Raghuram). New Delhi: SCON-ING.
Deshpande, P. A., Hegde, M. S., & Madras, G. (2010). *Hydrogen from biomass*. ING Bulletins on Regional Assessment of Reactive Nitrogen, No. 17 (Ed. Bijay-Singh). New Delhi: SCON-ING.
Dhar, D. W., Prasanna, R., & Swarnalakshmi, K. (2010). *Harnessing the nitrogen fixing potential of cyanobacteria in integrated nutrient management strategies for sustainable agriculture*. ING Bulletins on Regional Assessment of Reactive Nitrogen, No. 8 (Ed. N. Raghuram). New Delhi: SCON-ING.
Goyal, S. K., & Chalapati Rao, C. V. (2010). *Contribution of energy sector to NOx emissions*. ING Bulletins on Regional Assessment of Reactive Nitrogen, No. 15 (Ed. H. Pathak). New Delhi: SCON-ING.
Gupta, S., Gupta, R. C., Gupta, A. B., Prakasa Rao, E. V. S., Puttanna, K., & Singhvi A. (2010). *Pathophysiology of nitrate toxicity in human and its mitigation measures*. ING Bulletins on Regional Assessment of Reactive Nitrogen, Bulletin No. 20 (Ed. N. Raghuram). New Delhi: SCON-ING.

Kumar, D. M. (2010). *Nitrogen in North Indian Ocean*. ING Bulletins on Regional Assessment of Reactive Nitrogen, No. 10, (Ed. N. Raghuram). New Delhi: SCON-ING.

Pathak, H., Bhatia, A., Jain, N., & Aggarwal, P. K. (2010). *Greenhouse gas emission and mitigation in Indian agriculture—a review*. ING Bulletins on Regional Assessment of Reactive Nitrogen, No. 19 (Ed. Bijay-Singh). New Delhi: SCON-ING.

Prasad, R., Pathak, H., & Ladha, J. K. (2010). *Nitrogen management in rice-wheat cropping systems of Asia and its environmental and health implications*. ING Bulletins on Regional Assessment of Reactive Nitrogen, No. 1 (Ed. Bijay-Singh). New Delhi: SCON-ING.

Raghuram, N., Sangar, S., & Abrol, Y. P. (2011). Nitrogen management for sustainable development. *Current Science, 101*(9), 1126–1127.

Ramakrishnan, P. S. (2010). *Nutrient dynamics under shifting agricultural (Jhum) landscape in north- east India and linked sustainability issues*. ING Bulletins on Regional Assessment of Reactive Nitrogen. No. 12 (Ed. Bijay-Singh). New Delhi: SCON-ING.

Rengarajan, R, Srinivas, B., Sarin, M. M. (2010). *Atmospheric deposition of reactive nitrogen over continental sites and oceanic regions of India: A review*. ING Bulletins on Regional Assessment of Reactive Nitrogen, No. 4 (Ed. H. Pathak). New Delhi: SCON-ING.

Sharma, K. L., Maheswari, M., Venkateswarlu, B., Abrol, Y. P. (2010). *Nitrogen use in agriculture in arid and semiarid regions—Indian perspective*. ING Bulletins on Regional Assessment of Reactive Nitrogen, No. 9 (Ed. Bijay Singh). New Delhi: SCON-ING.

Singh, R., Adholeya, A., & Pant, D. (2010). *Management of solid wastes in agricultural systems for nitrogen conservation in India*. ING Bulletins on Regional Assessment of Reactive Nitrogen, No. 7, (Ed. N. Raghuram). New Delhi: SCON-ING.

Subbarao, G. V., Sahrawat, K. L., Nakahara, K., Kishii, M., Rao, I. M., Hash, C. T., George, T. S., et al. (2010). *Nitrification: is it a strategic point of intervention to limit nitrogen losses from agricultural systems?* ING Bulletins on Regional Assessment of Reactive Nitrogen, No. 13. (Ed. Bijay-Singh). New Delhi: SCON-ING.

Varadachari, C., & Goertz, H. M. (2010). *Slow-release and controlled-release nitrogen fertilizers*. ING Bulletins on Regional Assessment of Reactive Nitrogen, No. 11 (Ed. Bijay Singh). New Delhi: SCON-ING.

Chapter 33
The INI East Asia Regional Nitrogen Centre: Balancing Food Production and Environment—Nitrogen-Related Research and Management in East Asia

Xiaoyuan Yan, Chaopu Ti, and Kentaro Hayashi

Abstract East Asia is one of the most densely populated regions in the world, and is also a region with intensive nitrogen (N) loading. Nitrogen-related environmental problems, such as surface water eutrophication and atmospheric pollution including smog are of common concern. Efforts are made to balance the effect of N on food security and its environmental impacts. In this chapter, we summarize the relevant research and management activities in China and Japan, as representatives of East Asia.

Keywords Nitrogen management · Air pollution · Water pollution · Food security · Nitrogen policy

33.1 Introduction

The East Asia Centre of the International Nitrogen Initiative (INI) provides a focus for a science-led analysis of the challenges arising mainly from the problem of too much nitrogen (N) in the environment. The approach taken gives special attention to improving N management in the food system, but also considers other N losses, such as from combustion sources, especially as these are currently increasing at unprecedented rates across much of East Asia. In the present chapter, we focus on making the comparison between the situations in China and Japan.

X. Yan (✉) · C. Ti
State Key Laboratory of Soil and Sustainable Agriculture, Institute of Soil Science, Chinese Academy of Sciences, Nanjing 210008, China
e-mail: yanxy@issas.ac.cn

C. Ti
e-mail: cpti@issas.ac.cn

K. Hayashi
Institute for Agro-Environmental Sciences, National Agriculture and Food Research Organization (NARO), 3–1–3, Kannondai, Tsukuba, Ibaraki 305–8604, Japan
e-mail: kentaroh@affrc.go.jp

M. A. Sutton et al. (eds.), *Just Enough Nitrogen*,
https://doi.org/10.1007/978-3-030-58065-0_33

33.2 Nitrogen Management in China

To meet the food demands of the 22% of the world's population, China became the world's largest producer and consumer of synthetic N fertilizer. The total synthetic N input to cropland increased from 9.4 Tg to 29.5 Tg from 1980 to 2010 in China (Yan et al. 2014). Per-hectare N addition was 209, 210 and 220 kg N ha^{-1} for rice, wheat and corn in recent years (Chen et al. 2014), with a much higher rate for intensive double-cropping systems (550–600 kg N ha^{-1}) and some vegetables grown in greenhouses (> 1000 kg N ha^{-1}) as reported by Ju et al. (2009) and Shi et al. (2009).

Although a large amount of fertilizer N application plays a vital role in ensuring food security, it has contributed to low fertilizer recovery and environmental impacts. Many field experiments showed that the in-season fertilizer N recovery was 30–35% in the 1990s (Zhu and Chen 2002), 26–28% in 2001–2005 for major cereal crops (Zhang et al. 2007), relative to 52% in America and 68% in Europe (Ladha et al. 2005). While some of the fertilizer is used by crops in subsequent seasons, much of the remainder N can be considered as an expensive and environmentally damaging waste, leading to emissions of greenhouse gases, loss of biodiversity, and degradation of soil and freshwater. For instance, a survey of eutrophication of the major lakes in China revealed that about 50% of lakes investigated were eutrophic (Jin et al. 2005), while 49% of groundwater samples had nitrate N content exceeding 11.3 mg L^{-1} (the criterion for drinking water supplies in Europe) in a typical vegetable-growing area (Chen et al. 2000; Zhu 2002). In addition, over-application of fertilizers also contributes to increased N emissions and soil acidification. Guo et al. (2010) reported that soil pH declined significantly from the 1980s to 2000s in the major Chinese crop-production areas, with N fertilizer being the major contributor.

China has developed a series of programs, projects and regulations to enhance nutrient use efficiency and reduce environmental risk, in the search for sustainable agriculture development. For example, the Ministry of Agriculture (MOA) has implemented a soil testing and fertilizer recommendation (STFR) program to reduce the over-usage of fertilizer N on cereal crops since the late 1990s. The central government invested approximately 2.3 million US dollars (US$) on the program to implement STFR in all the 2498 agricultural counties in 2009 and the implementation area consequently increased from 16.7 million hectares in 2005 to 66.7 million hectares in 2009. A campaign on "zero growth of the fertilizer and pesticide consumption by 2020" was launched by the MOA on 18th March 2015, to explore a road to modern agricultural development with high efficiency, high product security, while being resource-saving and having an environmental-friendly target. According to the Zero Growth Program, China will promote precision agriculture, adjust fertilizer consumption structure, and use more organic fertilizer to replace chemical fertilizers.

Scientific research to improve nutrient use efficiency has been conducted widely in China, mainly with financial support from the MOA and the Ministry of Science and Technology (MOST). A successful example is the integrated soil-crop system management (ISSM) project led by China Agriculture University. With practices based on modern understanding of crop ecophysiology and soil biogeochemistry, farmers were able to increase average yields for rice, wheat and maize from 7.2 million grams per hectare ($Mg\ ha^{-1}$), 7.2 $Mg\ ha^{-1}$ and 10.5 $Mg\ ha^{-1}$ to 8.5 $Mg\ ha^{-1}$, 8.9 $Mg\ ha^{-1}$ and 14.2 $Mg\ ha^{-1}$, respectively, without any increase in N fertilizer (Zhang et al. 2011; Chen et al. 2014).

To clarify the atmospheric chemistry behavior and the global change effects of reactive N, and to improve the reactive N utilization efficiency in the cascade and reduce its emissions, a project entitled "Reactive nitrogen sources of China and their impacts on air quality and climate change" was supported by the National Key Basic Research Program of the MOST, which started in 2014. With this project, it is anticipated to answer questions such as:

(1) What is the N use efficiency during the flow from fertilizer to food? What are the key driving factors?
(2) Which roles does reactive N play in the aerosol formation? What is its contribution?
(3) What is the influence of atmospheric N deposition on the balance of greenhouse gases in the terrestrial ecosystem?

To gain insight into the agronomy effect and the environmental impact of N, various activities relevant for N assessment have been started in China. For instance, early researchers such as Xing and Zhu (2002) estimated N budgets in terrestrial ecosystems for a single year for all of China and its three major watersheds. Ti et al. (2012) evaluated N input and output in mainland China using updated data with temporal and spatial resolution. Ma et al. (2008) reported on N flows, losses, and N use efficiency in the production and utilization of three major grain crops. They pointed out that mean N surpluses of crop fields were 144 $kg\ ha^{-1}$ for wheat, 184 $kg\ ha^{-1}$ for rice, and 120 $kg\ ha^{-1}$ for maize in 2004, and they estimated that between 50% and 85% of N harvested as grain is lost during the pathway of utilization by humans and animals. Gu et al. (2012) presented an assessment of ammonia (NH_3), nitrogen oxides (NO_x), and nitrous oxide (N_2O) emissions in China based on a full N cycle analysis. Their results showed that the total health damage related to atmospheric reactive N reached US$19–62 billion in 2008, accounting for 0.4–1.4% of China's gross domestic product. A more comprehensive N assessment is still in progress in China.

33.3 Nitrogen Management in Japan

A framework for the comprehensive management of environmental N loads due to human activities such as food production and consumption and energy use is still absent in Japan. Although a strategic report (CRDS 2013) pointed out the necessity of research for understanding the nature of N issues and developing a N management framework in Japan, awareness of N issues has not yet sufficiently penetrated into Japanese policy sections and funding agencies.

In Japan, the main N issues generally recognized are the pollution of air and water. A number of studies have been conducted in Japan in relation to pollution control. Nitrogen-related studies on the basis of both process-oriented and regulatory sciences within each sector, e.g., agriculture, industry, and transport, have also been conducted. Meanwhile, multidisciplinary studies of N are currently limited (e.g., so far no nationwide N assessment published). Recognizing these limitations, the current Japanese status is shown below.

An important feature of the Japanese N flows is that Japan largely depends on imported food and animal feed. Their self-sufficiency ratios are fairly small, i.e., 39% for food and 26% for feed in 2013 (MAFF, 2014). This strong import dependency results in excess N inflow, and simultaneously contributes to environmental N loads in the countries exporting food and feed to Japan. In Japan at present, a large amount of food is wasted without consumption, so-called "food loss", which is estimated to be a range of 5–8 Tg of food in 2010 (MAFF 2013). Food loss is a direct N loss, while even more input N used to produce the food is also wasted as a result of the food loss. In this regard, the concept of N footprint is a good indicator of individual N loads, and is also effective in raising public awareness of the N issues. Shibata et al. (2014) estimated the current N footprint of Japan was 28.1 kg N cap^{-1} $year^{-1}$.

Air pollution by NO_x has been well controlled in Japan. However, the status of other air pollutants in relation to reactive N such as oxidants (mainly tropospheric ozone) and fine particles (regulated as $PM_{2.5}$) is worse. Exceedance of the environmental standard for photochemical oxidants was nearly 100%, and that for $PM_{2.5}$ was approximately half in 2012 (MOE 2014a).

Nationwide monitoring of atmospheric deposition (originally with respect to acid rain) has been conducted since 1980s in Japan, though there is no legislation to directly control N deposition. Japanese N deposition (wet and dry deposition) is generally up to 20 kg N ha^{-1} $year^{-1}$ except hot-spots such as in intensive livestock farming areas (Hayashi and Yan 2010) where larger rates are estimated. Nitrate NO_3^- wet deposition increased from 1989 to 2008 with a rate of 2–5% $year^{-1}$ especially in southwestern Japan, and currently NO_3^- and ammonium NH_4^+ make similar contributions to wet N deposition across Japan (Morino et al. 2011). Recently, N saturation of Japanese forest ecosystems were reported (e.g., Nakahara et al. 2010; Chiwa et al. 2012), to which long-term N deposition might contribute.

Water pollution by NO_3^- is problematic particularly for lakes and groundwater, in which the environmental standard for NO_3^- is exceeded at many monitoring plots (MOE 2013, 2014b). The source of NO_3^- is attributed to application of fertilizer and manure, discharge from livestock facilities, and miscellaneous drainage. However, it is noted that submerged paddy fields induce complete denitrification owing to their anaerobic conditions, in which NO_3^- is reduced to molecular N with less N_2O emissions. A topographical chain of uplands and paddies results in reduction of NO_3^- and N_2O loads in a watershed scale through the complete denitrification at the paddies which consume the leaching NO_3^- from the uplands (Eguchi et al. 2009).

Nitrogen surplus (difference between N input and crop N demand) in Japanese croplands is very high (153 kg N ha^{-1} in 2005), in which N input has been decreasing for chemical fertilizer but increasing for manure (Shindo 2012). Uplands, orchards, and tea fields can also be a remarkable source of NO_3^- and N_2O to the environment in Japan. As mentioned above, N loads from paddies are relatively small but not zero.

The following illustrate recent examples of N research in agriculture:

- A mechanistic model with good reproducibility of the N dynamics at an upland was developed (Asada et al. 2013, 2015).
- A shallow groundwater table was found to increase indirect N_2O emissions from uplands (Minamikawa et al. 2013).
- A review concluded that soil–plant interactions strongly affect cropland emissions of N_2O, in which functions of plant roots affecting biogeochemical factors in the rhizosphere and phenological changes are particularly important (Hayashi et al. 2015).

In addition to these findings, perturbation of N cycle due to elevated atmospheric CO_2 levels and climate change has been found to cause large uncertainties of future food production. This is especially linked to unknown responses of the N cycle to the perturbation including interactions between the N and carbon C cycles, e.g., Hayashi et al. 2020 (Chap. 22, this volume) for paddy ecosystems.

33.4 Conclusions

East Asia faces major challenges associated with excess N, which results from both fertilizer use in the food system and from fossil fuel combustions sources. Rapidly increasing economic development has been associated with substantial increases in N inputs and emissions to the environment, worsening pollution levels over the last decades. Although policies are being brought into address parts of this challenge in both China and Japan, these have a long way to go, also in making the links between source sectors and impacts across the N cycle. As part of ongoing work, both China and Japan are contributing with other countries to developing an East Asian Nitrogen Assessment as part of the regional demonstration activities of the International Nitrogen Management System.

Acknowledgements This study was financially supported by the National Natural Science Foundation of China (41425005).

References

Asada, K., Eguchi, S., Urakawa, R., Itahashi, S., Matsumaru, T., Nagasawa, T., et al. (2013). Modifying the LEACHM model for process-based prediction of nitrate leaching from cropped Andosols. *Plant and Soil, 373*(1), 609–625.

Asada, K., Eguchi, S., Tsunekawa, A., Tsuji, M., Itahashi, S., & Katou, H. (2015). Predicting nitrogen leaching with the modified LEACHM model: Validation in soils receiving long-term application of animal manure composts. *Nutrient Cycling in Agroecosystems, 102*(2), 209–225.

Chen, X. P., Ji, H. J., & Zhang, F. S. (2000). The integrated evaluation on effect of excess fertilization application on nitrate content of vegetable in Beijing. In X. L. Li (Ed.), *Fertilizing for sustainable production of high quality vegetables* (pp. 270–277). Beijing: Chinese Agriculture Publication. (In Chinese with English abstract).

Chen, X. P., Cui, Z. L., Fan, M. S., Vitousek, P., Zhao, M., Ma, W. Q., et al. (2014). Producing more grain with lower environmental costs. *Nature, 514*(7523), 486–489.

Chiwa, M., Onikura, N., Ide, J., & Kume, A. (2012). Impact of N-saturated upland forests on downstream N pollution in the Tatara river basin. *Japan Ecosystems, 15*(2), 230–241.

CRDS. (2013). *Strategic proposal: consolidated research initiative for sustainable nitrogen cycle.* CRDS-FY2012-SP-01. Japan: Center for Research and Development Strategy (CRDS), Japan Science and Technology Agency (JST). (In Japanese with English abstract).

Eguchi, S., Nakajima, Y., Yabusaki, S., Kasuya, M., Shibayama, H., & Imai, K. (2009). Denitrification during vertical upwelling at an alluvium-diluvium interface below the upland perimeter of a riparian paddy. *Journal of Environmental Quality, 38*(6), 2198–2209.

Gu, B. J., Ge, Y., Ren, Y., Xu, B., Luo, W. D., Jiang, H., et al. (2012). Atmospheric reactive nitrogen in China: Sources, recent trends, and damage costs. *Environmental Science and Technology, 46*(17), 9420–9427.

Guo, J. H., Liu, X. J., Zhang, Y., Shen, J. L., Han, W. X., Zhang, W. F., et al. (2010). Significant acidification in major Chinese croplands. *Science, 327*(5968), 1008–1010.

Hayashi, K., & Yan, X. Y. (2010). Airborne nitrogen load in Japanese and Chinese agroecosystems. *Soil Science and Plant Nutrition, 56,* 2–18.

Hayashi, K., Tokida, T., Kajiura, M., et al. (2015). Cropland soil–plant systems control production and consumption of methane and nitrous oxide and their emissions to the atmosphere. *Soil Science and Plant Nutrition, 61,* 2–33.

Hayashi, K., Ono, K., Tokida, T., Matsushima, M. Y., Yano, M., Toyada, S., et al. (2020). Nitrogen aspects of the free-air CO2 enrichment (FACE) study for paddy rice ecosystems. In M. A. Sutton, K. E. Mason, A. Bleeker, W. K. Hicks, C. Masso, N. Raghuram, S. Reis, & M. Bekunda (Eds.), *Just enough nitrogen: Perspectives on how to get there for regions with too much and too little nitrogen (Chapter 22).* The Netherlands: Springer.

Jin, X. C., Xu, Q. J., & Huang, C. Z. (2005). Current status and future tendency of lake eutrophication in China. *Science in China, Series C: Life Sciences, 48*(S2), 948–954.

Ju, X. T., Xing, G. X., Chen, X. P., Zhang, S. L., Zhang, L. J., Liu, X. J., Cui, Z. L., et al,. (2009). Reducing environmental risk by improving N management in intensive Chinese agricultural systems. *Proceedings of the National Academy of the Sciences of the United States of America (PNAS), 106*(9), 3041–3046; and Correction *106*(19), 8077.

Ladha, J. K., Pathak, H., Krupnik, T. J., Six, J., & van Kessel, C. (2005). Efficiency of fertilizer nitrogen in cereal production: Retrospects and prospects. *Advances in Agronomy, 87,* 85–156.

Ma, W. Q., Li, J. H., Ma, L., Wang, F. H., Sisák, I., Cushman, G., et al. (2008). Nitrogen flow and use efficiency in production and utilization of wheat, rice, and maize in China. *Agricultural Systems, 99*(1), 53–63.

MAFF. (2013). *For Reduction of Food Loss.* Japan: Ministry of Agriculture, Forestry and Fisheries, Government of Japan (MAFF). (In Japanese).

MAFF. (2014). *Reference of the food self-sufficiency in Japan as of 2013.* Japan: Ministry of Agriculture, Forestry and Fisheries, Government of Japan (MAFF). (In Japanese).

Minamikawa, K., Eguchi, S., Nishimura, S., Ihara, H., Maeda, M., Yagi, K., et al. (2013). Groundwater-induced emissions of nitrous oxide through the soil surface and from subsurface drainage in an Andosol upland field: A monolith lysimeter study. *Soil Science and Plant Nutrition, 59*(1), 87–95.

MOE. (2013). *Report of Water Quality in FY2012.* Japan: Ministry of the Environment (MOE). (In Japanese).

MOE. (2014a). *Report Of Air Quality in FY2012*. Japan: Ministry of the Environment (MOE). (In Japanese).

MOE. (2014b). *Report of Groundwater Quality in FY2012*. Japan: Ministry of the Environment (MOE). (In Japanese).

Morino, Y., Ohara, T., Kurokawa, J., Kuribayashi, M., Uno, I., & Hara, H. (2011). Temporal variations of nitrogen wet deposition across Japan from 1989 to 2008. *Journal of Geophysical Research, 116*(D6), D06307. https://doi.org/10.1029/2010JD015205.

Nakahara, O., Takahashi, T., Sase, H., Yamada, T., Matsuda, K., Ohizumi, T., et al. (2010). Soil and stream water acidification in a forested catchment in central Japan. *Biogeochemistry, 97*(2), 141–158.

Shi, W. M., Yao, J., & Feng, Y. (2009). Vegetable cultivation under greenhouse conditions leads to rapid accumulation of nutrients, acidification and salinity of soils and groundwater contamination in South-Eastern China. *Nutrient Cycling in Agroecosystems, 83,* 78–84.

Shibata, H., Cattaneo, L. R., Leach, A. M., & Galloway, J. N. (2014). First approach to the Japanese nitrogen footprint model to predict the loss of nitrogen to the environment. *Environmental Research Letters, 9*(11), 115013. https://doi.org/10.1088/1748-9326/9/11/115013.

Shindo, J. (2012). Changes in the nitrogen balance in agricultural land in Japan and 12 other Asian Countries based on a nitrogen-flow model. *Nutrient Cycling in Agroecosystems, 94,* 47–61.

Ti, C. P., Pan, J. J., Xia, Y. Q., & Yan, X. Y. (2012). A nitrogen budget of mainland China with spatial and temporal variation. *Biogeochemistry, 108*(1), 381–394.

Xing, G. X., & Zhu, Z. L. (2002). Regional nitrogen budgets for China and its major watersheds. *Biogeochemistry, 57*(58), 405–427.

Yan, X. Y., Ti, C. P., Vitousek, P., Chen, D., Leip, A., Cai, Z. C., et al. (2014). Fertilizer nitrogen recovery efficiencies in crop production systems of China with and without consideration of the residual effect of nitrogen. *Environmental Research Letters, 9*(9), 095002. https://doi.org/10.1088/1748-9326/9/9/095002.

Zhang, F. S., Cui, Z. L., Fan, M. S., Zhang, W. F., Chen, X. P., & Jiang, R. F. (2011). Integrated soil-crop system management: reducing environmental risk while increasing crop productivity and improving nutrient use efficiency in China. *Journal of Environmental Quality, 40*(4), 1051–1057.

Zhang, F. S., Cui, Z. L., Wang, J. Q., Li, C. J., & Chen, X. P. (2007). Current status of soil and plant nutrient management in China and improvement strategies. *Chinese Bulletin of Botany, 24,* 687–694. (In Chinese with English abstract).

Zhu, J. H. (2002). Study on fate and utilization of nitrogen in protected vegetable fields Ph.D. Dissertation. Beijing: China Agriculture University. (In Chinese with English abstract).

Zhu, Z. L., & Chen, D. L. (2002). Nitrogen fertilizer use in China—contributions to food production, impacts on the environment and best management strategies. *Nutrient Cycling in Agroecosystems, 63*(2), 117–127.

Chapter 34
The INI North American Regional Nitrogen Center: 2011–2015 Nitrogen Activities in North America

Jill S. Baron and Eric A. Davidson

Abstract The North American Nitrogen Center (NANC) carries out three main charges: (1) conducting assessments on nitrogen (N) flows within North America and the consequences for human health, water resources, biodiversity, and greenhouse gas emissions; (2) facilitating efforts to develop solutions to the problem of excess nitrogen in agricultural, institutional, and natural resource management sectors; and (3) presenting these results to policy makers. There are formidable challenges in reducing N loss from all parts of the North American food production and supply chain, including altering consumer behavior. The NANC is working with producers, trade groups, universities, and supply chains to develop effective practices for minimizing loss of reactive nitrogen (N_r) to the environment. The NANC is also helping public land management and regulatory agencies prepare effective policy approaches toward minimizing ecological damage from atmospheric N_r deposition.

Keywords Nitrogen · Agriculture · Biodiversity · Water resources · Health · Climate change · Nitrogen use efficiency · Sustainable agriculture · Critical loads · Ecosystem services

34.1 Introduction

The nitrogen (N) challenge in North America is one of excess. Our technologically advanced modern lifestyle and protein-rich diets cause more reactive nitrogen (N_r) to be lost to the North American environment than is produced in usable goods and services (Houlton et al. 2013). The North American Nitrogen Center carries out three main charges: (1) conducting assessments on nitrogen flows within North America and the consequences for human health, water resources, biodiversity, and

J. S. Baron (✉)
U.S. Geological Survey, Colorado State University, Fort Collins, CO 80523–1499, USA
e-mail: jill.baron@colostate.edu

E. A. Davidson
University of Maryland Center for Environmental Science, Appalachian Laboratory,
301 Braddock Road, Frostburg, MD 21532, USA
e-mail: edavidson@umces.edu

M. A. Sutton et al. (eds.), *Just Enough Nitrogen*,
https://doi.org/10.1007/978-3-030-58065-0_34

greenhouse gas emissions; (2) facilitating efforts to develop solutions to the problem of excess nitrogen in agricultural, institutional, and natural resource management sectors; and (3) presenting these results to policy makers.

34.2 Activities

34.2.1 North American Nitrogen Assessments

A number of highly productive assessment activities were launched by a National Science Foundation (NSF)-funded Research Coordination Network led by Eric A. Davidson and Alan Townsend during 2011–2015. Among the first was the publication of an Ecological Society of America *Issues in Ecology* in 2012: "Excess Nitrogen in the U.S. Environment: Trends, Risks, and Solutions." A team of authors led by Eric A. Davidson reported that there have been "important successes in reducing nitrogen emissions to the atmosphere, and this has improved air quality." They also noted that effective options have been identified for reducing nitrogen losses from agriculture, "although political and economic impediments to their adoption remain (Davidson et al. 2012)." Available in both Spanish and English, the *Issues in Ecology* is written for general audiences and is downloadable (see Davidson et al. 2012 for the web-link).

A workshop on climate-nitrogen interactions held at the John Wesley Powell Center for Analysis and Synthesis in 2011 was designed to acknowledge the interactive role of N_r and climate change on natural resources and health. With an emphasis on the United States (US), the main product was a technical report submitted to the National Climate Assessment (NCA): "The Role of Nitrogen in Climate Change and the Impacts of Nitrogen-Climate Interactions on Terrestrial and Aquatic Ecosystems, Agriculture and Human Health in the United States" (Suddick and Davidson 2012). The technical report was cited widely in the Biogeochemical Cycles Sector Report of the 2014 National Climate Assessment (Galloway et al. 2014).

Each chapter of the NCA technical report was published separately in a 2013 special issue of *Biogeochemistry,* and they have since been cited extensively. An overview of this special issue and summary of major climate-nitrogen interactions and their consequences was presented by Suddick et al. (2013). Other contributions to the *Biogeochemistry* special issue include the following:

- Using mass-balance principles and pre-existing data on long-term trends in N_r creation, Houlton et al. (2013) examine N_r use efficiencies within the food, fiber, energy and industrial sectors of the continental US.
- More detail on the impacts of human alteration of the nitrogen cycle in the US on the atmosphere was provided by Pinder et al. (2013), who report that the net effect of all of the many N cycle processes on radiative forcing in the US may be a modest cooling effect for a 20-year time frame, followed by and a modest warming for a 100-year time frame.

- Robertson et al. (2013) addressed the impacts of climate change—nitrogen interactions in agricultural systems, and identified a set of compelling research needs: (1) an improved understanding of agricultural N cycle responses to changing climate; (2) a systems-level understanding of important crop and animal systems sufficient to identify key interactions and feedbacks; (3) the further development and testing of quantitative models capable of predicting N-climate interactions with confidence across a wide variety of crop-soil-climate combinations; and (4) socio-ecological research to better understand the incentives necessary to achieve meaningful deployment of realistic solutions.
- Baron et al. (2013) noted the loading of N_r from watersheds and atmospheric deposition has greatly increased the flux of N_r to estuaries and coastal oceans, in some cases by more than an order of magnitude. Nearly all freshwaters and coastal zones of the US, including remote alpine lakes, are degraded by air pollution. Interactions with climate change will be felt most strongly through alterations of the hydrologic cycle, influencing both rates of N_r inputs to aquatic ecosystems and groundwater and the water residence times that affect N_r removal within aquatic systems.
- Climate change and N_r from anthropogenic activities independently are causing some of the most rapid changes in terrestrial and aquatic biodiversity. Porter et al. (2013) conclude from evaluation of empirical studies and modeling that N_r and climate change can interact to drive losses in biodiversity greater than those caused by either stressor alone.
- Peel et al. (2013) examine the potential human health implications of climate change and N cycle interactions related to ambient air pollution. Among their findings are that while it is known that nitrogen oxides and other compounds cause a variety of serious health effects and premature deaths, changes in temperature and precipitation may exacerbate these effects by lengthening the ozone season or intensifying ozone concentrations through a series of direct effects and feedbacks from wildfires and societal responses to warming.

The Sixth annual International Nitrogen Initiative (INI) meeting in Kampala, Uganda featured a talk "Perspectives on interactions and feedbacks between regulation and scientific investigation of nitrogen pollutants affecting human health," by Emma Suddick and Eric Davidson that summarized results from a workshop on human health impacts of excess nitrogen in air and water, held in Bethesda, MD, in 2012.

Finally, a paper by Clair et al. (2014) calculated a reactive nitrogen budget for Canada, finding that Canada receives a significant amount of N_r from the US, and is a major exporter of N as hydrocarbons, fertilizers, and food.

34.2.2 Sustainable Agriculture

In August 2013, about 160 agronomists, scientists, extension agents, crop advisors, economists, social scientists, farmers, representatives of regulatory agencies and non-governmental organizations (NGOs), and other agricultural experts gathered in Kansas City, Missouri to discuss the vexing challenge of how to produce more food to nourish a growing population while minimizing pollution to the environment. The conference, jointly sponsored by INI, the Soil Science Society of America, the American Geophysical Union, The International Plant Nutrition Institute, and The Fertilizer Institute, focused on technical, economic, and social impediments and opportunities for improving nitrogen use efficiency (NUE) in agriculture. A consensus statement was developed (NANC 2013a), and a follow-up brochure on NUE is available (NANC 2013b). A briefing to the National Academy of Science is available on YouTube (Davidson 2013). Briefings on the conference were given in Washington, DC, to a number of organizations and agencies, including the US Department of Agriculture, Environmental Protection Agency, staff members of the US Congress House and Senate Agriculture Committees, the Farm Bureau, and non-governmental organizations that were convened by the World Resources Institute. Dr. Jean Brender, who participated in the sustainable agriculture workshop, presented a keynote address at the 6th International Nitrogen Conference in Kampala, Uganda (see Brender 2020, Chap. 18, this volume). A paper in *Current Opinion on Environmental Sustainability* describes the need to foster partnerships to promote NUE research, extension, implementation, and performance indicators that encompass technical, social, and economic drivers of nutrient management (Davidson et al. 2014). Recent innovative approaches include market trading, supply chain incentives, and consumer awareness. A special issue of the *Journal of Environmental Quality* was published in 2015 with 14 papers from the conference, with an introduction by Davidson et al. (2015) under the title "More Food, Low Pollution (Mo Fo Lo Po): A Grand Challenge for the twenty first Century." A few success stories chronicled in the special issue have a common theme of tailoring regulations, incentives, and outreach to local conditions, administered and enforced by local entities, and where local "buy-in" has been obtained (Davidson et al. 2015).

34.2.3 Establishing and Using Critical Loads for Resource Management

In the US and Canada, atmospheric N_r deposition has caused changes ranging from eutrophication and changes in biodiversity to acidification in remote protected national parks and forests. Adopting practices commonly applied in Europe, critical loads for deposition based on scientifically-derived values for the amount below which there are no known ecological effects are coming into use in the US (CDPHE 2013). Empirical and calculated critical loads data for N_r (as well as for acidic deposition) in the US were synthesized from dozens of regional and national-scale monitoring networks, research projects and publically available databases following an approach similar to that used in Europe (Pardo et al. 2011; Baron et al. 2011). These national critical loads are now being used in a tool to guide natural resource managers making management and policy decisions through the implementation of critical loads for land management planning, as described in the Air Quality Portal for Forest Planning (O'Dea and Huber 2012). This tool is still in development and requires testing and analysis by the US Forest Service, National Park Service, and other agencies to ensure that the process is clear, efficient, and accurately conveys the uncertainty associated with the components of the tool.

The ecosystem components that directly change with N_r are difficult to identify, so a workshop to link critical loads to ecosystem services (Air Quality and Ecosystem Services) for communication, natural resources management planning, and policy consideration took place in February 2015 in Thousand Oaks, CA (Blett et al. 2016). Participants included 25 invited subject matter experts, including social scientists, economists, and ecologists from academia and public land management agencies. Attendees identified the links between ecosystem components that change when a critical load is exceeded and the ecosystem goods and services that are impacted. The workshop resulted in a 2017 Special Feature on Air Quality and Ecosystem Services published in the journal *Ecosphere*, accessible at https://esajournals.onlinelibrary.wiley.com/doi/toc/10.1002/(ISSN)2150–8925.SF-AQ (accessed September 2019).

34.2.4 The Nitrogen Footprint Calculator (N-PRINT)

Dr. James Galloway (University of Virginia) and colleagues developed a personal N_r calculator to allow individuals to examine their personal N_r budgets and use the results to take steps to reduce their footprint (www.N-Print.org). The first university nitrogen footprint was calculated for the University of Virginia following the development of the institution-level Nitrogen Footprint Tool (Leach et al. 2013). The footprint calculation takes into account the food purchased and consumed at university dining venues, energy used for electricity and heating, fuel used by university vehicles and commuters, research animals, and fertilizer application on University grounds. With funding provided by the US Environmental Protection Agency Sustainable and Healthy Communities program the institution-level N footprint program has been extended to other universities, secondary schools, residents of the Chesapeake Bay Watershed, and an urban area (Baltimore, Maryland). Other universities, including Brown, Colorado State, Dickinson College, Eastern Mennonite Universities, the Marine Biological Lab, and the University of New Hampshire are calculating their university footprint (Castner et al. 2017). In a panel discussion "Measuring and Reducing Campus Nitrogen Footprints," presented at the 2015 American Association of Sustainability in Higher Education (AASHE) meeting in Milwaukee, MN, panelists described the tool, research methods, and roles of students in the research, presented results for their respective institutions, discussed the implications, and invited the audience to share ideas for reducing nitrogen footprints. A second cohort of participating universities began developing their nitrogen footprints in 2015.

34.3 Current Challenges Being Addressed

The North American Nitrogen Center (NANC) has made great progress in assessing the causes and consequences of excess N_r in North America, although research still remains to be done. Current challenges rest with facilitating efforts to develop solutions and implementing effective policy.

More than half of the N_r produced in the US goes toward food, and here there are large challenges. Real progress will require changing the entire food supply chain from producer to consumer (Davidson et al. 2015). Recent studies suggest that while knowledge and techniques are available to advance the dual goals of making agriculture more productive and environmentally sustainable, in North America economic and social barriers stand in the way of widespread adoption of these practices by farmers. Economic signals revolve around the perception that risk of applying too little N is high; abundant N application is insurance of maximum yields that outweigh off-farm environmental impacts or the added cost of buying and applying too much fertilizer. Sociological research shows that most US farmers now obtain the majority of their information about management from family members, retailers, and private sector crop advisors instead of traditional university extension services. The current political aversion in the US and Canada to command-and-control regulation requires alternative approaches, including new or different communication strategies, and creating effective partnerships among private and public sectors and multiple stakeholders.

As with much of the northern hemisphere, increasing human population, increasing per capita meat and dairy consumption, and increasing demand for energy and goods creates the ultimate challenge of reducing unintentional losses of N_r to the environment. A large challenge is changing behaviors of individual consumers. Education to raise awareness of sensible choices in diet ("meatless Mondays"), transportation (mass transit, bicycles), and energy consumption (unplugging "vampire" electronics) can all reduce release of unwanted N_r to the environment, yet these are difficult patterns to change.

There are currently no air quality standards in the US related to atmospheric deposition of N_r to land and water resources, although an Integrated Science Assessment is underway in preparation of the scientific grounds for establishing such standards. Reviews of the literature are taking place, with INI involvement. In the absence of standards, the development and use of critical loads of air pollutant deposition in the US is gaining momentum. Critical loads are used to quantify the levels of air pollutants that are expected to impact forest health, soil fertility, aquatic biota condition, and other ecosystem responses. In addition, model refinements for improving critical loads estimates, and maps for illustrating critical loads for acidification and nitrogen saturation and eutrophication resulting from excess nutrient nitrogen have been developed at various scales (Blett et al. 2014).

34.4 Further Ambitions

Other activities include planned workshops, along with increased outreach to the public and policy makers. A workshop took place in 2016 at the 16th National Conference of the National Council on Science and the Environment (NCSE) that addressed what success in N_r management might look like (Davidson et al. 2016). Another workshop held in 2016 in Bellingham Washington to develop a Nooksack Basin Nitrogen Assessment and Management Program launched the North American Demonstration Project exploring N budgets and management across the international boundary of a shared watershed between Canada and the US. A growing concern is that optimum N_r management might still fall far short of achieving natural resource and human health protection goals. A third workshop on N_r management as related to nutrient management in agriculture took place in 2017 in New Orleans, Louisiana. "Regional and Global N Input Datasets and Global N_2O Modeling" was jointly sponsored by the Global Carbon Project and the International Nitrogen Initiative Global N_2O Budget Workshop.

Communication of N_r reduction goals and solutions continues to be facilitated by working through professional scientific societies with histories of effective communications campaigns: the Ecological Society of America and with the American Geophysical Union. We anticipate expanding the N footprint programs to more universities and institutions.

Acknowledgments This work has been supported by a National Science Foundation Research Coordination Network grant award DEB-1049744 to EA Davidson since 2011, by the Environmental Protection Agency, and by the U.S. Geological Survey John Wesley Powell Center for Analysis and Synthesis.

References

Baron, J. S., Driscoll, C. T., Stoddard, J. L., & Richer, E. E. (2011). Empirical critical loads of atmospheric nitrogen deposition for nutrient enrichment and acidification of sensitive US lakes. *BioScience, 661*(8), 602–613.

Baron, J. S., Hall, E. K., Nolan, B. T., Finlay, J. C., Bernhardt, E. S., Harrison, J. A., et al. (2013). The interactive effects of excess reactive nitrogen and climate change on aquatic ecosystems and water resources of the United States. *Biogeochemistry, 114*(1), 71–92.

Blett, T. F., Lynch, J. A., Pardo, L. H., Huber, C., Haeuber, R., & Pouyat, R. (2014). FOCUS: A pilot study for national-scale critical loads development in the United States. *Environmental Science & Policy, 38,* 225–236.

Blett, T. F., Bell, M. D., Clark, C. M., Bingham D., Phelan J. N., Nahlik, A. M., Landers, D. H., et al. (2016). *Air quality and ecosystem services workshop report: Santa Monica Mountains*

National Recreation Area, Thousand Oaks, CA—February 24–26, 2015. Natural Resource Report. NPS/NRSS/ARD/NRR—2016/1107. USA, Fort Collins, Colorado: National Park Service.

Brender, J. D. (2020). Human health effects of exposure to nitrate, nitrite, and nitrogen dioxide. In M. A. Sutton, K. E. Mason, A. Bleeker, W. K. Hicks, C. Masso, N. Raghuram, S. Reis, & M. Bekunda (Eds.), *Just enough nitrogen: Perspectives on how to get there for regions with too much and too little nitrogen (Chapter 18)*. The Netherlands: Springer.

Castner, E. A., Leach, A. M., Leary, N., Baron, J., Compton, J. E., Galloway, J. N., Hastings, M. G., et al,. (2017). The nitrogen footprint tool network: A multi-institution program to reduce nitrogen pollution. *Sustainability: The Journal of Record, 10*(2), 79–88.

CDPHE. (2013). *Rocky mountain national park initiative*. Colorado Department of Health and Environment. Website at: https://www.colorado.gov/pacific/cdphe/rocky-mountain-national-park-initiative. Accessed August 2019.

Clair, T. A., Pelletier, N., Bittman, S., Leip, A., Arp, P., Moran, M. D., et al. (2014). Interactions between reactive nitrogen and the Canadian landscape: A budget approach. *Global Biogeochemical Cycles, 28*(11), 1343–1357.

Davidson, E. A. (2013). *Solutions for a nitrogen-soaked world*. Briefing to the National Academy of Sciences, for Distinctive Voices at the Jonsson Centre. Video available on YouTube at: https://www.youtube.com/watch?v=ICURE43uHGM. Accessed August 2019.

Davidson, E. A., David, M. B., Galloway, J. N., Goodale, C. L., Haeuber, R., Harrison, J. A., Howarth, R. W., et al,. (2012). *Excess nitrogen in the U.S. environment: trends, risks, and solutions*. Issues in Ecology, Report Number 15. Ecological Society of America. Available online at: http://www.esa.org/esa/wp-content/uploads/2013/03/issuesinecology15.pdf. Accessed August 2019.

Davidson, E. A., Galloway, J. N., Millar, N., & Leach, A. M. (2014). N-related greenhouse gases in North America: Innovations for a sustainable future. *Current opinion on environmental sustainability, 9–10*, 1–8.

Davidson, E. A., Suddick, E. C., Rice, C. W., & Prokopy, L. S. (2015). More food, low pollution (Mo Fo Lo Po): A grand challenge for the twenty-first century. *Journal of Environmental Quality, 44*(2), 305–311.

Davidson, E. A., Nifong, R. L., Ferguson, R. B., Palm, C., Osmond, D. L., & Baron, J. S. (2016). Nutrients in the nexus. *Journal of Environmental Studies and Sciences, 6*(1), 25–38.

Galloway, J. N., Schlesinger, W. H., Clark, C. M., Grimm, N. B., Jackson, R. B., Law, B. E., Thornton, P. E., et al,. (2014). Ch. 15: Biogeochemical Cycles. In J. M. Melillo, Terese (T.C.) Richmond, G. W. Yohe, (Eds.), *Climate change impacts in the united states: The third national climate assessment* (pp. 350-368). USA: U.S. Global Change Research Program. Available online at: https://nca2014.globalchange.gov/report/sectors/biogeochemical-cycles. Accessed August 2019.

Houlton, B. Z., Boyer, E., Finzi, A., Galloway, J., Leach, A., Liptzin, D., et al. (2013). Intentional versus unintentional nitrogen use in the United States: Trends, efficiency and implications. *Biogeochemistry, 114*(1), 11–23.

Leach, A. M., Majidi, A. N., Galloway, J. N., Greene, A. J. (2013). Toward institutional sustainability: A nitrogen footprint model for a university. *Sustainability: The Journal of Record, 6*(4), 211–219.

NANC. (2013a). *The Kansas city consensus—2013*. Cornell, USA: North American Nitrogen Center (NANC). Available online: http://nitrogen.ipni.net/article/NNN-5018. Accessed August 2019.

NANC. (2013b). *The Kansas city consensus on nitrogen use efficiency 2013*. Follow-up Brochure to the Kansas City Consensus. Cornell, USA: North American Nitrogen Center (NANC). Available online at: http://nitrogen.ipni.net/article/NNN-5018. Accessed August 2019).

O'Dea, C. & Huber, C. (2012). Workshop summary. *Air quality portal for forest planning: Critical loads component*. Available online at: http://www.fs.fed.us/air/documents/CLs%20Workshop_version%20to%20post.pdf. Accessed August 2019.

Pardo, L. H., Fenn, M. E., Goodale, C. L., Geiser, L. H., Driscoll, C. T., Allen, E. B., et al. (2011). Effects of nitrogen deposition and empirical nitrogen critical loads for ecoregions of the United States. *Ecological Applications, 21*(8), 3049–3082.

Peel, J. L., Haeuber, R., Garcia, V., Russell, A. G., & Neas, L. (2013). Impact of nitrogen and climate change interactions on ambient air pollution and human health. *Biogeochemistry, 114*(1), 121–134.

Pinder, R. W., Bettez, N. D., Bonan, G. B., Greaver, T. L., Wieder, W. J., Schlesinger, W. H., et al. (2013). Impacts of human alteration of the nitrogen cycle in the US on radiative forcing. *Biogeochemistry, 114*(1), 25–40.

Porter, E. M., Bowman, W. D., Clark, C. M., Compton, J. E., Pardo, L. H., & Soong, J. L. (2013). Interactive effects of anthropogenic nitrogen enrichment and climate change on terrestrial and aquatic biodiversity. *Biogeochemistry, 114*(1), 93–120.

Robertson, G. P., Bruulsema, T. W., Gehl, R. J., Kanter, D., Mauzerall, D. L., Rotz, C. A., et al. (2013). Nitrogen–climate interactions in US agriculture. *Biogeochemistry, 114*(1), 41–70.

Suddick, E. C., & Davidson, E. A. (2012). *The role of nitrogen in climate change and the impacts of nitrogen-climate interactions on terrestrial and aquatic ecosystems, agriculture, and human health in the united states: a technical report submitted to the us national climate assessment.* USA: North American Nitrogen Center of the International Nitrogen Initiative (NANC-INI), Woods Hole Research Center. Available online at: http://whrc.org/wp-content/uploads/2015/09/SuddicketalWHRC.12.pdf. Accessed August 2019.

Suddick, E. C., Whitney, P., Townsend, A. R., & Davidson, E. A. (2013). The role of nitrogen in climate change and the impacts of nitrogen–climate interactions in the United States: Foreword to thematic issue. *Biogeochemistry, 114*(1), 1–10.

Chapter 35
The Latin America Regional Nitrogen Centre: Concepts and Recent Activities

Jean P. Ometto, Nataly L. Ascarrunz, Amy T. Austin, Mercedes M. Bustamante, Gisleine Cunha-Zeri, Maria C. Forti, Judith Hoelzemann, Victor J. Jaramillo, Luiz A. Martinelli, Felipe Pacheco, Cecilia Perez, Tibisay Perez, and Ariel Stein

Abstract The activities related to the International Nitrogen Initiative (INI) Latin America Regional Centre started in the early 2000s, in association with the international actions within the nitrogen (N) field. The office in Latin America was proposed in the context of broad global and local environmental changes and the agricultural

J. P. Ometto (✉) · G. Cunha-Zeri · M. C. Forti · F. Pacheco
Earth System Science Center (CCST), National Institute for Space Research (INPE), Av. dos Astronautas, 1758, São José dos Campos, SP 12227-010, Brazil
e-mail: jean.ometto@inpe.br

G. Cunha-Zeri
e-mail: gisleine.zeri@inpe.br

M. C. Forti
e-mail: crisforti@gmail.com

F. Pacheco
e-mail: felipe.pacheco@inpe.br

N. L. Ascarrunz
Instituto Boliviano de Investigación Forestal (IBIF), Casilla Postal 6204, Santa Cruz de La Sierra, Bolivia
e-mail: nascarrunz@gmail.com

A. T. Austin
University of Buenos Aires and IFEVA, Avenida San Martín (C1417DSE), Buenos Aires, Argentina
e-mail: austin@ifeva.edu.ar

M. M. Bustamante
Departamento de Ecologia, Universidade de Brasília, Brasília, CEP 70910-900, Brazil
e-mail: mercedesmcb@gmail.com

J. Hoelzemann
Universidade Federal do Rio Grande do Norte, Av. Senador Salgado Filho, s/n Natal, RN 59078-000, Brazil
e-mail: judith.hoelzemann@gmail.com

V. J. Jaramillo
Universidad Nacional Autónoma de México, Centro de Investigaciones en Ecosistemas, A.P. 27-3 Sta. María de Guido, C.P. 58090 Morelia, Michoacán, Mexico
e-mail: luque@cieco.unam.mx

M. A. Sutton et al. (eds.), *Just Enough Nitrogen*,
https://doi.org/10.1007/978-3-030-58065-0_35

need for nutrient addition, looking into processes that would horizontally permeate between these two important elements for human survival and for the Earth's system. However, the lack of information on the nitrogen cycle in Latin America is a serious impediment to evaluate properly how human activity is altering nitrogen pools and turnover at regional and global scales. Empirical measurements of N deposition and other N cycle processes are extremely scarce in Latin America, and data feeding to global and regional circulation models lack spatial distribution information in this region.

Keywords Atmospheric deposition · Biological nitrogen fixation · Nitrogen cycle · Water pollution · Food security · Transboundary pollution · Social dimension · Policy responses

35.1 Introduction

The Latin America Regional Centre of the International Nitrogen Initiative (INI-LA) has catalyzed a broad integrative research and outreach network across multiple ecoregions and socio-economic backgrounds in Latin America. The INI-LA works with the scientific community on synthesizing scientific information, producing new data and informing the policy processes concerning the nitrogen (N) budget and nutrient management in this large region. Within the INI-LA the Nitrogen Human Environment Network (Nnet) aims to examine human impact in natural and modified ecosystems across a wide range of climates, ranging from direct measurements to regional modelling. It aspires to build a greater understanding of how nitrogen excess or shortage affects ecosystem processes and related biodiversity, food production and environmental pollution. With a common framework of experimental design and sample collection in sites distributed along a regional precipitation gradient, the project is providing original, innovative, and integrative results related to ecosystem

L. A. Martinelli
CENA-Universidade de São Paulo, Avenida Centenario 303, 13416-000 Piracicaba, SP, Brazil
e-mail: lamartinelli50@gmail.com

C. Perez
Instituto de Ecología Y Biodiversidad (IEB), Universidad de Chile, Las Palmeras, 3425 Ñuñoa, Santiago, Chile
e-mail: cperez@bio.puc.cl

T. Perez
Instituto Venezolano de Investigaciones Cientificas, Lab. Química Atmosférica, Aptdo. 20632, Caracas 1020A, Venezuela
e-mail: tibisay.j.perez@gmail.com

A. Stein
Earth Resources and Technology on assignment to NOAA's Air Resources Laboratory, 1315 East West Hwy., SSMC3/Room 3316, Silver Spring, MD 20910, USA
e-mail: ariel.stein@noaa.gov

functioning and nitrogen dynamics in the region. In the study sites, defined according to physiographic and/or socio-economic attributes, the following inputs and outputs of nitrogen are being reviewed and analyzed:

(i) Inputs: natural and cultivation induced biological nitrogen fixation (BNF), fertilizer use, atmospheric deposition;
(ii) Outputs: net exports of agricultural products at regional level and estimates at site scale of gaseous emissions from land-use (fertilizer volatilization, biogenic soil emissions and burning) and export of N to groundwater and surface waste (domestic, agricultural and industrial).

The up-scaling of local nitrogen studies aims to constrain regional atmospheric chemistry and transport modelling, providing input for global models, and greatly enhancing the understanding of global patterns of alterations to the nitrogen cycle. The transboundary level of analysis refers to the social component, which is framed to examine the nitrogen emissions drivers through its sources and the socio-economic and political factors, as well as the policy responses at both local and regional levels. The network's ultimate goal is to have an integrated view of nitrogen management in the environment, maximizing N's essential role in sustaining life and minimizing its environmental effects for Latin America.

35.2 Conceptual Approach

The Nnet project has established a scientific cooperation network across Latin American countries (Argentina, Bolivia, Brazil, Chile, Mexico, and Venezuela) to investigate the processes that modify different aspects of the nitrogen cycle. The Nnet activity consists of working groups focusing on the following themes:

Dynamics of Biological Nitrogen Fixation (BNF) BNF is an essential source of N to terrestrial ecosystems, especially for natural systems, and is also an important input of N to the production of agricultural crops. In general, understanding of the rates and controls on BNF in natural tropical ecosystems is relatively scarce. Much of this knowledge gap stems from the lack of data and lack of standardization of analytical methodologies, making comparative studies challenging. In addition, there is great uncertainty about how BNF will respond to multiple aspects of global environmental change and how feedback processes will redefine the dynamics of nitrogen cycling on different scales. This component aims to reduce the uncertainty associated with the quantification of N input to terrestrial and aquatic systems in the studied regions, which will feed back into the regional nitrogen balance (RNB) to be produced by the network.

Nitrogen Dynamics and Ecological Processes This topic encompasses studies of N cycle dynamics and associated ecological processes in terrestrial and aquatic ecosystems. Excess N may change the equilibrium and rates of nitrogen processes including nitrification, denitrification, anaerobic ammonium oxidation and dissimilatory reduction of nitrate to ammonia (RDNA), influencing NO, N_2O and NH_3 emissions from aquatic and terrestrial ecosystems. The positive feedbacks concerning the N cycle are poorly understood in temperate environments and are not known in tropical ecosystems. Some results on these topics have been already been produced, based on the most recent scientific literature and on the data production at several sampling sites.

Nitrogen Deposition in Latin America Measurements of nitrogen deposition in Latin America are very scarce and limited. This information is fundamental for evaluating the implications of N deposition on ecosystem functioning and crop productivity, as well as on atmospheric chemistry. Observations are also crucial as a foundation to extrapolate regionally and globally and ultimately determine the magnitude of N deposition and the causes of the most important alterations of the N present in the atmosphere of the Latin American region.

Modelling the Atmospheric Part of the Nitrogen Cycle in Latin America This comprises studies of the main atmospheric chemistry and transport processes related to the N cycle, aiming to contribute to the development of an atmospheric model suitable to integrate consistently the observations of N fluxes. This component will produce maps for the Latin American region with stocks and fluxes of nitrogen, as well as highlighting the most critical variables for N cycling in the region.

Integrative Science and Results Modelling is being used as an integrative tool, e.g., the INLAND model, as surface modeling, the EURAD-IM model (EURopean Air Pollution Dispersion model), as atmospheric transport modelling, LUCC-ME, as land-use change data platform; sites in the region are part of the World Meteorological Organization (WMO) Global Atmospheric Watch (GAW) world observation network. This network aims to offer a unique opportunity to increase the global database on the nitrogen cycle and feed regional and global models.

Social Dimension and Policy Context The outcomes of this activity include the delivery of: (i) a literature review on nitrogen cycle in Latin America through a social-ecological system perspective; (ii) a conceptual framework of nitrogen emissions drivers to provide insights on socio-economic dynamics and political/institutional arrangements concerning the subject; and (iii) the production of resourceful concepts to support the policy-making process.

Training and Education Activities Expected results include training students within the different institutions participating in the Nnet initiative. These students will also have an opportunity to increase their exposure to international experts from diverse fields.

35.3 Background

Nitrogen availability is essential to the functioning of aquatic and terrestrial ecosystems, and N is often a limiting element for food production (Vitousek et al. 1997). The deposition of reactive nitrogen (N_r) has increased in the last three decades and it has been identified as an important factor related to global environmental change, particularly due to its impacts on biodiversity and human health (Galloway et al. 2004). The overall increase in N_r derived from anthropogenic activities has surpassed by far the rates of BNF in all natural terrestrial systems and the rate of increase of the introduction of N_r has accelerated markedly since 1960 (Galloway et al. 2004). The positive aspect of this growth is that about 48% of the world's population is now fed by crops fertilized with reactive nitrogen derived from human activities (Erisman et al. 2008).

35.3.1 Evidence from Latin America

Outside North America and Europe, little is known about the magnitude of N deposition or its impacts on ecosystems (Lara et al. 2001; Filoso et al. 2003; Trebs et al. 2006; Filoso et al. 2006). There are a number of reasons for this deficiency, including a lack of long-term observations and large unknowns regarding the responses of diverse tropical and subtropical ecosystems to the N_r input. Accordingly, the impacts of excess/deficit in N deposition in Latin America and the interaction with other human impacts, such as changes in other nutrient cycles, agricultural intensification and expansion, urbanization, and climate change, are still open issues (Martinelli et al. 2006; Austin et al. 2013).

Additionally, while most research has focused on the effects of excess nitrogen and its impacts on biotic systems (Galloway et al. 2008; Gruber and Galloway 2008), many regions outside North America and Europe are suffering from a substantial deficit of N_r for agricultural production, highlighting the need to establish a research agenda emphasizing the complex, and coupled, socio-ecological dimensions of N cycle alterations (Martinelli et al. 2006; Galloway et al. 2008). There is an open niche for identifying and producing detailed spatial information regarding the production and the fate of N_r to formulate regionally appropriate policies.

Due to economic and social factors the use of nitrogen fertilizers is uneven among different Latin American countries, being highest in more developed countries and below the nitrogen levels needed in less developed countries (Martinelli et al. 2006). This observed large variation leads to diverse inputs to the atmosphere. For instance, Brazil's N_r inputs through nitrogen fertilizer and atmospheric deposition derived from fossil fuel combustion are relatively high in the country's Southeast (high population centres) compared to that of the Amazonian region; whereas in the central area, dominated by the "Cerrado" biome, the nitrogen budget indicates an increase of anthropogenically derived N atmospheric deposition (Filoso et al. 2006). On the other hand, Bolivia may be experiencing high rates of N deposition from heavy biomass burning, with hotspots concentrating in the eastern lowlands and Andean foothills (Ulke et al. 2001).

35.3.2 Interactions with Biodiversity Change

Sala et al. (2000) have shown that, according to projections, nitrogen atmospheric deposition would be the third-largest determinant of biodiversity loss throughout the twenty-first century behind land-use and climate changes. The availability of nutrients is a key factor in determining the plant community composition and for various ecosystems, the stability of community processes is dependent on the low fertility of these ecosystems; therefore, an increase in N deposition is a major ecological threat. For instance, declines in plant diversity have been found as a result of high N availability because some nitrophilous species may grow faster than those adapted to low N availability. Secondary factors associated with the larger N supply, such as soil acidification and plants susceptibility to herbivores and drought, also may lead to biodiversity loss through competitive exclusion (Bobbink and Lamers 2002).

Most of the world´s biodiversity hotspots are located in tropical or subtropical areas from developing countries, which underline the importance of understanding the nitrogen deposition patterns in these regions. Actually, atmospheric N deposition and its potential negative consequences for biodiversity have not yet been fully quantified, but the potential impacts could be very large (Phoenix et al. 2006; Souza et al. 2020). Additionally, the incomplete understanding of the relationships between nutrient availability and carbon (C) cycling hampers the prediction of productivity responses of tropical ecosystems to global environmental changes. A recent meta-analysis conducted by Cleveland and O'Connor (2011) indicated that interactions between the N and phosphorus (P) cycles in lowland tropical forests may also have implications for C uptake and loss. The links between N and P cycles may emerge because of an underlying control by P that is expressed in two ways. Firstly, N fixation rates often increase with greater P availability and relatively high N inputs via N fixation could result in both higher soil and foliar N content. Secondly, P regulation of the N cycle goes beyond its effects on N fixation, being, for instance, an important factor in the photosynthetic process. In addition, P exerts a broader level of control over soil N turnover.

35.3.3 Interactions with Atmospheric Composition and Freshwaters

Another concern is the possible transboundary and long-range transport of N compounds derived from the periods of biomass burning (July–September) in central

Brazil and western Bolivia, as shown through plume rise and smoke observations and transport modelling (Pereira et al. 2009). The Andean mountain range acts as a geographic barrier to atmospheric circulation, thus redistributing the chemicals produced in central Brazil and western Bolivia towards the east face of the Andean range and down to the southern part of South America. Evidence of this transport was found on rainfall samples collected in a rural area of Uruguay (Zunckel et al. 2003). According to Butler et al. (2008), the plume transport and the chemical changes in the presence of nitrogen oxides (NO_x) and hydrocarbons are critical factors in photochemical oxidation reactions because high NO_x concentrations may lead to ozone (O_3) and reactive radical species production. Studies developed by Artaxo et al. (2009) regarding the elemental composition of aerosol and gases have shown that there is a significant deposition of nutrients and diverse organic compounds, but the partitioning between the different chemical species, especially between inorganic and organic nitrogen compounds, is still unresolved.

Assessing the effects of a changing N cycle in South America is also critical for understanding the state of the continent's considerable freshwater resources (Martinelli et al. 2010). For the Northern Hemisphere, knowledge of the extent, timing, and impact of N deposition on lakes is increasing rapidly. For example, Holtgrieve et al. (2011) used a geographically distributed set of lake sediment cores to show that inputs of anthropogenic perturbations of the N cycle occurred more or less in synchrony around the Northern Hemisphere with impacts observed even in very remote environments (e.g., the high Arctic). Various studies of lake plankton have shown that these changes have induced major changes in nutrient cycling, nutrient availability, and ecosystem function (Elser et al. 2009, 2010; McCrackin and Elser 2010, 2011).

35.3.4 Key Challenges for Latin America

Latin America is now at a crossroads where a balance needs to be found between production of the major agricultural commodities, reasonable and planned urbanization, and conservation of its natural ecosystems and associated goods and services. Most of the global natural biological fixation occurs in forests of Latin America.

On the other hand, despite recent reductions, Latin America has one of the highest rates of deforestation in the world, and one of the highest increases in the use of nitrogen fertilizers. A better understanding of the responses of the N cycle to human impacts will allow better conservation of biodiversity and natural resources, with an improvement in food security and more effective land-use choices in biofuel development. Latin America is a unique region in multiple aspects, and particularly relevant are the broad climatic gradient and economic patterns that include a diverse range of natural ecosystems and socio-economic development pathways. Additionally, the region is impaired by the lack of observational information on actual impacts of human activity on N cycling across this diverse range of ecosystems. Finally, the large expanse of tropical ecosystems and reservoirs of biodiversity juxtaposed with an intense economic incentive for development make the comprehension of human impacts in this context particularly important for global change research in the region. An evaluation of current and predicted changes in climate and land-use on nitrogen stocks and fluxes in the region is a comprehensive product being developed by the network.

35.4 Description of the Network Activities

35.4.1 Study Sites

The study sites in the network are distributed in Latin America according to Fig. 35.1. The factors considered in choosing the sites were (i) infrastructure and related ongoing work; (ii) scientific representativeness and history of information; (iii)

Fig. 35.1 Study sites across Latin America. The map especially highlights the locations and ecosystems of the Brazilian study sites

climatic and ecological characteristics; (iv) human activities, e.g., presence of agriculture, urban areas, deforestation; and (v) relevance to the regional understanding of the N cycle.

35.4.2 Distribution of Study Sites and Sampling for Atmospheric Nitrogen Deposition

The selection of the sites considered for the measurement network of atmospheric N deposition follows a precipitation gradient and model simulation on atmospheric long-distance transport and deposition for South America based on information from Pereira et al. (2009).

Rainfall sampling in some selected sites is being conducted following the procedure described in the GAW Manual for the GAW Precipitation Chemistry Programme (GAW/WMO 2004) using a wet-only collector, while for gas and particle sampling, a denuder based system developed by Sutton et al. (2001) is applied. This system is implemented in the UK and other parts of Europe (Tang et al. 2009). Although the target chemicals are total nitrogen and N_r species (nitrate and ammonium), all major anions (chloride, nitrate, sulphate) and cations (proton, sodium, potassium, magnesium, ammonium and calcium) in the rain samples and water extracts are being analyzed to address the chemistry of these solutions.

35.4.3 Distribution of Study Sites for Nitrogen Dynamics in Terrestrial Ecosystems

This component includes environmental gradients (precipitation and altitude) and land-use intensity (natural, rural and urban areas) gradients. Measurements are being conducted at the local level applying widely used and recognized methods.

35.4.4 Social Dimension and Policy Context

Land-use and land cover changes are brought about by social and institutional arrangements that need to be better understood and incorporated into models relating to the N cycle and its impacts. Regardless of the undeniable importance of curbing nitrogen emissions from agriculture and biomass burning in Latin American countries, N_r is essential for producing sufficient food and energy for their population. The main questions that emerge from this statement are: What factors are responsible for the increasing of emissions from nitrogen use? Is population growth the ultimate cause for the continuous expansion of N_r emissions? Are there other factors influencing N_r emissions in addition to agriculture and biomass burning? What is the role of government institutions in provoking this situation and ultimately in controlling and reversing this trend?

Based on a model designed for overviewing deforestation drivers (Herold et al. 2008), a conceptual framework has been developed for helping answer the above questions (Fig. 35.2). A careful analysis based on this scheme will contribute to a better understanding of the potential interactions between the different levels of direct and/or indirect drivers of nitrogen emissions (in virtually all its reactive forms), and for a more visible discernment of the associated impacts and risks. In fact, the ultimate goal will be to provide consistent information to support the formulation of adequate policies for Latin American countries dealing with the trade-offs of nitrogen emissions by comprehending their sources, dynamics and impacts in an integrated way.

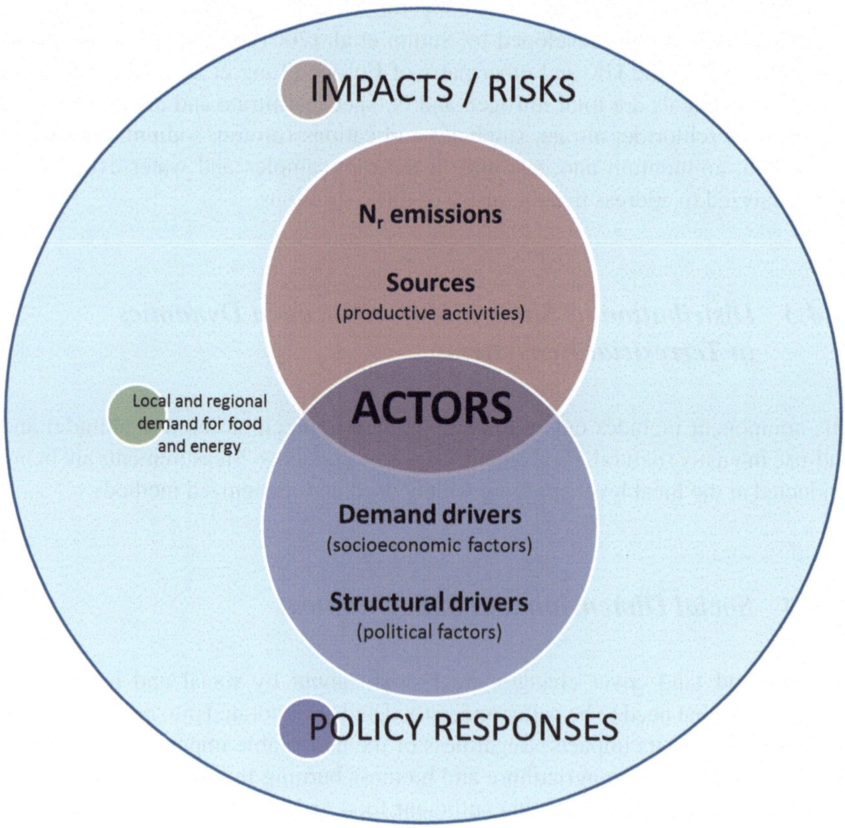

Fig. 35.2 Conceptual framework of nitrogen emissions drivers in Latin America

At the stage of conceiving a policy framework, it will be essential to take into account the fact that Latin America countries depend heavily on the agricultural and livestock sectors to boost their economies (ECLAC/FAO/IICA 2015; FAO 2014; IMF 2016); therefore, drastic measures may not have overall support. Considering the political sensitivity of the agricultural sector, the case for reducing nitrogen emissions is likely to rely upon emphasizing the wide and positive impacts of better nitrogen management practices, rather than focusing on the negative effects, along with a proper quantification in monetary terms of the benefits of nitrogen emissions mitigation.

35.5 Policy Relevance

The information produced will enable the production of a comprehensive understanding of the most important alterations of N_r in the Latin American region. In particular, it will provide information to support the design and implementation of effective and integrated policies to deal with the effects of excess or lack of nitrogen. The formulation of policies must take into account concerns associated with the environmental and economic impacts, in a way that matters to society.

Other interesting and potential synergies in this project are associated with the mapping of the spatial disconnection between production and consumption in the current accounting of greenhouse gas emissions from land-use and land-use change activities. This is highly relevant to Latin American countries where the exports of agriculture and biomass commodities constitute an important portion of the gross domestic product (GDP) and where countries struggle to maintain a balance between sustainable use practices and the heavy demands of socio-economic development.

Understanding the social and institutional forces related to the N cycle is an important element for evaluating and proposing policies that are better articulated in relation to food security and ecosystem protection. Many participants of the Nnet project are involved in several actions that range from establishing a regional assessment process for nitrogen to increasing public awareness of its benefits and adverse impacts on humans.

35.6 Some Results

The Nnet activity is providing a combination of direct measurement activities and a modelling exercise, associated with a social dimension analysis. At the present phase of the project, the observational work is still running, as well as the modelling simulations. In this section, some of the general results achieved are presented.

The results concerning wet deposition measurements showed relatively similar amounts of nitrogen within years and non-significant differences among sites (Argentina, Brazil, Venezuela), except for the most populated sites (e.g., Buenos

Aires), which showed higher ammonium depositions. In Venezuela, the results suggested that dry and gaseous deposition might be a more significant contributors to the nitrogen deposition due to the decrease in precipitation rates observed in recent years. On the other side, anthropogenic sources (biomass and fossil fuel combustion) are the main sources of nitrogenous species in rainy periods.

The atmospheric concentration measuring activities have been developing through a regional collaboration in many Brazilian locations. An important message emerging from gaseous measurements is the confirmation that longer sampling periods are essential in order to minimize uncertainties concerning results. The statistical details indicate higher concentrations in larger urban centres (e.g., São Paulo), but without influencing the results of smaller cities nearby. For some samples, there were mostly no significant differences between sites.

The work on BNF was reported by country and biome (Reis et al. 2020), and the interpolation technique presented provides a general overview of nitrogen fixation in the Latin American region. For instance:

(i) Results in Argentina suggests that soil organic matter formation was similar between grasses and legumes, although they differ in nitrogen inputs. The conclusion is that appropriate selection of legume crops is a key issue for increasing nitrogen fixation in crop rotations;
(ii) The Brazilian experiment concludes that the large variety of environmental problems associated with the balance of N_r in the environments tested, including adverse impacts, requires an integrated nitrogen management approach that would allow for the creation and closure of N budgets;
(iii) An adverse climate situation (e.g., decline in precipitation) may have a detrimental effect on nitrogen cycling in Chile, with consequences on nitrogen availability in soils on the Mediterranean forests;
(iv) The Mexican results show the high uncertainty of estimating nitrogen inputs from BNF for tropical forests (dry and moist/humid). The results suggest a modelling approach would be useful to further evaluate the constant deforestation threats in these forests.

The early results on atmospheric modelling show the largest concentrations throughout eastern Brazil, with higher population index and, consequently, more gaseous emissions (NO_2) (Souza et al. 2020). On the other hand, other gases (NH_3 and HNO_3) were found to have a more dispersed distribution over all regions, but with different levels of concentration. Future modelling exercises will include updated observational data in order to validate the modelling results.

An overall outcome from observational activities indicates that the assessment of nitrogen budgets and the improvement of nitrogen management use can be effective mechanisms to prevent undesirable losses in the region, both concerning agricultural production and threats to biodiversity (Tôsto et al. 2019). On the social dimension analysis, the results suggest that a precise definition of the interactions between demand and structural drivers of nitrogen emissions (Fig. 35.2) are vital in order to define the most efficient management strategies for the countries.

Finally, a review of the peer-reviewed literature indicated that science on nitrogen management approaches is still in the diagnostic phase of the problem, with limited understanding of the social feedbacks involved in this interaction. It also indicated the low participation of social scientists in publications related to benefits/threats of nitrogen in Latin America. The lack of a clear communication strategy involving a close interaction among social, political and environmental scientists was also identified.

35.7 Related Work of Interest to the International Nitrogen Initiative

Brazil The activities of the Nnet in Brazil are being conducted in association with several ongoing research projects related to biogeochemical cycling and land-use change in different Brazilian biomes (e.g., Cerrado, Caatinga and Atlantic Forest).

Mexico Research in the tropical dry forest region of Chamela involves a long-term project in which budgets, pools and fluxes of nutrients, water and energy have been monitored with variable degree of detail since 1982. The biogeochemical consequences of land-use change and woody plant recovery during secondary succession have also been under investigation for a number of years. More recently, the interplay between ecological processes and stakeholders in the region and ecosystem services of the tropical dry forest have been incorporated into a number of research projects. A recent group project has focused in identifying the vulnerability of the socio-ecological system in Chamela to the impacts of land-use practices and climate change.

Bolivia Collaborations are being developed with existing regional projects to build databases and enhance the level of productivity. So far, there is a collaboration with a European Commission funded network between Latin American and European partners, which aims to examine the role of biodiversity in mitigating climate change (ROBIN) . This network forms a partner activity to Nnet.

Argentina Ongoing research in the Pampa region of Argentina shows great promise for expansion into Nnet activities both in the region and in other ecoregions. Nitrogen deposition and gaseous N emissions from various situations of land-use change are being addressed in several projects. These projects are evaluating both baseline emissions and deposition in natural ecosystems, as well as quantifying the impact of human activities.

Venezuela Since 2008, the Atmospheric Chemistry Lab at the Instituto Venezolano de Investigaciones Cientificas (IVIC) has been collecting a time series of monthly sample collections and analysis of physicochemical parameters at the river mouth of four Venezuelan rivers that drain to the northern coast (NSF project OCE-0928941). The overall objective is to quantify the effect of different stages of development in the carbon and nitrogen cycling of these rivers and coastal areas. By using the data produced from this ongoing research, the goal is to generate the first Venezuelan nitrogen deposition flux estimates, contributing to the compilation effort the Latin American Nitrogen network.

Acknowledgements The 'Nitrogen Cycling in Latin America: Drivers, Impacts and Vulnerabilities (Nnet Project)', is supported by the Inter-American Institute for Global Change Research, IAI/CRN 3005, and by São Paulo Foundation Research, FAPESP process 2012/06416-1. The authors gratefully acknowledge preparatory support from the Global Environment Facility, through UN Environment and INI, for the establishment of the Latin American demonstration activity of the International Nitrogen Management System (INMS).

References

Artaxo, P., Rizzo, L. V., Paixão, M., Lucca, S. de, Oliveira, P. H., Lara, L. L., Wiedemann, K. T., Andreae, M. O., Holben, B., Schafer, J., Correia, A. L., & Pauliquevis, T. M. (2009). Aerosol particles in Amazonia: Their composition, role in the radiation balance, cloud formation, and nutrient cycles. In M. Keller, M. Bustamante, J. Gash & P. Silva Dias (Eds.), *Amazonia and global change* (pp. 233–250). Geophysical Monograph Series, 186. Washington DC: American Geophysical Union.

Austin, A. T., Bustamante, M. M. C., Nardoto, G. B., Mitre, S. K., Pérez, T., Ometto, J. P. H. B., et al. (2013). Latin America's nitrogen challenge. *Science, 340*(6129), 149. https://doi.org/10.1126/science.1231679.

Bobbink, R., & Lamers, L. (2002). Effects of increased nitrogen deposition. In J. Bell, & M. Treshow (Eds.), *Air pollution and plant life* (pp. 201–235). 2nd Edition. Chichester, England: John Wiley & Sons, Ltd.

Butler, T. M., Taraborrelli, D., Brühl, C., Fischer, H., Harder, H., Martinez, M., et al. (2008). Improved simulation of isoprene oxidation chemistry with the ECHAM5/MESSy chemistry-climate model: Lessons from the GABRIEL airborne field campaign. *Atmospheric Chemistry and Physics, 8,* 4529–4546.

Cleveland, C. J., & O'Connor, P. A. (2011). Energy return on investment (EROI) of oil shale. *Sustainability, 3,* 2307–2322.

Elser, J. J., Kyle, M., Steger, L., Nydick, K. R., & Baron, J. S. (2009). Nutrient availability and phytoplankton nutrient limitation across a gradient of atmospheric nitrogen deposition. *Ecology, 90,* 3062–3073.

Elser, J. J., Peace, A. L., Kyle, M., Wojewodzic, M., McCrackin, M. L., Andersen, T., et al. (2010). Atmospheric nitrogen deposition is associated with elevated phosphorus limitation of lake zooplankton. *Ecology Letters, 13*(10), 1256–1261.

Erisman, J. W., Sutton, M. A., Galloway, J., Klimont, Z., & Winiwarter, W. (2008). How a century of ammonia synthesis changed the world. *Nature Geoscience, 1,* 636–639. https://doi.org/10.1038/ngeo325.

ECLAC/FAO/IICA. (2015). *The outlook for agriculture and rural development in the Americas: a perspective on Latin America and the Caribbean 2015–2016*. San José, Costa Rica: ECLAC/FAO/IICA. http://www.iica.int/sites/default/files/publications/files/2015/b3696i.pdf. Accessed January 2017.

FAO. (2014). *FAO Statistical Yearbook 2014—Latin America and the Caribbean food and agriculture.* Santiago: Food and Agriculture Organization (FAO) of the United Nations Regional Office for Latin America and the Caribbean.

Filoso, S., Martinelli, L. A., Howarth, R. W., Boyer, E. W., & Dentener, F. (2006). Human activities changing the nitrogen cycle in Brazil. *Biogeochemistry, 79*(1/2), 61–89.

Filoso, S., Martinelli, L. A., Williams, M. R., Lara, L. B., Krusche, A., Ballester, M. V., et al. (2003). Land use and nitrogen export in the Piracicaba River basin, Southeast Brazil. *Biogeochemistry, 65*(3), 275–294.

Galloway, J. N., Dentener, F. J., Capone, D. G., Boyer, E. W., Howarth, R. W., Seitzinger, S. P., et al. (2004). Nitrogen cycles: Past, present, and future. *Biogeochemistry, 70*(2), 153–226.

Galloway, J. N., Townsend, A. R., Erisman, J. W., Bekunda, M., Cai, Z., Freney, J. R., et al. (2008). Transformation of the nitrogen cycle: Recent trends, questions, and potential solutions. *Science, 320*(5878), 889–892.

GAW/WMO. (2004). *Manual for the GAW Precipitation Chemistry Programme: Guidelines, data quality objectives and standard operating procedures.* GAW Report 160. Prepared by the GAW Precipitation Chemistry Advisory Group. World Meteorological Organisation, Geneva.

Gruber, N., & Galloway, J. N. (2008). An Earth-system perspective of the global nitrogen cycle. *Nature, 451,* 293–296.

Herold, A., Cunha-Zeri, G., Dietz, J., Freibauer, A., Hüttner, M., Jung, M., Mollicone, D., & Scheffler, M. (2008). *Berücksichtigung von Treibhausgasemissionen und -festlegungen durch Landnutzungsmaßnahmen (LULUCF) im Post-Kioto-Regime—quantitative Analyse zur Einbeziehung von reduzierter Entwaldung in ein künftiges Klimaregime* (Emissions and removals from land-use, land-use change and forestry activities in a post-Kyoto regime—quantitative analysis of a framework for reducing deforestation). Climate Change 16/08. Berlin: Umweltbundesamt. (In German).

Holtgrieve, G. W., Schindler, D. E., Hobbs, W. O., Leavitt, P. R., Ward, E. J., Bunting, L., et al. (2011). A coherent signature of anthropogenic nitrogen deposition to remote watersheds of the Northern Hemisphere. *Science, 334*(6062), 1545–1548.

IMF. (2016). *World Economic Outlook, April 2016: Too slow for too long.* Washington DC: International Monetary Fund (IMF).

Lara, L., Artaxo, P., Martinelli, L. A., Victoria, R. L., Camargo, P. B., Krusche, A., et al. (2001). Chemical composition of rainwater and anthropogenic influences in the Piracicaba River Basin, Southeast Brazil. *Atmospheric Environment, 35,* 4937–4945.

Martinelli, L. A., Howarth, R. W., Cuevas, E., Filoso, S., Austin, A. T., Donoso, L., et al. (2006). Sources of reactive nitrogen affecting ecosystems in Latin America and the Caribbean: Current trends and future perspectives. *Biogeochemistry, 79,* 3–24.

Martinelli, L. A., Joly, C. A., Nobre, C. A., & Sparovek, G. (2010). A falsa dicotomia entre a preservação da vegetação natural e a produção agropecuária. *Biota Neotropica, 10*(4), 324–330.

McCrackin, M. L., & Elser, J. J. (2010). Atmospheric nitrogen deposition influences denitrification and nitrous oxide production in lakes. *Ecology, 91,* 528–539.

McCrackin, M. L., & Elser, J. J. (2011). Greenhouse gas dynamics in lakes receiving atmospheric nitrogen deposition. *Global Biogeochemical Cycles, 25*(4), GB4005. https://doi.org/10.1029/2010gb003897.

Pereira, G., Freitas, S. R., Moraes, E. C., Ferreira, N. J., Shimabukuro, Y. E., Rao, V. B., et al. (2009). Estimating trace gas and aerosol emissions over South America: Relationship between fire radiative energy released and aerosol optical depth observations. *Atmospheric Environment, 43,* 6388–6397.

Phoenix, G. K., Hicks, W. K., Cinderby, S., Kuylenstierna, J. C. I., Stock, W. D., Dentener, F. J., et al. (2006). Atmospheric nitrogen deposition in world biodiversity hotspots: The need for a greater global perspective in assessing N deposition impacts. *Global Change Biology, 12*(3), 470–476.

Reis, C. R. G., Pacheco, F. S., Reed, S. C., Tejada, G., Nardoto, G. B., Forti, M. C., & Ometto, J. P. (2020). Biological nitrogen fixation across major biomes in Latin America: Patterns and global change effects. *Science of the Total Environment, 746,* 140998. https://doi.org/10.1016/j.scitotenv.2020.140998.

Sala, O. E., Chapin, F. S., III, Armesto, J. J., Berlow, E., Bloomfield, J., Dirzo, R., et al. (2000). Global biodiversity scenarios for the year 2100. *Science, 287*(5459), 1770–1774.

Souza, M. d. A., Pacheco, F. S., Palandi, J. A. d. L., Forti, M. C., Campos, M. L. A. M., Ometto, J. P. H. B., Reis, D. C. O., & de Carvalho Junior, J. A. (2020), Atmospheric concentrations and dry deposition of reactive nitrogen in the state of São Paulo, Brazil. *Atmospheric Environment, 230*, 117502. https://doi.org/10.1016/j.atmosenv.2020.117502.

Sutton, M. A., Tang, Y. S., Miners, B., & Fowler, D. (2001). A new diffusion denuder system for long-term, regional monitoring of atmospheric ammonia and ammonium. *Water, Air and Soil Pollution: Focus, 1*(5), 145–156.

Tang, Y. S., Simmons, I., van Dijk, N., Di Marco, C., Nemitz, E., Dämmgen, U., et al. (2009). European scale application of atmospheric reactive nitrogen measurements in a low-cost approach to infer dry deposition fluxes. *Agriculture, Ecosystems and Environment, 133,* 183–195.

Tôsto, K. L., Pacheco, F. S., Gomes, L., Miranda, M., Oliveira, S. M. L., Pessi, D. D., Cox, C., & Ometto, J. P. H. B. (2019). Nitrogen use efficiency: a local and regional approach for Brazilian agriculture. *Raega - O Espaço Geográfico em Análise, 46*(3), 125–139. https://doi.org/10.5380/raega.v46i3.67095.

Trebs, I., Lara, L. L., Zeri, L. M. M., Gatti, L. V., Artaxo, P., Dlugi, R., et al. (2006). Dry and wet deposition of inorganic nitrogen compounds to a tropical pasture site (Rondônia, Brazil). *Atmospheric Chemistry and Physics, 6,* 447–469.

Ulke, A. G., Longo, K. M., & de Freitas, S. R. (2001). Biomass burning in South America: Transport patterns and impacts. In D. Matovic (Ed.), *Biomass—detection, production and usage (Chapter 19* (pp. 387–408). Rijeka, Croatia: InTech.

Vitousek, P. M., Mooney, H. A., Lubchenco, J., & Melillo, J. M. (1997). Human domination of Earth"s ecosystems. *Science, 277*(5325), 494–499.

Zunckel, M., Saizar, C., & Zarauz, J. (2003). Rainwater composition in northeast Uruguay. *Atmospheric Environment, 37,* 1601–1611.

Part XI
Conclusions and Outlook

Chapter 36
Global Challenges for Nitrogen Science-Policy Interactions: Towards the International Nitrogen Management System (INMS) and Improved Coordination Between Multi-lateral Environmental Agreements

Mark A. Sutton, Clare M. Howard, Will J. Brownlie, David R. Kanter, Wim de Vries, T. K. Adhya, Jean P. Ometto, Jill S. Baron, Wilfried Winiwarter, Xiaotang Ju, Cargele Masso, Oene Oenema, N. Raghuram, Hans J. M. van Grinsven, Isabelle Van der Beck, Christopher Cox, Steffen C. B. Hansen, Ramesh Ramachandran, and W. Kevin Hicks

M. A. Sutton (✉) · C. M. Howard · W. J. Brownlie
UK Centre for Ecology & Hydrology, Edinburgh Research Station, Bush Estate, Penicuik, Midlothian, EH26 0QB, UK
e-mail: ms@ceh.ac.uk

C. M. Howard
e-mail: cbritt@ceh.ac.uk

W. J. Brownlie
e-mail: wilown@ceh.ac.uk

W. J. Brownlie
School of Geosciences, University of Edinburgh, King's Buildings, West Mains Road, Edinburgh, EH9 3JW, UK

D. R. Kanter
New York University, Department of Environmental Studies, 285 Mercer St, New York, NY 10003, USA
e-mail: david.kanter@nyu.edu

W. de Vries
Wageningen University and Research (WUR), Droevendaalsesteeg 3, 6708 PB Wageningen, The Netherlands
e-mail: wim.devries@wur.nl

T. K. Adhya · N. Raghuram
South Asian Nitrogen Centre, Indian Nitrogen Group, Sustainable India Trust and Society for the Conservation of Nature, F-4, A-Block, NASc Complex, DPS Marg, Pusa, New Delhi 110012, India
e-mail: adhyas@yahoo.com

N. Raghuram
e-mail: raghuram@ipu.ac.in; raghuram98@hotmail.com

M. A. Sutton et al. (eds.), *Just Enough Nitrogen*,
https://doi.org/10.1007/978-3-030-58065-0_36

Abstract Human interference with the nitrogen cycle has doubled reactive nitrogen inputs to the global biosphere over the past century, leading to changes across multiple environmental issues that require urgent action. Nitrogen fertilizers and biological nitrogen fixation have allowed benefits of increased crop harvest and livestock production, while in some areas there is insufficient nitrogen to fertilize crops. Whether in excess or deficit, nitrogen losses from its inefficient use are causing a combination of freshwater and marine pollution, air pollution, alteration of climate balance, stratospheric ozone loss, biodiversity loss and reduction of soil quality. The resulting nitrogen pollution affects human health, well-being and livelihoods. Scientific efforts have begun to bring these issues together. However, there is still a high degree of fragmentation between research on the different benefits and threats of reactive nitrogen and between the respective policy frameworks, especially at the global scale. We argue that a more joined-up approach to managing the global nitrogen cycle is needed to develop the 'gravity of common cause' between nitrogen issues and to avoid policy trade-offs. We describe how a coherent system for science evidence provision is being developed to support policy development through the 'International Nitrogen Management System' (INMS). There is now a matching challenge to bring together the multiple policy agreements relevant for nitrogen as a foundation to address synergies/trade-offs and to set priorities. Based on review of

T. K. Adhya
School of Biotechnology, Kalinga Institute of Industrial Technology (KIIT), Bhubaneswar, Odisha 751024, India

J. P. Ometto
Earth System Science Center (CCST), National Institute for Space Research (INPE), Av. dos Astronautas, 1758, São José dos Campos, SP 12227-010, Brazil
e-mail: jean.ometto@inpe.br

J. S. Baron
U.S. Geological Survey, Colorado State University, Fort Collins, CO 80523-1499, USA
e-mail: jill.baron@colostate.edu

W. Winiwarter
International Institute for Applied Systems Analysis (IIASA), Laxenburg, Austria
e-mail: winiwarter@iiasa.ac.at

X. Ju
Chinese Agricultural University, College of Resources and Environmental Sciences, Beijing, China
e-mail: juxt@cau.edu.cn

C. Masso
International Institute of Tropical Agriculture (IITA), 1st Main IRAD Road, Nkolbisson, P.O. Box 2008 (Messa), Yaoundé, Cameroon
e-mail: C.Masso@cgiar.org

O. Oenema
Alterra, Wageningen University and Research Centre (WUR), P.O. Box 47, 6700 AA Wageningen, The Netherlands
e-mail: oene.oenema@wur.nl

existing frameworks, we outline the concept for an Interconvention nitrogen coordination mechanism. This could make a major contribution to multiple Sustainable Development Goals by stimulating the next generation of international nitrogen strategies: maximizing the benefits of efficient nitrogen use, while minimizing its many environmental threats.

Keywords Nitrogen cascade · Policy · Intergovernmental agreements · Pollution · Interconvention coordination

36.1 Introduction

Human perturbation of the global nitrogen cycle in the twenty-first century is leading both to massive benefits for food and energy production and to multiple environmental threats (e.g., Fowler et al. 2013; Sutton et al. 2013). Although nitrogen (N) is abundant in the atmosphere in its unreactive form (as N_2), this is unavailable for most organisms, with the supply of reactive nitrogen (N_r) compounds limited under natural conditions. Anthropogenic sources have massively increased N_r formation

N. Raghuram
University School of Biotechnology, Guru Gobind Singh Indraprastha University, Dwarka Sec. 16C, New Delhi 110078, India

H. J. M. van Grinsven
PBL Netherlands Environmental Assessment Agency, Bezuidenhoutseweg 30, 2594 AV The Hague, The Netherlands
e-mail: Hans.vanGrinsven@pbl.nl

I. Van der Beck
United Nations Environment Programme, Ecosystems Division (International Waters), 900 17th St NW, Washington, DC 20006, USA
e-mail: isabelle.vanderbeck@un.org

C. Cox
United Nations Environment Programme, P.O. Box 30552, Nairobi, Kenya
e-mail: christopher.cox@un.org

S. C. B. Hansen
Global Environment Facility, GEF Secretariat—Programs Unit, Washington, DC, USA
e-mail: shansen@thegef.org

R. Ramachandran
National Centre for Sustainable Coastal Management, Ministry of Environment Forest & Climate Change, Government of India, Anna University Campus, Chennai 600025, India
e-mail: rramesh_au@yahoo.com

W. K. Hicks
Stockholm Environment Institute (SEI), Department of Environment and Geography, University of York, York, YO10 5DD, UK
e-mail: kevin.hicks@york.ac.uk

over the last century. These include fertilizer production, biological nitrogen fixation, and the unintended formation of nitrogen oxides (NO_x) from combustion. As a result of these activities, humans have more than doubled global terrestrial rates of N_r formation (Galloway et al. 2008).

The benefits of increased fertilizer N production (and biological nitrogen fixation) have been huge. Fertilizer N_r from the Haber-Bosch process is estimated to sustain nearly 50% of the global human population according to current diets, without which there would be massive problems of hunger and malnutrition in many parts of the world (Erisman et al. 2008). While there are opportunities for improved adoption of organic agriculture, thereby reducing the need for chemical fertilizers (Van Grinsven et al. 2015), the rise in crop and animal production over the last century have enriched human diets. Agricultural N_r inputs also provide a foundation for bioenergy production, offering an option for replacing fossil fuels.

Against these benefits, the unintended and often negative environmental consequences of anthropogenic transformation of N_2 to N_r have been correspondingly large. The overall global doubling of N_r flows has led to a web of pollution problems, often described in terms of the 'nitrogen cascade' (Galloway et al. 2003), where N_r converts between many chemical forms, resulting in multiple environmental impacts. This process is driven by the dissipation of energy contained in the N_r until it is eventually 'denitrified' back to atmospheric N_2. The consequences include water pollution of both freshwater and coastal marine systems (including drinking water contamination, harmful algal blooms and hypoxia), air pollution, greenhouse gas emissions, stratospheric ozone depletion, with negative effects on ecosystems, biodiversity, soil quality and human health (Sutton et al. 2011a; UNEP 2013). The end result is an array of adverse impacts on health, environment and livelihoods. This situation has been described as 'too much' nitrogen, while in areas with 'too little' nitrogen, such as in much of sub-Saharan Africa, agricultural soils are mined of nutrients, leading to reduced soil quality and erosion, making farming practices unsustainable.

The primary goal of intentional N_r fixation is to increase plant and animal growth, forming N compounds essential to life such as amino acids, proteins, enzymes and DNA. Key losses of N_r to the environment include ammonia (NH_3), nitric oxide (NO), nitrous oxide (N_2O) and nitrate (NO_3^-). Even denitrification losses to form N_2 are indirectly polluting, since they represent a waste of the substantial resources (2% of world energy) used to make N_r. With global efficiency of N use at around 20% (Sutton et al. 2013), approaches aimed at increasing nitrogen use efficiency (NUE) throughout the economy hold the simultaneous prospect to reduce N_r pollution and improve outcomes for food and energy production (Lassaletta et al. 2014; Zhang et al. 2015).

In recognition of these challenges, many researchers are investigating different parts of the nitrogen cycle. Some focus primarily on improving the benefits of intentional N_r inputs for agricultural productivity or on improving access to N_r in agriculture (e.g., Raun and Johnson 1999; Herridge et al. 2008; Vanlauwe and Giller 2006). However, assessment of loss pathways and impacts to air, water, atmosphere, and human health arising from excess nitrogen use or inadvertent N_r production is typically conducted by separate sets of researchers with different disciplinary backgrounds. The outcome of all this science, however, has surely not been used to best advantage. The historical specialization of science means that many scientists addressing different parts of the nitrogen cycle find themselves speaking a different language. For example, few individual scientists would be able to integrate understanding of satellite approaches for atmospheric ammonia measurement (Clarisse et al. 2009), mechanisms of denitrification in agricultural soils (Selbie et al. 2015), the role of amines in aerosol pre-nucleation (Nadykto et al. 2011) and nitrogen flows related to toxicity of nitrates or algal blooms (Powlson et al. 2008; Cochlan et al. 2008). Developing the linkages between these issues, requires scientists with a broad training.

It is even harder when it comes to developing the scientific evidence for linking the benefits of nitrogen use to minimizing environmental threats. The lack of scientists competent to cover the entire nitrogen cycle in all its details provides a significant barrier to the provision of sound scientific evidence to underpin future policies. This is further compounded by the general absence across many global regions, of intermediary science-to-policy processes. These are needed to draw on scientific evidence where it may exist, to interweave with national policy approaches.

We address the need for a more integrated nitrogen science-policy interface in three stages. First, we summarize the kind of scientific evidence that policy makers need to inform on options for better nitrogen management. This draws on the experience of nitrogen scientists and policy analysts involved in a wide array of threats and benefits of N_r. Secondly, we review the status and nitrogen science needs of different international policy frameworks relevant to nitrogen. Specifically, we ask to what extent these frameworks offer potential to act as a primary ‘policy home’ for the global nitrogen challenge. Thirdly, we illustrate how the architecture is being developed to link expertise and information towards a framework of science support for international nitrogen policy, the ‘International Nitrogen Management System’ (INMS). In the process, we reflect on the challenge to work towards a more coherent policy approach. Rather than concluding on a single ‘policy home’ for nitrogen, we instead identify an alternative model that emphasizes the need to bring governments and other actors together to improve understanding and coordination among existing policy processes. We argue that a joined-up (i.e., an integrated) approach for nitrogen management should help reframe reduction in pollution as a positive contribution toward improving nitrogen use efficiency, with benefits for jobs, while spurring innovation across multiple sectors.

36.2 Science Evidence to Inform Nitrogen Policies

36.2.1 Why Is a Joined-Up Approach to Nitrogen Necessary?

The first question that needs to be asked is why special attention is needed on the nitrogen cycle. The key to this is the multiplicity of nitrogen in our world, leading to many benefits and threats as summarized above. In addition to the prime benefits of N_r for food and bioenergy production, plant and animal products also have many other uses, while a plethora of chemical products require nitrogen compounds in their manufacture, including artificial fibres, dyes and most explosives. It is, however, the unintended consequences of nitrogen use that mean nitrogen is of special interest. The mobility of N_r and its ability to convert into so many different chemical forms—and their importance in biological metabolism—mean that human alteration of the nitrogen cycle is having *systemic consequences* across all compartments of the environment. This is clearly illustrated by Fig. 36.1, showing the 'nitrogen cascade'. Even though this is given in a highly simplified form—for example, only a few of the major inorganic forms of nitrogen are noted, while the many organic nitrogen compounds are not shown—it powerfully demonstrates the cross-cutting impact of nitrogen on global systems.

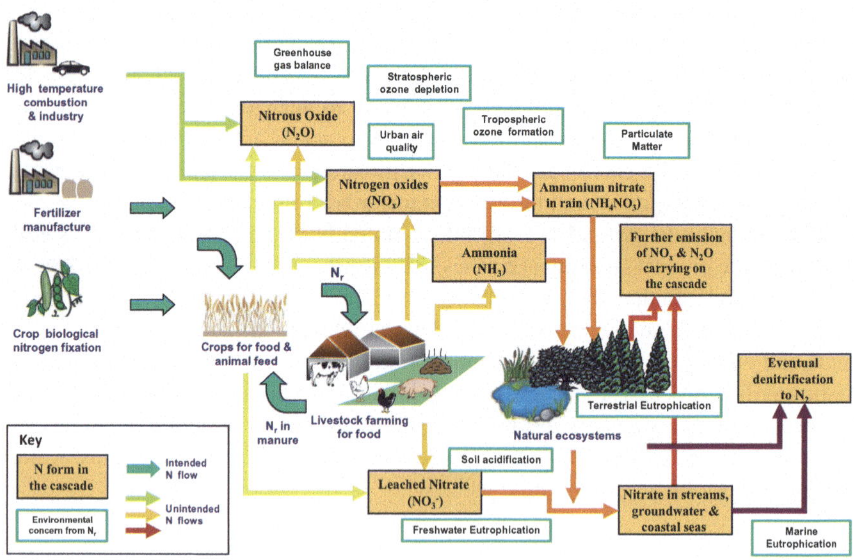

Fig. 36.1 Simplified view of the nitrogen cascade. Nitrogen (N) is present in low energy state as N_2 gas, so conversion to form reactive nitrogen (N_r) requires substantial energy, which is eventually dissipated in the cascade as N_r components react to make many other compounds, before eventually being denitrified back to N_2. In the process, the same nitrogen atom can contribute to several N compounds with multiple effects on the food, energy and environmental systems (redrawn from Sutton et al. 2011a)

Taking an integrated approach to the nitrogen cycle can therefore be catalytic in two ways. Firstly, we expect that linking of existing nitrogen science and policy responses will help to overcome the barriers to better nitrogen management. Secondly, such a 'joined-up' approach becomes illustrative as a case study of 'partial integration', showing how marine, freshwater, terrestrial and atmospheric scientists can work together with the matching policy domains to find an appropriate balance between focus (the nitrogen cycle) and integration (multiple sources, sectors, systems and effects).

Concerning the term 'joined-up', this can be considered as broadly synonymous with 'integrated', meaning that efforts are placed to connect the different benefits and threats, synergies and trade-offs, linking sectors, science areas and solutions. However, it is recognized that that the term 'integrated' can be considered as off-putting to some audiences, linked to concerns of increasing complexity and difficulty of implementation. The term 'joined up' therefore emphasizes importance of identifying and linking the most important issues, especially given the urgent need to overcome barriers to multiple Sustainable Development Goals through the wide range of nitrogen effects.

We emphasize that a focus on nitrogen does not imply the exclusion of interactions with other element cycles. Clearly, a balance needs to be found to allow optimal progress to be made. On one hand, addressing all issues simultaneously could lead to failure, while on the other adopting too narrow a focus will not allow all the key interactions to be addressed. The capacity for integrating research and policy areas is also expected to change over time. This means that the optimal degree of integration from a pragmatic standpoint needs to consider both the precedents and the extent of institutional capacity to take the next steps (cf. Jordan and Lenschow 2010).

Overall, the philosophy of developing an integrated nitrogen approach aims to ensure that:

(a) the multiple benefits and threats are considered,
(b) the synergies/trade-offs and opportunity for co-benefits are identified,
(c) the primary interactions with other element cycles are recognized, and
(d) the priorities for action can be identified, between sources, sectors, impacts and solutions.

The nature of the primary interactions between nitrogen and other element cycles differs across environmental media (air, soil, water). For example, in aquatic systems nitrogen interactions occur with nutrient limitation of phosphorus and silica influencing primary productivity. In considering greenhouse gas emissions from terrestrial systems, the primary nitrogen interactions are with carbon compounds, including both carbon dioxide and methane. In relation to air quality threats, nitrogen interacts with sulphur emissions, as well as with volatile organic compounds, ozone and particulate matter. Conversely, in crop production, balanced fertilization means accounting for nitrogen and in relation to other limiting nutrients in order to maximize nitrogen use efficiency. These examples illustrate the systemic consequences of altering the global nitrogen cycle, while pointing to the need to optimize the level of integration with the key aspects of these other issues and element cycles.

36.2.2 What Kind of Nitrogen Science Is Needed?

A comprehensive and broad-based approach to understanding the nitrogen cycle is needed to inform policy development, both to maximize the benefits of intended N_r use and to minimize the unintended threats. One condition for robust policies is that they are based on a sound scientific understanding of mechanisms, processes and interactions across the nitrogen cycle. If a key process is not understood or even missing, this could lead to the provision of misleading science advice for policy makers. Scientific information on nitrogen must also be communicated appropriately, emphasizing the need for engagement with stakeholders including policy makers, the public and private sector, civil society and non-government organizations among others, both to mobilize public understanding and to ensure well-informed policy decisions. This means learning to speak the language of the private agents, including highlighting the potential for short-term cost savings, lowered risk, increased productivity etc., as well as the environmental benefits of taking action.

The needs of policy makers are often very practical, so that the *application* of existing scientific knowledge is often a more urgent priority than improving fundamental understanding. For example, policy makers need to make decisions where a forward-look is necessary (requiring scenarios to address 'what if' questions), while cost-benefit analysis is central to the decision-making process. In the case of nitrogen, such cost-benefit analyses must be based on a chain of prior scientific information that starts with the magnitude of N_r flows, considers their fate and consequences, and eventually associates economic or social value with the different consequences of these flows and impacts. For example, Van Grinsven et al. (2013) applied this approach at the European scale, while Keeler et al. (2016) have considered the social costs of nitrogen in North America. At the same time, the management and mitigation options of what could be done better need to be clearly outlined and demonstrated, with strong scientific and technical underpinning to demonstrate the economic, societal, and environmental benefits.

In painting this picture, it is worth recognizing that the global nitrogen challenge points to a rather different science need than the last decades of science to underpin climate policies. In the **first stage** of science for climate policy, the question was whether there is a problem, i.e., to show whether there is human-driven climate change, and if so, by how much, where and when. As consensus on this central question was gradually reached, the science agenda then turned to emphasize a **second stage** focused on quantifying the present and anticipated impacts of climate change. Subsequently a **third stage** focused on exploring the possible solutions, either through mitigation or adaptation, as illustrated by the approach of UNEP (2013) to addressing nitrous oxide emissions.

By contrast, there is little sign of any 'nitrogen sceptics' who would argue that there is no such thing as nitrogen-induced water or air pollution. The problems of nitrogen in the environment are already widely acknowledged. This means that research on the nitrogen challenge has long focused on the second and third stages outlined above. Challenges for **stage-two** nitrogen research include investigation of whether

N pollution problems are getting worse or better (across both space and time), and how the benefits of nitrogen to the global food system can be linked to costs to the environment and human health. For nitrogen, key challenges now focus on **stage three**: to find approaches to reduce nitrogen pollution problems while meeting food and energy goals, investigating the appropriate measures, techniques and practices.

In making this comparison, the challenge for nitrogen also approaches a **fourth stage**: the search for optimized 'joined-up' solutions to a highly complex problem. Here the focus is on approaches that can help overcome the barriers to change—in businesses, national economies and across world regions. These include exploring how a joined-up 'nitrogen cycle approach' may better help achieve desired outcomes than fragmented approaches each focusing on different nitrogen impacts.

The 'Our Nutrient World' report prepared for the United Nations Environment Programme (Sutton et al. 2013) examined the possible elements of a future policy approach for nutrients including nitrogen, and the science needs resulting from this. The following priorities were identified, here applied to the nitrogen case:

(a) To establish a global assessment process for nitrogen between air, land, water, climate and biodiversity, considering the main driving forces, the interactions with food and energy security, the costs and benefits and the opportunities for the Green Economy.
(b) To develop consensus on the operational indicators, with benchmarking to record progress on improving nitrogen use efficiency and reducing the adverse environmental impacts.
(c) To investigate options for improvement of nitrogen use efficiency, demonstrating benefits for health, environment, and the supply of food and energy.
(d) To address barriers to change, fostering education, multi-stakeholder discourse and public awareness.
(e) To establish internationally agreed targets for improved N_r management at regional and planetary scales.
(f) To quantify the multiple benefits of meeting the nitrogen management targets for marine, freshwater and terrestrial ecosystems, mitigation of greenhouse gases and other climate threats, and improvement of human health/welfare.
(g) To develop and implement an approach for monitoring time-bound achievement of the nitrogen management targets, and for sharing and diffusing new technologies and practices that would help to achieve the targets.

Of these goals, points (a), (b), (c) and (d) match specific science requirements. Additional science challenges are included in both points (f) and (g), especially in relation to innovation and sharing technologies. By contrast, the setting of internationally agreed goals (e), while informed by science, was concluded to be the task of governments and policy makers, which needs to be addressed using relevant policy frameworks.

36.2.3 On What Time-Scales Is Nitrogen Science Needed to Support Policy?

A classic debate between scientists and policy makers concerns the timescale of evidence provision. Science is a slow process, which takes many years to come to fruition as measurements are made, models are built and fundamental understanding deepens. By contrast, policy makers often operate within much shorter timeframes, where science advice is often needed in a few weeks' time-horizon. Timing difference also applies to the private sector, which needs well-timed solutions to comply with policy decisions.

It is also important to distinguish between the *environmental time-scales of interest* to policy makers and *the time-scales of when they require essential information* from the science community. For example, policymakers may wish to see trends over the past century ("How much and when did the problem worsen?"), trends over the past decade ("Were our policies successful?") and projections over future decades or centuries ("What are the consequences of action versus no action?"). In each case, scientists need several years to decades to collect and process data, deepen understanding, and then build and integrate models to be able to deliver answers to such questions. Against this recognition, policy makers may assume that the answers to their questions are already available, and that the science community can quickly deliver the answers within a matter of weeks or even days, to support the policy discussion of the moment.

The answer to this dichotomy is actually well established. It means that policy makers cannot hope to make progress with scientific underpinning of their questions by ad hoc or short-term policy interventions. Rather, an ongoing process of dialogue between policy makers, policy institutions, and scientists is needed that deepens mutual understanding of each-others' needs and the feasible role of science to support policy development. This means that a viable science support process for any policy area has to be long-term. It must gradually build capacity to be able to answer policy makers' questions, and include a broad spectrum of disciplines, relevant to the questions being asked (e.g., natural sciences, social sciences, political sciences). Not only that, but to be truly successful, through sustained engagement, the science community must develop sufficient understanding to be able to anticipate which questions the policy makers are going to have in the next days, months and years, even before they have asked them, as well as consider the implications for the private sector (e.g., economic growth) and for environmental quality. This requires the establishment of a long-term partnership that builds mutual understanding of the science and policy needs, the likely priorities and the operational realities for data collection, management and interpretation. Several international conventions give good examples of such practice, as illustrated in Sect. 36.3 (this chapter).

These reflections point to the conditions of good policy making itself. They highlight the need for policy processes to be long-term (i.e., over several cycles of elected governments), where the policy makers know well in advance the questions that they want to address and the anticipated timescales of evidence provision that they will require. It is critical here that a policy process is not seen as a series of isolated initiatives, but is a coherent approach towards a longer-term goal, which is also required to monitor policy implementation (see item (g) above). This highlights the essential role of intersessional meetings by the bureaus of policy frameworks. At the same time, it points to the need to involve representatives of the science community to contribute to such intersessional meetings. This is vital to ensure that the science evidence will be available when required.

36.3 Nitrogen Science Needs and International Policy Frameworks

In this section, we provide an overview of major international policy frameworks relevant to nitrogen. This overview can serve as a basis to consider how nitrogen science can best support these processes and to understand better the character of the main relevant frameworks.

36.3.1 The Global Programme of Action for the Protection of the Marine Environment from Land-Based Activities (GPA) and Other International Water Conventions

The GPA is the only global intergovernmental mechanism directly addressing the connectivity between terrestrial, freshwater, coastal and marine ecosystems. The programme represents a non-legally binding framework that is designed to address marine ecosystem degradation by encouraging governments and regional organizations to develop and implement national plans of action (NPAs) to protect their marine environments from land-based pollution. It includes efforts to identify and assess the nature and severity of marine water pollution problems in relation to food security and poverty alleviation, public health, coastal and marine resources, ecosystem health including biological diversity, and economic and social benefits.

To date just over 90 countries have developed NPAs or other relevant action plans, and are in various stages of their implementation of such plans. The GPA's Third Intergovernmental Review (IGR-3) in 2012, in Manila, identified nutrient pollution and nutrient management as one of the core priorities for the GPA. Countries committed to step up their "*efforts to develop guidance, strategies or policies on the sustainable use of nutrients so as to improve nutrient use efficiency with attendant economic benefits for all stakeholders, including farmers, and to mitigate negative environmental impacts through the development and implementation of national goals and plans over the period 2012-2016, as necessary*" (UNEP 2012).

At the IGR-3, the International Nitrogen Initiative (INI) was requested by UNEP to support the process of garnering country consensus around a possible global target for nutrient management that would benefit marine environmental conservation, delivering estimates of what a possible goal to improve nitrogen use efficiency (NUE) by 20% could mean for countries and at the global level (UNEP 2011a; Bleeker et al. 2011). A proposal was made by some governments to use this as a basis for an aspirational goal, although such a goal was not ultimately adopted in the resulting Manila Declaration. The lack of agreement to include such an aspirational goal for 20% increase in NUE is perhaps not surprising as the proposal was only introduced a few weeks before the IGR-3. A much longer process of consensus building with countries would be needed before such an agreement could be reached.

In fact, substantial consensus development has since taken place, at least at a technical level, in agreeing on definitions and metrics related to NUE. Building on the foundations provided by nitrogen budget analysis of the Organisation for Economic Co-operation and Development (OECD) and the Task Force on Reactive Nitrogen (see Sects. 36.3.2 and 36.3.8, this chapter), this work has been taken forward in parallel by the EU Nitrogen Expert Panel (EU-NEP) (Oenema et al. 2015) and the Global Partnership on Nutrient Management (GPNM) (Norton et al. 2015). The GPNM itself is a multi-actor partnership which was established in 2009 as a specific contribution to the GPA (UNEP 2010). The GPNM secretariat is served by the GPA.

The Intergovernmental reviews of GPA take place approximately every five years, with IGR-4 held in October 2018 in Bali. In the absence of a strong intersessional process to show how improved nitrogen management can strengthen GPA's approach to meet its goals, the IGR-4 meeting focused on core operational modalities of the GPA rather than on reviewing scientific evidence or developing substantive goals (UNEP 2018). This means that such global goals still need to be considered for nitrogen in the future, such as improving nitrogen use efficiency, or reducing nitrogen waste (such as presented to the adjacent Our Ocean Conference in Bali by INI offering to support a global goal to 'halve nitrogen waste', CEH 2018). The contrast between IGR-4 and the Our Ocean Conference in Bali was especially instructive. The former represented a coherent intergovernmental framework focused mainly around governments; however with little intersessional involvement of governments since 2012, there appeared to be little momentum to drive future change. Conversely, the Our Oceans Conference represents a high-profile informal arrangement of voluntary action by governments, business and civil society. Many of the commitments made are ambitious, yet, reflecting the informal approach, there is little coherency of the goals or monitoring of the commitments made. This comparison may be contrasted with more formal approaches adopted in other frameworks, as summarized in the following sections.

It should be noted that the GPA is linked to a series of Regional Seas Conventions and Action Plans (RSCAPs), including legally binding instruments covering 18 regional sea areas. The Regional Seas programmes supported under these frameworks generally include provisions to address marine pollution from nutrient loading. The United Nations Environment Programme (UNEP) is mandated to coordinate the activities of these regional seas programmes through the GPA. These include the

Helsinki Commission (Baltic), the Cartagena Convention (Caribbean), the Black Sea Commission and the Oslo and Paris Commission (Atlantic) among others. For freshwater, the main international convention is the UNECE Convention on transboundary Water Courses (Water Convention), which has in the last years been opened for signatory by additional parties outside of the UNECE region.

36.3.2 Convention on Long-Range Transboundary Air Pollution (LRTAP)

The UNECE Convention on Long-range Transboundary Air Pollution was established in 1979 and is now the main international framework for science and policies related to transboundary air pollution (UNECE 2016; Maas and Grennfelt 2016).

The LRTAP Convention works by enabling national Parties to agree to legally binding protocols for reducing air pollutant emissions and their transboundary consequences. Most relevant for nitrogen is the Gothenburg Protocol, which was signed in 1999 and revised in 2012 (Reis et al. 2012; UNECE 2013a). This protocol includes emission ceilings for nitrogen oxides (NO_x) and ammonia (NH_3), makes mandatory requirements for emission-related practices in combustion, transport and agriculture, and includes guidance on how to achieve the requirements. Emissions data must be reported annually by Parties to the convention, which are then used by the convention's EMEP activity (European Monitoring and Evaluation Programme) as input to modelling and other assessment activities.

The Task Force on Reactive Nitrogen (TFRN) was established under the LRTAP Convention in 2007. It has the twin aims of providing necessary information to support revision of regional air pollution policies for nitrogen (e.g., Gothenburg Protocol Revision) and developing the vision and scientific basis to implement an integrated approach to reactive nitrogen management, counting the multiple co-benefits of taking action. The TFRN has developed the UNECE guidance documents on NH_3 abatement (UNECE 2014, 2015) and on national nitrogen budget approaches adopted by LRTAP (UNECE 2013b). It has also examined the relationship between nitrogen and climate, nitrogen and food (Westhoek et al. 2014, 2015), and developed the nitrogen links between the LRTAP and the UNECE Transboundary Water Convention.

A major output of TFRN and LRTAP was the European Nitrogen Assessment (ENA) (Sutton et al. 2011a, b). A key conclusion of the ENA was that the environmental impact of N_r emissions in Europe, at around 70 billion to 320 billion Euro per year, is of similar magnitude to the direct agricultural benefits of nitrogen use (not including the downstream benefits in the food chain). In addition, through the ENA, the TFRN has developed the thinking around counting the multiple benefits of improved N use.

The TFRN has benefited significantly from (and fed-back into) what is a mature science policy support process in the LRTAP, with well-established science and policy groups, and a strong intersessional process (Reis et al. 2012). Finally, the TFRN and ENA have played an important role in raising public awareness of the nitrogen challenge, including developing links with business communities, civil society, communication tools (e.g., ENA video on YouTube, viewed by 18,000 people, see NitrogenScientists 2011) and public awareness through press interventions (e.g., working in partnership with the London-based Science-Media Centre).

These experiences from LRTAP and TFRN have provided lessons that can be applied to developing a global science support process for international N policy. Key points include:

(a) the benefit of closely linking science and policy engagement to strengthen *international agreement* (Reis et al. 2012),
(b) the benefit of a long-term process with regular intersessional Working Group and Task Force meetings (e.g., twice yearly) in-between the high-level meetings of the Parties,
(c) the benefit of improving *mutual understanding* between scientists and policy makers through regular a regular process of exchange,
(d) the importance of developing key metrics of interest to policy makers, including critical thresholds (from critical loads to planetary boundaries), economic valuation of nitrogen flows and impacts, nitrogen budgets and nitrogen use efficiency (NUE), and
(e) the importance of strengthening engagement with international news media.

36.3.3 UN Framework Convention on Climate Change (UNFCCC) and the Link to the Intergovernmental Panel on Climate Change (IPCC)

The United Nations Framework Convention on Climate Change (UNFCCC) is relevant for nitrogen as the Kyoto basket of greenhouse gases includes nitrous oxide (N_2O), while perturbation of the global nitrogen cycle also alters the radiative budget in other ways. These include increasing carbon sinks through fertilization by atmospheric N_r deposition to forests, formation of tropospheric ozone (O_3) which reduces carbon sinks and is itself a greenhouse gas, and formation of particulate matter that has both direct and indirect cooling effects of climate (Butterbach-Bahl et al. 2011). In addition, a further industrially produced nitrogen compound, nitrogen trifluoride (NF_3) is a powerful greenhouse gas that is included under the Kyoto Protocol.

The consolidation of science evidence to UNFCCC is provided through the Intergovernmental Panel on Climate Change (IPCC), which is a legally separate body. However, in certain cases, the UNFCCC may request specific action from IPCC, such as on emission calculation methodologies. The major Assessment Reports of IPCC have delivered evidence on the science understanding of climate change, as well as the mitigation and adaptation opportunities. The clear separation between

IPCC and UNFCCC is a notable contrast to the LRTAP model that limits the opportunity for close interaction between the science community and policy makers (Reis et al. 2012). An exception is the finalization of the Summaries for Policy-Makers of the IPCC Assessment Reports, which are subject to negotiations with participating national governments. However, the focus of this engagement is on policy makers scrutinizing scientific outputs, rather than on developing mutual understanding of the wider science and policy goals. It is notable that IPCC was established in 1988, four years in advance of the UNFCCC, in the same way that INMS has also been established ahead of any global policy process for nitrogen.

36.3.4 UN Convention on Biological Diversity (CBD)

The CBD provides a broad framework for developing international cooperation and agreements on biodiversity protection. It includes twenty targets under the Aichi process of which one of the indicators is focused on reducing nutrient pollution (CBD 2016). As part of this action, INI provides support as a lead partner for the nitrogen related indicator within the Biodiversity Indicators Partnership (Bleeker et al. 2012).

One of the advantages of the broad approach of CBD is that it is naturally able to link all different threats of alteration of the nitrogen cycle on biodiversity, including air, land and water. At the same time, this exceptional breadth makes the CBD a highly complex and busy market place within which to set an agenda towards better global management of the nitrogen cycle.

Similar to the climate process, CBD is now closely associated with an independent scientifically-oriented process, the Intergovernmental Platform on Biodiversity and Ecosystem Services (IPBES). With the CBD signed in 1992, establishment of IPBES (in 2012) responded to a perceived gap for scientific support on biodiversity and ecosystem services. IPBES develops assessment reports for specific regions and for the globe.

36.3.5 Vienna Convention and the Montreal Protocol

The Vienna Convention of the Protection of the Ozone layer is a global agreement with a secretariat based at UNEP. It was adopted in 1985 and came into force in 1988, achieving universal ratification in 2009 (UNEP 2016). Its objectives are to promote cooperation between its Parties though observations, research and information exchange on how human activities are affecting the ozone layer and to put in place measures against activities that adversely affect stratospheric ozone concentrations.

The Vienna Convention is highly relevant for nitrogen, since with the effective control of most ozone depleting substances (most notably CFCs and HCFCs), nitrous oxide (N_2O) is the main ozone depleting substance, and will likely continue to be for the rest of the twenty-first century and beyond (Ravishankara et al. 2009). Nitrous oxide is already listed under the Vienna Convention as a substance that modifies "the chemical and physical properties of the ozone layer" (Article 3). Nevertheless, at present, the Montreal Protocol, which is the main legal instrument of the Vienna Convention, does not include N_2O in its list of regulated ozone depleting substances. It has been suggested therefore that N_2O be included (Kanter et al. 2013; UNEP 2013).

The Montreal Protocol has been recognized as being highly effective in achieving its aims of phasing out the production and consumption of 97 ozone depleting substances. This has made it attractive as a potential 'policy home' for N_2O, and with it, potentially an integrated approach to nitrogen management. Against this attractiveness is the fact that around 70% of global N_2O emissions results from agriculture, which is a sector where the Montreal Protocol has experienced the most challenges—namely the phase-out of methyl bromide, a soil fumigant (Gareau 2010).

It is likely that the high effectiveness of the Montreal Protocol results from its dealing with a small number of large companies that were able to produce replacement compounds without economic disruption, combined with appropriate financing and control of the trade and movement of Ozone Depleting Substances (ODS) to achieve the proposed changes. This contrasts with the challenges of dealing with a large number of diverse stakeholders (farmers, citizens etc.), whose actions all influence the nitrogen cycle. Nevertheless, the same dynamic that existed with the major CFC companies (where they supported the Montreal Protocol because they could capitalize on the market for CFC alternatives) could potentially exist with major fertilizer companies. For example, such companies increasingly provide fertilizer services to help farmers improve nitrogen management practices and now give increased attention to producing more efficient fertilizer products that can reduce nitrogen pollution.

In addition to the specific conventions noted above, the following sections on UN and OECD activities give examples of other forums where nitrogen may be addressed.

36.3.6 The UN Environment Programme (UNEP) and the UN Environment Assembly (UNEA)

The United Nations Environment Programme (UNEP) hosts a large and diverse set of international actions. For example, it provides the secretariat to the Montreal Protocol, to the GPA and GPNM and to the Climate and Clean Air Coalition (CCAC). The breadth of experience of UNEP, and close link to its mandate, could therefore make it attractive as an organization to host a more joined-up approach to nitrogen.

The GPA and GPNM have already been noted above. The CCAC is a voluntary group where countries and other stakeholders commit to reduce short-lived climate pollutants (SLCP) concentrations in the atmosphere, especially methane, tropospheric ozone and black carbon (CCAC 2016). The CCAC work to implement an identified set of measures for reducing emissions of black carbon and methane that were determined by a global integrated assessment of black carbon and tropospheric ozone (UNEP 2011b). The CCAC promotes funding from an international trust fund for actions to reduce these emissions as a contribution to meeting both climate and air pollution goals. As part of its agriculture programme, there is a direct connection with nitrogen through manure management and the transport related SLCP measures reduce NO_x as an air pollutant co-emitted with black carbon (Shindell et al. 2012). Like the GPA, the CCAC works with national governments to implement national plans for action to reduce these pollutants.

The UN Environment Programme also fosters and coordinates actions with many other regional environmental programmes. These include the 18 Regional Seas Programmes, seven of which are administered directly by UNEP while the others are managed through independent intergovernmental mechanisms. For example, the South Asia Seas Programme is coordinated by the South Asia Cooperative Environment Programme (SACEP). Such programmes provide opportunities to consider regional priorities in more detail and to support progress on regional implementation.

In 2014, UNEP was upgraded to achieve universal membership of all UN countries, thereby substantially strengthening its role in the UN system. This has been reflected in the change from the regular (annual to 2-yearly) meetings of the UNEP Governing Council and Global Ministerial Environmental Forum to be replaced by the United Nations Environment Assembly (UNEA). The first three sessions of UNEA took place in 2014, 2016 and 2017. UNEA is advertised as the highest level UN decision-making body on the environment. It considers both governance of the UNEP programme of work and serves as a platform to draw attention to specific issues of importance to countries, for example, by countries proposing resolutions for agreement by UNEA. For these reasons, the UNEA presents a promising mechanism to develop a joined-up approach to nitrogen management (Sutton et al. 2019).

Significant progress has already been achieved, facilitated by support from the INMS community to member states and UNEP. For example, the resolution on "Preventing and reducing air pollution to improve air quality globally" at UNEA-3 encouraged governments "to take advantage of synergistic effects of efficient nitrogen management on reducing air, marine and water pollution." (UNEP/EA.3/Res.8.). A major step has recently been achieved in following this up at UNEA-4 (March 2019), with the first ever resolution on 'Sustainable Nitrogen Management', as noted at the conclusion of this chapter.

36.3.7 UN Sustainable Development Goals and Other UN Programmes

An emerging focus in global sustainability has been the establishment of the Sustainable Development Goals (SDGs), which now replace the Millennium Development

Goals, originally established under the 1992 UN Conference on Environment and Sustainable Development (Rio Earth Summit). Since 2013, the SDGs have been addressed under the High-level Political Forum on Sustainable Development (HLPF), which has its secretariat at the UN Division for Sustainable Development (UN DSD 2016). Technical support in the development of SDG indicators is provided through the UN Statistical Commission. Of the 17 SDGs, most are relevant to nitrogen, and imply a requirement for better N_r management. However, despite being almost everywhere relevant, nitrogen is virtually invisible in the SDG process (see Table 36.1).

Indicators to assess progress towards these SDGs were agreed upon in 2015 (UN Statistical Commission 2016). Nitrogen appeared in only one place in the proposed indicators list as a 'nitrogen use efficiency composite indicator' for protection of the marine environment. However, this was not adopted in the finalized text (UN Statistical Commission 2016, compare Annex III, 14.1.1 with Annex IV, 14.1.1). The procedural difficulty of negotiating such indicators between UN agencies emphasizes the difficulty to address the interactions between SDGs. By contrast, Table 36.1 highlights the systemic relevance of nitrogen across most of the SDGs.

The final agreed SDG indicator list points to a highly complex process with 230 indicators overall. The failure to include nitrogen-specific indicators (despite efforts by the science and stakeholder community including by the GPNM and the Sustainable Development Solutions Network), reflects the SDG process being a negotiation among many actors according to existing sustainability paradigms and headlines. This is likely to have the general result that emerging or cross-cutting issues, such as nitrogen, are not included. By contrast, the very absence of nitrogen in the SDGs becomes an opportunity for a headline message. The concept of #EverywhereAndInvisible reflects the position of nitrogen both across the environment and across the SDGs. Narratives such as this highlight the potential to work with countries in showing how nitrogen management could help deliver multiple SDGs.

The catch-all nature of the SDG process means that in most cases delivery of the SDGs will only be achievable through separate, focused efforts taking place in parallel, such as by the international conventions and programmes described above.

Similarly, many intergovernmental organizations also have a role to play in improving management of the global nitrogen cycle. These include the United Nations Development Programme (UNDP), the United Nations Industrial Development Organization (UNIDO), the UN Food and Agriculture Organization (FAO), the World Meteorological Organization (WMO), the World Health Organization (WHO) and the International Oceanographic Commission of the United Nations Educational, Scientific and Cultural Organization (IOC-UNESCO). The wide relevance of nitrogen to these many organizations highlights the complexity of the challenge faced and the need to coordinate efforts.

Table 36.1 Relevance of nitrogen to the UN Sustainable Development Goals with a comparison of the most relevant SDG indicators agreed by the (UN Statistical Commission 2016), Annex IV

Goals especially relevant for nitrogen	Indicators agreed by UN	Comment
SDG 2. Zero Hunger: End hunger, achieve food security and improved nutrition and promote sustainable agriculture 2.1 By 2030, end hunger and ensure access by all people, in particular the poor and people in vulnerable situations, including infants, to safe, nutritious and sufficient food all year round 2.3 By 2030, double the agricultural productivity and incomes of small-scale food producers 2.4 By 2030, ensure sustainable food production systems and implement resilient agricultural practices that increase productivity and production	2.1.1 Prevalence of undernourishment 2.3.1 Volume of production per labour unit by classes of farming/pastoral/forestry enterprise size	Access to sufficient N_r resources through fertilizers and biological nitrogen fixation in farming will be essential to meet this goal Also relevant for SDG1: Zero poverty
SDG 3. Ensure healthy lives and promote well-being for all at all ages 3.9 By 2030, substantially reduce the number of deaths and illnesses from hazardous chemicals and air, water and soil pollution and contamination	3.9.1 Mortality rate attributed to household and ambient air pollution	N_r especially as NH_4^+ and NO_3^- is a major fraction of fine particulate air pollution while NO_x contributes to tropospheric O_3 pollution
SDG 6. Clean Water and Sanitation: Ensure availability and sustainable management of water and sanitation for all 6.3 By 2030, improve water quality by reducing pollution, eliminating dumping and minimizing release of hazardous chemicals and materials, halving the proportion of untreated wastewater and substantially increasing recycling and safe reuse globally	6.3.1 Proportion of wastewater safely treated 6.3.2 Proportion of bodies of water with good ambient water quality	Concentrations of NO_3^- in surface waters and aquifers are often above internationally agreed standards for human health

(continued)

Table 36.1 (continued)

Goals especially relevant for nitrogen	Indicators agreed by UN	Comment
SDG 7. Ensure access to affordable, reliable, sustainable and modern energy 7.2 By 2030, increase substantially the share of renewable energy in the global energy mix	7.2.1 Renewable energy share in the total final energy consumption	Producing bioenergy is necessarily associated with major N_r flows, while NO_x emissions are closely linked to total energy production by fossil fuels
SDG 9. Industry, Innovation and Infrastructure: Build resilient infrastructure, promote inclusive and sustainable industrialization and foster innovation 9.4 By 2030, upgrade infrastructure and retrofit industries to make them sustainable, with increased resource-use efficiency and greater adoption of clean and environmentally sound technologies	9.4.1 CO_2 emission per unit of value added	There has so far been little focus on innovation in the nitrogen cycle, which offers major opportunities in developing the circular economy (e.g., NO_x recapture & utilization technology)
SDG 11. Sustainable Cities and Communities: Make cities and human settlements inclusive, safe, resilient and sustainable 11.6 By 2030, reduce the adverse per capita environmental impact of cities, including by paying special attention to air quality and municipal and other waste management	11.6.2 Annual mean levels of fine particulate matter (e.g., $PM_{2.5}$ and PM_{10}) in cities (population weighted)	N_r air pollution is a key threat to sustainable cities, including from particulate matter and NO_2 concentrations
SDG 12. Responsible Consumption and Production: Ensure sustainable consumption and production patterns 12.3 By 2030, halve per capita global food waste at the retail and consumer levels and reduce food losse s along production and supply chains, including post-harvest losses	12.3.1 Global food loss index	In addition to food waste, excess consumption of livestock products is critical for N_r pollution and not currently addressed by the SDG indicators

(continued)

Table 36.1 (continued)

Goals especially relevant for nitrogen	Indicators agreed by UN	Comment
SDG 13. Climate Action: Take urgent action to combat climate change and its impacts 13.2 Integrate climate change measures into national policies, strategies and planning	13.2.1 Several countries have established integrated policy/strategy/plans to increases their ability to adapt to the adverse impacts of climate change, and foster climate resilience and low greenhouse gas emissions development in a manner that does not threaten food production	Nitrous oxide is one of the main global greenhouse gases, while N_r also affects climate in several other ways. Improving NUE is often one of the most cost-beneficial approaches to reduce GHG emission s from agriculture
SDG 14. Life Below Water: Conserve and sustainably use the oceans, seas and marine resources for sustainable development 14.1 By 2025, prevent and significantly reduce marine pollution of all kinds, in particular from land-based activities, including marine debris and nutrient pollution	14.1.1 Index of coastal eutrophication and floating plastic debris density. (Annex III proposal for 'Nitrogen use efficiency indicator' not adopted)	Nutrient pollution is one of the key threats identified by the GPA. Data on N_r is needed to calculate the index of coastal eutrophication potential
SDG 15. Life on Land: Protect, restore and promote sustainable use of terrestrial ecosystems, sustainably manage forests, combat desertification, and halt and reverse land degradation and halt biodiversity loss 15.1 By 2020, ensure the conservation, restoration and sustainable use of terrestrial and inland freshwater ecosystems and their services, in particular forests, wetlands, mountains and drylands, in line with obligations under international agreements 15.5 Take urgent and significant action to reduce the degradation of natural habitats, halt the loss of biodiversity and, by 2020, protect and prevent the extinction of threatened species	15.1.2 Proportion of important sites for terrestrial and freshwater biodiversity that are covered by protected areas, by ecosystem type 15.5.1 Red List Index	Insufficient N_r availability and soil-mining is a key threat in least-developed countries, while excess atmospheric N_r deposition threatens terrestrial biodiversity in many parts of the world

36.3.8 Organization for Economic Cooperation and Development (OECD)

The OECD represents a major global partnership consisting of 34 countries, representing much of the world economy. The OECD hosts a well-established approach to calculating national nitrogen balances in agricultural soils (OECD 2001). This represents a key baseline that, through partnership with the Expert Panel on Nitrogen Budgets (EPNB) of the TFRN (UNECE 2013b), offers a starting position in the construction of full nitrogen budget approaches. In parallel, the OECD has been exploring the concept of 'Economy-wide Nitrogen Use Efficiency' (Bleeker et al. 2013) as a high-level indicator to complement the nitrogen budgets approaches.

Engagement of INMS with the OECD during the preparation phase of INMS has led to the nitrogen challenge being presented to the OECD's Environmental Policy Committee (EPOC), including at a Ministerial level, and its Working Party on Water Biodiversity and Ecosystems (WPWBE). This is building the links with member countries to support approaches to address the nitrogen challenge, especially through developing country case studies. As such, the OECD provides an important venue for further exploration of approaches to joined-up nitrogen management, as well as raising the profile of the challenge at a high level (OECD 2018).

Since it is the country governments that ultimately have the responsibility to implement recommendations from all these international conventions and programmes, plans for improved nitrogen management need to include building national capacity to implement recommendations. Many of the processes noted above are developing case studies, tools and training programmes to achieve this.

36.4 Science to Support Nitrogen Policy Development

Recognizing the multi-dimensional nature of the nitrogen challenge, it is clear that there are several existing policy processes of high relevance. The consequence is that few individuals—science experts or policy makers—are competent to comment on the full diversity of relevant frameworks. In the case of nitrogen, there needs to be a clear process that develops mutual understanding among communities, that sets reasonable expectations on achievable timescales, and that communicates effectively between science provision and policy needs.

Before addressing the question of possible policy homes for nitrogen, we first map out in more detail how a process of science evidence support for the global nitrogen cycle is being established through the 'International Nitrogen Management System' (INMS).

36.4.1 Towards the International Nitrogen Management System

The vision for the 'International Nitrogen Management System' (INMS) is to establish a coordinated approach to science provision for policies on the global nitrogen cycle. In broad terms it might be considered as something like an 'IPCC for nitrogen', but with substantial differences in the model being developed. For example, it is establishing a science-led multi-actor partnership, with much closer engagement between science and policy. In this regard, INMS is more similar to the model established under the LRTAP Convention. Its activities are supported by finance from the Global Environment Facility (GEF), UNEP, INI and around 80 partner organizations through the 'Towards INMS' project (INMS 2016).

INMS recognizes that the present lack of a coherency across the nitrogen cycle contributes substantially to the barriers-to-change that prevent the implementation of a more optimized global nitrogen cycle. This means that maximizing the benefits for one policy domain (such as aquatic ecosystems and the coastal zone) requires taking account of the other benefits to which other possible actions could contribute. A particular emphasis is given to valuing the N_r resource, both as its equivalent cash value as fertilizer and the wider societal value of N_r pollution. By emphasizing the profits to be made by better N_r management, a stronger case for action may be expected, helping to overcome the barriers-to-change. For example, based on the flows reported in 'Our Nutrient World' (Sutton et al. 2013), total losses of nitrogen in the environment (including both N_r and denitrification to N_2) amount to around 200 million tonnes annually. With a nominal fertilizer price of $1 USD per kg N, this is equivalent to a resource loss of $200 billion USD. Put another way, an aspirational goal to #HalveNitrogenWaste would provide an opportunity to save $100 billion USD per year (CEH 2018; Sutton et al. 2019). The example illustrates the need for much greater emphasis on integrating nitrogen science, supported by appropriate tools and options.

Considering this rationale, INMS addresses the hypothesis that joined-up management of the nitrogen cycle will offer many co-benefits that strengthen the case for action for cleaner water, cleaner air, reduced greenhouse gas emissions, ozone layer recovery, better soil and biodiversity protection, while at the same time helping to meet food and energy goals.

This leads to a broad approach where the challenges of one issue become linked to the challenges and opportunities of the interacting issues. For example, when actions needed to reduce the effects of N_r on transboundary waters can be shown simultaneously to deliver quantified co-benefits for air, climate, food, energy, this will more strongly motivate the necessary changes for water protection. The same applies for each of the other threat and benefit policy domains (food, air, climate, soil etc.). By acting together through the nitrogen cycle, there is the potential to transform efforts for a cleaner and healthier environment.

An initial concept for INMS, developed in 2013, illustrates the kinds of information that are expected to be needed (Fig. 36.2). Each of the light blue boxes represents a

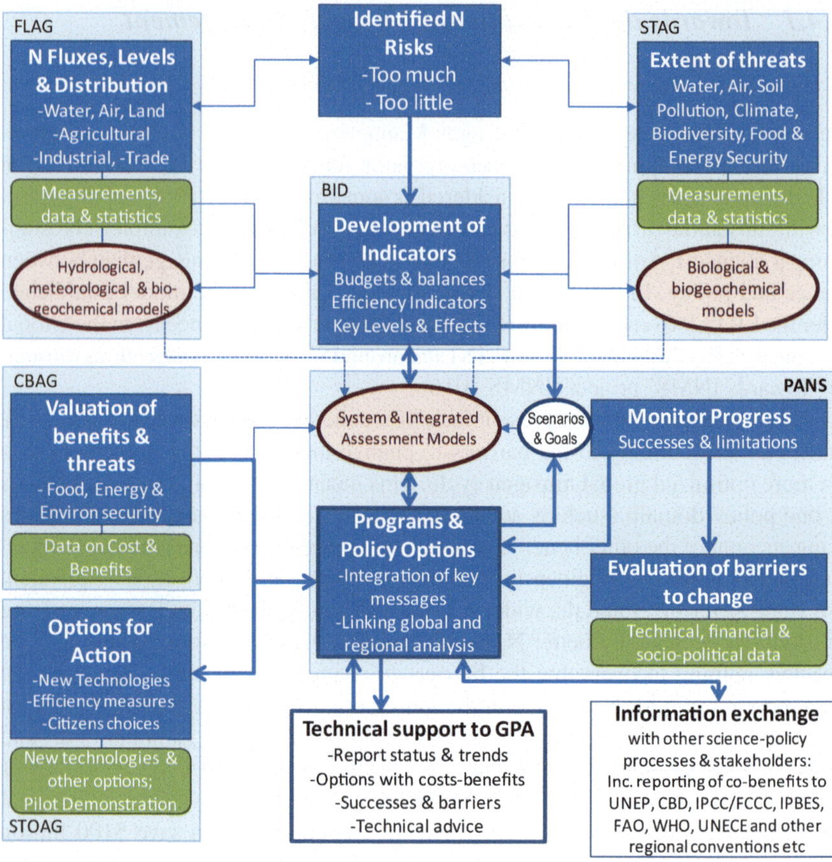

Fig. 36.2 Initial Concept for how an International Nitrogen Management System (INMS) could operate. This visualization (from 2013) illustrates how INMS could support the GPA process. It should be evident that the INMS outputs are equally relevant to other processes, as illustrated in the 'nitrogen policy arena' concept (see Fig. 36.4, this chapter). The acronyms refer to concept advisory groups i.e., FLAG: Fluxes & Levels Assessment Group; STAG: Sustainability and Threats Assessment Group; BID: Budgets and Indicators Development; CBAG: Costs & Benefits Assessment Group; PANS: Policies and Analysis of Nitrogen/Nutrient Synergies; STOAG: Societal & Technical Options Assessment Group

concept working or assessment group, which addresses issues (dark blue), supported by information (green) and models (brown). This original visualization was targeted to illustrate evidence support to the GPA. However, the same principles apply where such an evidence system is provided to support wider challenges, for example for CBD, UNFCCC, LRTAP, Vienna Convention, UNEA, OECD, FAO, WMO, WHO etc.

Such an approach as outlined in Fig. 36.2 is currently largely missing from the GPA, which has only included ad hoc science support until now. At the same time the

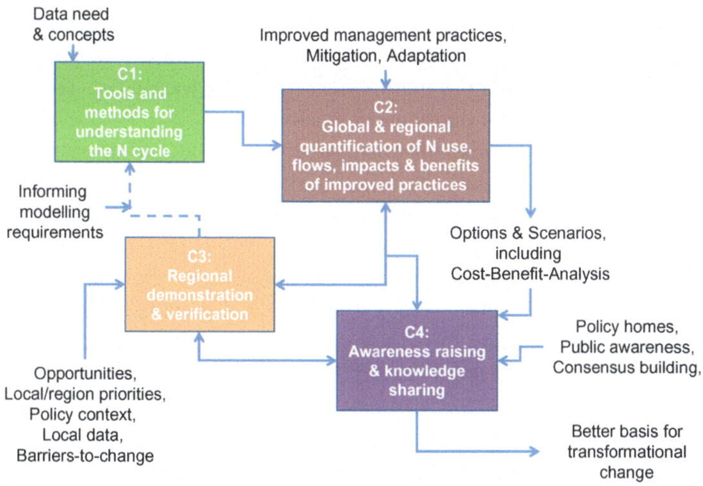

Fig. 36.3 Simplified overview of the four main components the "Towards INMS" project, as financed with the support of GEF, UNEP, INI and partners (INMS 2016)

model indicates a much closer degree of cooperation than current between IPCC and UNFCCC, which has been shown in the LRTAP process to offer substantial benefits in improving mutual understanding (Reis et al. 2012).

Since Fig. 36.2 was drafted, substantial progress has been made in bringing INMS to fruition through funded cooperation between UNEP, INI and GEF. For this purpose, the project design has been developed consisting of four main components (C1–C4), for which the tasks and linkages are summarized in Fig. 36.3.

36.4.2 Clarifying the Relationship Between Science, Policy and Practice

In developing the concept of what INMS should look like, a wide range of views has been encountered among different stakeholders about the kind of information and approach that is needed. For example, some stakeholders have encouraged that INMS should itself deliver a policy process. Conversely, other stakeholders have expressed concerns that INMS might become a policy process. Such ends of the spectrum appear to reflect different stakeholder views on the desirability of further developing governance concerning the global nitrogen cycle. At the same time, even the process of discussing a science-based support process has stimulated other stakeholders to reflect on whether they want any form of regional or global governance for nitrogen, and, if so, what form it should take.

These kinds of reactions show that the process of developing INMS is itself serving a useful role in stimulating thinking at the interface of science and policy. However, they also point to the need to clarify the exact role of the INMS process, both about what it is and what it is not. In responding to such questions, four parallel tracks have been identified. The reality is that INMS is centrally focused on Track 2, while stimulating wider engagement on the other tracks:

Track 1: Policy Development for Nitrogen: This is the role of governments in cooperation with all stakeholders. Negotiation of agreements needs to be based on robust scientific evidence, while also requiring appropriate indicators for monitoring success, which should be based on sound science. Agreeing on new policies is not the role of the science community, though it should generate the science required to inform effective policy decisions (Track 2, below) and provide the knowledge that can feed into capacity building to help governments and their agencies respond to the nitrogen challenge effectively (Track 3, below).

Track 2: Scientific Support for Nitrogen Policy Development: This role is necessarily under the lead of the science community and needs to be organized in such a way that the full range of stakeholder inputs are included, while developing an effective approach that is responsive to the needs of policy makers. Key elements of this track include providing the evidence of the multiple threats and benefits of nitrogen management, the provision of scenarios demonstrating cost-benefit of particular policy choices based on operational realities, the harmonization and benchmarking of performance indicators, the sharing and dissemination of best practices, and the synthesis of indicator monitoring.

Track 3: Practices and Technologies for Better Nitrogen Management: Through INMS, the science community can play a key role in identification of the most suitable options that maximize the nitrogen co-benefits, while profiling the potential of success stories for wider dissemination and adoption. Implementing wide-scale adoption of better practices and technologies is especially the role of governments and agencies, with engagement from a wide range of actors.

Track 4: Public Engagement about the Nitrogen Threats and Opportunities: Without significant public engagement little substantive progress can be expected in the exchange between policy making, scientific support and practice development. The key actors benefiting from N_r use and contributing to N_r pollution would have insufficient information on how to improve, while governments would not be empowered to take action by their citizens. It is therefore vital that the science process of INMS (Track 2) focuses on developing clear public messages and actively engages with industry, business, media and civil society.

36.5 Possible Models for Nitrogen Policy Homes

With this clearer view of the kinds of science needed and ways of envisaging international nitrogen science support for policy, the next question is how to join up international nitrogen policy development (Track 1) with the scientific support process (Track 2).

This raises the central question: what would be the most suitable policy home that INMS should eventually support? Here we consider four contrasting options.

36.5.1 Option 1: The Status Quo, with Fragmentation Across Policy Processes

It should be clear that there is currently no single international policy framework that addresses all the issues relevant for nitrogen. Similarly, each of the existing frameworks, such as the GPA, LRTAP, UNFCCC, CBD, Vienna Convention and the Regional Seas Conventions, as well as other groups such as OECD and the UN High-Level Political Forum for Sustainable Development (incorporating the SDGs), face many challenges to make progress in meeting their goals. In the case of nitrogen, it is evident that these different domains hardly work together at present, with many missed policy opportunities, for example to avoid trade-offs and identify multiple win-wins.

The *status quo* is also unattractive from a perspective of efficiency in providing science support to policy processes. For example, it means that the INMS community needs to engage one-by-one in multiple policy processes, which is extremely time consuming.

36.5.2 Option 2: Work Under the Lead of One Existing Convention or Programme

Following the advice of policy makers, discussions on a policy approach for nitrogen have continued at length in the wings of numerous meetings of the processes outlined in Sect. 36.3 of this chapter. Examples include with GPA, UNEP, UNEA, CBD, UNECE (TFRN, LRTAP and the Transboundary Water Convention), OECD, European Commission and with representatives of many national governments. Could one of these existing processes realistically take the lead in an integrated policy approach on the challenges of global nitrogen cycle?

Here we reflect briefly on the character, strengths and limitations of the different existing processes as regards their possible further application to develop an integrated nitrogen policy approach. The key opportunities and limitations of each framework are summarized in Table 36.2.

(1) **WATER: Global Programme of Action to Protect the Marine Environment from Land-based Activities** (GPA). The GPA is the only international programme to address the connection between land-based pollution and the marine environment. Since the Manila Declaration (UNEP 2012), nutrient pollution of the marine environment is considered as one of its three core challenges

Table 36.2 Opportunities and limitations of different conventions and programmes as potential policy homes for an integrated nitrogen approach

Framework	Opportunities	Limitations
GPA	The only global international agreement addressing the land-water-sea continuum Strong engagement with regional programmes and financing mechanisms including GEF Global approach with close partnership with UNEP and INMS	Mandate limited to the marine environment No quantitative aspirational goals or legally binding agreements Weak intercessional process with no formal process for sustained science support
Geneva Air Convention (LRTAP)	The main international agreement on transboundary air pollution Strong intersessional mechanisms and science support processes allowing sustained long-term development Closely structured engagement between science and policy communities allowing improved mutual understanding Interactions across the nitrogen cycle are addressed (e.g., Task Force on Reactive Nitrogen) Establishes clear legally binding agreements with quantitative goals	Mandate limited to air pollution, restricting the extent to which nitrogen co-benefits for climate and water are addressed in policy development Limited regional extent, albeit with significant partnerships and dialogue beyond UNECE region
UNFCCC and IPCC	Together, the leading global environmental agreement and science support processes for climate Strong intersessional processes allowing sustained long-term development Well-established long-term science support process through IPCC Establishes clear legally-binding agreements with quantitative goals	Strong separation between policy (UNFCC) and science support process (IPCC) limiting close engagement and mutual understanding Negotiating mandate limited to climate, limiting the extent to which nitrogen co-benefits for air quality and water are addressed in policy development Exceptionally busy process with well-established priorities, which provides a barrier to innovative consideration across nitrogen issues

(continued)

Table 36.2 (continued)

Framework	Opportunities	Limitations
CBD	The main international agreement on biodiversity, with science support from the Intergovernmental Platform on Biodiversity & Ecosystem Services (IPBES) Strong engagement with regional programmes and financing mechanisms including GEF Strong intersessional processes allowing sustained long-term development Clear indicator process with relevant established goals (e.g., Aichi Target for nitrogen and nutrient pollution)	Mandate highly diverse, yet limited to biodiversity interactions, so constraining the ability to address climate, air pollution and water quality goals Many diverse biodiversity-related challenges would make it operationally difficult to establish legally-binding agreements with quantitative goals for nitrogen
Vienna Convention and Montreal Protocol	The main international agreement on stratospheric ozone depletion Strong intersessional mechanisms and science support processes allowing sustained long-term development Establishes clear legally binding agreements with quantitative goals Strong financing mechanism can help facilitate regional adoption of measures Global approach with close partnership with UNEP	Mandate limited to stratospheric ozone, so constraining ability to address climate, air pollution and water quality goals Although nitrous oxide is recognized in the Vienna Convention, it is excluded from the Montreal Protocol Experience with the financing mechanism is mainly with industrial sources, and does not cover the wide diversity of agricultural, wastewater and traffic sources relevant to nitrogen

(together with waste water and marine litter). A strength of the GPA its established experience of working with countries through Regional Seas Programmes at the global level. Conversely, a limitation for nitrogen is that the focus of GPA is specifically on the marine environment. Issues of wider nitrogen management are therefore not automatically a priority, unless it can be demonstrated how joined-up nitrogen management strengthens the opportunity to meet the marine goals of the GPA. There is a clear need for science evidence provision to GPA, as shown by experience at the IGR-3. However, a present limitation of GPA is its undeveloped intersessional process. This means that it is not currently easy to connect science efforts between the IGR meetings (every 4–5 years) to support advance planning by the countries of their desired outcomes.

(2) **AIR: Convention on Long-range Transboundary Air Pollution** (LRTAP). Substantial progress has been made by the LRTAP convention in addressing the nitrogen issue and pioneering thinking connected with the wider nitrogen cycle. The LRTAP convention has a strong intersessional process, allowing the development of both long-term science capacity and a mutual understanding of the needs between the policy and science communities. In particular, through its Working Group on Strategies and Review, the architecture of the LRTAP Convention allows a close interaction between policy and science expertise. Apart from its substantive commitments on N_r emissions reductions to the atmosphere, the Gothenburg Protocol (UNECE 2013a) took a significant step in introducing voluntary reporting of national nitrogen budgets. The limitations of the LRTAP convention for an integrated approach on the global nitrogen cycle are two-fold: First, the convention is restricted to goals related to air pollution, and second, it only covers the geographic scope of the UNECE region. Although the UNECE Transboundary Waters Convention has shown that it is possible to include Convention parties beyond this region, it has so far not proved possible to agree to this within LRTAP. There is also the potential for much stronger cooperation between the UNECE LRTAP and Transboundary Waters conventions. However, these have different modes of operation, which provides a barrier to stronger linkage.

(3) **GREENHOUSE GAS: UN Framework Convention on Climate Change** (UNFCCC) and the **Intergovernmental Panel on Climate Change** (IPCC). At present the UNFCCC is one of the largest international agreements linked to the environment. The IPCC is also one of the world's leading science assessment processes. These are key strengths of the UNFCCC as a potential policy home for nitrogen, which naturally emphasize the links between N_r and climate change. Against this opportunity is the complexity of dealing with an extremely large organization that is already over-busy with its own challenges. N_2O is only one of the wider basket of Kyoto Protocol gases. Similarly, as a specific industrially chemical, the challenges for nitrogen trifluoride (NF_3) are rather separate. In practice the prospect of embedding a multi-threat nitrogen perspective within UNFCCC appears unlikely at present. Nevertheless, the strength of the climate issue can provide an entry-point for governments to recognize the wider benefits of joined-up nitrogen management. The strict separation of the UNFCCC-IPCC

model may not offer the most suitable approach for nitrogen, given the need for a closer engagement between science and policy. As shown by the LRTAP-TFRN approach, there are substantial benefits to be found from developing a close interface between these communities. Finally, it is unclear how the Paris Climate Accord could help or hinder a more joined-up approach. While N_2O has been included in many of the proposed country plans (being mentioned in 94 of 119 submitted 'Intended Nationally Determined Contributions', (UNFCCC 2015), the major focus still appears to be on carbon dioxide and the energy sector.

(4) **ECOSYSTEMS AND BIODIVERSITY: UN Convention on Biological Diversity** (CBD). The INI already works closely with the CBD, acting as the delivery partner for its nitrogen deposition indicator under the Aichi Targets process. This has led to INI contributing to several CBD meetings, building understanding of the CBD process. At the same time, the CBD secretariat has been similarly active in supporting the development of the INMS process. CBD, however, represents a highly diverse set of biodiversity interests. As a very busy 'market-place', an issue like nitrogen is under strong competition for attention with many other topics in the CBD. Although N_r is multi-source, multi-impact (matching to CBD), as the challenge of nitrogen is fundamentally biogeochemical, it naturally has high commonality with conventions dealing specifically with material flows (like GPA, LRTAP, UNFCCC).

(5) **STRATOSPHERIC OZONE: Vienna Convention and Montreal Protocol.** It has already been noted that N_2O now represents the main source of stratospheric ozone depletion. Given this point, it remains an open question whether N_2O control should become part of the group of pollutants that are addressed under the Montreal Protocol. Advocates of its inclusion emphasize the success of the Montreal Protocol in decreasing CFC and HCFC emissions substantially over the last 20 years. Conversely, critics have emphasized that the success of the Montreal Protocol was connected with the availability of financially supported alternatives to CFCs and HCFCs, while being focused on a few large well-organized companies producing CFCs and HCFCs. Although some N_2O arises from large industrial operations, over 70% arises from agricultural sources, implying the need for the Montreal Protocol to deal with a much wider and more diverse set of stakeholders than it has in the past (e.g., its work on methyl bromide has given it some experience of dealing with the agricultural sector). Irrespective of this debate, it remains an open question whether the Montreal Protocol would be ready to make a double leap to address all the main polluting and beneficial effects of reactive nitrogen.

While each of these policy frameworks is highly relevant for nitrogen, a clear message emerges that none of them (as they stand at present) appears to be well suited to act as a main policy home for nitrogen. This is not surprising. If the solution were easy, it would have already presented itself at an earlier stage. In addition to the challenges facing these environmental agreements, as summarized in Table 36.2, attention also needs to be given to integrate the opportunities for resource mobilization. This implies developing the policy and practice links with agriculture, wastewater and industrial sources, pointing to the need for engagement with WHO, FAO, UNDP, OECD and the International Resources Panel (IRP) for health, livelihoods and development of the circular economy.

36.5.3 Option 3: Establish an Intergovernmental Convention on Nitrogen

Already in 1990s the fragmentation of science and policy of the nitrogen cycle was identified (e.g., Cowling et al. 1998; Bull and Sutton 1998). This growing recognition led to the establishment of the INI in 2003 as a focal point to bring science evidence more closely together. Associated with this, scientists have often suggested that an intergovernmental convention on nitrogen issues was needed. Such calls (e.g., Saltsjöbaden-3 workshop, Erisman et al. 2007) led to the establishment of the UNECE Task Force on Reactive Nitrogen by the Executive Body of the LRTAP Convention (UNECE 2007). Nevertheless, although the TFRN has a mandate to address the full nitrogen cycle from a technical perspective, this mandate is set within a negotiating context of a specific threat (in this case air pollution).

Many have called for a new international 'Nitrogen Convention'. Such a position has strong public appeal because of its simplicity (e.g., Bull et al. 2011; Sutton et al. 2011b), with a clear focus on bringing together the different issues connected to nitrogen. At the same time, the response of many policy makers to this suggestion has been equally interesting. Through the 'corridor discussions' of many intergovernmental meetings, Mark Sutton has posed this question to numerous government officials. The response seems to be almost universally: "*we already have enough intergovernmental conventions; we don't need more. Do your best to work with the existing policy processes.*" There is thus a case in-principle for a joined-up Nitrogen Convention, but simultaneously major political barriers to such an approach. Reflection on this policy response also needs to recognize the timing of the discussion (e.g., after the economic crisis of 2007) and the tough challenges already experienced in achieving the goals of existing international agreements.

36.5.4 Option 4: Establish an Interconvention Coordination Mechanism on Nitrogen

The comparison of the existing frameworks prepares the way for a fourth option. This originated during discussions in the margin of the United Nations Environment Assembly (UNEA-1, Nairobi, 2014). It was subsequently refined following discussions at the Environmental Policy Committee (EPOC) of the OECD (February 2015) and then extended on the basis of stakeholder feedback at the first INMS plenary meeting (INMS-1, Lisbon, April 2015). Here the approach is intermediate between the second option ('work with existing conventions') and the third option (a 'nitrogen convention'). Under this approach, the importance is recognized of the 'policy arena for nitrogen', which links each of the main environmental and other international frameworks. Under this approach an intergovernmental mechanism would be established that would help coordinate the efforts on nitrogen across the international conventions. Such a coordination mechanism would provide a framework under the lead of governments that makes the links to ensure better-informed policy coordination minimizing overall environmental pollution, while maximizing the benefits of nitrogen use for food, energy and livelihoods.

As can be seen from Fig. 36.4, the Interconvention Nitrogen Coordination Mechanism would be served with scientific support from INMS, while establishing the connections with each of the other international frameworks. In this way, the nitrogen policy arena can be developed as an achievable goal. It meets both objectives of working with existing frameworks and addresses the present lack of policy coordination. Precedents also exist for such an Interconvention Nitrogen Coordination Mechanism. Examples include the 'Major Economies Forum on Energy and Climate' and the 'Global Landscapes Forum', which are activities to support progress

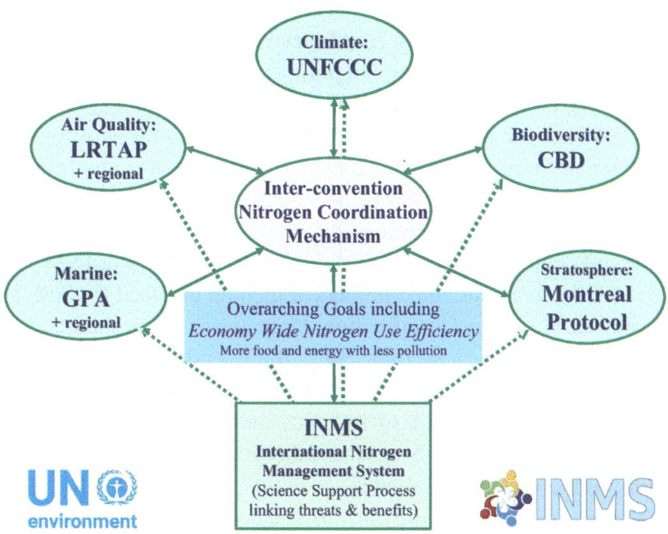

Fig. 36.4 Concept of the Policy Arena for Nitrogen showing how it could connect science support from INMS with the major effect-based multi-lateral environmental agreements. Currently, the international agreements operate largely in isolation from each other, failing to exploit synergies that operate across the global nitrogen cycle. This could be addressed by establishing an Interconvention Nitrogen Coordination Mechanism, e.g., under the auspices of the Committee of Permanent Representatives of the UN Environment Programme

in UN climate negotiations, the WHO Global Coordination Mechanism on Non-Communicable Diseases (GCM/NCD), and the UNEP Environmental Management Group that promotes coordination across the UN system in the field of environment. While the mode of operation in each case is different, the key challenges for the nitrogen policy arena must be to:

- build long-term understanding between issues among UN member states with global coverage, while recognizing regional differences,
- identify the synergies and trade-offs between issues across the nitrogen cycle,
- identify win-wins that could help overcome the barriers to achieving SDGs and other existing commitments,
- harmonize common approaches, building on the efforts of existing intergovernmental conventions and programmes,
- share experiences and success stories between countries and between intergovernmental conventions and programmes, and
- consider shared goals that could help mobilize change to meet multiple environmental, economic and social goals.

As regards a possible home for the nitrogen policy arena, this must be a question for further discussion by countries. Given the broad scope of their mandates both UNEA and OECD can serve as important forums to further refine the concept and build support with countries for the approach. At a regional scale, frameworks such as UNECE and other regional bodies could serve to support and further develop the approach in cooperation with the nitrogen policy area at a global scale. For example, the South Asia Cooperative Environment Programme (SACEP) covers all environmental issues relevant for nitrogen making it highly relevant as a regional policy home for nitrogen in South Asia. The exact form and design of the Interconvention Nitrogen Coordination Mechanism must be a matter for further discussion as INMS engages with the relevant policy frameworks.

Overall, the stakeholder discussions during plenary meetings of INMS have given strong support for this concept. Stakeholders agreed on the need for both INMS and the Interconvention Nitrogen Coordination Mechanism at the heart of Fig. 34.6. Starting with the bodies noted in Table 36.2, the overall message of stakeholders at INMS-1 (Lisbon, 2015) was one of high ambition to strengthen and extend the concept by increasing the number of linkages and goals. If the outcome of this consultation (as shown in Fig. 36.5) seems rather daunting, it clearly highlights the systemic nature of the nitrogen challenge.

Substantial progress in developing the nitrogen coordination mechanism has since been made. As part of stakeholder engagement at INMS-3 (Edinburgh, 2018) it was highlighted and agreed that the approach needs to emphasize the Interconvention nature of the mechanism. Most significantly UNEA-4 (Nairobi, March 2019) has adopted a resolution on Sustainable Nitrogen Management (UNEP/EA.4/L.16) (UNEP 2019), which encapsulates many of the ideas developed in this chapter. The background to the adoption of this resolution indicates the process of developing thinking and partnership building. During a meeting on Air Quality and Agriculture under the Working Group on Strategies and Review of LRTAP (Geneva, May 2017)

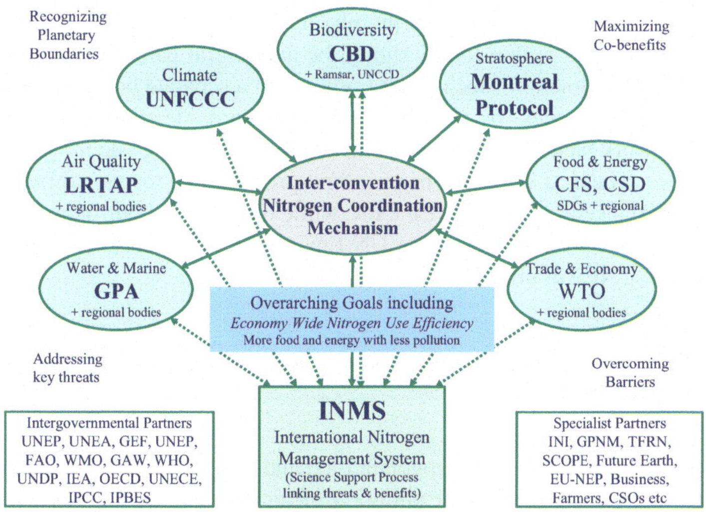

Fig. 36.5 Extended view of the Nitrogen Policy Arena showing how it could connect science support from INMS with major international and other agreements. This version is the result of stakeholder review at INMS-1 of the original version (Fig. 36.4) as developed during UNEA-1 and at OECD-EPOC

involvement of SACEP allowed agreement to be reached on holding a joint workshop between INMS and SACEP on prioritizing nitrogen threats in South Asia (Malé, Maldives, September 2017), in partnership with the South Asian Seas Programme. That meeting agreed that all nitrogen threats must be addressed, and accordingly drafted and agreed a Nitrogen Resolution to be submitted to UNEA.

The resolution was subsequently endorsed by the SACEP Governing Council, representing the Environment Ministers of all eight countries of South Asia (March 2018), agreeing that India would lead the submission on behalf of the region. Work through 2018 focused on reaching detailed agreement on the submission process with support of the Indian Minister of Environment, engagement with the UNEP Committee of Permanent Representatives (CPR, October 2018), including active support from Sri Lanka, and the drafting of a supporting concept note explaining the resolution, including the four options for improved coordination across the nitrogen cycle (September 2018). Following submission with the support of both Bangladesh and India, the resolution was formally presented by India for negotiation at the Open Ended Meeting of the CPR in March 2019, allowing its subsequent adoption at UNEA-4.

Mobilization of the draft nitrogen resolution was significantly helped by the preparation of a chapter for the UNEP 2018/2019 Frontiers report (Sutton et al. 2019), launched during the Open-Ended CPR with a Foreword by the Acting Executive Director of UNEP. Finally, holding of a high-level segment of INMS-4 (Nairobi, April 2019), including government members of the UNEP CPR and representatives of international conventions and programmes, has allowed rapid follow up of the

UNEA-4 Nitrogen Resolution further mobilizing government awareness. One of the emerging conclusions is that INMS should be an integral part of the Interconvention Nitrogen Coordination Mechanism (rather than a separate body), which would both improve effectiveness by allowing countries to provide guidance on priorities of the INMS work plan, and improve resilience by establishing one integrated body instead of two. The next steps will be to submit proposals on the way forward to the October (2019) meeting of the UNEP CPR, while the UNEA-4 Nitrogen Resolution requests the Executive Director of the UN Environment Programme to report back on progress to UNEA-6 in 2022.

36.6 Conclusions

Considering the crucial role of nitrogen in food and energy production and the associated multiple threats in view of pollution of air, freshwater and coastal waters, alteration of climate balance, stratospheric ozone loss, loss of biodiversity and soil quality, we argue that a more integrative or 'joined-up' approach to management of the global nitrogen cycle is needed. This is now being supported by the International Nitrogen Management System (INMS).

INMS is being supported by the Global Environment Facility and its project partners through the 'Towards INMS' project. At the same time, it is important to make the distinction between 'Towards INMS' is a funded GEF/UNEP project (2016–2022), and the INMS process, which needs to develop into the future.

The very establishment of INMS also encourages governments, business and civil society to consider whether the current set of policy frameworks provides the optimal way to address the nitrogen challenge. As shown here, the current nitrogen policy landscape is highly fragmented according to major threats.

In reviewing possible 'policy homes' to tackle the global nitrogen challenge, we have noted four options (Sect. 36.5, this chapter). The first is simply the *status quo* of fragmentation between intergovernmental programmes and policies. Secondly, the potential for individual existing policy frameworks to provide a lead policy home for nitrogen was considered. These include the Global Program of Action for the Protection of the Marine Environment from Land-based Activities (GPA), the UNECE Convention on Long-range Transboundary Air Pollution (LRTAP), the UN Framework Convention on Climate Change, the UN Convention on Biological Diversity (CBD) and the Vienna Convention on Protection of the Ozone Layer. In each case, institutional limitations linked to the particular mandates of these frameworks restrict their scope. The result is that there is no current international policy process that is ideally suited to develop a joined-up response to the global nitrogen challenge.

The third option is the establishment of a focused 'Nitrogen Convention'. While such an approach may be attractive to build coherency in policies relevant to the nitrogen cycle, this was not attractive to policy makers, who have instead emphasized working with existing conventions and frameworks (Option 2).

Considering these limitations, we propose a fourth option, intermediate between Options 2 and 3. This envisages developing gravity around the nitrogen policy area, leading to the establishment of an Interconvention Nitrogen Coordination Mechanism. Under this approach each of the key international frameworks (GPA, LRTAP, UNFCCC, CBD, Vienna Convention etc.), UN programmes and organizations (UNEP, UN Habitat, UNDP, FAO, WHO, WMO etc.), and stakeholders come together under the lead of governments to develop shared understanding and common goals.

One of the purposes of this chapter has been to stimulate governments and other actors to consider which of these directions they wish to take. Here we emphasize that nitrogen is not 'just another problem'. Rather, nitrogen is part of the solution, as it links together many existing challenges. In this way a joined-up approach to nitrogen offers many win-wins with economic opportunities to help overcome the barriers to change.

Estimates from the European Nitrogen Assessment (ENA) (Sutton et al. 2011a) illustrate the economic argument. Nitrogen pollution has societal costs to health, ecosystems and climate estimated at 70–320 billion Euro per year, with these values later updated to 75–485 billion Euro by Van Grinsven et al. (2013). Even if one is sceptical about the 'willingness-to-pay' approach underlying these numbers, the cash value of lost nitrogen illustrates the opportunity even more clearly. Based on the ENA, total EU emissions of nitrogen are around 22 M tonnes per year, with agriculture contributing a 78% share. Considering a fertilizer price of 0.80 Euro per kg N, this amounts to a lost fertilizer value from agriculture worth 14 billion Euro per year. The cost of the EU Common Agricultural Policy (CAP) is 57 billion Euro per year, which means that agricultural nitrogen emissions represent a cash loss worth 25% of the EU CAP budget. Such a calculation highlights how reducing N_r losses can become a key driver for developing the circular economy.

Estimates from 'Our Nutrient World', (Sutton et al. 2013) allow equally compelling numbers to be estimated at the global scale. On this basis a global goal to 'halve nitrogen waste' (reducing all N losses) would offer a potential cash resource-saving worth $100 billion USD annually (excluding the societal costs).

From a science and practice perspective, INMS is now addressing such issues by conducting cost-benefit and scenario analyses of better N management. These will inform stakeholders of the economic implications of the societal choices of nitrogen management.

While the exact form that the nitrogen policy arena should take is still open, experience from existing science engagement with policy leads to several conclusions concerning the nature of the evidence needed to support nitrogen policy development:

(1) The more specific and focused the agreement that policy makers propose to establish, the more specific and robust the science evidence needs to be in order to support that agreement.
(2) A broad combination of evidence is needed, including information on agreed indicators, scenarios and methods to achieve the desired outcomes (technologies, practices etc.). Moreover, assessment of the costs and impacts of action

on other policy areas is required, including on other environmental, economic or social aspects, allowing cost-benefit analysis and stronger engagement with the general public.

(3) Governments are likely to be interested in science to assist with policy setting in relation to, e.g., fiscal support measures, subsidies, extension and support services, investments in pollution control. Science outcomes and options need to be framed with these in mind.

(4) Long-term policy processes with strong intersessional activity appear to provide the foundation for the most robust, specific and ambitious agreements. One of the reasons for this is that sustained science input allows the parties of a proposed agreement to draw upon a robust long-term body of science. This can then build confidence in the science evidence, while enabling policy makers to request tasks of the science community to address their concerns. Together with an improved technical underpinning of the possible practices, it gives the countries confidence to know that their agreement is both achievable and that the benefits outweigh the costs.

(5) The evidence needed by policy processes varies between simple to highly complex approaches. On the one hand, a simple analysis can have great power in policy context (e.g., planetary or regional boundaries), while conversely, where there are objections, there may be calls for more and more detail. This reflects the interface between political negotiation and scientific evidence. It also emphasizes how the science must go beyond technical approaches to understand the opportunities and the barriers-to-change, based on the diversity of stakeholders involved and the operational realities.

(6) Global policy frameworks need evidence of varying detail, especially to allow data-poor areas of the world to engage fully in the process. This calls for the science community to deliver a range of approaches to satisfy all needs, from those countries and regions where only basic evidence is possible (implying the requirement for simple indicators and results from global models etc.) to those developed regions where there is the call for more-sophisticated approaches; consequently, regional cooperation and capacity building on a case by case basis will be crucial to achieve better nitrogen management at the global scale.

The list illustrates the opportunity for INMS to engage with governments and other actors in developing a more effective interaction between the worlds of science, policy and improvement of nitrogen management practices, from local and national to regional and global scales.

Finally, in responding to this developing agenda, we envisage that INMS should engage with policy makers over the next five to ten years at three scales:

(a) Strengthening science support to individual multilateral agreements, according to their topic focus (e.g., GPA, CBD, UNECE/LRTAP, UNFCCC, Vienna Convention, FAO, WHO etc.).

(b) Continuing to work with relevant global and regional multi-stakeholder partnerships to build deeper understanding of the cross-cutting issues (e.g., GPNM,

CCAC, Future Earth, International Long-term Ecological Research), while working with countries keen to champion the nitrogen challenge.

(c) Engaging in discussions towards establishment of an interconvention nitrogen coordination mechanism, cooperating with overarching frameworks to stimulate thinking by governments and other actors, especially within the frame of UNEA.

INMS has made substantial progress by engaging with the UN Environment Programme 'Committee of Permanent Representatives' (CPR) and the South Asian Cooperative Environment Programme (SACEP), under the leadership of country champions from South Asia. This has allowed UNEA-4 to adopt a first ever resolution on Sustainable Nitrogen Management (March 2019, UNEP/EA.4/L.19) (UNEP 2019), which is now pointing the way to develop the next steps.

Acknowledgements This chapter is a contribution from the International Nitrogen Initiative (INI), a joint activity of SCOPE and Future Earth. The authors gratefully acknowledge support from the Global Environment Facility (GEF), through the UN Environment Programme (UNEP), in funding 'Towards INMS', and from the UK Natural Environment Research Council (NERC), International Opportunities Fund for its funding of the 'INMS Pump Priming' project. We thank the UK Biological and Biotechnological Sciences Research Council (BBSRC) and Indian Department of Biotechnology (DBT) for funding the NEWS India-UK virtual joint centre on agricultural nitrogen, the UK Research and Innovation (UKRI) for support of the Global Challenges Research Fund (GCRF) South Asian Nitrogen Hub, and the UK Centre for Ecology & Hydrology for support through its SUNRISE and wider National Capability programs. We express our thanks to the South Asia Cooperative Environment Programme (SACEP) and the governments of Afghanistan, Bangladesh, Bhutan, India, Maldives, Nepal, Pakistan and Sri Lanka for their support in developing and championing the resolution on Sustainable Nitrogen Management adopted by the UN Environment Assembly (UNEA) and in establishment of the South Asian Nitrogen Hub. We thank David Hooper and Jana Compton for their detailed reviews of this chapter.

References

Bleeker, A., Erisman, J. W., Howard, C., & Sutton, M. (2011). *Potential targets for nutrient management to be included in the Manilla declaration.* Informal Document in support of the 3rd Intergovernmental Review of the Global Program of Action for the Protection of the Marine Environment from Land-based Activities. Included as Annex I to UNEP/GPA/IGR3/INF/7 (see UNEP, 2011).

Bleeker, A., Sutton, M. A., Leach, A. Erisman, J. W., & Galloway J. (2012). *How to meet the CBD N target for reducing critical load exceedance.* Report to the Secretariat of the Convention on Biological Diversity. Netherlands: ECN.

Bleeker, A., Sutton, M., Winiwarter, W., & Leip A. (2013). *Economy-wide nitrogen balances and indicators: Concept and methodology.* ENV/EPOC/WPEI(2012)4/REV1. Paris: Organisation for Economic Co-operation and Development (OECD) (Working Party on Environmental Information).

Bull, K. R., & Sutton, M. A. (1998). Critical loads and the relevance of ammonia to an effects-based nitrogen protocol. *Atmospheric Environment, 32*(3), 565–572.

Bull, K. R., Hoft R., & Sutton, M. A. (2011). Coordinating European nitrogen policies between international conventions and intergovernmental organizations. In M. A. Sutton, C. M. Howard, J. W. Erisman, G. Billen, A. Bleeker, P. Grennfelt, H. van Grinsven, & B. Grizzetti (Eds.), *The*

European nitrogen assessment: Sources, effects and policy perspectives (Chapter 25, pp. 570–584). Cambridge University Press.

Butterbach-Bahl, K., Nemitz, E., Zaehle, S., Billen, B., Boeckx, P., Erisman, J. W., Garnier, J., Upstill-Goddard, R., Kreuzer, M., Oenema, O., Reis, S., Schaap, M., Simpson, D., de Vries, W., Winiwarter, W., & Sutton M. A. (2011). Effect of reactive nitrogen on the European greenhouse balance. In M. A. Sutton, C. M. Howard, J. W. Erisman, G. Billen, A. Bleeker, P. Grennfelt, H. van Grinsven, & B. Grizzetti (Eds.), *The European nitrogen assessment: Sources, effects and policy perspectives* (Chapter 19, pp. 434–462). Cambridge University Press.

CBD. (2016). *Aichi indicators of the UN convention on biological diversity*. https://www.cbd.int/sp/targets. Accessed December 2016.

CCAC. (2016). *Climate and clean air coalition*. http://www.ccacoalition.org/en/content/about-us. Accessed December 2016.

CEH. (2018). *INI commits to support a global goal to halve nitrogen waste by 2030. Prof Mark Sutton, CEH, announces commitment at Our Ocean Conference*. UK: Centre for Ecology & Hydrology. https://www.ceh.ac.uk/news-and-media/news/ini-commits-support-global-goal-halve-nitrogen-waste-2030. Accessed December 2018.

Clarisse, L., Clerbaux, C., Dentener, F., Hurtmans, D., & Coheur, P. F. (2009). Global ammonia distribution derived from infrared satellite observations. *Nature Geoscience, 2*, 479–483.

Cochlan, W. P., Herndon, J., & Kudela, R. M. (2008). Inorganic and organic nitrogen uptake by the toxigenic diatom *Pseudo-nitzschia australis* (Bacillariophyceae). *Harmful Algae, 8*(1), 111–118.

Cowling, E. B., Erisman, J. W., Smeulders, S. M., Holman, S. C., & Nicholson, B. M. (1998). Optimizing air quality management in Europe and North America: Justification for integrated management of both oxidized and reduced forms of nitrogen. *Environmental Pollution*, 102(1). *Supplement, 1*, 599–608.

Erisman, J. W., Spranger, T., Sutton, M. A., Akselsson, C., Amin-Hanjani, S., Anderson, H. V., ApSimon, H., Fagerli, H., Harmens, H., Havlikova, M., Hetteling, J. P., Hicks, K., Horvath, L., Hutchings, N., Maasikimets, M., Maione, M., Reis, R., & Stenby, C. (2007) Working group 5: Nitrogen—integrated environmental policies. *Proceedings of the UNECE/EC Saltsjobaden 3 Workshop: Air pollution and its relationship to climate change and sustainable development. Linking immediate needs with long term challenges*. Available online at: http://asta.ivl.se/Workshops/Saltsjobaden3/Conclusions/WG5.pdf. Accessed December 2016.

Erisman, J. W., Sutton, M. A., Galloway, J. N., Klimont, Z., & Winiwarter, W. (2008). How a century of ammonia synthesis changed the world. *Nature Geoscience, 1*, 636–639.

Fowler, D., Pyle, J. A., Raven, J. A., & Sutton, M. A. (2013). *The global nitrogen cycle of the twenty-first century*. Special Issue. Philosophical Transactions of the Royal Society, Series B, 368 (1621).

Galloway, J. N., Aber, J. D., Erisman, J. W., Seitzinger, S. P., Howarth, R. W., Cowling, E. B., et al. (2003). The nitrogen cascade. *BioScience, 53*(4), 341–356.

Galloway, J. N., Townsend, A. R., Erisman, J. W., Bekunda, M., Cai, Z., Freney, J. R., et al. (2008). Transformation of the nitrogen cycle: Recent trends, questions and potential solutions. *Science, 320*(5878), 889–892.

Gareau, B. (2010). A critical review of the successful CFC phase-out versus the delayed methyl bromide phase-out in the Montreal Protocol. *International Environmental Agreements: Politics, Law and Economics, 10*, 209–231.

Herridge, D. F., Peoples, M. B., & Boddey, R. M. (2008). Global inputs of biological nitrogen fixation in agricultural systems. *Plant and Soil, 311*, 1–18.

INMS. (2016). *Towards the establishment of an International Nitrogen Management System*. http://www.inms.international. Accessed December 2016.

Jordan, A., & Lenschow, A. (2010). Environmental policy integration: a state of the art review. *Environmental Policy and Governance, 20*(3), 147–158.

Kanter, D., Mauzerall, D. L., Ravishankara, A. R., Daniel, J. S., Portmann, R. W., Grabiel, P. M., et al. (2013). A post-Kyoto partner: Considering the stratospheric ozone regime as a tool to manage nitrous oxide. *Proceedings of the National Academy of Sciences of the United States of America (PNAS), 110*(12), 4451–4457.

Keeler, B. L., Gourevitch, J. D., Polasky, S., Isbell, F., Tessum, C. W., Hill, J. D., et al. (2016). The social costs of nitrogen. *Science Advances, 2*(10), e1600219.

Lassaletta, L., Billen, G., Grizzetti, B., Anglade, J., & Garnier, J. (2014). 50 year trends in nitrogen use efficiency of world cropping systems: the relationship between yield and nitrogen input to cropland. *Environmental Research Letters, 9*(10), 105011. https://doi.org/10.1088/1748-9326/9/10/105011.

Maas, R., & Grennfelt, P. (Eds.). (2016). *Towards Cleaner Air. Scientific Assessment Report 2016*. Oslo: EMEP Steering Body and Working Group on Effects of the Convention on Long-Range Transboundary Air Pollution. Available at: https://www.unece.org/fileadmin/DAM/env/lrtap/ExecutiveBody/35th_session/CLRTAP_Scientific_Assessment_Report_-_Final_20-5-2016.pdf. Accessed December 2016.

Nadykto, A. B., Fangqun, Y., Jakovleva, M. V., Herb, J., & Xu, Y. (2011). Amines in the Earth's atmosphere: A density functional theory study of the thermochemistry of pre-nucleation clusters. *Entropy, 13*(2), 554–569.

NitrogenScientists. (2011) *European Nitrogen Assessment*. Video on YouTube. https://www.youtube.com/watch?v=uuwN6qxM7BU. Accessed December 2016.

Norton, R., Davidson, E., & Roberts, T. (2015). *Position paper—Nitrogen use efficiency and nutrient performance indicators*. Technical Paper 01/2015. Global Partnership on Nutrient Management (GPNM). Available at: http://www.nutrientchallenge.org/sites/default/files/documents/files/NUE%20and%20nutrient%20performance%20indicators_GPNM2015_FINAL.pdf. Accessed March 2020.

OECD. (2001). *OECD National soil surface nitrogen balances. Explanatory notes*. Paris: Organisation for Economic Co-operation and Development (OECD). Available at: http://www.oecd.org/tad/sustainable-agriculture/1916652.pdf. Accessed December 2016.

OECD. (2018). *Human acceleration of the nitrogen cycle. Managing risks and uncertainty*. Paris: Organisation for Economic Co-operation and Development (OECD). Available at: http://www.oecd.org/environment/human-acceleration-of-the-nitrogen-cycle-9789264307438-en.htm. Accessed December 2018.

Oenema, O., Brentrup, F., Lammel, J., Bascou, P., Billen, G., Dobermann, A., Erisman, J. W., Garnett, T., Hammel, M., Haniotis, T., Hillier, J., Hoxha, A., Jensen, L. S., Oleszek, W., Pallière, C., Powlson, D., Quemada, M., Schulman, M., Sutton, M. A., Van Grinsven, H. J. M., & Winiwarter, W. (2015). *Nitrogen use efficiency (NUE)—An indicator for the utilization of nitrogen in food systems*. Available at: http://www.eunep.com/wp-content/uploads/2017/03/Report-NUE-Indicator-Nitrogen-Expert-Panel-18-12-2015.pdf. Accessed March 2020.

Powlson, D. S., Addiscott, T. M., Benjamin, N., Cassman, K. G., De Kok, T. M., Van Grinsven, H., L'hirondel, J. L., Avery, A. A., & Van Kessel, C. (2008). When does nitrate become a risk for humans? *Journal of Environmental Quality, 37*(2), 291–295.

Raun, W. R., & Johnson, G. V. (1999). Improving nitrogen use efficiency for cereal production. *Agronomy Journal, 91*, 357–363.

Ravishankara, A. R., Daniel, J. S., & Portmann, R. W. (2009). Nitrous oxide (N_2O): The dominant ozone-depleting substance emitted in the 21st Century. *Science, 326*(5949), 123–125.

Reis, S., Grennfelt, P., Klimont, Z., Amann, M., ApSimon, H., Hettelingh, J. P., et al. (2012). From acid rain to climate change. *Science, 338*(6111), 1153–1154.

Selbie, D. R., Lanigan, G. J., Laughlin, R. J., Di, H. J., Moir, J. L., Cameron, K. C., et al. (2015). Confirmation of co-denitrification in grazed grassland. *Nature Scientific Reports, 5*, 17361.

Shindell, D., Kuylenstierna, J. C. I., Faluvegi, G., Milly, G., Emberson, L., Hicks, K., et al. (2012). Simultaneously mitigating near-term climate change and improving human health and food security. *Science, 335*(6065), 183–189.

Sutton, M. A., Howard, C., Erisman J. W., Billen, G., Bleeker, A., Grennfelt, P., van Grinsven, H., & Grizzetti, B. (Eds.). (2011a). *The European Nitrogen Assessment: Sources, effects and policy perspectives*. Cambridge University Press.

Sutton, M. A., Oenema, O., Erisman, J. W., Leip, A., van Grinsven, H., & Winiwarter, W. (2011b). Too much of a good thing. *Nature, 472,* 159–161.

Sutton, M. A., Bleeker, A., Howard, C. M., Bekunda, M., Grizzetti, B., de Vries, W., van Grinsven, H. J. M., Abrol, Y. P., Adhya, T. K., Billen, G.,. Davidson, E. A, Datta, A., Diaz, R., Erisman, J. W., Liu, X. J., Oenema, O., Palm, C., Raghuram, N., Reis, S., Scholz, R.W., Sims, T., Westhoek, H., Zhang, F. S., with contributions from Ayyappan, S., Bouwman, A.F., Bustamante, M., Fowler, D., Galloway, J. N., Gavito, M. E., Garnier, J., Greenwood, S., Hellums, D. T., Holland, M., Hoysall, C., Jaramillo, V. J., Klimont, Z., Ometto, J. P., Pathak, H., Plocq-Fichelet, V., Powlson, D., Ramakrishna, K., Roy, A., Sanders, K., Sharma, C., Singh, B., Singh, U., Yan, X. Y., & Zhang Y. (2013) *Our Nutrient World: The challenge to produce more food and energy with less pollution.* Global Overview of Nutrient Management. Edinburgh: CEH on behalf of the GPNM and INI for the United Nations Environment Programme.

Sutton, M., Raghuram, N., Adhya, T. K., Baron, J., Cox, C., de Vries, W., Hicks, K., Howard, C., Ju, X., Kanter, D., Masso, C., Ometto, J. P., Ramachandran, R., van Grinsven, H., & Winiwarter, W. (2019). The nitrogen fix: From nitrogen cycle pollution to nitrogen circular economy. *Frontiers 2018/2019: Emerging Issues of Environmental Concern* (pp. 52–65). Nairobi: United Nations Environment Programme. Available at: www.unenvironment.org/resources/frontiers-201819-emerging-issues-environmental-concern. Accessed March 2019.

UN DSD. (2016). *Sustainable development knowledge platform.* United Nations Division for Sustainable Development. Available at: https://sustainabledevelopment.un.org/about. Accessed December 2016.

UN Statistical Commission. (2016). *Report of the inter-agency and expert group on sustainable development goal indicators* [including proposed indicators in Annex III and agreed indicators in Annex IV for the 2030 sustainable development agenda]. UN Economic and Social Council (19 February 2016) E/CN.3/2016/2/Rev.1. Available at: http://unstats.un.org/unsd/statcom/47th-session/documents/2016-2-IAEG-SDGs-Rev1-E.pdf. Accessed December 2016.

UNECE. (2007). *Decision 2007/1. Establishment of a task force on reactive nitrogen.* Executive Body of the Convention on Long-range Transboundary Air Pollution (ECE/EB.AIR/91/Add.1). Geneva, Switzerland: United Nations Economic Commission for Europe (UNECE). Available at: https://www.unece.org/fileadmin/DAM/env/documents/2007/eb/EB/EB%20Decisions/Decision%202007.1.pdf. Accessed December 2016.

UNECE. (2013a). *1999 Protocol to abate acidification, eutrophication and ground-level ozone to the Convention on Long-range Transboundary Air Pollution, as amended on 4 May 2012.* Executive Body for the Convention on Long-range Transboundary Air Pollution (ECE/EB.AIR/114, 6 May 2013). Geneva, Switzerland: United Nations Economic Commission for Europe (UNECE).

UNECE. (2013b). *Guidance document on national nitrogen budgets.* Executive Body for the Convention on Long-range Transboundary Air Pollution (ECE.EB/AIR/119). Geneva, Switzerland: United Nations Economic Commission for Europe (UNECE). Available at: http://www.unece.org/environmental-policy/conventions/envlrtapwelcome/guidance-documents-and-other-methodological-materials/gothenburg-protocol.html. Accessed March 2020.

UNECE. (2014). *Guidance document on preventing and abating ammonia emissions from agricultural sources.* Executive Body for the Convention on Long-range Transboundary Air Pollution (ECE.EB/AIR/120). Geneva, Switzerland: United Nations Economic Commission for Europe (UNECE).

UNECE. (2015). *Framework code for good agricultural practice for reducing ammonia emissions.* Geneva, Switzerland: United Nations Economic Commission for Europe (UNECE).

UNECE. (2016). *The Convention and its achievements. A common framework for transboundary cooperation on air pollution.* Geneva, Switzerland: United Nations Economic Commission for Europe (UNECE). Available at: https://www.unece.org/environmental-policy/conventions/envlrtapwelcome/the-air-convention-and-its-protocols/the-convention-and-its-achievements.html. Accessed December 2016.

UNEP. (2010). *Building the foundations for sustainable nutrient management*. A publication of the Global Partnership on Nutrient Management. Nairobi: United Nations Environment Programme (UNEP). Available at: http://www.unep.org/pdf/Building_the_foundations-2.pdf. Accessed December 2016.

UNEP. (2011a) *Addressing the nutrient challenge: Where we are, what we need to know, and what we need to do?* (UNEP/GPA/IGR3/INF/7, Dec 2011). Third Intergovernmental Review Meeting on the Implementation of the Global Programme of Action for the Protection of the Marine Environment from Land-based Activities. Third Session. Nairobi: United Nations Environment Programme (UNEP). Available at: http://www.nutrientchallenge.org/sites/default/files/documents/files/Addressing%20Nutrient%20Challenge%20-%20UNEP-GPA-IGR3-INF.%20doc7.pdf. Accessed December 2016.

UNEP. (2011b). *Integrated assessment of black carbon and tropospheric Ozone*. Nairobi: United Nations Environment Programme (UNEP) and World Meteorological Organization (WMO).

UNEP. (2012). *Manila Declaration on furthering the implementation of the GPA* (UNEP/GPA/IGR.3/CRP.1/Rev.1). Intergovernmental Review Meeting of the Global Programme of Action for the Protection of the Marine Environment from Land-based Activities. UN Environment Program. Third Session. Nairobi: United Nations Environment Programme (UNEP). Available at: http://www.unep.org/ecosystemmanagement/water/regionalseas40/Portals/50221/WBRS18_INF10_ManilaDeclaration.pdf. Accessed December 2016.

UNEP. (2013). *Drawing down N_2O to protect climate and the ozone layer*. A UNEP Synthesis Report. Nairobi: United Nations Environment Programme (UNEP).

UNEP. (2016). *Vienna Convention on the protection of the ozone layer*. Nairobi: United Nations Environment Programme (UNEP). Available at: http://ozone.unep.org/en/treaties-and-decisions/vienna-convention-protection-ozone-layer. Accessed December 2016.

UNEP. (2018). *Intergovernmental Review Meeting on the Implementation of the Global Programme of Action for the Protection of the Marine Environment from Land-based Activities. Fourth session, Bali, Indonesia, 31 October and 1 November 2018. Draft Bali Declaration* (UNEP/GPA/IGR.4/6). Nairobi: United Nations Environment Programme (UNEP). Available at: https://papersmart.unon.org/igr-meeting/sites/default/files/unep_gpa_igr.4_6_draft_bali_declaration.pdf. Accessed December 2018.

UNEP. (2019). *United Nations Environment Assembly of the United Nations Environment Programme, Fourth session, Nairobi, 11–15 March 2019. Resolution on Sustainable Nitrogen Management* (UNEP/EA.4/L.16). Available at: http://web.unep.org/environmentassembly/ministerial-declaration-resolutions-and-decisions-unea-4. Accessed March 2019.

UNFCCC. (2015). *Synthesis report on the aggregate effect of the intended nationally determined contributions*. Conference of Parties, 21st session, Paris (FCCC/CP/2015/7). United Nations Framework Convention on Climate Change (UNFCCC). Available at: http://unfccc.int/resource/docs/2015/cop21/eng/07.pdf and http://unfccc.int/focus/indc_portal/items/9240.php. Accessed December 2016.

Van Grinsven, H. J. M., Holland, M., Jacobsen, B. H., Klimont, Z., Sutton, M. A., & Willems, W. J. (2013). Costs and benefits of nitrogen for Europe and implications for mitigation. *Environmental Science and Technology, 47*(8), 3571–3579.

Van Grinsven, H. J. M., Erisman, J. W., de Vries, W., & Westhoek, H. (2015). Potential of extensification of European agriculture for a more sustainable food system, focusing on nitrogen. *Environmental Research Letters, 10*(2), 025002. https://doi.org/10.1088/1748-9326/10/2/025002.

Vanlauwe, B., & Giller, K. E. (2006). Popular myths around soil fertility management in sub-Saharan Africa. *Agriculture, Ecosystems & Environment, 116*(1–2), 34–46.

Westhoek, H., Lesschen, J. P., Rood, T., Wagner, S., De Marco, A., Murphy-Bockern, D., et al. (2014). Food choices, health and environment: Effects of cutting Europe's meat and dairy intake. *Global Environmental Change, 26,* 196–205.

Westhoek, H., Lesschen, J. P., Rood, T., Leip, A., Wagner, S., De Marco, A., Murphy-Bokern, D., Pallière C., Howard, C. M., Oenema, O., & Sutton M. A. (2015). *Nitrogen on the table: The influence of food choices on nitrogen emissions and the European environment.* (European Nitrogen Assessment special report on nitrogen and food). UK: Centre for Ecology & Hydrology.

Zhang, X., Davidson, E. A., Mauzerall, D. L., Searchinger, T. D., Dumas, P., & Shen, Y. (2015). Managing nitrogen for sustainable development. *Nature, 528,* 51–59.

Chapter 37
Pre-informed Consumers on a Pre-adjusted Menu Had Smaller Nitrogen Footprints During the N2013 Conference, Kampala, Than Those on a Conventional Menu

Trust Tumwesigye, Giregon Olupot, Patrick Musinguzi, Adrian Leip, Mateete Bekunda, and Mark A. Sutton

Abstract International conferences are hotspots of food wastage and release of reactive nitrogen (N_r) into the environment, but there is limited data about extent of food wastage and food product-specific Nitrogen (N) Footprints of consumers from such conferences. This study was aimed at evaluating the impact of pre-information and pre-adjusted menu on food-product specific N Footprints of the 6th International Nitrogen (N2013) conference held in Kampala, Uganda (average of 140 participants). For comparison, we also computed N Footprints for a baseline conference held at the same venue (average of 180 participants). At N2013, the delegates, hotel management and chefs had been pre-informed about a pre-adjusted menu designed to substitute half of animal-based sources of protein with plant sources (demitarian diet). Average meat consumption (excluding eggs) during the N2013 conference

T. Tumwesigye · G. Olupot (✉) · P. Musinguzi
Department of Agricultural Production, Makerere University, Kampala, Uganda
e-mail: giregono@gmail.com

T. Tumwesigye
e-mail: trusttumwesigye@gmail.com

P. Musinguzi
e-mail: musipato@yahoo.com

A. Leip
European Commission, Joint Research Centre (JRC), Directorate D—Sustainable Resources, Ispra, VA, Italy
e-mail: Adrian.LEIP@ec.europa.eu

M. Bekunda
International Institute of Tropical Agriculture (IITA), Arusha, Tanzania
e-mail: M.Bekunda@cgiar.org

M. A. Sutton
UK Centre for Ecology & Hydrology, Edinburgh Research Station, Bush Estate, Penicuik, Midlothian, EH26 0QB, UK
e-mail: ms@ceh.ac.uk

M. A. Sutton et al. (eds.), *Just Enough Nitrogen*,
https://doi.org/10.1007/978-3-030-58065-0_37

was 118 g capita^{-1} day^{-1} on dry weight basis, while milk consumption (excluding powdered milk) was 75 g capita^{-1} day^{-1} (fresh weight basis). These values were smaller than those of the baseline conference where meat consumption (excluding eggs) averaged 234 g capita^{-1} day^{-1} on dry weight basis and milk consumption (excluding powdered milk) averaged 159 g capita^{-1} day^{-1} (fresh weight basis). The reduction in meat consumption during the N2013 conference was compensated for by eating more fruits (102 g capita^{-1} day^{-1}) and vegetables (45 g capita^{-1} day^{-1}) than during the baseline conference (69 and 33 g capita^{-1} day^{-1}, respectively). Overall, the Nitrogen Footprint for the N2013 conference was 97 g N capita^{-1} day^{-1}, representing a reduction of 40% compared with the baseline conference of 160 g N capita^{-1} day^{-1}. The Nitrogen Footprint for the N2013 conference would have been even lower, had it not been for over-supply beyond demand that left a substantial amount of food wasted. We conclude that pre-information and a pre-adjusted menu with clear guidelines to actors in food procurement, preparation and consumption are critical to mitigating food wastage from international conferences. The experience demonstrates how a conference approach to cutting down N_r consumption simultaneously helps raise awareness, while allowing delegates to reduce their N footprints with environmental and health benefits.

Keywords Food choice · Nitrogen footprint · Demitarian · Conference planning · Environment · Health

37.1 Introduction

Reactive nitrogen (N_r) is an essential element for life needed for nearly all metabolic functions (proteins, enzymes, hormones, etc.) and its deficiency can severely hamper growth (Galloway et al. 2003; Sutton et al. 2011, 2013). Crop productivity on which humans depend (directly or indirectly) for food and fibre is most limited by N_r especially in low external input agro-ecosystems (Sanchez and Swaminathan 2005; Musinguzi et al. 2016). However, the 14-fold N_r increase from 15 Tg year^{-1} in 1860 (1 Tg = 10^{12} g) to 210 Tg year^{-1} in 2010 presents one of the most challenging human and environmental health hazards of our times (Galloway et al. 2003; Sutton et al. 2011; 2013), involving complex exposure pathways and processes (Wetzel 2001) and cascading negative impacts (Galloway et al. 2013; Sutton et al. 2013; Stevens et al. 2014). Reactive nitrogen is a major constituent of particulate matter and smog in the atmosphere associated with stratospheric ozone depletion and global warming (Stevens et al. 2014; Pierer et al. 2014). In terrestrial ecosystems, excessive N_r fuels loss of biodiversity, forest dieback and soil acidification (Stevens et al. 2014) as well as global warming (Springman et al. 2016). Livestock and poultry consumption is responsible for 14.5–18% of greenhouse gas emissions (Steinfeld et al. 2006; Cribb 2010; Stehfest et al. 2013). In aquatic ecosystems, N_r causes acidification of marine ecosystems and eutrophication of freshwater bodies (Xue and Landis 2010; Meier and Christen 2013; Stevens et al. 2014) with negative consequences on aquatic life and biodiversity (Tukker et al. 2006).

Over-consumption of N_r-rich foods especially of animal origin increases blood levels of low-density and very low-density lipoproteins, a predisposing factor to obesity, diabetes, hypertension, chronic heart disease, strokes and death (Willis 1994). Overweight and obesity killed 3.4 million people, and caused loss of 3.9% of years of life and 3.8% of disability-adjusted life-years globally in 2010 (Marie et al. 2014).

Furthermore, about one-third of the food produced annually is wasted (Higher Level Panel of Experts—HLPE 2014); with about 1.3 billion megagram (Mg) of food wasted in 2010 alone globally (FAO 2011b). In America and Europe, food loss and wastage peaks at 280–300 Mg per capita per year (Mg cap^{-1} $year^{-1}$) compared with 120–170 Mg cap^{-1} $year^{-1}$ in sub-Saharan Africa and South East Asia (HLPE 2014). At the consumption stage, food wastage and avoidable N_r losses into the environment manifest in the form of preparation of unnecessarily large quantities and consumption of N_r-rich foods way above normal body requirements (Stevens et al. 2014; Vanham et al. 2015; HLPE 2014).

Tackling the damaging effects of N_r on human and environmental health and the global economy requires reversing a global trend towards unsustainable animal-based diets (Cribb 2010; Springman et al. 2016), in line with Sustainable Development Goal (SDG) No.12: Achieving sustainable consumption and production patterns. Recent evidence indicates that a shift to more vegetarian (low-N_r) diets globally could reduce greenhouse gas (GHG) emissions by two-thirds (29–70% of the 2050 projections); mortality by 6–10% (contributing to a healthier population); with savings to the global economy of US$ 1 trillion–3 trillion (0.4–13% of global GDP) by 2050 (Springman et al. 2016). If China adopts the newly proposed guideline of consuming 14.6–27.4 kg meat cap^{-1} $year^{-1}$ down from the 93 kg meat cap^{-1} $year^{-1}$ projected by 2030 based on the current consumption pattern of 63 kg meat cap^{-1} $year^{-1}$, it will cut CO_2 emissions from the livestock industry down from 1.8 to 0.8 Tg $year^{-1}$.

However, there is limited awareness in the public of the impact of dietary decisions associated with releases of N_r on human and environmental health. Bridging this awareness gap through the engagement of key actors along the entire food value chain is critical to reversing the negative human and environmental health impacts of N_r. Scientific conferences which relate directly to management of N_r present a unique entry point for generating empirical evidence about how contributors to, perpetrators and victims of N_r pollution can be helped to make choices that mitigate avoidable food wastage and N_r losses into the environment.

The importance of modifying food intake at conferences was highlighted by the Barsac Declaration (Sutton et al. 2009), a commitment by over 200 scientists linked to the International Nitrogen Initiative (INI) to provide a 'demitarian' or half-meat option at their conferences. This approach was first tested at the 'Nitrogen Deposition, Critical Loads and Biodiversity Workshop', 16–18 November 2009 in Edinburgh (Sutton et al. 2014) and the 'NitroEurope Open Science Conference', 1–5 February 2010, Solothurn, Switzerland (Ambus et al. 2011). The dietary experiences from these conferences were not published in the scientific volumes resulting from these conferences, but can be summarized as follows. In the Edinburgh 2009 workshop, clear labelling of buffet material allowed delegates to self-select a demitarian option, but no change was made to the menu itself. It provided a talking point, but appeared not to alter overall intake substantially, while the same amount of food was prepared. At Solothurn, the menu was adjusted to reduce the meat serving, including at the conference dinner. At this event food was served in groups of 50, leading to some of those being served last to have a feeling of 'missing out', which appeared to be connected to a combination of both less meat, less overall food to normal and the 'pure-meat' nature of the dishes served.

Based on these experiences, a different approach was adopted at the 'Nitrogen and Climate Change' conference 11–15 April 2011 of around 350 delegates (Sutton et al. 2012), where a pre-arrangement was made with the catering team to prepare meals with half the amount of meat to normal. Feedback from the catering manager showed that they enjoyed the challenge, that they focused on preparing mixed meat and vegetable dishes that had a lower fraction of meat, that they used the money saved to diversify and improve the vegetable ingredients (improving taste) and that the chef had (privately) chosen to extend the approach, actually reducing intake to one-third (reducing from 180 g meat fresh weight (FW) per person per meal down to 60 g per person). The 'demitarian' approach provided a talking point, raising awareness of delegates to the discussion. A survey at the end of the conference showed that 92% of respondents (around 250) "did not miss out on the amount of meat served", while 4% "missed out" and the remaining 4% were "not sure" (Sutton and Howard 2011).

These examples provided a suitable starting point to trial dietary change in conferences linked to the INI network. However, they also pointed to further challenges. Firstly, work was needed to make a scientific evaluation of the outcomes of the changes. Secondly, each of the conferences noted above took place in Europe. Would it be acceptable to adopt such an approach in a future INI conference in sub-Saharan Africa, where the intake of many citizens is already below dietary guidelines?

Concerning tools for generating the necessary data, these already exist, although with some challenges. Xue and Landis (2010) established a Nitrogen (N) Footprint by systematically investigating the eutrophication potential of food consumption (considering both N_r and phosphorus), whereas Lassaletta et al. (2013) explored N_r use efficiencies of diets from a N_r budget approach at country-level scale, analyzing N_r flows. Leach et al. (2012) developed a coherent and systematic methodology for computing N Footprints for the USA that is being adapted to other countries worldwide, with slight modifications. However, the above studies are based on national-scale estimates of N Footprint (NFP). "Leip et al. (2013) assessed the NFP of a broad range of food products but for EU and at 'farm-gate'". Similarly, Pierer et al. (2014), Shibata et al. (2014) and Stevens et al. (2014) established Nitrogen Footprints for a broad range of products for Austria, the UK and Japan, respectively. However, country- or region-based N Footprints do not pinpoint hotspots of food wastage and N_r release into the environment such as international conferences. Even where N Footprints of a broad range of food products have been established, wheat, maize, rice and other cereals have been lumped together by taking their averages yet production practices, yields and consumption patterns differ considerably among these cereals.

Besides, none of these studies have considered other food types like *matooke* (cooking type of bananas), cassava and sweet potatoes which are Uganda's staple food crops, or considered local foods such as vegetables and fruits. In addition, all these studies have been conducted for the affluent and temperate north with high external-input agro-ecosystems antithetical of the low external-input agro-ecosystems that typify developing countries like Uganda.

To address these issues, we combined a decision to modify the dietary intake of the 6th International Nitrogen Conference (N2013) held in Kampala, Uganda in November 2013, with collection of information on the foods typical in this part of Sub-Saharan Africa. We established product-specific N Footprints including staple foods indigenous to Uganda like *matooke*, cassava, sweet potatoes and some vegetables on the basis of Nitrogen Loss Factors developed for Tanzania, a country with similar agricultural production systems to Uganda's by Hutton et al. (2017). In parallel, we recorded observations of routine practices with regard to food supplies, preparation, service, consumption, recycling and wastage at Speke Resort, Munyonyo Hotel (SRMH) in Kampala, which hosted the N2013 conference. This was the first time that the N2013 International Conference was held in Uganda and indeed, in Africa. In the N2013 conference, consumers, hotel management and chefs had been pre-informed about the health, economic and environmental benefits of a pre-adjusted menu with 50% of animal-based sources of protein being replaced by vegetable sources, thereby catering for a demitarian diet on average among the delegates. The N2013 conference thus allowed this study to evaluate the impact of pre-information and pre-adjusted menu on food-product specific N Footprints.

37.2 Materials and Methods

This study was conducted at Speke Resort, Munyonyo Hotel (SRMH), located 12 km to the southeast of Kampala, the capital and business city of Uganda on an area spanning approximately 20.41 ha and extending 400 m along the northern shores of Lake Victoria. With 59 presidential suites, 355 rooms, a 1000-seat ballroom, nine multifunctional meeting rooms each with a capacity of 10–300 people (http://www.spekeresort.com/), SRMH is a suitable hotel for international conferences, including the 2013 6th International Nitrogen Conference (N2013) that was held for the first time in Africa.

37.2.1 Involvement of Speke Resort, Munyonyo Hotel

Four visits were made to SRMH to seek permission to conduct this study, to make the hotel management aware about the purpose of the study, including its potential to cut down on the costs associated with the purchases of more expensive N_r-rich foods, and to enlist the support and cooperation of SRMH management and staff during data collection. In this study, avoidable food losses include food damage or spoilage from poor handling during purchase and/or preparation that renders food unsuitable for human consumption, and also preparation of food in excess of what could be eaten (HLPE 2014). 'Food recycled' refers to non-consumed food items served to conference participants that were either retained to be served later (re-use) or left-over food fed to livestock. 'Food not eaten' combines food items purchased that were unfit to be prepared for consumption because of spoilage and waste and food items served in excess of what could be eaten. Uneaten food was either recycled or destined for disposal or dumping.

37.2.2 Data Collection

A two-fold survey was conducted at Speke Resort to collect data on the number of meals served per day, menu (food items served) per meal, number of guests per meal per day, quantity of each food item purchased, quantity of each food item served consumed, recycled and not eaten per day. The first survey was conducted for the baseline conference whereas the second survey was for the 2013 International Nitrogen Conference (N2013). The survey for the baseline conference was conducted from 15th to 21st July 2013. It was organized by the ICSB and was conducted to gain insights into the N footprints of food items from routine practices at Speke Resort with regard to food purchases, preparation, service, consumption, recycling and wastage with which, the N Footprints of the N2013 conference would be compared. The chefs of Speke Resort Munyonyo and the N2013 conference delegates were sensitized about the human and environmental health impacts of N_r as well as the economic, human and environmental health benefits of shifting to low-N_r diets. The N2013 conference delegates comprising Ugandans and non-Ugandans were also encouraged to voluntarily participate in an N_r impact compensation scheme (Leip et al. 2014). This decision was made to be consistent with the theme for the N2013 conference: "*Just Enough N: Perspectives on how to get there for 'too much' and 'too little' Regions*" under four sub-themes: (i) Nitrogen in the context of food production; (ii) Nitrogen impacts on human and ecosystem health; (iii) Management approaches in too much and too little N regions, and (iv) Integrated assessment of N dynamics both spatially and temporally (e.g., Leip et al. 2014).

Data on daily total supply, service, consumption, recycling and wastage were collected for all meals for each of the 28 raw, largely unprocessed basic food items from both animal and plant origin for either of the baseline event or the N2013 conference. The 28 food items were: *matooke* (cooking type of bananas, a major staple food in Uganda), cassava, sweet potatoes, potatoes, rice, corn, wheat, cereal (a cocktail of processed cereals), peas (mean of field pea, green gram, cowpea, pigeon pea), beans (average of bush bean and climbing bean), groundnuts, banana (average of exotic and indigenous yellow banana served as dessert), onions, fruits, vegetable (average of indigenous and exotic vegetables), oil, fish, beef, pork, goat meat, mutton, chicken, eggs, coffee, tea, milk (fresh), powder milk and sugar. Data

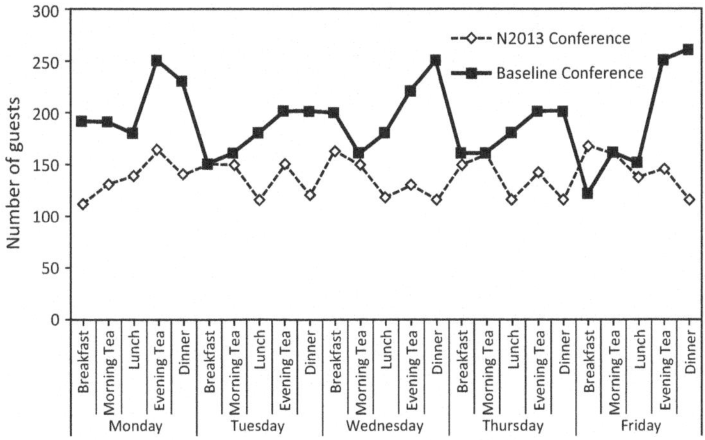

Fig. 37.1 Daily variation in the number of delegates for meals served for the baseline conference and the 6th International Nitrogen conference (N2013) at Speke Resort Munyonyo Hotel, Uganda in 2013. Although the baseline conference continued for seven days, results are here shown for five days to illustrate the comparison with N2013

for the baseline event were collected for seven days whereas data collection for the N2013 event lasted five days (during the duration of the conference). Data were also concomitantly collected on number of consumers per meal per day (Fig. 37.1). Total daily per capita supply, service, consumption, recycling and wastage for each of the 28 food items was computed by dividing the total quantity of each food item under its respective fate: supply/purchase, served, consumed, recycled or wasted, by the total number of consumers per meal. To compute per capita fate of food items per day, all quantities of each food item were summed up for all meals per day. To trace the origin (farming system) of these food items, the two main markets where Speke Resort purchases its food (Nakasero Market in Kampala City and Kabira Country Club in Ntinda, one of the main suburbs of Kampala City), were visited and necessary information collected. The majority of the food items were sourced from upcountry (in different agro-ecological zones), but some processed foods had been imported.

37.2.3 Determination of Dry Weight of Each Fresh Food Item

Individual fresh weight (*fwi*) of each of the 28 food items (*FI*) was taken and four representative replicate subsamples of each food item (*FI*) immediately collected, weighed and sealed in pre-weighed and pre-labeled sampling bags. The subsamples were oven-dried at 70 °C to constant weight (48 h for grains, cereals, vegetables; and 72 h for animal products). The dry weight of individual food item (*FIdwi*) was computed from its fresh weight (*FIfwi*) from Eq. 37.1:

$$FIdwi = FIfwi * \left(\frac{MFI\,sub\,sample\,fwi - MFI\,sub\,sample\,dwi}{MFI\,sub\,sample\,dwi} \right) \quad (37.1)$$

where: *MFIsubsamplefwi* is the mean fresh weight of the four subsamples per *FI* and *MFIsubsampledwi* is the mean dry weight of the four subsamples per *FI*.

37.2.4 Quantification of N_r in Individual Food Items for Each Fate of the Food Item

Because samples of the individual food items collected were not analyzed for N_r contents ($FIN_{r\,conti}$), we used published literature (Table 37.1) to convert *FIdwi* into its respective N_r stock ($FIN_{r\,stock}$) for all the food items using Eq. 37.2:

$$FIN_{r\,stock} = FI_{dwiI} * FIN_{r\,conti} \quad (37.2)$$

37.2.5 Computation of Partial Nitrogen Footprints of Food Items

Nitrogen Footprints were calculated according to Leip et al. (2014). Total N Footprint associated with food consumption was computed by summing up all the food item-specific total Footprints for the 28 food items. For technical difficulties and also for the purpose of this study, we did not consider N Footprints associated with energy consumption along the food chain from production.

37.3 Results and Discussion

37.3.1 Quantities of the Major Food Items Purchased for the Baseline Conference and the N2013 Conference

Milk, wheat, sugar, rice, *matooke* and fruits were the six food items purchased in largest quantities in that order for both conferences. Generally, higher quantities of food items were purchased for the baseline conference than for the N2013 conference, except for fruits, eggs, groundnuts, cereal, beans and vegetables (Fig. 37.2). For example, an equivalent of 176 g milk cap^{-1} day^{-1} were purchased for the baseline conference, 61 g milk cap^{-1} day^{-1} in excess of the milk for the N2013 conference

Table 37.1 Reactive nitrogen contents (kg N_r kg^{-1} of food item) used to convert dry weights of the food items for the baseline and N2013 conferences into N_r stocks at Speke Resort Munyonyo Hotel, Uganda in 2013

Food item	N_r content (kg^{-1})	Data source
Matooke	0.00112	a
Cassava	0.00144	a
Sweet potatoes	0.00112	a
Potatoes	0.00256	a
Rice	0.012	a
Corn	0.0152	a
Wheat	0.01952	a
Cereals	0.0128	a
Peas	0.00336	a
Bean	0.0048	a
Groundnuts	0.04112	a
Bananas	0.0032	a
Onions	0.00272	a
Fruits	0.00144	a
Vegetables	0.00224	a
Oil	0.00528	a
Fish	0.02581	b
Beef	0.0296	a
Pork	0.02144	a
Goat meat	0.0224	a
Mutton	0.0224	a
Chicken	0.01968	a
Eggs	0.01712	a
Milk	0.00528	a
Milk powder	0.04208	a
Coffee	0.0128	a
Tea	0.016	a
Sugar	0	a
Millet	0.01552	a

Data Sources: a = FAO (2011a) as compiled by Lassaletta et al. 2014, supplementary information; b = Ramseyer 2002

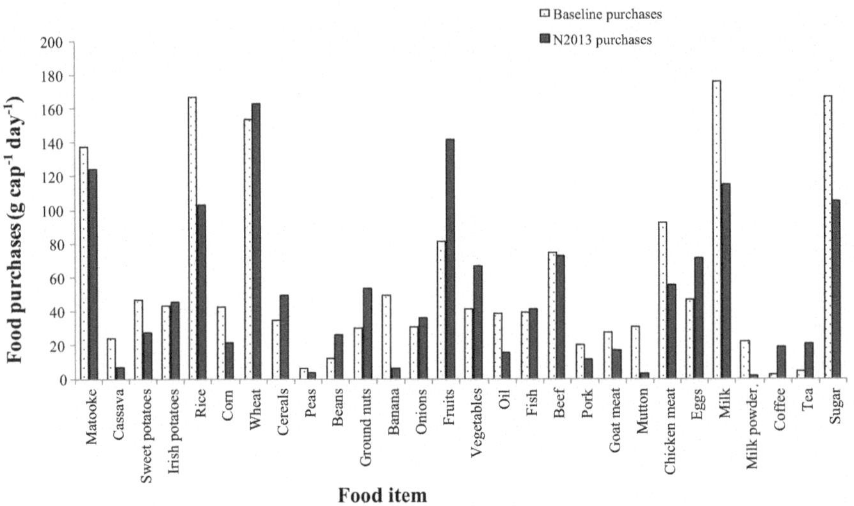

Fig. 37.2 Purchases for each of the food items for the baseline conference and the 6th International Nitrogen conference (N2013) that were hosted at Speke Resort Munyonyo Hotel, Uganda in 2013. All items are expressed as g dry matter cap^{-1} day^{-1}, with the exception of milk which is expressed g fresh weight cap^{-1} day^{-1} (equivalent to ml cap^{-1} day^{-1})

(115 g milk cap^{-1} day^{-1}) on fresh weight (FW) basis. Similarly, the quantities of sugar and rice (either at 166 g cap^{-1} day^{-1}) for the baseline conference were 62 g cap^{-1} day^{-1} in excess of the supplies for the N2013 conference (104 g cap^{-1} day^{-1} year^{-1}) on dry matter (DM) basis. In contrast, 147 g (fruits + banana) cap^{-1} day^{-1} were purchased for the N2013 conference in excess of 17 g (fruits + banana) cap^{-1} day^{-1} for the baseline conference. Peas were the food item purchased in the smallest quantity for both the baseline conference (6 g cap^{-1} day^{-1}) and N2013 conference (4 g cap^{-1} day^{-1}). *Matooke* is Uganda's staple food crop and purchase was 137 and 124 g cap^{-1} day^{-1} for baseline conference and N2013, respectively. Wheat purchases were comparable for the baseline conference (154 g cap^{-1} day^{-1}) and the N2013 conference (162 g cap^{-1} day^{-1}). The generally low purchases for groundnuts, beans and vegetables, could imply that these alternatives to animal-based sources of proteins are not yet well appreciated. Reversing this trend is not a matter of choice but a pre-condition to reduction of N Footprints. The beneficial impacts of this reversal on economic, human and environmental health are widely published (Stevens et al. 2014; Springman et al. 2016).

37.3.2 Consumption Patterns for the Different Food Items

As expected, the food items purchased in largest quantities were also the most consumed (Fig. 37.3), implying that consumer preferences inform how much of

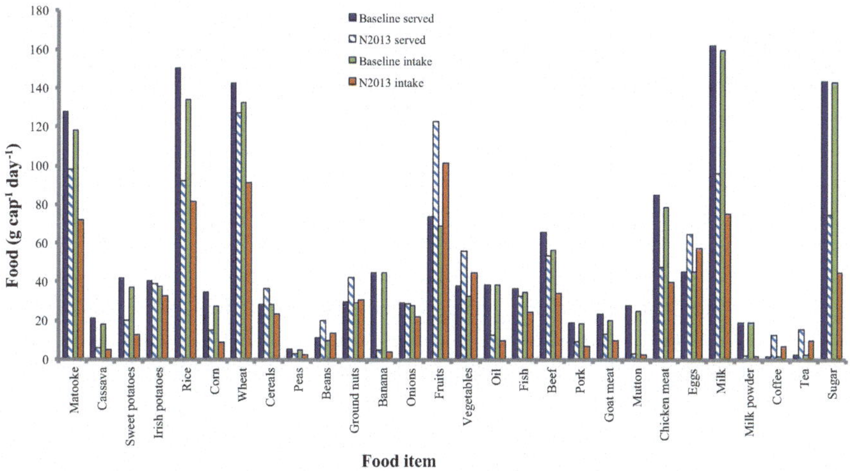

Fig. 37.3 Food items served and consumed for the baseline conference and the 6th International Nitrogen (N2013) conference, that were hosted at Speke Resort Munyonyo Hotel, Uganda in November 2013. All items are expressed as kg dry matter cap^{-1} day^{-1}, with the exception of milk which is expressed g fresh weight cap−1 day−1 (equivalent to mg cap^{-1} day^{-1})

which food item is purchased. The quantity of each food item consumed was generally higher for the baseline conference than it was for the N2013 conference with the exception of groundnuts which was at about 31 g cap^{-1} day^{-1} for both of the conferences. Higher quantities of beans, fruits, vegetables and eggs were consumed during the N2013 conference than during the baseline conference. The consumption of milk during the baseline conference was more than twice (159 g milk cap^{-1} day^{-1}) that of the N2013 conference (75 g milk cap^{-1} day^{-1}, FW basis), with over three times as much sugar consumed during the baseline conference (143 g sugar cap^{-1} day^{-1}) than it was during the N2013 conference (45 g sugar cap^{-1} day^{-1}). Particularly worth noting was the marked reduction in the consumption of high N_r foods mostly the meats during the N2013 conference relative to the baseline conference. Meat, fish and eggs consumption totaled 279 g cap^{-1} day^{-1} for the baseline conference but only 176 g cap^{-1} day^{-1} for the N2013 conference.

The marked reduction in meat consumption during the N2013 conference was compensated for by eating more fruits (102 g cap^{-1} year^{-1}, excluding banana) and vegetables (45 g cap^{-1} day^{-1}, excluding onions) than during the baseline conference (69 and 33 g cap^{-1} day^{-1}) for fruits and vegetables, respectively low N_r foods. It was also interesting to note that within the high-N_r (animal-based) foods, there was a shift towards consumption of more eggs during the N2013 conference, with nearly twice as many eggs (32%) consumed compared to 17% during the baseline conference (Fig. 37.4). Per capita tea and coffee consumption was also higher during N2013 than during the baseline conference.

Speke Resort Munyonyo Hotel hosts international conferences of this nature almost on daily basis; with even more guests than those hosted during the two conferences considered in this study. The reduction in the consumption of high-N_r foods especially the meats and milk during the N2013 conference relative to the baseline conference in preference to more fruits, legumes and vegetables lends support to the view by Pierer et al. (2014) that people can make healthy food choices if they have correctly packaged information and sustainable alternatives. However, the rate of meat consumption for either the baseline conference (over 282 g meat cap^{-1} day^{-1}) or the N2013 conference (equivalent to 178 g meat cap^{-1} day^{-1}) is still way above the 14.6–27.4 kg meat cap^{-1} $year^{-1}$ recently proposed guidelines for China to tackle surging obesity and global warming (Springman et al. 2016). Consumers at international conferences and hosts of such conferences need to be mindful of the fact that consumption of such quantities of meat especially in Uganda, a country blighted by severe soil degradation blamed for the at least 40% of children under the age of five, pregnant and breast-feeding mothers who are the hardest hit by malnutrition (Sanchez and Swaminathan 2005), is not consistent with SDG No. 12: Achieving sustainable consumption and production patterns.

37.3.3 Other Pathways for the Food Items

There are several food loss pathways during such conferences, including leftover food (i.e., balance of the food served that was not eaten), food wastage (i.e., a combination of food purchased that could not be prepared and served, but destined for disposal as waste), as well as recycled food. For example, Fig. 37.3 shows how the difference between food purchase and intake (i.e., food wastage) was generally higher for the N2013 conference than for the baseline. The same can also be seen for leftover food and recycled food (Table 37.2). Among animal-based food, leftover items and wastage were particularly high for beef, chicken and milk, with more food wasted

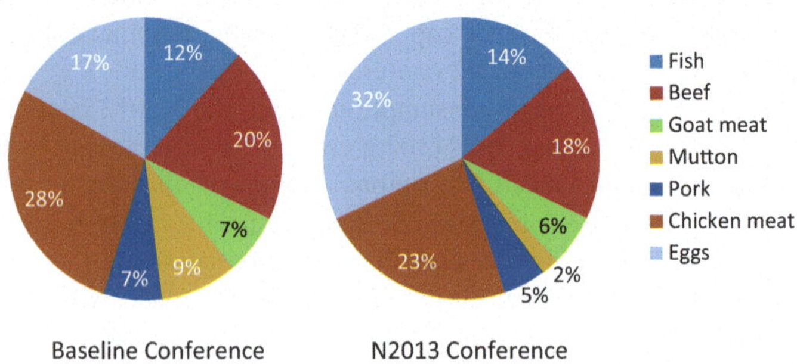

Fig. 37.4 Comparison of meat consumption patterns for the baseline (left) and the 6th International Nitrogen (N2013) conferences (right) that were hosted at Speke Resort Munyonyo Hotel, Uganda in November 2013

Table 37.2 Fate of each of the 28 food items (quantified in g dry matter (DM) cap^{-1} day^{-1}) purchased for the baseline conference and the 6th International nitrogen conference (N2013) used to calculate Nitrogen Footprints at Speke Resort Munyonyo Hotel, Uganda in 2013

	Food supply (g cap^{-1} day^{-1})		Food left over (g cap^{-1} day^{-1})		Food wasted (g cap^{-1} day^{-1})		Food recycled (%)	
Food item	Baseline	N2013	Baseline	N2013	Baseline	N2013	Baseline	N2013
Matooke	137	124	10	26	19	52	0	26
Cassava	24	7	3	1	6	2	0	13
Sweet potatoes	47	27	5	7	9	15	0	32
Irish potatoes	44	45	3	6	6	13	0	12
Rice	166	103	16	11	33	22	0	12
Corn	42	22	8	6	15	13	0	8
Wheat	154	162	10	36	21	71	0	18
Cereals	35	49	0	13	6	26	21	32
Peas	6	4	0	1	1	1	0	12
Beans	12	26	1	6	2	13	0	25
Ground nuts	30	53	0	11	0	22	0	25
Banana	49	6	0	1	4	2	10	10
Onions	31	36	1	7	3	14	0	10
Fruits	81	141	5	21	12	40	2	13
Vegetables	42	67	5	11	9	22	0	9
Oil	38	15	0	3	0	5	0	20
Fish	39	42	2	8	4	17	2	23
Beef	75	72	9	19	18	38	0	36
Pork	20	11	1	2	1	4	0	19
Goat meat	27	17	4	4	7	7	0	27
Mutton	30	3	3	0	6	1	0	10
Chicken meat	92	55	6	8	13	15	1	16
Eggs	47	71	0	7	1	14	2	10
*Milk	176	115	3	20	16	40	12	20
Milk powder	22	2	0	0	3	0	16	0
Coffee	2	19	0	6	1	12	44	45
Tea	4	21	0	5	1	11	36	36
Sugar	166	105	0	30	23	60	16	37
Total	1639	1423	96	275	243	551	162	555

* Milk quantities are fresh weight and not in dry matter

during the N2013 conference than during the baseline, which was paradoxical given that there were more purchases of these food items for the baseline conference than for the N2013 conference. For example, leftover and waste for beef during the N2013 conference were 18 and 38 g beef cap^{-1} day^{-1}, respectively which were at least twice the left-over and waste for beef during the baseline conference. We attributed this to the impact of sensitization of the N2013 participants about the negative impact of over-consumption of high N_r foods, which was not done for the baseline conference delegates, suggesting that the purchased amounts could have been reduced even further. Fortunately, food recycling was more associated with the N2013 conference than it was with the baseline conference.

In developing countries like Uganda, the bulk of food loss and wastage occurs at the production stage, mainly because of pest damage and poor post-harvest handling whereas in industrialized countries, food loss and wastage are mainly at the level of consumption (HLPE 2014). However, based on this study, food loss and wastage at the consumption stage seem to be as important in developing countries as it is in industrialized ones. For example, out of a total of 1423 g food supplied cap^{-1} day^{-1} for all the food items during the N2013 conference, at least 551 g cap^{-1} day^{-1} (39%) was neither consumed nor recycled into the production system. This was more than twice the 243 g cap^{-1} day^{-1} (15%) of the food loss and wastage for the baseline conference despite the fact that total food supply was higher for the baseline conference (1639 g cap^{-1} day^{-1}). These results are suggestive of luxurious consumption during baseline conference, a predisposing factor to avoidable N_r loss into the environment (HLPE 2014) and human and environmental health (Willis 1994; Springman et al. 2016). It is beyond the scope of this study whether this important commitment to reducing excessive food consumption by informed delegates becomes part of their daily lifestyle beyond the N2013 conference. What cannot be over-emphasized is the potential contribution of awareness raising (i.e., 'sensitization') of all actors along the consumption and waste management part of the food chain during international conferences to reduce over-consumption and wastage of food. Similar sensitization strategies in all countries would significantly benefit both the local producers (the farmers) and the consumers, especially in sub-Saharan Africa where pressure on soil resources is worrying, and food insecurity, malnutrition, chronic poverty and hunger still prevalent. Food loss and waste can mean a wasted investment for poor farmers reducing their income, but can also mean an increase in the consumer's expenditure. The sensitization efforts at all levels are critical for boosting a sustainable food and environment systems. This, however, must be done without comprising the human nutrition requirements particularly for people from sub-Saharan Africa.

37.3.4 Contribution of the Different Food Items to N_r Pathways

Chicken, beef, wheat, fish, rice and mutton were the six most important sources of N_r intake in that order, accounting for about 17.9 g N cap^{-1} day^{-1} (61.2% of total N_r intake) for the baseline conference. This N_r was higher than the 11.4 g N cap^{-1} day^{-1} (61.8% of total N_r intake) for the N2013 conference (Table 37.3—see values in bold), where groundnuts replaced mutton amongst the most important N-sources, and beef and chicken swapped ranks. With the exception of groundnuts, intake of N_r through peas and legumes which are potential substitutes to animal sources of proteins was very low for both conferences, even lower than N_r intakes from coffee and tea (for the N2013 conference) that are not naturally rich sources of N_r. Overall, total N_r intake was 29.2 g N cap^{-1} day^{-1} for the baseline conference which was higher than 11.4 g N cap^{-1} $year^{-1}$ for the N2013 conference.

Beef, chicken, wheat, rice, goat meat and fish were the six most important sources of N_r in food waste in that order, for the baseline conference. The six food items alone accounted for about 3.9 g N cap^{-1} day^{-1} (69% of total N_r in food wasted). In contrast, for the N2013 conference, beef, wheat, fish, groundnuts, chicken and onions were the most important sources of N_r in food waste in that order, accounting for 8.2 g N cap^{-1} day^{-1} (68% of total N_r in food waste). It is important to note that the amount of N_r in food waste during the N2013 conference (12.1 g N cap^{-1} day^{-1}) was more than twice the 5.6 g N cap^{-1} day^{-1} for the baseline conference.

Food recycling during the N2013 conference was at least 10-fold (5.1 g N cap^{-1} day^{-1}) that of the baseline conference (0.5 g N cap^{-1} day^{-1}). Beef, wheat, fish, groundnuts, chicken and onions were the six most important sources of recycled N_r for the N2013 conference, whereas powdered milk, fresh milk, cereal, fish, banana and chicken were the most important for the baseline conference. The amount of N_r in food served at Speke Resort Munyonyo Hotel for the baseline conference (32 g N cap^{-1} day^{-1}) was higher than the about 25 g N cap^{-1} day^{-1} for the N2013 conference (Table 37.3). About 92% of the N_r in total food served during the baseline conference was consumed compared to only about 76% of the N_r in total food served during the N2013 conference.

Keeping N_r intake preferably within dietary recommendations, optimizing the recycling of N_r in unconsumed food and food waste constitute key strategies for mitigating avoidable loss of N_r into the environment. The sum of N_r in food consumed and food wasted for the baseline conference totaled about 35 g N cap^{-1} day^{-1} (29.2 consumed, 5.6 wasted, Table 37.3). This compared with total for the N2013 conference of 31 g N cap^{-1} day^{-1} (18.5 consumed, 12.1 wasted, Table 37.3). The N_r in crop and livestock food products consumed or wasted is withdrawn from the soil and must be replenished to sustain production especially in low external input agro-ecosystems. This becomes critical in Uganda where average per capita fertilizer use is less than 2.0 kg fertilizer per farmer per year. World-over, cities including Kampala where these conferences were held are choking with food waste (HLPE 2014) with increasing difficulty to locate new landfill sites for disposal as old sites fill-up very

Table 37.3 Fate of N_r in the 28 the food items (in g N cap^{-1} day^{-1}) for the baseline conference and the 6th International Nitrogen Conference (N2013) used to calculate per capita Nitrogen Footprint at Speke Resort Munyonyo Hotel, Uganda in 2013

	N_r in food served (g N cap^{-1} day^{-1})		N_r in food consumed (g N cap^{-1} day^{-1})		N_r in food recycled (g N cap^{-1} day^{-1})		N_r in food wasted (g N cap^{-1} day^{-1})	
Food item	Baseline	N2013	Baseline	N2013	Baseline	N2013	Baseline	N2013
Matooke	0.61	0.47	0.57	0.35	0.00	0.12	0.09	0.25
Cassava	0.07	0.02	0.06	0.02	0.00	0.00	0.02	0.01
Sweet potatoes	0.14	0.07	0.13	0.04	0.00	0.02	0.03	0.05
Irish potatoes	0.50	0.48	0.46	0.40	0.00	0.06	0.08	0.16
Rice	2.20	1.35	**1.96***	**1.19***	0.00	0.16	0.48	0.32
Corn	0.65	0.28	0.51	0.16	0.00	0.02	0.28	0.24
Wheat	3.39	3.02	**3.15***	**2.17***	0.01	0.54	0.50	1.69
Cereals	0.36	0.46	0.36	0.30	0.08	0.15	0.08	0.33
Peas	0.11	0.06	0.10	0.05	0.00	0.01	0.01	0.02
Beans	0.11	0.20	0.10	0.14	0.00	0.05	0.02	0.13
Ground nuts	1.22	1.73	1.22	**1.27***	0.00	0.43	0.01	0.92
Banana	0.50	0.06	0.50	0.05	0.05	0.01	0.05	0.02
Onions	1.03	1.00	0.98	0.77	0.00	0.10	0.10	0.50
Fruits	0.69	1.15	0.65	0.96	0.01	0.14	0.11	0.37
Vegetables	0.79	1.15	0.68	0.92	0.00	0.11	0.18	0.45
Oil	0.20	0.07	0.20	0.05	0.00	0.01	0.00	0.03
Fish	3.15	2.79	**2.99***	**2.09***	0.06	0.65	0.37	1.47
Beef	4.82	3.92	**4.15***	**2.51***	0.00	1.40	1.33	2.81
Pork	1.20	0.57	1.15	0.43	0.00	0.11	0.08	0.27
Goat meat	1.43	0.80	1.21	0.59	0.00	0.21	0.44	0.43
Mutton	1.40	0.15	**1.25***	0.13	0.00	0.01	0.29	0.03
Chicken meat	4.71	2.64	**4.37***	**2.21***	0.03	0.42	0.75	0.85
Eggs	0.78	1.10	0.78	0.99	0.02	0.11	0.02	0.23
Milk	0.86	0.51	0.84	0.40	0.10	0.10	0.09	0.21
Milk powder	0.80	0.08	0.80	0.08	0.13	0.00	0.13	0.00
Coffee	0.02	0.16	0.02	0.09	0.01	0.07	0.01	0.15
Tea	0.05	0.25	0.05	0.16	0.02	0.09	0.02	0.18
Sugar	0.00	0.00	0.00	0.00	0.00	0.00	0.00	0.00

(continued)

Table 37.3 (continued)

	N_r in food served ($g\ N\ cap^{-1}\ day^{-1}$)		N_r in food consumed ($g\ N\ cap^{-1}\ day^{-1}$)		N_r in food recycled ($g\ N\ cap^{-1}\ day^{-1}$)		N_r in food wasted ($g\ N\ cap^{-1}\ day^{-1}$)	
Food item	Baseline	N2013	Baseline	N2013	Baseline	N2013	Baseline	N2013
Total	31.8	24.5	29.2	18.5	0.5	5.1	5.6	12.1

*For explanation of numbers in **bold** please refer to Sect. 37.3.4 in this chapter

fast. Loss of N_r this way not only constitutes an environmental burden to society in terms of the pollution associated with such waste (HLPE 2014; Stevens et al. 2014); it is also robbing arable lands of the much needed soil organic matter, a reservoir for 99% N_r (Schmidt et al. 2011) that gets buried in landfills irretrievably. The use of the phrase food recycling here should not be interpreted literally, since we used it to define re-use mostly by livestock farmers who collect the leftover and food waste for feeding their livestock. This clarification is important because there is no guarantee that the waste from livestock is actually ploughed back into the agricultural fields, let alone the fields far away in rural areas, from where the bulk of the food originates that was used in these two conferences.

37.3.5 Computation of Food Item-Specific Nitrogen Footprints

The N Footprint for the baseline was 160 g N cap^{-1} day^{-1} which was higher than 97 g N cap^{-1} day^{-1} for the N2013 conference. For specific food products, N Footprints were dominated by beef (38.6 g N cap^{-1} day^{-1}), wheat (23.7 g N cap^{-1} day^{-1}), rice (17.6 g N cap^{-1} day^{-1}), goat meat (11.2 g N cap^{-1} day^{-1}), milk (11.2 g N cap^{-1} day^{-1}) and onion (5.9 g N cap^{-1} day^{-1}), amounting to 114 g N cap^{-1} day^{-1} (about 71.3% of total N Footprint) for the baseline conference. These N Footprints were higher than for the N2013 conference: 23.7, 22.2, 10.7, 5.7, 5.6 and 5.4 g N cap^{-1} day^{-1} for beef, wheat, rice, fruits, vegetables and goat meat, respectively, amounting to 73.3 g N cap^{-1} day^{-1} (about 76.6% of total N Footprint, Fig. 37.5). Animal products accounted for about 79.5 g N cap^{-1} day^{-1} (49.7% of total N Footprint) for the baseline conference and about 38.7 g N cap^{-1} day^{-1} (39.9% of total N Footprint) for the N2013 conference.

Matooke, cassava and sweet potatoes which are staple food crops in Uganda had smaller N Footprints than the cereals ranging from 0.20 g N cap^{-1} day^{-1} (cassava) to 1.87 g N cap^{-1} day^{-1} (*matooke*) for the baseline and 0.06 g N cap^{-1} day^{-1} (cassava) to 1.14 g N cap^{-1} day^{-1} (*matooke*) for the N2013 conference. For cereals, the N Footprints ranged from 3.29 g N cap^{-1} day^{-1} (corn) to 32.14 g N cap^{-1} day^{-1} (wheat) for the baseline and 1.05 g N cap^{-1} day^{-1} (corn) to 22.16 g N cap^{-1} day^{-1} (wheat) for N2013. Legumes had the smallest N Footprints that is, 0.03 g N cap^{-1} day^{-1}

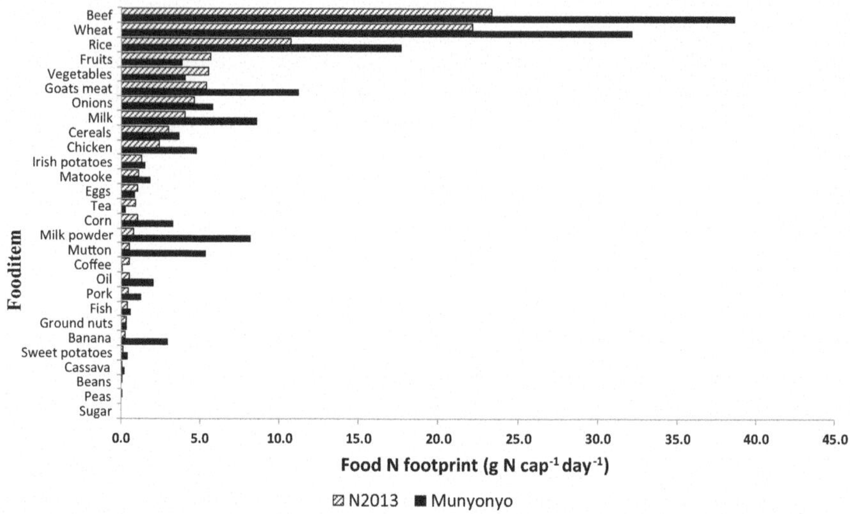

Fig. 37.5 Food product-specific reactive nitrogen footprints for the baseline conference (labelled 'Munyonyo') and the N2013 conference (g N cap^{-1} day^{-1}) held in Speke Resort Munyonyo Hotel, Uganda in 2013

(either peas or beans) for the baseline or 0.02 g N cap^{-1} day^{-1} (peas) and 0.04 g N cap^{-1} day^{-1} (beans) for the N2013 conference. Interestingly, the N Footprints for legumes were even smaller than those for coffee and tea. Moreover, even fruits and milk that were consumed in largest quantities had typically low N Footprints.

We estimated the N Footprints for 28 food items that constitute the menu in Speke Resort Munyonyo Hotel. Legumes (peas and beans) are the most N_r-efficient, followed by sweet potatoes, cassava and corn, which are major staple foods in Uganda as well as fruits and vegetables. Given that the total N Footprint for either the baseline or N2013 conference is higher than those reported for the N Footprint of an average citizen in industrialized countries including Austria (Pierer et al. 2014), United Kingdom (Stevens et al. 2014) and Japan (Shibata et al. 2014), a shift to these more N_r-efficient foods is critical to cutting down N_r release into the environment and sustaining productivity especially in low external input agro-ecosystems. The beneficial effects of shifting to more vegetarian diets on human and environmental health have been documented (Westhoek et al. 2014, 2015; Springman, et al. 2016). The challenge now is how to convince and wean consumers from addiction to and over-reliance on animal-based sources of proteins especially given that these alone accounted for 50% and 40% of the total N Footprints for the baseline and N2013 conferences.

37.4 Conclusions

Per capita food purchases, food served and consumed were generally lower for the N2013 conference than for the baseline conference, except for fruits, vegetables and eggs, suggesting an increased shift to more plant-based foods and eggs in preference to meats during the N2013 conference. Food recycling was also associated more with the N2013 conference than it was with the baseline, an important step towards mitigating avoidable food wastage and N_r loss. The surprisingly higher food wastage during the N2013 conference than during the baseline conference, despite the fact that more food items were purchased and served in the baseline conference, is attributable to reduced intake of high N_r foods such as chicken, mutton, milk and beef in the N2013 conference, which is an important step towards increased human and environmental health. Consequently, the Nitrogen Footprint of the N2013 conference was nearly half that of the baseline. This implies that pre-information and a pre-adjusted menu with clear guidelines for food preparation and consumption are critical in reducing avoidable over-consumption and food wastage usually associated with increased N_r release into the environment. However, even the Nitrogen Footprint of the N2013 conference was still way above that of an average citizen in industrialized countries including United Kingdom, Austria and Japan. The implication is that more aggressive strategies are needed to further reduce over-consumption and food wastage at international conferences. This is especially so in developing countries where the food wasted in such conferences originates from severely degraded soils, where ordinary citizens' lives are wrecked by hunger and malnutrition, and where cities struggle to cope with the burden of food waste. In particular, consumers at such international conferences should be encouraged and provided with incentives to trim their appetites to eat just enough of the right food for normal body requirements, with associated planning of menus and food purchases that allow food wastage to be reduced.

The N2013 conference has provided an important test in implementing the goals of the Barsac Declaration (Sutton et al. 2009). N2013 is the first INI conference to have been organized with a comprehensive collection of information on diets, pre-informing those purchasing, preparing and eating, and analyzing the results in comparison to a baseline conference. The fact that this achievement was accomplished in in sub-Saharan Africa is even more significant. It shows that it is possible to develop a conversation about excess intake of meat and dairy products in a region where many have insufficient food. The outcomes demonstrate that it is not just average intake levels that need to be adjusted, but especially situations of high intake where there is also most freedom of choice to allow changes to be made. It is also important to note that Speke Resort Munyonyo is only one among many similar hotels in Kampala Capital City and indeed many other capital cities in Africa. The findings of this study are therefore, only a snapshot of what could be happening and calls for further studies to paint a consistent picture on the Nitrogen Footprints from international conferences similar to the ones investigated in this study.

Acknowledgements We are grateful to the administrators and chefs of Speke Resort Munyonyo Hotel for their positive cooperation while collecting data for this research. We would also like to acknowledge the EU for the financial support without which this work would not have been conducted. This paper is a contribution to the work of the International Nitrogen Initiative.

References

Ambus, P., Skiba, U., Butterbach-Bahl, K., Sutton, M. A. (Eds.). (2011). Reactive nitrogen and greenhouse gas flux interactions in terrestrial ecosystems. (Third NitroEurope Special Issue) *Plant and Soil, 343*(1–2), 1–185.

Cribb, J. (2010). *The coming famine: The global food crisis and what can be done to avoid it*. Collingwood VIC, Australia: CSIRO Publishing.

Dalgaard, T., Olesen, J. E., Petersen, S. O., Petersen, B. M., Jørgensen, U., Kristensen, T., et al. (2011). Developments in greenhouse gas emissions and the net energy use in Danish agriculture—how to achieve substantial CO_2 reductions? *Environmental Pollution, 159*(11), 3193–3203.

FAO. (2011a). *Food balance sheets. A handbook*. Rome: FAO.

FAO. (2011b). *Global food losses and food waste—extent, causes and prevention*. In J. Gustavsson, C. Cederberg, U. Sonesson, R. van Otterdijk, A. Meybeck (Eds.). Rome: FAO. Available online at: http://www.fao.org/docrep/014/mb060e/mb060e00.pdf. Accessed September, 2019.

FAO. (2015). *World fertilizer trends and outlook to 2018*. Rome: FAO. Available online at: http://www.fao.org/3/a-i4324e.pdf. Accessed September, 2019.

Galloway, J. N., Aber, J. D., Erisman, J. W., Seitzinger, S. P., Howarth, R. W., Cowling, E. B., et al. (2003). The nitrogen cascade. *BioScience, 53*(4), 341–356.

Galloway, J. N., Leach, A. M., Bleeker, A., & Erisman, J. W. (2013). A chronology of human understanding of the nitrogen cycle. *Philosophical Transactions of the Royal Society B, 368*(1621), 20130120.

HLPE (Higher Level Panel of Experts). (2014). *Food loss and waste in the context of sustainable food systems. A report by the Higher Level Panel of Experts on Food Security and Nutrition of the Committee on World Food Security*. Rome: HLPE.

Hutton, M. O., Leach, A. M., Leip, A., Galloway, J. N., Bekunda, M., Sullivan, C., et al. (2017). Toward a nitrogen footprint calculator for Tanzania. *Environmental Research Letters, 12*(3), 034016.

Lassaletta, L., Billen, G., Romero, E., Garnier, J., & Aguilera, E. (2013). How changes in diet and trade patterns have shaped the N cycle at the national scale: Spain (1961–2009). *Regional Environmental Change, 14*(2), 785–797.

Lassaletta, L., Billen, G., Grizzetti, B., Garnier, J., Leach, A. M., & Galloway, J. N. (2014). Food and feed trade as a driver in the global nitrogen cycle: 50-year trends. *Biogeochemistry, 118*, 225–241.

Leach, A. M., Galloway, J. N., Bleeker, A., Erisman, J. W., Kohn, R., & Kitzes, J. (2012). A nitrogen footprint model to help consumers understand their role in nitrogen losses to the environment. *Environmental Development, 1*(1), 40–66.

Leip, A., Weiss, F., Lesschen, J. P., Westhoek, H. (2013). The nitrogen footprint of food products in the European Union. *Journal of Agricultural Science, 152*(S1, Special Issue from the 17th International Nitrogen Workshop), 20–33.

Leip, A., Leach, A., Musinguzi, P., Tumwesigye, T., Olupot, G., Tenywa, J. S., et al. (2014). Nitrogen-neutrality: A step towards sustainability. *Environmental Research Letters, 9*(11), 115001. https://doi.org/10.1088/1748-9326/9/11/115001.

Marie, Ng, Fleming, T., Robinson, M., Thomson, B., Graetz, N., Margono, C., et al. (2014). Global, regional and national prevalence of overweight and obesity in children and adults during the

1980–2013: A systematic analysis for the global burden of disease study. *The Lancet, 384*(9945), 766–784.
Meir, T., & Christen, O. (2013). Environmental impacts of dietary recommendations and dietary styles: Germany as an example. *Environmental Science Technology, 47*(2), 877–888.
Musinguzi, P., Ebanyat, P., Tenywa, J. S., Basamba, T. A., Tenywa, M. M., & Mubiru, D. N. (2016). Critical soil organic carbon range for optimal crop response to mineral fertiliser nitrogen on a Ferralsol. *Experimental Agriculture, 52*(4), 635–653.
Pierer, M., Winiwarter, W., Leach, A. M., & Galloway, J. N. (2014). The nitrogen footprint of food products and general consumption patterns in Austria. *Food Policy, 49*(1), 128–136.
Ramseyer, L. J. (2002). Predicting whole-fish nitrogen content from fish wet weight using regression analysis. *North American Journal of Aquaculture, 64*(3), 195–204.
Sanchez, P. A., & Swaminathan, M. S. (2005). Cutting world hunger in half. *Science, 307*(5708), 357–359.
Schmidt, M. W. I., Torn, M. S., Abiven, M., Dittmar, T., Guggenberger, G., Jamssens, I. A., et al. (2011). Persistence of soil organic matter as an ecosystem property. *Nature, 478,* 48–56.
Shibata, H., Cattaneo, L. R., Leach, A. M., & Galloway, J. N. (2014). First approach to the Japanese nitrogen footprint model to predict the loss of nitrogen to the environment. *Environmental Research Letters, 9*(11), 115013.
Springman, M. H., Godfray, H. C. J., Rayner, M., & Scarborough, P. (2016). Analysis and Valuation of the health and climate change co-benefits of dietary change. *Proceedings of the National Academy of Sciences of the United States of America (PNAS), 113*(15), 4146–4151.
Stehfest, E., van den Berg, M., Woltjer, G., Msangi, S., & Westhoek, H. (2013). Options to reduce the environmental effects of livestock production—comparison of two economic models. *Agricultural Systems, 114,* 38–53.
Steinfeld, H., Gerber, P., Wassenaar, T., Castel, V., Rosales, M., & de Haan, C. (2006). *Livestock's long shadow. Environmental issues and options.* Rome: FAO.
Stevens, C. J., Leach, A. M., Dale, S., & Galloway, J. N. (2014). Personal nitrogen footprint tool for the United Kingdom. *Environmental Science: Processes & Impacts, 16,* 1563–1569.
Sutton M. A., Bealey W. J., Stevens C., Bleeker A., & over 200 signatories. (2009). *The Barsac Declaration: Environmental sustainability and the demitarian diet.* NinE/BEGIN/INI. Available online at: http://www.nine-esf.org/node/281/index.html. Accessed September 2019.
Sutton, M., & Howard, C. (2011) Distilling nitrogen science. *Planet Earth* (NERC Magazine), Winter 2011, 18–19. Available online at: http://www.nerc.ac.uk/latest/publications/planetearth/archive/planet-earth-win11. Accessed September 2019.
Sutton, M. A., Howard, C. M., Erisman, J. W., Billen, G., Bleeker, A., Grennfelt, P., et al. (Eds.). (2011). *The European Nitrogen Assessment. Sources, effects and policy perspectives.* Cambridge: Cambridge University Press.
Sutton, M. A., Billen, G., Cellier, P., Erisman J. W., Mosier A. R., Nemitz, E., van Grinsven, H., et al. (Eds.) (2012). Nitrogen and global change (Fifth NitroEurope Special Issue), *Biogeosciences, 9.* Available online at: http://www.biogeosciences.net/special_issue93.html. Accessed September 2019.
Sutton, M. A., Bleeker, A., Howard, C. M., Bekunda, M., Grizzetti, B., de Vries, W., van Grinsven, H. J. M., et al. (2013). *Our Nutrient World: The challenge to produce more food and energy with less pollution.* Global Overview of Nutrient Management. Edinburgh: Centre for Ecology and Hydrology, Edinburgh, on behalf of the Global Partnership on Nutrient Management and the International Nitrogen Initiative.
Sutton, M. A., Mason, K. E., Sheppard, L. J., Sverdrup, H., Haeuber, R., Hicks, W. K. (Eds.) (2014). *Nitrogen deposition, critical loads and biodiversity (Proceedings of the INI Workshop, linking experts of the Convention on Long-range Transboundary Air Pollution and the Convention on Biological Diversity).* Dordercht: Springer.
Tukker, A., Huppes, G., Guinée, J., Heijungs, R., de Koning, A., van Oers, L., Suh, S., et al. (2006). *Environmental impact of products (EIPRO). Analysis of the lifecycle environmental impacts related to the final consumption of the EU-25.* Technical Report EUR 22284 EN. European

Commission. http://ec.europa.eu/environment/ipp/pdf/eipro_report.pdf. Accessed 28 December 2016.
Vanham, D., Bouraoui, F., Leip, A., Grizzetti, B., & Bidoglio, G. (2015). Lost water and nitrogen resources due to EU consumer food waste. *Environmental Research Letters, 10*(8), 084008.
Westhoek, H., Lesschen, J. P., Rood, T., Wagner, S., De Marco, A., Murphy-Bokern, D., et al. (2014). Food choices, health and environment: Effects of cutting Europe's meat and dairy intake'. *Global Environmental Change, 26,* 196–205.
Westhoek, H., Lesschen J. P., Leip, A., Rood, T., Wagner. S., De Marco, A., Murphy-Bokern, D., et al. (2015). *Nitrogen on the table: The influence of food choices on nitrogen emissions and the European environment.* (European Nitrogen Assessment Special Report on Nitrogen and Food). Edinburgh: Centre for Ecology and Hydrology.
Wetzel, R. G. (Ed.). (2001). *Limnology. Lake and river ecosystems* (3rd ed.). San Diego: Academic Press.
Willis, R. (1994). *Your health is in your hands.* Lincolnshire, UK: The Stanborough Press.
Xue, X., & Landis, A. E. (2010). Eutrophication potential of food consumption patterns. *Environmental Science and Technology, 44*(16), 6450–6456.

Chapter 38
The Kampala Statement-for-Action on Reactive Nitrogen in Africa and Globally

Mark A. Sutton, Peter Ebanyat, N. Raghuram, Mateete Bekunda, John S. Tenywa, Wilfried Winiwarter, Albert Bleeker, Eric A. Davidson, Jan Willem Erisman, Wim de Vries, James N. Galloway, Patrick Heffer, W. Kevin Hicks, Cargele Masso, Cheryl A. Palm, Clifford S. Snyder, Bernard Vanlauwe, Shamie Zingore, and Delegates of the 6th International Nitrogen Conference, Kampala

Abstract Following discussions at the 6th International Nitrogen Conference, Kampala (24th–27th November 2013), the delegates agreed the *Kampala Statement-for-Action on Reactive Nitrogen in Africa and Globally.* The Statement-for-Action highlights the global challenge of aiming for just enough nitrogen: enough to meet human needs for food, fuel and fibre, while avoiding excess that contributes to air and water pollution, climate change and ecosystem degradation. The following priorities for Africa are highlighted: (i) Sub-Saharan African agriculture needs to be part of the solution to regional and world food security. This will require restoring and sustaining

---

M. A. Sutton (✉) · Delegates of the 6th International Nitrogen Conference, Kampala
UK Centre for Ecology & Hydrology, Edinburgh Research Station, Bush Estate, Penicuik, Midlothian, EH26 0QB, UK
e-mail: ms@ceh.ac.uk

P. Ebanyat · J. S. Tenywa
Department of Agricultural Production, Makerere University, Kampala, Uganda
e-mail: peter.ebanyat@gmail.com

J. S. Tenywa
e-mail: johntenywa@gmail.com

P. Ebanyat
International Institute of Tropical Agriculture (IITA), Kampala, Uganda

N. Raghuram
University School of Biotechnology, Guru Gobind Singh Indraprastha University, Dwarka Sec. 16C, New Delhi 110078, India
e-mail: raghuram@ipu.ac.in; raghuram98@hotmail.com

South Asian Nitrogen Centre, Indian Nitrogen Group, Sustainable India Trust and Society for Conservation of Nature, F-4, A-Block, NASc Complex, DPS Marg, Pusa, New Delhi 110 012, India

M. Bekunda
International Institute of Tropical Agriculture (IITA), Arusha, Tanzania
e-mail: M.Bekunda@cgiar.org

M. A. Sutton et al. (eds.), *Just Enough Nitrogen*,
https://doi.org/10.1007/978-3-030-58065-0_38

the fertility of Sub-Saharan African soils through better use of fertilizers in combination with other nutrient sources, such as organic matter and biological fixation. (ii) Smart input subsidy schemes helping to trigger profitable nutrient use should take care that increases in fertilizer find a balance between food security and reducing environmental effects, while recognizing that, (iii) Africa is already facing environmental challenges of nitrogenous air and water pollution even with current levels of nitrogen input. Current nitrogen inputs are thought to be mainly from sources other than chemical nitrogen fertilizers (e.g., biological nitrogen fixation, organic nitrogen inputs, wastewater), though further study is needed to demonstrate apportionment between nitrogen sources and sinks. The following global priorities are highlighted: (i) Improving nitrogen management should be incorporated as a critical component across the Sustainable Development Goals. (ii) The benefits of reducing nitrogen losses from agriculture, industry, transport and energy, of improved waste treatment and of better-informed individuals and institutions should be highlighted, including an emphasis on innovative nutrient recycling and on equitable diet and energy choices. (iii) There is a need for innovation and increased awareness on the nitrogen challenge, including through better communication, education and training. (iv) Solutions to the nitrogen issue should be tuned to regional conditions and require cross-ministerial, trans-disciplinary, multi-sectoral cooperation to create effective policies that fulfill regional and global commitments.

Keywords Nitrogen strategies · Nitrogen declarations · Nitrogen policy · International development · Sustainable development

---

W. Winiwarter
International Institute for Applied Systems Analysis (IIASA), Laxenburg, Austria
e-mail: winiwarter@iiasa.ac.at

A. Bleeker
National Institute of Public Health and Environment (RIVM), P.O. Box 1,
3720 BA Bilthoven, The Netherlands
e-mail: albert.bleeker@rivm.nl

E. A. Davidson
University of Maryland Center for Environmental Science, Appalachian Laboratory,
301 Braddock Road, Frostburg, MD 21532, USA
e-mail: edavidson@umces.edu

J. W. Erisman
Louis Bolk Institute for International Advice and Research on Sustainable Agriculture,
Nutrition and Health Care, Driebergen, The Netherlands
e-mail: j.w.erisman@cml.leidenuniv.nl

VU University, Amsterdam, The Netherlands

Institute of Environmental Studies (CML), Leiden University, Leiden, The Netherlands

W. de Vries
Wageningen University and Research (WUR), Droevendaalsesteeg 3,
6708 PB Wageningen, The Netherlands
e-mail: wim.devries@wur.nl

38.1 Introduction

This chapter reports the *Kampala Statement-for-Action on Reactive Nitrogen in Africa and Globally* as it was agreed by the delegates of the 6th International Nitrogen Conference, Kampala (24th–27th November 2013). In order to give the context for the Statement-for-Action, a summary of prior declarations and other relevant international agreements was first prepared in advance of the conference. This provided a foundation for delegates to consider future priorities in the context of the past agreements. While the Kampala conference represents the first time that the International Nitrogen Initiative (INI) has hosted the International Nitrogen Conference in Africa, this background highlights the contribution of previous agreements both in Africa and in other parts of the world. A drafting committee formed during the conference included a range of scientific perspectives plus business representation. The group prepared a first draft of the Statement-for-Action, based on the messages emerging from the conference. This draft was then presented to the final plenary session of the conference, where the main text was revised by the conference participants. The main text (Sect. 38.2) represents the consensus document. It was also agreed that the background information be incorporated into a Technical Annex to the Statement-for-Action. This background is included as Sect. 38.3 and does not form part of the negotiated consensus text.

We here give the main text of the *Kampala Statement-for-Action*, followed by the Technical Annex that summarizes previous nitrogen agreements. References marked as [number] refer to further information that is provided in the Technical Annex.

J. N. Galloway
Environmental Sciences Department, University of Virginia, 291 McCornik Road, 400123, Charlottesville, VA 22904, USA
e-mail: jng@virginia.edu

P. Heffer
Agriculture Service, International Fertilizer Association, 49 avenue d'Iéna, 75116 Paris, France
e-mail: pheffer@fertilizer.org

W. K. Hicks
Stockholm Environment Institute (SEI), Department of Environment and Geography, University of York, York, YO10 5DD, UK
e-mail: kevin.hicks@york.ac.uk

C. Masso
International Institute of Tropical Agriculture (IITA), 1st Main IRAD Road, Nkolbisson, P.O. Box 2008 (Messa), Yaoundé, Cameroon
e-mail: C.Masso@cgiar.org

C. A. Palm
Institute for Sustainable Food Systems, University of Florida, Frazier Rogers Hall, 110570, Gainesville, FL 32611, USA
e-mail: cpalm@ufl.edu

38.2 The Kampala Statement-for-Action

The messages below are targeted to policy, industrial, agricultural and civil society fora that are active in addressing issues of reactive nitrogen in food security, energy, health, environment, biodiversity and climate change, either directly or indirectly around the world.

38.2.1 The Situation

Recognizing that Africa is entering a new green revolution, the International Nitrogen Initiative (INI) held its tri-annual conference in Kampala, Uganda, which was the first time in Africa. 161 delegates from 37 countries represented disciplines ranging from agricultural science and atmospheric science to medical science, and included private, public sector and civil society representatives.

Nitrogen (N) is an essential nutrient to sustain life. The theme of the conference—"Let us aim for just enough N"—addresses both the crucial need for enough nitrogen input to grow crops and livestock and also the potential that too much, too little or poorly managed nitrogen inputs can result in environmental degradation, such as water and air pollution, climate change, stratospheric ozone depletion, human health risks, and biodiversity loss.

After five previous INI conferences in the last fifteen years that brought attention to the urgency of improving nitrogen management (see notes [1] and [2] in the Technical Annex), we note considerable recent momentum, including:

- the Rio+20 summit [3] in 2012, where the role of nitrogen in the green economy and for advancing the three equally important pillars for an "economically, socially and environmentally sustainable future" was emphasized;
- the UN Framework Convention on Climate Change, the UN Convention on Biological Diversity, the UN Convention to Combat Desertification, the Vienna

C. S. Snyder
Conway, AR 72034, USA
e-mail: cssnyder@conwaycorp.net

B. Vanlauwe
International Institute of Tropical Agriculture (IITA), PMB 5320, Oyo Road, Ibadan 200001, Oyo State, Nigeria
e-mail: B.Vanlauwe@cgiar.org

S. Zingore
International Plant Nutrition Institute (IPNI), c/o ICIPE Duduville Campus, Kasarani, P.O. Box 30772–00100, Nairobi, Kenya
e-mail: szingore@ipni.net

Convention for the Protection of the Ozone Layer, and the Convention on Long-range Transboundary Air Pollution, where recognition of the role of nitrogen in these environmental challenges has grown;

- the UNEP Global Program of Action for the Protection of the Marine Environment from Land-based Activities (GPA), committing to sustainable management of nutrients (including N) and supporting Global Partnership on Nutrient Management [4, 5];
- the Maputo Declaration [6] on Agriculture and Food Security (2003) and the Abuja Declaration [7] on Fertilizer for an African Green Revolution (2006) that highlighted the importance for investment in agriculture and the essential role of nutrient inputs for advancing a new green revolution in Africa.

These international efforts and other accords such as the Amsterdam declaration [8], the Edinburgh declaration [9] and the Aichi targets [10], have created a momentum whereby major new achievements are now possible. Considerable economic, social and environmental benefits could be obtained across the world by implementing the best practices for sustainable nitrogen management more widely. Specifically, we call for implementation of policies consistent with the following:

38.2.2 Specific Messages for Sub-Saharan Africa

I. **Improving Soil Fertility Status, Nutrient Use and Supply**[1]: Sub-Saharan African agriculture needs to be part of the solution to regional and world food security. Countries allocating 10% or more of their budget to agriculture, following their commitment spelled out in the Maputo Declaration, have been able to improve their food security status in recent years. However, continuing this positive trend will require restoring and sustaining the fertility of Sub-Saharan African soils through more and better use of fertilizers in combination with other nutrient sources, such as organic matter and biological fixation. This should be consistent with the concept of Integrated Soil Fertility Management. Management practices need to be tailored to the diversity of Sub-Saharan African soils, cropping and livestock systems. Improving reliable delivery of quality fertilizers to smallholder farmers is paramount to increase agricultural productivity per unit of land and water.

II. **Acting on Nutrient and Fertilizer Policy**: Available regional examples can motivate good practice and sustainable high yields with higher grain protein content. Sub-Saharan African nations would be well advised to continue their efforts to increase their level of sustainable use of fertilizer consistent with the Abuja Declaration.

[1] Nitrogen needs to be considered alongside other nutrients in balanced nutrient management because of multiple nutrient limitations.

- National authorities should foster enabling policies and a regulatory environment that encourage private sector investment in an effective value chain from fertilizer to products and farmer participation in value-added output markets.
- Implementing smart input subsidy schemes compatible with fair competition and long-term investments in rural transportation, market infrastructures, agricultural research and knowledge transfer helps trigger greater, efficient and profitable nutrient use. Such actions would need to take care that increases in fertilizer inputs are applied properly to balance food security and reduce environmental effects (see global messages, Sect. 38.2.3 below).

III. **Reducing Nitrogen's Contribution to Degradation of Water Bodies and Air Pollution**: Some regions in Africa are already facing the environmental challenges of nitrogenous air pollution and nutrient inputs to water bodies from agricultural run-off, atmospheric deposition, sewage and industrial discharge.[2]

38.2.3 Global Messages

I. **Improving Nitrogen Management**, including the 'just enough N' concept for policies to promote food security while avoiding environmental degradation, should be incorporated as critical components of the new international Sustainable Development Goals (SDGs).

II. **Many societal benefits** [11] can be obtained by reducing nitrogen losses to the environment [12]:

- **Reducing Nitrogen Losses from Agriculture**: Concepts are available and success stories exist, demonstrating that a decrease of environmental impacts and an increase in food production is possible. Science-based sustainable intensification of farming systems can help optimize nutrient inputs, by applying the right nutrient source, at the right rate, at the right time, in the right place.
- **Reducing Nitrogen Losses from Industry, Transport and Energy Sectors**: Nitrogen emissions from fossil fuel burning and industrial processes can be reduced by adopting existing technologies and innovation.

[2]The consensus text agreed at the Kampala conference did not specify the actions required to address this concern. These are now being explored under the East African Demonstration activity of the International Nitrogen Management System (www.inms.international). It may be considered as implicit that national and regional authorities and agencies should provide frameworks for implementing measures by land users and regulation of nitrogen-rich effluents from industry and cities, as well as gaseous emissions from industrial and transport systems to control air quality and water quality. Commitments in this direction have since been recognized in the resolutions of UNEA-3 (December 2017), https://papersmart.unon.org/resolution/ including UNEP/EA.3/L.23 and UNEP/EA.3/L.27 and at UNEA-4 (March 2019) through UNEP/EA.4/L.16.

- **Improving Treatment of Waste**: Sewage treatment and solid municipal waste (household wastes) are sources of nitrogen losses that could be reduced by treatment and/or recycling.
- **Informing Individuals and Institutions**: Enabling consumers to adjust lifestyle choices with equity, including diet [13], transportation and energy consumption will have important impacts on nitrogen losses to the environment [14].

III. **Innovation and awareness.** There is a clear need for innovation and increased awareness among the different stakeholders through information, communication, education, training and extension, modeling and providing policy briefs.

IV. **Policies and technologies** for solutions to the nitrogen issue should be tuned to regional and local conditions and require cross-ministerial, trans-disciplinary and multi-sectoral cooperation and coordination. This will enable countries to create effective policies and fulfill their regional and global commitments. Agencies such as the Global Environment Facility (GEF) can facilitate the regional/global linkages in this regard [15].

38.3 Technical Annex to the Kampala Statement-for-Action: Notes, Relevant Documents, Previous Activities

1. **The Nanjing Declaration** (2004) of the 3rd International Nitrogen Conference on Nitrogen Management (2004) understood reactive nitrogen as a critical nutrient for food, feed and fibre security. Its accumulation has negative effects on the environment and human health, as anthropogenic reactive nitrogen production exceeds natural rates of production in many regions of the world. Other areas including most of Africa and parts of South America and Asia suffer from the opposite problem of N deficiency in the soil, contributing to food insecurity and malnutrition. Therefore international efforts towards N assessment and efficient N management are urgently required for sustainable development.
2. **The Delhi Declaration** (2010) on Reactive Nitrogen Management for Sustainable Development stresses the leakage of reactive nitrogen from crop, animal, aquatic and industrial production systems into the environment as a cause for concern. Anthropogenic contribution of reactive nitrogen varies hugely between and within countries and economic sectors and accordingly the responsibility to mitigate the damage (due to excess N) varies proportionately. Sustainable nitrogen management should be built on five key pillars namely: food security, energy and industry, human health, ecosystem services and biodiversity, and climate, with efforts to integrate between these pillars. This would allow optimization for the efficient use of inorganic and organic fertilizers world-wide, and

facilitate enhanced access and sustainable use of N inputs in the predominantly N-deficient soils of Africa and parts of Latin America and Asia.

3. **The Rio+20 Declaration** (2012) titled *'The Future We Want'* included the principle of common but differentiated responsibilities, as set out in Principle 7 of the Rio Declaration. It emphasized the importance of the three Rio Conventions (UNFCCC, CBD, UNCCD) to advance sustainable development and sustainable use of natural resources and ecosystems. Moreover, it stated as a key priority for the international community to support Africa's sustainable development. It referred to green economy in the context of sustainable development and poverty eradication and emphasized the Global Environmental Outlook process for informed decision making. It included the concept to promote, enhance and support more sustainable agriculture, including crops, livestock, forestry, fisheries and aquaculture, that improves food security, eradicates hunger, and is economically viable, while conserving land, water, plant and animal genetic resources, biodiversity and ecosystems, and enhancing resilience to climate change and natural disasters. The Rio+20 Outcome document also noted "with concern that the health of oceans and marine biodiversity are negatively affected by marine pollution, including marine debris, especially plastic, persistent organic pollutants, heavy metals and nitrogen-based compounds, from a number of marine and land-based sources, including shipping and land run-off. We commit to take action to reduce the incidence and impacts of such pollution on marine ecosystems, including through the effective implementation of relevant conventions".
4. **The Global Partnership on Nutrient Management** (GPNM) was set up at the UN Commission on Sustainable Development in 2009 and is supported by a growing number of countries under the UNEP Global Programme of Action for the Protection of the Marine Environment from Land-based Activities, "to promote effective nutrient management, minimising negative impacts on the environment and human health, while maximising their contribution to global sustainable development and poverty reduction". Its scope is to promote sustainable use of nutrients, notably nitrogen, reduce nutrient losses, and improve overall nutrient use efficiency and effectiveness for enhanced food security, safer environment and greener economy, through global partnerships between countries and stakeholders, through its foundation document (2010), via key messages for the policy makers presented during the Rio+20 (2012), and most recently the Global Overview on Nutrient Management under the title *'Our Nutrient World'* (2013).
5. **The Manila Declaration** (2012) adopted by 65 participating national governments and the European Commission during the Third Intergovernmental Review Meeting (IGR3) of the UNEP 'Global Programme of Action for the Protection of the Marine Environment from Land-based Activities' in January 2012 committed to "develop guidance, strategies or policies on the sustainable use of nutrients so as to improve nutrient use efficiency" and to "support the further development of the Global Partnership on Nutrient Management and associated regional and national stakeholder partnerships."

6. **The Maputo Declaration on Agriculture and Food Security in Africa** (2003) committed to the allocation of at least 10 percent of national budgetary resources to agriculture and rural development policy implementation within five years.
7. **The Abuja Declaration** (2006) on Fertilizer for an African Green Revolution adopted by the Africa Fertilizer Summit of the African Union Ministers of Agriculture, emphasized the crucial role of fertilizers in restoring soil fertility and achieving the African Green Revolution for food security and other relevant Millennium Development Goals. It called for increasing the level of use of fertilizer in Sub-Saharan Africa from 8 kg/ha to an average of at least 50 kg/ha by 2015, declaring fertilizers (both organic and inorganic) as strategic commodities without borders. Moreover, it resolved to set up agro-dealer networks, agro-input credit mechanisms, smart subsidies, fertilizer manufacture, procurement and distribution systems, as well as an Africa Fertilizer Development Financing Mechanism through the Africa Development Bank by 2007.
8. **The Amsterdam Declaration** (2001) emerged from the Global Change Open Science Conference on the Challenges of a Changing Earth, held by the International Geosphere Biosphere Programme (IGBP), the International Human Dimensions Programme (IHDP), the World Climate Research Programme (WCRP) and Diversitas. It found that human activities are significantly influencing Earth's environment in many ways in addition to greenhouse gas emissions and climate change, well outside the range of the natural variability exhibited over the last half million years. It noted that an ethical framework for global stewardship and strategies for Earth System management is urgently needed, including a new system of global environmental science.
9. **The Edinburgh Declaration** (2011) on reactive nitrogen, adopted at the international conference '*Nitrogen and Global Change*', recognized the main messages of the European Nitrogen Assessment (ENA) launched at the same conference, specifically the ENA review of benefits and threats that provides options for improved N management. Nitrogen related risks and opportunities are well represented in the ENA for the geographical area of Europe. Five key threats of excess reactive nitrogen were identified in the ENA as Water quality, Air quality, Greenhouse gas balance, Ecosystems and biodiversity and Soil quality (WAGES). The Edinburgh declaration agreed that an overall mitigation strategy should focus on improving nitrogen use efficiency, particularly in agriculture.
10. Parties to the **Convention on Biological Diversity**, in 2010 in Nagoya, Japan, adopted the Strategic Plan for Biodiversity 2011–2020—Aichi Target 8: "By 2020, pollution, including from excess nutrients, has been brought to levels that are not detrimental to ecosystem function and biodiversity."
11. Specific benefits to society of mitigation action are:

    - Protection of inland and coastal waters from N outflow from agriculture and sewage can be expected to significantly reduce adverse impacts on human health and the environment (algal blooms and changes in biodiversity). The contribution of reactive N from sewage, industrial effluents etc., could be

very significant in some countries/regions and even most significant in poorly managed towns/cities and tourist sites, including many small island nations.
- Reduction of N emissions to the air using existing technologies can reduce impacts on human health (via particulate matter (PM2.5), nitrogen dioxide and tropospheric ozone), climate change and ozone layer depletion (via nitrous oxide) and the environment (soil and water acidification and eutrophication and effects on biodiversity).

12. '***Our Nutrient World****—the challenge to produce more food and energy with less pollution*', is a report published 2013 by the Global Partnership on Nutrient Management and the International Nitrogen Initiative (available at http://www.inms.international/sites/inms.international/files/ONW.pdf). This report calls for a global effort to address 'The Nutrient Nexus', where reduced nutrient losses and improved nutrient use efficiency across all sectors simultaneously provide the foundation for a Greener Economy to produce more food and energy while reducing environmental pollution. It calls for agreement between all relevant stakeholders of the global community on which existing inter-governmental process is considered best suited to take the lead in improving nutrient management for the twenty-first century, or whether a new policy process is needed. Nutrient use efficiency is named a key indicator to assess progress towards better nutrient management, with an aspirational goal for a 20% relative improvement in full-chain NUE by 2020, leading to an annual saving of around 20 million tonnes of nitrogen ('20:20 by 2020')
13. '***The Barsac Declaration*** *on Environmental Sustainability and the Demitarian Diet*' (2009) highlighted the importance of our own food choices impacting the environment by altering the requirements for different agricultural activities. In many developed countries and increasingly in some developing countries, individuals eat more animal products than is necessary for a healthy balanced diet. Reducing per capita consumption of animal products in such populations has the potential to improve nutrient use efficiency, reduce overall production costs and reduce environmental pollution, apart from significant health benefits. This can be achieved through promoting the 'demitarian' option, based on meals containing half the amount of meat or fish compared with the normal local amount of affluent diets, combined with a correspondingly larger amount of other food products, along with the normal vegetarian/non-vegetarian meal options.
14. An option to quantify such nitrogen losses could be linked to the concept of nitrogen neutrality. This concept requires to first minimize the nitrogen release associated with anthropogenic activities (nitrogen footprint) and to balance remaining emissions by achieving measured reductions of the reactive N release elsewhere and contributing to sustainable land management where this is not yet achieved.
15. **The Global Environment Facility** (GEF) has been successful in funding nutrient assessment & management activities under the Global Partnership on Nutrient Management that, e.g., have become visible in the report 'Our Nutrient

World'. Major progress is now being made through the recent establishment of the International Nitrogen Management System (INMS) as a joint activity of UN Environment and the International Nitrogen Initiative (INI), with funding from GEF (www.inms.international).

Acknowledgements We thank all participants for their contributions to the conference discussions and for theirs inputs, which led to this consensus document. The preparation of the Kampala Statement-for-Action was conducted under the overall supervision of MAS and MB in order to ensure a balanced approach with wide multi-actor involvement, both in the drafting team (WW, NR as co-chairs, AB, ED, JWE, WdV, JNG, PH, WKH, CP, CS, MAS, BV and SZ) and in the plenary discussion that revised and agreed the final text. The first version of the Technical Annex was drafted by NR. The final plenary discussion of the main consensus text of the Kamala Statement-for-Action was moderated by MAS and PE.

Appendix

The following additional papers resulting from the 6th International Nitrogen Conference, Kampala, Uganda (24–27th November 2013) have been published in an Environmental Research Letters Special Issue (*Focus on Nitrogen Management Challenges: From Global to Local Scales*). All papers are open access and are listed below in the order they appear in the special issue.

Reis, R., Bekunda, M., Howard, C. M., Karanja, N., Winiwarter, W., Yan, X., Bleeker, A., Sutton, M. A. (2016). Synthesis and review: Tackling the nitrogen management challenge: from global to local scales. *Environmental Research Letters*, *11*, 120205.

Bodirsky, B. L., & Müller, C. (2014). Robust relationship between yields and nitrogen inputs indicates three ways to reduce nitrogen pollution. *Environmental Research Letters*, *9*, 111005.

Schullehner, J., & Hansen, B. (2014). Nitrate exposure from drinking water in Denmark over the last 35 years. *Environmental Research Letters*, *9*, 095001.

Yan, X., Ti, C., Vitousek, P., Chen, D., Leip, A., Cai, Z., Zhu, Z. (2014). Fertilizer nitrogen recovery efficiencies in crop production systems of China with and without consideration of the residual effect of nitrogen. *Environmental Research Letters*, *9*, 095002.

Fitton, N., Datta, A., Hastings, A., Kuhnert, M., Topp, C. F. E., Cloy, J. M., Rees, R. M., Cardenas, L. M., Williams, J. R., Smith, K., Chadwick, D., Smith, P. (2014). The challenge of modelling nitrogen management at the field scale: simulation and sensitivity analysis of N_2O fluxes across nine experimental sites using DailyDayCent. *Environmental Research Letters*, *9*, 095003.

Du, E., de Vries, W., Galloway, J. N., Hu, X., Fang, J. (2014). Changes in wet nitrogen deposition in the United States between 1985 and 2012. *Environmental Research Letters*, *9*, 095004.

Luo, X. S., Tang, A. H., Shi, K., Wu, L. H., Li, W. Q., Shi, W. Q., Shi, X. K., Erisman, J. W., Zhang, F. S., Liu, X. J. (2014). Chinese coastal seas are facing heavy atmospheric nitrogen deposition. *Environmental Research Letters*, *9*, 095007.

Glibert, P. M., Maranger, R., Sobota, D. J., Bouwman, L. (2014). The Haber Bosch–harmful algal bloom (HB–HAB) link. *Environmental Research Letters*, *9*, 105001.

Rufino, M. C., Brandt, P., Herrero, M., Butterbach-Bahl, K. (2014). Reducing uncertainty in nitrogen budgets for African livestock systems. *Environmental Research Letters*, *9*, 105008.

Zhou, M., Brandt, P., Pelster, D., Rufino, M. C., Robinson, T., Butterbach-Bahl, K. (2014). Regional nitrogen budget of the Lake Victoria Basin, East Africa: syntheses, uncertainties and perspectives. *Environmental Research Letters*, *9*, 105009.

Lassaletta, L., Billen, G., Grizzetti, G., Anglade, J., Garnier, J. (2014). 50 year trends in nitrogen use efficiency of world cropping systems: the relationship between yield and nitrogen input to cropland. *Environmental Research Letters*, *9*, 105011.

Davidson, E. A., & Kanter, D. (2014). Inventories and scenarios of nitrous oxide emissions. *Environmental Research Letters*, *9*, 105012.

M. A. Sutton et al. (eds.), *Just Enough Nitrogen*,
https://doi.org/10.1007/978-3-030-58065-0

Leip, A., Leach, A., Musinguzi, P., Tumwesigye, T., Olupot, G., Tenywa, J.T., Mudiope, J., Hutton, O., Cordovil, C. M. d. S., Bekunda, M., Galloway, G. (2014). Nitrogen-neutrality: A step towards sustainability. *Environmental Research Letters*, *9*, 115001.

Dalgaard, D., Hansen, B., Hasler, B., Hertel, O., Hutchings, N.J., Jacobsen, B. H., Jensen, L. S., Kronvang, B., Olesen, J. E., Schjørring, J. K., Kristensen, I. S., Graversgaard, M., Termansen, M., Vejre, H. (2014). Policies for agricultural nitrogen management—trends, challenges and prospects for improved efficiency in Denmark. *Environmental Research Letters*, *9*, 115002.

Galloway, J. N., Winiwarter, W., Leip, A., Leach, A. M., Bleeker, A., Erisman, J. W. (2014). Nitrogen footprints: Past, present and future. *Environmental Research Letters*, *9*, 115003.

Powell, J. M. (2014). Feed and manure use in low-N and high-N-input dairy cattle production systems. *Environmental Research Letters*, *9*, 115004.

Wang, Q. X, Koshikawa, H., Liu, C., Otsubo, K. (2014). 30-year changes in the nitrogen inputs to the Yangtze River Basin. *Environmental Research Letters*, *9*. 115005.

Misselbrook, T. H., Cardenas, L. M., Camp, V., Thorman, R. E., Williams, J. R., Rollett, A. J., Chambers, B. J. (2014). An assessment of nitrification inhibitors to reduce nitrous oxide emissions from UK agriculture. *Environmental Research Letters*, *9*, 115006.

Xue, L. H., Yu, Y. L, Yang, L. Z. (2014). Maintaining yields and reducing nitrogen loss in rice–wheat rotation system in Taihu Lake region with proper fertilizer management. *Environmental Research Letters*, *9*, 115010.

Hutchings, N. J., Nielsen, O.-K., Dalgaard, T., Mikkelsen, M. H., Børgesen, C. D., Thomsen, M., Ellermann, T., Højberg, A. L., Mogensen, L., Winther, M. (2014). A nitrogen budget for Denmark; developments between 1990 and 2010, and prospects for the future. *Environmental Research Letters*, *9*, 115012.

Shibata, H., Cattaneo, L. R., Leach, A. M., Galloway, J. N. (2014). First approach to the Japanese nitrogen footprint model to predict the loss of nitrogen to the environment. *Environmental Research Letters*, *9*, 115013.

Bealey, W. J., Loubet, B., Braban, C. F., Famulari, D., Theobald, M. R., Reis, S., Reay, D. S., Sutton, M. A. (2014). Modelling agro-forestry scenarios for ammonia abatement in the landscape. *Environmental Research Letters*, *9*, 125001.

Galy-Lacaux, C., & C Delon, C. (2014). Nitrogen emission and deposition budget in West and Central Africa. *Environmental Research Letters*, 9, 125002.

Singh, S., & Kulshrestha, U. C. (2014). Rural versus urban gaseous inorganic reactive nitrogen in the Indo-Gangetic plains (IGP) of India. *Environmental Research Letters*, *9*, 125004.

Sanz-Cobena, A., Lassaletta, L., Estellés, F., Del Prado, A., Guardia, G., Abalos, D., Aguilera, E., Pardo, G., Vallejo, A., Sutton, M. A., Garnier, J., Billen, G. (2014). Yield-scaled mitigation of ammonia emission from N fertilization: the Spanish case. *Environmental Research Letters*, *9*, 125005.

Bouraoui, F., Thieu, V., Grizzetti, B., Britz, W., Bidoglio, G. (2014). Scenario analysis for nutrient emission reduction in the European inland waters. *Environmental Research Letters*, *9*, 125007.

Billen, G., Lassaletta, L., Garnier, J. (2015). A vast range of opportunities for feeding the world in 2050: Trade-off between diet, N contamination and international trade. *Environmental Research Letters*, *10*, 025001.

Van Grinsven, H. J. M., Erisman, J. W., de Vries, W., Westhoek, H. (2015). Potential of extensification of European agriculture for a more sustainable food system, focusing on nitrogen. *Environmental Research Letters*, *10*, 025002.

Sobota, D. J., Compton, J. E., McCrackin, M. L., Singh, S. (2015). Cost of reactive nitrogen release from human activities to the environment in the United States. *Environmental Research Letters*, *10*, 025006.

Zhang, Y., Bleeker, B., Liu, J. G. (2015) Nutrient discharge from China's aquaculture industry and associated environmental impacts. *Environmental Research Letters*, 10, 045002.

Bustamante, M. M. C., Martinelli, L. A., Pérez, T., Rasse, R., Ometto, J. P. H. B., Pacheco, F. S., Lins, S. R. M., Marquina, S. (2015). Nitrogen management challenges in major watersheds of South America. *Environmental Research Letters*, *10*, 065007.

Grizzetti, B., Passy, P., Billen, G., Bouraoui, F., Garnier, J., Lassaletta, L. (2015). The role of water nitrogen retention in integrated nutrient management: assessment in a large basin using different modelling approaches. *Environmental Research Letters, 10*, 065008.

Wu, L., Chen, X. P., Cui, Z. L., Wang, G. L., Zhang, W. F. (2015). Improving nitrogen management via a regional management plan for Chinese rice production. *Environmental Research Letters, 10*, 095011.

Leip, A., Billen, B., Garnier, J., Grizzetti, B., Lassaletta, L., Reis, S., Simpson, D., Sutton, M. A., de Vries, W., Weiss, F., Westhoek, H. (2015). Impacts of European livestock production: nitrogen, sulphur, phosphorus and greenhouse gas emissions, land-use, water eutrophication and biodiversity. *Environmental Research Letters, 10*, 115004.

Van Meter, K. J., Basu, N. B., Veenstra, J. J., Burras, C. L. (2016). The nitrogen legacy: Emerging evidence of nitrogen accumulation in anthropogenic landscapes. *Environmental Research Letters, 11*, 035014.

Index

M. A. Sutton et al. (eds.), *Just Enough Nitrogen*,
https://doi.org/10.1007/978-3-030-58065-0

The manufacturer's authorised representative in the EU is Springer
Nature Customer Service Centre GmbH, Europaplatz 3, 69115 Heidelberg,
Germany. If you have any concerns regarding our products, please
contact ProductSafety@springernature.com

Printed and bound by CPI Group (UK) Ltd, Croydon, CR0 4YY
30/09/2026
02225506-0018